Atomic Physics of Highly Charged Ions

E. Salzborn P. H. Mokler A. Müller (Eds.)

Atomic Physics of Highly Charged Ions

Proceedings of the Fifth International Conference
on the Physics of Highly Charged Ions
Justus-Liebig-Universität Giessen
Giessen, Federal Republic of Germany,
10–14 September 1990

With 318 Figures

Springer-Verlag
Berlin Heidelberg New York
London Paris Tokyo
Hong Kong Barcelona
Budapest

Professor Dr. Erhard Salzborn
Institut für Kernphysik, Strahlenzentrum der Justus-Liebig-Universität Giessen,
Leihgesterner Weg 217, W-6300 Giessen, FRG

Professor Dr. Paul H. Mokler
Gesellschaft für Schwerionenforschung mbH, Planckstrasse 1, Postfach 11 05 52
W-6100 Darmstadt 11, FRG

Professor Dr. Alfred Müller
Institut für Kernphysik, Strahlenzentrum der Justus-Liebig-Universität Giessen,
Leihgesterner Weg 217, W-6300 Giessen, FRG

Cover page shows charge state spectrum of Xe^{q+} recoil ions
produced in collisions of 1.4 MeV/u Fe^{15+} projectile ions with Xe target atoms

This book originally appeared as a supplement to Vol. 21 (1991) of the journal
Zeitschrift für Physik D - Atoms, Molecules and Clusters
(ISSN 0178-7683) © Springer-Verlag Berlin Heidelberg 1991

International Scientific Committee

Chairman: E. Salzborn, FRG

A. Bárány, Sweden	P. Mokler, FRG	S. Ohtani, Japan
S. Bliman, France	R. Morgenstern, The Netherlands	R. Olson, USA
F. Brouillard, Belgium	R. Phaneuf, USA	H. Schmidt-Böcking, FRG
B. Gilbody, UK	L. Presnyakov, USSR	H. Winter, Austria
P. Hvelplund, Denmark		

Local Organization

E. Salzborn	R. Schulze	J. Haselbauer	R. Traßl
A. Müller	W. Spies	J. Krüger	O. Uwira
F. Melchert	R. Völpel	J. Neumann	B. Weißbecker
G. Hofmann	M. Wagner	S. Petri	S. Pfaff
S. Krüdener	A. Frank	U. Pracht	
M. Liehr	T. Grewe	M. Schlapp	
S. Schennach	M. Hagmann	M. Stenke	

Secretaries: U. Rittinger, M. Sterlepper, G. Bernhardt

ISBN-13: 978-3-642-76660-2 **e-ISBN-13: 978-3-642-76658-9**
DOI: 10.1007/ 978-3-642-76658-9

Softcover reprint of the hardcover 1st edition 1991

2156/3140-5 4 3 2 1 0 - Printed on acid-free paper

Preface

This book contains the invited lectures and contributed papers presented at the *V International Conference on the Physics of Highly Charged Ions,* which was held at the Justus-Liebig-Universität Giessen, 10–14 September 1990. This conference was the fifth in a series - after Stockholm (1982), Oxford (1984), Groningen (1986) and Grenoble (1988) - to deal with a rapidly growing field, which comprises the spectroscopy of highly charged ions and their interactions with photons, electrons, atoms, ions, and solids.

Most of the matter of the universe is in the ionized state. Investigations dealing with hot plasmas on earth have been greatly furthered by thermonuclear-fusion research. The increasing maturity of this programme has revealed the fundamental role of highly charged ions in fusion plasmas. Today, it is clear that a detailed knowledge of the production mechanisms of highly charged ions and their interactions with other plasma constituents is an important prerequisite for a better understanding of the microscopic and macroscopic plasma properties.

The study of highly charged ions involves various branches of physics. It was the aim of the conference to bring together physicists working in atomic collisions and spectroscopy, in plasma physics and astrophysics, as well as in solid-state and ion-source physics. About 220 scientists from 20 nations attended the conference, indicating the strong worldwide interest and the vitality of research in this field.

We are grateful to all invited speakers and authors of contributed papers reporting on the rapid progress achieved in the last two years, both experimentally and theoretically. Many thanks are due to all colleagues who - as members of the scientific or organizing committees and/or chairmen - have contributed to making the conference lively and fruitful. Finally, we gratefully acknowledge the financial support given to the conference by sponsors.

Giessen
March 1991

Erhard Salzborn

Sponsors

Justus-Liebig-Universität Giessen

Giessener Hochschulgesellschaft e.V.

DPG-Regionalverband Hessen-Mittelrhein-Saar e.V.

Gesellschaft für Schwerionenforschung (GSI)

Hessisches Ministerium für Wissenschaft und Kunst

Deutsche Forschungsgemeinschaft (DFG)

Bundesministerium
für Forschung und Technologie (BMFT)

Contents

Invited Papers

Contributed Papers

Supplement to Z. Phys. D - Atoms, Molecules and Clusters 21, S 1-5 (1991)

Atoms, Molecules
Zeitschrift für Physik D and Clusters

Role of highly-charged ions in astrophysical and laboratory plasmas

B. C. Fawcett.
Rutherford Appleton Laboratory, Science and Engineering Research Council, Chilton, Didcot, Oxfordshire, OX11 0QX, England

Abstract

This presentation considers how the interaction between physical conditions in plasmas and atomic-structure properties affects the role played by highly-charged ions. It is emphasised that reliable and accurate atomic data are required to determine plasma conditions from the analysis of spectral line intensities. Atomic-structure problems which limit the quality of available atomic data are discussed. Spectroscopic observations quoted provide information on the behaviour of ions in plasmas.

I. Introduction

Sometimes at low density or temperature, sometimes at high density or temperature, sometimes in equilibrium, sometimes transient; plasmas have many forms in the universe and within them each ion plays different roles. The behaviour of an ion in some remote plasma can tell us about conditions in which it exists provided we can understand the message it sends. We have in recent years become expert at recording information sent by far away ions in stars or from closer laboratory plasmas with sophisticated spectrometers, telescopes or satellite borne instrumentation. The ion radiates spectral lines and we measure their intensities along with the continuum which may be modified by the ions. Much of the information about plasma conditions contained in these spectra however remains locked up as securely as that in Egyptian hieroglyphics was to early explorers. True we can interpret the spectra of simple ions and Doppler widths, but most spectral lines found in astrophysical spectra belong to complex ions. Limitations in our knowledge of the atomic physics of complex ions prevents us deciphering the language of spectral line intensities. The ions role is now for convenience discussed under two headings. Firstly the role due to plasma conditions and secondly the role governed by its atomic-structure properties. Finally spectroscopic observations are discussed which show how plasma conditions can be revealed with the aid of a knowledge of atomic-structure.

II. Plasma conditions and the ion's role.

One part of the ion's role we can describe best in physical terms is its movement while our knowledge of atomic-structure properties is somewhat limited. Thus we can describe the kinetics with a theoretical description of the state of equilibrium of the plasma, its electrons and ions and determine whether the plasma components have time to relax to a statistical distribution of velocities following a change in plasma conditions induced by external means such as magnetic or electric interaction or, alternatively, radiation or conduction. In many plasmas the electrons do have time to relax and hence their temperature is meaningful, but the plasma may not have a unique temperature since the relaxation time between ions and electrons is 1800 times longer than electron-electron relaxation times while the the relaxation time for protons is about 40 times longer than that for electrons. In laboratory plasmas ion and electron temperatures are usually different and occasionally the rate of energy input is so fast or the electron density so low that the electrons escape from the plasma before they relax to a statistical distribution of velocities.

What role does the ion play at different plasma conditions? Well at the low densities, such as in the solar corona, ionisation is caused predominantly by electron impact and recombination by radiative decay of a free electron from the continuum. The ratio of the population of a lower ion state to the next stage above it is simply the ratio of the radiative recombination rate coefficient to the ionisation rate coefficient, and each coefficient is a function of temperature. The population of an excited energy level which can decay to emit allowed spectral lines is usually governed by collisional excitation from the ground level and spontaneous radiative decay, but since the decay is the faster process the population is mainly in the ground level. The intensity of a spectral line is then given by the integral over the line of sight of the instrument:

$$I(u,l)=\frac{1}{4\pi}\int n_e n(z,g)X(T_e,g,u)\,\frac{A(u,l)}{\sum A(u,k)}\,ds$$

where n_e is the electron density and n(z,g) the population density of the ground level g for the charge state z; X (T_e,g,u) is the excitation rate coefficient at electron temperature; A is the spontaneous transition probability and u and l signify the upper and lower energy levels for the transition producing the spectral line.

In low density laboratory plasmas conditions are usually changing faster than ionisation and recombination relaxation times and successive stages of ionisation must be determined by a set of time dependent equations. The above equation may still apply for line intensities because the excitation and spontaneous decay rates are much faster than the ionisation and recombination rates but step-wise excitation via metastable levels has also to be taken into account.

Transitions from metastable levels can be exploited to determine electron density. At very low density the metastable levels are depopulated by spontaneous decay with the emission of either forbidden lines (electric quadrupole or magnetic dipole transitions) or intercombination lines. As the electron density is increased more electrons are excited into the levels and whereas the electrons arriving in the upper levels of the permitted line decay quickly to radiate a more intense line, the metastable levels may not have time to decay and emit radiation before the other electron collisions depopulate them. They can therefore acquire a substantial population in the interval between excitation and the depopulating collision.

As the density is increased, particularly above 10^{17} cm^{-3}, collisional processes which depopulate the excited levels may compete with or become faster than radiative decay. Consequently, step wise excitation processes and cascade become more significant in the population of levels. For level population distribution in an optically thin plasma, collisional dominate radiative processes when the electron density exceeds approximately $1.6 \times 10^{12} T_e^{1/2} \Delta E$ cm^{-3}, where ΔE is the energy separation in eV of two levels and T_e the electron temperature in K. At still higher densities, all the level populations become dominated by collision processes and level and ion populations are described by the Boltzman and Saha equations. An intermediate situation occurs in some dense plasmas where the upper energy levels are statistically populated and the population distributions of many lower levels mainly influenced by collisional excitation and spontaneous radiative decay.

The widths of Doppler-broadened spectral lines are due to the combined effects of the ion temperature and mass motion of the ions and plasma density dependent Stark widths are a function of the atomic properties of the ion.

The ion also plays a role in shaping continuum in the emission of recombination continuum and through absorption.

III. The role governed by the ion's atomic structure.

Astrophysicists and plasma physicists usually need to know electron and ion temperature and density, chemical composition, density of each charge state, particle and energy transport, magnetic and electric fields and the temporal and spatial dependence of these parameters. The intensities of the spectral lines can be used to determine the various parameters of these plasmas, only if the relevant atomic processes are understood at the particular temperature and density under consideration. Several fundamental atomic values must be reliably known. These are the excitation rate coefficient and the transition probability which are a function of collision strength and oscillator strength respectively. In addition ionisation and recombination rates will be required to determine the ionisation balance. To be more specific it is our inability to determine these fundamental atomic quantities and their error-bars which frustrates us from reading the content of the spectra. We simply cannot determine the oscillator strengths and collision strengths accurately enough to adequately enable the interpretation of most spectral line intensities radiated from complex ions. This arises because of difficulties in dealing with atomic-structure and collisional problems. In simple ions the atomic quantities we require for the interpretation of spectral line intensities can often be calculated through the adoption of sophisticated wavefunctions and comprehensive treatment of configuration interaction in advanced atomic-structure(1-10) or collision codes (11-14) but for complex ions a number of factors, such as the inadequate modelling of configuration interaction, make the determination of error bars on the calculated data very uncertain. For complex ions it may not be presently possible to include all the configurations required for reliable atomic-structure calculations so forced approximations are usually made. Further perturbations may come from inner shells or levels lying in the continuum. Cancellation effects(1) can lead to serious errors and others can occur in the radial dipole matrix elements. The calculation of line strengths and oscillator strengths involves a double summation over amplitudes whose correct relative phases must be known. This involves the signs of the eigenvectors and the signs of the radial dipole integrals and can result in destructive interference effects. This cancellation if not adequately taken into account can leave gross errors on individual oscillator strength values.

Cancellation factors are printed by some codes(1). If they are small, errors in corresponding *gf* values may be large.

A practical way to improve the accuracy of the data is to enhance the quality of eigenvectors. Eigenvectors computed *ab initio* can be seriously inaccurate but they can be improved with semi-empirical techniques which involve the adjustment of Slater parameters(15) by least-squares optimisation to obtain agreement between observed and calculated energy level differences. This can provide a better model for level mixing and cancellation. Small oscillator strengths can be subject to gross errors if one of the levels is mixed with a level associated with a large oscillator strength. Imperfections in computed eigenvectors give rise to predicted differences in the level's composition and therefore greatly overestimate the smaller oscillator strength. On the other hand large oscillator strengths are relatively insensitive. It was also demonstrated, with collision strength calculations for Fe XIII and Fe IX(16,17), that the procedure involving the adjustment of Slater parameters to improve energy levels and eigenvectors can be applied to improve collision strengths as well since the collision target is described by the same eigenvectors. The atomic-structure model was developed with the Cowan atomic-structure codes(1) which have Hartree-Fock and optional Slater parameter optimisation routines. Resulting Slater parameters were inserted into the SUPERSTRUCTURE(9) atomic-structure code. A program JAJOM (18,19) used term coupling coefficients from SUPERSTRUCTURE and reactance matrices from Distorted wave collision codes(12,13) to make the transformation to collision strengths between fine structure levels.

The ion's relativistic effects also have to be taken into account. Non-relativistic calculations break down at high atomic numbers so fully relativistic Multi-Configuration-Dirac-Fock codes(2-5) are necessary. As the Dirac-Fock equations involve multi-component wavefunctions, array sizes in the computations are much larger than for Hartree-Fock(1,6,7). It can therefore be very important to use these relativistic and non-relativistic codes in conjunction as the one type gives the advantage of fully relativistic calculations and the other gives the ability to include more configurations in the model due to the much smaller arrays inherent in the matrices.

The calculation of atomic data for simple atomic structures is much more satisfactory. Consequently it is frequently necessary to restrict plasma diagnostics to methods which utilise such data. An example of accurate calculations of oscillator strengths in simple ions is given by Froese-Fischer(20) who used her multi-configuration Hartree-Fock method (MCHF) for calculations in Li I, Na I and K I which included correlation in the core. These were compared with other calculations which applied Hylleraas type wavefunctions or relativistic many-body perturbation theory and also with accurate experimental values obtained from laser beam experiments. Agreements are impressive with deviations only in the third decimal place. The MCHF results were also supported the matching of length and velocity formulations.

IV. The role of ions through spectroscopic observations.

The role of ions in plasmas can be further highlighted with spectroscopic observations of astrophysical and laboratory plasmas. Examples are now cited to underline several of the physical phenomena discussed in previous sections.

The first time resolved intensity profiles of spectral lines of highly ionised argon recorded were emitted from the Zeta torroidal plasma [21]. What does this observation tell us? Well clearly it established the presence of a given ion at a particular time. If however one wishes to obtain an accurate ionisation balance from the data one is limited by the size of error bars on ionisation and recombination rates. This emphasises a problem which is common to many aspects of the analysis of spectral line intensities. The trouble is that our knowledge of the ion's role is restricted by the accuracy of the atomic data at out disposal.

For our next next example we recall that the first solar extreme ultraviolet records[22-24] were obtained lines back in 1960 and revealed a group of intense lines in wavelength region from 171 to 220 Å which for a long time remained a mystery. Even the basic line identification problems were not solved until our analysis[25-28] in 1965. Although interpretation of the intensities of these spectral lines could tell us much about the ions role in the sun, it is only quite recently that the necessary atomic data has become available to analyse the information confined in some of these line intensities. Many of the line ratios are for instance sensitive to electron density. It was pointed in 1965[25] when the lines were identified that for Fe X the line intensity ratio $3s^23p^5\ ^2P_{1/2}$- $3s^23p^43d\ ^2D_{1 1/2}$ / $3s^23p^5\ ^2P_{1 1/2}$- $3s^23p^43d\ ^2D_{2 1/2}$ was reduced due to electron density effects. This occurs because the population of the $3s^23p^5\ ^2P_{1/2}$ level is lower in the coronal plasma than when in Boltzmann equilibrium with the ground level because of its decay by a "forbidden" transition. The term forbidden is generally used for transitions "forbidden" by electric-dipole which are allowed via magnetic-dipole, electric-quadrupole or higher order transitions. In Fe XI to Fe XIII the ground term is of different multiplicity to other terms belonging to the ground configuration and the latter again decay via the forbidden transitions identified in the classic work of Edlén[29]. Excitation from these levels and subsequent decay results in the emission of many spectral lines whose relative intensity depends on electron density. Some have been applied by Dere et al.[30] to determine solar flare electron densities which are in the regime where line ratios are particularly sensitive. One problem which limits the accuracy of such density determinations is the difficulty in calculating atomic data such as collision strengths and oscillator strengths and determining their error bars for such complex ions as Fe XI and FeXII. *Ab initio* methods tend to be unreliable for these ions as they predict levels too widely spaced and thus give the wrong eigenvectors and level mixing. Semi-empirical methods which improve eigenvectors have therefore been adopted for oscillator strengths [31-36] but there remains a need for better collision strength calculations. Some progress has recently been made as semi-empirical methods have also been incorporated into the ion-targets used in collision calculations for Fe XIII and Fe IX [16,17]. For Fe IX the density sensitive lines are $3s^23p^6\ ^1S_0$-$3s^23p^53d\ ^3P_1$ and $3s^23p^6\ ^1S_0$-$3s^23p^53d\ ^3P_2$. The second is a forbidden transition and population of the $3s^23p^53d\ ^3P_2$ level is sensitive to electron density. The recent calculations [17] have improved the collision cross sections affecting excitation to the $3s^23p^53d\ ^3P_1$ and $3s^23p^53d\ ^3P_2$ levels from the $3s^23p^6\ ^1S_0$ ground level and thereby also the precision of the density diagnostics.

Spectra of solar flares near 100 Å were acquired by Goddard Space-flight Centre experiment on the OSO 5 satellite[37]. The lines were identified as Fe XVIII to Fe XXIII δn=2-2 transitions[38] with the aid of isoelectronic studies[39,40]. They can be used for diagnostic purposes. The Fe XXI $2s^22p^2\ ^3P_0$-$2s2p^3\ ^3D_1$ and $2s^22p^2\ ^3P_1$-$2s2p^3\ ^3D_2$ are for example density sensitive for the same reasons as described for Fe X with one line only having an electric-dipole transition to the ground.

More recently solar flare spectra of Fe XVII and higher stages of ionisation were recorded[41] with the Solar Maximum Mission (SMM) X-ray Polychromator flat beryl crystal spectrometer and by the ASAF P78-1 satellite[47]. Once again the lines were identified with the aid of isoelectronic studies[42-45] as Fe XVII to Fe XXIV δn=2-3 transitions which can be emitted from solar flares in this spectral region. In fact the highest ionisation stages are not present on the SMM spectra due to the existing temperature conditions but are on ASAF P78-1 satellite spectra[47]: which highlights another diagnostic method. As a given stage of ionisation is dominant in near equilibrium plasmas in a restricted range of temperatures, its observation can in certain circumstances be applied to estimate electron temperature limits. For this an emission measure[30] for a line of a specific ion is applied to analyse the plasma in line of sight of the instrument and this can often define the maximum electron temperature present.

Several of the Fe XVII to Fe XXIV δn=2-3 transitions, as well as δn=2-4,5 transitions[45] at shorter wavelengths, have been applied to electron density and temperature diagnostics [41,46,47]. Some of the considerations involved are almost identical to those described for the Fe XVIII to Fe XXIII δn=2-2 transitions. The subject of recent studies in Fe XIX[46] involved line intensity analysis aided by new complicated atomic-structure and collision calculations. In fact the latter Fe XIX study was partly made to resolve Fe XIX lines from Ne IX lines. He-like Ne IX has a much simpler atomic-structure which introduces the next subject.

For solar studies transitions between simple structures can often be chosen [47]. The atomic data is much more reliable for these ions which may have between one and four electrons outside a closed shell. An important example is found on a solar soft-Xray spectrum acquired[48] with a rocket borne grazing incidence spectrograph. The OVII, NVI and CV $1s^2\ ^1S_0$- $1s2s\ ^3S_1$ forbidden lines were identified on this spectrum. The He-like spectrum of transitions $1s^2\ ^1S_0$- $1s2p\ ^1P_1$, $1s^2\ ^1S_0$- $1s2p\ ^3P_1$ and $1s^2\ ^1S_0$- $1s2s\ ^3S_1$, which correspond to an allowed dipole transition, an intercombination line, and a forbidden line were all present on this record and are the basis for a classic diagnostic method[49]. Isoelectronic lines can be used to measure a wide range of electron densities. For example in O VII the ratio of $1s^2\ ^1S_0$- $1s2p\ ^3P_1$ to $1s^2\ ^1S_0$- $1s2s\ ^3S_1$ is sensitive to electron densities near 10^{11} cm^{-1} while the ratio of $1s^2\ ^1S_0$- $1s2p\ ^3P_1$ to $1s^2\ ^1S_0$- $1s2p\ ^1P_1$ is sensitive to electron densities near 10^{14} cm^{-1}. Corresponding lines have since been found in other solar rocket [50] and satellite spectra [41,51] for isoelectronic solar abundant ions including Fe XXV[51]. Many including Fe XXV were recorded with the bent crystal spectrometer on the SMM satellite[41,51] and by the USAF P78-1 satellite[47]. The Fe XXV lines are useful not only for diagnostics of the solar and stellar plasmas but also for laboratory plasmas such as those contained in TOKAMAK[52] fusion research machines. The emission measure[47] for Fe XXV can indicate the high electron temperature of the plasma.

There are furthermore satellite lines [53] to the long wavelength side of the $1s^2\ ^1S_0$- $1s2p\ ^1P_1$ He-like resonance lines which have been applied[47,52] to electron temperature and density diagnostics of solar flares and TOKAMAK[52] fusion research plasmas. Some of the satellite lines are Li-like[53] and these are the most useful as they have been resolved as satellites of several ions in the isoelectronic sequence between C V and Fe XXV. Both the theory and observation of these lines present more difficulties than the aforementioned He-like lines. For instance the autoionisation transition probabilities required for

their analysis are sensitive to the square of the Slater configuration interaction integrals (R^k) and as different atomic-structure codes may calculate significantly different values for the R^k integrals, then the autionisation rates are not always in agreement leaving uncertainties in the analysis.
As the subject of this presentation is the "role of highly-charged ions in astrophysical and laboratory plasmas" and highly-charged ions mainly emit their lines at short wavelengths most useful observations occur in the far ultraviolet soft-Xray region. There are however a number of emission lines from fairly high ionisation stages present in the ultraviolet spectrum between 300 Å and 2000 Å[55,56]. At the shorter limit near 330 Å are the $3s^23p^n$ - $3s\,3p^{n+1}$ Fe X to Fe XIV lines which were first identified[28] in the theta-pinch spectra. The solar spectra of these transitions have similar diagnostic potential as for the Fe X to Fe XIV lines between 170 Å and 220 Å.
In the spectral range between 1100 Å and 2000 Å important solar spectra were recorded by rocket[55] and SKYLAB[56] borne spectrographs. This data have been applied by Feldman[47] for diagnostic purposes. For this analysis he has cautiously chosen lines belonging to ions with relatively simple structures for which the required atomic data are most reliable. Thus he has used Li-like and Na-like lines of C IV, N V, Al III and Si IV for relating line emission to electron temperature. He has also utilised Be-like C III, N IV and O V lines and Mg-like lines of S V for electron density determination taking advantage of similar dependencies as discussed above for C V to Fe XV. Electron densities were also estimated from B-like N III and O IV $2s^2\,2p\ {}^2P$- $2s\,2p^2\ {}^4P$ Al-like S IV $3s^2\,3p\ {}^2P$- $3s\,3p^2\ {}^4P$ transitions as well as C-like O III $2s^2\,2p^2\ {}^3P_1$- $2s\,2p^3\ {}^5S_2$ transitions using the density dependence of intercombination lines.
One of the most important features in this spectral region, and indeed throughout the visible spectrum, are the forbidden lines which appear in solar and stellar spectra. Their intensities can infer limits on electron temperatures but perhaps they are of most value for Doppler width measurements. At these longer wavelengths widths can be measured accurately enough to place small errors on ion temperatures or mass motion of ions in plasmas.
Stark broadening in astrophysical plasmas is perhaps best represented by hydrogen Balmer lines, which are not in the context of highly-charged ions, but Stark widths in highly-charged ions are very prominent features in laser-produced plasmas. Stark broadened O VI and K IX lines have been applied [57] to measure electron density.
The same paper[57] also reported laser-plasma continuum studies. The electron temperature was obtained by directly measuring the frequency interval over which the intensity of the recombination continua for C V and CVI fell to e^{-1}. Merging of the Lyman series in carbon ions was also reported.
In conditions where the instrumental width is less than broadening the precise point of merging can be applied to obtain electron density using the Inglis-Teller theory.
The first laser produced plasma of highly-charged ions ever recorded[58] were of carbon spectra, spectra of Fe XV and Fe XVI and of isoelectronic nickel. In these iron and nickel spectra another aspect of the role of highly charged ions in dense plasmas was demonstrated. Levels of high principal quantum number are populated at the high electron densities and long Rydberg series are observed. At very high densities the upper levels can be statistically populated and related to the population of the next highest stage of ionisation. They may therefore be applied to determine the densities of different ionisation states provided the Saha equation is obeyed. In such spectra of dense plasmas line to line or line to continuum methods of electron temperature determination are valid if the Saha-Boltzman equation describes the population of atomic energy levels.

Another topical role of ions in plasmas is for X-ray lasers. Highly-charged ions feature in new proposals for resonant photo-pumping. Lee et al.[58] propose pumping the 5p $^2P_{1/2}$ and 5p $^2P_{1 1/2}$ levels in Fe XXIV using aluminium Lyman α radiation. This would, if successful, give X-ray lasing at 70 Å due to population inversion of the n=4 and n=5 levels.

V. Discussion

At present advanced spectroscopic satellite borne instruments are recording spectra of stellar plasmas but the interpretation of the spectral line intensities is restricted because much of the atomic data required for the analysis in not known with sufficient accuracy. Consequently knowledge of the role of many highly-charged ions in these plasmas will remain a mystery until the necessary atomic data calculations or measurements are completed reliably. For solar spectra it has been possible to work with simple atomic structures to perform the analysis but with stellar plasmas this choice is not always available. Even with laboratory plasmas it is often desirable to perform diagnostics with complex ions (ie: for impurity determination) and ground based astrophysical records present many problems involving complex ions for which there are unreliable data. The need for more good atomic data measurements and calculations is therefore stressed. New methods are now available which can greatly improve the situation. For instance lifetimes of atomic energy levels can be measured accurately after excitation with tuned lasers. Ion traps also enable valid measures of long lifetimes. Single and double electron captures are being achieved through focussing high energy ion beams into gas cells[59] thereby enabling the investigation of lines which decay for levels of high principal quantum number. The R-matrix method is being applied[60] to calculate collision strengths and hence derive excitation rate coefficients for electron temperature and density diagnostics. More powerful new computers allow improved calculations. The role of highly-charged ions in astrophysical and laboratory plasmas will therefore be better understood in the light of such future advances.

References

1. Cowan, R. D. 'The Theory of Atomic Structure and Spectra' (Univ. California Press, Berkeley and Los Angeles, 1981).
2. Grant, I. P. Mckenzie B. J. Norrington P. H. Computer Physics Comm. **21**, 207 (1980) ibid **21**, 233.
3. Dyall, K. G. Grant, I. P. Johnson, C. T. Parpia F. A. and Plummer, E. P., Comput. Phys. Commun. **55**, 425 (1989)
4. Parpia, F. A. Grant, I. P. Dyall, K. G. and Fischer, C. F. (In preparation)
5. Desclaux, J. P., Computer Phys. Commun. **9**, 31 (1975)
6. Froese Fischer, C, Comput Phys Commun. **4**, 107 (1972) and Ibid **7**, 236 (1974)
7. Froese Fischer, C, 'The Hartree-Fock method for Atoms', (Publ. J. Wiley, NY. 1977)
8. Klapisch, M, C. R. Acad. Sci. B **265**, 914 (1967)
9. W.Eissner, M. Jones and H. Nussbaumer, Comput. Phys. Commun. **8**, 230 (1974)
10. Hibbert, A., Computer Phys. Commun. **9**, 141-172 (1975)
11. W.Eissner and M. J. Seaton, J. Phys. B. Atom. Molec Phys. **5**, 2187 (1972)
12. P. G. Burke and W. E. Eissner, "Low Energy Collisions in Complex Atoms and Ions" Chapter I of *Atoms in Astrophysics* (Publ. Plenum Press NY. 1983)

13. K. A. Berrington, P. G. Burke, J. J. Chang, A. T. Chivers, W. D. Robb and K. T. Taylor, Computer Phys. Commun. **8**, 1491 (1974)
14. K. A. Berrington, P. G. Burke, M. Le. Dourneuf, W. D. Robb, K. T. Taylor and Vo Ky Lan, Computer Phys. Commun. **14**, 367 (1978)
15. Fawcett, B. C. Atomic Data and Nuclear Data Tables **41**,182-255 (1989)
16. Fawcett, B. C. and Mason, H.E. Atomic Data and Nuclear Data Tables **43**, 245-258, (1989).
17. Fawcett, B. C. and Mason, H. E. (In the press 1990) Atomic Data and Nuclear Data Tables **47**.
18. H. E. Saraph, Computer Phys. Commun. **1**, 232 (1970)
19. H. E. Saraph, Computer Phys. Commun. **3**, 256 (1972)
20. Froese Fischer, C., Nuclear Instruments and Methods in Physics Research **B31**, 265 (1988).
21. Fawcett, B. C., Jones, B.B., and Wilson, R., Proc. Phys. Soc. **78**, 1223 (1961)
22. Hinteregger, H. E., Astrophys. J. **132**, 801 (1960)
23. Hinteregger, H. E., J. Geophys. Res. **66**, 175 (1961)
24. Austin, W. E. Purcell, J. D. and Tousey, R., Astron. J. **67**, 110 (1962)
25. Gabriel, A. H., Fawcett, B. C., Jordan, C., Nature Vol. **206**, 390 (1965)
26. Gabriel, A. H., Fawcett, B. C., Jordan, C., Proc. Phys. Soc. **87**, 825 (1966)
27. Fawcett, B. C., Gabriel, A.H., Saunders, P.A.H., Proc. Phys. Soc. **90**, 863 (1967)
28. Fawcett, B. C., J. Phys. B. Atom. Molec. Phys. **4**, 1577 (1971)
29. Edlén, B., Z. Astrophys. **20**, 30 (1942)
30. Dere, K. P., Mason, H. E., Widing, K.G. and Batia, A. K., Astrophys. J. Suppl. **40**, 341 (1979).
31. Bromage, G.E., Cowan, R.D., and Fawcett, B. C., Physica Scripta **15**, 177 (1977)
32. Bromage, G. E., Cowan, R. D., Fawcett, B. C., Mon. Not. R. astr. Soc. **183**, 19 (1978).
33. Fawcett, B. C., Atomic Data and Nuclear Data Tables **35**, 185 (1986)
34. Fawcett, B. C., Atomic Data and Nuclear Data Tables **35**, 203 (1986)
35. Fawcett, B. C., Atomic Data and Nuclear Data Tables **36**, 129 (1987)
36. Fawcett, B. C., (1987) Atomic Data and Nuclear Data Tables **36**, 151 (1987)
37. Kastner, S. O. Neupert, W. M. and Swartz, M., Astrophys. J. **191**, 261 (1974)
38. Fawcett, B. C., and Cowan, R. D., Mon. Not. R. astr. Soc. **171**, 1 (1975)
39. Fawcett, B. C., and Hayes, R. W., Mon. Not. R. astr. Soc., **170**, 187 (1975)
40. Fawcett, B. C., J. Phys. B. Atom. Molec. Phys. **3**. 1732 (1970)
41. Phillips, K. J. H., Leibacher, J. W., Wolfson, C .H., Parkinson, J. H., Fawcett, B. C., Kent, B. J., Mason, H. E., Acton, L. W., Culhane, J. L., Gabriel, A. H., Astrophysical Journal **256**, 774 (1982)
42. Fawcett, B. C., Hayes, R.W., Physica Scripta, **8**. 244 (1973)
43. Fawcett, B. C., Cowan, R. D., and Hayes, R. W., Astrophys. J., **187**, 377 (1974)
44. Fawcett, B. C.,Ridgeley, A., Hughes, T. P., Mon. Not. R. astr. Soc. **188**, 363 (1979)
45. Fawcett, B. C., Jordan, C., Lemen, J. R., and Phillips, K. J. H., Mon. Not. R. astr. Soc. **225**, 1013 (1987)
46. Bhatia, A. K., Fawcett, B. C., Lemen,J. R, Mason, H. E. and Phillips, K. J. H., Mon. Not. R. astr. Soc. **242**, 421 (1989)
47. Feldman, U., Physica Scripta **24**, 681-711 (1981).
48. Freeman, F. F. and Jones, B. B., Solar Physics, **15**, 288 (1970)
49. Freeman, F. F. Gabriel, A. H., Jones, B. B.and Jordan, C., Phil. trans. Roy. Soc. Lond. A. **270**, 127 (1970)
50. Acton, L. W., Bruner, M. E., Brown, W. A., Fawcett, B. C., Schweizer, W., Speer, R. J., Astrophys. Jour. **291**, 865 (1984)
51. Gabriel, A. H., Culhane, J. L., Acton, L.W.,Antonucci, E., bently, R. D. Jordon, C., Phillips, K. J. H., Leibacher, J. W., Parmer, A. N., Rapley, C. G., Wolfson, C.H., and Strong, K. T.,Proc. of COSPAR, **E2.1.6**. (1980)
52. Hill, K. W, von Goeler, S., Bitter, M., Cambell, L, Cowan, R. D., Freankel, B., Greenburger, A. Horton, R, Hovey, J., Roney, W., Sauthoff, N. R. and Stodiek, W., Phys. Rev. A. **19**, 1770 (1979)
53. Gabriel, A. H., Mon. Not. Roy. Astr. Soc. **189**, 319 (1972)
54. Burton, W. M and Ridgeley, A., Solar Physics, **14**, 3 (1970)
55. Doschek, G. A., Feldman, U. Van Hoosier, M. E. and Bartoe, J. D. F., Astrophys. J. Suppl. **31**, 417 (1976)
56. Burgess, D. D., Fawcett, B. C. and Peacock, N. J., Proc. Phys. Soc. **92**, 805 (1967).
57. Fawcett, B. C., Gabriel, A. H., Irons, F. E., Peacock, N. J., Saunders, P. A. H., Proc. Phys. Soc., **88**, 1051 (1966)
58. Lee, Y. T., Howard, W. M. and Nash, J. K., J. Quant. Radiat. Transfer. **43**, 335 (1990)
59. Boduch, P. Chantepie, M. Fawcett, B. C. Hennecart, D. Husson, X. Kucal, H. Leclar, D. and Stolterfoht, N. (1990) 'Single and Double Electron Captures in 160Kev Ar^{8+} - $He(H_2)$ collisions' (Presented at the 21st. EGAS Conference Uppsala)
60. Kato, T. Lang, J. and Berrington, K. E. Atomic Data and Nuclear Data Tables **44**, 133 (1990)

Supplement to Z. Phys. D - Atoms, Molecules and Clusters 21, S 7-11 (1991)

Atoms, Molecules
Zeitschrift für Physik D and Clusters

Theory of highly-ionized few-electron systems

G. Soff
Gesellschaft für Schwerionenforschung (GSI), Planckstraße 1, Postfach 110552, W-6100 Darmstadt, FRG

Abstract

Quantum electrodynamical corrections in hydrogen-like atoms are reviewed. In particular the Wichmann-Kroll contribution to the total vacuum polarization charge density is investigated. Limits for precision tests of quantum electrodynamics in strong Coulomb fields are given. A significant parity violation effect in helium-like uranium is pointed out.

I. Introduction

A precision spectroscopy of high-Z hydrogen-like atoms provides an unique opportunity to test quantum electrodynamics of strong fields. The fundamental QED processes causing energy corrections of bound electrons are depicted by the Feynman diagrams in figure 1. Double lines signify electron wavefunctions and propagators in the nuclear Coulomb potential. Part a) visualizes vacuum polarization. Here an electron interacts with a photon which virtually creates an electron-positron pair indicated by the loop. Part b) displays the electron self energy. It represents a self-interaction of the electron by emitting and absorbing again a photon.

The most accurate determination of radiative corrections in quantum electrodynamics is achieved in measurements of the anomalous magnetic moment of electrons and muons as well as in Lamb-shift studies in hydrogen. In contrast to these QED phenomena in relatively weak external fields the Lamb-shift in high-Z systems can be utilized to verify theoretical predictions for quantum electrodynamical corrections in strong fields. For heavy elements the coupling constant $Z\alpha$ of the external field is not small compared with unity in order to allow for any reliable perturbation expansion in $Z\alpha$. Furthermore relativistic bound states with high electron densities are entangled in the calculations. This implies a different region of involved momenta compared with QED processes in weak external fields being characterized by an expansion in α.

Various peculiar features can be attributed to strong external fields. For example the process [1,2] of spontaneous electron-positron pair creation in a homogeneous electric field will occur only if one goes beyond the bounds of a critical electric field strength. This phase transition of the QED vacuum remains completely unobservable even in utmost precise experimental studies in weak fields. It has also been demonstrated [3] that possible nonlinear extensions of ordinary quantum electrodynamics would be most pronounced in level shifts of bound electrons in very heavy atoms. E.g., one may consider the Lagrangian

$$\mathcal{L} = \mathcal{L}_{QED} + \mathcal{L}_{cor} \tag{1}$$

including the nonlinear correction term

$$\mathcal{L}_{cor} = \lambda(\overline{\Psi}\Gamma\Psi)^n \tag{2}$$

with a general vertex function Γ and parameter n. The most stringent boundary for the coupling constant λ is derived from a Lamb-shift analysis in high-Z atoms.

Quantum electrodynamical energy shifts for K- and L-shell electrons in hydrogen-like atoms have been tabulated [4,5]. The biggest uncertainty in table [5] results from vacuum polarization corrections of higher order in $Z\alpha$. Meanwhile these terms have been evaluated with a much better accuracy [6,7,8]. Furthermore we can now employ additional data [9] for nuclear radii. Also the influence of the vacuum polarization correction of order $\alpha^2(Z\alpha)$ denoted as Källén-Sabry contribution has been computed recently [10]. Moreover nuclear polarization corrections [11,12,13] can be taken into account. For a compilation of energy eigenvalues in helium-like atoms we may refer to ref. [14]. As a particular example we briefly discuss in the next section the evaluation of the Wichmann-Kroll contribution [15] to the Lamb shift.

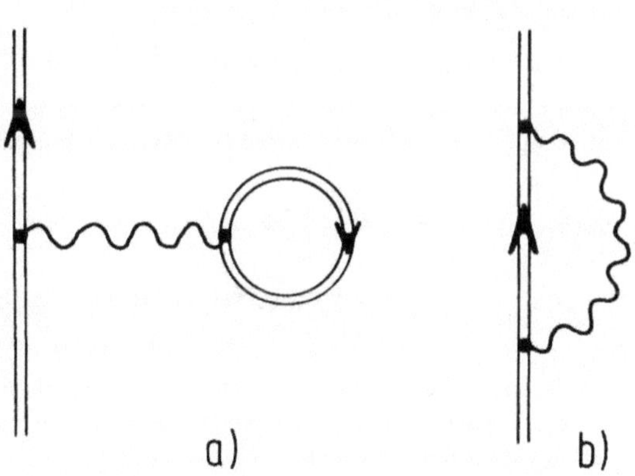

Figure 1: Feynman diagrams for the lowest-order (a) vacuum polarization and (b) self energy.

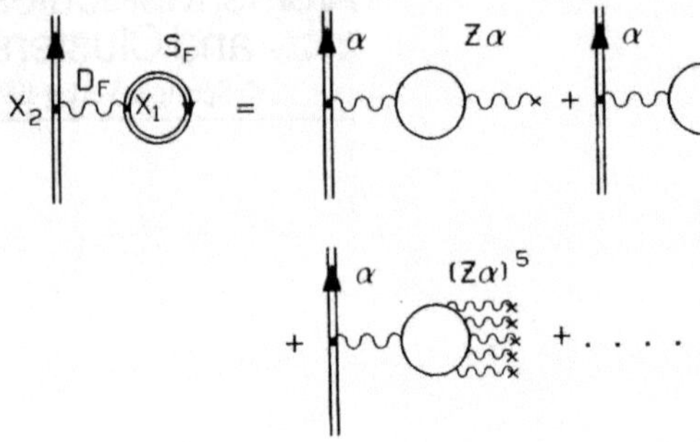

Figure 2: $Z\alpha$-expansion of the vacuum polarization diagram.

II. Vacuum polarization in a strong Coulomb field

The Feynman diagram illustrating the vacuum polarization correction in a strong Coulomb field is depicted in figure 2. The double lines indicate wavefunctions and propagators taking into account the external field to all orders while single lines signify free propagators. An expansion in the coupling constant $Z\alpha$ of the nuclear Coulomb field is displayed on the right hand side of figure 2. The Uehling term of order $\alpha(Z\alpha)$ can be calculated employing standard techniques [5,16]. The Wichmann-Kroll terms of order $\alpha(Z\alpha)^n$ with $n = 3, 5, 7, \ldots$ are more difficult to evaluate. Their influence on binding energies of electrons, muons and pions can be deduced from refs. [6,7,8].

According to the Feynman rules [17] the energy shift for an electron state $\Phi_n(\vec{x}_2)$ caused by the vacuum polarization diagram in figure 2 is given by

$$\Delta E = \int d\vec{x}_2\, \Phi_n^+(\vec{x}_2)\, \Phi_n(\vec{x}_2)\, U(\vec{x}_2) \tag{3}$$

with the effective potential

$$U(\vec{x}_2) = 4\pi i\alpha \int d(t_2 - t_1) \int d\vec{x}_1\, D_F(x_2 - x_1)\, \mathrm{Tr}[\gamma_0 S_F(x_1, x_1)]. \tag{4}$$

$D_F(x_2 - x_1)$ denotes the free photon propagator

$$D_F(x_2 - x_1) = \frac{-i}{(2\pi)^4} \int d^4k\, \frac{\exp\{-ik(x_2 - x_1)\}}{k^2 + i\epsilon}. \tag{5}$$

The Feynman propagator $S_F(x_1, x_2)$ of the electron results from the solution of the inhomogeneous Dirac equation. To account for the Coulomb field to all orders in $Z\alpha$ we start from the Dirac equation already incorporating the external potential A_μ. Thus we have as defining equation

$$[\gamma^\mu (i\partial_\mu - eA_\mu(x_1)) - m] S_F(x_1, x_2) = \delta^4(x_1 - x_2). \tag{6}$$

A representation of the full electron propagator $S_F(x_2, x_1)$ in terms of occupied ($n+$) and unoccupied ($n-$) states is provided by

$$S_F(x_1, x_2) = \Theta(t_2 - t_1) \sum_{n+} \Phi_n(x_2)\overline{\Phi}_n(x_1) - \Theta(t_1 - t_2) \sum_{n-} \Phi_n(x_2)\overline{\Phi}_n(x_1). \tag{7}$$

After regularization [6] the finite vacuum polarization potential can be expressed as a Coulomb integral over the vacuum polarization charge density $\rho(\vec{x}_1)$

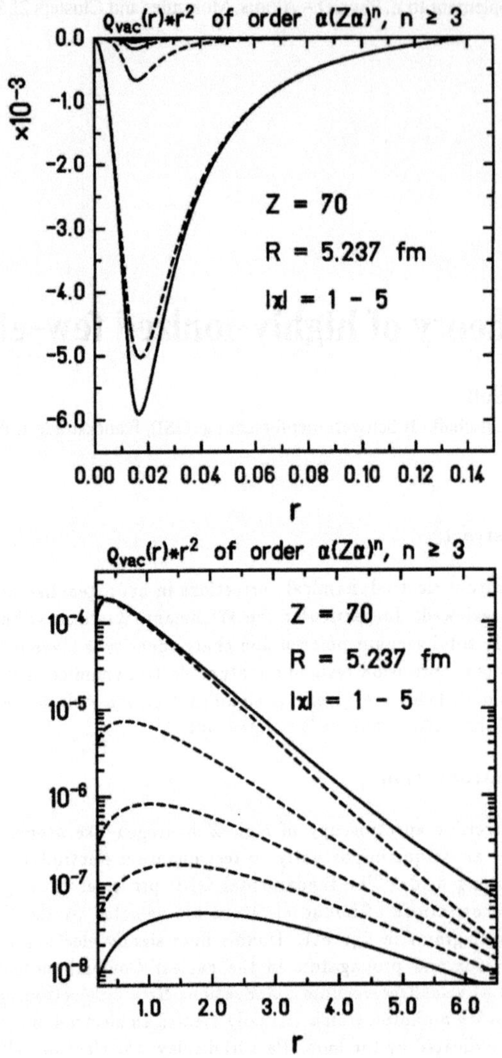

Figure 3: Radial vacuum polarization charge density $\rho_{|\kappa|} \cdot r^2$ of order $\alpha(Z\alpha)^n$ with $n \geq 3$ for the system $Z = 70$ with a nuclear radius $R = 5.237$ fm versus the radial coordinate r in natural units. The various contributions for $|\kappa| = 1$, 2, 3, 4 and 5 are shown separately by the dashed lines. $\rho_{|\kappa|} \cdot r^2$ is given in units of the elementary charge e. The solid line indicates the sum $\rho \cdot r^2$ of the various angular momentum components. a) Linear scale for the range in which the charge density is negative. b) Logarithmic scale to demonstrate the large distance behaviour of $\rho_{|\kappa|} \cdot r^2$. Here the vacuum polarization charge density is positive.

$$U(\vec{x}_2) = -e \int d\vec{x}_1\, \frac{\rho(\vec{x}_1)}{|\vec{x}_2 - \vec{x}_1|}. \tag{8}$$

In figure 3 we show the computed vacuum polarization charge density of order $\alpha(Z\alpha)^n$ with $n \geq 3$ for a nuclear charge number $Z = 70$. A spherical shell with radius $R = 5.237$ fm is

n	l	j	$\Delta E_{VP}[\alpha(Z\alpha)]$	$\Delta E_{VP}[\alpha(Z\alpha)^3]$
6	0	1/2	-6.053E-01	2.665E-02
6	1	1/2	-1.316E-01	7.985E-03
6	1	3/2	-6.768E-03	1.030E-03
6	2	3/2	-3.202E-04	7.747E-05
6	2	5/2	-5.949E-05	2.876E-05
6	3	5/2	-1.059E-06	1.701E-06
6	3	7/2	-2.787E-07	1.164E-06
6	4	7/2	-2.080E-09	1.336E-07
6	4	9/2	-6.467E-10	1.200E-07
6	5	9/2	-1.707E-12	2.045E-08
6	5	11/2	-5.861E-13	1.963E-08
7	0	1/2	-3.764E-01	1.657E-02
7	1	1/2	-8.231E-02	4.989E-03
7	1	3/2	-4.289E-03	6.514E-04
7	2	3/2	-2.094E-04	5.040E-05
7	2	5/2	-3.909E-05	1.873E-05
7	3	5/2	-7.573E-07	1.166E-06
7	3	7/2	-1.998E-07	7.910E-07
7	4	7/2	-1.807E-09	9.682E-08
7	4	9/2	-5.631E-10	8.645E-08
7	5	9/2	-2.383E-12	1.651E-08
7	5	11/2	-8.194E-13	1.576E-08
7	6	11/2	-1.310E-15	3.627E-09
7	6	13/2	-4.806E-16	3.547E-09
8	0	1/2	-2.497E-01	1.098E-02
8	1	1/2	-5.480E-02	3.319E-03
8	1	3/2	-2.883E-03	4.372E-04
8	2	3/2	-1.436E-04	3.444E-05
8	2	5/2	-2.689E-05	1.281E-05
8	3	5/2	-5.482E-07	8.223E-07
8	3	7/2	-1.449E-07	5.552E-07
8	4	7/2	-1.459E-09	7.048E-08
8	4	9/2	-4.554E-10	6.268E-08
8	5	9/2	-2.398E-12	1.267E-08
8	5	11/2	-8.253E-13	1.205E-08
8	6	11/2	-2.176E-15	3.054E-09
8	6	13/2	-7.991E-16	2.978E-09
8	7	13/2	-8.323E-19	8.394E-10
8	7	15/2	-3.199E-19	8.280E-10

n	l	j	$\Delta E_{VP}[\alpha(Z\alpha)]$	$\Delta E_{VP}[\alpha(Z\alpha)^3]$
9	0	1/2	-1.739E-01	7.650E-03
9	1	1/2	-3.827E-02	2.317E-03
9	1	3/2	-2.028E-03	3.073E-04
9	2	3/2	-1.024E-04	2.450E-05
9	2	5/2	-1.922E-05	9.118E-06
9	3	5/2	-4.051E-07	5.975E-07
9	3	7/2	-1.072E-07	4.021E-07
9	4	7/2	-1.155E-09	5.221E-08
9	4	9/2	-3.609E-10	4.632E-08
9	5	9/2	-2.156E-12	9.672E-09
9	5	11/2	-7.427E-13	9.188E-09
9	6	11/2	-2.487E-15	2.449E-09
9	6	13/2	-9.138E-16	2.384E-09
9	7	13/2	-1.604E-18	7.296E-10
9	7	15/2	-6.171E-19	7.184E-10
9	8	15/2	-4.420E-22	2.353E-10
9	8	17/2	-1.759E-22	2.332E-10
10	0	1/2	-1.259E-01	5.538E-03
10	1	1/2	-2.776E-02	1.680E-03
10	1	3/2	-1.479E-03	2.240E-04
10	2	3/2	-7.539E-05	1.801E-05
10	2	5/2	-1.418E-05	6.710E-06
10	3	5/2	-3.059E-07	4.459E-07
10	3	7/2	-8.108E-08	2.995E-07
10	4	7/2	-9.142E-10	3.949E-08
10	4	9/2	-2.858E-10	3.498E-08
10	5	9/2	-1.852E-12	7.460E-09
10	5	11/2	-6.384E-13	7.078E-09
10	6	11/2	-2.463E-15	1.944E-09
10	6	13/2	-9.054E-16	1.890E-09
10	7	13/2	-2.052E-18	6.054E-10
10	7	15/2	-7.896E-19	5.954E-10
10	8	15/2	-9.702E-22	2.094E-10
10	8	17/2	-3.862E-22	2.073E-10
10	9	17/2	-1.984E-25	7.632E-11
10	9	19/2	-8.107E-26	7.584E-11

Table 1: Energy shifts ΔE_{VP} in units of eV for electron states with quantum numbers n, l and j in hydrogen-like uranium caused by vacuum polarization corrections of order $\alpha(Z\alpha)$ and $\alpha(Z\alpha)^3$. (1.0E+01 = $1.0 \cdot 10^1$).

assumed to describe the extended nucleus. The dashed lines indicate the various contributions arising from the expansion of the electron Green function in terms of the Dirac angular momentum quantum number κ. Obviously the term with $\kappa = \pm 1$ dominates. The minimum of the induced charge density is located at around the nuclear radius. Close to the positively charged nucleus the computed vacuum polarization charge density is negative as expected intuitively. For distances $r \gtrsim \lambda_e$ with the electron Compton wavelength $\lambda_e \simeq 386$ fm the charge density assumes positive values.

As required by charge conservation the total induced vacuum polarization charge vanishes

$$\int_0^\infty \rho(r)\, r^2\, dr = 0\,. \tag{9}$$

Numerically this could be achieved with a relative accuracy of about $10^{-5} - 10^{-6}$. For hydrogen-like uranium the 1s binding energy amounts to 131 823.2 eV. The total 1s Lamb shift is predicted to be 457.6 eV [5,6,10,11]. The Wichmann-Kroll contribution of the vacuum polarization effect yields an 1s energy correction of 5 eV [6]. In view of the aspired experimental precision of about 0.1 eV the Wichmann-Kroll term can be identified with a relative accuracy of about 2 %.

III. Radiative corrections for Rydberg electrons

From fig. 3 we can deduce that the vacuum polarization charge density of order $\alpha(Z\alpha)^n$ with $n \geq 3$ does not decline exponentially for $r \to \infty$. The contribution of order $\alpha(Z\alpha)^3$ displays a r^{-7}-dependence at large distances. The corresponding vacuum polarization potential $V_{13}(r)$ diminishes as [18,19]

$$V_{13}(r) \overset{r\to\infty}{=} \frac{\alpha(Z\alpha)^3}{\pi r}\frac{32}{225}\frac{1}{(2r)^4}\,. \tag{10}$$

For large r this causes the interesting consequence that V_{13} exceeds the Uehling potential V_{11} of order $\alpha(Z\alpha)$ which falls off exponentially for radial distances being much larger than the electron Compton wavelength. We evaluated [20] the energy shift of Rydberg electrons produced by the vacuum polarization potential of order $\alpha(Z\alpha)$ and $\alpha(Z\alpha)^3$. The results for electron states with principal quantum numbers between $n = 6$ and $n = 10$ in hydrogen-like uranium ($Z = 92$) are presented in table 1. Already for the $6d$-state the striking long-range dependence (10) leads to a dominance of the Wichmann - Kroll correction of order $\alpha(Z\alpha)^3$ over the Uehling correction of order $\alpha(Z\alpha)$. However, the total energy shifts are rather tiny which may hinder an immediate experimental verification.

IV. Comparison with experimental data

In figure 4 the theoretical values for the 1s Lamb shift in hydrogen-like ions are compared with available experimental data [21-28]. The energy shift is given in units of $\alpha/\pi\,(Z\alpha)^4\,mc^2$. This scaling displays the major advantage that the total Lamb shift is just a number between 1 and 10 for all elements in our periodic system. The finger points to precise determination [27] of the 1s Lamb shift in hydrogen-like argon ($Z = 18$). Also the measurements of the Lamb shift in hydrogen-like nickel ($Z = 28$) [28] and krypton ($Z = 36$) [26] are in agreement with the theoretical result (solid line).

The ultimate aim of this experimental project is the measurement of the 1s Lamb shift in hydrogen-like lead ($Z = 82$) or uranium ($Z = 92$) with maximum precision. However, for very heavy elements internal degrees of freedom of the nucleus provide severe limitations for precision tests of QED. Besides large error bars for nuclear radii the nuclear polarization causes major uncertainties in level shifts of strongly bound electrons. The influence of nuclear polarization is extensively discussed in refs. [11-13]. In particular $^{238}_{92}$U is a strongly deformed nucleus with a low-lying first excited state. Consequently this nucleus is easy to polarize. For hydrogen-like uranium we obtained an associated energy shift of 1 eV. The corresponding effect for the spherical double-magic nucleus $^{208}_{82}$Pb is by an order of magnitude smaller. Here we found $\Delta E_{1s}^{nucpol} = 0.11$ eV. In conclusion measurements of radiative corrections in strong Coulomb fields should be performed with hydrogen-like lead rather than with uranium.

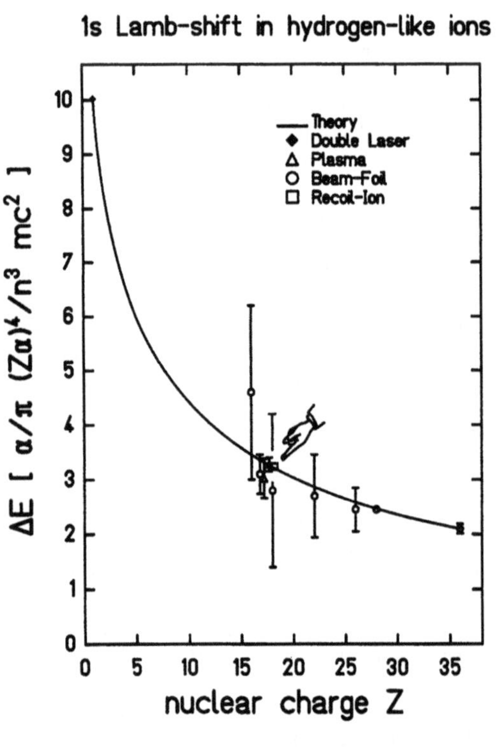

Figure 4: Comparison of theoretical results for the total 1s Lamb shift with available experimental data. The finger points to a precise experimental result of ref. 27.

V. Parity violation in helium-like uranium

To observe parity violation effects in atomic systems two essential conditions have to be fulfilled. Firstly, electron wavefunctions should exhibit a considerable overlap with the nuclear interior. Parity violation is caused by the exchange of intermediate vector bosons. Due to their large mass the weak interaction is of extreme short-range nature. Secondly, to allow for any parity admixture two almost degenerated states with opposite parity should exist. By chance these prerequisites are accomplished in helium-like uranium [29].

To calculate the energy eigenvalues of bound electrons we employed the multiconfiguration Dirac-Fock method. One starts from the Hartree-Fock description of a many-electron system in which the many-electron wavefunction $\Psi(\vec{r}_1,\vec{r}_2,\ldots,\vec{r}_N)$ is approximated by an antisymmetric product of single-particle wavefunctions $\Phi(\vec{r}_i)$ [30], i.e.

$$\Psi(\vec{r}_1,\vec{r}_2,\ldots,\vec{r}_N) \;\cong\; \mathcal{A}\,\Phi(\vec{r}_1)\,\Phi(\vec{r}_2)\ldots\Phi(\vec{r}_N)\,. \tag{11}$$

The single-particle states in their spherical representation read

$$\Phi_{n\kappa m}(r,\theta,\phi) \;=\; \frac{1}{r}\begin{pmatrix} G_{n\kappa}(r) & \chi_{\kappa m}(\theta,\phi) \\ i\,F_{n\kappa}(r) & \chi_{-\kappa m}(\theta,\phi)\end{pmatrix}\,. \tag{12}$$

The variational principle is employed in the multiconfiguration method. The variational space consists of linear combinations of configuration state functions which are classified by parity Π, total angular momentum J and z-projection M

$$\Psi(\Pi JM) \;=\; \sum_{\nu=1}^{NCF} c_\nu\,|\nu\Pi JM>\,. \tag{13}$$

The sum is performed over the number of configurations NCF being characterized by ν. c_ν denotes the mixing coefficients. The total Hamiltonian contains a sum over single-particle Hamiltonians as well as a sum over electron-electron interactions $g(i,j)$.

$$H \;=\; \sum_i h_D(i) + \sum_{i,j} g(i,j) \tag{14}$$

with

$$h_D \;=\; c\vec{\alpha}\cdot\vec{p} + \beta mc^2 + V_{nuc}(r)\,. \tag{15}$$

In lowest order the electron-electron interaction incorporates the $1/R$ Coulomb interaction as well as the Breit interaction which accounts for retardation and magnetic interaction. In Coulomb gauge it reads [30]

$$g(1,2) \;=\; \frac{1}{R} - \frac{\vec{\alpha}_1\cdot\vec{\alpha}_2}{R}\cos(\omega R) + (\vec{\alpha}_1\cdot\vec{\partial}_1)(\vec{\alpha}_2\cdot\vec{\partial}_2)\frac{\cos(\omega R)-1}{\omega^2 R}\,. \tag{16}$$

There are published computer codes [31,32] to evaluate relativistic atomic structure properties within the multiconfiguration Dirac-Fock method. Another successful approach to treat theoretically few-electron systems is the relativistic many-body perturbation theory [33-35] which can be derived directly from bound-state QED.

As binding energy for the $^1S_0(1s,2s)$-state in helium-like uranium taking into account also QED corrections we obtained -165114.406 eV, for the $^3P_0(1s,2p)$-state with opposite parity it results -165113.5 eV. The energy difference amounts to

$\Delta E = -0.906$ eV. According to Drake [14] one finds $\Delta E = 1.816$ eV. However, one has to emphasize that the uncertainty in this energy difference is comparable to its magnitude. Due to the presence of the 1s-state in these configurations a large overlap with the nuclear interior is guaranteed.

In lowest order perturbation theory the parity admixture is given by

$$\eta = \frac{\langle 2\,^3P_0 \left| \frac{G_F}{2\sqrt{2}} \left(1 - 4\sin^2\theta_W - \frac{N}{Z}\right) \rho_{el}\gamma_5 \right| 2\,^1S_0 \rangle}{E(2\,^3P_0) - E(2\,^1S_0)}. \qquad (17)$$

G_F is the Fermi constant of the electroweak theory, θ_W the Weinberg angle, N the neutron number of the nucleus, Z the proton number and ρ_{el} the electric charge density. As final result we computed for the parity admixture

$$|\eta| = \frac{4.62 \cdot 10^{-6}\ \text{eV}}{|\Delta E|}. \qquad (18)$$

For the discussion of experimental prospects of a parity violation experiment we refer to ref. [29]. Similar crossings of the $^3P_0$ and the $^1S_0$-state occur for $Z \simeq 64$ and $Z \simeq 6$.

VI. Summary

The major insights can be summarized by the follwing items:

(a) Precise measurements of the Lamb-shift in high-Z hydrogen-like systems represent a critical test of quantum electrodynamics of strong Coulomb fields [36].

(b) However, precision tests of quantum electrodynamics are bounded by nuclear degrees of freedom. E.g., for very heavy elements the uncertainty of nuclear radii and internal nuclear excitations represent severe limitations for Lamb-shift measurements with relative accuracies better than 10^{-3}.

(c) Theoretical predictions for radiative corrections are provided. Future improvements should imply a more accurate determination of the nuclear-size correction on the self energy as well as self-energy corrections of higher order incorporating the exchange of two or more virtual photons.

(d) There is a significant parity violation effect in helium-like uranium.

(e) Precise measurements of transition energies in few-electron systems can be untilized to test the theory of relativistic atomic structure. Highly charged ions are an ideal testing ground for relativistic many-body theories.

Acknowledgement: The author is very grateful for the fruitful cooperation with Walter Johnson (Notre Dame University) and with Peter Mohr (National Institute of Standards and Technology, Gaithersburg).

References

1. W. Greiner, B. Müller, J. Rafelski, Quantum electrodynamics of strong fields, (Springer, Berlin, 1985)
2. C. Itzykson, J.-B. Zuber, Quantum Field Theory, (McGraw-Hill, New York, 1980)
3. D.C. Ionescu, J. Reinhardt, B. Müller, W. Greiner, G. Soff, Phys. Rev. A38, 616 (1988)
4. P.J. Mohr, At. Data Nucl. Data Tables 29, 453 (1983)
5. W.R. Johnson, G. Soff, At. Data Nucl. Data Tables 33, 405 (1985)
6. G. Soff, P. Mohr, Phys. Rev. A38, 5066 (1988)
7. G. Soff, Z. Physik D11, 29 (1989)
8. J.M. Schmidt, G. Soff, P.J. Mohr, Phys. Rev. A40, 2176 (1989)
9. H. de Vries, C.W. de Jager, C. de Vries, At. Data Nucl. Data Tables 36, 495 (1987)
10. T. Beier, G. Soff, Z. Physik D8, 129 (1988)
11. G. Plunien, B. Müller, W. Greiner, G. Soff, Phys. Rev. A39, 5428 (1989)
12. G. Plunien, Kernpolarisation in schweren Atomen und superschweren Quasiatomen, Report GSI-89-17
13. G. Plunien, B. Müller, W. Greiner, G. Soff, Nuclear Polarization in heavy atoms and superheavy quasiatoms, Preprint GSI-90-41
14. G.W.F. Drake, Can. J. Phys. 66, 586 (1988)
15. E.H. Wichmann, N.M. Kroll, Phys. Rev. 101, 843 (1956)
16. S. Klarsfeld, Phys. Lett. 66B, 86 (1977)
17. S.S. Schweber, An Introduction to Relativistic Quantum Field Theory, (Harper & Row, New York, 1961)
18. J. Blomqvist, Nucl. Phys. B48, 95 (1972)
19. K.-N. Huang, Phys. Rev. A14, 1311 (1976)
20. G. Soff, P.J. Mohr, Phys. Rev. A40, 2174 (1989)
21. C. Wieman, T. Hänsch, Phys. Rev. A22, 192 (1980)
22. L. Schleinkofer, F. Bell, H.D. Betz, G. Trollmann, J. Rothermel, Physica Scripta 25, 917 (1982)
23. J.P. Briand, J.P. Mosse, P. Indelicato, P. Chevallier, D. Girard-Vernet, A. Chetioui, M.T. Ramos, J.P. Desclaux, Phys. Rev. A28, 1413 (1983)
24. E. Källne, J. Källne, P. Richard, M. Stöckli, J. Phys. B17, L115 (1984)
25. P. Richard, M. Stöckli, R.D. Deslattes, P. Cowan, R.E. LaVilla, B. Johnson, K. Jones, M. Meron, R. Mann, K. Schartner, Phys. Rev. A29, 2939 (1984)
26. J.P. Briand, P. Indelicato, M. Tavernier, O. Gorceix, D. Liesen, H.F. Beyer, B. Liu, A. Warczak, J.P. Desclaux, Z. Physik A318, 1 (1984)
27. H.F. Beyer, R.D. Deslattes, F. Folkmann, R.E. LaVilla, J. Phys. B18, 207 (1985)
28. H.F. Beyer, P. Indelicato, K.D. Finlayson, D. Liesen, R.D. Deslattes, Measurement of the 1s Lamb shift in hydrogen-like nickel, preprint 1990
29. A. Schäfer, G. Soff, P. Indelicato, B. Müller, W. Greiner, Phys. Rev. A40, 7362 (1989)
30. J.P. Desclaux, in: Relativistic, quantum electrodynamic and weak interaction effects in atoms, AIP conference proceedings 189, eds.: W. Johnson, P. Mohr, J. Sucher, p. 265, (American Institute of Physics, New York, 1989)
31. J.P. Desclaux, Comput. Phys. Commun. 9, 31 (1975)
32. K.G. Dyall, I.P. Grant, C.T. Johnson, F.A. Parpia, E.P. Plumer, Comput. Phys. Commun. 55, 425 (1989)
33. J. Sapirstein, Physica Scripta 36, 801 (1987)
34. W.R. Johnson, S.A. Blundell, J. Sapirstein, Phys. Rev. A38 2699, 2764 (1988)
35. S.A. Blundell, W.R. Johnson, J. Sapirstein, Phys. Rev. A41, 1698 (1990)
36. G. Soff, Quantum electrodynamical corrections in critical fields, Preprint GSI-90-51

Supplement to Z. Phys. D - Atoms, Molecules and Clusters 21, S13-16 (1991)

Atoms, Molecules
Zeitschrift für Physik D and Clusters

Precision spectroscopic measurements in few-electron ions

R. W. Dunford, H. G. Berry, D. A. Church*, T. P. Dinneen, M. Hass+, C. J. Liu, N. Berrah-Mansour, R. C. Pardo, M. L. A. Raphaelian, L. Young, and B. J. Zabransky
Physics Division, Argonne National Laboratory, Argonne, IL 60439, USA

L. J. Curtis
Department of Physics and Astronomy, University of Toledo, Toledo, OH 43606, USA

Abstract

We describe recent precision experiments in few-electron ions including measurements of the lifetimes of two-photon-emitting levels in Ni^{26+} and Ni^{27+}, a measurement of the lifetime of the 2^3S_1 level in Br^{33+} and measurements of the $2^3S_1 \rightarrow 2^3P_{0,1,2}$ transition energies in B^{3+}.

I. Introduction

Few-electron systems, particularly hydrogen-like and helium-like ions provide an important testing ground for our understanding of atomic structure. In the one-electron systems, there are no calculational uncertainties, so experiments in these systems provide tests of the underlying theory of Quantum Electrodynamics (QED). Current interest is in the question of whether there is a breakdown of QED in highly-charged one-electron atoms. The theory of the two-electron system is not as well understood. Here one has the additional problem of treating correlations between the two electrons and a complete Hamiltonian for the system does not exist [1]. The importance of the helium-like systems is that they are the simplest atoms in which the electron-electron correlations must be taken into account.

In this paper we describe four recent experiments done by our group which study the structure of one- and two-electron ions. The first two experiments are measurements of the lifetimes of the $2^2S_{1/2}$ level in one-electron Ni^{27+} and the 2^1S_0 level in two-electron Ni^{26+}, both of which decay primarily by two-photon emission. The third experiment is a measurement of the lifetime of the 2^3S_1 level in helium-like Br^{33+}, which decays to the ground state via a M1 transition. The final experiment is a precision measurement of the $2^3S_1 - 2^3P_{0,1,2}$ transition energies in two- electron B^{3+}. This experiment was done by driving resonances in a fast beam of metastable B^{3+} using a frequency doubled dye laser.

*Present address: Texas A&M University, College Station, TX 77843.

+Present address: University of Rochester, Rochester, NY 14627.

II. Two-Photon Decays in One- and Two-Electron Nickel

The 2^1S_0 level (see Fig. 1) in helium-like ions is forbidden to decay to the 1^1S_0 ground state by the emission of a single photon and so it decays via two-photon emission. The $2^2S_{1/2}$ level in hydrogen-like ions also decays to the ground state via two-photon emission although there is also a small M1 branch (17% in Ni^{27+}). We have measured the lifetimes of these two-photon emitting levels in one- and two-electron nickel [2]. Our technique is to measure the rate of two-photon coincidences as the distance between our detectors and the beam-foil is varied.

ANL-P-19,734

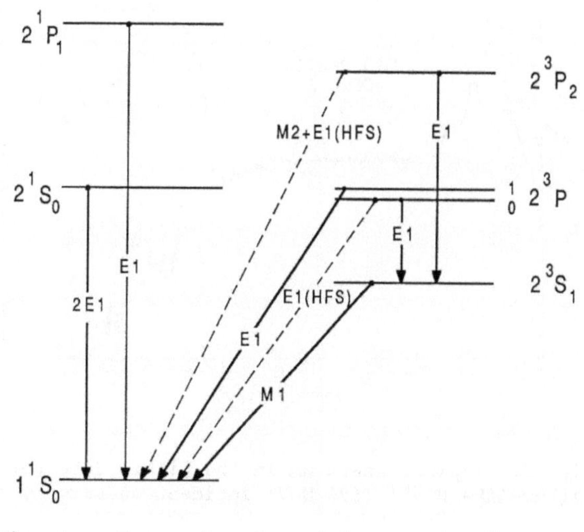

Fig. 1. Energy levels and decay modes for helium-like ions.

The data were taken using nickel beams from the Argonne Tandem-Linac (ATLAS). After acceleration, the ions were stripped in a thick carbon foil and the 26+ or the 28+ charge state was directed to the target area. The beam energy was measured by a time-of-flight energy analyzer and corrected for energy loss in the stripper foils [3]. In the target area, a thin carbon target (12 $\mu g/cm^2$) is moved relative to three fixed Si(Li) detectors using a precision translation stage. Two of the Si(Li) detectors are collimated so they observe a region of length 5 mm along the beam path and subtend a solid angle of 0.1% at the beam. The third detector is not highly collimated and accepts photons into a solid angle of 1%. There is also a lower resolution silicon X-ray detector attached to the target holder which is used for normalization.

Events were recorded for coincidences between any two of the three Si(Li) detectors, but most of the coincidences are between one of the collimated detectors and the large solid-angle detector. The collimators serve to localize the events in position along the beam while the large solid-angle detector provides a good efficiency for detecting the second photon. In Fig. 2 we show a typical singles spectrum for one of the Si(Li) detectors for the case of Ni^{26+} incident on the target. The broad continuum extending out to 7.8 keV is mostly due to the decay of the 2^1S_0 level in helium-like Ni^{26+}. In Fig. 3 we show the sum-energy spectrum for true coincidences. In this case the sum of the energies of the two photons forms a peak at the transition energy. This peak is almost entirely from the helium-like two-photon decay. The spectrum taken with the 28+ beam is predominantly from the two-photon decay of hydrogen-like nickel but there is a small contamination from the helium-like sum energy peak which lies 300 eV lower in energy.

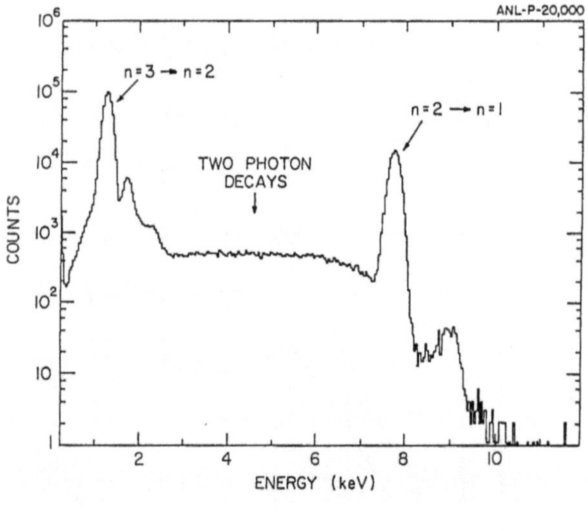

Fig. 2. Typical spectrum in the Si(Li) detector. Helium-like Ni^{26+} (670 MeV) incident on target.

The data are normalized using the count from the normalization detector and the decay curve is fit to a function with two parameters: a) the amplitude at the closest position and b) the lifetime. The lifetimes from these two component fits for each of the three detector pairs are combined and then corrected for time dilation to obtain the final result. A number of additional systematic effects were considered but found to be unimportant for these measurements including gain

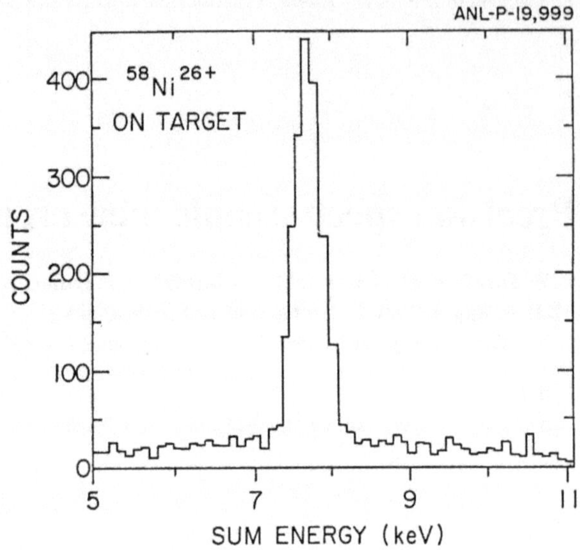

Fig. 3. Sum-energy spectrum for true coincidences with 670 MeV Ni^{26+} incident on target.

shifts, quenching in electric and magnetic fields, contributions from ions with high-lying spectator electrons [4] and backgrounds from coincidences caused by Compton scattering of Lyman-α photons between two of the Si(Li) detectors.

Two separate measurements were made for each of the two lifetimes. In each case, the measurements were consistent within errors and we simply combined them. For the helium-like decay the combined result is 156.1(1.6) ps which is in substantial agreement with the theoretical [5] value of 154.3(0.5) ps. For the decay of the $2^2S_{1/2}$ state in hydrogen-like Ni^{27+} we find a lifetime of 217.1(1.8) ps which agrees with the theoretical value [6] of 215.45 ps.

III. M1 Decay of the 2^3S_1 level $^{79}Br^{33+}$

The same apparatus has been used to measure the lifetime [7] of the 2^3S_1 level in helium-like $^{79}Br^{33+}$. This level decays to the 1^1S_0 ground state by a forbidden M1 transition. Bromine-79 is a good choice for the measurement of the lifetime of the helium-like 2^3S_1 level because, in this isotope, all of the other n=2 levels which decay by single-photon emission are much shorter lived and so there is a minimum of complication from the decay of other levels. An important factor here is that the $^3P_0$ level is quenched to the ground state by hyperfine mixing induced by the magnetic moment of ^{79}Br (I=3/2). This reduces the lifetime of this level from 1.48 ns (unquenched) to 5.6 ps (quenched).

The bromine beam was provided by the Argonne Tandem-Linac (ATLAS). The helium-like 33+ charge state is magnetically analyzed and directed to the experimental area. Two detectors were used for this experiment and they were collimated so that they observe a region 2 mm along the beam.

At each foil-detector separation, peaks near 12 keV in the Si(Li) detectors, which correspond to single-photon decays from the n=2 levels of Br^{33+} were fitted to a Gaussian line shape. The peak areas at each distance were divided by the count for the normalization detector and these data were fitted to a decay curve to obtain the lifetime. Our final result is $\tau(2^3S_1)$=224.1(7.1) ps, which agrees with the calculation by Drake [8] which gives 230(2) ps and the calculation by Johnson and Lin [9] which gives 228.2 ps.

IV. Laser Spectroscopy

Spectroscopic measurements in helium-like ions include high-precision measurements in helium [10] and lithium [11] which test the nonrelativistic parts of the interaction, and lower precision measurements in heavy ions which test the relativistic and QED corrections. A complete list of recent experiments is given by Drake [12] along with his most recent calculations. The measurements in lithium test relativistic corrections to higher precision than measurements in helium but are limited at present by uncertainties in the nonrelativistic calculations of the energy. We have recently completed precision laser measurements of the 2s to 2p transitions in B^{3+} where the relativistic and QED corrections are a larger fraction of the transition energy so they can be tested with less uncertainty arising from the nonrelativistic contribution to the calculations.

Metastable B^{3+} ions were produced in the Positive Ion Injector Electron Cyclotron Resonance (PII-ECR) source at ATLAS. The source was operated at 10 kV extraction voltage. Beam currents in the interaction region varied from 0.5 to 1.5 μA. Transitions from 2^3S_1 to $2^3P_{0,1,2}$ were driven by 282 nm laser light produced by intra-cavity frequency doubling of the 564 nm fundamental of a Coherent 699 ring dye laser. The maximum doubled output obtained from the dye laser was 7 mW.

The ion beam and laser beam were overlapped to take advantage of the increased interaction time and the velocity compression provided by the accelerated beam. The Doppler width (FWHM) was 1 GHz, corresponding to a beam energy spread of 21 eV. The transitions were monitored by detecting the fluorescence radiation emitted as the 2^3P levels decay back to the 2^3S_1 level. The sensitivity of the experiment was sufficient to allow most of the hyperfine components of $^{11}B^{3+}$ (nuclear spin I=3/2) to be observed. In order to make the Doppler correction, the laser light was reflected back through the interaction region and the laser was scanned over both the red and the blue Doppler-shifted components while simultaneously measuring an iodine absorption spectrum with the fundamental output of the dye laser. After making an absolute wavelength determination for one of the 2^3P_2 hyperfine transitions, all other transitions were measured relative to it.

To obtain hyperfine-free wavelengths it is necessary to measure all of the hyperfine components. Since the laser power stability was poor over large frequency scan ranges, a different technique was used to measure the hyperfine structure. This involved using the interaction region as a post acceleration section, and, keeping the laser frequency fixed, the ion-beam velocity was tuned through the resonances by varying the interaction region voltage. In this way all hyperfine components were obtained in a single spectrum as shown in Fig. 4.

The experiment resulted in precision laser measurements of the fine structure, the hyperfine structure and absolute wavelengths of the $(1s2s)^3S_1$ to $(1s2p)^3P_{0,1,2}$ transitions in helium-like boron. These results are being prepared for publication [13]. The measurements provide a precision at the level of 0.1% of the one-electron Lamb shift for the 2s → 2p transition.

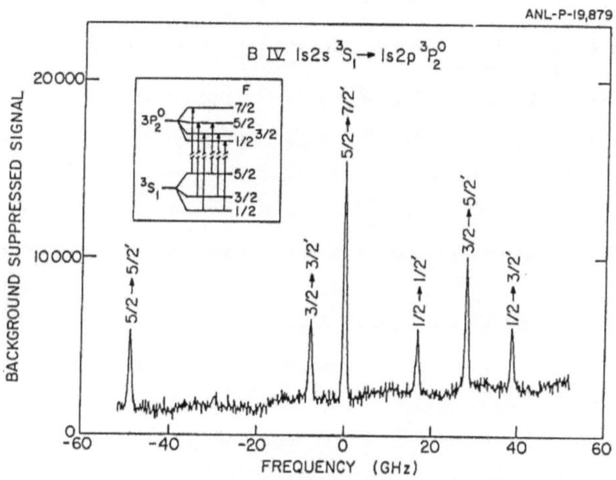

Fig. 4. Nearly complete hyperfine spectrum of the 2^3S_1 → 2^3P_2 transition in B^{3+}. The resonances were scanned by varying the potential in the light collection region from 0 to -800 volts.

V. Conclusion

We have discussed a number of experiments that are sensitive to the relativistic and QED effects in one- and two-electron ions. The measurements of the lifetimes of the two- photon emitting states in Ni^{26+} and Ni^{27+} used the coincidence technique which resulted in simple-decay curves that could be fit with only two parameters. In the future we will apply the coincidence technique to more highly-charged one- and two-electron ions. The experiment to measure the lifetime of the 2^3S_1 level in $^{79}Br^{33+}$ illustrates the potential of utilizing the hyperfine quenching effect to remove unwanted components from a blended line. We also described the first laser measurement of the 2^3S_1 → $2^3P_{0,1,2}$ transition in the B^{3+} system. With improved ion-beam intensity and metastable yield, there is potential for significant improvement in the precision of this experiment. We will also utilize this method to make measurements on other metastable helium-like ions which can be produced by our ECR ion source.

This work was supported by the U. S. Department of Energy, Office of Basic Energy Sciences, under Contract No. W-31-109-ENG-38.

References:

1. Lindgren, I., Nuc. Inst. Meth. Phys. Res. 31, 102 (1988).

2. Dunford, R. W., M. Hass, E. Bakke, H. G. Berry, C. J. Liu, and M. L. A. Raphaelian, L. J. Curtis, Phys. Rev. Lett. 62, 2809 (1989).

3. Ziegler, J. F., Handbook of Stopping Cross Sections for Energetic Ions in all Elements (Pergamon, New York, 1980).

4. Lin, D. L. and L. Armstrong, Phys. Rev. A 16, 791 (1977).

5. Drake, G. F. W., Phys. Rev. A 34, 2871 (1986).

6. Parpia, F. A. and W. R. Johnson, Phys. Rev. A 26, 1142 (1982).

7. Dunford, R. W., D. A. Church, C. J. Liu, H. G. Berry, M. L. A. Raphaelian, M. Hass, L. J. Curtis, Phys. Rev. A. 41, 4109 (1990).

8. Drake, G. W. F., Phys. Rev. A 3, 908 (1971); G. W. F. Drake (private communication).

9. Johnson, W. R. and C. P. Lin, Phys. Rev. A 9, 1486 (1974).

10. Zhao, P., J. R. Lawall, A. W. Kam, M. D. Lindsay, F. M. Pipkin, W. Lichten Phys. Rev. Lett. 63, 1593 (1989).

11. Riis, E., H. G. Berry, O. Poulsen, S. A. Lee, S. Y. Tang, Phys. Rev. A 33, 3023 (1986).

12. Drake, G. F. W., Can. J. Phys. 66, 586 (1988).

13. Dineen, T. P., N. B. Mansour, H. G. Berry, L. Young and R. C. Pardo, in preparation.

Supplement to Z. Phys. D - Atoms, Molecules and Clusters 21, S 17-21 (1991)

Atoms, Molecules
Zeitschrift für Physik D and Clusters

Atomic spectroscopy in highly ionised plasmas

H. P. Summers, P. Thomas, R. Giannella, M. von Hellermann, W. Dickson, K. Lawson, W. Mandl, P. Briden and members of Experimental Division II, JET.
JET Joint undertaking, Abingdon, Oxon, U.K.

Abstract

Observations of the JET tokamak are used to illustrate contrasting spectral emission from a range of plasma environments differing in temperature, density, interaction with boundary surfaces, dynamic state and excitation mechanisms. Some spectral features of diagnostic value are identified. Models based on collisional radiative theory are used to investigate these features. The generality of this approach to modelling the radiating properties of atoms in arbitrary dynamic plasmas is indicated.

I. Introduction

The magnetic confinement fusion research machine, JET (Joint European Torus), sustains a plasma composed normally of deuterium, minor impurities such as beryllium, carbon and oxygen and traces of metals such as chromium and nickel at central electron temperatures $\lesssim$12 keV (typically 8 keV) and central electron densities $\lesssim 2\times10^{14}\mathrm{cm}^{-3}$ (typically 5×10^{13} cm^{-3}). On occasions further gaseous species such as helium are introduced and solid species are (laser) ablated into the plasma. The plasma is optically thin in all impurity radiation. It has strong similarities to astrophysical plasmas such as the solar corona together with some special environments, active control and diagnostic systems which make it a valid and very fruitful source for the study and exploitation of atomic spectroscopy in highly ionised plasmas[1]. This paper rests on a basis of JET spectroscopic observations, however it is believed that the methods used, reliance on and impact on fundamental experimental and theoretical studies of atomic cross-sections etc. are of general relevance.

Figure 1 shows the geometrical layout of JET. The existence of qualitatively distinct radiating zones in the JET plasma arises because of the magnetic field structure and because of the plasma limiting surfaces. Thus, although the bulk plasma, with particle confinement times of order 1 sec radiates in a manner typical of near equilibrium moderate density plasma (albeit at very high temperature), the behaviour of atoms and ions near the edge or in the shadow of limiters is much more dynamic influenced by sources, diffusion and precise plasma positioning. Then again, neutral deuterium heating beams cause profound modification of the radiating behaviour of impurities along or in the locality of the beam lines. Ion cyclotron resonance heating, solid deuterium pellet injection, lower hybrid current drive etc. all alter the radiation emission.

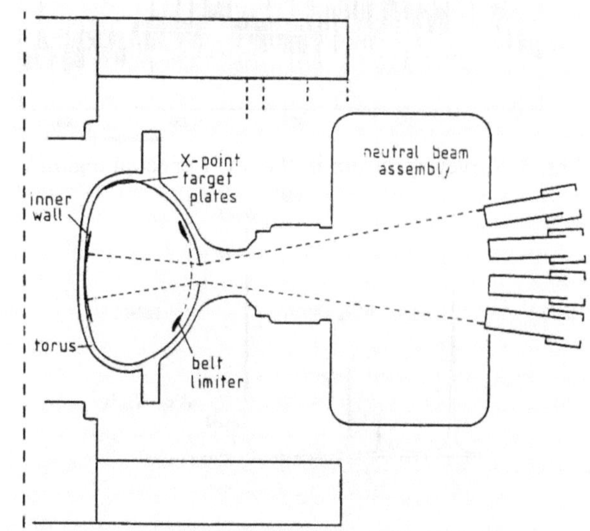

Fig. 1. A vertical section of JET. Note the graphite tile protected inner wall, beryllium belt limiters and the neutral beam injector (NBI) assembly.

II. Plasma and impurity spectral emission in JET

A VUV survey spectrum is shown in figure 2. It is a radial view across the plasma in the horizontal mid-plane. In this case molybdenum was introduced into a plasma of central electron temperature ~8 keV by laser ablation. Molybdenum is ionised to approximately the stage Mo^{+40} at this temperature. Lower ionisation stages occur in shells at progressively larger radii from the plasma axis and therefore lower temperature where a balance between ionisation and recombination processes with some influence of diffusion occurs. The deduction of the species concentration from measurement of the line of sight integrated emission in such spectrum lines is the *standard analysis* [2]. Beryllium by contrast is fully ionised throughout virtually the whole plasma. The BeIII emission at 100.26A occurs only at the periphery of the plasma close to the limiting surfaces which are the source of the beryllium. The electron temperature quite close to these surfaces is high compared with the ionisation potential of Be^{+2} and it is probable that Be^{+2} penetrates to such a highly ionising environment before ionisation. Thus the localisation of such an ion is very different resulting from a balance of ionisation and transport processes.

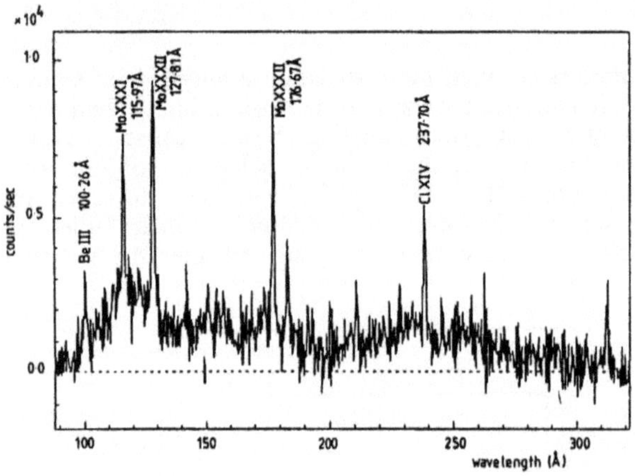

Fig. 2. Survey spectrum in the VUV spectral region.

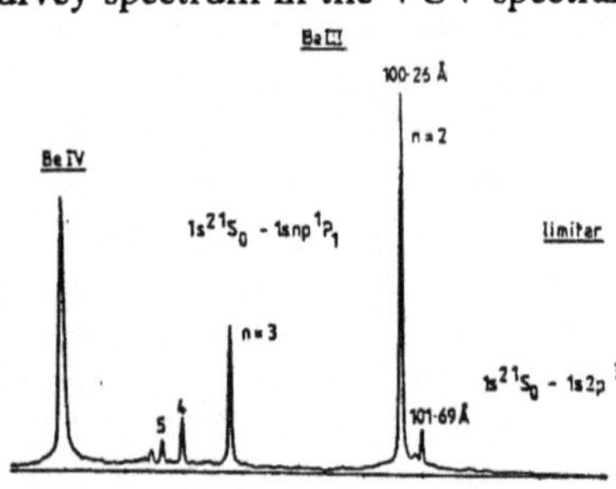

Fig. 3. XUV spectrum along a line of sight directed at the inner wall: (a) Plasma in contact with the outer belt limiters.(b) Plasma in contact with the inner wall.

The flux of impurities into the plasma and the plasma parameters local to the surfaces which influence impurity release are important matters [3]. Figures 3a and 3b show a high resolution XUV spectrum of BeIII when the plasma was out of and then in contact with the inner wall. Note the singlet series and the intercombination line in figure 3a. The Lyman series of BeIV extends to shorter wavelength, only the first member being present on the spectra. Dynamic state and temperature influence these spectral intensities. Figure 3b introduces a new phenomemon. Disturbance of the series decrement probably indicates the presence of charge transfer reactions from neutral deuterium released from the inner wall by the outward plasma ion flux to it [4].

Visible spectra taken along a horizontal line of sight intersecting and inclined to the neutral heating beam path are shown in figures 4a and 4b. The narrow spectral emission lines in the n = 6-5 principal quantum shell transition of BeIV and in BeII in figure 4b are edge localised emissions similar to those of BeIII in figure 2. The pronounced change in figure 4a in the BeIV lines arises from emission at the beam/viewing line intersection and is driven by charge exchange reactions of Be^{+4} in the core plasma and deuterium in the beams. This is diagnostic of conditions in the plasma centre [5].

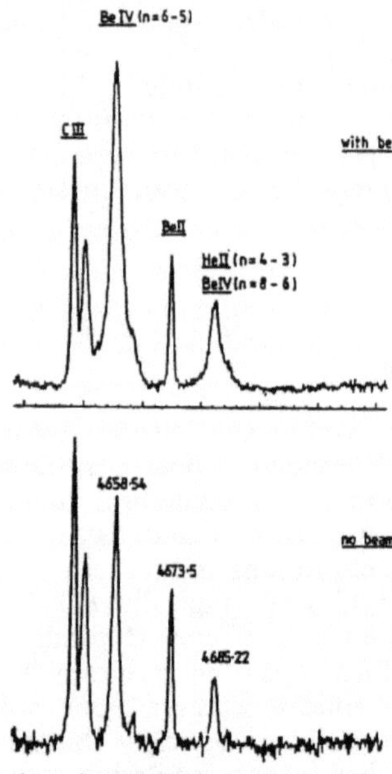

Fig. 4. Visible spectrum along a line of sight intersecting the neutral deuterium beams: (a) Beams active, (b) Beams inactive.

Finally figure 5 shows spectra along the same line of sight, but in the wavelength vicinity of Dα at times spanning beam switch on. The charge exchange driven feature at 6561A is similar to that shown in figure 4a above but note the Doppler shifted beam emission split into Stark components by the **vxB** electric field. The principal energy fraction of the beam is at 40 keV/amu, giving electric fields ~100kV/cm. There are also half and third energy components [6]. Excitation of deuterium in the beam is predominantly by plasma ion collisions. The narrow component at 6561A is plasma edge emission in the line of sight. For completeness it should be mentioned that observations of other impurity neutral atom emission, such as BeI and CrI, are also made. Such emission is strongly localised at source surfaces.

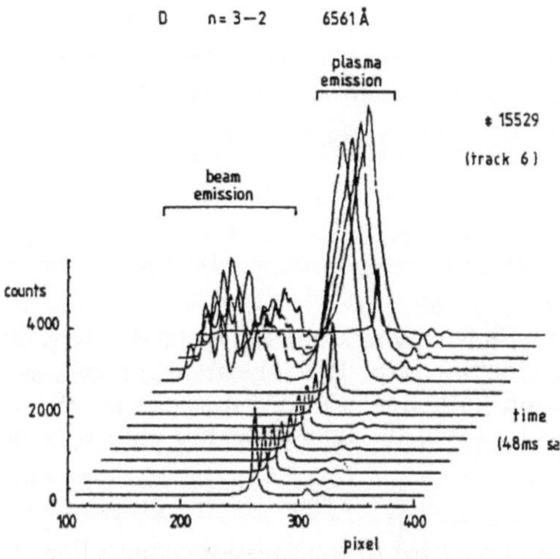

Fig. 5. Visible spectrum along a line of sight intersecting the neutral deuterium beams

The above observations form a connected set linked by atomic reaction processes and by diagnostic usage. Consider firstly the neutral deuterium heating beams. Their penetration into the plasma is a critical parameter. The attenuation is by ionising and charge exchanging collisions with plasma ions (deuterons and impurities) and to a lesser extent by electron collisions. Since it is the simultaneous excitation of the deuterium in the beams which leads to the Stark features, observations such as figure 5 chart the attenuation experimentally. With knowledge of the attenuation, the charge exchange features of figures 5 and 4a allow deduction of absolute radially resolved deuteron and impurity densities. These of course are responsible for the beam attenuation in the first place - a cyclical connection.

The charge exchange features must be separated from superposed cold edge features (figure 4a and 4b). The latter, which are of high principal quantum shell transitions in the visible, are from a complementary cascade path to the resonance lines in the XUV such as shown in figure 3. Excitation of the lower levels of the edge species is principally by electron impact. So the singlet series decrement in figure 3a gives electron temperature. For neutral species, such as Be^0, visible emission is from excited principal quantum shells whose populations are disturbed by further collisions at JET edge densities. So edge electron density is a spectroscopically addressable parameter. The enhancement of the high series members when the plasma is in contact with the inner wall is due to charge transfer from excited states of thermal deuterium near the walls. The excited deuterium presence is revealed in the Balmer alpha edge emission (figure 5) although the excitation is by electrons rather than ions as was the case for the Stark features. The Stark emission, visible wavelength beam driven charge exchange emission and neutral edge emission are all influenced by higher order collisions with excited states. Such processes also modify the effective attenuation rates for the beams, dielectronic recombination etc. under JET conditions and so ultimately influence ionisation equilibria of impurity ions and the *standard analysis* of eg. Mo^{+30} (figure 2)

III. Interpretative modelling

Formally the line of sight emissivity in the n --> n′ principal quantum shell transition following charge transfer from neutral deuterium in its ground state in the beams to a fully ionised plasma impurity of charge z_0 is

$$I(z,n\text{-->}n') = q_{eff}(z,n\text{-->}n') \int N^{(+z_0)} N_D^{(beam)} dl \quad (1)$$

where q_{eff} denotes the effective rate coefficient for the particular photon emission and the integral is called the emission measure. z ($=z_0$-1) simply denotes the final charge state. q_{eff} is the derived theoretical quantity which allows the physically useful emission measure to be obtained. That it is not simply the direct charge transfer rate coefficient to the level n times a branching ratio is due to cascade, redistributing secondary collisions etc. It therefore depends on local electron and ion temperature and density as well as beam particle energy. The effective coefficient can appear outside the integral because of the short line-of-sight/beam line intersection. If the attenuated neutral deuterium density at the observed volume is known by calculation or measurement then the emission measure can be resolved to give the impurity density.

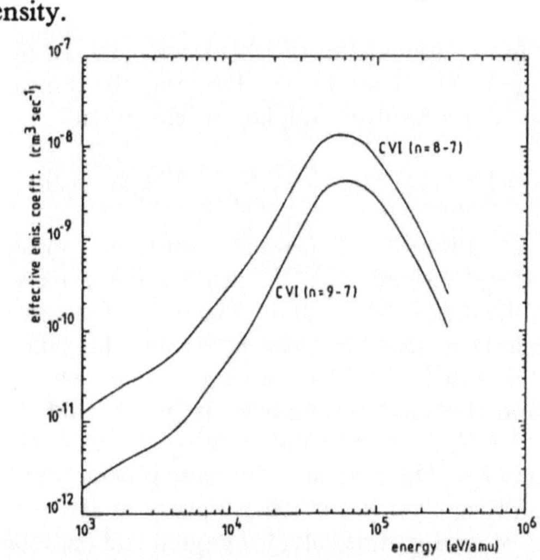

Fig. 6. Charge exchange effective emission coefficients for CVI(n=8-7) & CVI(n=9-7) at T_I=5000eV, $N_I = 2.0\times10^{13}$ cm^{-3}, Z_{eff}=2.

Incidentally, the emission measure in equation (1) is almost the same as that involved in formation of the beam Stark emission although in the latter case there is a sum of similar terms for each plasma ion type with appropriate effective coefficients (driven by ion impact excitation rather than charge exchange) resolved into Stark components (see figure 7).

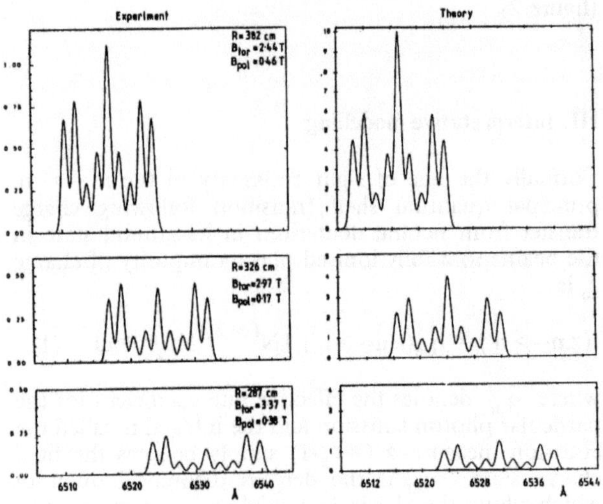

Fig. 7. Comparison of the measured primary energy Stark features, isolated by multiple Gaussian fiiting procedures with the modelled features normalised to each other at r/a = 0.7, where a is the horizontal semi-major axis.

For emission from a complex (more than one electron) species, X^{+z} (such as Be^{+2}), near the plasma edge, the situation is formally similar to that of equation (1), except that the fundamental driving process is electron impact excitation and the ion species may have significantly populated metastable states, indexed by σ, ($1s^2\ {}^1S$ and $1s2s\ {}^3S$ for Be^{+2}) . Denoting the transition by i--> j for a resolved coupling scheme, then

$$I(z,\sigma,i\text{-->}j) = \int q_{eff}(z,\sigma,i\text{-->}j)N_\sigma^{(+z)}N_e dl \quad (2)$$

with a different differential emission measure $N_\sigma^{(+z)}N_e dl$ where $N_\sigma^{(+z)}$ denotes the density of ions in the metastable state σ. The view of metastable populations as quasi-independent is somewhat different from the usual astrophysical approach in which an ionisation balance assumption is made and then $N_\sigma^{(+z)} / N_{tot}$ incorporated with q_{eff} giving a '$g(T_e)$' function [8]. The present is the more general case and is certainly appropriate at the periphery of JET where collisional mixing timescales for ground and metastable populations are comparable with diffusion and dynamical timescales. This is illustrated in figure 8. The JET data (see figure 3a) matches the transient case.

For inflowing atoms or ions from localised surfaces ionising fully near the surface without dispersion , the flux can be related to the emission along a line of sight orthogonal to the surface from any one of its ionisation stages as

$$\Gamma = \Sigma_\sigma \Gamma(z,\sigma) = \Sigma_\sigma E(z,\sigma,i\text{-->}j)\ I(z,\sigma,i\text{-->}j) \quad (3)$$

where $\Gamma(z,\sigma)$ is the partial flux in metastable state σ of the impurity ion of charge state z. E is the number of ionisations per photon or the reciprocal of the *photon efficiency* and is the relevant derived theoretical quantity needed for interpretation. An observed spectral intensity is associated with each metastable. In general the photon efficiency also depends on density as well as temperature particularly for species in neutral or near neutral stages of ionisation.

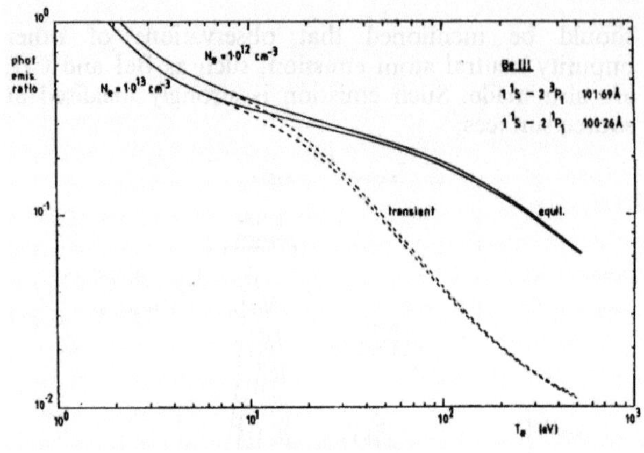

Fig. 8. Emissivity ratio for BeIII ($1s^2\ {}^1S$ - $1s2p\ {}^1P$) and BeIII ($1s^2\ {}^1S$ - $1s2p\ {}^3P$). The transient case assumes evolution from Be^{+1} to Be^{+4} in a plasma of specified temperature and the time integrated emissivity ratio is given.

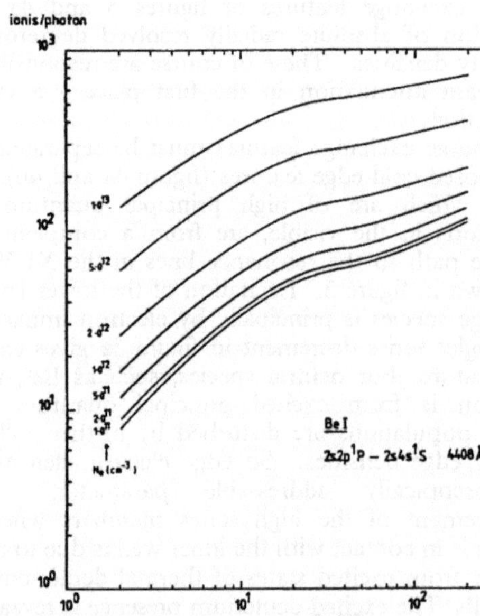

Fig. 9. Ionisations per photon for BeI 4408A. corresponding to the transition $2s4s\ {}^1S$ - $2s2p\ {}^1P$.

IV. The unified picture

The effective coefficients used in the above studies were obtained by collisional radiative population calculations[9]. In these it was necessary to compute both large sets of principal quantum shell ('bundle-n') [10] populations and large sets of ('resolved') low level populations of ions - the latter with precision. In some cases, high level bundle-n calculations were coupled to low level resolved calculations. Complex ions had to be addressed and metastable states distinguished.
The theoretical picture can be completed by considering ionisation and recombination. Then the evolution of ionisation stage populations into each other can be modelled. It is at this point that usages for general plasma simulations and detailed interpretative spectroscopy tend to be most discordant. Properly, each stage must be represented by its ground and metastable populations and the effective ('generalised collisional dielectronic') coefficients established between them, namely $\alpha_{cd}(z+1,\rho';z,\rho)$, $S_{cd}(z,\rho;z+1,\rho')$ and the cross-coupling coefficients $q_{cd}(z,\rho;z,\sigma)$. They depend on density as well as temperature. It was these coefficients which were used for figure 8. Supplemented with power loss coefficients they provide the effective atomic source terms for arbitrary finite density dynamic plasma theoretical simulations matching spectroscopy[11].

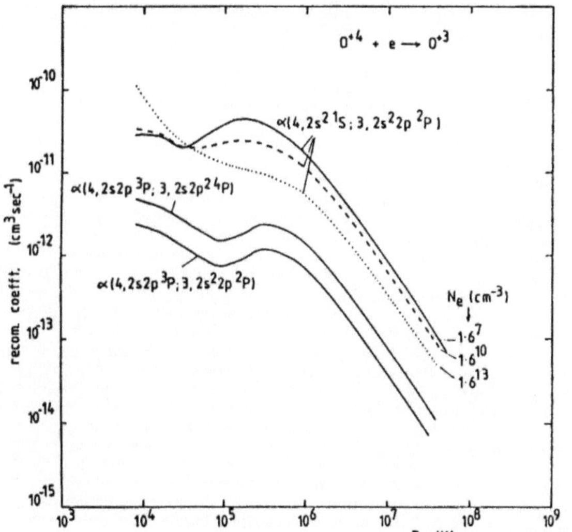

Fig. 10. Generalised collisional-dielectronic recombination coefficients for O^{+4} + e ---> O^{+3}

On a theoretical note, the detailed effective emission coefficients and the generalised collisional dielectronic coefficients can be computed simultaneously. The procedures upon which this work draws organise recombining ion metastable (parent) and recombined ion spin system resolved bundle-n calculations to provide projection matrices which are expanded over resolved low level manifolds and combined with high quality low level rates from assessed atomic data bases. Subsequent matrix manipulations yield all relevant quantities.

V. Conclusion

In this paper spectrocopic observation and detailed interpretative atomic modelling have been closely linked. It has served to clarify a framework and quality of derived atomic data structures appropriate both for diagnostic spectroscopy and general plasma simulation. The starting point is however electron/atom, atom/ion and atom/atom collision data of good reliability. This paper can be seen as one view of how to make best use of spectroscopic observations and fundamental data in a fusion plasma context.

References

1. JET Progress Report, EUR 12808 EN, EUR-JET-PR7 (1989)
2. Behringer K., Engelhardt W.W., Horton L. et al.: Proc XIII Symposium on Physics of Ionised Gases, Yugoslavia (1986)
3. Behringer K.H., Summers H.P., Denne B., Forrest M. & Stamp M.F.: Plasma Phys. & Contr. Fusion **31**, 2059 (1989)
4. Mattioli M., Peacock N.J., Summers H.P., Denne B. & Hawkes N.C.: Phys. Rev. **A40**, 3886 (1989)
5. Boileau A., von Hellermann M., Horton L.D. & Summers H.P.: Plasma Phys & Contr. Fusion **31**, 779 (1989)
6. Boileau A., von Hellermann M., Mandl W., Summers H.P., Weisen H. & Zinoviev A.: J Phys B **22**, 1145 (1989)
7. Summers H.P., Behringer K.H. & Wood L.: Physica Scripta **35**, 303 (1987)
8. Lang J., McWhirter R.W.P. & Mason H.: Solar Physics - in press (1990)
9. Bates D.R., Kingston A.E. & McWhirter R.W.P.: Proc. Roy. Soc. **A276**,297 (1962)
10. Burgess A. & Summers H.P.: Mon. Not. R. Astr. Soc. **226**,257 (1987)
11. Summers H.P. & Hooper M.B.: Plasma Physics **25**,1311 (1983)

Supplement to Z. Phys. D - Atoms, Molecules and Clusters 21, S 23-27 (1991)

Atoms, Molecules
Zeitschrift für Physik D and Clusters

Electron scattering theory for highly charged ions

M. S. Pindzola and N. R. Badnell
Department of Physics, Auburn University, Auburn, Alabama, USA

D. L. Moores
Department of Physics and Astronomy, University College London, London, UK

D. C. Griffin
Department of Physics, Rollins College, Winter Park, Florida, USA

Abstract

Lowest-order QED theory using distorted wave states provides a convenient description for many electron scattering processes involving highly charged atomic ions. We apply QED theory to calculate ionization, excitation, and dielectronic recombination cross sections for electron scattering from highly charged uranium ions. For ionization and dielectronic recombination we compare theory with recent heavy ion accelerator measurements. We also apply lowest-order electroweak gauge theory using distorted wave states to examine the Z-scaling of parity-violating neutral current contributions to the KLL Auger decay of atomic ions.

I. Introduction

Electron scattering processes involving highly charged ions provide information on the quantum mechanical dynamics of few-body relativistic systems under a known force law. Dynamical information so gleaned may then prove useful in studies of other relativistic systems, like quarks inside hadrons[(1)], for which the force laws are more complicated. Lowest-order QED theory[(2)] provides a convenient description for many electron-ion scattering processes. For near threshold ionization and excitation cross sections, as well as dielectronic recombination cross sections, the use of plane wave states in QED is no longer valid. In this paper we first review the application of distorted wave theory to the calculation of the lowest-order Feynman diagrams for electron-ion scattering[(3)]. We then apply QED theory using distorted-wave states to calculate ionization, excitation, and dielectronic recombination cross sections for electron scattering from highly charged uranium ions. For ionization and dielectronic recombination we compare theory with recent heavy ion accelerator measurements[(4),(5)]. As a challenge to future heavy ion accelerator and storage ring experiments, we apply electroweak gauge theory[(6)] using distorted wave states to calculate parity-violating neutral current contributions to the KLL Auger decay of uranium

II. Theory

The electron-impact ionization cross section may be derived starting with the two Feynman diagrams shown in Fig. 1. The direct scattering amplitude (in atomic units) is given by[(2)]

$$S_d = i\int d^4x \int d^4y \int d^4q \ \frac{1}{(2\pi)^4} \ \bar{\psi}_f(x)\gamma_u\psi_i(x) \ \frac{e^{-iq(x-y)}}{q^2 + i\alpha} \ \bar{\psi}_e(y)\gamma^u\psi_b(y) \ , \qquad (1)$$

while the exchange scattering amplitude is given by

$$S_e = -i\int d^4x \int d^4y \int d^4q \ \frac{1}{(2\pi)^4} \ \bar{\psi}_e(x)\gamma_u\psi_i(x) \ \frac{e^{-iq(x-y)}}{q^2 + i\alpha} \ \bar{\psi}_f(y)\gamma^u\psi_b(y) \ , \qquad (2)$$

where the γ_u are 4x4 matrices, q is the 4 momentum transfer, and the Lorentz gauge has been chosen for the photon propagator. The

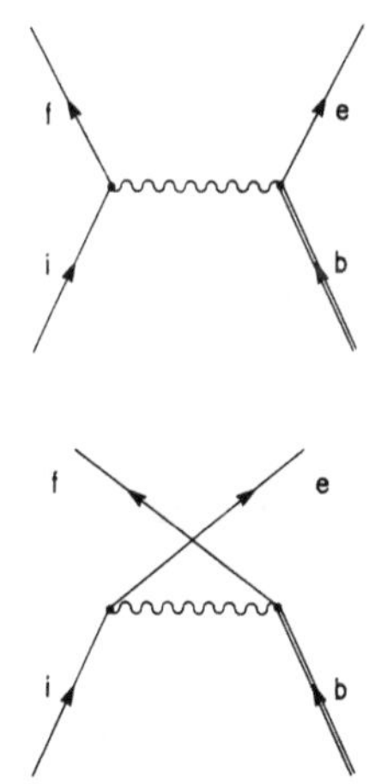

Fig. 1 Lowest-order Feynman diagrams for electron scattering with a bound state electron

Dirac bispinor wavefunctions, $\psi(x)$, are labeled as to the target bound state and the incident, final, and ejected scattering states. The 12 dimensional integrals found in Eqs. (1) and (2) may be reduced to 2 dimensional radial integrals through the introduction of partial wave expansions for the incident, final, and ejected wavefunctions, as well as the photon propagator.

For an electron ionization process of the form:

$$e^- + (n_b\ell_b j_b)^{w} \rightarrow (n_b\ell_b j_b)^{w-1} + 2e^- \quad , \qquad (3)$$

where w is the occupation number of the relativistic subconfiguration $(n\ell j)$, the subconfiguration-average ionization cross section is given by[3]

$$\sigma_i = \int_0^{E/2} d\varepsilon_e \, \frac{16w}{p_i^3 p_e p_f} \sum_{\ell_i j_i} \sum_{\ell_f j_f} \sum_{\ell_e j_e} [j_i, j_e, j_f] \, |V|^2_{avg} \quad , \qquad (4)$$

where ε is the total energy of the electron minus its rest energy, $E = \varepsilon_e + \varepsilon_f$, the linear momentum $p = (2\varepsilon + \varepsilon^2/c^2)^{1/2}$, c is the speed of light, and the notation $[j] = 2j + 1$ is used.

The average of the square of the interaction matrix element found in Eq. (4) is given by[3]

$$|V|^2_{avg} = \sum_{\lambda,\lambda',L} (-1)^{\lambda+\lambda'} [j_i, j_f, j_e, j_b]^{-1} \times \Bigg(\frac{F(\lambda,L)F^*(\lambda',L) + G(\lambda,L)G^*(\lambda',L)}{[L]} + 2\sum_{L'} \begin{Bmatrix} j_f & j_i & L \\ j_e & j_b & L' \end{Bmatrix} \text{Real } F(\lambda,L)G^*(\lambda',L') \Bigg), \qquad (5)$$

where the standard notation is used for the 6j symbol and the "natural" phase approximation has been chosen. The direct multipole function is given by[3]

$$F(\lambda,L) = R_\lambda(fe;ib)\,\delta_{\lambda,L} \langle\kappa_f||C^{(\lambda)}||\kappa_i\rangle\langle\kappa_e||C^{(\lambda)}||\kappa_b\rangle + \sum_{\{a_f a_i a_e a_b\}} g(a_f a_i a_e a_b)\, S_\lambda^{a_f a_i a_e a_b}(fe;ib)$$

$$\delta(1\lambda L)\langle a_f\kappa_f||V^{(L)}||a_i\kappa_i\rangle\langle a_e\kappa_e||V^{(L)}||a_b\kappa_b\rangle, \qquad (6)$$

where the set $\{a_f a_i a_e a_b\}$ is summed over (+-+-), (+--+), (-++-), (-+-+), and g = -1,+1,+1,-1 for each of the 4 cases respectively. In Eq. (6), $\langle\kappa_1||C^{(\lambda)}||\kappa_2\rangle$ and $\langle a_1\kappa_1||V^{(L)}||a_2\kappa_2\rangle$ are previously defined[3] reduced matrix elements of tensor operators and $\kappa = -2(j-\ell)(j + 1/2)$. After the interchange $f \leftrightarrow e$, the exchange multipole function, $G(\lambda,L)$, has the same form as $F(\lambda,L)$. The 2d retarded static integral is given by

$$R_\lambda(fe;ib) = \int_0^\infty dr \int_0^\infty dr' (2\lambda+1)\xi j_\lambda(\xi r_<) h_\lambda^+(\xi r_>) \times (P_f(r)P_i(r) + Q_f(r)Q_i(r)) (P_e(r')P_b(r') + Q_e(r')Q_b(r')), \qquad (7)$$

and the 2d retarded magnetic integral is given by

$$S_\lambda^{a_f a_i a_e a_b}(fe;ib) = \int_0^\infty dr \int_0^\infty dr' (2\lambda+1)\xi j_\lambda(\xi r_<) h_\lambda^+(\xi r_>) F_{a_f}(r)F_{a_i}(r)F_{a_e}(r')F_{a_b}(r') \quad , \qquad (8)$$

where $j_\lambda(x)$ and $h_\lambda^+(x)$ are spherical Bessel functions, $\xi = (\varepsilon_f - \varepsilon_i)/c$, $F_{a_i}(r) = P_i(r)$ if $a_i = +$, and $F_{a_i}(r) = Q_i(r)$ if $a_i = -$. The continuum radial orbitals found in the 2d integrals of Eqs. (7) and (8) are solutions to the single-channel Dirac equation

$$\left(\frac{d}{dr} + \frac{\kappa}{r}\right) P_{\varepsilon\kappa}(r) + \frac{1}{c}\left[V_n(r) - \varepsilon - 2c^2 + V_{DF}\right] Q_{\varepsilon\kappa}(r) = 0 \; ,$$

$$\left(\frac{d}{dr} - \frac{\kappa}{r}\right) Q_{\varepsilon\kappa}(r) - \frac{1}{c}\left[V_n(r) - \varepsilon + V_{DF}\right] P_{\varepsilon\kappa}(r) = 0 \; , \qquad (9)$$

where $V_n(r)$ is the nuclear potential, the distorting potential operator V_{DF} is constructed from previously calculated Dirac-Fock target orbitals, and the continuum normalization is $(1 + \varepsilon/2c^2)^{1/2}$ times a sine function.

For an electron excitation process of the form:

$$e^- + (n_1\ell_1 j_1)^{w_1}(n_2\ell_2 j_2)^{w_2} \rightarrow (n_1\ell_1 j_1)^{w_1-1}(n_2\ell_2 j_2)^{w_2+1} + e^- \quad , \qquad (10)$$

the subconfiguration-average excitation cross section is given by

$$\sigma_x = \frac{8\pi\, w_1(2j_2 + 1 - w_2)}{p_i^3 p_f} \sum_{\ell_i, j_i} \sum_{\ell_f, j_f} [j_i, j_f] \, |V|^2_{avg} \quad , \qquad (11)$$

where the substitutions $b \rightarrow 1$ and $e \rightarrow 2$ are made in the evaluation of $|V|^2_{avg}$ from Eq. (5).

By the principle of detailed balancing, the dielectronic capture cross section is related to the corresponding autoionization rate. For an autoionization process of the form:

$$(n_1\ell_1 j_1)^{w_1}(n_2\ell_2 j_2)^{w_2} \rightarrow (n_1\ell_1 j_1)^{w_1+1}(n_2\ell_2 j_2)^{w_2-2} + e^- \quad , \qquad (12)$$

the subconfiguration-average Auger rate is given by[3]

$$A_a = \frac{4}{p_f} \, \frac{(2j_1+1-w_1)w_2(w_2-1)(2j_2+1)}{4j_2} \sum_{\ell_f, j_f} [j_f] \, |V|^2_{avg} \quad , \qquad (13)$$

where the substitutions $i \rightarrow 2$, $b \rightarrow 2$, and $e \rightarrow 1$ are made in the evaluation of $|V|^2_{avg}$ from Eq. (5). For an autoionization process of the form:

$$(n_1\ell_1 j_1)^{w_1}(n_2\ell_2 j_2)^{w_2}(n_3\ell_3 j_3)^{w_3} \rightarrow \qquad (14)$$

$$(n_1\ell_1 j_1)^{w_1+1}(n_2\ell_2 j_2)^{w_2-1}(n_3\ell_3 j_3)^{w_3-1} + e^- \; ,$$

the subconfiguration-average Auger rate is given by

$$A_a = \frac{4}{p_f}(2j_1+1-w_1)w_2\, w_3 \sum_{\ell_f j_f} [j_f]\, |V|^2_{avg} , \quad (15)$$

where the substitutions $i \to 3$, $b \to 2$, and $e \to 1$ are made in the evaluation of $|V|^2_{avg}$ from Eq. (5). The dielectronic recombination cross section is obtained by multiplying the dielectronic capture cross section by the branching ratio for radiative stabilization.

Beyond the subconfiguration-average approximation, more detailed level specific cross sections including electron correlation effects may be calculated using the multiconfiguration Dirac-Fock (MCDF) approximation for atomic structures[7]. In the last few years the MCDF method has been successively applied to the calculation of dielectronic recombination cross sections and rate coefficients for many atomic ions[8-10]. In particular Chen has extended the MCDF method to include the Breit interaction and has found significant effects on Auger transitions of highly charged ions[11]. The MCDF method has also been applied to the calculation of electron-impact excitation[12-17] and ionization[18] cross sections for atomic ions.

Electron-impact scattering cross sections may also be derived starting with the two Feynman diagrams of Fig. 1 where the wavy line represents the exchange of the heavy Z_o boson instead of the photon[6]. The direct scattering amplitude is given by

$$\bar{S}_d = \frac{i}{4\sin^2\theta}\int d^4x \int d^4y \int d^4q \frac{1}{(2\pi)^4}$$

$$\bar{\psi}_f(x)\gamma_u(c_V - c_A\gamma^5)\psi_i(x)\, \frac{e^{-iq(x-y)}}{q^2 - M^2c^2}$$

$$\bar{\psi}_e(y)\gamma^u(c_V - c_A\,\gamma^5)\psi_b(y) , \quad (16)$$

while the exchange scattering amplitude is given by

$$\bar{S}_e = \frac{-i}{4\sin^2\theta}\int d^4x \int d^4y \int d^4q \frac{1}{(2\pi)^4}$$

$$\bar{\psi}_e(x)\gamma_u(c_V - c_A\gamma^5)\psi_i(x)\, \frac{e^{-iq(x-y)}}{q^2 - M^2c^2}$$

$$\bar{\psi}_f(y)\gamma^u(c_V - c_A\,\gamma^5)\psi_b(y) , \quad (17)$$

where $c_V = -1/2 + 2\sin^2\theta$, $C_A = -1/2$, M is the Z_o mass, and θ is the Weinberg angle. For energies relevant in atomic collisions ($q^2 << M^2c^2$), the Z_o boson propagator can be approximated as a 4 dimensional delta function. The 12 dimensional integrals found in Eqs. (16) and (17) may be reduced to a 1 dimensional radial integral through the introduction of partial wave expansions for the incident, final, and ejected wavefunctions, as well as the delta function.

Parity violating neutral current contributions to electron-ion scattering processes may be cast in almost the same form for the subconfiguration-average cross sections as derived in the preceeding paragraphs for the electromagnetic interaction. The average of the square of the interaction matrix element of Eq. (5) is now given by

$$\left\{|V|^2_{avg}\right\}_{PV} = \left(\frac{c_V c_A}{4\sin^2\theta M^2 c^2}\right)^2 \left\{|V|^2_{avg}\right\}_{EM} , \quad (18)$$

where the $1/M^4$ factor guarantees small cross sections. The forms of the multipole functions found in Eq. (5) are also different. The direct multipole function is now given by

$$F(\lambda,L) = \sum_{\{a_f a_i a_e a_b\}} g(a_f a_i a_e a_b)\, T_\lambda^{a_f a_i a_e a_b}(feib)$$

$$\times\ (\delta_{\lambda,L}\langle a_f\kappa_f||C^{(\lambda)}||a_i\kappa_i\rangle\langle a_e\kappa_e||C^{(\lambda)}||a_b\kappa_b\rangle \quad (19)$$

$$+\ \delta(1\lambda L)\langle a_f\kappa_f||V^{(L)}||a_i\kappa_i\rangle\langle a_e\kappa_e||V^{(L)}||a_b\kappa_b\rangle) ,$$

where the set $\{a_f a_i a_e a_b\}$ is summed over (+++-), (++-+), (--+-), (---+), (+-++), (+---), (-+++), (-+--) and g = +1,-1,+1,-1,+1,+1,-1,-1 for each of the 8 cases respectively. After the interchange $f \leftrightarrow e$, the exchange multipole function, $G(\lambda,L)$, has the same form as $F(\lambda,L)$. The odd number of like sign a's in any set for Eq. (19) guarantees an opposite parity selection rule from the even number of like sign a's found in Eq. (6). The 1d radial integral is given by

$$T_\lambda^{a_f a_i a_e a_b}(feib) = \int_o^\infty \frac{dr}{r^2}(2\lambda + 1)$$

$$F_{a_f}(r)\, F_{a_i}(r) F_{a_e}(r) F_{a_b}(r) . \quad (20)$$

III. Uranium Ion Results

To illustrate the application of lowest order QED theory using distorted wave states, we calculate various cross sections for electron scattering from highly charged uranium ions using the expressions found in the previous section. The near threshold electron-impact ionization cross section for U^{91+} is shown in Fig. 2.[3] The subconfiguration-average cross section calculation (solid curve) is obtained using the σ_i cross section expression of Eq. (4) and the direct term only of $|V|^2_{avg}$ from Eq. (5). As such the calculation contains contributions from the electrostatic, electromagnetic, and retardation parts of the full covariant two-body interaction. For comparison electrostatic only results are presented using both relativistic (dashed curve) and non-relativistic (dotted curve) wavefunctions. The QED distorted wave results (solid curve) agree to 10% with the QED plane-wave calculations of Scofield[19] (dot-dashed curve) at the higher energies. The plane-wave results for U^{91+} continue a steady rise with incident energy to a value of 11.6 barns at 1 GeV. This steady rise in the K shell ionization

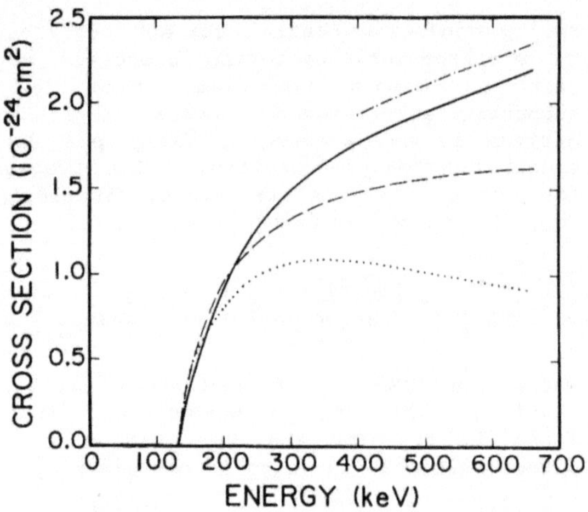

Fig. 2 Electron-impact ionization cross section for U^{91+}. Solid curve, QED theory; dashed curve, Dirac theory; dotted curve, Coulomb theory; dot-dashed curve, plane-wave theory of Scofield (Ref. 19).

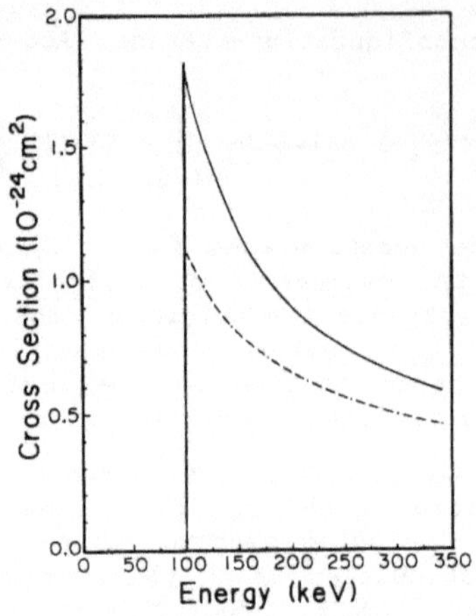

Fig. 3 Electron-impact excitation cross section for the 1s→2s transition in U^{91+}. Solid curve, QED theory; dashed curve, Dirac theory.

cross section has been confirmed experimentally using high-energy electron beams on fixed atomic targets(20). In the threshold region at 222 keV, recent heavy-ion channeling experiments(4) report a cross section value of 3.9 barns for U^{91+}, with an uncertainty in the measurement giving a range of values from 1.9 barns to 7.8 barns. The QED distorted wave calculation shown in Fig. 1 gives a value of 1.1 barns at 222 keV. Recently we have repeated our QED distorted wave calculations for the electron-impact ionization of U^{91+} including the direct, exchange, and interference terms of $|V|^2_{avg}$ from Eq. (5) and have obtained a value of 1.6 barns at 222 keV.

The near threshold electron-impact excitation cross section for the 1s→2s transition in U^{91+} is shown in Fig. 3. The subconfiguration-average cross section calculation (solid curve) is obtained using the σ_x cross section expression of Eq. (11) and the full expression for $|V|^2_{avg}$ from Eq. (5). For comparison electrostatic only results (dashed curve) are presented using relativistic wavefunctions. We thus observe a 60% increase in the threshold excitation cross section due to the electromagnetic and retardation parts of the full covariant two-body interaction. We have also calculated excitation cross sections for the 1s→2$\overline{p}$(j = 1/2) and 1s→2p(j = 3/2) transitions in U^{91+} and have found increases of 22% and 12% respectively at threshold.

Auger rates for the decay of the lowest autoionizing subconfigurations of U^{89+} are given in Table I. The subconfiguration-average (SCA) results are obtained using the A_a autoionizing rate expressions of Eqs. (13) and (15) and the full expression for $|V|^2_{avg}$ from Eq. (5). The multiconfiguration Dirac Fock-Breit (DFB) results are obtained using methods previously developed(10). The DFB results are level specific and based on configuration mixing within the even parity subconfigurations ($1s^22s$ + $1s2s^2$ + $1s2\overline{p}^2$ + $1s2\overline{p}2p$ + $1s2p^2$) and within the odd parity subconfigurations ($1s^22\overline{p}$ + $1s^22p$ + $1s2s2\overline{p}$ + $1s2s2p$). For those subconfigurations with more than one level, the DFB result is the (2J+1) weighted average over the allowed levels. The SCA results are obtained using exactly the same DFB orbitals, but of course configuration mixing is ignored. The difference in the rates found in Table I between the two methods can be largely attributed to correlation effects; for example the $1s2s^2$ and $1s2\overline{p}^2$ mix strongly due to their relatively small energy separation.

Table I: U^{89+} Autoionizing Rates

Subconfiguration	Energy (Ry)	SCA Rate (Hz)	DFB Rate (Hz)
$1s2s^2$	4635	5.64(14)	6.26(14)
$1s2s2\overline{p}$	4644	2.66(14)	2.75(14)
$1s2\overline{p}^2$	4663	3.85(13)	2.25(12)
1s2s2p	4962	5.12(13)	5.28(13)
$1s2\overline{p}2p$	4975	8.47(13)	7.73(13)
$1s2p^2$	5294	4.74(13)	4.47(13)

The dielectronic recombination cross section for U^{90+} in the energy range encompassing the lowest autoionizing levels of U^{89+} is shown in Fig. 4(10). Since the radiative rate for highly charged ions is much greater than the autoionizing rate for most doubly excited levels, the dielectronic recombination cross section approximately equals the dielectronic capture cross section which in turn is directly proportional to the various Auger rates. The dielectronic recombination cross section shown in Fig. 4 has been folded with the Compton profile for H_2 and thus represents the resonant transfer excitation (RTEX) cross section for U^{90+} + H_2 collisions in the rest frame of the

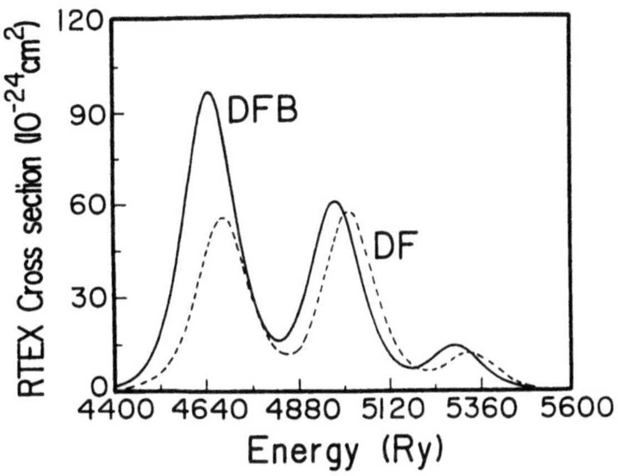

Fig. 4 RTEX Cross Section for U^{90+} collisions with H_2 as a function of projectile energy times m/M. Solid curve: Dirac-Fock method with Breit interaction corrections; dashed curve: Dirac-Fock method.

projectile. The multiconfiguration Dirac-Fock Breit (DFB) results differ substantially from the multiconfiguration Dirac-Fock (DF) results only for the lowest energy peak involving the Auger decay of the $1s2s^2$, $1s2s2\bar{p}$, and $1s2\bar{p}^2$ subconfigurations. Recent heavy-ion accelerator measurements[5] have confirmed both the three peak nature of the U^{90+} cross section and the enhancement of the lower energy peak due to the Breit interaction.

Parity conserving and parity violating Auger rates for the decay of the $2s^2$ ($j = 0$) state of U^{90+} are given in Table II. Both subconfiguration-average (SCA) results are obtained using the A_a autoionizing rate expression of Eq. (13) and the full expression for $|V|^2_{avg}$ from Eq. (5);

Table II: U^{90+} Autoionizing Rates

Subconfiguration	Energy (keV)	Parity Conserving SCA Rate (Hz)	Parity Violating SCA Rate (Hz)
$2s^2$	64.035	1.18(15)	3.04(-9)

except that the parity violating rate has the substitutions found in Eqs. (18) and (19). The parity conserving rate involves the even parity $1s\varepsilon s$ ($j = 0$) continuum, while the parity violating rate takes place through the odd parity $1s\varepsilon\bar{p}$ ($j = 0$) continuum. Based on other similar calculations, the parity violating Auger rate to fill a K shell hole in a singly charged ion scales roughly as Z^8, where Z is the atomic number. Work is in progress to find suitable candidates for analysis by high resolution angle dependent Auger spectroscopy.

IV. Summary

In conclusion lowest order QED theory employing distorted wave states may be applied to a wide range of electron-ion scattering processes. Electromagnetic and retardation corrections to the electrostatic two-body Coulomb interaction are found to be quite significant for highly charged ions. With further development of electron scattering experiments involving heavy-ion accelerators and storage rings it may be possible to not only provide accurate checks on QED theory for strong fields, but on other gauge theory forces as well.

Acknowledgments

This work was supported in part by the Office of Fusion Energy, U.S. Department of Energy, under Contract No. DE-FG05-86-ER53217 with Auburn University and in part by Lawrence Livermore National Laboratory under subcontract No. 8063670 with Auburn University.

References

1. F.E. Close, Proceedings of the 1989 Computational Atomic and Nuclear Physics Summer School (World Scientific, NY, 1990).
2. J.D. Bjorken and S.D. Drell, Relativistic Quantum Mechanics (McGraw-Hill, New York, 1964).
3. M.S. Pindzola, D.L. Moores, and D.C. Griffin, Phys. Rev. A40, 4941 (1989).
4. N. Claytor, B. Feinberg, H. Gould, C.E. Bemis, Jr., J. Gomez del Campo, C.A. Ludemann, and C.R. Vane, Phys. Rev. Lett. 61, 2081 (1988).
5. W.G. Graham, K.H. Berkner, E.M. Bernstein, M.W. Clark, B. Feinberg, M.A. McMahan, T.J. Morgan, W. Rathbun, A.S. Schlachter, and J.A. Tanis, Phys. Rev. Lett, submitted 1990.
6. C. Quigg, Gauge Theories of the Strong, Weak, and Electromagnetic Interactions, (Benjamin/Cummings, Reading, Mass., 1983).
7. I.P. Grant, Advan. Phys. 19, 747 (1970).
8. S. Dalhed, J. Nilsen, and P. Hagelstein, Phys. Rev. A33, 264 (1986).
9. M.H. Chen, Phys. Rev. A33, 994 (1986).
10. M.S. Pindzola and N.R. Badnell, Phys. Rev. A, submitted 1990.
11. M.H. Chen, Nucl. Instr. Methods in Phys. Res. B43, 366 (1989).
12. G.D. Carse and D.W. Walker, J. Phys. B6, 2529 (1973).
13. P.H. Norrington and I.P. Grant, J. Phys. B14, L261 (1981) and J. Phys. B20, 4869 (1987).
14. P.L. Hagelstein and R.K. Jung, At. Data and Nuc. Data Tables 37, 121 (1987).
15. K.J. Reed, M.H. Chen, and A.U. Hazi, Phys. Rev. A36, 3117 (1987).
16. W.J. Qian, Y.K. Kim, and J.P. Desclaux, Phys. Rev. A39, 4509 (1989).
17. H.L. Zhang, D.H. Sampson, and A.K. Mohanty, Phys. Rev. A40, 616 (1989).
18. D.L. Moores and M.S. Pindzola, Phys. Rev. A, submitted 1990.
19. J.H. Scofield, Phys. Rev. A18, 963 (1978).
20. B.L. Moiseiwitsch, Adv. At. and Mol. Phys. 16, 281 (1980).

Supplement to Z. Phys. D – Atoms, Molecules and Clusters 21, S 29–33 (1991)

Atoms, Molecules
Zeitschrift für Physik D and Clusters

Dielectronic and radiative recombination with highly charged ions

L. H. Andersen
Institute of Physic, Aarhus University, DK-8000 Aarhus C, Denmark

Abstract

Measurements of radiative recombination with bare ions as well as H-like ions are reported for relative energies between 0 and 1 eV. Further, measurements of dielectronic recombination with He-like and Li-like ions of fluorine are reported. The measurements have been performed with a single-pass, merged-beam apparatus.

I. Introduction

A free electron and a positive ion (atomic or molecular) may become attached to each other in several ways, depending on the physical conditions. As an electron approaches the positive ion, it is accelerated, and quite large kinetic energies may be obtained, especially in the vicinity of a highly charged ion. In order to become bound to the ion, the electron must exchange energy and momentum with a third 'particle'.

In the case of radiative recombination (RR), this exchange process takes place due to the coupling between matter and the electromagnetic field. In the vicinity of the ion, where the electron experiences a large acceleration, a photon may be created, and as a result, the electron has no longer sufficient energy to escape the Coulomb field of the positive ion. Predominantly, final states with low main quantum number n are populated. In the low energy range to be considered here, a large range of angular momentum states is populated.

The radiative-recombination process may be written as:

$$X^{Z+} + e^- \rightarrow X^{(Z-1)+}(n) + h\nu \; , \tag{1}$$

where X is the positive ion with total charge Z, hv the emitted photon, the energy of which equals the kinetic energy of the free electron plus the binding energy of the state n. The RR process may take place at any initial energy of the free electron and therefore the cross section for RR exhibits no resonant structure. When the ion is fully stripped, the mathematical description of RR involves hydrogenic wave functions and is considerably simpler than when non-bare ions are studied.

From a theoretical point of view, the RR process was the object of several papers starting as early as in the 1920s [1-5]. At that time, Kramers used semiclassical arguments to derive the cross section for RR for fully stripped ions. The ideas of Bohr's correspondence principle were used to obtain an approximate expression for the cross section. Later, several theoretical papers followed [6-9]. The main theme has been to find approximations to overcome the large numerical calculations involved for the excited states. To this end, scaling properties have been investigated [7,8].

When compared to the early theories for RR, the first attempted measurements of RR yielded rate coefficients which were orders of magnitude larger than that predicted by theory. It was realized by D'Angelo [10] that the discrepancy was in fact due to the dominance of another recombination process, ternary recombination (TR), in which the incoming electron interacts with another free electron, loses enough energy to be captured to one of the excited states of the ion, and finally ends up in the ground state after emission of one or several light quanta.

Radiative recombination has been observed in ion-storage rings [11,13]. These measurements were performed with a merged-beam technique in the electron cooler of the storage ring. Due to the fact that the electron beam in the cooler will accelerate the ion beam when the two beams have different velocities, observations were done only at zero relative energy. Due to a lack of precise knowledge about the electron distribution, no direct information on the RR process was derived. In fact, the observed rate for RR was used to estimate the "temperature" kT of the electron gas, and as a consequence, the theory for RR was not tested.

Despite the very early theoretical treatments of RR, it was not until recently that we were able to measure absolute rate coefficients for RR for bare ions of He, C, and F [14,15] as well as RR for H-like ions of C, O, and F and Li-like oxygen ions [16]. All measurements were performed for electron energies between 0 and 1 eV in a single-pass experiment, with no problems associated with acceleration of the ion beam. Low electron energies were chosen where the RR cross section has a magnitude which makes it possible to measure the RR process. For protons, the cross section for capture into the 1s orbit at 1 eV is about 2×10^{-21} cm^2.

The recombination process, known as dielectronic recombination (DR), is a resonant process in which the incoming electron excites an electron which is present on the ion. It thereby loses energy and will become bound to the ion, provided the initial kinetic energy is lower than the excitation energy. In this way, doubly excited ionic states are created. If these states decay radiatively, the electron cannot escape from the ion and the DR process is completed. If the doubly excited states decay via autoionization, the process results in a resonance in the elastic scattering cross section. The DR process can be represented by

$$X^{q+} + e^- \rightarrow (X^{(q-1)+})^{**} \rightarrow (X^{(q-1)+})^{*} + h\nu \; . \tag{2}$$

Unlike radiative recombination, dielectronic recombination occurs only when the ion X carries electrons. If the core-excitation energy is ΔE and the initial kinetic energy of the electron is E_r, then for capture to the state nl energy conservation yields

$$E_r + E_{nl} = \Delta E \; , \tag{3}$$

where E_{nl} is the binding energy of the state designated nl. In the case where ΔE is small compared to the total binding energy of the core electron high Rydberg states (nl) are populated. Such Rydberg electrons often act as a spectator as far as the stabilization of the core is concerned. However, there are important exceptions to this when more configurations can mix, either directly or through the continuum. For $\Delta n'=1$ core excitations such as 1s to 2p, the captured electron may end up in deeply bound states, like in the 2p state.

Historically, the DR process was first suggested as a recombination mechanism by Sayers in 1939 [17]. A few years later, Massey and Bates considered DR theoretically [18]. It was Burgess in 1964 [19], however, who recognized the importance of DR as a recombination process which in fact dominates over the radiative recombination process in the solar corona. After the discovery by Burgess, it was generally realized that DR plays a significant role in hot plasmas, affecting the ionization balance, emitted X-rays, and kinematics of the plasma. However, it was not until 1978 that the first direct observations of DR appeared from hot plasmas [20]. DR cross sections were measured directly with colliding beams of positive ions and free electrons for the first time in 1983 by several groups [21-23]. These early experiments were limited by either low count rates or relatively bad energy resolution.

With the development of electron coolers for ion-storage rings, a new era has begun. With these devices, it is possible to obtain good energy resolution and at the same time sufficiently high electron densities for merged-beam experiments to be performed. With this technique, measurements of DR with intrashell excitations ($\Delta n'=0$) have been obtained in Århus [24,25], and measurements with $\Delta n'>0$ have been performed at the Test Storage Ring in Heidelberg [26]. It is now possible to test the various theoretical approaches on a high level. This means that the magnitude of the cross section as well as spectroscopic information is available from the experiments. In this connection it should be pointed out that the spectroscopic information on the doubly excited resonances is of interest to test theories of atomic resonances which, again, is important for the understanding of correlated two-electron motion in the presence of a positive nuclear charge.

In electron-beam ion traps, ions of very highly charged states may be created. These ions may, in turn, recombine with the electrons inside the trap. By measuring the emitted X-ray spectrum [27], or simply the ion-charge state fractions [28], as a function of electron energy, one has been able to study DR with He-like nickel and argon.

Dielectronic recombination has an equivalent process in ion-atom collisions called resonant transfer and excitation (RTE) [29]. This process involves electron capture from a gas target, with simultaneous excitation of the ionic projectile via the electron-electron interaction. The situation is somewhat disturbed by the non-resonant transfer and excitation process (NTE), where the excitation is caused by constituents other than the captured electron, that may be the nuclear charge or other target electrons. In these experiments, the energy resolution is strongly limited by the target-electron momentum distribution which is to be folded onto the resonances.

Quite recently, the DR process has been studied using ion channeling through thin crystals [30,31]. In this case, the electron 'gas' is provided by the valence electrons in the crystal. The electron density is in the order of 10^{23} cm^{-3}, which is much more than that obtained in other experiments. As for RTE, the resonances are broadened by the initial electron-velocity distribution. In this connection it should be mentioned that it has been found quite recently [31] that the resonances obtained from channeled ions in Au are much narrower than that expected on basis of the Fermi velocity distribution. Thus, as an application,the DR process may be used to probe the electron-velocity distribution inside solids.

Before turning to a discussion of the experiment, it should be mentioned that with the electron densities present in electron coolers, the ternary recombination process can be neglected. Only at extremely high densities or at very low temperatures is the TR process significant [32]. Such conditions may be created in ion traps in the future.

II. Experiment

The measurements, which are reported here, were performed at the Tandem accelerator at the Institute of Physics, University of Aarhus. 1-2 MeV/amu ions were merged with an almost monoenergetic beam of electrons with essentially the same speed as the ion beam. Different relative energies were obtained by varying the electron-beam energy which was around 1 keV. The electron beam had a diameter of 1 cm and was guided by a magnetic field of about 200G which was applied to compensate for the space charge field. Both the ion and electron beam could be modulated at a frequency of 2 MHz, so that electrostatic pick-up plates could monitor the position of the two beams in the interaction region. Thereby the two beams were brought into full overlap. Figure 1 shows the experiment schematically. Details of the experimental apparatus may be found in previous publications [15,25].

The recombination process was identified by counting ions which had collected one electron after the passage through the interaction region. We used two different methods to distinguish between recombination (signal) and electron capture in the rest gas (background). For measurements of the DR process, the background contribution was found via a smooth fit to the yield of ions at energies outside the resonances. In the case of radiative recombination this technique was not applicable since the RR process exhibits no resonances. In order to distinguish between the relatively small RR signal and background, the electron beam was turned on and off by modulating the anode between -2kV and 0 kV at a frequency of 250Hz. At -2kV, the electron beam was turned off, and the true RR signal was obtained as the difference between the yield measured with (N) and without (N_0) electron beam being on.

The electron-velocity distribution was characterized by a product of two Maxwell distributions, one accounting for the transverse electron motion, and one for the longitudinal motion. It was found that the measured shape of the DR resonances contains information, which could easily yield the transverse energy spread $kT_\perp$ as well as the longitudinal energy spread $kT_\parallel$. It was found [14] that the two temperatures were $kT_\perp=0.15$ eV and $kT_\parallel=0.001\text{-}0.002$ eV.

Since very small relative energies were considered, the energy spread in the electron beam could not be neglected. Instead of the cross section, we determined the rate coefficient, which is defined as:

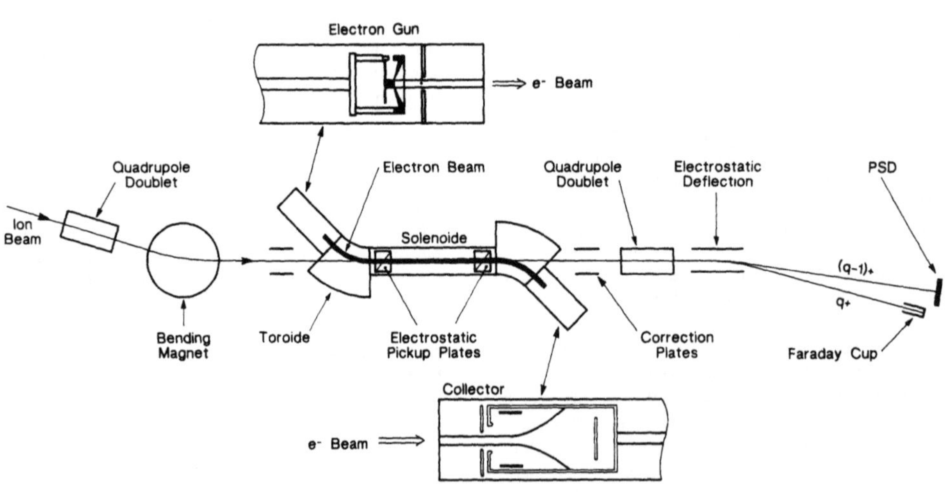

$$\langle v\sigma\rangle = \frac{N^{(Z-1)+} - N_0^{(Z-1)+}}{N^{Z+}} \frac{v_i}{L\rho\varepsilon}, \quad (4)$$

where Z is the total charge of the incoming ion, L is the length of the merge section, ε is the ion-detection efficiency, v_i is the ion velocity, and ρ is the electron density which typically was $5\times10^7/cm^3$.

III. Results

Radiative Recombination

In Fig.2 is shown $\langle v\sigma\rangle$ as a function of relative energy for F^{9+}. Negative energies correspond to the situation where the electrons are slower than the ions. In the case of fully stripped flourine ions, the signal is due to radiative recombination. Evidently, the rate coefficient exhibits a symmetric cusp which peaks at E=0.

In the case where the wave length of the photon is much larger than the extension of the bound state (dipole approximation), Bethe and Salpeter [5] obtained the following simple analytical expression for the cross section for radiative capture to a hydrogenic level with main quantum number n:

$$\sigma_n = 2.10x10^{-22}cm^2 \frac{Z^4E_0^2}{nE_e(Z^2E_0 + n^2E_e)}, \quad (5)$$

where E_e is the kinetic energy of the free electron, E_0 is the Rydberg energy, and Z is the charge of the fully stripped ion. The theory of Bethe and Salpeter was based on the assumption that the oscillator strength is continuous across the continuum limit. Formula (5) is thus expected to be most accurate for high values of n. Indeed, corrections must be applied for the lower states. For n=1, 2, and 3, where the largest cross sections are found, a more accurate calculation was given by Stobbe in 1930 [3]. At low energies, these corrections are 20% for n=1, 12% for n=2, and 9% for n=3, the cross sections of Stobbe being smaller than those of Bethe and Salpeter [15].

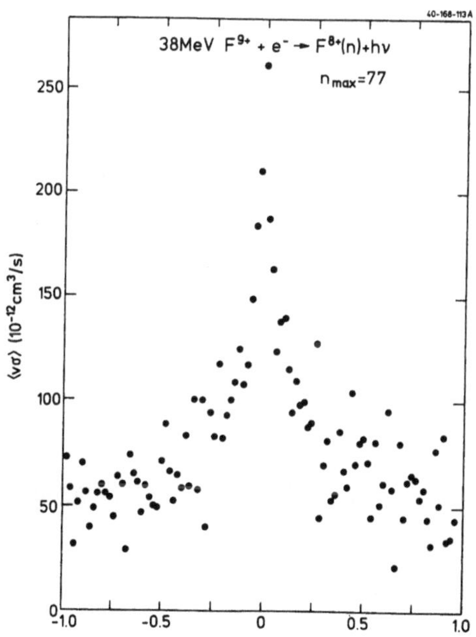

Fig 2: The rate coefficient for F^{9+} as a function of energy.

In Fig. 3 is presented the experimental results obtained with bare ions. The rate coefficient is shown as a function of kinetic energy of the initially free electron. The theoretical curve shown in Fig. 3 was obtained from

$$\langle v\sigma\rangle = \int \sum_n \sigma_n v f(v) dv, \quad (6)$$

with the velocity distribution f(v) given as a product of two Maxwell distributions. The theoretical cross section is given by Bethe and Salpeter, formula (5), corrected according to Stobbe for n=1, 2, and 3. Since the ion-charge-state separation field removes the outer Rydberg electrons, in the evaluation of $\langle v\sigma\rangle$ only values of n corresponding to the actual analyzer field were included. The uncorrected Bethe and Salpeter theory would result in a rate coefficient which is 8% higher than the curve shown. The uncertainty of the experimental rate coefficient is estimated to be 20%. This is mainly due to the uncertainty in the determination of the effective target thickness and the ion-detection efficiency. Evidently, for the bare ions, radiative recombination at low energy is well predicted theoretically.

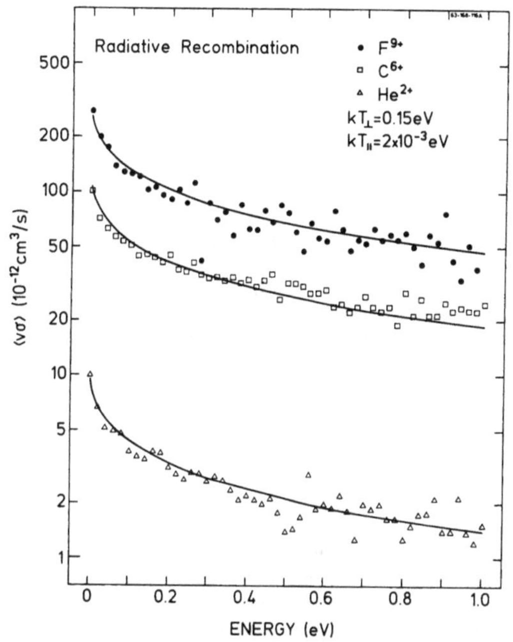

Fig. 3: The rate coefficient for radiative recombination for bare ions of He, C, and F.

Radiative recombination with non-bare ions is much more difficult to treat theoretically than radiative recombination with bare ions due to the much more complex wave functions involved in the calculation. As a first step on our way to gain knowledge about RR with non-bare ions, it is obvious to test how well the theory for bare ions can be used to predict the rate coefficient obtained with H-like ions. To this end, we measured the RR rate coefficient for C^{5+}, O^{7+}, and F^{8+} at relative energies between 0 and 1 eV.

The experimental results are shown in Fig. 4. The curves shown in the figure result from formula (5), corrected according to Stobbes theory for n=1,2, and 3. Further, the Pauli exclusive principle was taken into account by summing only over open shells in Eq. (6). We used an effective charge in Eq. (5) which was chosen to be the asymptotic (total) ion charge. The agreement between the calculated and experimental data is very good. As a comparison, we find that if the nuclear charge is used as the effective charge, the calculation for C^{5+} yields a rate coefficient which is about 50% higher than that obtained with the total charge.

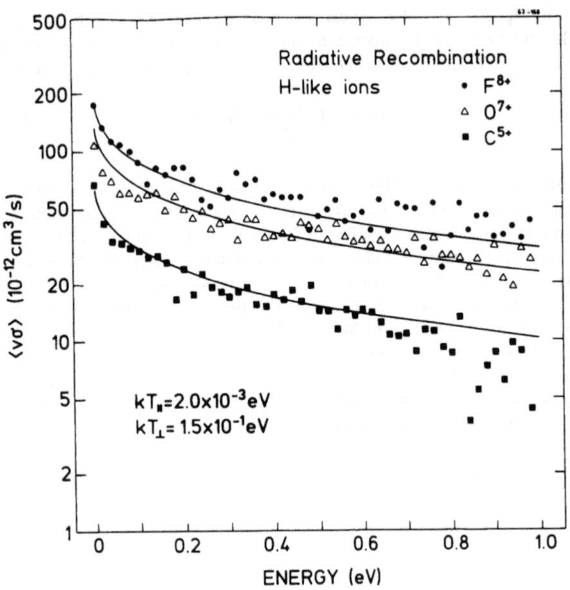

Fig. 4. The rate coefficient for radiative recombination for H-like ions of C, O, and F.

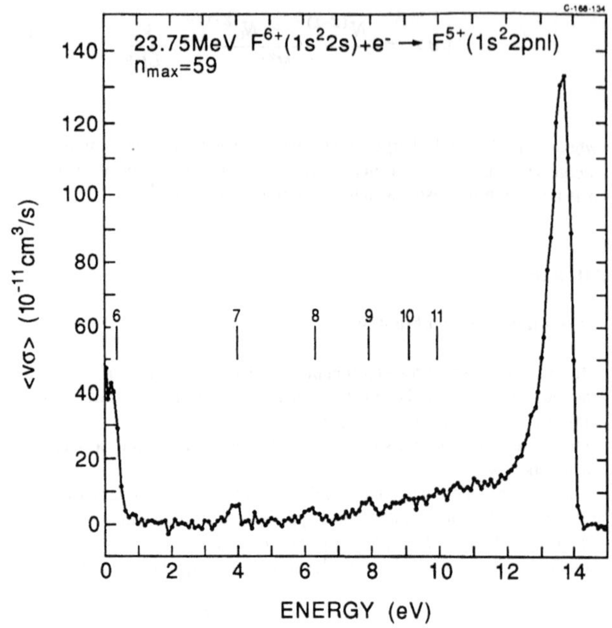

Fig. 5. The DR-rate coefficient for Li-like F^{6+}.

To qualitatively understand this, we note that at low energies ($E<<Z^2E_0/n^2$), σ_n scales as 1/n, meaning that a large part of the total cross section relates to the excited states. Further, the final states φ_{nl} mostly have angular momentum numbers l which are larger than zero [15]. Such states do not penetrate much into the core region, and consequently the matrix element $<\varphi_{nl}|r|\psi>$ is determined in the region of space outside the core. Hence, the total cross section is expected to be characterized by the total ion charge, in agreement with suggestions by Mclaughlin and Hahn [33].

Dielectronic Recombination.

We shall here be concerned with dielectronic recombination associated with the 2s - 2p excitation in Li-like fluorine and He-like fluorine. We have previously reported on the same transition in Li-like carbon and oxygen, and in He-like carbon and oxygen [24,25].

The simple initial state configuration $1s^22s$ of the Li-like isoelectronic sequence makes it suitable as a test case for theories. In Fig. 5 is shown the rate coefficient $<v\sigma>$ as a function of the kinetic energy of the electron in the ion frame for F^{6+}. Resonances associated with capture into n=6,7,8, and 9 are distinguishable. Capture into the high Rydberg states is seen as a large unresolved peak which ends at the excitation energy of the 2s - 2p transition (14 eV). This peak is large due to the large number of Rydberg states which contributes to the DR process. Again, the electrostatic-analyzer field sets an upper limit to the contributing states. Here n_{max} is 59. Further, the Rydberg states contribute much to the total DR rate due to the presence of a small electrostatic field in the interaction region. This field, perpendicular to the beam direction, is produced by the charged electrons. In the present experiment, with typically 10 mA electron beams, this space-charge potential, averaged over the size of a typical ion beam, is in about 2 V/cm.

The electrostatic field causes different angular momentum states to mix. The large number of Rydberg states with large angular momentum numbers l does not contribute much to the DR process due to the low Auger rate. In the presence of the electric field, however, these states mix with states with lower values of l. These have larger Auger rates and therefore contribute more to the DR rate [34]. In the case of C^{3+}, we found relatively good agreement between theory [35] and experiment [25] for fields in the order of 3 V/cm. For O^{5+}, the field had to be about 10 V/cm to make the theory fit the data. At 2 V/cm, theory and O^{5+} data disagreed by about a factor of two. The early measurements by the Oak Ridge group [36] showed a similar tendency. At present, no calculation is published for F^{6+}, yet a comparison would be interesting.

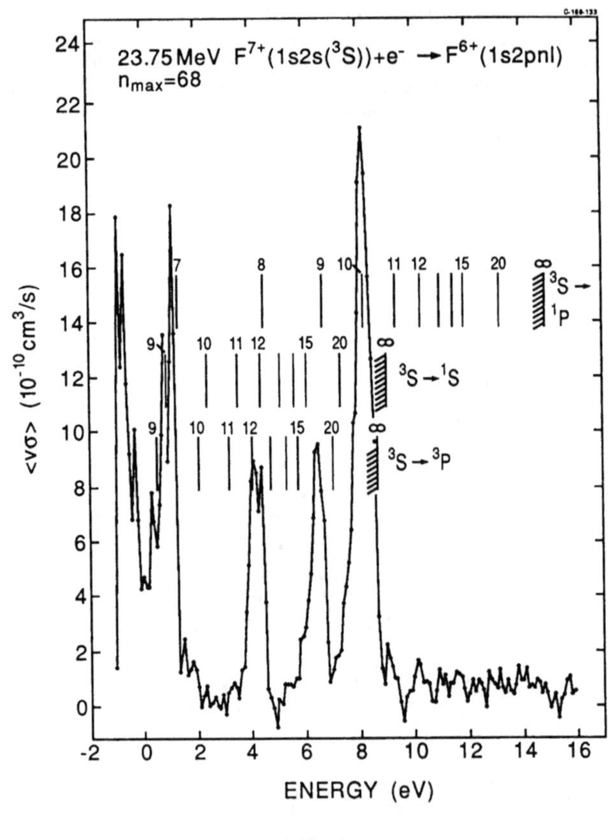

Fig. 6. The DR-rate coefficient for $F^{7+}(1s2s(^3S))$.

As a final example of DR, Fig. 6 shows the DR spectrum obtained with F^{7+} ions. In this case, the beam contains about 30% of the metastable $1s2s(^3S)$ [37], and about 0.07% of the metastable $1s2s(^1S)$ state, the remainder being in the ground state. As in the case of C^{4+} and O^{6+} [25], the spectrum exhibits some sharp resonances. In the case of F^{7+}, these are due to excitations of the ^3S state. Only one peak, at 0.3 eV, can be assigned to the ^1S initial state.

Members of the $(^1S)n'l'$ and $(^3P)n''l''$ intermediate configurations may mix with the $(^1P)nl$ configuration for n=8,9, and 10, either directly or through the continuum [24,25,38,39]. Whenever that happens, one would expect a large DR signal due to the strong radiative stabilization of the $1s2p(^1P)nl$ state to the $1s^2(^1S)nl$ state. The large signal observed at around 4.5 eV, 6.7 eV, and 8.2 eV may indeed be a result of this type of interaction. The data calls for theoretical calculations which include such effects.

IV. Conclusion

Examples of RR and DR measurements performed with the electron beam of an electron cooler have been given. The RR data provide a basis for tests of the theory for RR. The state selective DR data for Li-like F^{6+} and He-like F^{7+} is suitable for testing DR theories for their ability to take into account external electric fields and configuration mixing.

References

1. H. A. Kramers, Philos.Mag. **46**, 836 (1923)
2. J. R. Oppenheimer, Phys.Rev. **31**, 349 (1930)
3. M. Stobbe, Ann.Phys. **7**, 661 (1930)
4. W. Wessel, Ann.Phys. **5**, 611 (1930)
5. H. Bethe and E. Salpeter, 'Quantum Mechanics of One- and Two-electron Systems', Vol. **35** of Handbuch der Physik (Springer, Berlin, 1957).
6. C.M.Lee and R.H.Pratt, Phys.Rev. **A14**, 990 (1976)
7. Y.Hahn and D.W.Rule, J.Phys. B **10**, 2689 (1977)
8. Y. S. Kim and R. H. Pratt, Phys.Rev. A **27**, 2913 (1983)
9. A. I. Milstein, Phys.Lett. A **136**, 52 (1989)
10. N. D'Angelo, Phys.Rev. **121**, 505 (1961)
11. V. M. Kathov and V. M. Strakhovenko, Zh.Eksp.Teor.Fiz. **75**, 1269 (1978)
12. H. Poth et al., Z.Phys. A **332**, 171 (1989)
13. G. Kilgus et al. Phys.Rev.Lett. **64**, 737 (1990)
14. L. H. Andersen, J. Bolko, and P. Kvistgaard, Phys.Rev.Lett. **64**, 729 (1990)
15. L. H. Andersen and J. Bolko, Phys.Rev A. **42**, 1184 (1990)
16. L. H. Andersen and J. Bolko, J.Phys. B.(1990)
17. M. J. Seaton and P. J. Storey, in Atomic Processes and Applications, edited by P. G. Burke and B. L. Moiseiwitsch (North Holland, Amsterdam, 1976) p. 134
18. D.R. Bates and H. S. W. Massey,Phil. Trans.Roy.Soc. A **239**, 269 (1943)
19. A. Burgess, Astrophys.J. **139**, 766 (1964); ibid. **141**, 1588 (1965)
20. R. L. Brooks, R. U.Datla, and H. R.Griem, Phys.Rev.Lett. **41**, 107 (1978)
21. D.S.Belic, G.H.Dunn, T.J.Morgan, D.W. Mueller, and C. Timmer, Phys.Rev.Lett. **50**, 399 (1983)
22. J.B.A. Mitchell, C.T. Ng, J.L. Forand, D.P. Levac, R.E. Mitchell, A. Sen, D.B. Miko, and J.W. McGowan, Phys.Rev.Lett. **50**, 335 (1983)
23. P.F. Dittner, S. Datz, P.D. Miller, P.L. Pepmiller and C.M. Fou, Phys. Rev.Lett. **51**, 31 (1983)
24. L.H. Andersen, P. Hvelplund, H. Knudsen, and P. Kvistgaard, Phys. Rev.Lett. **23**, 2656 (1989)
25. L.H. Andersen, J. Bolko, and P. Kvistgaard, Phys.Rev. A **41**, 1293 (1990)
26. G. Kilgus et al. Phys.Rev.Lett. **64**, 737 (1990)
27. D.A. Knapp et al.,Phys. Rev. Lett.**62**, 2104 (1989)
28. R. Ali, C.P. Bhalla, C.L. Cocke, and M. Stockli, Phys.Rev.Lett. **64**, 633 (1990)
29. J.A. Tanis, Nucl.Instr. and Meth. Phys.Res. Sect. A **262**, 52 (1987)
30. S. Datz et al., Nucl.Instr. and Meth. **B48**, 114 (1990)
31. A.Belkacem et al. Phys.Rev.Lett. **64**, 380 (1990)
32. See. e.g., J. Stevefelt, J. Boulmer, and J-F. Delpech, Phys.Rev. A **12**, 1246 (1975).
33. D.J. McLaughlin and Y. Hahn,private communication.
34. A. Müller, et al., Phys.Rev. A. **36**, 599 (1987).
35. D.C. Griffin, M.S. Pindzola, and P. Krylstedt, Phys.Rev. A **40**, 6699 (1989).
36. D.F. Dittner, S. Datz, P.D. Miller, P.L. Pepmiller, and C.M. Fou, Phys. Rev. A **35**, 3668 (1987)
37. M. Terasawa et al., Phys.Rev. A **27**, 2868 (1983)
38. K. Taulbjerg and J. Macek, Phys.Rev.Lett. **62**, 2766 (1989)
39. N.R. Badnell, M.S. Pindzola, and D.C. Griffin, Phys.Rev. A **41**, 2422 (1990)

Supplement to Z. Phys. D - Atoms, Molecules and Clusters 21, S35-38 (1991)

Atoms, Molecules
Zeitschrift für Physik D and Clusters

Electron-impact excitation of multiply-charged ions using energy loss in merged beams: $e + Si^{3+}(3s^2S_{1/2}) \rightarrow e + Si^{3+}(3p^2P_{1/2,3/2})$

E. K. Wåhlin[a], J. S. Thompson, and G. H. Dunn[b]

Joint Institute for Laboratory Astrophysics
of the University of Colorado and the National Institute
of Standards and Technology,
Boulder, CO 80309, USA

R. A. Phaneuf and D. C. Gregory
Oak Ridge National Laboratory, Oak Ridge, TN 37831, USA

A. C. H. Smith[c]
Joint Institute for Laboratory Astrophysics,
Oak Ridge National Laboratory, and University College London,
Gower Street, London WC1E 6BT, U.K.

Abstract

For the first time absolute total cross sections for electron-impact excitation of a multiply-charged ion have been measured using an electron-energy-loss technique. Measurements were made near threshold for the process $e + Si^{3+}(3s^2S_{1/2}) \rightarrow e + Si^{3+}(3p^2P_{1/2,3/2})$ - 8.88 eV. The 10^{-15} cm^2 measured cross section agrees with results of 7-state close coupling calculations to better than the ±20% (90% CL) total uncertainty of the measurements. Convoluting the theoretical curve with a Gaussian energy distribution indicates an energy width of $0.15 \lesssim \Delta E \lesssim 0.20$ eV.

I. Introduction

The physics of multiply-charged ions has become the object of much attention recently, driven partly by the important role such ions play in fusion, astrophysical and other high-temperature plasmas and partly by the fact that new technology has become available which makes meaningful experimental investigations of multiply-charged ions possible and palatable. The collision physics associated with these "objects" has proven fascinating and challenging, particularly as the plethora of resonances and their roles in different collisions have been uncovered.

Despite this recent interest and activity, there is little experimental data on the electron-impact excitation of multiply charged ions. Using the technique of crossed beams and fluorescence detection, measurements of absolute cross sections have been made [1] only for C^{3+}, N^{4+}, Al^{2+}, and Hg^{2+}. The $\lesssim 10^{-4}$ detection efficiencies, the decreasing cross sections with increasing charge, the tenuous target densities, and increasingly difficult absolute radiometry with decreasing wavelength couple to make these measurements ever more difficult as the ionic charge increases. In addition, it is desireable to measure excitation of nonradiating states, since it is for such states that the fractional contribution of capture resonances becomes large and the test of the theory becomes most crucial. Thus, it has long been recognized [2] that new approaches must be found for the experimental study of electron-impact excitation of multiply charged ions.

The invention [3] of the electron beam ion trap (EBIT) and creative use of the electron beam ion source (EBIS) [4] have ushered in a significant new approach to the study of excitation. It is possible to obtain excitation cross sections for excitation to some states of highly charged ions, such as He-like argon and Ne-like barium or gold, with energy resolution characteristic of these devices (≈50 eV) by observing fluorescent X-rays and normalizing to theory for [3] radiative recombination or [4] dielectronic recombination. Though the energy resolution is large compared to state widths, it is narrow enough to clearly observe some excitation resonances in the cross sections measured.

To further fill the need for methods to study excitation, an energy-loss* merged-beams technique has been developed [5] and has now been installed on a beam line of the Oak Ridge ECR ion source and used to measure cross sections for excitation of Si^{3+}. The method achieves a detection efficiency of close to unity (≈0.7). The technique leads to absolute measurements requiring no normalization, has an energy

[a]Present address, Los Alamos National Laboratory, P-1, MS-E526, P.O. Box 1663, Los Alamos, NM 87544
[b]Staff Member, Quantum Physics Division, National Institute of Standards and Technology
[c]1989-90 JILA Visiting Fellow, on Leave from

* Others developing related trochoidal techniques for electron-ion excitation studies include: A. Chutjian *et al.* at Jet Propulsion Labs, D. Belic *et al.* at Belgrade University, D.W. Mueller *et al.* at North Texas State Univ., and I.D. Williams *et al.* (this volume) at Queen's University, Belfast. A. Chutjian has previously used conventional energy loss techniques to study relative differential cross sections of Zn^+, Cd^+, and Mg^+ over the angular range 6 to 17 degrees. A similar system for differential measurements on multiply charged ions is currently being developed in Grenoble by

resolution of about 0.2 eV, and gives cross sections for both radiating and nonradiating states in the important threshold region. The technique was designed for use with ion beams from ECR or EBIS sources, but should readily be adaptable for use with storage rings.

II. Experimental Approach

The experimental method is schematically illustrated in Fig. 1 and described in detail elsewhere [6]. The apparatus is immersed in a uniform magnetic field (for the present experiments ≈0.004 T) in the z direction. Sets of parallel plates labelled "merger" and "demerger" provide electric fields in the y direction. Charged particles undergo trochoidal motion in these regions with trajectories expressed by: $x=\rho/\omega_c(\omega_c t-\sin\omega_c t)$; $y=\rho/\omega_c(1-\cos\omega_c t)$; and $z=\dot{z}_0 t$. Particles reproduce their initial velocities after each integral number n of cyclotron periods $2\pi/\omega_c$, but with the particles displaced by $\Delta z = 2\pi n\dot{z}_0/\omega_c$ and $\Delta x = 2\pi n\rho/\omega_c$. Thus, in these two regions, charged particles undergo a drift in the x direction with drift velocity $\vec{\rho} = \vec{E}\times\vec{B}/B^2$. The fields are chosen such that there are exactly two cyclotron periods in the merger, so the electrons again have velocity $\dot{z}_0\vec{k}$ as they merge [7] with the ions coming along the z axis.

After the beams have been merged and allowed to collide over a nominal 8 cm (this path length is adjustable) path, the beams enter the demerger where the electrons separate from the ions. Because of the "competition" between the electrons' forward velocity and their drift velocity ρ, those electrons which have undergone scattering and reduction of their forward velocities will travel at a different angle in the demerger region. Thus, there is dispersion of the electrons here. The slow electrons, which have been inelastically scattered, impact a position sensitive detector (PSD) and are counted, while unscattered electrons are collected in the electron Faraday cup where their

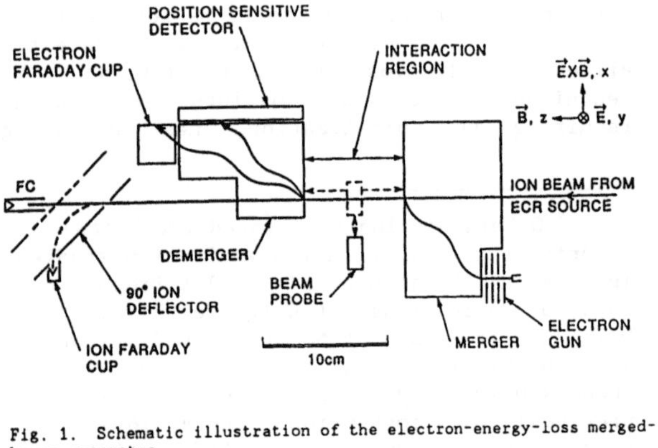

Fig. 1. Schematic illustration of the electron-energy-loss merged-beams apparatus.

current of about 100 nA is measured. It is clear that electrons which have elastically scattered at large angles may also impact the PSD, and these will be discussed later.

Ions are extracted from the electron-cyclotron-resonance (ECR) ion source to 10q keV, where q is the ion charge. Because of the ions' much larger mass, and hence much lower cyclotron frequency, they are not significantly deflected in these regions, and their trajectories are easily corrected with sets of deflector plates (not shown). The ions are bent by a 45° energy analyzer into a Faraday cup and their current measured. Despite beam intensities of several microamperes of Si^{3+} from the ion source, the high degree of collimation required for these experiments leads to about 30 nA of current in the collision region.

Cross sections for the inelastic scattering process can then be determined from the equation

$$\sigma = \frac{R}{\epsilon}\ \frac{v_e v_i}{(v_e - v_i)}\ \frac{q\ e^2}{I_e I_i}\ \frac{1}{\Omega} \qquad (1)$$

where R is the event count rate at the desired location on the PSD, ϵ is the detection efficiency for the scattered electrons and v_e and v_i and I_e and I_i are the velocities and currents of the electrons and ions respectively. The quantity $1/\Omega$ is the beam overlap factor determined from the beam density distributions G(x,y,z) and H(x,y,z) from the relationship

$$\Omega = \frac{\int G(x,y,z)H(x,y,z)\ dxdydz}{\int G(x,y,z)\ dxdy\ \bullet \int H(x,y,z)\ dxdydz}\ . \qquad (2)$$

The quantities G and H are determined by measuring two-dimensional distributions at a number of places (usually 7 - 9) along the interaction path using a fluorescent screen/ digital-video camera technique [8].

The efficiency ϵ is determined by measuring the current of a reduced (≈1 fA) electron beam at the Faraday cup (FC) and then deflecting it onto the PSD and counting the particles in the beam. Scanning the beam across the PSD affords the opportunity to determine that the efficiency of the detector is not spatially variant to any significant extent.

In practice, the counts arriving at the PSD from the reaction being studied constitute but a small fraction of the total, and measures must be taken to separate the signal counts from the "background." The vacuum in the cryo-pumped collision region is about 8 x 10^{-9} Pa. Still, there are about 50 s^{-1} counts/nA from electron scattering on the background gas, about 50 s^{-1}/nA from surface scattering of electrons, and about 30 s^{-1}/nA from gas scattering of the ion beam.

Overwhelming these sources is about 500 s^{-1}/nA from the ions due to the fact that *for Na-like ions* a fraction of the ions are in long-lived (1 -10 μs) autoionizing states [9]. Thus, there is a stream of electrons produced along the ion beam path, and in particular the electrons emitted by the ion beam while in the demerger constitute most of the background.

The above background rates, when combined with beam currents, lead to total background rates of 15 kHz from ions, 10 kHz from electrons, and about 100 Hz from "signal." Separation of signal from background is approached by beam chopping techniques, using a phased 4-way chopping scheme [6] at 1 kHz to minimize background neutral density modulation. The four gated rates for (electrons + dark), (ions + dark), (signal + ions + electrons + dark), and (dark) are stored according to location in four separate histogram memories, and appropriate arithmetic is done on the arrays on an updating basis during a run. This is straightforward, except that, because the backgrounds from ions and electrons are about the same, the total rate into the memory containing the signal is nearly twice that for either electrons or ions alone. This leads to the need to determine dead times to better than 1%. Dead times both of the electronics and of the micro-channel plates of the PSD were taken into account.

The electron-surface scattering at the entrance to and in the demerger was a problem which could lead to "false signals" due to micromodulation of the electron beam's position by the space charge of the ion beam. Great care was taken to ensure the absence of such effects.

The most serious limitation in the method is that a particle beam, upon entering the $\vec{E}\times\vec{B}$ region, experiences a potential gradient across the beam height; and this in turn makes it appear that the beam has a much greater energy spread -- the dispersion is greatly sacrificed to this so-called "shear effect" [10]. This leads to the most critical difficulty with the measurements: separation of inelastically scattered electrons from elastically scattered ones. Thus, the trochoidal demerger works basically by sending particles at an angle determined by the drift velocity $\rho = E/B$ and the longitudinal (along B) velocity, v , so $\theta = \tan^{-1}(\rho/v)$. Many electrons, elastically scattered at large angles from the Si^{3+} ions, will have the same v as the inelastically scattered electrons of interest and will thus appear at the detector over the same locus of positions. To partially obviate this problem, we introduced limiting apertures at the demerger entrance, keeping electrons elastically scattered at angles greater than 20° in the laboratory from entering the analyzer. This allows separation for a short range above threshold, but severely limits the energy range above threshold over which one can make measurements. As presently configured our modelling showed good separation for 2 eV above threshold, but we limited measurements to ±0.6 eV about the threshold energy of 8.88 eV.

Also associated with the potential drop across the demerger is the fact that low-energy electrons will be retarded and even rejected in the bottom half of the demerger. Electrons with zero energy in the center-of-mass frame (CM), resulting from excitation at threshold, will have about 0.86 eV in the laboratory frame, and if the potential at the location of the beam is more negative than this, these electrons will be rejected. The demerger is thus run in an "accelerating" mode so that all particles are slightly accelerated into the device. This, again, degrades the dispersion of the device. The beam distributions, G(x,y,z) and H(x,y,z), are used to determine a locus of origin of signal for the scattering events, and modelling calculations [6] are carried out to determine where the signal electrons will arrive on the detector. This allows one to look only at those coordinates where signal should arrive, thus blocking some background. It also determines a small correction for electrons not detected -- usually less than 6%. The modelling is carried out for each point for the full range of scattering angles in the CM.

III. Results

Measured cross sections are shown in Fig. 2 as solid points with error bars which represent 90% confidence limits (about 1.7σ) for relative uncertainties. These relative uncertainties are statistical counting uncertainties added in quadrature to uncertainties due to possible spurious signals, beam overlap, and incomplete signal collection. The larger circle at 9.2 eV has two sets of error bars: the inner represent the relative uncertainty described above, and the outer has added in quadrature uncertainties due to dead-time corrections, detector calibration and beam current measurements. The total absolute uncertainty near the maximum of the cross section is ±20% at 90% CL.

The solid curve in Fig. 2 has been generated by convoluting a hypothetical 0.2 eV Gaussian energy distribution with 7-state R-matrix close coupling calculations of Badnell *et al.* [11]. The theoretical curve has been shifted 0.17 eV in energy to bring the threshold into agreement with the spectroscopic value. Agreement with experiment is seen to be excellent and well within the ±20% uncertainty of the measurements. The theory did not take into account the 0.057 eV energy difference in the $3P_{1/2}$ and $3P_{3/2}$ levels. Thus, rather than a single step function at 8.88 eV, the hypothetical "true" cross section would be the sum of two step function cross sections: the $3P_{1/2}$ at 8.839 eV with a magnitude of about 3.7 x 10^{-16} cm^2 and the $3P_{3/2}$ at 8.896 eV with a magnitude of about 7.3 x 10^{-16} cm^2. If this is taken into account, convoluting this hypothetical curve with a 0.15 eV Gaussian energy distribution

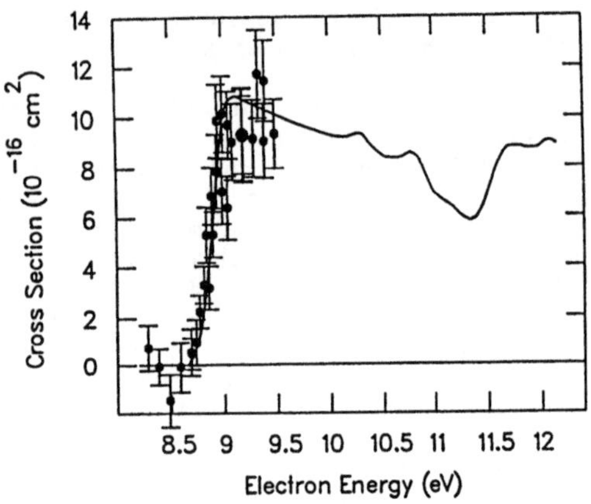

Fig. 2. Absolute total cross section for the electron impact excitation, e + $Si^{3+}(3s^2S_{1/2})$ e + $Si^{3+}(3p^2P_{1/2,3/2})$. Solid circles are present measurements; error bars represent total relative uncertainty at 90% CL; outer bar on the 9.2 eV measurement represents total absolute uncertainty. Curve represents 7-state close coupling calculations [10] convoluted with a 0.2 eV FWHM Gaussian as an approximation to the electron energy distribution.

(approximately the value expected as the thermal spread from the cathode) also gives good agreement with the experiment. One may conclude that the experimental energy distribution is $0.15 \leq \Delta E \leq 0.20$ eV, and that the theory accurately yields thetotal cross section near threshold for this ion and transition.

The very intense ion beam-induced background made diagnostics for spurious "signals" extremely difficult and time-consuming and thus made it impractical to extend the energy range to where the large "window" resonance is shown in the theory in Fig. 2. Testing of theory for resonance effects will be better carried out on other ion species without the overwhelming autoionization background. This is important, since testing of the theory must be the goal of the experiments, and for the present ion and transition, both the R-matrix and distorted wave [12] calculations give identical results at threshold.

Acknowledgments

Carl Timmer, David Swenson, Luc Forand, Brett DePaola, Alfred Müller, Dragan Belic, and Klaus Rinn were all heavily involved in the developoment and construction of the merged beams apparatus. Nigel Badnell, Don Griffin, Mitch Pindzola and Chris Bottcher suggested the choice of Si^{3+} as a test case for the first multiply charged ion to investigate. The work was supported in part by the Office of Fusion Energy of the U.S. Department of Energy under Contract No. DE-AI01-76PR06010 with the National Institute of Standards and Technology and Contract No. DE-AC05-840R21400 with Martin Marietta Energy Systems, Inc.

References

1. For a review, see Phaneuf, R.A.: Atomic processes in electron-ion and ion-ion collisions. pp. 117-156. New York: Plenum, 1986; Also see [2].
2. Dunn, G.H.: The Physics of Ionized Gases, pp. 49-95. ed. M. Matic. Belgrade, Boris Kidric, 1980.
3. Marrs, R.E., Levine, M.A., Knapp, D.A., Henderson, J.R.: Phys. Rev. Lett. **64**, 1715 (1988).
4. Ali, R., Bhalla, C.P., Cocke, C.L., Schulz, M., Stockli, M.: this issue, pp.
5. Timmer, C., Wahlin, E., DePaola, B., Rinn, K., Swenson, D., Belic, D., Dunn, G.: Abstracts, XV ICPEAC, Brighton, 1987, pp 805; Wahlin, E., Timmer, C., Forand, L., Swenson, D., DePaloa, B., Phaneuf, R., Belic, D., Rinn, K., Muller, A., Dunn, G.: Abstracts, XVI ICPEAC, New York, 1989, p 368.
6. Timmer, C.W.: PhD thesis, University of Colorado, 1988; E.K. Wahlin, PhD thesis, University of Colorado, 1990.
7. Auerbach, R., Cacak, R., Caudano, R., Gaily, T.D., Keyser, C.J., McGowan, J. Wm., Mitchell, J.B.A., Wilk, S.F.J.: J. Phys. **B10**, 3797 (1977).
8. Forand J.L., Timmer, C., Wahlin, E.K., DePaola, B.D., Dunn, G.H., Swenson, D.R., and Rinn, K.: Rev. Sci. Instrum. **61** Nov.1990.
9. Harris, S.E., Walker, D.J., Caro, R.G., Mendelsohn, A.J., Cowan, R.D.: Opt. Lett. **9**, 168 (1984).
10. McMillan, M.R., Moore, J.H.: Rev. Sci. Instrum. **51**, 944 (1980).
11. Badnell, N., Pindzola, M.S., Griffin, D.C.: Phys. Rev. A, submitted; private communication, 1990.
12. Mann, J.B.: in Los Alamos Scientific Laboratory report LA-8267-MS, March 1980.

Supplement to Z. Phys. D - Atoms, Molecules and Clusters 21, S 39–43 (1991)

Atoms, Molecules
Zeitschrift für Physik D and Clusters

Electron impact ionization of multiply charged ions

Alfred Müller
Institut für Kernphysik, Strahlenzentrum der Justus-Liebig-Universität, W-6300 Giessen, FRG

Abstract

This paper gives a brief review of the physics of electron impact ionization. Some of the latest accomplishments in both theory and experiment are discussed. The focus is on results for the isoelectronic sequences of H-like, Li-like and Na-like ions and in particular on the combination and relative importance of direct versus indirect ionization mechanisms. The discussion also includes complex many-electron systems.

Introduction

Ionization of ions by electron impact is one of the most fundamental atomic collision processes. Its importance for all kinds of plasmas has stimulated theoretical and experimental studies since the first systematic investigations of gas discharge phenomena. The field received new impetus in connection with the world wide efforts to accomplish controlled nuclear fusion and a longstanding interest in ionization has been kept up by the astrophysics community.

By now a substantial body of data as well as a substantially improved understanding of the process have been acquired. In particular, our knowledge about the importance of intermediate excited states formed during the collision has been greatly improved [1]. Beside direct "knock-on" ionization which is the conceptually simplest ionization mechanism there are a number of multi-step processes which can also significantly contribute to ionization cross sections. Such indirect ionization mechanisms give rise to distinct step and peak features in the cross section functions. By careful observation of these features in experiments with high energy resolution and good statistics a spectroscopy of intermediate multiply excited states of ions becomes feasible. By such a spectroscopy information can be obtained on level energies and on partial cross sections for direct ionization, excitation-autoionization (EA), resonant-excitation-double-autoionization (REDA) and resonant-excitation-auto-double-ionization (READI) contributing to net single ionization of the parent ion.

EA involves direct excitation of an inner-shell electron to a bound configuration of the ion A^{q+}

(1) $$e + A^{q+} \rightarrow A^{**q+} + e\ .$$

The multiply excited ion A^{**q+} can autoionize in a subsequent relaxation process and the net result is the production of an ionized ion $A^{(q+1)+}$

(2) $$A^{**q+} \rightarrow A^{(q+1)+} + e.$$

Due to a Coulomb focusing effect the cross section for direct (inner shell) excitation of an ion (in contrast to excitation of an atom) is non-zero at threshold which gives rise to a distinct step in the excitation function when the electron energy is scanned across the excitation threshold.

REDA and READI proceed via intermediate compound states of the projectile electron and the parent ion and thus are closely related to the process of dielectronic recombination. All these processes involve a radiationless (dielectronic) capture of the incident electron by the ion A^{q+}

(3) $$e + A^{q+} \rightarrow A^{**(q-1)+}.$$

This process is inverse to autoionization (or Auger electron emission) and thus it becomes clear that it can only occur when the energy of the free electron matches the resonance condition

(4) $$E_{res} = E(j) - E(i),$$

where E(j) and E(i) are the total energies of all bound electrons in the excited ion $A^{**(q-1)+}$ and in the parent ion A^{q+}, respectively.

The multiply excited product ion $A^{**(q-1)+}$ can follow several different decay routes. In case of the emission of photons the new charge state of the initial ion can be stabilized and thus dielectronic recombination completed. Trivially, also time-reverted dielectronic capture, i.e. autoionization, can occur and bring the ion back to its initial charge state. The net overall process is then a resonant (elastic or inelastic) scattering of an electron from the ion.

Contributions to net single or multiple ionization of the parent ion require emission of two or more electrons from the intermediate $A^{**(q-1)+}$ ion.

(5) $$A^{**(q-1)+} \rightarrow A^{(q+n)+} + (n+1)\,e \qquad n = 1,2,3,...$$

The emission of (n+1) electrons predominantly proceeds sequentially one by one, though the simultaneous emission of two electrons is also possible. For n=1, i.e. net single ionization of the parent ion, the sequential emission of two electrons (double-autoionization) completes the REDA process. Simultaneous emission of two electrons (auto-double-ionization) completes the READI process. While REDA requires the existence of an intermediate autoionizing state of the ion A^{q+}, in order to facilitate the ejection of the second electron, there is no such restriction for READI and hence in some cases the two processes can be experimentally distinguished.

Hydrogen-like ions: direct ionization

Obviously, one of the most important classes of experiments in electron-ion ionization is that dealing with hydrogen-like ions. With only one target electron the H-like systems seem ideal for making comparisons between experiment and theory and to test the understanding of the direct ionization process. Hence, it was no accident that He^+ was the object of the first crossed charged beams experiment carried out by Dolder et al. [2] in 1961. Measurements for the next member of the hydrogen isoelectronic sequence, Li^{2+}, were not reported until recently [3] . Already the first classical and quantum theoretical approaches [4,5] led to a scaling behaviour of electron-impact ionization cross sections

(6) $\sigma I^2 = f(E/I)$.

For ions along one particular isoelectronic sequence one has to expect a universal curve when the cross section multiplied by the square of the ionization potential I is plotted as a function f of the scaled electron energy E/I.

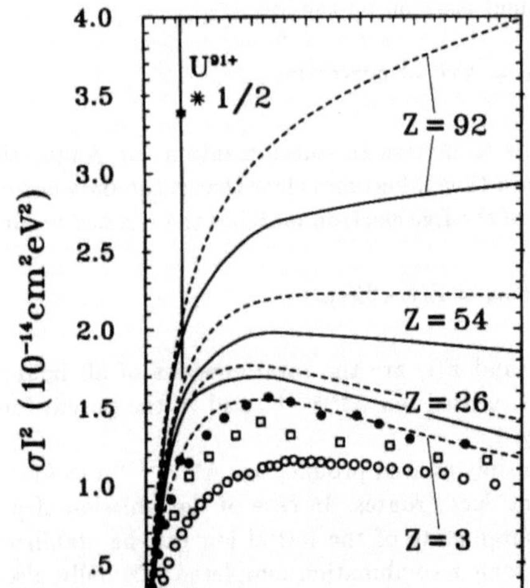

Fig.1 Scaled cross sections for electron-impact ionization of H-like ions (see Eq.6). Open circles are experimental for H atoms, open squares are for He^+ ions and solid circles for Li^{2+} ions (for references see [3]). The full star is experimental [12] for U^{91+} and multiplied by 1/2 to fit on the plot. Solid lines are relativistic calculations [11], dashed lines are estimated from the Lotz formula with a relativistic correction [3].

For the H-like atoms and ions which have been studied so far the scaling works well at sufficiently high energies. Close to the threshold, however, cross sections reflect the presence of a focusing effect on the incident electrons by the increasing target ion charge. At maximum the product σI^2 for Li^{2+} is about 20% higher than that for neutral H. Within this small deviation (which can be reproduced by theory using distorted waves [6] to describe the projectile electrons) the scaling, Eq.6, works remarkably well.

So, further experiments with ions of higher atomic number seemed like a waste of effort. It has been pointed out [3,7], however, that changes will occur when the atomic number increases to Z=50 or even to Z=92. For Z=92 , i.e. for U^{91+}, the electron binding energy is 131.8 keV [8] suggesting relativistic effects both on the bound electron and the projectile electron which needs relativistic velocities to remove the bound electron.

For a guess of relativistic effects on ionization cross sections Tinschert et al. [3] had used the Lotz formula [9] in connection with a relativistic correction suggested earlier by Gryziński [10]. Results of this simple calculation are compared in Fig. 1 with relativistic theory recently published by Moores et al. [11] and with experimental data for H, He^+, Li^{2+} and U^{91+}. The comparison obviates the necessity to perform more careful experiments possibly by using the colliding-beams technique in a heavy ion storage ring.

Lithium-like ions

The lithium iso-electronic sequence is the most extensively studied series both experimentally and theoretically. Experiments benefit from the absence of metastable ions in the parent beams so that well-defined collision systems can be investigated. The electronic structure is simple enough to facilitate fairly detailed theoretical calculations and to sort out individual ionization mechanisms. The presence of two inner-shell electrons on the other hand makes the system complex enough to allow a variety of indirect ionization mechanisms which can be studied in great detail provided the experimental technique is sensitive enough to distinguish small features due to inner-shell processes, on top of a dominant contribution from direct ionization of the comparatively weekly bound 2s electron.

More than a decade ago Crandall et al. [13] could clearly

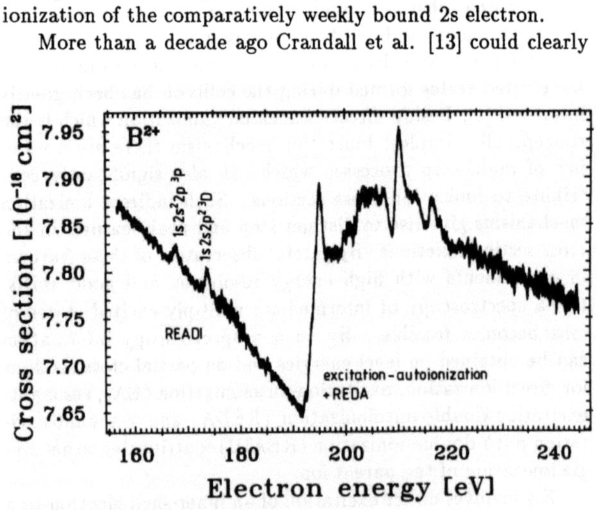

Fig.2 Cross sections for electron-impact ionization of B^{2+} ions [15]. Intermediate states and indirect mechanisms are indicated.

demonstrate the presence of indirect ionization mechanisms in Li-like ions of the elements C, N and O. The two step mechanism observed in these experiments was that of excitation-autoionization (see Eqs. 1 and 2).

In more recent experiments which employed a fast energy-scanning technique it was possible to perform high-precision experiments on the ionization of C^{3+} ions [14] and then also of B^{2+}, N^{4+}, O^{5+} and F^{6+} ions [15]. An example for these measurements is given in Fig. 2 which shows results for the ionization of B^{2+} ions in an energy range where 1s→2l inner-shell excitations become possible. Different from the older data the new experiments show structure in the excitation-autoionization step which is due to different terms in the Li-like 1s2s2l configurations. On top of the individual steps there are resonance peaks which are related to Be-like 1s2snln'l' configurations populated by dielectronic capture and subsequently decaying by two-electron emission. The two little peaks at about 168 eV and 174 eV are due to READI processes. From the areas of the READI peaks the resonance strengths S can be inferred. These numbers can be compared to predictions of Pindzola and Griffin [16] and can serve in guiding theory towards a proper description of double Auger processes. Fig. 3 presents a comparison of experimental values of S for the $1s2s^22p\ ^2P$ READI resonance with the range theoretically predicted by Pindzola and Griffin. The atomic numbers are between Z=5 and Z=8. The uncertainty of the theory is visualized by shading the area inside which the values of S are predicted (of course at discrete atomic numbers). The solid points with error bars represent the experimental data.

As one has to expect, the cross section features above the excitation-autoionization threshold are qualitatively the same for all members of the Li isoelectronic sequence and therefore can be discussed on the basis of one example. (Though indirect cross section contributions are more prominent for higher-Z members of the sequence.) For comparison with Fig. 2 a scan measurement [14] for C^{3+} parent ions is displayed in Fig.4 for energies at and above the excitation-autoionization threshold. Intermediate highly excited states have been assigned to the observed cross section features. Beside the excitation-autoionization steps there are a number of resonance contributions which have to be assigned to REDA processes. It is interesting to note the sizable contributions from a Rydberg series of intermediate excited states with configurations 1s2s3lnl' (n=3,4,5...) which are populated from the $1s^22s$ ground state by Δn=2 excitations with simultaneous capture of the incident electron. In a measurement on Li-like N^{4+} even resonant Δn=3 excitation could be experimentally observed [15].

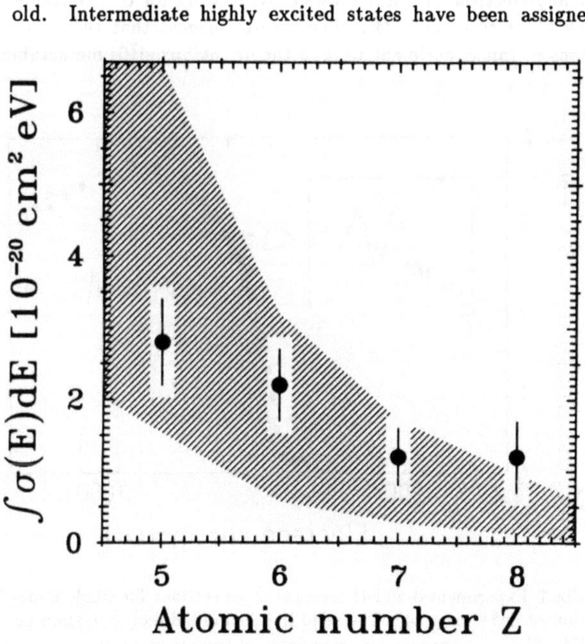

Fig.3 Comparison of experimental [14,15] and theoretical [16] $1s2s^22p\ ^2P$ READI resonance strengths S.

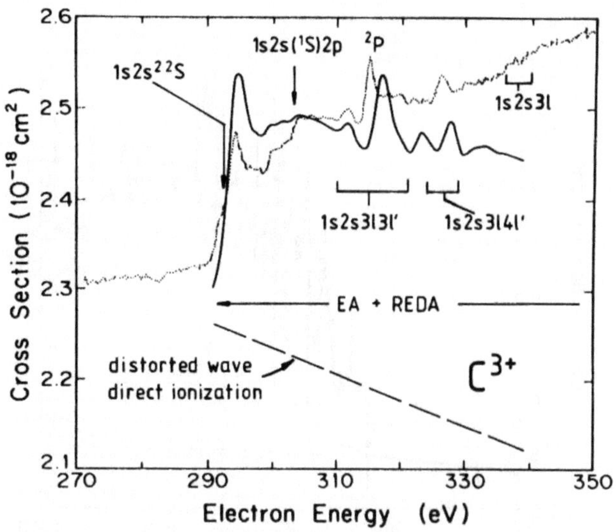

Fig.4 Comparison of experimental [14] and theoretical [17] cross sections for ionization of C^{3+} in the vicinity of the EA threshold.

Also shown in Fig. 4 is a 17-state close-coupling calculation by Tayal and Henry [17]. The theoretical results are convoluted with a 2-eV Gaussian. The agreement with the experimental data is truly remarkable.

Sodium-like ions

As the number of electrons in the outermost shell increases the ionization cross section goes up. Indirect contributions are based on inner-shell processes and thus do not change their magnitude much when outer electrons are attached. Hence, it becomes experimentally more difficult to identify indirect ionization mechanisms in isoelectronic sequences with several electrons in the outer shell. Cross sections are then predominantly determined by direct ionization. Thus, the next interesting isoelectronic sequence after the Li-like is the Na-like sequence. In the ground state there is one 3s electron outside filled K- and L-shells, a situation promising significant ionization contributions from intermediate autoionizing states.

Strong contributions to ionization of sodium-like Fe^{15+} from REDA had been estimated before by LaGattuta and Hahn [20] who initialized the search for resonances in the ionization of ions by their calculations. It was quite a step in the development of experimental techniques that led to the first measurement of electron-impact ionization of Fe^{15+} ions by Gregory et al. [21]. Their results are shown in Fig.5 together with recent theoretical results of Chen et al. [22].

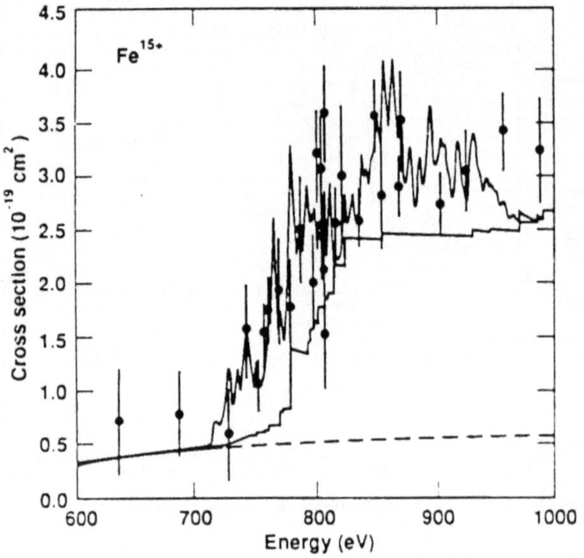

Fig.5 Experimental [21] and theoretical [22] cross sections for ionization of Na-like Fe^{15+} ions. The dashed line represents direct ionization, the lower solid line is direct ionization plus EA and the upper solid line comprises direct, EA and REDA contributions.

It is interesting to break down the cross section into the three main contributions: direct ionization, EA and REDA. These different contributions are indicated in Fig.5. Apparently, indirect ionization dominates the cross section at energies above about 750 eV and even the REDA processes give significant contributions reaching up to 50% of the total cross section at positions of strong resonances. These calculations provide a new challenge to experimentalists to further improve the experimental conditions and make the observation of REDA resonances in the electron-impact ionization of sodium-like Fe^{15+} feasible.

One step towards this goal was made by employing the fast energy-scanning technique to the ionization of sodium-like Mg^+ ions [23]. The result of this measurement is shown in Fig. 6.

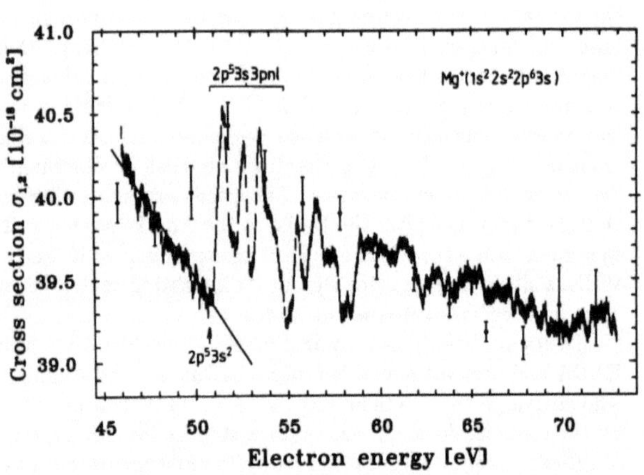

Fig.6 Experimental cross sections for ionization of Na-like Mg^+ [23].

The energy resolution in this experiment was 0.3 eV as evidenced by the width of the peak observed at 56 eV. A wealth of structure is observed above the excitation-autoionization threshold near 50 eV. The dominant resonance series below 55 eV is probably due to $2p^5 3s3pnl$ intermediate excited states of the Mg atom (compare the model calculations for Al^{2+} in Ref. [19]). The first three peaks were interpreted to be due to n=3, n=4 and n=5 and the shoulder on the high-energy side of the n=5 peak could be due to the unresolved resonance peaks with $n \geq 6$.

Complex ions: single and multiple ionization

Heavy atoms and ions with many electrons and especially the elements with atomic numbers Z = 54 to Z = 58 offer attractive opportunities to study electron-scattering processes in a region where the atomic structure is dominated by electron-electron interactions. In highly charged ions of these species it is again predominantly the Coulomb force which determines the static and dynamic behaviour, however, a new interesting aspect is then introduced by the increasing importance of relativistic effects both in atomic structure and in collision phenomena. An example for the latter has been discussed in the chapter on hydrogen-like ions. Large numbers of equivalent electrons in the outer shells of heavy elements also lead to enhanced effects on many-electron processes. Cross sections for multiple ionization may be nearly as big as for single ionization and strong contributions of multi-step processes to single and multiple-ionization are observed. In particular, REDA resonances may subtantially contribute to measured cross sections. Higher-order resonance processes like resonant-excitation-triple-auto-ionization (RETA) have been observed to make up for as much as 30% of the respective multiple ionization cross section.

As an example for ionization of a complex highly charged many-electron ion Fig.7 shows cross sections for U^{16+} obtained by Gregory et al. [24]. The results suggest that the parent ions in this experiment were in the $4f^{14}5s^2 5p^6 5d^7 6s$ metastable

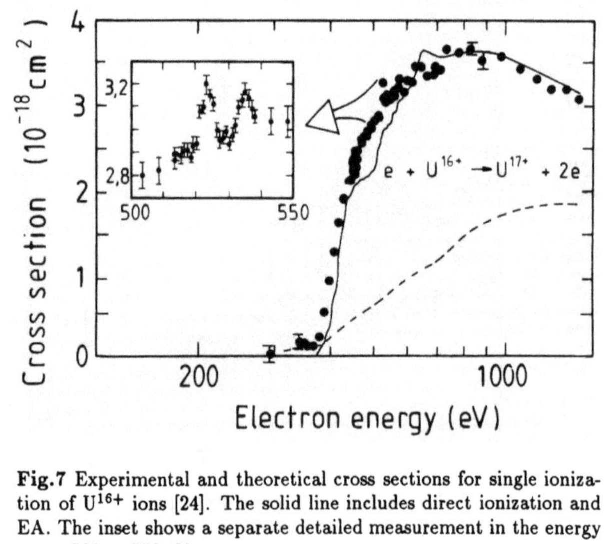

Fig.7 Experimental and theoretical cross sections for single ionization of U^{16+} ions [24]. The solid line includes direct ionization and EA. The inset shows a separate detailed measurement in the energy range 500 to 550 eV.

configuration. The dashed curve is a Lotz calculation for direct single ionization of electrons in that configuration. The solid curve is a distorted-wave calculation including excitation-autoionization effects. The discrepancy of theory and experiment at energies between about 400 and 600 eV is typical for such collision systems: theory without accounting for resonance phenomena may miss important cross section contributions. This explanation for the observed discrepancy is substantially supported by the observation of two very strong resonance peaks between 520 and 540 eV making up for nearly 15% of the measured cross section (see inset of Fig. 7).

In a series of detailed measurements on heavy metal ions [25-28] narrow capture resonances were found not only in single but also in double, triple and quadruple ionization channels. Fig. 8 shows an example for resonant contributions to double ionization of La ions. In this particular case, the intermediate multiply excited ion La^{+} *must* decay by the emission of three electrons and one of the emission processes *must* involve an auto-double-ionization. In spite of the complexity of the RETA mechanism the observed resonance feature makes up for about one third of the total double ionization cross section.

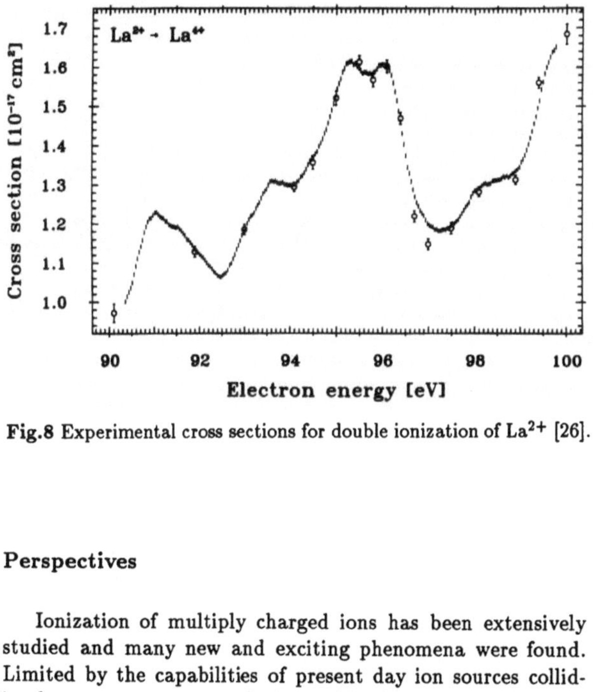

Fig.8 Experimental cross sections for double ionization of La^{2+} [26].

Perspectives

Ionization of multiply charged ions has been extensively studied and many new and exciting phenomena were found. Limited by the capabilities of present day ion sources colliding beams measurements for ions with charge states q beyond 10 are scarce, the record for ionization experiments is at q=16. The importance of multi-electron processes for very highly charged ions has not yet been studied, calculations including relativistic effects could not yet be tested sensitively for these ions. In the future, further development of ion source technology may push the experimental limits to higher ion charge states. Precise measurements on electron impact ionization of very highly charged ions, however, will devolve upon new developments employing electron-beam ion-trap sources or heavy ion storage rings. The experimental storage ring (ESR) of GSI in Darmstadt promises to provide stored U^{91+} ions within the next months and thus fulfills one necessary pre-condition for the ultimate ionization experiment: electron impact ionization of the heaviest naturally occuring element in a hydrogen-like state $e + U^{91+} \rightarrow U^{92+} + 2e$.

Acknowledgements

I want to thank all my colleagues who have contributed to our investigations of electron impact ionization of ions at E. Salzborn's laboratory. In particular, G.Hofmann and K.Tinschert deserve special recognition. Collaboration with G.H.Dunn, R.A. Phaneuf, D.C.Gregory, R.Becker, S.M.Younger, D.C.Griffin, M.S.Pindzola and many others is gratefully acknowledged. Part of the research reported here was supported by Deutsche Forschungsgemeinschaft (DFG), Bonn-Bad Godesberg. Collaborative work was made possible by NATO Collaborative Research Grant No. RG 86/0510.

References

1. Müller, A., in PHYSICS OF ION IMPACT PHENOMENA, ed. by D. Mathur, Springer, Berlin, Heidelberg, New York, 1991, in preparation
2. Dolder, K.T., Harrison, M.F.A., Thonemann, P.C., Proc.Roy.Soc.A 264, 367 (1961)
3. Tinschert, K., Müller, A., Hofmann, G., Huber, K., Becker, R., Gregory, D.C., Salzborn, E., J.Phys.B 22, 531 (1989)
4. Thomson, J.J., Phil.Mag. 23, 449 (1912)
5. Bethe, H., Ann.d.Phys. 5, 325 (1930)
6. Younger, S.M., in ELECTRON IMPACT IONIZATION, ed. by T.D.Märk and G.H.Dunn, Springer, Wien, New York, 1985, pp. 277-319
7. Pindzola, M.S., Moores, D.L., Griffin, D.C., Phys. Rev.A 40, 4941 (1989)
8. Johnson, W.R., Soff, G., At.Data Nucl.Data Tables 33, 405 (1985)
9. Lotz, W., Z.Phys. 232, 101 (1970); Z.Phys. 216, 241 (1968); Z.Phys. 206, 205 (1967)
10. Gryziński, M., Phys.Rev. 138, A305 (1965); Phys.Rev. 138, A322 (1965); Phys.Rev. 138, A336 (1965)
11. Moores, D.L., Pindzola, M.S., Phys.Rev.A 41, 3603 (1990)
12. Claytor, N., Feinberg, B., Gould, H., Bemis, C.E., del Campo, J.G., Ludemann, C.A., Vane, C.R., Phys.Rev.Lett. 61, 2081 (1988)
13. Crandall, D.H., Phaneuf, R.A., Hasselquist, B.E., Gregory, D.C., J.Phys.B 12, L249 (1979)
14. Müller, A., Hofmann, G., Tinschert, K., Salzborn, E., Phys.Rev. Lett. 61, 70 (1988)
15. Hofmann, G., Müller, A., Tinschert, K., Salzborn, E., Z.Phys.D - Atoms, Molecules and Clusters 16, 113 (1990)
16. Pindzola, M.S., Griffin, D.C., Phys.Rev.A 36, 2628 (1987)
17. Tayal, S., Henry, R.J.W., Phys.Rev.A 42, 1831 (1990)
18. Crandall, D.H., Phaneuf, R.A., Falk, R.A., Belić, D.S., Dunn, G.H., Phys.Rev.A 25, 143 (1982)
19. Henry, R.J.W., Msezane, A.Z., Phys.Rev.A 26, 2545 (1982)
20. LaGattuta, K.J., Hahn, Y., Phys.Rev.A 24, 2273 (1981)
21. Gregory, D.C., Wang, L.-J., Meyer, F.W., Rinn, K., Phys.Rev.A 35, 3256 (1987)
22. Chen, M.H., Reed, K.J., Moores, D.L., Phys. Rev. Lett. 64, 1350 (1990)
23. Müller, A., Hofmann, G., Tinschert, K., Weißbecker, B., Salzborn, E., Z.Phys.D - Atoms, Molecules and Clusters 15, 145 (1990)
24. Gregory, D.C., Huq, M.S., Meyer, F.W., Swenson, D.R., Sataka, M., Chantrenne, S.J., Phys. Rev.A 41, 140 (1990) and Pindzola, M.S., Buie, M.J., Phys.Rev.A 39, 1029 (1989)
25. Müller, A., Tinschert, K., Hofmann, G., Salzborn, E., Dunn, G.H., Phys.Rev. Lett. 61, 70 (1988)
26. Müller, A., Tinschert, K., Hofmann, G., Salzborn, E., Dunn, G.H., Younger, S.M., Pindzola, M.S., Phys.Rev.A 40, 3584 (1989)
27. Tinschert, K., Müller, A., Hofmann, G., Salzborn, E., Younger, S.M., submitted to Phys. Rev. A
28. Hofmann, G., Müller, A., Weißbecker, B., Stenke, M., Tinschert, K., Salzborn, E., this volume

Supplement to Z. Phys. D - Atoms, Molecules and Clusters 21, S 45-50 (1991)

Atoms, Molecules
Zeitschrift für Physik D and Clusters

Electron ion interactions in crystal channels: collisions in ultra-dense electron media

S. Datz, P. F. Dittner, J. Gomez del Campo, H. F. Krause, T. M. Rosseel and C. R. Vane
Oak Ridge National Laboratory, Oak Ridge, TN 37831-6377, USA

Abstract

Dielectronic excitation of H-like S, Ca and Ti is shown to occur in the dense electron gas of a crystal channel. Cross sections for collisional ionization of the short lived excited states can then be determined. Ionic excitation can also be achieved by resonant coherent excitation in which case specific m states can be excited for further study.

Introduction

In recent years considerable progress has been made in the study of collisions of multiply charged ions with electrons. Areas which have been opened to investigation include excitation, ionization, radiative recombination and dielectronic recombination. All of these processes are contributors to the dynamics of hot plasmas. However, in high density plasmas collisions of short lived excited ionic states with electrons also play a significant role and, until now, have not been amenable to study. We have utilized the channeling effect to create conditions under which such processes may be studied.

Ions entering a crystal at small angles with respect to low order crystaline axes and planes are constrained in their motion and interact only with weakly bound conduction or valence electrons [1]. Thus ions passing through a channel with velocity $v_i \gg v_f$, the Fermi velocity of the conduction electrons, may be viewed as being bombarded with a flux of electrons with velocity $v_e \simeq v_i$ with a distribution of energies determined by the Fermi distribution of momenta for the electrons in the channel. These electrons can participate in the formation of excited states by either direct excitation or by recombination processes [2]. Then, because of the great electron density ($> 10^{23}/\mathrm{cm}^3$), in the channel they can participate in secondary electronic collisions which further excite or ionize the short lived excited species.

* Presently visiting Tage Erlander Professor at Manne Siegbahn Institute of Physics, Frescativägen 24, S-104 05 STOCKHOLM, Sweden

Specific excited states of ions in channels may also be prepared by the process of resonant coherent excitation [3, 4, 5]. Here the ion is excited by the periodic potential of the crystal lattice. As it passes a row of atoms with atomic spacing, d, it experiences a perturbation with frequency (v_i/d). Excitation from state i to state j can occur when $hK(v_i/d) = \Delta E_{ij}$ where K is a harmonic $= 1, 2, 3\ldots$. Ions excited in this manner may be prepared in specific m states and ionization of these species by electronic collisions in channels may also be studied.

Experimental

The experiments were carried out using the 25-MV tandem accelerator of the Holifield Heavy Ion Research Facility (HHIRF) at ORNL. Ion beams were colimated and passed through channels in thin crystals and measurements were made of *a)* the charge state distribution of the emergent ions using either electrostatic or magnetc deflection and a solid state position sensitive detector and, *b)* the X-ray spectra using a Si(Li) detector. In some cases coincidences between charge capture and X-ray emission were measured.

In fig. 1 we show a measure of the "frozen" charge state effect using 315 MeV Ti^{q+}(q = 20,21,22) beams passing through a < 100 > channel in a 2.6 μm Si crystal and compare it to passage through the same crystal in a random direction. In the random direction, we obtain an equilibrium charge-state distribution independent of input charge state. For injection in the channeled direction, > 85% of the beam exits in its initial charge state.

The Si(Li) X-ray detector used had a resolution of 145 eV at 5.9 keV. For ions $Z \gtrsim 15$ we are easily able to resolve K-X rays from relaxation to a totally empty K shell from those arising from filling a K vacancy in an ion already containing a 1s electron. In fig. 2, we show X-ray spectra for 315 MeV Ti^{q+}(q = 20,21,22) channeled in < 100 > Si (2.6 μm) taken in coincidence with electron capture, $q \to q-1$. The peak at 5.1 KeV corresponds to a $(2p1s) \to 1s^2$ transition. The peak at 5.3 KeV comes from a $2p \to 1s$ transition in the one electron ion.

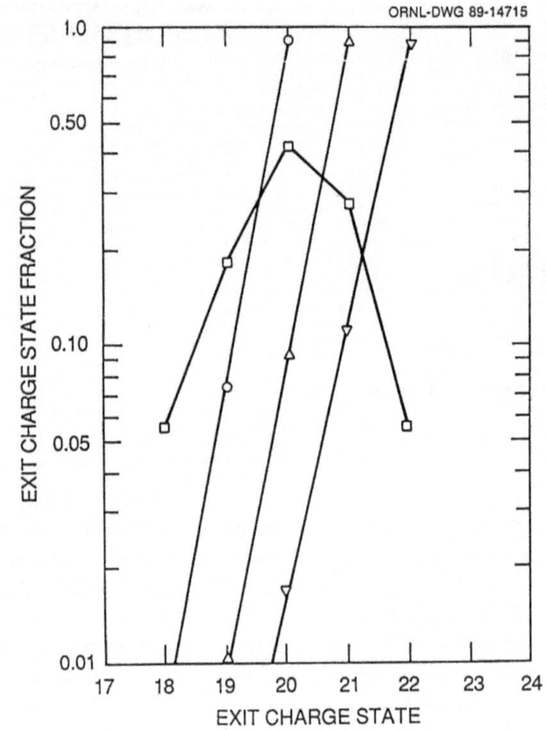

Fig. 1. Exit charge state distributions for injection of 315 MeV Ti^{20+} (o), Ti^{21+} (Δ), Ti^{22+} (∇) in a < 100 > channel in Si 2.6 μm thick, and ($\Box$) exit charge distribution for random direction.

Radiative Electron Capture

The radiative electron capture (REC) peaks shown in fig. 2 arise from recombination of Fermi electrons (e_f) with the penetrating ions. In the case of the bare ion $e_f + Ti^{22+} \to Ti^{21+}(1s) + h\nu$ (fig. 2, part III) and for the one electron ion $e_f + Ti^{21+}(1s) \to Ti^{20+}(1s^2) + h\nu$ (fig. 2, part II). The shapes of the peaks are determined by the Compton profile of the electrons in the channel and fit very well with a 10 eV Fermi distribution. Moreover, since the Fermi width is determined by the electron density and the radiative recombination cross sections are accurately known, we have a measure for expected profile for resonant collisional excitation processes, the electron target thickness and hence the cross section for production of other X-rays in the spectrum.

Results and Discussion

Consider, for example, a hydrogen-like ion passing through a crystal channel at a velocity $v_i \gg v_f$, with a relative velocity corresponding to that required to excite a K electron and to simultaneously capture the exciting electron into an L shell. The formation of this doubly excited state is the exact reverse of a KLL Auger transition.

A summary of most of the reactions which could occur under these conditions is given below, in equation 1.

$$I.\quad A^{(Z-1)+}(1s) + e \to \left[A^{(Z-2)+}2\ell 2\ell'\right]^{**}.$$

$$II.\ a)\quad \left[A^{(Z-2)+}(2\ell 2\ell')\right]^{**} \to A^{(Z-1)+}(1s) + e$$

$$b)\quad \left[A^{(Z-2)+}(2\ell 2\ell')\right]^{**} + e \to \left[A^{(Z-2)+}(2\ell' 2\ell'')\right]^{**} + + e.$$

$$III.\ a)\quad \left[A^{(Z-2)+}(2\ell 2\ell')\right]^{**} \to \left[A^{(Z-2)+}(1s2\ell)\right]^{*} + + \underline{h\nu_1}$$

$$b)\quad \left[A^{(Z-2)+}(2\ell 2\ell')\right]^{**} + e \to \left[A^{(Z-2)+}(2\ell 3\ell)\right]^{**} \xrightarrow{e} \left[A^{(Z-1)+}(2\ell)\right]^{*} + 2e.$$

$$IV.\ a)\quad \left[A^{(Z-2)+}(1s2\ell)\right]^{*} \to \underline{\left[A^{(Z-2)+}(1s^2)\right]} + \underline{h\nu_2}$$

$$b)\quad \left[A^{(Z-2)+}(1s2\ell)\right]^{*} + e \to \left[A^{(Z-2)+}(1s3\ell)\right]^{*} + e \to \left[A^{(Z-1)+}(1s)\right] + 2e.$$

$$V.\ a)\quad \left[A^{(Z-1)+}(2\ell)\right]^{*} \to \left[A^{(Z-1)+}(1s)\right] + \underline{h\nu_3}$$

$$b)\quad \left[A^{Z-1)+}(2\ell)\right]^{*} + e \to \left[A^{(Z-1)+}(3\ell)\right]^{*} + e \to \underline{A}^{Z+} + 2e.$$

Equation 1

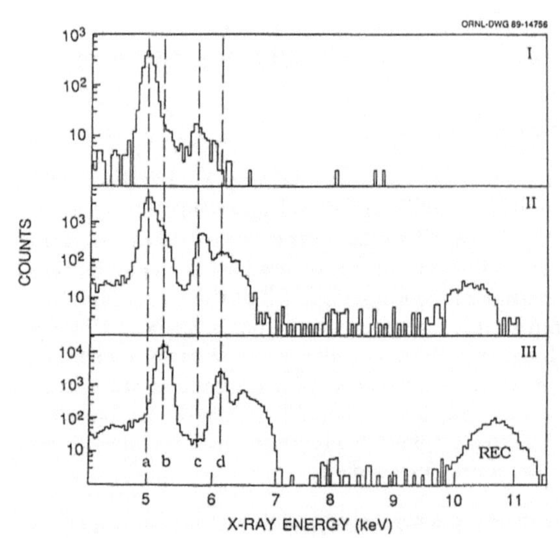

Fig. 2. X-ray spectra (singles) for 315 MeV Ti^{q+} channeled thorugh < 100 > Si 2.6 μm thick: (I) $q = 20$, (II) $q = 21$, and (III) $q = 22$ taken in coincidence with emergent charge state $(q-1)$. Specific features identified in the spectra are, respectively: (a) $1s2\ell \to 1s^2$, (b) $2p \to 1s$, (c) $1s3\ell \to 1s^2$, and (d) $3p \to 1s$. K-radiative electron capture X-rays appear at 10.3 KeV in II and 10.7 keV in III.

In a dilute electron target, only eqns 1-I, 1-III.a, and 1-IV.a are involved. The observable end products would be $h\nu_1$ from a $2\ell 2\ell' \to 1s2\ell$ transition and $h\nu_2$ from a $1s2\ell \to 1s^2$ transition and an ion of decreased charge $A^{(Z-2)+}$ all in coincidence. However, in a dense medium when collisional rates become comparable to radiative rates, the other channels come into play. The collisional path leads to either $h\nu_3$ from a $2\ell \to 1s$ transition (eqn. I-V.a) or to complete ionization (eqn. I-V.b).

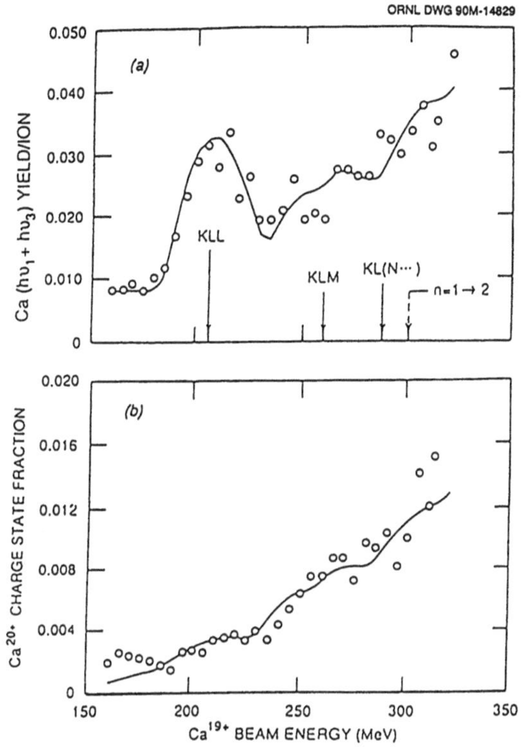

Fig. 3. (a) Yield of $h\nu_1(2\ell n\ell' \rightarrow 1sn\ell')$ plus $h\nu_3(2\ell \rightarrow 1s)$ calcium X-rays (see eqn. 1) and (b) charge fraction of Ca^{20+} ions as a function of Ca^{19+} ion energy incident on a < 100 > channel in Si (1.2 μm thick). The smooth curves are results of a computer simulation as described in the text.

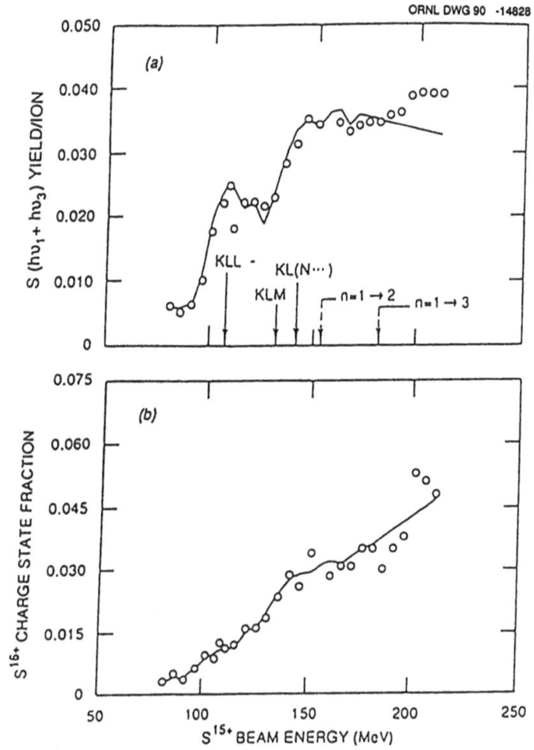

Fig. 4. (a) Yield of $h\nu_1(2\ell n\ell' \rightarrow 1sn\ell')$ plus $h\nu_3(2\ell \rightarrow 1s)$ sulphur X-rays (see eqn. 1) and (b) charge fraction of S^{16+} ions as a function of S^{15+} ion energy incident on a < 110 > channel in Si (1.2 μm thick). The smooth curves are results of a computer simulation as described in the text.

Magnitudes of these contributions can be estimated by assuming rapid collisional substate mixing [6], $n\ell \leftrightarrow n\ell'$ and then using the fastest available radiative and Auger rates. The radiative and Auger rates have been tabulated by Seely [7]. The ionization cross sections are estimated from the Lotz semi-empirical formula [8]. The excitation cross sections used are obtained from the Seaton formula [9].

No calculations exist for the cross sections for fine structure mixing (eqn. 1-II.b) for the ions in question. However, if we use the recent calculation [6] on, e.g., $(2s^2) \rightarrow (2s2p\,^3P_2)$ transition in Fe^{22+} we obtain a cross section of $\sim 3 \times 10^{-16}$ cm^2 at a velocity corresponding to KLL excitation of Fe^{23+}. Since the fine structure splitting decreases with decreasing Z, the corresponding cross sections for the ions in question should be even higher. The main point is that these cross sections are so large that complete mixing is assured. We have no direct calculation for the dielectronic excitation cross section (i.e., eqn. 1-I) instead, we estimate it from measurements of RTE cross sections on the same or similar systems. The collisional rates, r_i, are obtained from

$$r_i = \sigma_i \, \rho_e \, v_i$$

where ρ_e is the electron density taken to be 1.44×10^{23} cm^{-3} for the Si < 110 > channel[10] and σ_i is the cross section for the considered process. However, since the dielectronic excitation process (eqn. 1-I) is a resonant one, the actual rate is grossly reduced (factor of $\sim$ 1000) due to the Compton spread in relative energies. In our calculation, we use a 10 eV Fermi distribution for the electrons in a < 110 > Si channel. This value is taken from the work of Appleton et al. [10], and we derived a similar Fermi width from our measurement of the radiative electron capture (REC) peak (e.g., see fig. 2).

In fig. 3a, we show for channeled Ca^{19+} the yield per injected ion of $(h\nu_1 + h\nu_3)$ (eqn. 1-III.a and 1-V.a) as a function of ion energy. The features correspond to dielectronic excitation of KLL, KLM, KLN, etc., and, at 300 MeV, to direct $1s \rightarrow 2p$ excitation. The energies at which these contributions should appear are indicated by the arrows at the bottom of fig. 3a. The solid curves in fig. 3a are calculated from rate equations shown in eqn. 1, folded with the Compton profile of the 10 eV Fermi distribution. The populations of the ten lowest hydrogen- and helium-like configurations ($1s$ to 3ℓ and $1s^2$ to $3\ell 3\ell'$, respectively) are iteratively solved and summation bins for numbers of X-rays emitted are accumulated in 400 steps through the

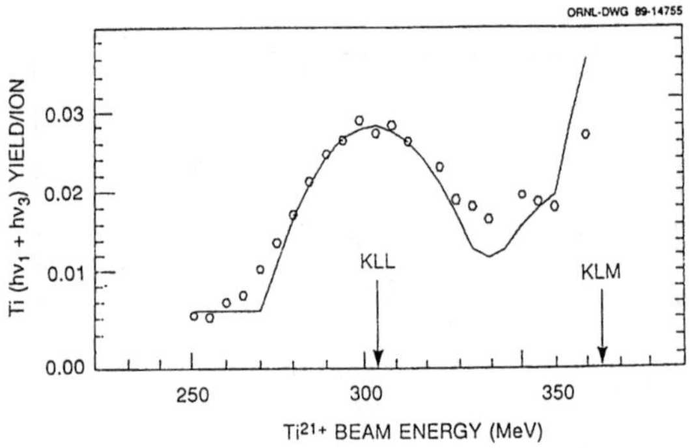

Fig. 5. (a) Yield of $h\nu_1(2\ell n\ell' \rightarrow 1sn\ell)$ plus $h\nu_3(2\ell \rightarrow 1s)$ titanium X-rays (see eqn. 1) as a function of Ti^{21+} ion energy incident on a < 110 > channel in Si (1.2 μm thick). The smooth curve is a result of a computer simulation as described in the text.

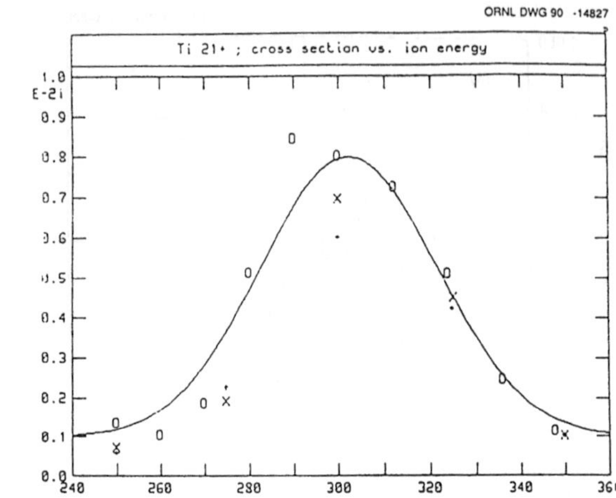

Fig. 6. Coincidence measurements for $Ti^{21+} + e \rightarrow Ti^{20+} + h\nu$ in < 110 > Si as a function of Ti ion energy. Yield in the KLL resonance region as a function of crystal thickness; 0–1.1 μm, + 2.1 μm, X 2.8 μm.

crystal length of 1.2 μm. Charge-state fractions and x-ray yields per ion are calculated, including effects of Auger and X-ray decay of excited ions downstream of the crystal. It should be noted that the only normalization is the addition of backgrounds linear in projectile energy and that the predicted values are absolute in X-rays per ion. Using the same rate coefficients, we can calculate, e.g., the yield of Ca^{20+} ions (cf. eqn. 1-V.b) and plot it together with the experimental data in fig. 3b.

In the case of Ca^{19+}, the radiative processes dominate, e.g., the ratio of rates for the first branch point in eqn. 1 (III.a/III.b) $\simeq$ 5. In the case of S^{15+}, the radiative rates are slower and the excitation and ionization cross sections are higher so that the collisional channel comes more into play, i.e., for S^{15+} (III.a/III.b) $\simeq$ 1. In fig. 4a, we show, as a function of S^{15+} ion energy, the data for the yields of $(h\nu_1 + h\nu_3)$. In fig. 4b we show the S^{16+} charge fraction which is correspondingly higher than the totally stripped ion fraction for Ca. The solid curves are calculated in the same manner as for Ca^{19+} above.

In the case of Ti^{21+}, (III.a/III.b) $\cong$ 9. Accordingly, the number of totally stripped ions becomes vanishingly small, but the effect is easily seen in the X-ray channel alone as demonstrated in fig. 5.

In all the above we have made many assumptions concerning rates. To validate these assumptions we take a simpler case where radiative processes should dominate (i.e. where secondary electron collisions are unimportant) and see whether we can derive a cross section for dielectronic recombination. For this we chose the case of the KLL resonance in Ti^{21+} and measured the X-ray spectra in coincidence with Ti^{20+}. This measures the dielectronic recombination rate. The importance of secondary collisions can be measured by varying the thickness of the Si crystal used i.e. in the absence of secondary collisions the yield will be linearly proportional to thickness. The results for the yield per unit thickness is shown for the three thicknesses used 1.1, 2.1, and 2.8 μm in fig. 6 and are sensibly independent of thickness. The curve in fig. 6 is for a 10 eV Fermi distribution fit to the points. From this we can derive a cross section of $0.70 \pm 0.10 \times 10^{-21}$ cm^2 per electron. To compare with RTE data [11] (where the width of the resonance is determined by the Compton profile of H_2) we multiply our values by 1.2 and, since the RTE cross sections are given per molecule, we divide those values by two. The value thus obtained is 0.72 ± 0.12.

Resonant Coherent Excitation (RCE)

In previous work [1, 3, 4, 5] we showed that hydrogenic ions passing through axial and planar channels could be excited from $n = 1$ to $n = 2$ when the frequency of perturbation by the atoms in the crystal spaced a distance d apart came into resonance when the spacing between eigenstates i and j,

$$\Delta E_{ij} = hK(v_i/d)$$

where K is a harmonic 1,2,3 ... of the (v_i/d) frequency. We also showed that the degeneracy in the $n = 2$ levels is removed; first by the assymetry in the crystal field and second by Stark mixing of $2s$ with $2p_z$ (z being the direction of ionic motion) which is caused by the wake field following the ion. The main point in the context of this paper is that the resonant frequency, and hence velocity, for excitation to $2p_{x,y}$ is different than that for $2p_z$ and they can be excited selectively.

In our previous work we used C, N, O and F ions available at EN Tandem energies. These light ions when excited to $n = 2$ had a high probability of ionizing by secondary collision in the channel (cf. eq. 1) and the resonances were observed as an increase in the population of totally stripped ions.

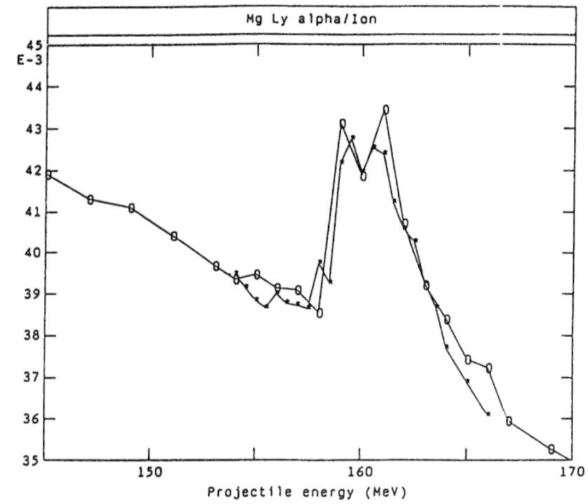

Fig. 7. Resonant coherent excitation of Mg^{11+} (4th harmonic) in < 100 > Au observed as Mg Ly α emission; Yield of X-rays as a function of Mg^{11+} energy.

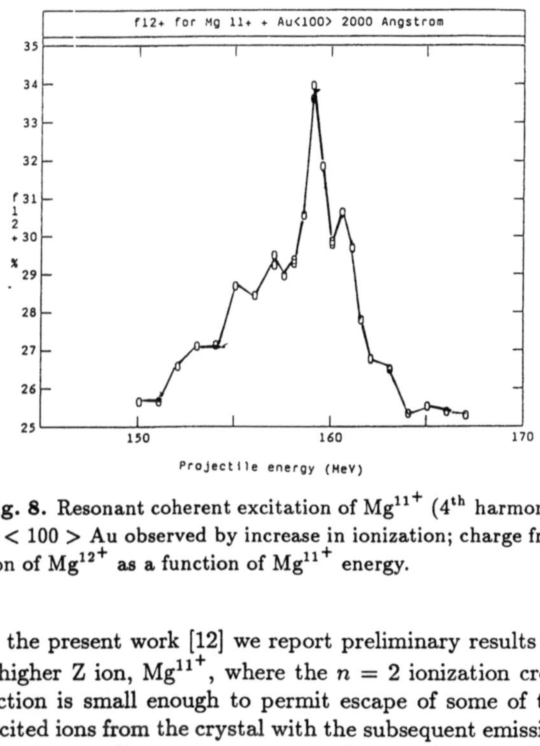

Fig. 8. Resonant coherent excitation of Mg^{11+} (4th harmonic) in < 100 > Au observed by increase in ionization; charge fraction of Mg^{12+} as a function of Mg^{11+} energy.

In the present work [12] we report preliminary results on a higher Z ion, Mg^{11+}, where the $n = 2$ ionization cross section is small enough to permit escape of some of the excited ions from the crystal with the subsequent emission of radiation. Since we can excite different orientations (m states) of the ion selectively by varying the velocity we can measure the separate ionization cross sections for these states by determining the yields of totally stripped ions compared to those which emit a Ly α X-ray.

We used the HHIRF to supply beams of Mg^{11+} at energies from 145 to 170 MeV. These were passed through an Au crystal 1200 Å thick in the < 100 > directions (d = 4.08 Å). The emergent ion charge distribution and two Mg K X-ray yields were measured as a function of ion energy. The results are shown in figs 7 and 8. The fourth harmonic of the $n = 1$ to 2 resonance (ΔE_{ij} = 1468 eV) should appear at 163 MeV if the transition were to occur in vacuum.

The crystal field effects shift the peaks to lower energies. The upper of the two peaks corresponds to excitation into a $2p_z$ state and the lower to a $2p_x$ or $2p_y$ state. A comparison of the two curves shows that the probability of escape from the crystal without ionization (X-ray curve) is greater for ions in the $2p_z$ state than those in the $2p_{x,y}$ state.

At this point these RCE data are not yet quantified and are presented as proof of principal for experiments which measure electron bombardment ionization cross sections for short lived excited states with specific polarization.

This research was sponsored by the U.S. Department of Energy, Office of Basic Energy Sciences, Division of Chemical Sciences, under Contract No. DE-AC05-840R21400 with Martin Marietta Energy Systems, Inc.

References

1. Datz, S., and Moak, C.D.: "Heavy-Ion Channeling," in Treatise on Heavy-Ion Science, ed. Bromley, D.A., Vol. **6**, Astrophysics, Chemistry, and Condensed Matter, Chapter 5 (1985) pp. 169–240.

2. Datz, S., Vane, C.R., Dittner, P.F., Giese, J.P., Gomez del Campo, J., Jones, N.C., Krause, H.F., Miller, P.D., Schulz, M., Schöne, H., and Rosseel, T.M.: Phys. Rev. Lett. **63**, 742 (1989); Nucl. Inst. Meth. **B48**, 114 (1990).

3. Datz, S., Moak, C.D., Crawford, O.H., Krause, H.F., Dittner, P.F.. Gomez del Campo, J., Biggerstaff, J.A., Miller, P.D., Hvelplund, P., and Knudsen, H.: Phys. Rev. Lett. **40**, 843 (1979).

4. Datz, S., Moak, C.D., Crawford, O.H., Krause, H.F., Miller, P.D., Dittner, P.F., Gomez del Campo, J., Biggerstaff, J.A., Knudsen, H., and Hvelplund, P.: Nucl. Inst. Meth. **170**, 18 (1980).

5. Miller, P.D., Krause, H.F., Biggerstaff, J.A., Crawford, O.H., Datz, S., Dittner, P.F., Gomez del Campo, J., Moak, C.D., Neskovic, N., Pepmiller, P.L., and Brown, M.D.: Nucl. Inst. Meth. **B13**, 56 (1986).

6. Kim, Y.-K., and Desclaux, J.-P.: Phys. Rev. **A38**, 1805 (1988).

7. Seely, F.: Atomic Data and Nuclear Data Tables **26**, 137 (1981).

8. Lotz, W.: Z. Physik **206**, 206 (1968).

9. Seaton, M.J.: "The Theory of Excitation and Ionization by Electron Impact," in Atomic and Molecular Processes, ed. Bates, D.R., p. 374 (New York, Academic Press 1962)

10. Appleton, B.R., Ritchie, R.H., Biggerstaff, J.A., Noggle, T.S., Datz, S., Moak, C.D., Verbeek, H., and

Neelavathi, V.N.: Phys. Rev. **B19**, 4347, see discussion on page 4356, (1979).

11. Presently no direct RTE measurement exists for H-like Ti^{21+}. To obtain a value for comparison we take the results for Li-like Ti^{19+} + H_2 given by Reusch et al. (Nucl. Inst. Meth. **B23**, 137 (1987)) and scale to Ti^{21+} using the values for Li-like, He-like and H-like Ca given by Tanis et al.(Phys. Rev. **A34**, 2543 (1986)).

12. Collaborating on this part of the work were Iwata, I., Komaki, M., Kimura, K., and Yamazaki, Y., University of Tokyo and Honda, F., and Fujimoto, F., University of Osaka.

Supplement to Z. Phys. D - Atoms, Molecules and Clusters 21, S 51-56 (1991)

Atoms, Molecules
Zeitschrift für Physik D and Clusters

Electron-electron interaction in fast ion-atom collisions*

A. Warczak
Institute of Physics, Jagiellonian University, Reymonta 4, 30-059 Kraków, Poland

Abstract

Electron correlation observed in ion-atom collisions involving inner-shell processes is discussed in brief. In particular, correlated double ionization of selected molecular orbitals is proposed in order to explain unusual behaviour of impact-parameter-dependent probabilities for vacancy production. Moreover, among transfer and excitation processes some evidence for a correlated three-electron process is emphasized.

I. Introduction

In the recent decades a considerable progress in our understanding of inner-shell processes, generated in heavy ion-atom collisions, has been achieved /1-4/. Essentially, this progress was based on a widely applied independent-electron approximation, where electron-electron interaction is ignored or replaced by central potentials. Within this approach it was possible to separate transitions of one active electron (excitation, ionization, capture), considering all the residual electrons as independent spectators.

Nothing prevents an ion-atom system that multiple combinations of one-electron transitions take place in a collision. Recently, growing interest in such processes was observed. In particular, this interest was focused on electron correlation manifested just in multiple-electron transitions /5-13/. Correlation may be understood as deviation from an independent-particle approximation and is an useful concept in the study of the many-body problem.

The aim of the paper is to present briefly two current aspects of correlation effects observed via inner-shell processes in collisions of heavy ions with atoms at MeV/u energies. The first problem, correlation in double inner-shell ionization, is discussed in Sect. II. Sect. III is concentrated on electron-electron interaction which is responsible for correlated transfer and excitation processes. In that section the process involving a two-electron transfer is emphasized.

* This work was supported in part by the Polish Research Project CPBP 01.09.

II. Correlated double inner-shell ionization

Two parameters, the asymmetry parameter (Z_1/Z_2) and adiabaticity parameter ($\eta=(v/v_e)^2$) are responsible for the classification of dynamic ion-atom systems relative to single-ionization mechanisms (Z_1 and Z_2 are the nuclear charges of projectile and target atom, respectively, v denotes projectile velocity and v_e the orbital velocity of the active electron). Such a rough classification, known as a Madison-Merzbacher map /14/, divides colliding systems into two regions (Fig.1). Region (1) with

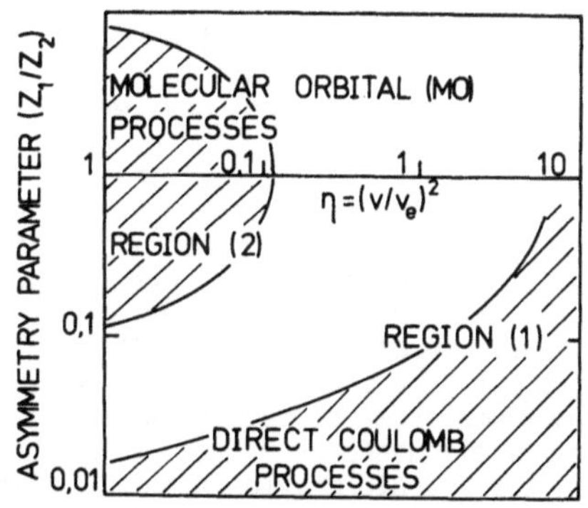

Fig. 1. Madison-Merzbacher map dividing ion-atom systems into two regions.

$Z_1/Z_2 \ll 1$ and (or) $\eta \gg 1$ corresponds to the situation where the perturbing influence of the nuclear motion on electrons is very weak due to a very small strength of the projectile-target interaction or due to a very short interaction time. In this case single-electron transitions are described in terms of processes acting direct on target or projectile states and induced by the changing Coulomb field. Region (2) with $Z_1/Z_2 \sim 1$ and $\eta < 1$ corresponds to a strong interaction between electrons and colliding nuclei which is responsible for the formation of a transient diatomic complex (quasimolecule). In this case excitation mechanisms influence the quasimolecular states (molecular orbital model).

It is well known that correlated two-electron processes in heavy atoms ($Z \geq 20$), observed in inner-shells in "static" situations, e.g. a two-electron one-photon transition into doubly ionized K shell /15/ or a double K-shell ionization due to internal conversion /16/, are very weak. There, transition rates for correlated two-electron jumps are of the order of 10^{-3}-10^{-5} in respect to single-electron transitions /15,16/. Therefore, it is quite obvious that the best experimental environment, for observation of correlation effects in collisions, exists in region (1) shown in Fig.1.

If one can calculate within the independent-particle model the probability for the corresponding single-electron ionization or one can measure it exclusively, then correlation effects can be readily examined by observing the probability for a two-electron process. Correlation means any deviation of this probability from the product of the probabilities for the one-electron transition. Collision dynamics in region (1) introduces a very small single-electron perturbation which, by turns, yields a very small probability for the uncorrelated double transition. Therefore, just here, any electron correlation should be identified much easier than in other regions of Fig.1. As, additionally, at high projectile velocities the physics of atomic collisions becomes relatively simple, a full advantage of region (1) has been taken, in in the past, to study correlation effects in double ionization of few-electron systems in collisions with fast projectiles /17,18/. However, at present, double ionization at high velocities is not well understood, even for the simplest two-electron targets /9,19/.

The MO picture (region (2) in Fig.1) has been widely applied in the last twenty years to analyse inner-shell ionization in heavy ion-atom systems /1,2,3,14/. In spite of the fact that very energetic heavy ions, up to several GeV, were very often used in experiments /2/, the projectile velocities were still smaller than the orbital velocities of inner-shell electrons considered. In the MO region (Fig.1) the probabilities for single K-shell vacancy production are rather large, typically, of the order 10^{-2}-10^{-1} or even larger /2,3/. Thus, if correlation between the two electrons in a shell in a colliding system would be still of the same order of magnitude as in the "static" cases discussed already, the required correlation tests would provide a hard challenge for the experimentalists.

Very recently, McGuire /6/ discussed extensively the problem of correlation in collisions of many-electron targets with charged projectiles. He separated, for the first time, two types of correlations (corrections to the independent-electron model): <u>static</u> and <u>scattering</u> correlation. Static correlation is that included in the asymptotic wave functions of the initial and final electron states. This part of correlation is obtained by switching off the electron-electron interaction for the collision time. Static correlation is already well known. It is the correlation that gives rise to the rearrangement processes (e.g. Auger effect). Scattering correlation is a new process which is intrinsic to the scattering. It is this part of correlation which occurs during the collision and influences the evolution operator. Thus, a hint has been given to look more widely for electron correlation in heavy ion-atom systems.

Recent experiments /19,20/ suggested that scattering correlation may be very important at large scattering angles (small impact parameters b), well beyond the region which mostly contributes to the total cross section for the particular electronic process. For inner-shell ionization this region corresponds to impact parameters below the so-called adiabatic radius r_{ad} /21/ ($r_{ad} = \hbar v/\Delta E$, where ΔE is the energy transfer to the removed electron). It became clear that measurements at small impact parameters, well inside r_{ad}, are required to look for possible scattering correlation while removing two electrons from an inner-shell.

Zoran et al. /13/ performed the first experiment aimed to look just for this effect in a system well placed in region (2) of Fig.1. In a triple-coincidence experiment the impact-parameter dependence was estimated for single- and double-vacancy production in a system close to K-L level-matching conditions (Ni + Pb at 1.5 MeV/u). In such systems K- and L-vacancy production in the lighter and heavier collision partner, respectively, is dominated by ionization of the 3dσ MO, thus, retaining the simplicity of the two-electron problem.
A significant enhancement, up to a factor of about 3, of the double ionization was observed in that experiment relative to the predictions of the independent-particle model. This enhancement extends in the region of reduced impact parameters between 0.1 and 0.3 ($b_{red} = b/r_{ad}$). The data suggest that the enhancement exhibits a peak-like structure with a maximum at $b_{red} \sim 0.15$. Moreover, the correlated electron process generates double-ionization probabilities in the 10^{-2} range /13/, i.e. of the order of probabilities expected from the independent-electron model. This findings provide a strong support of the suggestions drawn above that, indeed, collision dynamics is able to alter, even significantly, the electron correlation.

The impact-parameter dependence of inner-shell vacancy production was extensively investigated in the long-standing history of ion-atom collisions /2,3/. However, in most cases, inclusive measurements of excitation function, P(b), for single ionization are available, only. Thus , information on multiple-vacancy production is partially lost. P(b), the number of vacancies produced per collision in two-electron systems, may have a form

$$P(b) = P_{ex}(b) + 2P_d(b) , \qquad (1)$$

where $P_{ex}(b)$ means probability for exclusive single-vacancy production and $P_d(b)$ that for double-vacancy production. Previously, for $P(b) \ll 1$, the double-vacancy production has been considered as negligibly small. Such a treatment is correct within the independent-particle approximation. However, in the case of a strong correlation, as in /13/, P(b) must be influenced, even strongly, by multiple ionization. Therefore, the inclusive P(b) curves may exhibit unusual features which are not attributed to the single-vacancy production.

Let us discuss some examples of such an unusual behaviour of P(b). First, Fig.2 shows ionization of the 3dσ MO at 1.4 MeV/u projectile energy and for $Z_1+Z_2 \sim 130$ /22/, i.e. very close to the conditions of experiment /13/. It turns out from the figure that ionization probabilities in gas targets are significantly lower than in a solid. In a gas

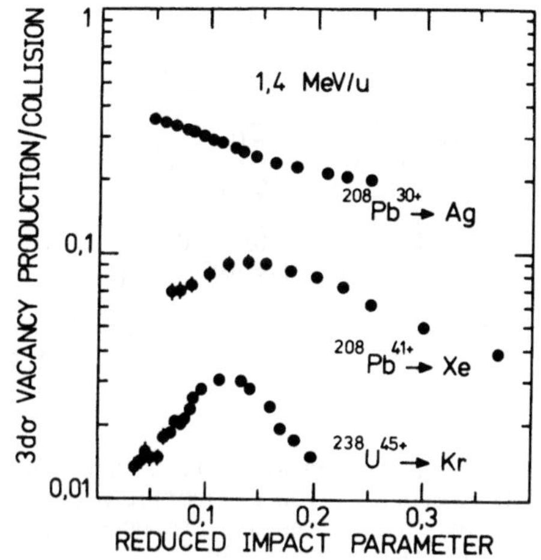

Fig. 2. $3d\sigma$ vacancy production per collision plotted versus reduced impact parameter. Data obtained in an inclusive measurement of single-vacancy production /22/.

target direct $3d\sigma$ excitation into continuum states dominates. Moreover, ionization probabilities in gas targets are just of the order of 10^{-2} (Fig.2), i.e. exactly of the order of correlated double $3d\sigma$ vacancy production attributed to direct ionization into continuum /13/. Thus, the distinct maxima in P(b) (gas-target data in Fig.2) are probably due to the onset of correlated double ionization (1). The position of the maxima which corresponds to that observed in /13/ strongly supports this suggestion. In a solid target, however, single-vacancy production increases by almost one order of magnitude (Fig.2), mainly ,due to the inverted character of the colliding Pb+Ag system /23/. Here, additional MO processes dominate the ionization /23/. Therefore, the correlated double-vacancy production does not emerge from the "background" (1). In the original paper /22/ the shape of gas-target data has been explained, controversially, by strong variation of fluorescence yield.

The second example (Fig.3) concerns a very similar but noninverted Kr+U system /25/. Here, again as in /13/, the direct $3d\sigma$ ionization plays an important role. A distinct structure in P(b) with the amplitude of the order of 10^{-2} at b_{red}~0.2 is seen, as well (Fig.3). The structure is not attributed to the known mechanisms of single-vacancy production.

Next, we consider structures observed in the $2p\sigma$ excitation curves, based on inclusive measurements of the single-vacancy production /24,26/. Several explanations of these structures have been proposed previously, running from state selective electron capture after a close collision to multiple resonant transfer and excitation (see Sect. III). It is very interesting to note that, similarly to the $3d\sigma$ case, the enhancement of P(b) with super-imposed structures rises significantly in the range 0.1-0.3 on the reduced impact-parameter scale. As, additionally, the absolute amplitudes of the structures on the probability scale, are of the order of 10^{-2}-10^{-3}, scattering correlation may be suggested as a common origin of all the P(b) features discussed already. Probably, the two $2p\sigma$ electrons undergo the same kind of scattering correlation as those occupying the $3d\sigma$ MO.

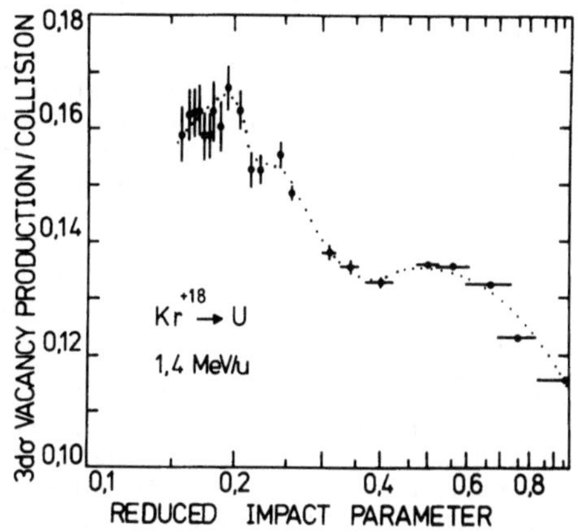

Fig. 3. $3d\sigma$ vacancy production per collision plotted versus reduced impact parameter. Data obtained in an inclusive measurement of single-vacancy production /25/. The logarithmic scale for impact parameters is used to expose the region of small impact parameters. The dotted line serves to guide the eye. Note that vertical scale does not start at zero.

Structures observed in P(b) can come about in collisions through interference of different contributions to the transition amplitude, whose relative phase depends on the parameters of the collision (e.g. impact parameter). In the cases discussed above we observe, most probably, interference between the probability amplitudes for correlated and uncorrelated double-vacancy production. Both components do not correspond to distinguishable processes. Consequently, it is appropriate to add the amplitudes which may interfere /6/. This effect should be pronounced in P(b) curves where scattering correlation provides double-ionization probabilities close to those for uncorrelated process. The above effect influences the $P_d(b)$ term in (1), only.

It seems that region (2) in Fig.1 contains much more surprising effects related to electron correlation, than expected. Therefore, theoretical and experimental work should be invested to proof suggestions given above.

III. Electron correlation in transfer and excitation processes

A very sensitive parameter for electron correlation effects are the rates of nonradiative vacancy decay. This vacancy decay is a two-electron process which occurs with the emission of an electron into continuum state (Auger effect). The inverse of Auger transition, the excitation part of dielectronic recombination (DR), is already well known /27/. In this process a free electron recombines with the projectile via the excitation of a projectile electron. This is a resonant process occuring only at special velocities of projectiles in respect to the free electron.

In ion-atom collisions , a combination of single-electron processes may involve a target-electron capture by the projectile and, simultaneously, an excitation of a projectile electron. This class of atomic reactions is known as transfer and excitation (TE) processes.

In collisions, the highly charged projectile may interact with weakly bound target electrons in a way similar to DR. Thus, a projectile electron becomes excited by an interaction with the captured quasi-free electron. This process (see inset of Fig.4), called resonant transfer and excitation (RTE), focused a considerable attention over the last decade /5,28/. First, due to electron correlation involved into the process which could be tested in collisions. Second, due to expected dominance of RTE among inner-shell processes at very high projectile energies /29/. The fact that in RTE the target electron is not exactly free leads to a broadening of the resonances in the cross sections by the momentum distribution of the target electron (Compton profile). In most experiments the RTE cross sections have been measured via registration of coincidences between projectile x rays, stabilizing the excited state, and the projectiles which have captured electrons /5,28/. An example of RTE excitation function is shown in Fig. 4.

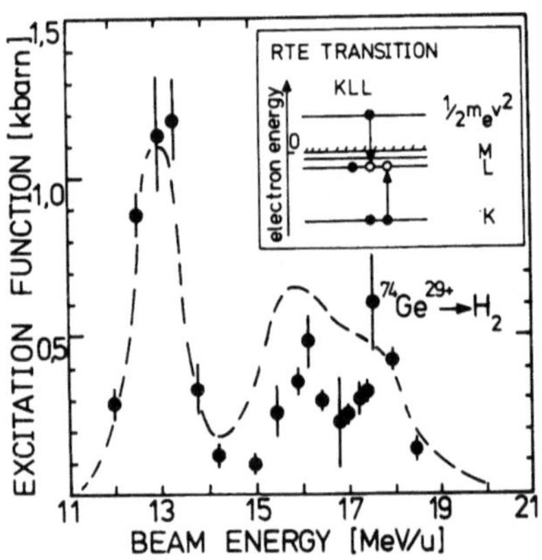

Fig. 4. RTE excitation function observed in Li-like Ge ions bombarding H_2 /36/. Dashed line - theoretical calculations. Inset shows schematically the mechanism of RTE (inversed Auger process).

It is possible to scale the RTE excitation function to the adiabaticity parameter η /30/. All the main resonances for various projectiles correspond, nearly, to the same η values shown in Fig.5 ($0.5 \lesssim \eta \lesssim 0.75$). Thus, the region of electron correlation, studied by means of RTE processes, may be localized in Fig.1.

Theoretical treatment of RTE observed via x-ray emission is based on DR /31/ and, at present, is able to handle the experimental data. This fact may be related to the static character of electron correlation involved into RTE (inversed Auger transition).

The presence of the target nucleus provides an additional process that can populate states indistinguishable from those populated by RTE. Here, the projectile-electron excitation by the target nucleus is independent of capture.

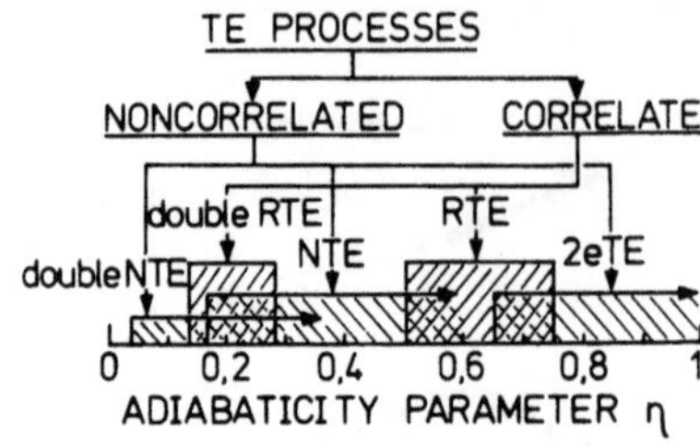

Fig. 5. TE processes shown on the axis of adiabaticity parameter η . For discussion see the text.

Nonresonant transfer and axcitation (NTE) is the dominant TE process at low projectile velocities /32/. At higher velocities NTE produces a background for RTE measurements. In Fig.5 the region for dealing with NTE is shown, as well.

Very recently, a new noncorrelated TE process has been observed /33-35/ with the treshold at $\eta \sim 0.65$. This process designated as two-electron transfer and excitation (2eTE) is a combination of the projectile-electron excitation in a collision with a target electron (electron-impact excitation) and, again as in NTE, of independent target-electron capture into projectile. The 2eTE overlaps with the upper edge of RTE (Fig.5) and may produce a considerable background for RTE resonances involving capture into higher projectile shells.

Capture of several electrons, alone, by a highly charged ion should manifest correlation effects, as well. There has recently been some discussion in respect to correlation in double electron capture /37-39/. The best environment for its observation would be provided by a process where energy released in a joint two-electron jump from a target into a highly ionized projectile is invested to excite another projectile electron. Such a resonant process, multiple (double) RTE, has been postulated /26/. The process should exhibit, again as in single RTE, a characteristic behaviour of the excitation function. Its observable signature is, for incoming hydrogenic projectiles, the coincidence between x rays stabilizing the intermediate multiply excited state and double charge exchange. This process may be considered as inversed double Auger transition (inset in Fig.6) which is a known and established phenomenon /40/. On the η scale double RTE should appear at values considerably lower than single RTE (Fig.5). Very recently, the experimental evidence for double RTE has been reported /11/. The experimental excitation function for this process is shown in Fig.6. The main background is provided by double NTE where two electrons are captured independently /41/.

At present, double RTE is open for theoretical as well as for further experimental study.

Surprisingly, the cross section for double RTE, a correlated three-electron process, is very large (~ 1 kbarn /11/) and may easily compete with that for single RTE (Fig.4). Therefore, exotic multiple combinations of correlated single-electron processes are expected to play a very important role in future experiments with highly charged heavy ions. Although, first results considering many-electron processes in relativistic heavy ion-atom collisions

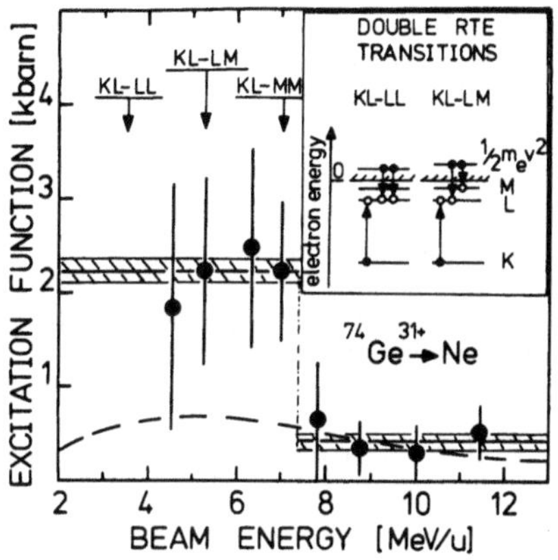

Fig. 6. Double RTE excitation function according to /11/. The arrows indicate the calculated positions of the expected maxima (see the inset). The shaded areas represent the errors of the fitted step-like function. The dashed line shows the double NTE contribution /41/.

show still importance of independent-electron approximation /42/, surprising phenomena may not be excluded,even in this region of projectile energies.

IV. Conclusions

Several experiments suggest that dynamic electron-electron correlation (scattering correlation) plays an important role, especially, in close collisions of very heavy ions with atoms. Thus, the MO region becomes very attractive for searching for signs of correlation effects. Molecular orbitals (3dσ, 2pσ), transiently formed in these collisions, offer simple quasi two-electron systems for probing electron correlation in much more complex objects. In this way, one may dose complications contained in the many-body problem.

Experiments related to TE processes provide additional evidence for electron correlation. A relatively large cross section for double RTE (three-electron correlation) shows that multiple-electron interaction may not be ignored in heavy ion-atom collisions. Perhaps, it will be possible in the future to observe the removal or capture of all electrons in a shell induced by a joint correlated process. The experimental base for performation of measurements of multiple-electron processes, including consideration of a full collision dynamics even at relativistic energies, is available /43-45/.

Acknowledgement

The author highly appreciates the fifteen years old collaboration with the Atomic Physics Group at GSI in Darmstadt where many of the experiments discussed in the paper were performed. Support in attending the Conference HCI V is also acknowledged.

References

1. Meyerhof, W.E., Taulbjerg, K.:Ann. Rev. Nucl. Sci. 27, 279(1977)
2. Mokler, P.H., Liesen, D.:in Progres in Atomic Spectroscopy. Part C., p. 321. New York:Plenum 1984
3. Wille, U., Hippler, R.:Phys. Rep. 132, 129(1986)
4. Cocke, C.L.:in Atomic Physics of Highly Ionized Atoms., p. 323. New York and London:Plenum 1983
5. Tanis, J.A.:Nucl. Instr. Meth. A262, 52(1987)
6. McGuire, J.H.:Phys. Rev. A36, 1114(1987)
7. Stolterfoht, N.:in Electronic and Atomic Collisions., p. 661. Amsterdam:North Holland 1988
8. Skutlartz, A., Hagmann, S., Schmidt-Böcking.,H: J. Phys. B21, 3609(1988)
9. Readig, J.F., Ford, A.L., Xushan Fang:Phys. Rev. Lett. 62, 245(1989)
10. Mack, M., Nijland, J.H., Straten, P., Niehaus, A., Morgenstern, R.:Phys. Rev. A39, 3846(1989)
11. Warczak, A., Stachura, Z., Szymański, A., Stöhlker, Th., Kozhuharov, C., Livingston, A.E., Mokler, P.H., Reusch, S.:Phys. Lett. A146,122 (1990)
12. Reading, J.F., Ford, A.L.:Comments At. Mol. Phys. 23, 301(1990)
13. Zoran, V., Enulescu, A., Piticu, I., Wintermeyer, G., Kambara, T., Gabr, M., Schuch, R.: Phys. Rev. Lett. 64, 527(1990)
14. Madison, D.H., Merzbacher, E.:in Atomic Inner-Shell Processes., p. 1. New York:Academic Press 1975
15. Wölfli, W., Stoller, Ch., Bonani, G., Suter, M., Stöckli, M.:Phys. Rev. Lett. 35, 656(1975)
16. Briand, J.P., Chevalier, P., Chetioui, A., Rozet, J.P., Tavernier, M., Touati, A.:Phys. Rev. A23, 39(1981)
17. McGuire, J.H.:Phys. Rev. Lett. 49, 1153(1982)
18. McGuire, J.H.:J. Phys. B17, L779(1984)
19. Giese, J.P., Horsdal, E.:Phys. Rev. Lett. 60, 2018(1988)
20. Kamber, E.Y., Cocke, C.L., Cheng, S., Varghese, S.L.:Phys. Rev. Lett. 60, 2026(1988)
21. Bang, J., Hansteen, J.M.:Phys. Scr. 22, 609(1981)
22. Liu, B., Liesen, D., Warczak, A.:J. Phys. B19, L241(1986)
23. Warczak, A.,Liesen, D., Maor, D., Schönfeldt, W.A.:J. Phys. B16, 1575(1983)
24. Liesen, D., Warczak, A., Armbruster, P., Bosch, F., Kozhuharov, Ch., Liu, B.:Z. Physik D5, 49 (1987)
25. Liu, B., Dohmann, H.D., Liesen, D., Warczak, A.: GSI Scientific Report 1982., p.150.
26. Warczak, A., Liesen, D., Liu, B.:J. Phys. B19, 3975(1986)
27. Burgess, A.:Astrophys. J. 139, 776(1964)
28. Mokler, P.H., Reusch, S., Stöhlker, Th., Schuch, R., Schulz, M., Wintermeyer, G., Stachura, Z., Warczak, A., Müller, A., Awaya, Y., Kambara, T.: Radiation Effects and Defects in Solids 110, 39 (1989)
29. Graham, W.G., Bernstein, E.M., Clark, M.W., Tanis, J.A.:Phys. Lett. A125, 134(1987)
30. Mokler, P.H., Reusch, S.:Z. Physik D8, 393(1988)
31. McLaughlin, Hahn, Y.:Phys. Lett. A88, 394(1982)
32. Clark, M., Brandt, D., Swenson, J.K., Shafroth, S.M.:Phys. Rev. Lett. 54, 544(1985)
33. Schulz, M., Schuch, R., Datz, S., Justiniano, E.L.B., Miller, P.D., Schöne, H.:Phys. Rev. A38, 5454(1988)

34. Schulz, M, Giese, J.P., Swenson, J.K., Datz, S., Dittner, P.F., Krause, H.F., Schöne, H., Vane, C.R., Benhenni, M., Shafroth, S.M.:Phys. Rev. Lett. 62, 1738(1989)
35. Zouros, T.J.M., Lee, D.M., Richard, P.:Phys. Rev. Lett. 62, 2261(1989)
36. Reusch, S., Awaya, Y., Kambara, T., Mokler, P.H., McLaughlin, D.J., Müller, A., Schuch, R., Schulz, M.:ICPEAC XV. Brighton 1987.Conference Abstracts., p.642.
37. Stolterfoht, N., Havener, C.O., Phaneuf, R.A., Swenson, J.K., Shafroth, S.M., Meyer, F.W.: Phys. Rev. Lett. 57, 74(1986)
38. Barat, M.:Comments At. Mol.Phys. 21,307(1988)
39. Mack, M., Nijland, J.H., Straten, P., Niehaus, A.:Phys. Rev. A39, 3846(1989)
40. Aberg, T.:in Atomic Inner-Shell Processes. Vol. 1., p. 353. New York:Academic Press 1975
41. Stöhlker, Th., Kozhuharov, Ch., Livingston, A. E., Mokler, P.H., Reusch, S., Stachura, Z., Warczak, A.:to be published.
42. Meyerhof, W.E., Anholt, R., Xiang Yuan Xu, Gould, H., Feinberg, B., McDonald, R.J., Wegner, H.E., Thieberger, P.:Nucl. Instr. Meth. A262,10 (1987)
43. Olson, R.E., Ullrich, J., Schmidt-Böcking, H.: Phys. Rev. A39, 5572(1989)
44. Ullrich, J., Schmidt-Böcking, H.:Phys. Lett. A125, 193(1987)
45. Ullrich, J., Schmidt-Böcking, H., Kelbch, C.: Nucl. Instr. Meth. A268, 216 (1988)

Supplement to Z. Phys. D - Atoms, Molecules and Clusters 21, S 57-61 (1991)

Atoms, Molecules
Zeitschrift für Physik D and Clusters

Dynamics of recoil-ion production*

R. Dörner[1], J. Ullrich[2], O. Jagutzki[1], S. Lencinas[1], H. Schmidt-Böcking[1], R. E. Olson[3]

[1] Institut für Kernphysik, Universität Frankfurt, W-6000 Frankfurt/M., FRG
[2] Gesellschaft für Schwerionenforschung, W-6100 Darmstadt, FRG
[3] Physics Department, University of Missouri-Rolla, Rolla, Missouri 65401, USA

*supported by BMFT and DFG

PACS: 34.10+x, 34.50 Fa

Abstract
The dynamics of the ionisation process of He by swift proton impact is discussed. Experimental results exploiting recoil-ion momentum spectroscopy (RIMS) are compared to results of nCTMC calculations. It is shown that the three body momentum exchange has explicitely to be taken into account to understand the scattering of the projectile in ionising collisions. For scattering angles around 0.55 mrad the scattering of the projectile at the target nucleus is of comparable strength to the scattering at the ionised electron. The electron emission is directed opposite to the scattered projectile for these scattering angles.

Introduction

Proton impact on helium is a fundamental collision system for our understanding of recoil ion-production by heavy particles. Nevertheless we are still far away from a comprehensive understanding of this most simple more electron collision system. In the last years great experimental efforts have been made to make highly differential data on this system available. Kamber et al [6,7] published differential cross sections for 3 and 6 MeV p-He for small scattering angles and found a shoulder at 0.55 mrad. Giese and Horsdal reported a peak in the ratio of double to total ionization of He by protons at a scattering angle of 0.9 mrad [8]. Schiwietz and Skogvall [12] measured for close collisions electron spectra in coincidence with projectile scattering angle and recoil charge state. In this paper we report on differential data exploiting recoil-ion momentum spectroscopy (RIMS [1]), an experimental technique specially suited for the investigation of the momentum exchange between projectile, recoil ion and electron.

Experiment

A novel recoil-ion momentum spectrometer was used to explore the momentum exchange between p, recoil ion and the ionized electron for the reaction

$$0.5\ \text{MeV p} + \text{He} \dashrightarrow \text{p} + \text{He}^{1+} + e^-$$

The momentum of the recoil-ion transverse to the beam direction was measured in coincidence with the transverse momentum transfer (=scattering angle and azimuthal angle) to the projectile (fig 1). For single ionization this means a complete determination of the transverse momentum transfer in the collission, since the momentum of the electron can be calculated from recoil-ion- and projectile momentum using momentum conservation.

The proton beam was collimated to a divergence less than 0.1 mrad. It was detected about 3 m downstreams the recoil-ion detector by a postion sensitive channel-plate detector with a resolution better than 0.4 mm. The direct beam was blocked by a 0.5mm wire in front of the channel-plate detector. Using a gas target pressure of about 10^{-3} hPa a ratio of protons scattered at the residual gas or the collimator slits to the protons scattered at the target gas of about 1:5 was achieved. The time of flight of the recoil-ions over field free drift path of 3.5mm (5mm or 2.5 mm in some cases) was measured to determine the recoil-ion momentum. A cylindrical tube of 10 mm diameter and 40 mm lengths is used as gas cell. The beam passes through the cylinder in 3.5 mm (5mm) distance to the cylinder wall. The recoil ions drift to the wall of the target cell and can exit through a slit of 20mm x 1mm parallel to the beam axis. Outside the gas cell they are accellerated in an electric field. After passing a focussing lens they are charge state separeted in a magnetic field and detected by a position sensitive channel-plate detector. To avoid a leakage of the acceleration field into the drift region and to assure a Faraday-cage free of electric fields even on the mV level in the target region the cylinder is covered by a copper mesh of 0.12 mm mesh-width and the gas cell is gold coated. The recoil-ion energies of interest are in the meV regime while the thermal energy at room temperature is about 40 meV. Therefore the spectrometer is mounted on a cold finger of a cryopump and the He gas is precooled. A gas temperatur of about 35 K was achieved. Besides thermal motion of the target atoms residual fringe fields in the target area are the limiting factor for the resolution of the spectrometer. From various test with different spectrometer materials and different lenght of the drift path we expect the systematical error of the measured recoil-ion energies due to fringe fields to be below 5 meV. For a more detailed discussion of the spectrometer see [1].

Theory

The experimental results are compared to calculations made using the n-body classical-trajectory Monte Carlo (n-CTMC) approach [2]. Two electrons are incorporated in the claculation. Both electrons are distributed within a microcanonical model, with a screening constant of 5/16 to the helium nucleus and a binding energy of 0.903 au equal for both electrons. Solving the classical equations of motion all interactions besides the electron electron interaction are taken into account. Doing this the momentum exchange between projectile and target nucleus, electron and target nucleus and projectile and electron is calculated. In a classical simulated atom in the center of mass system the nucleus is moving with a momentum which is opposite to the sum momentum of both electrons relative to the center of mass of the atom. While this internal motion is completely neglegible for calculation of total and differential cross sections and electron spectra, it has to be carefully taken into account to calculate recoil-ion energies in the meV regime. In a preceding paper [3 fig 2] we present data and first calculations of recoil-ion energies as function of scattering angle for 0.5 MeV p-He single ionisation. The theoretical results shown there are still influenced by the internal motion of the target nucleus wich adds an artificial mean energy of about 5 meV to all calculated recoil-ion energies. In this paper we present results of a revised version of the n-CTMC which is corrected for this effect.

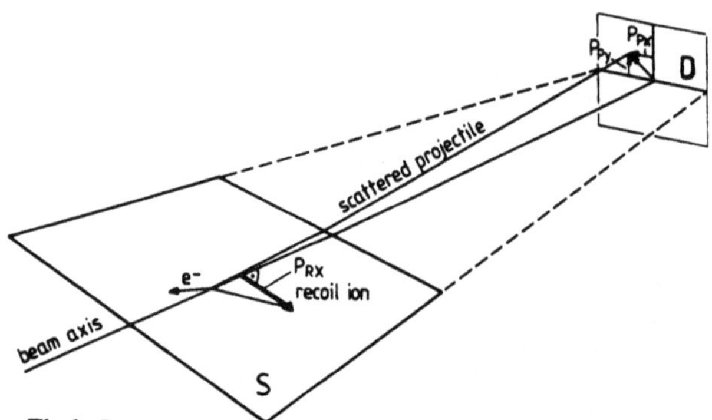

Fig 1: Scattering geometry. S: scattering plane, D: projectile detection plane, P_{Rx} recoil-ion transverse momentum which is measured in the experiment. $P_{px,py}$ projectile transverese momenta, measured with a two dimensional position senstive channel-plate detector.

Results and discussion

Fig. 2 shows the mean energy of the recoil ions perpendicular to the beam direction $\langle E_{R\perp} \rangle$ as function of the scattering angle. If the projectile would be scattered at the target nucleus only and the electron momentum could be neglected one would expect in good approximation

$$E_{R\perp} = E_0 * m_P/m_R * \vartheta^2$$

with $E_{R\perp}$ the transverse energy of the recoil ion, E_0 the energy of the projectile, $m_{P,R}$ the mass of projectile and recoil ion respectively and ϑ being the projectile scattering angle in radians. This is shown by the straight line (two body momentum exchange). The experimental results given by the open cycles are already corrected for the thermal motion of the target atoms (about 2.6 meV) and the calculated leakage of the acceleration voltage through the copper mesh (see [1]) which makes an effect of about 2 meV for the lowest recoil ion energies measured. The error bar shows the estimated systematical error due to fringe fields in the spectrometer. The scattering angle dependence of the mean recoil-ion transverse energy differs significantely from what could be expected for a pure two-body momentum exchange between projectile and He nucleus for all scattering angles below 1 mrad. This region of small scattering angles contributes the major part (97% for 0.5 MeV p-He) to the total cross section.

For very small scattering angles the calculated recoil energy does not drop to zero but has a saturation value of about 1.1 meV. Since for zero scattering angle the projectile does not carry any transverse momentum, the transverse momentum of the recoil ion is completely determined by the transverse momentum of the ionised electron. Due to momentum conservation for zero scattering angle electron and recoil nucleus are emitted back to back, seen in the plane perpendicular to the beam. The nCTMC calculations show for these small scattering angles a mean electron energy of 14 eV with a transverse component of 8 eV, which is in agreement with calculations in first Born approximation by Salin [4]. The mean electron energy for the total electron spectrum however is given by Rudd et al. [5] to be 54 eV. The calculated nCTMC result is in good agreement to this (56 eV). The nCTMC as well as first Born [4] predict a strong increase of the mean electron energy with scattering angle in the range from 0-1 mrad. The calculated recoil-ion energy for zero scattering angle does not reflect the

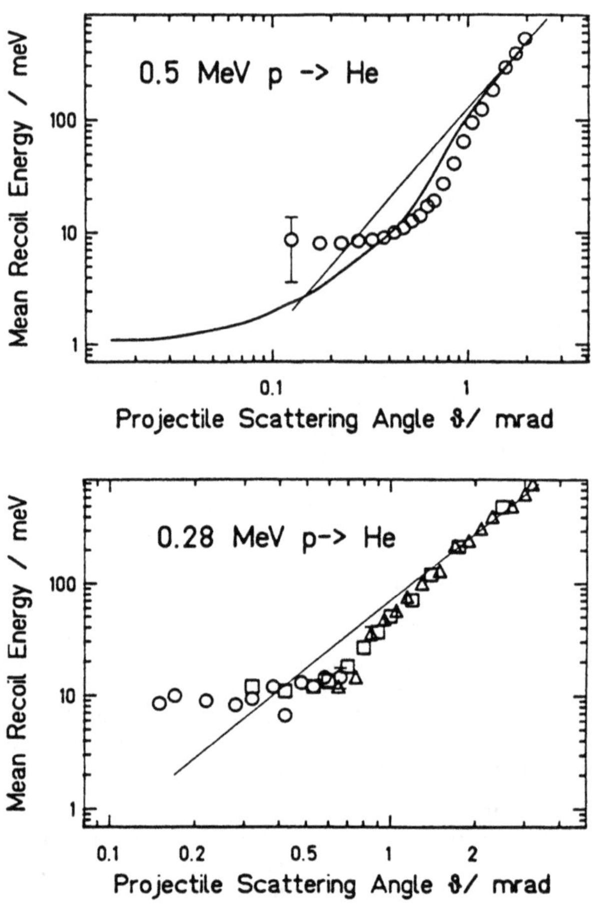

Fig. 2a,b Mean recoil-ion energy transverse to the beam direction as function of scattering angle for 0.5 and 0.28 MeV p-He single ionisation. Open symbols: experiment, the thermal motion at 30K has been substracted (see text), straight full line: two-body momentum exchange between projectile and target nucleus fat line: nCTMC calculation (without internal target motion, see text)

mean single electron energy (54eV) (as we claimed in [3]) but only the much lower electron energies coincident with this scattering angle. As disussed above in [3] we showed an early version of nCTMC calculations where all recoil-ion energies were contaminated by an error of about 5 meV. This lead to a wrong theoretical saturation recoil energy for zero scattering of 6.5 meV. By chance this is in agreement with a mean electron energy for zero scattering equal to the electron distribution integrated over all scattering angles.

For the smallest scattering angle measured the experimental value for the recoil-ion energy is 8±5meV compared to 2.6 meV predicted by the nCTMC calculations. This experimental value has been reproduced several times within the given error of ±5meV. In particular, with a gold coated and with a copper spectrometer and for three different field free flight pathes of 5, 3.5 and 2.5 mm. Nevertheless internal fields due to contact potential cannot be excluded completely. A more precise test of the spectrometer is in progress.

For scattering angles above 0.3 mrad experiment and theory show that the recoil-ion energy drops below the value expected for nucleus nuclues two-body momentum exchange. The transverse momentum gained by the projectile is much larger than the momentum transfered to the target nucleus. This difference in transverse momentum has been transfered to the ionised electron. In this scattering angle regime an important part of the projectile scattering is produced by a kind of binary collision between projectile and electron. For such a binary collision with one electron at rest the maximum scattering angle is given by the mass ratio of the two colliding particles m_p/m_e to be 0.55 mrad. However the target electrons have a compton profile which cannot be neglected at these impact energies so even larger scattering angles can be reached. The theoretical and

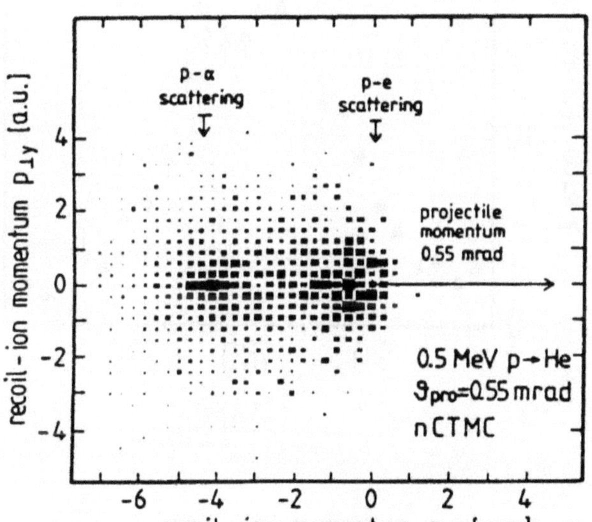

***Fig 3:** nCTMC calculation. recoil-ion momentum distribution perpendicular to the beam axis, for a fixed momentum transfer to the projectile of 4.5 a.u. (corresponding to 0.55mrad) and single ionisation*

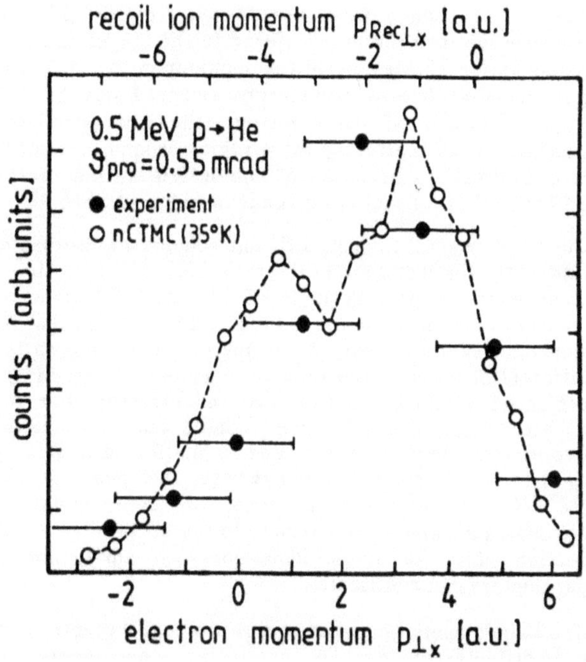

***Fig 4:** projection of fig 3 on the x-axis in comparison to the experiment. upper scale: recoil momentum, lower scale: electron momentum (s.text) open symbols: projection of nCTMC result of Fig3 folded with thermal motion of 35K. full symbols: experiment.*

experimental results demonstrate that in this intermediate region of scattering angles between 0.3 and 1 mrad two different mechanisms of projectil scattering are competing. The projectile can be scattered at the electron and at the target nucleus. The first mechanism (scattering at the electron) would leave the recoil ion with zero energy while the second mechanism would create a recoil ion with an energy idicated by the "two body kinematics" line. The experimental results clearly proof this for the mean recoil ion energy measured in this angular range are significantly lower than the expected two body value. For 0.28 MeV p impact the experiment shows the same features (Fig 2b). The different symbols in fig 2b show the results of three different runs with no beam dump for the total beam (open circles), and two different sizes of beam dumps (squares and triangles).

Kamber at al [6,7] found a sharp decrease in the differential single ionization cross sections for 3 and 6 MeV p-He at a scattering angle of about 0.55 mrad. They attributed this to the two different scattering mechanisms discussed above. Scattering at the electron reaching out only to 0.55 mrad leading always to an ionisation and the scattering at the nucleus with a much lower ionisation probability for all scattering angles. For a direct experimental test of this explanation there are two possibilities. One is to measure the electron spectra as function of scattering angle. An increase of the mean electron energy around 0.55 mrad due to the proton electron scattering should be found. This is predicted by quantum mechanical [4] as well as by the nCTMC calculations. The second possibility to test wether the shoulder in the differential single ionization crossection is due to the change from p-e to p-α scattering is to measure the momentum transfered to the target nuclues. Although for 0.5 MeV p-He the shoulder in the differential cross section is completely smeared out by the compton profile of the electrons, the low recoil-ion energies in the scattering angle region around 0.55 mrad give a direct confirmation of the interpretation of the differential cross sections given by Kamber et al [6,7].

Our experimental findings although support an interpretation of the distinct peak in the ratio $He^{2+}/(He^{1+}+He^{2+})$ at projectile scattering angle of 0.9 mrad [8] given by Olson et al [9] and Ullrich et al [10]. They argued that for single ionisation all impact parameters contribute to the differential cross section only up to about 0.55 mrad if the projectile would be scattered at the electron. For the double ionisation a subsequent double scattering at the two ionized electrons could lead to scattering angles up to 1.1 mrad. At 0.9 mrad where the peak in the $He^{2+}/(He^{1+}+He^{2+})$ ratio is found, for double ionisation all impact parameters contribute to the differential cross section while for single ionisation only small impact parameters can contribute.

The two competing mechanisms of projectile-electron and projectile-target nucleus scattering can nicely be recognized in Fig 3. It is shown the recoil-ion momentum distribution in the plane perpendicular to the beam for a fixed scattering angle of 0.5-0.6 mrad for 0.5 MeV p-He calculated in the nCTMC approach. The scattering angle of 0.5-0.6 mrad is equivalent to a transverse momentum transfer to the projectile of 4.1-4.9 au indicated by the arrow. Two regions of recoil-ion momenta can be distinguished. One of them around the negative projectile momentum resulting from a scattering of the projectile at the helium nucleus at small impact parameters. The recoil-ion momenta are broadend by the momentum of the ionized low energetic electron. The second maximum in the recoil ion momentum distribution is close to zero momentum. For these ionising events the projectile is scattered at the electron leaving a slow recoil ion behind. Fig.3 can although be read as an emission pattern of the ejected electron using momentum conservation

$$p_e = -(p_P + p_R) .$$

The p-α scattering leads to low energetic electrons wich are ejected to all azimuthal angles with respect to the projectile. The p-e scattering leads at 0.55 mrad to a transverse component of the electron energy of 275 eV. These high energetic electrons are emitted sharply peaked to the opposite side of the scattered projectile indicating a real binary encounter. As has been discussed above, the measured mean transverse recoil-ion energies directly support the intuitive picture of two different scattering mechanisms at intermediate scattering angles. Our experimental approach of RIMS does not only allow to look at mean recoil-ion energies but also more differential at all transverse momentum components (fig. 1). Fig 4 shows a projection of fig. 3 on the x axis together with the experimental results. In addition the nCTMC results in Fig 4 have been folded with a thermal motion of 35 K which smears out the double peak structure. The experimental results do not show two peaks for p-α and p-e scattering, but one broad peak. For we did not measure absolute differential cross sections, the absolute height of experiment is adjusted to the theortical curve. The total experimental resolution as indicated by the error bars in fig.4 (±1.2 **au**) is limited by the beam divergence, projectile detector resolution and systematical error on the recoil-ion energy measurement. The measured azimuthal distribution of the emitted electrons for a projectile scattering of 0.5-0.6 mrad is strongly peaked opposite to direction of the scattered projectile (Fig. 5a) again underlining the contribution of the scattering at the electron to the total scattering of the

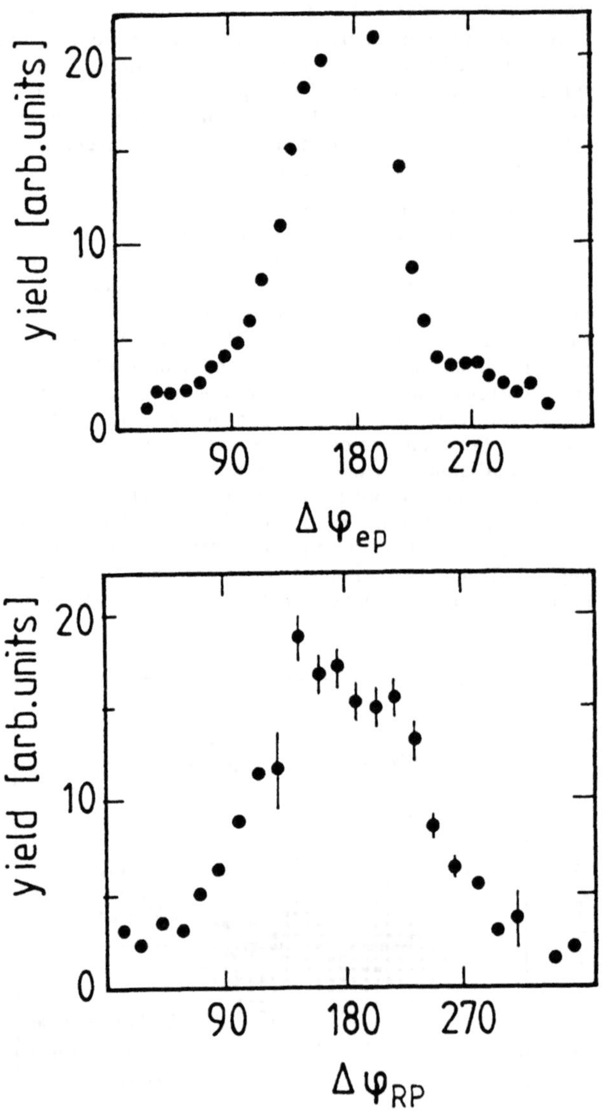

***Fig. 5:** a)Azimuthal angular distribution (plane perpendicular to the beam) of electron emission for 0.5 MeV p-He single ionisation and a scattering angle of 0.55 mrad. 0 degree direction of the scattered particle. The electrons are predominatly ejected opposite to the scattered projectile, indicating a binary type of collision between p and electron. b) same as a) for recoil ions.*

projectile. We note that the shown coincident emission pattern of the electrons has been experimentally achieved without detecting any electron by measuring the recoil ion and projectile transverse momentum and exploiting momentum conservation law. The azimuthal distribution of the recoil ions show a similar behavior (fig 5 b). If we go now to large scattering angles above 1 mrad wich only contribute below 3 % to the total single ionization cross section mean recoil ion energy is due to a hard two-body collision of the projectile with the target nucleus (fig2) and the recoil ions are found to bee emitted sharply back to back (in the plane pependicular to the beam) to the projectile But even for these violent collisions the emerging electron gives a measurable momentum contribution to the projectile scattering [3,11].

Conclusions

Using recoil-ion momentum spectroscopy (RIMS, for a review of this experimental technique see [1]) an insight in the dynamics of the single ionisation process of helium by swift proton impact was given. It was demonstrated that for scattering angles below 1 mrad which contribute the dominant part to the total cross section the ejected electron has an important influence on the scattering of the projectile. For all these ionising events the scattering of the projectile cannot be described by a one center potential (for example not by a screened potential). For these events there is also no correlation between impact parameter and scattering angle. For scattering angles around 0.55 mrad it was shown that the process of projectile-electron and projectile-target nucleus scattering are competing leading to an electron emission peaked opposite to the scattered projectile and a recoil ion transverse momentum much lower than the momentum necessary to balance the projectile momentum.

The experimental work was supported by the Bundesministerium für Forschung und Technologie (BMFT: 060F/73)

References

1. Ullrich, J., Dörner, R., Lencinas, S., Jagutzki, O., Schmidt-Böcking, H
in: Proceedings of the 11th Conference on the Application of Accelerators in Research and Industry, Denton 1990

2. Olson, R.E.,
in Electronic and Atomic Collisions, edited by Gilbody, H.B., Newell, W.R., Read, F.H., and Smith, A.C.H. (Elsevier, Amsterdam, 1988) pp. 271-285

3. Dörner, R., Ullrich, J., Schmidt-Böcking, H. and Olson, R.E.
Phys.Rev.Lett. **63**,147(1989)

4. Salin, A.,
J.Phys. B: 22 (1989) 3901

5. Rudd, M.E. , Toburen, L.H., Stolterfoht, N.
At. Data and Nucl. Data Tables 18, (1976) 413

6. Kamber, E.Y., Cocke, C.L., Cheng, S., Varghese, S.L.
Phys.Rev.Lett. 60 (1988) 2026

7. Kamber, E.Y., Cocke, C.L., Cheng, S., McGuire, J.H., Varghese, S.L.
J.Phys.B 21 (1988) L455

8. Giese, J.P., Horsdal, E.,
Phys.Rev.Lett. 60 (1988) 2018

9. Olson, R.E., Ullrich, J., Dörner, R., Schmidt-Böcking, H.
Phys.Rev.A 40 (1989) 2843

10. Ullrich, J. , Olson, R.E., Dörner, R., Schmidt-Böcking, H.
in: The Physics of Electronic and Atomic Collisions, ed. Dalgrano, A., Freuns, R.S., Koch, P.M., Lubell, M.S., Lucatorto, T.B. New York 1989

11. Horbatsch, M. J.Phys. B 22 (1989) L639

12. Schiewitz, G., Skogvall, B., Stolterfoht, N., Schneider, D., Montemayor, V.
in: The Physics of Electronic and Atomic Collisions, ed. Dalgrano, A., Freuns, R.S., Koch, P.M., Lubell, M.S., Lucatorto, T.B. New York 1989

Supplement to Z. Phys. D - Atoms, Molecules and Clusters 21, S 63-67 (1991)

Atoms, Molecules and Clusters
Zeitschrift für Physik D

Theory of multiple ionization and capture in energetic ion-atom collisions

Marko Horbatsch
Department of Physics, York University, Toronto, Ontario, Canada, M3J 1P3

Abstract

This short review includes classical, semiclassical and quantal methods. Information about charge state population as a function of scattering angle and about the distribution of ionized electrons in energy and solid angle can be extracted for projectile energies in the range of 50 keV/amu up to 1 GeV/amu.

1. Introduction

The subject of this review concerns many electron processes in heavy ion collisions, mainly capture and multiple ionization (MI). We discuss the following problem: a bare nucleus of charge i impinges on a neutral target atom with N electrons and produces a q-fold charged recoil ion by capturing m and ionizing n electrons:

$$P^{i+} + T \rightarrow P^{(i-m)+} + T^{q+} + ne^{-}. \quad (1)$$

In the early work only the recoil ions were detected [2] (recoil ion production cross sections) while a coincidence count of projectile ions with known charge state with the recoils yields charge state correlated cross sections [3]. Most theories dealing with this problem do not rely on a quantum mechanical (QM) description, but on semiclassical or classical approximations. This works well for the prediction of the population of charge states on projectile and target after the collision and for information differential in projectile (and target) scattering angle and for electron emission spectra. Detailed QM information, such as quasiresonant K-K shell exchange can be obtained in density functional theory (DFT).

The problem of N electrons and the two colliding nuclei is usually reduced to an explicitly time dependent (TD) N-body problem with a fixed nuclear trajectory. This elimination of two degrees of freedom is convenient as long as one is interested in total charge state production cross sections. For information differential in the projectile or target scattering angle it poses a problem insofar as the information about deflection is contained in the phase of the nuclear scattering wave function. A further common simplification is to reduce the remaining explicitly TD many body problem to an independent particle model (IPM), such as e.g. the TD Hartree-Fock approximation or simplifications thereof. For many purposes it is sufficient to further simplify and give up on shell structure. At this level the model becomes statistical in nature, is based on the $\hbar = 0$ limit and is related either to the TD Thomas-Fermi or the Vlasov model. To regain QM information it is possible to use the mean field potential from such a calculation in an effective single-particle TD Schrödinger equation (SE), which amounts to a DFT approach.

Three other approaches that do not rely on this 'standard' sequence of approximation have been applied: (i) an IPM based on perturbative expansions from stationary scattering theory, (ii) classical N+2-body calculations, and (iii) purely statistical models (energy deposition model) which do not attempt to follow the detailed evolution of the collision [1]. This review does not contain material about heavy ion collisions at low energies, where highly charged recoil ions are predominantly produced by capture.

2. Theory

2.1 Stationary scattering theory methods

The application of perturbative Born-type methods based on the stationary scattering theory approach to MI is relatively recent [3-5]. It is so far limited to collision systems for which simultaneous capture and ionization of electrons does not reach the unitarity limit at small impact parameters. The impact energy of 35 MeV/amu is high enough to exclude a significant amount of target electron capture by the projectile. The ion velocity of about 0.25c on the other hand is low enough to avoid a relativistic formalism. An IPM is constructed based on the Plane Wave Born Approximation (PWBA). In this model a single-particle (SP) ionization probability is calculated within each shell from the PWBA amplitude:

$$T_{fi} = < e^{i\mathbf{k}_f \dot{\mathbf{R}}} \Phi_f(\mathbf{r}) | \frac{Z_P}{R} - \frac{Z_P}{|\mathbf{r}-\mathbf{R}|} | e^{i\mathbf{k}_i \dot{\mathbf{R}}} \Phi_i(\mathbf{r}) > . \quad (2)$$

Here Z_P is the projectile charge, $\mathbf{R}$ the internuclear separation, $\mathbf{k}_i$ and $\mathbf{k}_f$ the initial and final relative momentum vectors between the two nuclei and $\Phi_{i/f}(\mathbf{r})$ the electronic wave function before and after the collision. Several approximations are being made: screening effects are static only, i.e. changes in electron binding and polarization effects induced by the projectile are ignored; in the partial wave analysis of (2) it is assumed that the electronic wave function does not depend on R, justified by the sudden approximation.

q-fold electron ionization probabilities are calculated by combining the results from individual shells according to the shell-IPM [6]. Using the SP ionization probabilities for each active shell one forms products of these and sums all combinations that lead to a given charge state q of the recoiling target ion. Total recoil ion production cross sections are then obtained from an integration over the impact parameter b, namely

$$\sigma_q = 2\pi \int_0^\infty P_q(b) b db, \quad P_q(b) = \sum_{n_1=0}^{N_1} \sum_{n_2=0}^{N_2} .. \sum_{n_m=0}^{N_m} \prod_{j=1}^{m} P_{n_j}(b). \quad (3)$$

Here m is the number of active electron shells, N_i the number of electrons in each shell, P_{n_j} the impact parameter dependent statistical distribution to remove n_j of the N_j electrons in this shell given by (4) below. For the purpose of calculating recoil production the information about individual shells is almost irrelevant. One can form a statistical SP ionization probability for the whole atom by first combining the individual shell ionization probabilities with weighting factors for the number of electrons in each shell and calculate the q-fold ionization rate from the binomial distribution:

$$P_q(b) = \frac{N!}{q!(N-q)!} (P_s)^q (1-P_s)^{N-q}. \quad (4)$$

Significant deviations between the more sophisticated evaluation employed in [3,4] and the statistical prescription (4) appear only for charge states q that are close to N, even though the individual shell probabilities may be quite different (see Figs. 1 and 3).

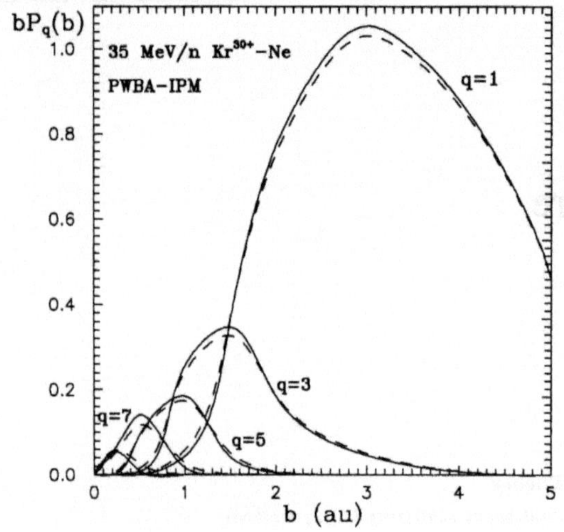

Fig. 1 MI probabilities using shell-IPM (3) (solid) vs. statistical atom (4) (dashed) evaluation for 35 MeV/n Kr^{30+} - Ne [3].

2.2 TDSE-based methods

In the semiclassical approximation (SCA) the QM $(N+2)$–body problem is reduced to Newton equations for the two nuclei

$$M_P \ddot{\mathbf{R}}_P = -\nabla_{\mathbf{R}_P} U_{eff}[\rho_{el}, \mathbf{R}_{P/T}], \quad M_T \ddot{\mathbf{R}}_T = -\nabla_{\mathbf{R}_T} U_{eff}[\rho_{el}, \mathbf{R}_{P/T}] \tag{5}$$

and the N–electron TDSE for the electronic motion:

$$i\partial_t \Psi(\mathbf{r}_1, \mathbf{r}_2, ..\mathbf{r}_N, t) = \hat{H}(t)\Psi(\mathbf{r}_1, \mathbf{r}_2, ..\mathbf{r}_N, t) \tag{6}$$

Atomic units are used throughout, $M_{P/T}$ are the masses of the two respective nuclei, $\mathbf{R}_{P/T}$ their position vectors and $\rho_{el}(t)$ is the electronic density calculated from the solution of (6) in each time-step. $\hat{H}(t)$ contains the full target Hamiltonian with inter- electronic forces as well as the explicitly time-dependent Coulomb interactions with both nuclei.

The standard approximation to (6) is to introduce an IPM, i.e. to assume a determinantal form for the many-electron wave function Ψ. Such a wave function is correct only for non-interacting fermions, but it samples the inter-electronic forces. By this assumption (6) reduces to a coupled set of SP TDSE's for the orbitals that make up the Slater determinant:

$$i\partial_t \psi_j(\mathbf{r}, t) = \hat{h}(t)\psi_j(\mathbf{r}, t), \hat{h}(t) = -1/2\nabla^2 + v_{eff}[\rho_{el}(t), \mathbf{r}, t]. \tag{7}$$

The index j runs from 1 up to some number j_0 required by the model to accomodate the N electrons. The potential can be taken either from the TDHF prescription, from TD DFT, or from a semiclassical ($\hbar-$ expanded) calculation such as the Vlasov-IPM.

A partial solution to the problem of the temporal oscillation of TDHF- projections [7] is given by the semiclassical DFT approach. The densities from the Vlasov-IPM settle to some average electronic configuration on projectile and target after the collision and are free from QM oscillations (see Fig. 2). Use of these electronic densities removes the consistency problem in the HFSlater potential

$$v_{eff}(\mathbf{r}, t) = -\frac{Z_T}{r} - \frac{Z_P}{|\mathbf{r} - \mathbf{R}(t)|} + \int \frac{\rho_{el}(\mathbf{r}', t)}{|\mathbf{r} - \mathbf{r}'|} d^3r' - c_X \rho_{el}(\mathbf{r}, t)^{1/3}, \tag{8}$$

and thus linearizes the QM problem. Therefore, transition probabilities to bound states can be calculated without additional averaging, as documented in the calculation of quasi-resonant K-K transfer [8]. A current investigation of MI shows, however, that the question of the appropriate asymptotic Hamiltonian onto which the TDHFS-like solution should be projected remains.

Lüdde et al. [9] have further developed a Feshbach projection formalism to avoid the technical problem of integrating continuum channels. Within this formalism the single-particle Hilbert space is split up into a P- and Q-part, whereby the P-part is calculated explicitly by means of, e.g., an atomic orbital (AO) expansion and the Q-part is defined as the rest, which cannot be spanned by the expansion. One- and two-center versions were developed and used for single electron problems. The one-center version has been applied to heavy ion collisions for the calculation of total recoil ion production [10]. In this work the SP model included a frozen target HFS-potential, and the following equations were solved:

$$\dot{\mathbf{a}} = \frac{1}{f}(\mathbf{1} - \bar{\mathbf{G}}^{-1} - if\epsilon)\mathbf{a}, \tag{9}$$

where

$$G_{kj} = < k|\frac{1}{1 - ifZ_P/|\mathbf{r} - \mathbf{R}(t)|}|j > . \tag{10}$$

$\mathbf{R}(t)$ represents the inter-nuclear separation, $f = (b+1)/Z_P$ is the scaling factor and ϵ is a diagonal matrix containing the HFS orbital energies and four states $(1s, 2s, 2p_0, 2p_1)$ were kept. The calculation provides Q-space occupation probabilities for the individual shells, from which total recoil ion production can be calculated as in the PWBA-IPM. To obtain projectile charge state correlated cross sections a two-center AO-Feshbach calculation would have to be performed. A calculation with a semiclassical TD potential of form (8) in the one-center AO-F formalism is under way.

2.3 Semiclassical methods

Classical three-body calculations were introduced as a means to obtain reliable capture and ionization cross sections for $p - H$ collisions [11] and were termed the Classical Trajectory Monte-Carlo (CTMC) method. Olson extended this method to an IPM [12,13] and later to a many body method (N-CTMC) [14-16]. Scheid and coll. [17] have looked for the connection of this method to QM. The semiclassical ($\hbar = 0$) Liouville equation can be derived as a lowest order in $\hbar$ limit of (7). This approach differs from the CTMC method not only in that it relies on the IPA, but also the question of an initial distribution to describe the stationary atom is somewhat open [18].

For the many-electron case an IPM with consistent TD mean field can be defined and is termed the Vlasov-IPM. The equation which replaces the TDHF orbital equations is given by

$$\partial_t f(\mathbf{r}, \mathbf{p}, t) + \mathbf{p} \cdot \nabla_{\mathbf{r}} f(\mathbf{r}, \mathbf{p}, t) - \nabla_{\mathbf{r}} v_{eff}(\mathbf{r}, t) \cdot \nabla_{\mathbf{p}} f(\mathbf{r}, \mathbf{p}, t) = 0. \tag{11}$$

At $t = 0$ an atomic distribution function can be defined as a product of two step functions distributing the binding energy equally between the highest and lowest occupied SP energies as obtained in an IPM for atomic structure. For the many electron case it is quite important to work with an initial distribution, which contains a wide range of binding energies. In the early versions of the semiclassical Vlasov-calculations a somewhat different form was employed for the initial distribution [19], but it also had the property of a broad distribution of binding energies. The numerical integration of (11) is performed by a test-particle discretization. At the level of neglect of exchange interactions the correct Vlasov potential takes the form (8) with $c_X = 0$. In the early versions [19] this was replaced by a simple model potential:

$$v_{eff}(\mathbf{r}, t) = -\frac{Z_T^{eff}(t)}{r} - \frac{Z_P}{|\mathbf{r} - \mathbf{R}(t)|}, \tag{12}$$

where the TD model potential parameter Z_T^{eff} was chosen for a given target atom at $t = 0$ and depended on time according to the TD average target charge state determined from the electronic density. It is possible to incorporate the correct Vlasov Hamiltonian (8) by summing the forces between all test-particles [8]. We have, now, adopted an aproach whereby a fit is done to the axially symmetric part of the interaction potential w.r.t. the internuclear axis. This part of the electronic potential w.r.t. the azimuthal angle contains the bulk part of the information about charge transfer. Also shown is at which time the reduction of this potential at the target nucleus sets in due to MI (Fig. 2).

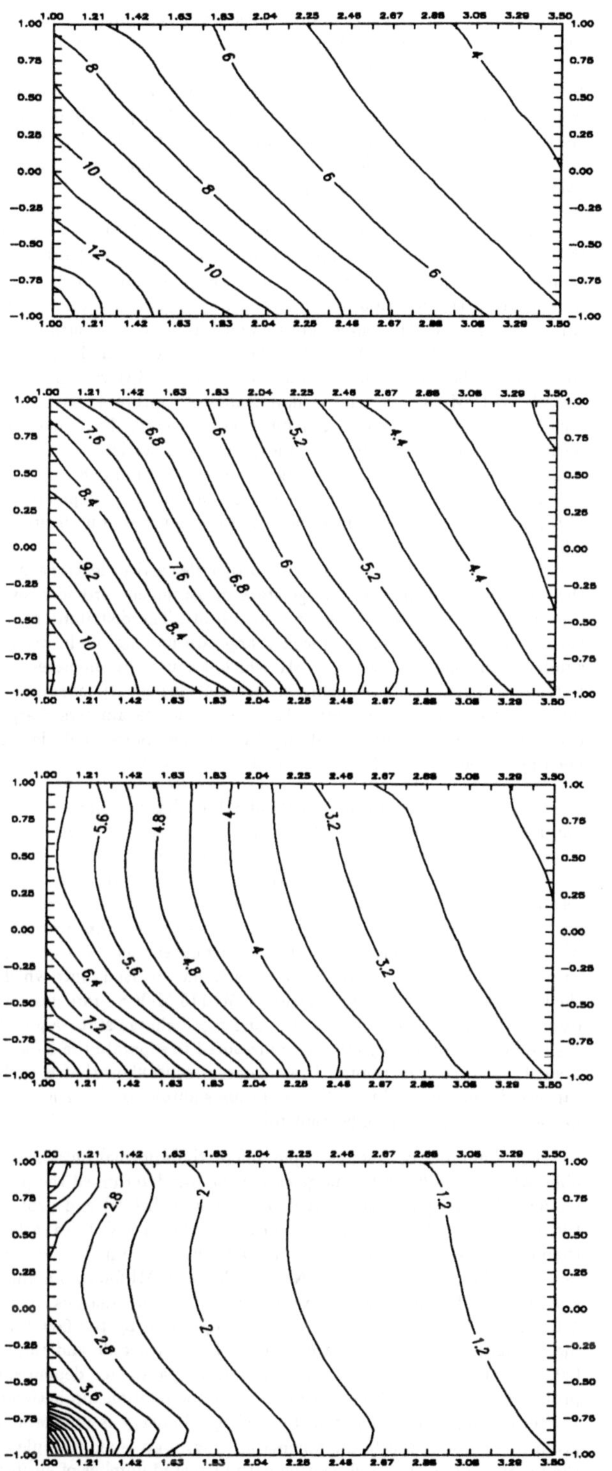

Fig. 2 Contour plots of the axially symmetric part of the electronic potential for several internuclear distances in the 100 keV/n C^{6+} - Ne system for $b = 1au$. The abscissa displays $\xi = (r_1 + r_2)/2R$ and the ordinate $\eta = (r_1 - r_2)/2R$ (elliptic coordinates, $1 \le \xi \le 3.5$, $-1 \le \eta \le 1$), the target nucleus is always in the lower left-hand corner and the projectile nucleus in the upper one. The distances shown are (in au) R=1.4 (before closest approach), R=1.4 (after cl.ap.), R=2.2 and R=5.14. The average charge states after the collision are $q_P = 3.1$ and $q_T = 5.8$.

Capture and ionization probabilities are obtained from the trinomial distribution, which results as a consequence of the separation of configuration space into three parts: target (T), projectile (P) and continuum (C) [19]. In terms of those three SP probabilities the probability to capture m and to ionize simultaneously n electrons out of a total of N is:

$$P_{(m,n)} = \frac{N!}{m!n!(N-m-n)!}(P_s^P)^m(P_s^C)^n(P_s^T)^{(N-m-n)}. \qquad (13)$$

It is possible to extend this analysis to include electrons carried by the projectile into the collision [20]. A detailed comparison of SP emission distributions in energy and scattering angle for fixed b was given in [21]. MI into a window in phase space can also be calculated in the statistical model [22].

In the past three years investigations have concentrated on the question of impact parameter- or scattering angle-dependent multiple capture and ionization probabilities [14,23-25]. For close collisions a direct test of the statistical IPM is possible as impact parameter and projectile scattering angle or target recoil momentum are uniquely related in this limit. However, for peripheral collisions the nuclear kinematics in ionizing collisions are significantly perturbed by the total transverse momentum balance, i.e. the nuclear transverse momenta become comparable to the momenta of the ionized electrons. How serious this problem is was understood first in the classical N-CTMC model [25] and incorporated later in the Vlasov-IPM [21,26].

The average projectile deflection calculated from the interaction with the electronic density distribution and the nuclei (5) is small. Thus, in an ionizing collision at large impact parameters the momentum transfer is dominated by the momentum of the emitted electrons. It is possible in the sense of an IPM to generate distributions of transverse momenta for target and projectile nuclei corresponding to the various MI channels by using the SP distributions of emitted electrons. Corresponding average deflection angles can be calculated and used in the classical expression for the differential cross section [26]. Differential cross sections calculated in this way agree better with experiment at small scattering angles than the use of a screened Rutherford trajectory. It is ironic that the inclusion of electron momentum transfer annihilates the effect of screening and increases average scattering angles up to Rutherford values even at large b.

To account for the detailed structure in the ratio of MI probabilites for small scattering angles [27] one has to collect the deflections into a given scattering angle window from all impact parameters for each charge state achieved. This can be conveniently performed based on the observed Gaussian distribution of transverse momenta, which is independent of b over a wide range and one obtains [21]:

$$\left(\frac{d\sigma}{d\theta}\right)_{(m,n)} \approx 2\pi\int_0^\infty bdbP_{(m,n)}(b_{av}(\theta))\exp[-(\frac{\theta-\theta_{av,n}(b)}{\Delta\theta})^2]. \qquad (14)$$

2.4 Classical trajectory methods

An extension of the classical trajectory method [11] to many electron targets was developed by Olson [1,12,13]. This IPM distinguished between K- and L-shell and used a hydrogen-like model Hamiltonian. The charge parameters were determined from the requirement to reproduce HF-expectation values for the potential energy. The initial distribution was chosen to be microcanonical with binding energies taken from atomic structure.

For the calculation of target recoil momentum and projectile deflection it is important to note that the CTMC model describes the relative motion of the target nucleus with the test-electron, which is absent in the IPA-based calculations, where the whole electron cloud interacts with the target nucleus and thus, at $t = 0$, the nucleus has a stable equilibrium point. This leads to some discrepancies in the width of the projectile and target distributions between the N-CTMC and Vlasov-IPM calculations [26,28].

The early CTMC-IPM calculations described the production of low charge recoils well, but suffered from an overestimation of the high charge states. A possible reason for this overestimation is the equal binding of all electrons within a given shell with the same orbital energy corresponding to a neutral IPM-atom.

In the context of single- and double-ionization of He by proton impact some refinements of the CTMC-IPM have been proposed. These refinements concern the question of a proper mean field potential instead of a simple Coulomb potential with effective charge parameter [29,30]. It is possible to construct an appropriate microcanonical distribution and to determine a static screening potential self-consistently in which the coordinate space distribution is quite different [31]. These calculations do not incorporate dynamical (TD) screening effects. The CTMC-IPM has been extended recently to treat molecular targets and differences between ionization of atomic and molecular hydrogen have been studied [32]. Another study investigates the difference between proton-atom and antiproton-atom collisions [33].

To gain access to correlations between final charge states and target and projectile deflections Olson developed an $N-$particle CTMC model [15]. In the N-CTMC model all electrons are treated separately in a hydrogen-like fashion with a model Hamiltonian of the form

$$H = T_{nucl} + V_{nucl} + \sum_{i=1}^{N} \frac{p_i^2}{2} + \sum_{i=1}^{N} \frac{Z_T z_i}{|\mathbf{r}_i - \mathbf{R}_T|} + \sum_{i=1}^{N} \frac{Z_P z_i}{|\mathbf{r}_i - \mathbf{R}_P|}, \quad (15)$$

where the individual electronic charge parameters z_i are obtained using experimentally known ionization potentials for the various ionic charge states. The binding energies for each individual electron that have been constructed are used in the microcanonical distribution for each electron and due to this construction a significant variation of these energies is achieved. The electron-electron interaction is not included in the Hamiltonian (except for a calculation in fast p-He collisions [16]) and thus the model is similar to an IPM. The classical character of the N-CTMC model is evident from the distinguishability of the electrons.

3. Results

Reasonable agreement with experiment for the distribution of recoil charge state production cross sections as a function of recoil charge can be obtained from most models with some restrictions given in brackets: energy deposition model [1] (higher charge states only); CTMC-IPM [1,12,13] (high charge states overestimated); PWBA-IPM (fast collisions only); AO-F ($q = 1$ problematic); Vlasov-IPM; N-CTMC.

An interesting comparison between two different theoretical approaches can be made concerning the contribution of various subshells towards recoil ion production. This information is available from quantal calculations: the PWBA-IPM calculations for $Kr^{30+} - Ne$ at 35 MeV/n and the AO-F results with static screening for 0.1 - 1 MeV/n $C^{6+} - Ne$ collisions. Both sets of calculations have in common in their $b-$ dependent SP ionization probabilities the difference between K- and L-shell contributions as well as the size of discrepancies within the L-shell. However, while the PWBA-IPM probabilities for the L-shell are ascending in order $(2s, 2p_0, 2p_1)$ for all b [3], in the AO-F results ionization from the $2p_1$ subshell dominates at large b over $2p_0$, but lies below for $b < 2a.u.$ [10].

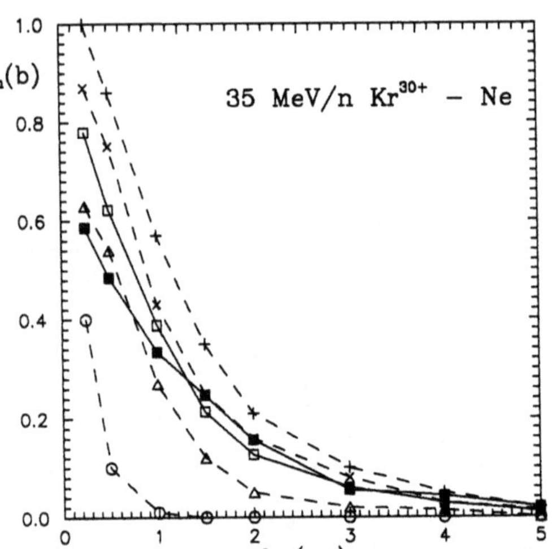

Fig. 3 SP ionization probabilities for 35 MeV/n Kr^{30+} -Ne collisions from PWBA-IPM (open sq.) and Vlasov-IPM (full sq.). PWBA-shell results: o (1s), Δ (2s), × $2p_0$, + $2p_1$.

Charge state correlated cross sections were available since 1980 [2] and usually analyzed by forming products of independently obtained single-particle capture and ionization probabilities [13]. The separation of the contributions to recoil production into direct MI, single capture and MI, double capture and MI provides a much more stringent test for theory. A quantitative explanation of multiple capture and MI based on the trinomial distribution (13) was given in the Vlasov-IPM and the relative importance of capture + MI vs. direct MI channels was given as a function of energy. Experimental results [20] show, however, that these quantum statistical calculations somewhat underestimate single capture + MI and overestimate direct ionization. It should also be noted that the Z_P- dependence of capture + MI distributions in fast collisions is not well understood [13].

For ion-atom collisions where the projectile carries electrons into the collision the coincident charge state measurement provides also information about multiple electron loss + MI. The distributions of these cross sections over the recoil charge are looking qualitatively similar to the ones for capture + MI, which is caused by the necessity to remove tightly bound projectile electrons occuring only for small to intermediate impact parameters. At these impact parameters charge exchange between projectile and target is possible. Some analysis has been performed for the F-Ne system at 4-15 MeV [20].

To provide more information about MI one has determined other parameters in coincidence with the projectile and target charge states. Recordings of target K-Auger coincidences with recoil charge states were made to study vacancy sharing in $F^{8+/9+}$ - Ne and quasi-resonant K-K transfer [34]. Quasiresonant K-K transfer is one example where DFT can be used to describe single electron processes within the mean field created by the ionizing outer-shell electrons [8]. Triple coincidence measurements were performed to study into which shells capture occurs in single capture + MI 115 MeV S - rare gas collisions [35]. In addition to projectile and recoil charge state K_α and K_β radiation on the projectile were measured to separate the K- and L-shell capture contributions. These measurements show that K-shell capture dominates in this system and causes autoionizing transitions in the recoil to play an important role.

For a number of systems triple differential information is available, where in addition to charge states one has determined the projectile deflection [20,23,28], target recoil motion [24,25] and projectile inelasticity [14]. One wanted to determine MI probabilities rather than cross sections in order to test the validity of the statistical distributions. The 1.4 MeV/n U^{32+} - Ne data for direct MI follow binomial statistics at all angles measured quite well [24]. The analysis of the 4-15 MeV F^{6+} - Ne data, however, is complicated by the fact that many charge state channels are open for both projectile and target [20,23]. These measurements provide impact parameter dependent probabilities only for close collisions, where the relationship between scattering angle, impact parameter and recoil energy is unique and given by two-body scattering kinematics. Fig. 4 gives an example of the broadening in the $b - \theta$ relation with increased number of ejected electrons according to the Vlasov-IPM.

This effect is deemed responsible for the observed peaks in the ratio of double to single ionization in MeV p-He collsions below 1 mrad scattering angle [27]. While the absolute height of this ratio has not been reproduced yet by any calculation, the peak has been found in N-CTMC [16] as well as Vlasov-IPM calculations based on (14) [21]. Further experimental work would be welcome to study the effect of thermal motion in projectile beam and target on the peak position.

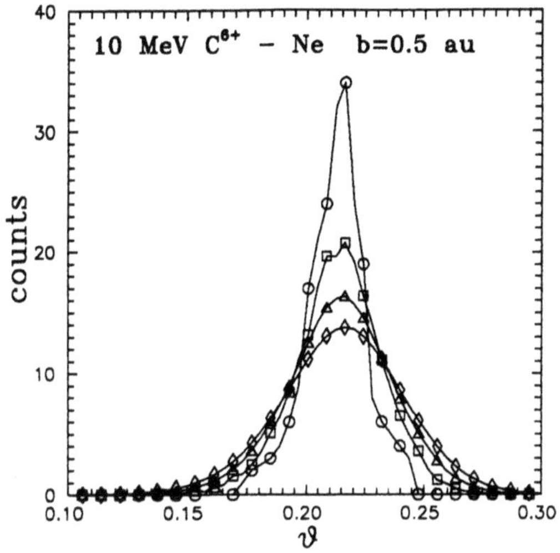

Fig. 4 Distribution of projectile scattering angles (in mrad) in one- (circles) to four-fold (diamonds) ionizing collisions with fixed impact parameter due to balancing of momenta in the Vlasov-IPM.

An experiment performed on the $0.53 MeV/n F^{8+} - Ne$ system has found anomalously large projectile azimuthal scattering angles [36] for double capture events. Charge states on projectile and target and transverse momenta of both nuclei were determined in coincidence. Calculations based on the Vlasov-IPM show a broadening of distributions with final target charge states, but their width is much smaller (Fig. 5). We conclude that the reported results on the average electronic and recoil momenta appear to be strongly affected by Auger processes.

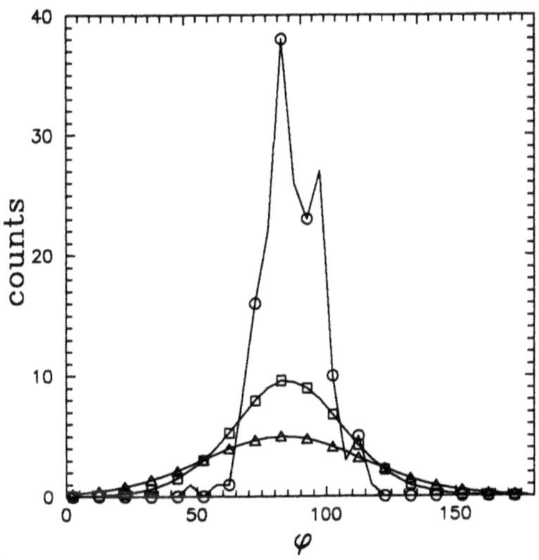

Fig. 5 Distribution of azimuthal projectile scattering angles in $b = 1$ 10 MeV C^{6+} - Ne collisions for one-, three- and five-fold electron emission, labelled by circles, squares and triangles respectively. The scattering plane is at $\phi = 90$ deg.

Some information about $\delta-$electron emission in heavy ion collisions was available [37], which was compared to SCA calculations. Recent experiments performed at GSI on the U-Ne,Ar systems at 1.4 MeV/n provide detailed energy and scattering angle dependencies and are described well by the N-CTMC model [25].

For the p-He system electron spectra as a function of emission angle and energy were taken with fixed, large projectile scattering angle [38]. The statistics of these data are limited, but the data have been compared to CTMC-, Born-, and Vlasov-IPM calculations [21].

4. Conclusions

Experimentally available information about electronic processes in heavy ion collisions has been described with a variety of methods, including a statistical 'black box' approach, the quantum statistical ($\hbar = 0$) Vlasov IPM, CTMC shell model and $N-$ particle calculations. Experimental data are often 'obscured' by post-collision effects like autoionization, which require correction of theoretical data. From a theorist's point of view the situation is not fully satisfying yet: a large amount of data cannot be calculated with QM methods. This represents a mainly technical and computational problem and will certainly keep another generation of physicists occupied.

Acknowledgment

This work would not have been possible without discussions and collaboration with many theoretical and experimental colleagues, all of whom I wish to thank. Financial assistance through NSERC, Canada and the NATO Exchange Programme is gratefully acknowledged.

References

[1] Cocke C.L., Phys Rev. **A 20**, 749 (1979).
[2] Gray T.J. et al, Phys Rev. **A 22**, 849 (1980).
[3] Lesteven-Vaïsse I. et al, J. Phys. France, **49**, 1529 (1988).
[4] Gayet R., J. Phys. France **50**, C1-53 (1989).
[5] Salin A., J. Phys. **B 22**, 3901 (1989).
[6] Hansteen J.M., Mosebekk O.P., Phys. Rev. Lett. **29**, 1361 (1972).
[7] Gramlich K. et al, J. Phys. **B 19**, 1457 (1986).
[8] Toepfer A. et al, Phys. Lett. **A 126**, 11 (1987).
[9] Lüdde H.J. et al, Phys. Lett. **A 125**, 197 (1987); J. Phys. **B 21**, 4131 (1988); J. Phys. **B 22**, 3243 (1989); Z. Phys. **D 15**, 141 (1989).
[10] Lüdde H.J. et al, Phys. Lett. **A 145**, 173 (1990).
[11] Abrines R., Percival I.C., Proc. Phys. Soc. **88**, 861 (1966)
[12] Olson R.E., J. Phys. **B 12**, 1843 (1979).
[13] Müller A. et al, Phys. Rev. **A 33**, 3010 (1986).
[14] Schuch R. et al, Phys. Rev. Lett. **60**, 925 (1988).
[15] Olson R.E., in *Electronic and Atomic Collisions*, ed. by Gilbody H.B. et al, North Holland, Amsterdam (1988), pp. 271-85.
[16] Olson R.E. et al, Phys. Rev. **A 40**, 2843 (1989).
[17] Eichenauer D. et al., J. Phys **B 14**, 3929 (1981).
[18] Schmidt A. et al., J.Phys. **B 23**, 2327 (1990).
[19] Horbatsch M., et al Phys. Lett. **A 113**, 251 (1985); Z. Phys **D 1**, 337 (1985); Z. Phys **D 2**, 183 (1986); J. Phys **B 19**, L193 (1986).
[20] Kelbch, S. et al, J. Phys. **B 23**, 1277 (1990).
[21] Horbatsch M., Comp. Phys. Comm., in press (1990).
[22] Horbatsch M., Phys. Lett. **A 137**, 466 (1989).
[23] Kelbch, S. et al, Phys. Lett. **A 127**, 92 (1988).
[24] Ullrich J. et al, J. Phys. **B 21**, 611 (1988).
[25] Olson R.E. et al, J. Phys. **B 20**, L 809 (1987); Phys. Rev. **A 39**, 5572 (1989).
[26] Horbatsch M., J. Phys. **B 22**, L639 (1989).
[27] Giese J.P., Horsdal E., Phys. Rev. Lett. **60**, 2018 (1988).
[28] Dörner, R. et al, Phys. Rev. Lett. **63**, 147 (1989).
[29] Peach G. et al, J. Phys. **B 18**, 3921 (1985).
[30] Reinhold C.O., Falcón C.A., Phys. Rev. **A 33**, 3859 (1986).
[31] Montemayor V.J., Schiwietz G., J. Phys. **B 22**, 2555 (1989).
[32] Meng L. et al, Phys. Rev. **A 40**, 3637 (1989).
[33] Reinhold C.O., Olson R.E., Phys. Rev. **A 39**, 3861 (1989).
[34] Hagmann S., et al., Phys. Rev. **A 34**, 2897 (1986).
[35] Konrad J., Ph.D. Dissertation, Universität Heidelberg, 1986.
[36] González A.D., et al., J. Phys. **B 23**, L 303 (1990).
[37] Skutlartz A., Hagmann S., Phys. Rev. **A 28**, 2863 (1983)
[38] Schiwietz G., Phys. Rev. **A 37**, 370 (1988).
[39] Datz S. et al, Phys. Rev. **A 41**, 3559 (1990).

Supplement to Z. Phys. D - Atoms, Molecules and Clusters 21, S 69-75 (1991)

Atoms, Molecules
Zeitschrift für Physik D and Clusters

Experiments with highly-charged ions in the storage ring TSR

A. Wolf, J. Berger, M. Bock,
D. Habs, B. Hochadel, G. Kilgus, G. Neureither,
U. Schramm, D. Schwalm, and E. Szmola
Physikalisches Institut der Universität Heidelberg
and Max-Planck-Institut für Kernphysik, Heidelberg, FRG

A. Müller and M. Wagner
Institut für Kernphysik, Justus-Liebig-Universität, Giessen, FRG

R. Schuch
Manne-Siegbahn Institute, Stockholm, Sweden

Abstract

Experiments with merged electron and ion beams at the Heidelberg Test Storage Ring (TSR) are described. The experimental conditions of radiative and dielectronic recombination measurements are discussed. Results are presented for the radiative recombination rates of bare C and Cl nuclei, and for the dielectronic recombination of H-like, He-like, and Li-like ions up to charge 26.

I. Introduction

In the early discussions of ion storage rings it was recognized (see, e.g., [1]) that merged electron–ion beam experiments were particularly well suited to these facilities and could take advantage of them in several respects. First, an electron-beam 'target' inside the storage ring presents only a small perturbation to the circulating ion beam, and the possiblity to reach a high ion current by recirculation and by stacking techniques can be exploited. Next, the detection of ions changing their charge state in the electron beam can be readily implemented; all atomic processes leading to a change of the ion charge can thus be detected with a high efficiency. Moreover, the ion storage rings offer very low residual gas pressure, as required for the suppression of background, and a powerful electron 'target' is already present in most of these facilities, where an intense and cool electron beam is used for electron cooling of the ion beam.

Consequently, the first experiments with highly-charged ions at the Test Storage Ring (TSR) were centered on the investigation of recombination processes of ions with free electrons in the electron cooling device. Roughly, two experimental situations can be distinguished.

In the *equilibrium* state of the electron-cooled ion beam, the ion velocity adjusts itself to the electron velocity and, seen from a reference frame moving with the beam velocity, the ions are effectively at rest in a cold, thin electron gas. In this situation, only a single constant quantity, the recombination rate coefficient, is measured, which represents mainly *radiative recombination* and possibly small additional contributions due to ternary recombination [2].

On the other hand, the average relative energy between electrons and ions can be varied and the recombination rate be measured *before* a new equilibrium of electron and ion velocities is reached. Then, for ions with one or more attached electrons, reconances due to *dielectronic recombination* can be observed and spectroscopic investigations of the doubly excited states corresponding to these resonances become possible.

Even though single-pass merged-beam experiments recently were greatly improved and applied to radiative and dielectronic recombination measurements [3], the high current and the relatively high energy of the stored ions in the TSR are of great advantage for the measurement of weak dielectronic resonances. Thus, dielectronic recombination of hydrogen-like ions was observed for the first time at high energy resolution. In addition, measurements of weaker high-lying resonances of helium-like and lithium-like ions were performed.

The purpose of the present paper is to give an overview of these results and to explain the experimental procedure of radiative and dielectronic recombination measurements at the TSR. Perspectives for the extension of these experiments in thė future will be discussed shortly in the conclusion.

II. Operation of the TSR with highly-charged ions

The heavy-ion storage ring TSR [4] at the Max-Planck-Institut für Kernphysik in Heidelberg was built to store and cool ions in a wide range of energies and charge states, injected by a 13-MV tandem followed by a radiofrequency linear accelerator. The ring can accomodate ions of a momentum up to $q \cdot 510$ MeV/c, corresponding to its maximum magnetic field strength, where q indicates the ion charge number and c the velocity of light. Over the ring circumference of 55.4 m, the ions circulate in an average vacuum of presently 3×10^{-11} mbar. In the electron cooling device [5] an intense electron beam travelling at the same veloc-

Table 1. Present performance of the TSR with highly-charged ions. Dominant loss processes are R: recombination in the electron cooler, EC: electron capture, and EL: electron loss in the residual gas.

Ion	Energy MeV/amu	Current mA	Mean life	Loss process
Bare nuclei:				
C^{6+}	6.1	18	1 h	R
Cl^{17+}	5.8	0.1	3 min	EC
Few-electron ions:				
O^{7+}	8.9	0.04	7 min	EL
Cu^{26+}	8.1	0.1	1 min	EC

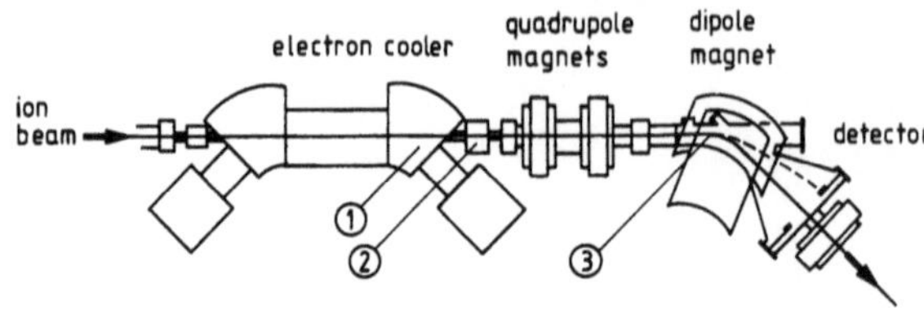

Fig. 1. Part of the TSR with the electron–ion interaction region and the detector for recombined ions. The numbers indicate the principal regions where a transverse magnetic field is present, which is listed below for operation with bare light nuclei at 6 MeV/amu. 1: Toroid section of the electron guide field (< 27 mT); 2: correction dipole (61 mT); 3: deflection dipole (0.6 T). The straight section of the electron cooler is 1.5 m long.

ity as the ions is merged with the ion beam. The electron energy ranges up to 16 keV and the electron current up to 2 A. By elastic collisions of the ions with the electrons, phase-space cooling of the ion beam is accomplished and the transverse and longitudinal velocity spread of the ion beam, as well as its size, are reduced.

The operational performance of the TSR and its experimental program were discussed earlier [6,7]. Some recent, typical numbers for ion energies, currents, and lifetimes relevant for the operation with highly-charged ions are listed in Table 1. Usually with the help of electron cooling, the ions are accumulated in the ring by adding newly injected ions in small portions to a continuously circulating and cooled stack [8]. Thus, the ion current in the steady state is given by the beam lifetime and the refilling rate. The main mechanisms causing loss of circulating ions, as listed in Table 1, differ according to the ion type and its energy [6,7].

III. Radiative recombination

The setup of the merged-beam experiment formed by the electron cooler, the ring magnets, and the ion detector is shown in Fig. 1. The ion beam of a diameter of $\approx 1 \ldots 3$ mm travels in the center of the electron beam of 5.1 cm diameter over a nominal length of 1.5 m. A magnetic field of ≈ 50 mT, collinear to the electron beam, prevents the electrons from diverging because of their space charge. Behind the dipole magnet, recombined ions form a spot of a size close to that of the circulating beam. At the indicated detector position, the distance of the recombined ions from the stored beam depends on the initial charge state q and amounts, e.g., to 140 mm for $q = 6$ and to 30 mm for $q = 26$. A multichannel plate (MCP) of 20 mm diameter was used for the measurements; a 10×10 mm^2 solid-state detector could be moved to the position of the MCP for verifying its efficiency, which was found to be higher than 95% in most situations.

The typical count rate observed during electron cooling, when the ion beam had reached its equilibrium state, was several 10^4 s^{-1}. The ratio of the recombination signal to the background due to electron capture in the residual gas varied between ≈ 3 and 0.5. The background was measured after increasing the electron energy (typically 5 keV during cooling) by ≈ 1 keV. Under these conditions the relative energy between electrons and ions was so large that the expected recombination signal practically vanished.

From the electron density, the ion current, and the (nominal) length of the interaction region the recombination rate coefficient was determined after background subtraction. Results obtained for bare ions are listed in Table 2. In addition to the given statistical error, a maximum systematic error of $\pm 15\%$ is estimated, which is mainly attributed to uncertainties in the measured ion current and in the effective overlap length between electrons and ions. According to Table 2 the rate coefficient of C^{6+} was measured for different values of the electron density; a variation, which would indicate a contribution from ternary recombination, was not observed.

The calculation of rate coefficients for radiative recombination in merged beams was carefully considered by Bell and Bell [9]. They analytically obtained the sum of the recombination cross-section over all principal quantum numbers n. The theoretical results listed in Table 2 were calculated using equation 28 of [9] for a transverse energy spread of the electrons corresponding to the cathode temperature ($k_B T = 0.10$ eV). The longitudinal energy spread in the center-of-mass frame, which is strongly reduced during the acceleration of the electrons, was assumed to vanish ("flattened" velocity distribution). This choice of temperatures for the comparison between theory and experiment is justified by the analysis of the transverse and longitudinal energy spread in the electron cooler during dielectronic recombination measurements (see below).

The good agreement between theory and experiment documented by Table 2 would be satisfying if there were not some inconsistencies with respect to the expected field ionization of highly excited recombined ions. On one hand, one can estimate from the calculated rate coefficient the limit n_{pop} up to which excited states are *initially* populated with relative probabilities decreasing not faster than $\propto n^{-1}$, and hence contribute significantly to the recombination rate. This theoretical limit depends on the ion charge,

Table 2. Recombination rate coefficient α measured for bare nuclei; also listed are the electron density n_e and theoretical results [9] for radiative recombination, obtained for a transverse electron temperature of 0.1 eV/k_B.

Ion	Energy MeV/amu	n_e 10^7 cm^{-3}	$\alpha/(10^{-10}$ s^{-1}cm$^3)$ measured[a]	calculated
C^{6+}	6.1	3.5	1.82(10)	1.85
"	"	1.9	1.77(10)	"
"	"	1.2	1.81(20)	"
Cl^{17+}	5.8	3.2	17.2(10)	17.7

[a] Statistical errors only; systematic errors are of the order of ±15%

Table 3. Critical quantum numbers for the initial population after radiative recombination (n_{pop}) and for ionization in motional electric fields (n_{sp}) at a beam velocity of 0.11·c; numbers indicate zones marked in Fig. 1. Lower bounds are given for the limiting quantum number n_{exp} obtained by comparing the measured rate coefficient to theory [9].

Ion	n_{pop}	n_{sp} (1)	(2)	(3)	n_{exp}
C^{6+}	190	53	43	24	>69
Cl^{17+}	530	115	94	53	>140

the electron temperature, and also on the shape of the electron velocity distribution; it is given for a "flattened" distribution and for the relevant ions in Table 3. On the other hand, recombined ions in highly excited states can be field-ionized on their way to the detector by motional electric fields due to transverse magnetic fields. For the three most important regions with transverse magnetic fields, reached by an ion from the center of the overlap region after a typical flight time of 30 ns, 80 ns, and 150 ns, respectively, the 'saddle point' quantum numbers n_{sp} are also listed in Table 3. These are the principal quantum numbers of unshifted levels with an energy corresponding to the saddle point formed by the superposition of the Coulomb and the external potential; slightly above these limits field ionization should occur. Since all values of n_{sp} are clearly below n_{pop}, field ionization of highly excited states should reduce the measured recombination rate unless these ions decay to lower states before reaching the critical field regions.

The average angular momentum $\ell \approx n/3$ of the states populated by radiative recombination is known theoretically [10,11]. In field-free regions, the radiative lifetime of states with this ℓ exceeds the time of flight to the dipole magnet (150 ns) already for $n \gtrsim 30$ ($q = 6$) or for $n \gtrsim 70$ ($q = 17$). With these quantum numbers as limits for field ionization, the theoretical prediction of the rate coefficient is reduced to values of 1.2×10^{-10} s^{-1}cm^3 for C^{6+} and 12×10^{-10} s^{-1}cm^3 for Cl^{17+}. The lower bounds of the field ionization limit n_{exp} which can be estimated from the *measured* rate coefficients (considering the statistical and systematic error) are listed in the last column of Table 3.

A conclusive explanation of the experimental results cannot be given at present. As one possibility, the fast passage of the highly excited ions through spatially varying external fields might modify the ℓ or even the n distribution of the Rydberg ions and let them decay more rapidly to lower states which cannot be field ionized in the dipole magnet. As another possibility, environmental conditions in the overlap region, such as the space-charge field of the electron beam and the ≈ 50-mT magnetic field in beam direction, might yield values higher than presently calculated for the recombination rate itself.

The rate coefficient measured at the TSR for C^{6+} ions [$1.80(30) \times 10^{-10}$ s^{-1}cm^3, including the systematic error] also differs significantly from the result obtained by Andersen *et al.* [11] under quite similar experimental conditions [$1.00(15) \times 10^{-10}$ s^{-1}cm^3]. For the latter experiment, which was performed at a beam velocity and a magnetic field in beam direction both lower by roughly a factor of 2, satisfactory agreement is found with theory for the applicable values of the field ionization limit ($n = 60$) and the transverse temperature (0.15 eV/k_B). In contrast, the TSR result obviously reveals discrepancies which call for further experimental and theoretical studies.

IV. Dielectronic recombination

Experimental procedure

The dielectronic recombination (DR) measurements at the TSR were performed by recording the recombination rate as a function of the laboratory electron energy E_e, while the ion energy remained essentially constant. Approximately, the center-of-mass (c.m.) electron energy E is given by the non-linear relation

$$\sqrt{E} \approx \sqrt{E_e} - \sqrt{E_e^{cool}} \quad , \qquad (1)$$

where E_e^{cool} is the electron energy used to cool the ions, which determines their velocity. A typical value for the cooling energy is $E_e^{cool} \approx 5$ keV, which leads to measuring energies $E_e \approx 6.5$ keV for resonances close to $E = 100$ eV, and to energies $E_e \approx 10.5$ keV for resonances at $E = 1000$ eV. Slightly different experimental procedures were used for c.m. energies in the higher range $E > 100$ eV and in the lower range ≈ 0.01 eV $< E < 100$ eV.

For high c.m. energy, the electron acceleration voltage was stepped up after cooling to a base level close to the resonances. Then it was switched at a high rate (≈ 15 s^{-1}) between this level, where the background rate was measured, and a variable second level, for which the signal was measured. By varying slowly the second level a scan over the resonances was performed. The ions were slowly accelerated because of Coulomb collisions with the electrons. However, since the acceleration rate is inversely proportional to the c.m. energy, it was small enough that the measuring phase could be extended over ≈ 1 min before stepping down to the cooling energy again.

For resonances at c.m. energies in the lower range a similar square-modulation scheme was used, now keeping the lower level at the cooling energy. Here, the modulation was essential for maintaining a constant ion energy during the measurement, since the acceleration rate by Coulomb collisions was much higher in the lower range of E. The modulation period was short enough to prevent the acceleration of the ions towards the 'detuned' electron energy.

Both the electron and the ion energy in the laboratory have to be known accurately in order to determine the absolute c.m. energies of the DR resonances. As far as the ion energy is concerned, it is known within 10^{-3} from the calibration of an analyzing magnet between the accelerator and the storage ring. The energy of the *circulating* ions can be monitored very accurately by recording the noise spectrum of the ion beam, which reflects the distribution function of the revolution frequency. Relative changes of the average energy at a level of 10^{-4} or less can be evaluated quantitatively.

The electron energy is derived from the electric potential difference between the cathode and the interaction region of electrons and ions. Basically the acceleration voltage of the electrons, applied between the cathode and the beam tube, is measured. However, the space-charge potential of the electrons introduces a substantial potential difference between the beam tube and the interaction region, which must be subtracted from the acceleration voltage. The electrostatic potential is lowest in the center of the electron beam; as a function of the distance from the center, r, it increases $\propto r^2$ up to the electron beam edge and $\propto \ln r$ beyond the beam edge. For the typical electron density of 5×10^7 cm^{-3} the potential difference is -146 V between the center and the beam edge ($r = 25.4$ mm), and -546 V between the center and the grounded beam tube ($r = 100$ mm).

A reliable correction for the space-charge potential is possible for the following reasons. First, the geometry can be controlled by adjusting the relative position of electron and ion beam until the lowest revolution frequency of the stored ion beam is observed in the beam noise spectrum. The ion and the electron beam should then be coaxial. Second, the electron beam diameter equals the known cathode diameter within 0.3 mm since the electrons are immersed in a homogeneous magnetic guiding field; hence, the electron density can be determined accurately from the measured electron current. Third, the space-charge compensation by positive ions generated by collisions of the electrons with residual gas molecules can be kept below 10^{-3} of the electron charge. These slow ions are prevented from accumulating in the electron beam by clearing electrodes, located in the toroid magnet downstream of the interaction region. After the space-charge correction, the overall systematic error of the c.m. energy scale is estimated to be below ± 1 eV at $E \approx 500$ eV.

Of particular interest is the energy resolution of the DR spectra, given by the c.m. energy deviation of the electrons.

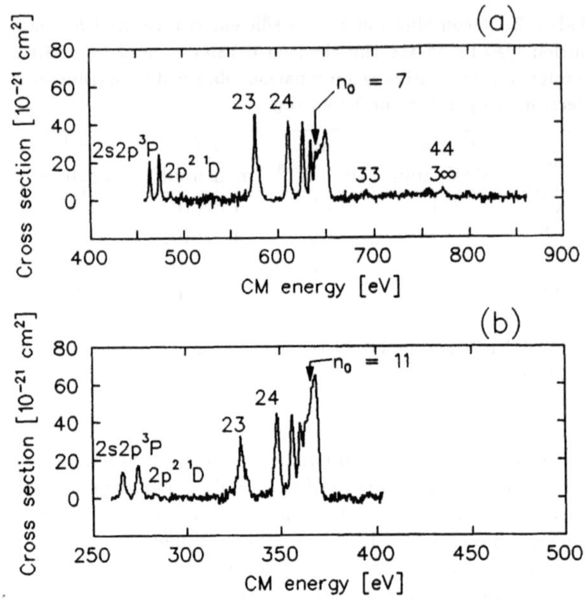

Fig. 2. DR spectra of H-like ions. **a:** O^{7+} ions; **b:** C^{5+} ions. Above the $2\ell 2\ell'$ levels, the resonances are labeled by the principal quantum numbers of the inner and the outer electron. The arrows indicate the energy at the critical outer quantum number n_0.

The relevant electron beam properties are the longitudinal and the transverse energy spread $\delta E_{\parallel}$ and $\delta E_{\perp}$, defined in the electron rest-frame, which enter into the c.m. energy deviation δE at a resonance energy E_{res} according to

$$\delta E = \delta E_{\parallel} + \delta E_{\perp} \pm 2\sqrt{E_{res}\delta E_{\parallel}} \quad . \tag{2}$$

A longitudinal energy spread $\delta E_{\parallel}$ between 1×10^{-3} eV and 5×10^{-4} eV and a transverse spread compatible with the cathode temperature of 0.105 eV/k_B (1220 K) is consistently indicated by the observed shape of DR resonances. At higher $E_{res} \approx 300$ eV the resonances are symmetric and have a width of ± 1 eV, determined by the last term in (2). At $E_{res} \lesssim 10$ eV the resonance signal becomes asymmetric (see also [12]) and its low-energy tail (corresponding to positive electron energy deviations δE) assumes a constant width of $\delta E_{\perp}$.

Hydrogen-like ions

The dielectronic resonances investigated with these ions are of the type

$$O^{7+}(1s) + e^- \rightarrow \begin{cases} O^{6+}(2\ell n\ell') & (n = 2, \ldots, \infty) \\ O^{6+}(3\ell n\ell') & (n = 3, \ldots, \infty) \end{cases} .$$

For the first DR measurement [13] at the TSR, using O^{7+} ions of 8.9 MeV/amu, and for a later measurement with C^{5+} ions of 9.4 MeV/amu, the DR cross-section as a function of the c.m. energy is shown in Fig. 2. For a general understanding of the pattern formed by all $2\ell n\ell'$ resonances it is useful to recall that the radiative stabilization, mainly due to the transition ($2p\,n\ell \rightarrow 1s\,n\ell$), has an approximately constant rate for all n, which, however, varies with the ion charge $\propto q^4$. On the other hand, the resonance formation rate (inverse Auger rate) scales $\propto n^{-3}$ for $n \gg 1$, but is roughly independent of q. Furthermore, the number of sublevels contributing to the DR signal for a given n is limited by the existence of a maximum angular momentum $\ell \approx 4$ (see, e.g., [14]). Hence the DR rate, determined by the smaller one of the two rates, decreases $\propto n^{-3}$ above the critical value n_0, at which the radiative and Auger rates are equal to each other. The quantity n_0 determines the relative contribution of high Rydberg states in the DR spectrum, which is larger for smaller q (see Fig. 2).

A detailed comparison [13,15] of the DR spectrum of O^{7+} to theoretical calculations [15] shows a remarkable agreement, in particular with respect to the resonance strengths. The comparison also served to verify the degree of refinement required in the theoretical analysis [15].

Presently, the resonances $3\ell n\ell'$ are observed only at a rather low signal-to-noise ratio (see Fig. 2). Improved measurements in this region are intended. For O^{7+}, the observed energies of the series limits $2\ell n\ell'$ and $3\ell n\ell'$, $n \rightarrow \infty$, agree within 0.3 eV with the theoretical excitation energies of the hydrogen-like recombining ion. Thus, the accuracy of the c.m. energy scale is confirmed.

Helium-like ions

Because of the fine structure of the first excited states of helium-like ions, the dielectronic resonances are grouped in overlapping series corresponding to the reactions

$$C^{4+}(1s^2) + e^- \rightarrow \begin{cases} C^{3+}[1s2s(^{1,3}S)\,n\ell'] \\ C^{3+}[1s2p(^{1,3}P)\,n\ell'] \end{cases} \quad (n = 2, \ldots, \infty) \ .$$

The DR spectrum obtained at the TSR with C^{4+} ions of 6.0 MeV/amu is shown in Fig. 3. Owing to the strong stabilizing transition $2\,^1P \rightarrow 1\,^1S$, the series of the $1s2p\,^1P$ core dominates the spectrum. In comparison to the spectra for hydrogen-like ions the regularity of the resonance strength along the series is disturbed by autoionization into degenerate continuum states for $n \geq 5$. Correspondingly, much more processes must be taken into account for the interpretation of this spectrum [16].

Similar overlapping series of DR resonances at much lower c.m. energies, corresponding to innershell excitation from the $2\,^3S$ and $2\,^1S$ states, were observed in a single-pass experiment [12] using incident metastable helium-like ions. Compared to these spectra, the TSR measurements allow the series to be followed down to the lowest principal quantum numbers; moreover, complications of the analysis of the single-pass spectra, due to the incomplete knowledge of the metastable ion fractions in the beam, are avoided.

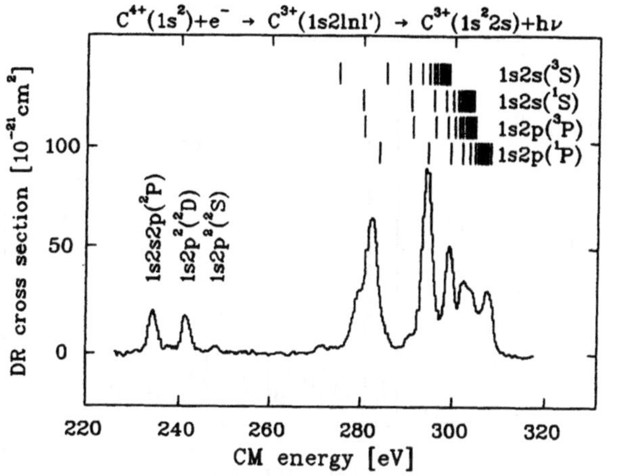

Fig. 3. DR spectrum of He-like carbon ions in their ground state. Rydberg ladders attached to the first excitation thresholds are shown.

Lithium-like ions

Even at very high charge states, lithium-like ions produce DR resonances in the c.m. energy range presently accessible at the TSR ($\lesssim 3000$ eV). Hence, it was possible to investigate the DR of highly ionized ions with a nuclear charge close to that of Fe. Data on atomic processes of these ions, which are typical impurity ions in fusion plasmas, are of particular practical interest. Using a Cu^{26+} beam of 8.1 MeV/amu, resonances of the following types were observed:

$$Cu^{26+}(1s^22s) + e^- \rightarrow$$
$$\rightarrow \begin{cases} Cu^{25+}(1s^2\,2p_{1/2}\,n\ell) & (n = 13, \ldots, \infty) \\ Cu^{25+}(1s^2\,2p_{3/2}\,n\ell) & (n = 11, \ldots, \infty) \\ Cu^{25+}(1s^2\,3\ell\,3\ell') & \\ Cu^{25+}(1s^2\,3\ell\,n\ell') & (n = 4, \ldots, \infty) \ . \end{cases}$$

The first two series at c.m. energies below 85 eV (see Fig. 4a) include very strong resonances close to the ionization threshold of Cu^{25+}. The maximum observed cross-section at the first resonance close to 0.7 eV amounts to $\approx 1.6 \times 10^{-16}$ cm^2. The integrated resonance cross-sections as a function of n will provide a sensitive test of DR calculations for highly-charged lithium-like ions, as they were published for Fe^{23+} [17].

The much weaker DR resonances of the $1s^2\,3\ell3\ell'$ intrashell states are shown in Fig. 4b. Preliminary level assignments [18] for the strongest resonances are indicated. The magnetic fine structure of the $3s3p\,^3P$ term is partially resolved.

A more detailed evaluation of these recent data, including a comparison to theory, is in progress. This comparison

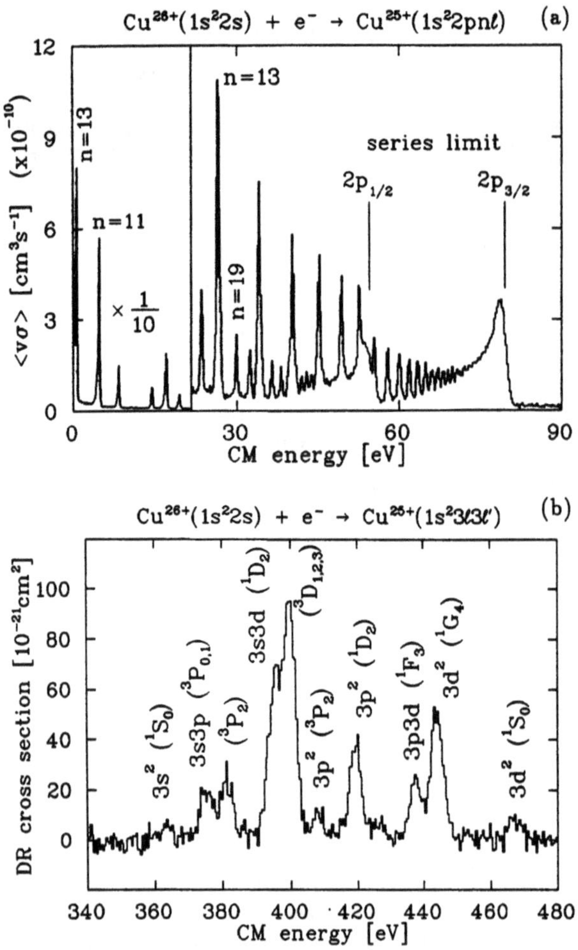

Fig. 4. DR spectra of Li-like Cu ions. **a:** Low-lying resonances corresponding to innershell excitation of Cu^{26+}. **b:** LMM resonances with term assignments according to the text.

already now suggests that, contrary to the DR measurements presented so far, the contribution of high Rydberg states (see Fig. 4a) is affected, and indeed reduced, by external fields. In particular, the observed cutoff of $n \approx 90$ corresponds to field ionization in the deflection dipole (position 3 in Fig. 1) but, for the expected ℓ distribution of the Rydberg ions after DR ($\ell \lesssim 12$ [18]), only ions in states with $n > 150$ would reach the dipole before their radiative decay to lower states. Thus, simple field-ionization and radiative lifetime estimates seem to be inconsistent with the experimental results.

V. Outlook

The DR spectra of few-electron ions obtained at the TSR provide a useful testing ground for the calculational methods applied to describe complex atomic processes, which are of great importance for plasma physics and astrophysics. In this context, an extension of the DR measurements to more complicated ions in the range of Be-like to Na-like systems is envisaged. Similar to the Cu^{26+} experiment, it can be expected that considerable currents of such ions can be accumulated in the TSR at high charge states. With respect to few-electron ions, the aim of studying DR resonances of higher-lying intrashell states has already been pointed out.

In radiative and dielectronic recombination measurements, which frequently are sensitive to the contribution of high Rydberg states, models describing the initial population of these states and their survival in external fields are put to a test. Most of the DR spectra presented above show no field effects. On the other hand, the few cases where such effects appear indicate that the simple models used in the interpretation up to now are not adequate and should be refined. More insight is expected from additional systematic measurements on the effect of external fields.

New possibilities of spectroscopic studies in merged-beam experiments are opened up by laser-induced processes, such as laser-induced radiative recombination [19] and laser-induced dielectronic recombination (see discussion in [1]). In particular, the energy resolution for the observation of these processes is determined by the spectral width of the laser radiation in the c.m. frame ($\lesssim 10^{-4}$ eV), rather than by the electron energy spread which is relevant for spontaneous processes. Laser-induced radiative recombination is presently investigated at the TSR. At the time of writing, laser-induced recombination of protons and electrons into the $n = 2$ state of the hydrogen atom has been observed in this experiment and its dependence on the laser wavelength has been studied; detailed results will be published. It can be expected that these measurements, which will be extended to highly-charged ions, yield additional information about the influence of external fields and highly excited states on the recombination process.

Acknowledgements

We thank the TSR group, in particular E. Jaeschke, M. Grieser, and M. Steck, for the efficient support during our experiments. This work has been funded by the German Minister for Research and Technology (BMFT) under contract number 06 HD 133 I, and by the Gesellschaft für Schwerionenforschung (GSI), Darmstadt.

References

1. Datz, S.: Nucl. Instrum. Methods **B24/25**, 3 (1987).
2. Beyer, H.F., Liesen, D., Guzman, O.: Part. Accel. **24**, 163 (1989).
3. Andersen, L.H.: These Proceedings, and references therein.

4. Krämer, D., Bisoffi, G., Blum, M., Friedrich, A., Geyer, Ch., Holzer, B., Heyng, H.W., Habs, D., Jaeschke, E., Jung, M., Ott, W., Pollock, R.E., Repnow, R., Schmitt, F.: Nucl. Instrum. Methods **A287**, 268 (1990).
5. Steck, M., Blum, M., Grieser, M., Jaeschke, E., Kleffner, C.M., Krämer, D., Ott, W., Repnow, R.: Electron cooling of heavy ions at the Heidelberg cooler storage ring TSR, *in:* Proc. Workshop on Electron Cooling and New Cooling Techniques, Legnaro, 1990 (World Scientific, Singapore, to be publ.)
6. Habs, D. *et al.*: Nucl. Instrum. Methods **B43**, 390 (1989).
7. Wolf, A. *et al.*: Proc. XVI Int. Conf. on the Physics of Electronic and Atomic Collisions, New York, 1989 (AIP Conf. Proc. 205, New York, 1990), p. 378.
8. Bisoffi, G., Grieser, M., Jaeschke, E., Krämer, D., Noda, A.: Nucl. Instrum. Methods **A287**, 320 (1990); Grieser, M., Blum, M., Habs, D., v. Hahn, R., Hochadel, B., Jaeschke, E., Kleffner, C.M., Stampfer, M., Steck, M., Noda, A.: Advanced stacking methods using electron cooling at the TSR Heidelberg *in:* Proc. 19th INS Symp. on Cooler Rings and their Applications, Tokyo, 1990 (World Scientific, Singapore, to be publ.)
9. Bell, M., Bell, J.S.: Part. Accel. **12**, 49 (1982).
10. Omidvar, K., Guimaraes, P.T.: Astrophys. J. Suppl. Ser. **73**, 555 (1990).
11. Andersen, L.H., Bolko, J.: Phys. Rev. A **42**, 1184 (1990).
12. Andersen, L.H., Bolko, J., Kvistgaard, P.: Phys. Rev. A **41**, 1293 (1990).
13. Kilgus, G., Berger, J., Blatt, P., Grieser, M., Habs, D., Hochadel, B., Jaeschke, E., Krämer, D., Neumann, R., Neureither, G., Ott, W., Schwalm, D., Steck, M., Stokstad, R., Szmola, E., Wolf, A., Schuch, R., Müller, A., Wagner, M.: Phys. Rev. Lett. **64**, 737 (1990).
14. Müller, A., Belić, D.S., DePaola, B.D., Djurić, N., Dunn, G.H., Mueller, D.W., Timmer, C.: Phys. Rev. A **36**, 599 (1987).
15. Pindzola, M.S., Badnell, N.R., Griffin, D.C.: Phys. Rev. A **42**, 282 (1990).
16. Badnell, N.R., Pindzola, M.S., Griffin, D.C.: Phys. Rev. A **41**, 2422 (1990).
17. Griffin, D.C., Pindzola, M.S.: Phys. Rev. A **35**, 2821 (1987).
18. Badnell N.R., Pindzola, M.S.: Priv. comm.
19. Neumann, R., Poth, H., Winnacker, A., Wolf, A.: Z. Phys. A **313**, 253 (1983).

Supplement to Z. Phys. D - Atoms, Molecules and Clusters 21, S 77-80 (1991)

Atoms, Molecules
Zeitschrift für Physik D and Clusters

Status of the theory of electron capture in ion-atom collisions at low and intermediate energies

Knud Taulbjerg
Institute of Physics, Aarhus University, DK-8000 Århus C, Denmark

Abstract

The theory is reviewed in a general perspective without reference to particular systems. Analytic work, exploiting the special symmetry property of the Coulomb field is emphasized. Coupled-channel methods are discussed with respect to Galilei-invariance and convergence. Our final remarks concern the question about electron correlation in the double capture process.

I. Introduction

In the context of the present conference, there is no need to go into any details to introduce the basic physical mechanisms involved in electron capture processes in collisions of highly charged ions with atoms or simple molecules. We all are familiar with the endothermic near-resonant reaction, mediated by series of curve crossings between the perturbed energy levels of the initial and final arrangement channels. It is also well known that the population amplitudes subsequently can be redistributed by rotational coupling due to the angular motion of the internuclear line.

This is not to say that all theoretical problems are solved in this field. There are practical difficulties with the choice of a limited basis and with convergence of coupled-channel calculations; and a more fundamental problem appears to remain in connection with a consistent account of momentum transfer to ensure proper Galilei-invariance in slow collisions. Further, it is desirable to develop simpler models and analytic methods to allow flexibility and readiness in qualitative predictions. Finally, it should be recalled that it is only processes involving a single active electron that are fairly well understood. Multielectron capture by highly charged ions generally produce multiply excited states, and there are so many of them that traditional coupled-channel calculations are overwhelming in size. Electron correlation is known to play an essential role for the description of such final states but the influence of the electron correlation potential *during* the collision has not been clarified.

This report consists of three parts: First, we discuss simple models which allow analytic solutions and a corresponding insight into the physical process. The second part is concerned with numerical methods using large basis expansions. Finally, we briefly discuss some of the difficulties that arise in the description of the two-electron capture process.

II. Models

The application of the Landau-Zener model in the description of electron capture by highly-charged ions is well established. Since the relative slope of the intersecting potential-energy curves is readily expressed in this case, there is only one parameter to be determined, namely the coupling strength $H_{12}(R_c)$ at the intersection. An analytic approximation $H_{12}^{OS}(R_c)$ for this parameter, valid for fully stripped ions, was derived by Olson and Salop [1] many years ago.

The capture cross section in the Landau-Zener model depends in a characteristic way on the position of the crossing radius: there is a typical *reaction window* for electron capture by a curve crossing [2].

The coupling strength is to be modified when the degeneracy in principal shells (n) is lifted in case of partly stripped ions. This is readily done when it is realized that the coupling in the fully stripped ion case is totally dominated by the interaction with the most strongly polarized Stark-state in the manifold of degenerate states. The interaction with individual subsells (nl) is therefore obtained by a corresponding projection. Due to the O(4) symmetry property of the Coulomb field, the resulting branching of the coupling strength appears to be given by simple Clebsch-Gordan coefficients

$$H_{12}^{OST}(R_c) = H_{12}^{OS}(R_c)\langle jjj-j \mid l0\rangle ,$$

where $j = (n-1)/2$. When this effect is properly included, there is a reaction window for each subshell, in better agreement with data [2].

A reaction window analysis has proven to provide a reasonable estimate of the subshell distribution of electron capture by partly stripped ions [3]. A detailed account is, however, not expected since multiple interactions and rotational coupling are ignored completely. Different groups have worked to expand the simple Landau-Zener formulation to include such effects. The multichannel Landau-Zener formulation represents a simple probabilistic approach with branching at each curve crossing [4].

Rotational coupling can be included in a similar way in case of fully stripped ions by an application [5] of theory due to Demkov et al.

[6] and developed for highly charged ions by others [7,8]. Demkov's approach is based on the dipole approximation for the interaction of the captured electron with the residual target ion, and exploits the O(4) symmetry of the Coulomb field in great details.

To give a brief overview, we recall that the energy eigenstates of a hydrogen-like ion (charge Z) in the Stark representation can be taken as simultaneous eigenstates of the z components of the angular momentum operator $\vec{l}$ and the Runge-Lenz vector

$$\vec{a} = -(2H)^{-1/2}\left[\frac{1}{2}(\vec{p}\times\vec{l}-\vec{l}\times\vec{p})-2\vec{r}/r\right] \quad . \qquad (1)$$

Specifically, we have

$$\begin{aligned} l_z|nkm\rangle^{st} &= m|nkm\rangle^{st} \\ a_z|nkm\rangle^{st} &= k|nkm\rangle^{st} \end{aligned} \qquad (2)$$

where k is the polarization quantum number ($k=n_2-n_1$ in terms of parabolic node numbers).

In O(4) applications Stark states are identified as product states in pseudo-spin space

$$|j_1m_1\rangle|j_2m_2\rangle = |nkm\rangle^{st} \quad , \qquad (3)$$

where

$$\begin{aligned} j_1 = j_2 = j &= (n-1)/2 \\ m_1 &= (m-k)/2 \\ m_2 &= (m+k)/2 \end{aligned} \quad . \qquad (4)$$

The corresponding spin operators are given by

$$\begin{aligned} \vec{j}_1 &= \frac{1}{2}(\vec{l}-\vec{a}) \\ \vec{j}_2 &= \frac{1}{2}(\vec{l}+\vec{a}) \end{aligned} \quad . \qquad (5)$$

The key element in Demkov's application is the operator replacement relation [9]

$$\vec{r} = -\frac{3n}{2Z}\vec{a} \quad , \qquad (6)$$

which is an identity when restricted to the manifold of the considered principal shell. This allows to write the interaction with the residual target ion in dipole approximation as

$$-\frac{1}{|\vec{R}-\vec{r}|} \approx -\frac{1}{R}+\frac{1}{R^3}\vec{R}\cdot\vec{r} = -\frac{1}{R}-\frac{3n}{2Z}\frac{1}{R^3}\vec{R}\cdot\vec{a} \quad . \qquad (7)$$

Writing the Hamiltonian in the co-rotating frame, we then have

$$H_1' = H_0' - \frac{1}{R^2}\left[\frac{3n}{2Z}\vec{e}_R\cdot\vec{a}+(vb)\,\vec{e}_N\cdot\vec{l}\right] \quad , \qquad (8)$$

where H_0' is the unperturbed part including the monopole term. The last term is the Coriolis coupling due to the rotation of the system about the unit vector $\vec{e}_N$, perpendicular to the scattering plane which is spanned by the velocity $\vec{v}$ and the impact parameter $\vec{b}$.

Equation (5) is now used to replace $\vec{a}$ and $\vec{l}$ by $\vec{j}_1$ and $\vec{j}_2$ to obtain

$$H_1' = H_0' - \frac{1}{R^2}\left[\vec{\omega}_1\cdot\vec{j}_1+\vec{\omega}_2\cdot\vec{j}_2\right] \quad , \qquad (9)$$

where

$$\vec{\omega}_q = \frac{3n}{2Z}\vec{e}_R-(-1)^q\,vb\,\vec{e}_N \quad . \qquad (10)$$

It is now clear that H_1' is diagonal in the pseudo-spin representation

$$|j_1m_1\rangle_{\vec{e}_1}|j_2m_2\rangle_{\vec{e}_2} \quad , \qquad (11)$$

where the z-axis is determined by unit vectors $\vec{e}_1$ and $\vec{e}_2$, directed along $\vec{\omega}_1$ and $\vec{\omega}_2$, respectively. The corresponding eigenenergies are

$$\varepsilon_{m_1m_2}(R) = -\frac{Z^2}{2n^2}+\frac{Z-1}{R}-\frac{\omega}{R^2}(m_1+m_2) \quad , \qquad (12)$$

where $\omega=|\vec{\omega}_1|=|\vec{\omega}_2|$.

It is, of course, trivial to solve the time-dependent Schrödinger equation in the eigen representation. A practical application therefore merely involves a projection of a given initial state in the manifold of the considered principal shell onto the representation given by eq. (11). This is accomplished by simple rotations in pseudo-spin space. After propagation in time in the eigen representation, inverse rotations can be applied to get back to the co-rotating Stark representation where transition amplitudes can be analyzed.

In a multichannel Landau-Zener model with rotational coupling, amplitudes can be propagated through a sequence of curve crossings with intermediate allowance for rotational coupling by such a procedure. Results for total capture are simplest to derive [7] since it suffices to keep track of the amplitude in the initial state (all other states are capture channels). Individual cross sections for the l-distribution have been obtained recently [8] and compare quite well with extensive multistate calculations [10].

In closing this section, it should be emphasized that the rotational coupling model just discussed only applies in case of fully stripped ions. Since rotational coupling is non-localized, it appears unlikely that the method can be suitably amended to apply in cases where core interaction and subshell splitting become important. In such cases we still have to rely on the more forcefull coupled-channel method.

III. Coupled-channel methods

The coupled-channel method is based on an expansion in a *suitable* finite basis and a numerical integration of the corresponding matrix representation of the Schrödinger equation. The method has been described in details on many occasions over the years. Here, we shall only consider two aspects, namely the significance of translation factors and the question of convergence.

The translation factor principally represents the Galilei transformation from one inertial frame to another. Translation factors are accordingly objective devices when atomic expansions are used. There, the factor appears in the general form

$$\exp(i\vec{v}_c\cdot\vec{r})\exp(-i\tfrac{1}{2}v_c^2t) \qquad (13)$$

and its purpose is plainly to refer target (C=T) and projectile (C=P) centred basis functions to the common inertial frame where the Schrödinger equation is to be solved. Coupling elements between projectile and target centred basis functions therefore generally appear in the form

$$M_{ij}^{TP}(t) = e^{i\frac{1}{2}(v_T^2-v_P^2)t}\langle\varphi_i^T(t)|e^{i\vec{v}\cdot\vec{r}}H|\varphi_j^P(t)\rangle \quad , \tag{14}$$

where $\vec{v}=\vec{v}_P-\vec{v}_T$ is the usual impact velocity.

One could get the impression from eq. (14) that the effect of the plane wave factor $\exp(i\vec{v}\cdot\vec{r})$ is of first order in velocity. This is, however, not the case as is seen by expanding this factor about the point $\vec{r}=\vec{r}_0(t)$, defined such that the first-order term disappears [11]. Discarding terms of *second* order, we then obtain

$$M_{ij}^{TP}(t) = \exp\left[i\frac{1}{2}(v_T^2-v_P^2)t+i\vec{v}\cdot\vec{r}_0(t)\right]\langle\varphi_i^T|H|\varphi_j^P\rangle \tag{15}$$

Now (choosing, as is always possible, the origin in a point on the internuclear axis), it is easy to see that $\vec{r}_0 = g(R)\vec{R}$, where $g(R)$ typically is a slowly varying function. The full time-dependent phase factor in eq. (15) is accordingly seen to be of the order v^2.

The expansion in eq. (15) is useful in practical calculations at low velocities since it allows that the matrix element can be computed independently of the trajectory parameters $\vec{v}$ and $\vec{b}$.

The coupled-channel method is exactly Galilei invariant in the general form of the atomic expansion as well as in the first-order approximation described in eq. (15). This is not the case when a molecular expansion is used. In fact, results of molecular-based calculations can change very significantly when the reference frame is changed. However, there *are* ad hoc methods, some more plausible than others, to cure the problem:

The original idea of Bates and McCarroll [12] was to satisfy Galilei invariance by attaching *asymptotic* plane-wave factors to the molecular basis functions, i.e., to use eq. (13) throughout, with the centre C determined by the dissociation limit. There are two problems with this formulation. First, the ortho-normality of the molecular basis is lost. This is annoying but certainly tractable in practice. It is more serious, conceptually, that the plane-wave translation factor represents a momentum-wise attachment to one of the two centres while the basic philosophy of the molecular expansion is to allow the adiabatic relaxation of the electron cloud in the combined two-centre field.

Both these objections are eliminated when a common switching function is introduced [13,14]. The general form is

$$\exp(if(\vec{r},\vec{R})\vec{v}\cdot\vec{r})\exp(-i\frac{1}{2}(f(\vec{r},\vec{R})v)^2t) \quad . \tag{16}$$

The only remaining problem is that there is no theory to provide a unique form of the switching function, $f(\vec{r},\vec{R})$. There *are* physical conditions to be satisfied at large separations and, if relaxation is considered to be complete, also in the united atom limit, but this is not sufficient to determine the switching function. The remaining flexibility in the choice of switching function appears at the same fundamental level as the choice of basis type and basis size: it relies on physical intuition and should be tested for stability against reasonable variations.

Various forms of switching functions have been suggested, and some were tested in a very careful study [15] of the prototype collision system He^{++}-H. The results indicate that the simple parameter-free form derived in Ref. 14, by a consideration of the electrostatic forces, is quite adequate.

For numerical ease, simpler switching functions have been chosen in other calculations. Considering that "united atom phenomena" are unimportant in highly charged ion collisions, it is plausible that the plane-wave form should provide a sufficiently accurate representation. Then it is a good approximation at low energies to drop the translation factor *after* the proper coordinate origin has been identified for each individual matrix element. Then, direct elements on the target, direct elements on the projectile and exchange-type matrix elements appear to refer to different centres. Unfortunately, however, many workers have gone one step further, namely to the "prescribed common origin method" which is difficult to justify.

Our final remark in connection with the coupled-channel method is concerned with convergence. We have often argued [11,3] that the atomic expansion is more appropriate than the molecular expansion, even at low energies, because it avoids the problems with translation factors and since the processes under consideration in highly charged ion collisions are governed by interactions at intermediate and large separations where the LCAO representation, after all, is adequate to describe molecular effects. There is, however, a price to pay for this: We have found [3] that it is important to include complete principal shells on the highly charged ion even if only a single subshell is populated at the end of the collision. The remaining states are populated *transiently* during the collision and this affects the final result. In retrospect, this is easy to understand since complete shells are needed to model the adiabatic relaxation but it was surprising and a bit annoying to us in the beginning that we had to include 5d, 5f, and 5g states to compute the important 5s and 5p capture cross sections in Ar^{+6}-H collisions. This should be contrasted with the molecular expansion method where it is expected that a minimum basis, determined by the important final channels, would be sufficient.

IV. Double capture

In most theoretical work until now, it has been assumed that only a single electron is active in the considered process. The reason for this is not that multielectron processes are uninteresting. They are just too complicated to handle with quantitative methods unless the independent particle picture is assumed to be valid.

Qualitative considerations, based on potential-energy diagrams, may of course proceed as in the single electron case; but there are different types of curve crossings, involving one- as well as two-step processes, and the competition between these has not yet been fully understood. It is clear, though, that electron correlation plays an important role.

It is generally appreciated that electron correlation is extremely difficult to incorporate in dynamic processes. The situation is probably further complicated in the case of highly-charged ion collisions since final-state interaction due to the residual target ion is likely to be essential for a correct description of the distribution over final states. To understand this, we recall that doubly-excited states of highly-charged ions are strongly correlated - primarily in their angular motion. The physical reason is that two electrons in a strong central Coulomb field tend to stabilize each other in strongly polarized Stark-type states on opposite sides of the nucleus to reduce the interelectronic repulsion energy [16]. Referring to the discussion in Section II, it is noted that these correlated states are well described by so-called DESB states [17], obtained by a suitable recoupling of the four pseudo-spins pertaining to the two electrons. The DESB states are characterized by correlation quantum numbers K and T. What is important in the present connection is that K is a measure of angular correlation

$$\langle \cos\theta_{12} \rangle \propto -K \tag{17}$$

and that K accordingly is large and positive for the more strongly bound states within a given configuration.
Considering that a typical impact parameter in collisions of highly-charged ions with simple atoms is significantly larger than the radius of the target, it becomes clear that the part of Hilbert space, that is needed in the initial phase of the collision, must represent two electrons on the *same* side of the projectile ion. In the DESB language such states have large but negative values of the correlation quantum number K. Doubly excited states of this kind are also quite stable if isolated from external perturbations but it can be expected that a strong mixing with other states in the manifold will be introduced by the field of the residual target ion. A proper theory to account for this mixing is not available.

References

1. Olson, A. and Salop, A., Phys.Rev. A **14**, 579 (1976).
2. Taulbjerg, K., J.Phys. B **19**, L367 (1986).
3. Hansen, J.P. and Taulbjerg, K., J.Phys. B **21**, 2459 (1988).
4. Salop, A. and Olson, R.E., Phys.Rev. A **13**, 1312 (1976).
5. Janev, R.K., Belic, D.S., and Bransden, B., Phys.Rev. A **28**, 1293 (1983).
6. Demkov, Yu.N., Ostrovskii, V.N., and Solov'ev, E.A., Sov.Phys. JETP **39**, 57 (1974).
7. Abramov, V.A., Baryshnikov, F.F., and Lisitsa, V.S., Sov. Phys. JETP **47**, 469 (1978).
8. Macek, J. and Dong, X.Y., Phys.Rev. A **40**, 95 (1989).
9. Englefield, M.J.: *Group Theory and the Coulomb Problem* (Wiley, New York, 1972).
10. Green, T.A., Shipsey, E.J., and Browne, J.C., Phys. Rev. A **25**, 1364 (1982).
11. Larsen, O.G. and Taulbjerg, K., J.Phys. B **17**, 4523 (1984).
12. Bates, D.R. and McCarroll, R., Proc.Roy.Soc. (London) **245**, 175 (1958).
13. Schneiderman, S.B. and Russek, A., Phys.Rev. **181**, 311 (1969).
14. Vaaben, J. and Taulbjerg, K., J.Phys. B **14**, 1815 (1981).
15. van Hemert, M.C. van Dieshoeck, E.F., van der Hart, J.A., and Kioke, F., Phys.Rev. A **31**, 2227 (1985).
16. Mølmer, K. and Taulbjerg, K., J.Phys. B **21**, 1739 (1988).
17. Sinanoglu, O. and Herrick, D.R., J.Chem.Phys. **62**, 886 (1975); Lin, C.D., Phys.Rev. A **29**, 1019 (1984).

Supplement to Z. Phys. D - Atoms, Molecules and Clusters 21, S 81-85 (1991)

Atoms, Molecules and Clusters
Zeitschrift für Physik D

Photons shedding light upon basic charge exchange processes

R. Hoekstra[1,2], F. J. de Heer[2] and R. Morgenstern[1]

[1] Kernfysisch Versneller Instituut, Zernikelaan 25, 9747 AA Groningen, The Netherlands
[2] FOM-Institute for Atomic and Molecular Physics, PO Box 41883, 1009 DB Amsterdam, The Netherlands

Abstract

Here we present recent progress made in our Photon Emission Spectroscopy studies of state selective electron capture in fundamental collision systems, like highly charged ions colliding on H.
In hydrogenic ions the *l* states are degenerate and therefore light emission from different *l* levels can not be resolved spectroscopically. However we succeeded to determine $He^+(4l)$ electron capture cross sections from the $HeII(n = 4 \rightarrow n = 3)$ line emission by exploiting the relatively long and different lifetimes of the $HeII(4l)$ states.
A further recent progress has been the extension of the experiments down to energies well below 1 keV/amu. Results for C^{4+} - H collisions will be presented and discussed on basis of the classical overbarrier model.
Finally we discuss the *m* state distribution of capture into $C^{3+}(3p)$ in collisions on molecular hydrogen. The distribution is deduced from the angular dependence of the related vuv radiation.

1. Introduction

Electron capture processes in collisions of highly charged ions on neutral particles are intensively studied both theoretically and experimentally (see e.g. [1-3] and these proceedings). Aside from pure scientific aspects this interest arises from the importance of these charge exchange processes for various other fields of physics, like controlled magnetically confined fusion research and astrophysics. Generally, for the charge transfer processes a lot of reaction channels are possible due to the many available electronic vacancies in the highly charged ion. Here we will discuss the one-electrons reactions

$$A^{q+} + H(H_2) \rightarrow A^{(q-1)+*}(nl) + H^+(H_2^+) \quad (1)$$
$$\hookrightarrow A^{(q-1)+*}(n'l') + h\nu$$

leading to electron capture into specified excited *nl* states of the $A^{(q-1)+}$ ion which decays under photon emission. At low velocities v « 1 a.u. only a limited number of *nl* states is predominantly populated by (near-) resonant charge transfer of the target electron at potential curve crossings.

By means of Photon Emission Spectroscopy we have extensively studied the above mentioned reactions, especially for the impact of He-like and fully stripped ions (see e.g. Dijkkamp *et al* [4] and Hoekstra *et al* [5] and references therein). In this paper we will review three recent developments in our PES studies:

1) determination of contributions from degenerate states in hydrogenic ions,
2) measurements at impact energies well below 1 keV/amu and
3) measurements of *m* state selective electron capture cross sections from vuv radiation.

Before discussing these developments on basis of fundamental collision processes, the experimental set-up will be described briefly.

2. Experiment

The highly charged ions have been produced by the ECR ion source installed at the KVI Groningen. In the experimental set-up, schematically shown in fig. 1 the ions cross a pure H_2 or a mixed H/H_2 target beam produced by a radio-frequency source. The absolute density profiles of the atomic and molecular components of the target beam are determined by observation of electron impact induced atomic (Balmer-ß) and molecular radiation [6]. The photons are observed with a monochromator for visible light (300-600 nm). In this way we found an effective fraction of atomic hydrogen of 75%, see e.g. Čirič *et al* [6] and Hoekstra [7].

The photon emission resulting from the decay of the excited product ions has been observed not only with the above mentioned spectrometer but also with a spectrometer for vuv radiation (10 - 80 nm). Generally this spectrometer is positioned under the magic angle of 54.7° with respect to the ion beam and it is tilted by 45° since under these angles influences of polarization effects cancel, see e.g. [7]. This monochromator is equipped with a position sensitive microchannelplate detector which enables simultaneous detection of emission lines within ranges of about 20 nm. The wavelength dependent sensitivity of the photon detection systems has been determined absolutely by various electron and ion impact processes with known cross sections [4,7].

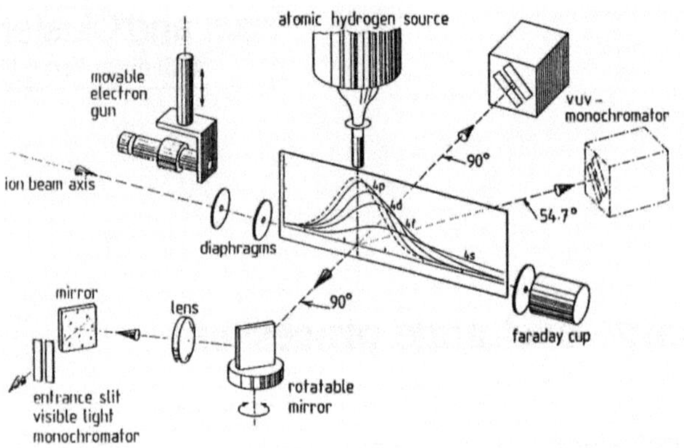

Fig. 1. Schematic view of the experimental set-up. Typical emission profiles of the $He^{+}(4l)$ states are included (see text). The target profile is indicated by the dashed line.

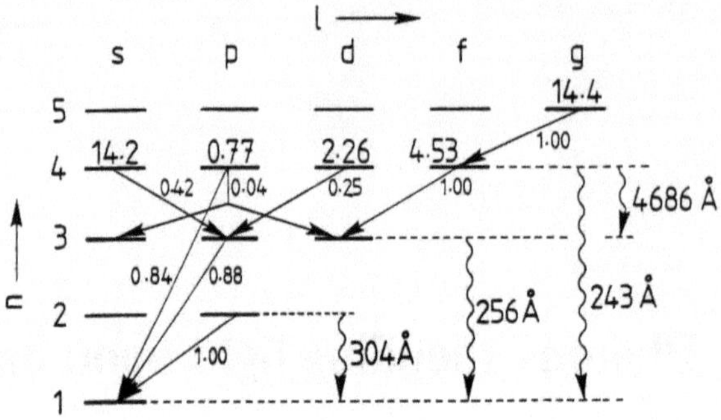

Fig. 2. Energy level scheme of HeII. The relevant transitions are shown together with their branching ratios and the lifetimes (in nsec) of the upper states.

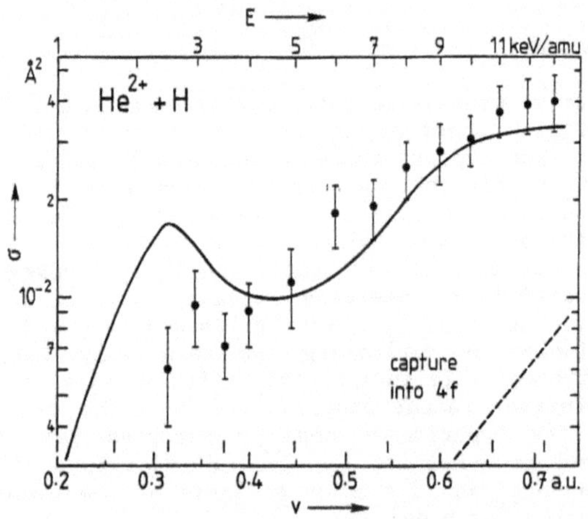

Fig. 3. Absolute electron capture cross sections for charge transfer into $He^{+}(4f)$ in collisions of He^{2+} on H. Theory: —— Fritsch [11] and - - - Ryufuku [12].

3. Determination of contributions from degenerate states in hydrogenic ions

In hydrogenic ions the l levels within one n shell are quasi-degenerate and therefore light emission from different l levels can not be resolved spectroscopically. So measurements of transitions in H-like ions have to be compared with the sum of the various theoretical contributions. Comparing in this way experiments for C^{6+} - H collisions with the most sophisticated calculations it was found that there is good agreement between theory and experiment for the dominant capture channels, whereas for the non-dominant high-n states there were considerable differences [5]. However these transitions between high-n states, yielding light in the visible spectral range, are particularly important for future plasma diagnostics [8]. Due to the strong fields in a tokamak the l states are mixed [8] and hence it is essential to know all the l state electron capture cross sections separately. For the case of He^{2+} colliding on atomic hydrogen we have deduced the separate $4l$ cross sections from the measurement of the HeII($n = 4 \rightarrow n = 3$) transition at 468.6 nm [7,9]. To that end we have exploited the fact that the lifetimes of the states are different. For these HeII($4l$) states relevant lifetimes and branching ratios are shown in fig. 2. Radiation from shortlived states, e.g. $4p$ is mainly concentrated on the target area, whereas radiation from longlived states, like $4s$ is still emitted downstream the ion-beam axis (cf. fig. 1). Therefore the measurement of the emission profiles along the ion beam gives information on the l state population. This method, often used in collision experiments on static targets and in beam foil experiments, has only once before been used in a crossed beam experiment, namely by Aumayr *et al* [10] for H^{+} - Li collisions.

Neglecting cascades the emission profiles along the ion beam axis, P_{4l} are described by

$$P_{4l}(z) = \frac{1}{v\tau_{4l}}\int_0^z T(z')e^{-(z' - z)/v\tau_{4l}}dz' \qquad (2)$$

with v the velocity of the ions and τ_{4l} the lifetime of state $4l$, z the position along the ion beam axis and T(z′) the target density profile. The measured signals, S(z) are equal to

$$S(z) = K \sum_l \beta(4l \rightarrow 3)\, P_{4l}(z)\, \sigma_{4l} \qquad (3)$$

with K an absolute calibration constant, $\beta(4l \rightarrow 3)$ the branching ratio for decay to $n = 3$ and σ_{4l} the electron capture cross sections. By measuring the signals at some 25-30 positions along the ion-beam axis, it is possible to deduce the cross sections from a least squares fit. To increase the accuracy of the fit results the contribution of the $4p$, which is relatively small due to the small branching ratio (fig. 2) has first been substracted from the signals. The σ_{4p} could be determined directly from the $4p \rightarrow 1s$ transition in the vuv. As a typical example of the experimental results obtained in this way fig. 3 shows the data for the $4f$ state, together with the results of the large basis set Atomic Orbital (AO) calculation by Fritsch [11] and the Unitarized Distorted Wave Approximation (UDWA) by Ryufuku [12]. It can be seen that the experimental results are almost 10 times larger than the UDWA results which demonstrates the known inappropriateness of this theory at $v < 1$ a.u. for calculating capture into nondominant states. Although the experimental results increase smoothly with energy, whereas the AO theory shows a structure around 4 keV/amu the agreement is good, especially if we note that capture into the *4l* states is two to three orders of magnitude smaller than capture into the dominantly populated *2p* state. Recently these experiments have been extended into the impact energy range of 30 -130 keV/amu [13].

4. Measurements at impact energies « 1 keV/amu

Generally the PES measurements are limited to the impact range of roughly 1 - 10 keV/amu, which is the range in which (ECR) ion sources produce beams routinely. We have extended our impact energy down to lower energy by means of a five-element lens system [14] which replaces the set of collimating diaphragms, cf. fig. 1. The last element of the lens encloses the collision region. Hence the potential difference between the last element and the ECR ion source defines the kinetic energy of the ion beam. For impact energies well below 1 keV/amu a lot of discrepancies exist between semi-classical calculations and total one-electron capture measurements. To find the reason for these disagreements and to obtain more insight in low-energy charge transfer state selective cross sections are needed.

Some of the experimental results for C^{4+} - H collisions are shown in figs. 4 and 5 which also include the theoretical results [15,16,17] and related measurements [4,18]. In this collision system electron capture populates predominantly the $C^{3+}(3l)$ states and at energies around 100 eV/amu almost only the $C^{3+}(3p)$ state is populated, cf fig. 4. From the figures it is seen that our results are a logical extension down to lower energies of the data by Dijkkamp *et al* [4] and that there is generally a fair agreement with the theoretical predictions. The calculations by Gargaud *et al* [15] (MO7) and by Gargaud and McCarroll [16] (MO4) use molecular orbitals as basis set and are essentially the same, differing only in the number of molecular states included. The older calculations, MO4 include only the σ orbitals correlated to the $n = 3$ manifold of CIV, the more recent calculations MO7 include also the π and δ orbitals. Therefore MO7 includes both radial and rotational coupling whereas MO4 only includes radial coupling. The AO+ calculations by Fritsch and Lin [17] use atomic orbitals, namely all $n = 3$, *4* and *5* CIV orbitals, the H(*1s*) orbital and some united atom orbitals to account for molecular binding effects at the lower velocities.

Comparing in more detail theory and experiment it is seen from fig. 4 that for energies above 100 eV/amu there is good agreement between experiments and MO7 and AO+ calculations. The MO4 results are at variance with the experiments, which might indicate the importance of rotational couplings which have not been included in the MO4 calculation. In that respect it should be noted that whereas in the energy range of 100 to 200 eV/amu the relative *3l* results of the MO7 work are in agreement with experiment, they differ in absolute value, see fig. 5 and Hoekstra *et al* [14]. Since the *3p* is the by far

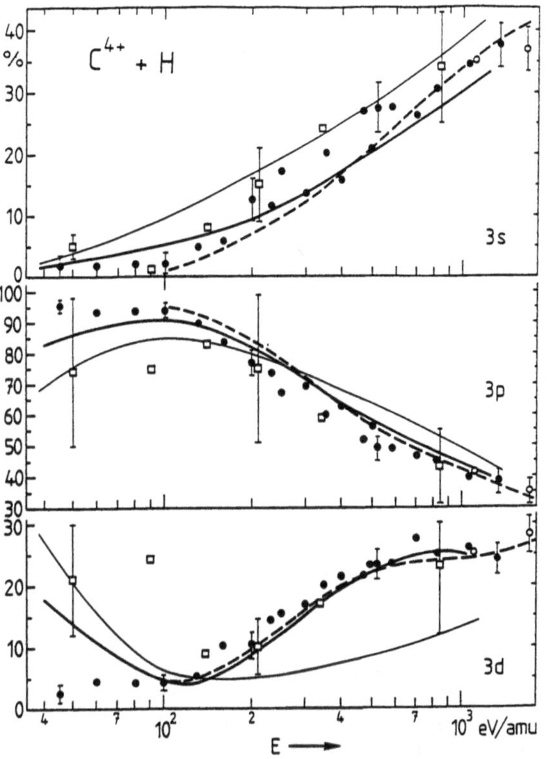

Fig. 4. Relative electron capture cross sections ($\sigma(3l)/\sigma(n = 3)$) for C^{4+} - H collisions. Theory: —— MO7 Gargaud *et al* [15], —— MO4 Gargaud and McCarroll [16] and – – – AO+ Fritsch and Lin [17]. Experiments: □ - Baptist *et al* [18], ○ - Dijkkamp *et al* [4] and ● - Hoekstra *et al* [14].

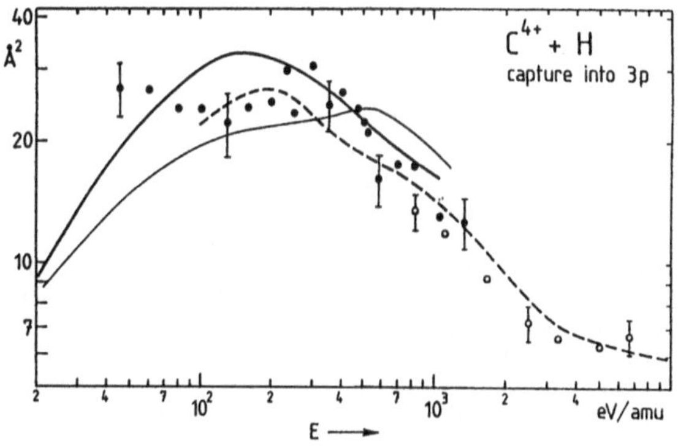

Fig. 5. Electron capture into $C^{3+}(3p)$ in C^{4+} - H collisions. Theory and experiment as in fig. 4.

most populated state (cf. fig. 4) the theoretical overestimation of this state is reponsible for the long standing discrepancy between theory and total charge transfer measurements by Phaneuf *et al* [19]. Below 100 eV/amu the theory predicts a decrease in the *3p* cross section and a strong increase in *3d* cross section. This effect is not yet observed in our measurements which are therefore also in disagreement with the results of Baptist *et al* [18]. The reason for this discrepancy is yet unclear, however with different lens settings (changing the overlap of target and primary ion beam) we could reproduce the results

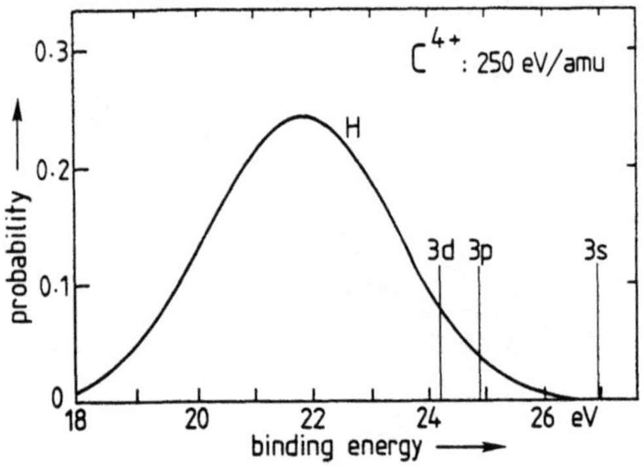

Fig. 6. Reaction window for electron capture in collisions of 250 eV/amu C^{4+} colliding on H.

at 60 eV/amu within error bars. At last it may be stated that below 100 eV/amu the total charge transfer cross sections are in agreement with the MO7 calculations (see Hoekstra *et al* [14]), however the l distribution is clearly different (see fig. 4).

The general tendencies in charge transfer in C^{4+} - H collisions may be understood classically with the "dynamic" version (Niehaus [20]) of the classical over-barrier model. With this model we have calculated (as described by Hoekstra [7]) the reaction window for 250 eV/amu C^{4+} colliding with H. Fig. 6 shows the reaction window together with the energy levels of the populated $C^{3+}(3l)$ states. From this figure it can be seen that the $3l$ levels are situated on the edge of the reaction window. From these positions we expect $\sigma(3s)$ to be relatively small and $\sigma(3p)$ and $\sigma(3d)$ to be of the same order of magnitude. Whereas the first expectation is confirmed by the experiments the latter one is not, since $\sigma(3p)$ is much larger than $\sigma(3d)$, cf. fig. 4. The importance of the $3p$ state may be understood from angular momentum arguments [21]. In the frame of the C^{4+} ion, the target electron has an apparent angular momentum of the order of bv with b the impact parameter and v the velocity of the ion. The maximum impact parameter at which the electron still can transfer (classically) to the ion is 10 a.u., cf. [14]. Therefore the angular momentum for 250 eV/amu C^{4+} (v = 0.1 a.u.) ranges between 0 and 1.0. These values fit obviously better to the p than to the d state. It should be noted that this argument is not totally complete since there is also a contribution to the angular momentum of the intrinsic motion of the target electron. So from this angular momentum argument the population of the d state will become more important with increasing energy, which is indeed seen from figs. 4 and 5. It has been found by Dijkkamp *et al* [4] that at impact energies of approximately 6 keV/amu l distributions become more or less statistical. It should be noted that a statistical population of the l states does not imply that the m levels within one l state are equally populated as will be shown in the next section.

5. Measurements of *m* state selective electron capture cross sections from vuv radiation

Whereas experimental results for electron capture into nl states are becoming more and more available, results for m states are scarce. However, the m distribution gives the most direct information on the primary capture processes since in the first place molecular σ- ($m = 0$) and π- ($m = 1$) orbitals are generated by radial and rotational coupling, respectively. A nonuniform population of nlm levels with different $|m|$ manifests itself by anisotropic and polarized line emission [7] during decay of these levels. Here we describe measurements of the polarization of the $3p \rightarrow 2s$ transition (at 31.2 nm) of $C^{3+}(3p)$ ions which have been produced in slow collisions on molecular hydrogen. The procedure for evaluating the polarization of the $3p \rightarrow 2s$ transition is to measure the relative change in the intensity at two spectrometer orientations, namely at 90° and 54.7° with respect to the ion beam, cf. fig. 1. At 54.7°, the *magic angle*, there is no influence of polarization. The intensity of light emitted at a certain angle θ, I(θ) is related to the intensity, I(90°) and polarization, Π(90°) at 90° in the following way

$$I(\theta) = [1 - \Pi(90^\circ)\cos^2(\theta)]\, I(90^\circ) \qquad (4)$$

The measured signals at 54.7° and 90°, S(54.7°) and S(90°), respectively have to be normalized since the effective target pressure observed under these angles is different. This can be done by relating them to the signals of the $3s \rightarrow 2p$ transition, since this transition is unpolarized and hence isotropic. Using this normalization we can deduce the polarization of the $3p \rightarrow 2s$ transition from the measured signals by applying the following relation deduced from eq. 4

$$S(54.7^\circ) = S(90^\circ)\,\frac{S_{3s-2p}(54.7^\circ)}{S_{3s-2p}(90^\circ)}\left[1 - \frac{\Pi(90^\circ)}{3}\right] \qquad (5)$$

For an $np \rightarrow ns$ *doublet* line the cross sections σ_0 and σ_1 for populating the $m = 0$ and $|m| = 1$ magnetic substates of the np state respectively are related to the polarization at 90° by

$$\Pi(np \rightarrow ns) = \frac{3\sigma_0 - 3\sigma_1}{7\sigma_0 + 11\sigma_1} \qquad (6)$$

From fig. 7 it can be seen that the polarization varies from Π = 0.05 at v = 0.22 a.u. to Π = 0.34 at v = 0.5 a.u. Therefore the population of the $m = 0$ substate represents between 56% (v = 0.22 a.u.) and 92% (v = 0.5 a.u.) of the total $3p$-state population. To some extent a preference for populating $m = 0$ at higher velocities can be understood from the following classical picture: In the frame of the C^{4+} ion the target electron initially has two velocity components, namely one due to its intrinsic velocity distribution and one equal but opposite to the ion velocity v. The latter one obviously increases with increasing impact energy and thus the angle between the total velocity vector and the ion beam direction becomes smaller. Therefore the angular momentum vector becomes more and more perpendicular to the ion

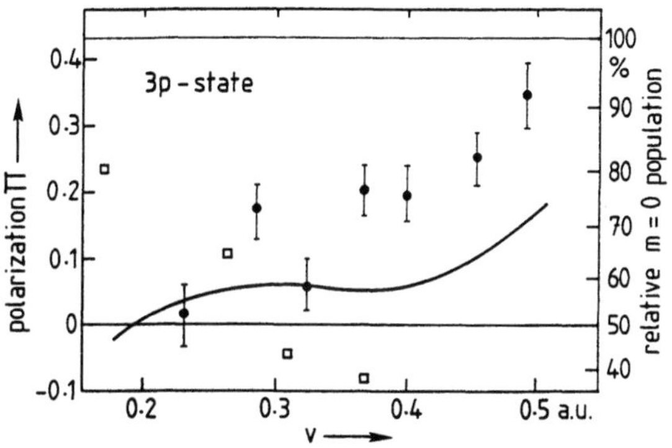

Fig. 7. The polarization Π and the relative $m = 0$ population of the *3p* state of CIV, produced in collisions of C^{4+} on molecular hydrogen. Theory: —— Harel and Jouin [24] (C^{4+} - H). Experiment: □ Baptist *et al* [23] and • Hoekstra *et al* [22].

beam axis (z axis). Hence the $m = 0$ states are increasingly populated. Although our data cannot directly be compared with the theoretical predictions of Harel and Jouin [24] (C^{4+} - H) it is interesting to note that both their predictions and our experimental results show a comparable velocity dependence and a predominant population of the $m = 0$ substate. It should be noted that the predominant population of the $m = 0$ state is at variance with the equal population of m states implicitly suggested by the approximately statistical distribution of l state selective cross sections, at roughly $v = 0.5$ a.u. (see e.g. Dijkkamp *et al* [4]).

Conclusion

We have discussed three recent developments in our Photon Emission Spectroscopy studies of basic charge exchange processes:

1) determination of contributions from degenerate states in hydrogenic ions,
2) measurements at impact energies well below 1 keV/amu and
3) measurements of m state selective electron capture cross sections from vuv radiation.

The first point was achieved by measuring emission profiles along the ion beam axis and thereafter deconvoluting these profiles to the specific contributions. For $He^{+}(4l)$ population in collisions on H it was shown that there is good agreement with large basis set AO+ calculations [11]. At impact energies well below 1 keV/amu it was found for C^{4+} colliding on atomic hydrogen that generally there is a fair agreement with theory, however some discrepancies persist. The common tendencies in the measurements could be understood on basis of the "dynamic" version of the classical overbarrier model. Finally we discussed m state selective measurements in the vuv for C^{4+} colliding on molecular hydrogen. The predominant population of the $m = 0$ state at higher velocities shows that the statistical l distribution experimentally found [4] does not imply that the m states are equally populated.

Acknowledgement

The authors gratefully acknowledge the technical support by J. Eilander and J. Sijbring and the participation of J.P.M. Beijers and A.R. Schlatmann in the low energy measurements and of M.G. Suraud in the m state selective measurements. This work was performed as part of the research program of the "Stichting voor Fundamenteel Onderzoek der Materie" (FOM), which is financially supported by the "Stichting voor Nederlands Wetenschappelijk Onderzoek" (NWO).

References

1. Janev, R.K., Presnyakov, L.P.: *Physics of highly charged ions* (Springer, Berlin 1985)
2. Janev, R.K., Winter, H.,: Phys.Rep. **117** 265 (1985)
3. Gilbody, H.B: *Advances in Atomic and Molecular Physics* (eds. Bates, D., Bederson, B.) **22**, 143 (Academic Press, London 1986)
4. Dijkkamp, D., Čirič, D., Vlieg, E., de Boer, A., de Heer, F.J.: J.Phys. **B18**, 4763 (1985)
5. Hoekstra, R., Čirič, D., de Heer, F.J., Morgenstern, R.: Phys.Scr. **T28**, 81 (1989)
6. Čirič, D., Dijkkamp, D., Vlieg, E., de Boer, A., de Heer, F.J.: J.Phys. **B18**, 4745 (1985)
7. Hoekstra, R.: thesis R.U. Groningen, available on request (1990)
8. Fonck, R.J., Darrow, D.S., Jähnig, K.P.: Phys. Rev. **A29**, 3288 (1984)
9. Hoekstra, R., de Heer, F.J., Morgenstern, R.: *Electronic and Atomic collisions* (edited by Dalgarno, A. et al) 390 (American Institute of Physics, New York 1990)
10. Aumayr, F., Fehringer, M., Winter, H.:J.Phys. **B17**, 4201 (1984)
11. Fritsch, W.: Phys.Rev. **A38**, 2664 (1988) and private communication.
12. Ryufuku, H.:Rep.JAERI-M-82-031 (1982)
13. Frieling, G.J., Hoekstra, R., de Heer, F.J.: these proceedings
14. Hoekstra, R., Beijers, J.P.M., Schlatmann, A.R., Morgenstern, R., de Heer, F.J.: Phys. Rev. **A41**, 4800 (1990)
15. Gargaud, M.,McCarroll, R.,Valiron, P.: J.Phys. **B20**, 1555 (1987)
16. Gargaud, M., McCarroll, M.:J.Phys. **B18**, 463 (1985)
17. Fritsch, W. Lin, C.D.: J.Phys. **B17**, 3271 (1984)
18. Baptist, R., Bonnet, J.J., Bonnefoy, M., Boursey, E., Brenac, A., Chassevent, M., Chauvet, G., Dousson, S., Le Duff, Y., Fleury, A., Gargaud, M., Hitz, D.:Nucl.Instrum.Methods **B23**, 123 (1987)
19. Phaneuf, R.A., Alvarez, I., Meyer, F.W., Crandall, D.H.: Phys.Rev. **A26**, 1892 (1982)
20. Niehaus, A.: J.Phys. **B19**, 2925 (1986)
21. Burgdörfer, J., Morgenstern, R., Niehaus, A.: Nucl.Instrum.Methods **B23**, 120 (1987)
22. Hoekstra, R., Suraud, M.G., de Heer, F.J. Morgenstern, R.: J.Physique Coll. **50**, 387 (1989)
23. Baptist, R., Bonnet, J.J. Chauvet, G., Desclaux, J.P., Dousson, S., Hitz, D.: J.Phys. **B17**, L417 (1984)
24. Harel, C., Jouin, H.: J.Phys. **B21**, 859 (1988)

Supplement to Z. Phys. D - Atoms, Molecules and Clusters 21, S 87-91 (1991)

Atoms, Molecules
Zeitschrift für Physik D and Clusters

Calculations for doubly-excited states populated in two-electron processes

J. E. Hansen, H. van der Hart, M. Landtman, Y.-T. Shen and N. Vaeck*
Zeeman Laboratory, University of Amsterdam, Plantage Muidergracht 4, 1018 TV Amsterdam, The Netherlands

* Research Assistant of the Belgian National Fund for Scientific Research (FNRS)

Abstract

Recent calculations of doubly-excited states for two and four-electron states in neutral atoms and ions are described. In particular the radiative properties of these states are considered. It is pointed out that some of these states live so long that triply-excited states perhaps can be formed in neutralisation processes at surfaces.

I. Introduction

The interest in doubly-excited states has been growing the last 20 years, for reasons partly connected with the fact that some of these states represents a breakdown of the conventional single-particle picture and thus to some extent a new type of state. On the other hand these states although first noticed in He has now been observed in highly-ionized systems also. Some of them form apparently readily for example in collision experiments.

Doubly-excited states can have rather different properties from singly excited ones. Let us consider He as an example. The ground state is $1s^2$ and if we excite one electron we have configurations like $1sn\ell$ where n perhaps can be large. The interaction between 1s and $n\ell$ will then be weak. There will be other configurations $1sn\ell'$ nearby, because $E(n\ell)=E(n\ell')$ in a hydrogenic approximation, and if the $1sn\ell'$ configuration has the same parity as $1sn\ell$, there could be an interaction between the two which potentially could be large because of the energy degeneracy but in practise is weak or absent because the interacting states have different LS values.

However, if both 1s electrons are excited we can get a different situation. If, for example, both electrons are excited to n=4 we have a rather large number of configurations with several terms of the same type. The configurations are, of even parity $4s^2$, 4s4d, $4p^2$, 4p4f, $4d^2$ and $4f^2$ and of odd parity 4s4p, 4s4f, 4p4d and 4d4f. Again all these configurations are degenerate in energy at least in the hydrogenic approximation and this time also the interactions are large since all four electrons are located in the same area of space. Thus we expect all the configurations of the same parity to interact strongly with each other. The importance of this interaction was predicted by D. Layzer in 1959 and he called the set of configurations "the complex" [1]. Since then it has become standard practise to include at least these configurations in configuration interaction calculations also for groundstates. One example is Be $1s^22s^2$ for which the $1s^22p^2$ configuration belongs to the same complex and is a strong perturber. Also if we have a configuration like $3\ell4\ell'$ we can construct a strongly interacting complex. The even parity complex comprises 3s4s, 3s4d, 3p4p, 3p4f, 3d4s and 3d4d while 3s4p, 3s4f, 3p4s, 3p4d, 3d4p and 3d4f belong to the odd parity complex. Also in this case we have degeneracy of the (hydrogenic) basis states and a strong interaction although weaker than for the $4\ell4\ell'$ complex because the n=3 and n=4 electrons have a smaller overlap with each other than the n=4 electrons.

II. Other quantum numbers

The fact that we have this strong interaction means that the states belonging to the complex can be very mixed and thus not well described by single configuration quantum numbers. One of the earliest examples is the $1s^2\ ^1S$ - $2\ell n\ell\ ^1P^o$ series in He which were observed for the first time by Madden and Codling in 1963 [2]. We would expect to see three series; 2snp, 2pns and 2pnd but experimentally only one was observed at first, later a few weak lines of another series was found. The interpretation by Cooper, Fano and Prats [3] was the observation that for a particular n (>2), the three series form a Layzer complex and thus interact strongly. This leads to the formation of three new series, of which the two observed ones are termed + (for symmetric in the mixing of 2snp and 2pns) and - (for antisymmetric) in which the two electrons can be interpreted as performing oscillations in and out of phase with each other. The + series in which the two electrons are close to the nucleus at the same time, has the larger probability for excitation from the ground state and is responsible for the strong observed series, while the - series can be identified with the weak series. The third even weaker series can be interpreted as primarily 2pnd.

The only terms that can be observed in photoabsorption from the ground state in He are of $^1P^o$ character but there are of course many other terms in particular when we go to higher n values like $3\ell 3\ell'$. Fig.1 shows the terms in the $1s^2 3\ell 3\ell'$ configurations in N IV.

Fig.1. Doubly-excited energy levels in N IV

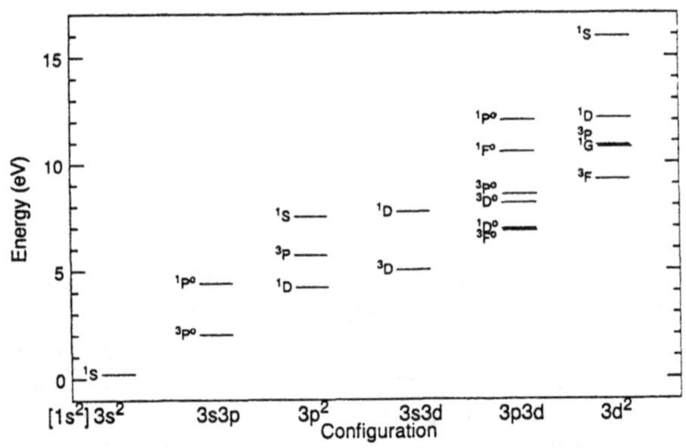

A glance at this figure shows the strength of the interaction between the different configurations. In the $3p^2$ configuration one would normally have the three terms in the order ^3P, ^1D and ^1S. The reason that this is not the case here is that there is a strong mixing of $3p^2\ ^1D$ with in particular $3s3d\ ^1D$ and $3d^2\ ^1D$. There are several other strong mixings especially for the low L values and this and other considerations has led to an interest in whether other quantum numbers can be found which can give a better description of the real states. It is not the purpose of this paper to discuss this subject but Fig. 2 shows the terms from Fig.1 arranged according to the so called linear rotor-vibrator picture in which the two electrons and the nucleus are pictured as a tri-atomic molecule performing collective, that is to say correlated, motions [4]. This picture uses the $[K,T]^A$ quantum numbers, described for example in ref.4, to classify the correlation patterns of the doubly-excited states.

Fig.2. *d* supermultiplet classification of energy levels in N IV, see ref.4.

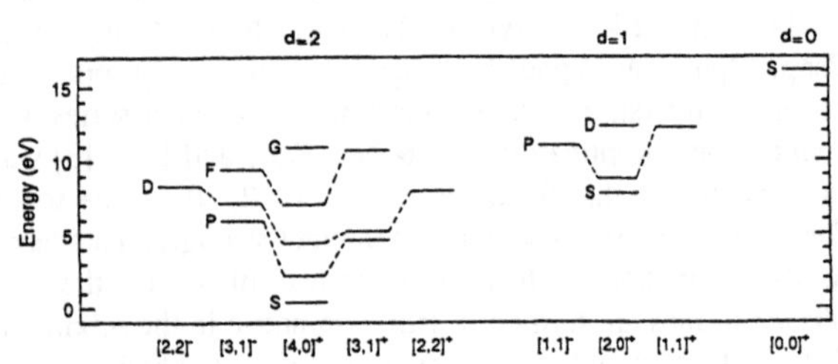

It should be pointed out that the theory predicts a near degeneracy (T-doubling) between terms on opposite sides of the middle row with quantum numbers [K,0]. This degeneracy is to a fair degree present in the He isoelectronic sequence but not in evidence in Fig.2 in the presence of a $1s^2$ core [5].

III. Calculations

It is fair to say that although the molecular analogy just described has had a large influence on our understanding of the doubly-excited states, most of the actual calculations have not used this concept except to classify the results and the actual calculations has been carried out by a variety of methods, most of which does not use the molecular picture. In fact since this is the simplest non-trivial many-body problem practically every method of calculation has been used to try to solve it. Most work has been carried out for He and H^- and Table 1 shows a collection of some of methods which has been used to calculate two-electron states in He or in the He isoelectronic sequence.

Table 1:

A selection of methods used to solve the two-electron problem in the He isoelectronic sequence.

Truncated diagonalisation	
Hydrogenic basis	ref.7
B-spline basis	present work
Feshbach projection	
Conventional	ref.24
Model potential	ref's 11 and 12
Complex coordinate rotations	ref.25
Close-coupling	ref.26
MBPT	ref.27

The essential differences between the different methods can perhaps best be understood from Fig.3, which shows some of the bound states and continua associated with the three lowest ionization limits 1s, 2s and 2p in He. Let us assume that we are concerned with the determination of the properties of the lowest states of the type $2\ell n\ell'$. These states belonging to the so called closed channels are degenerate with continuum states of the same symmetry of the type $1s\epsilon\ell$ (belonging to the open channels) and the interaction with the nominally bound $2\ell n\ell'$ states leads to the non-radiative decay of the latter.

The interaction between bound and continuum states [6] is the reason for the complicated lineshape of the $1s^2\ ^1S \rightarrow 2\ell n\ell'\ ^1P^o$ absorption lines in He and is one of the basic difficulties in the calculation.

The truncated diagonalisation method is one of the oldest theorical approaches [7] and it has provided a very substantial part of the calculated energy values obtained. It consists of choosing a large enough (hydrogenic) basis set of bound states belonging to the closed channels and diagonalising the matrix consisting of the matrix elements of the (electrostatic) Hamiltonian between the basis states. The autoionization decay rate is calculated by second order perturbation theory

Fig.3. Some open and closed channels for $2\ell n\ell'$ $^1P^o$ terms in He-like systems.

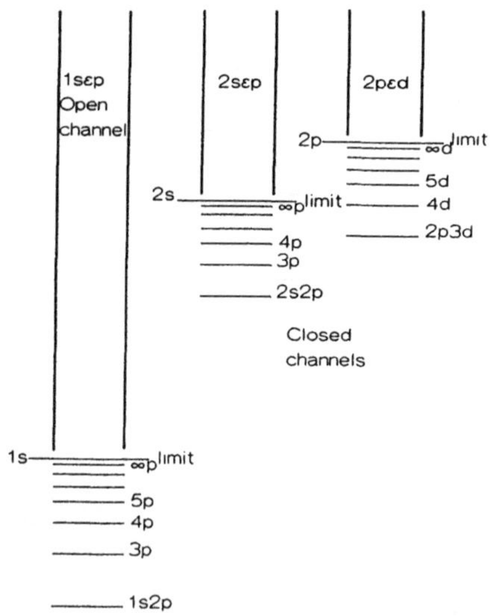

between the bound wavefunction and a continuum $1s\epsilon\ell$ wavefunction of the proper energy. There are several problems with this method. One problem is that the diagonalization for the bound spectrum can lead to very long eigenvectors which makes it difficult to characterise the states in a simple way. Another is that the interaction with the open channel leads to a shift of the resonance energy, since for example, the interaction with the bound part of the $1sn\ell$ series is neglected, as well as to a broadening. The first problem was one of the reasons for the interest in other types of (collective) quantum numbers. However, the computational problem is that the use of hydrogenic basis functions requires very many functions. The reason being in physical terms that the $n\ell$ hydrogenic function is a bad approximation to the helium $n\ell^2$ or $n\ell n\ell'$ configuration wave function when $n \sim n'$ due to the large screening by the other electron. In principle this problem can be solved by including in the calculation a complete $n\ell/\epsilon\ell$ basis but to do this using hydrogenic wavefunctions (or HF wavefunctions for problems with more than two electrons) requires an infinite set including continuum functions for the closed channels. This has been one reason for trying to invent other methods which like the close-coupling method, for example, is variational and thus can pick up the effect of a large number of basis states in one function. Recently an alternative basis set has been introduced which can represent an infinite basis set with a finite set of functions. We have used these functions, which are called B-splines [8], to make truncated diagonalisations effectively in the same manner as Conneely and Lipsky [9], for example, but as Table 2 shows with even better results for the energies as could be expected since we effectively include the continuum belonging to the closed channels.

We compare with the results of Ho [10] who has used the complex coordinate rotation method which avoids the explicit use of continuum wavefunctions from the open channels but in principle gives a solution which includes the interaction between open and closed channels. The B-spline results include only s, p, d and f orbitals at present and neglects as mentioned the interaction with the open channel completely. With this in mind the improvement compared to the hydrogenic basis is very satisfactory.

Most of the methods mentioned above are most easily applied to two-electron systems and extension to more electrons are not always obvious. One reason being that Hylleraas type wavefunctions often are used in order to circumvent the hydrogenic basis set problem, mentioned earlier. Recently the Feshbach projection method has been extended to many-electron systems by use of a model potential [11] which also simplifies the calculations in cases when several open channels are present (even in the two-electron case). The method has been applied to two-electrons outside a $1s^2$ shell [12] with good results [13]. It is also easy to extend the truncated-diagonalisation method to many-electron systems [11]. If we consider two-electrons outside closed shells the calculations only have to be based on another basis set which takes the screening by the core into accound in order to include the interaction between the two outer electrons accurately but neglecting core polarization which can be expected to be a good approximation when the outer electrons are far away from the $1s^2$ core. We are presently working on extending the B-spline technique to such cases.

Table 2:

Results using the truncated diagonalization method with a B-spline basis compared to results using a hydrogenic basis as well as results obtained from the complex rotation method. The results are for the lowest $2\ell n\ell'$ ^1S and $^1P^o$ terms with $n \geq 2$ in He. Energies are given in a.u.

B-spline basis	Hydrogenic basis ref.9	Complex rotation ref.10
^1S		
-0.778756	-0.7752447	-0.77787
-0.621992	-0.6151335	-0.621928
-0.590086	-0.5881421	-0.589925
-0.548153	-0.5469404	-0.54809
$^1P^o$		
-0.6925150	-0.6883642	-0.693135
-0.5970934	-0.5965637	-0.5970745
-0.5639601	-0.5629202	-0.5640
-0.5470360	-0.5468116	-0.54693

IV. Application to experiment

Let us now turn to some of the applications to experiments. Doubly-excited states have been observed for example in charge exchange experiments [15] and it is particularly interesting that the PCI effect allows the lifetimes of the states to be probed experimentally so that the experiments not only provide a check on the calculated energies but also on the calculated autoionization rates [13]. Two questions have been of particular interest, one is the question of which states are populated and whether a connection to the collective picture can be found. The other is the question of the stability of the resulting ion. We will consider the latter here.

If we consider the decay of the states in the $1s^23\ell3\ell'$ configurations, then we have possibilities for radiative decay to $1s^22\ell3\ell'$ configurations and possibilities for non-radiative decay to $1s^22\ell$ configurations. Due to the dependence of the radiative rate on transition energy which increases regularly from ion to ion in the Be isoelectronic sequence we expect the total radiative decay to increase with Z in fact as Z^4 [16]. On the other hand, since the free electron energy in the non-radiative decay to $1s^22\ell$ is increasing with Z, the non-radiative decay rates which depend on the overlap between the continuum electron and the bound electrons, can be expected to decrease with Z. In a hydrogenic approximation the autoionization rate is independent of Z [16] and in reality, the autoionization rates for the Be isoelectronic sequence have been found to be fairly independent of Z over the range C^{2+} to Xe^{50+} [14]. Thus we expect the end result to be a degree of radiative stabilization with increasing Z.

However, the models of two-electron transfer in collisions between ions and atoms predict that the electrons with increasing Z are picked up in states with higher and higher n values [17]. Thus although we predict radiative stabilization in the $1s^23\ell3\ell'$ configuration with increasing Z, the actual experiments would lead to configurations $1s^2n\ell n'\ell'$ with higher and higher n when Z increases and the question is whether we would obtain radiative stabilization in this case. At first sight this might seem likely, because if we consider the transitions to the same states and limits as above we would in fact find the same behaviour as described above. However, there is one complicating factor. We do in fact have an increase in the number of limits available for non-radiative decay, not only $1s^22\ell$ but also higher $1s^2n\ell$ limits can be reached [18]. Every time we cross over a limit, the free electron energy in the Auger decay to the last limit is reduced and thus the overlap with the bound orbitals increased. This leads to an increase in the Auger decay probabilities and we can therefore expect "windows" in the probabilities for radiative decay relative to Auger which perhaps can be used in the unraveling of the precise pick-up mechanism.

There is recent evidence for radiative stabilization in collisions between highly ionized Ar [20] and Xe [21] and neutrals. However, the precise explanation for this is not yet known. In the case of Ar a fairly thorough calculation of branching ratios have been carried out [22] which do not appear to be able to explain the stabilization observed.

Fig.4. Position of the doubly-excited $1s^2n\ell n'\ell$ (n=7-9; n'=6-9) in Ar^{14+} relative to the $1s^2n\ell$ (n=4-7) limits in Ar^{15+} from ref.18.

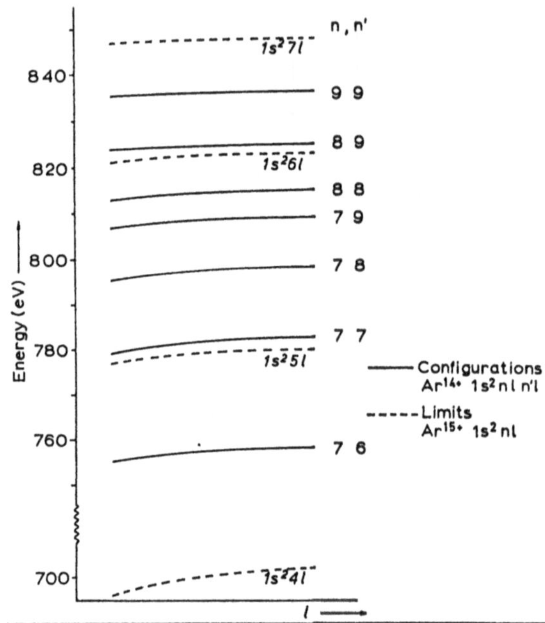

V. Neutralisation at surfaces

There is another experiment where we expect that highly-excited states can be formed, namely neutralisation of ions at surfaces. Here it is believed that the ion attracts electrons from the surface during the last 10^{-14} sec or so before impact and that these electrons although originally attracted into states with high n values have managed to reach the n=1 and 2 shells before the impact. This means that in the best case radiative and in the worst case non-radiative decay probabilities must lie around 10^{-15} sec. If we consider the case of completely stripped Na ions, Table 3 shows non-

Table 3:

Average autoionization decay rates for 16s15p and 16s17p states in Na^{9+} assuming a statistical population. See also ref.19.

$15s^2$	$\rightarrow n\ell\epsilon\ell$	n=1,10 ; ℓ=s,p,d,f,g	$A^a=2.8\cdot10^{12}$
15s16s		n=1,11	$A^a=2.1\cdot10^{12}$
15p16s		n=1,11	$A^a=2.8\cdot10^{12}$
$16s^2$		n=1,11	$A^a=3.0\cdot10^{12}$
16s16p		n=1,11	$A^a=2.0\cdot10^{12}$
16s17p		n=1,12	$A^a=1.4\cdot10^{12}$
$17s^2$		n=1,12	$A^a=3.1\cdot10^{12}$

radiative decay rates for some 16snp states which clearly shows that these states will not have decayed before impact.

This raises the possibility that a third electron is picked up before the doubly-excited state can decay and if the non-radiative decay of the triply-excited state can compete with the decay rates of the doubly-excited state the decay of the triply-excited state will be a competitor to the radiative decay of the doubly-excited states. In a separate contribution to these proceedings [23] we report calculations admittedly for a system with much lower n values, namely $1s^2 3\ell 3\ell' 3\ell''$, and we find in fact that the average non-radiative lifetime for this configuration, assuming a statistical population, against decay to $1s^2 2\ell 3\ell'$ is 40 times shorter than the average lifetime for the $1s^2 3\ell 3\ell'$ states. Thus the triply-excited states can in fact be one of the underlying mechanisms for the neutralisation of ions at surfaces.

Acknowledgments

This work is part of the research program of the stichting voor Fundamenteel Onderzoek der Materie (FOM), which is financially supported by NWO

References

1. Layzer, D.: Ann. Phys., NY **8** 271 (1959)
2. Madden, R.P., Codling, K.: Phys. Rev. Lett. **10**, 516 (1963)
3. Cooper, J.W., Fano, U., Prats, F.: Phys. Rev. Lett. **10,** 518 (1963)
4. Herrick, D.R., Kellman, M.E.: Phys. Rev. A **21**, 418 (1980)
5. Chen, Z., Lin C.D.: J. Phys. B. **22**, 2875 (1989)
6. Fano, U.: Phys. Rev. **124**, 1866 (1961)
7. Altick, P.L., Moore, E.N.: Phys. Rev. Lett. **15**, 100 (1965)
8. Johnson, W.R., Sapirstein, J.: Phys. Rev. Lett. **57**, 1126 (1986)
9. Conneely, M.J. and Lipsky, L.: J. Phys. B **24**, 4135 (1978)
10. Ho, Y.K.: Phys. Rev. **A 23**, 2137 (1981)
11. Martín, F., Mó, O., Riera, A., Yáñez, M.: Europhys. Lett. **4**, 799 (1987)
12. Bachau, H., Galan, P., Martín, F., Riera, A., Yáñez, M.:At. Data Nucl. Data Tables, **44**, 305 (1990)
13. Posthumus, J.H., Morgenstern, R. : J. Phys. B **23**, 2293 (1990)
14. Vaeck, N., Hansen, J.E.: J. Phys. B **22**, 3137 (1989)
15. Mack, M.: Thesis, Groningen (1987)
16. Cowan, R.D.: The Theory of Atomic Structure and Spectra. Berkeley: University of Calif. Press (1981) Ch. 19
17. Niehaus, A.,: J. Phys. B **19**, 2925 (1986)
18. Hansen, J.E.: J. Phys. (Paris) **50**, C1-603 (1989)
19. Andrä, H.J.: Atomic Physics of Highly Charged Ions, Editor: R. Marrus. New York: Plenum Press (1989)
20. Guillemot, L., Roncin, P.: Private communication in ref 22 and Roncin, P.: Contribution to this conference
21. Cederquist, H., Anderson, H., Astner, G., Hvelplund, P., Pedersen, J.O.: Phys. Rev. Lett. **62**, 1465 (1989) and Cederquist, H.: Contribution to this conference
22. Luc-Koenig, E., Bauche, J.: J. Phys. B **23**, 1763 (1990)
23. Vaeck, N., Hansen, J.E.: Contribution to this conference
24. Feshbach, H.: Ann. Phys. **5**, 357 (1958); **19**, 287 (1962)
25. Ho, Y.K.: Phys. Rep. **99**, 1 (1983)
26. Callaway, J.: Phys. Rep. **45**, 89 (1978)
27. Salomonson, S., Carter, S.L., Kelly, H.P.: Phys. Rev. **A 39,** 5111 (1989)

Supplement to Z. Phys. D - Atoms, Molecules and Clusters 21, S 93-98 (1991)

Atoms, Molecules
Zeitschrift für Physik D and Clusters

Two electron processes in charge exchange reactions involving multiply charged ions.

P. Roncin, H. Laurent*, L. Guillemot, M. N. Gaboriaud, and M. Barat
Laboratoire des Collisions Atomiques et Moléculaires, Université Paris-Sud bât. 351, 91405 Orsay Cedex, France
*I.P.N.O.-Bat. 100, BP No. 1 F-91406 Orsay, France

Experiments performed at LAGRIPPA CENG BP 85X 38041 Grenoble Cedex, France.

Abstract:
Multiple electron capture mechanisms are discussed in light of differential energy gain measurements. Two electron processes are examined in terms of one electron processes. Special emphasis is brought to transfer excitation processes

Introduction

Much experimental work has been devoted to the study of electron capture processes by multiply charged ions at few keV collision energies. Writing the general reactions as $A^{q+} + B \rightarrow A^{(q-s)+} + B^{r+} + (r-s)\ e^- + h\nu('s)$, it is realized that only joint efforts from all possible spectroscopies may lead to substantial progress,(q is the projectile A ionic charge, r the number of electrons removed from target B and s the number of electron actually stabilized onto the projectile). While the electron, X-`ray , U.V. and visible photon spectroscopy collect refined informations on the decay of the excited states populated by the collision, differential energy gain spectroscopy gives insights on the population mechanisms. We have specialized in measuring differential cross sections associated with the various electron capture processes identified by their energy gain and by both projectile and target final charge state. As the obvious characteristic of multiply charged ions is to carry a large ionic charge, the scattering profile will be dictated by the coulombic repulsion between the charge exchange projectile and the ionized target. For a given process, the careful analysis of the differential cross section may answer questions such as where (at which internuclear distance) was the electron captured? what was the "collision path"? Since much is now known on the mechanism for single electron capture, we have mainly tried to analyse our scattering data in terms of independent one electron capture processes.

I One electron processes

It is now well established that the classical coulombic barrier model[1] (C.B.M.) very well describes the gross features of single electron capture (S.E.C.) processes. In particular, such model straightforwardly predicts the principal quantum number n_1 of the dominant electron capture level, the internuclear distance R_{cbm} at which capture takes place and the order of magnitude of the total cross section. The quasimolecular description sets up another approach in which electron captures occur in the vicinity of well separated curve crossings. Using an atomic wave function basis set the one electron "exchange" interaction can be pictured as mediated by the tiny overlap between the ground state target electron wave function and the one of the projectile capture level (either on top or tunneling through the coulombic barrier). In addition Olson and Salop[2] have derived a useful semi empirical formula to evaluate this interaction $\mathcal{V}_{01}$ at the relevant curve crossing. In this simple "*valence bond*" type approach, transition probabilities can be evaluated in the Landau Zener formalism which predicts internuclear distances ("reaction window"[3]) at which one electron capture is favored. This approach easily confirms the n *selectivity* but also provides the velocity dependance that lacks in the barrier model. Higher collision energies will favor transitions at shorter internuclear distances and thus capture into lower lying states. Since we will not focus on this particular aspect and for sake of simplicity we will suppose here that R_{cbm} indeed coincides with the maximum of the Landau Zener reaction window and that capture occurs dominantly into the n_1 shell. At this stage only the channels for S.E.C. into n_1 and n_1-1 are needed in addition to the entrance channel (fig 1.).

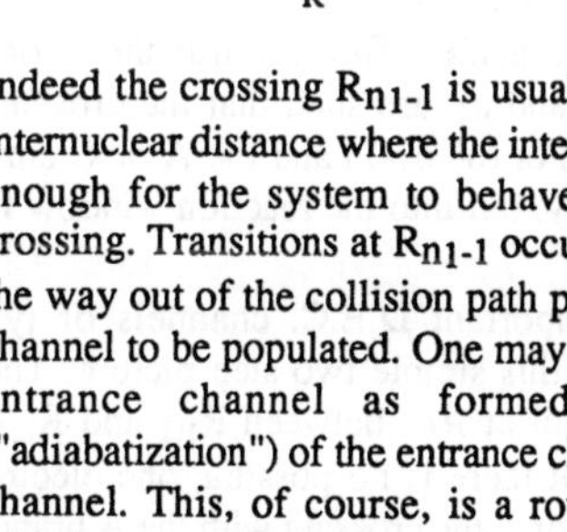

Figure.1
Schematic potential energy curves for one electron capture into the shell n_1 and n_1-1
The dashed line results from the diagonalization of the entrance channel with the S.E.C. channel for capture into n_1-1

Indeed the crossing R_{n_1-1} is usually located at much shorter internuclear distance where the interaction $\mathcal{V}_{01}$ is usually large enough for the system to behave very adiabatically at the crossing. Transitions at R_{n_1-1} occur both in the way in and in the way out of the collision path preventing the (n_1-1) S.E.C. channel to be populated. One may then consider the effective entrance channel as formed from diagonalization ("adiabatization") of the entrance channel with the n_1-1 S.E.C. channel. This, of course, is a rough picture. The crossing with the n_1-1 S.E.C. channel is not fully adiabatic and S.E.C. into n_1-1 will be favored by increasing energy. More interesting is the population mechanism of the "n_1+1" shell. The direct crossing with the incident channel is often located at very large internuclear distance making a direct transition at such R_{n1+1} crossing completely negligible. This channel may however be populated and photon[4] and energy gain[5,26] spectroscopies indicate that the corresponding cross section increases with collision energy. In addition, angular differential cross sections[5] show population of n_1+1 only at "large" scattering angle. Recently, the mechanism responsible for feeding this channel has been identified[5] as due to transition at inner crossings. The existence of these crossings was predicted by the "electron promotion model"[6]. It can be shown that the relative ordering of the crossings is such that this excitation process is limited to $\Delta n=+1$ transitions. The part of the flux which did transit on the S.E.C. into n_1 at the Rn_1 crossing (A branch in fig 1.) will be redistributed between n_1 and n_1+1 at shorter internuclear distances whereas the flux which did not transit at Rn_1 will only feed the n_1 channel via the n_1-1 channel (C branch in fig 1.).

Concerning the angular momenta l of the capture states, the situation is more intricate. For projectiles of moderate charge state and with a large enough ionic core the energy separation between the different possible l values may allow each channel to be treated separately with a l dependant formulation of the exchange interaction as given by Taulbjerg[7]. On the other hand, when the l states are almost degenerated the manyfold of l states is easily coupled over a wide range of internuclear distances. Rotational coupling as well as post collisional effects may then be essential. Photon data however indicate a strong tendency towards population of high l for collision velocity[8] close to one a.u.. Finally, a recent experiment[9] has demonstrated strong orientation propensity[10] in the $B^{3+}+He \rightarrow B^{2+}(2p)+He^+$ system even at a collision velocity as small as 0.07 a.u.

II Double electron capture

The coulombic barrier model has been extended[11,12] to multiple electron capture. It then allows determination of the internuclear distance R_{2cbm} where double electron capture (D.E.C.) will take place. Similarly using the formula by Olson and Salop[2] to describe the interaction $\mathcal{V}_{12}$ between a S.E.C. and a D.E.C. channel, one derives a reaction window where crossings between S.E.C and D.E.C. channel (if any) would be effective. Of course this requires that only states which differ by one spin-orbital are coupled. From there on, D.E.C. is easily explained by "one electron" crossing between S.E.C. and D.E.C. channels. D.E.C. will take place into (n_2,n_1) for crossings on the A branch in fig 2 whereas crossings on the C branch will feed (n_2',n_1-1). Which situation is to be favored is very dependant on the dynamical behavior of the R_{n1} crossing which directs most of the incident flux towards the A or C branch in fig 2. Here n_2 and n_2' are such that the crossings between the states (n_2,n_1) or (n_2',n_1-1) and the A or C curve (fig 1. and 2a) respectively, fall into the reaction window for D.E.C.

However, some important D.E.C. channels of type (n_2',n_1-1) do not fall into this simple two step picture. They cross the entrance channel at $\mathcal{R}_c$ between R_{n1} and R_{n1-1} (fig 2), and $n_2'\neq n_1$ so that there is no possible one electron interaction neither at $\mathcal{R}_c$ nor at the crossing with the A branch. Differential energy gain measurements[13] clearly indicate a direct population mechanism at the $\mathcal{R}_c$ crossing. One must

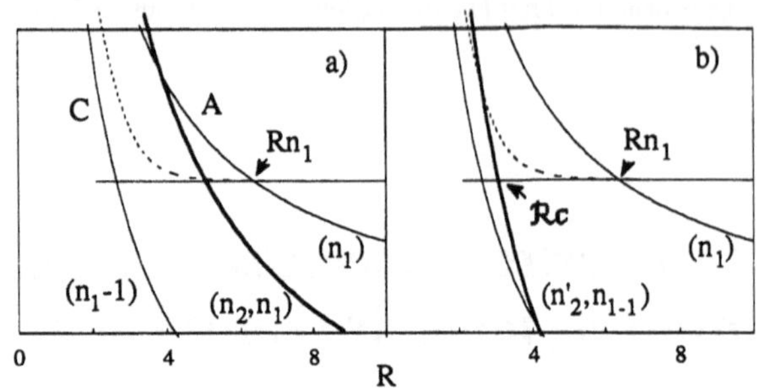

Figure 2 Two curve crossing schemes for D.E.C.
a) D.E.C. occurs by two successive one electron captures at well separated crossings.
b) The D.E.C. channel crosses the entrance channel in the region where the first electron is captured.

remember however that the crossing with the n_1-1 channel is usually very adiabatic. In such adiabatic regime the distance over which the transition to n_1-1 actually takes place is given[14] in the Landau-Zener model as $\Delta R=2\mathcal{V}_{01}/\Delta F$ where $\Delta F=2q/(Rn_1\text{-}1)^2$ is the slope difference at the Rn_1-1 crossing. Within this width the wave function describing the system already has non zero projection over the n_1-1 S.E.C. state, from there on the capture of the second electron appears as a one electron transition. This can be turned quantitative in an adiabatic two state approximation. The projection $\mathcal{P}$ of the adiabatic state onto the diabatic state is given by $\mathcal{P}_{01}=\mathrm{Sin}(\chi/2)^2$ where the "mixing angle" $\chi=\mathrm{Tan}^{-1}(2\mathcal{V}_{01}/\Delta E)$ and $\Delta E(R)$ is the diabatic energy separation, $\mathcal{P}_{01}$ is then interpreted as the probability to make a "virtual" transition to the diabatic state considered. At the crossing $\mathcal{R}c$ with the D.E.C. channel we end up with a transition probability P_{02} given by **(1)** $P_{02}=\mathcal{P}_{01}.P_{12}$ with P_{12} the Landau-Zener one electron transition probability $P_{ij}=1-\exp[-2\pi\mathcal{V}_{ij}^2/v\Delta F]$ (v the radial velocity). At this stage it is interesting to refer to the notion of "correlated" two electron transition which was a subject of wide discussion during the past years. As usually defined uncorrelated processes should write as probability products for individual events as in eq.(1). On the other hand, using a *three states* second order perturbation approximation Laurent *et al*[13] derived an expression for the effective interaction $\mathcal{V}_{02}$ coupling the entrance channel to the D.E.C. channel at the $\mathcal{R}c$ crossing **(2)** $\mathcal{V}_{02}=\mathcal{V}_{01}\mathcal{V}_{12}/\Delta E$. Unlike expression (1) this interaction does not lead to transition probability which factorizes, however both expressions have the same physical ground and give comparable transition probabilities. In fact expression (1) and (2) can be shown to be equivalent if both $\mathcal{V}_{01}/\Delta E$ and $\mathcal{V}_{12}$ are small enough, then $\mathcal{P}_{01}\approx(\mathcal{V}_{01}/\Delta E)^2$, $P_{12}\approx\alpha\mathcal{V}_{12}^2$ whereas $P_{02}\approx\alpha(\mathcal{V}_{01}\mathcal{V}_{12}/\Delta E)^2$ ($\alpha=2\pi/v\Delta F$). The present situation might appear like a topological accident where the three entrance, S.E.C and D.E.C. channels just happened to cross at the same internuclear distance. However considering the width associated to the Rn_1-1 crossing, this situation turns out to be rather general for the part of the flux which passes the dominant S.E.C. channel diabatically.In the two state approximation the transition will be said to derive from virtual transitions to the n_1-1 S.E.C. channel whereas in the three state model the transition appears as a second order effect.

Both approaches require that the $\mathcal{R}c$ crossing lies in the vicinity of R_{n1-1} i.e. that both crossing regions more or less overlap. For this reason the distinction between one step and two step processes which relates to scattering pattern measurement (differential cross section) and therefore to collision path, may not be completely meaningful if interpreted in term of interactions. In high energy atomic collisionswhere an important additional width arises from the Heisenberg uncertainty principle, the terms "one or two step" refer much more to the nature of the interactions than to the curve crossing scheme, so that the presently called "one step" transition would be called "two step" at higher collision energy.

In some tricky situations the one electron exchange interactions $\mathcal{V}_{01}$ and $\mathcal{V}_{12}$ may be so large (and therefore the atomic states so mixed) that four state perturbation has to be considered. This is the case for the C^{4+} He system where the dominant capture process is the D.E.C. into $C^{2+}(1s^2\ 2s^2)$. Here no crossing exists with S.E.C. channel around R_{cbm} and the D.E.C. occurs in "one step" at relatively small internuclear distance. The entrance channel is there strongly coupled to the neighbouring $C^{3+}(1s^2\ 2p)$ and $C^{3+}(1s^2\ 2s)$ S.E.C. states which have been shown[15] to mediate, at least partly, the D.E.C. into $C^{2+}(1s^2\ 2s^2)$.

Since the underlying hypotheses which govern the D.E.C. into (n_2,n_1) on the A branch are more or less the same as those of the C.B.M. which assume complete "molecularisation"[11,12], we may here again assume that n_2 is the value which may be evaluated by these models. On the other hand, the paths which lead to $(n_2',n_1\text{-}1)$ population are not taken into account. From fig 2b. it is easily seen that these $(n_2',n_1\text{-}1)$ D.E.C. channels give rise to larger energy gains and paradoxically to comparatively smaller scattering angles. These two characteristics already explain most of the discrepancies between the predictions of the molecular coulombic barrier model[12] and the experimental coincident differential energy gain measurements[16] on multiple electron capture processes in reactions $Ar^{q+}+B \rightarrow Ar^{(q-s)+}+B^{r+}+(r-s)\ e^-$ (q=7..11 r=1..7 B=Ne,Ar,Kr,Xe).

Including now the possibility for each captured electron to suffer the one electron promotion mechanism described previously[5], then the *one electron transition hypothesis* developed here easily explains (and evaluates) most of the D.E.C. processes.

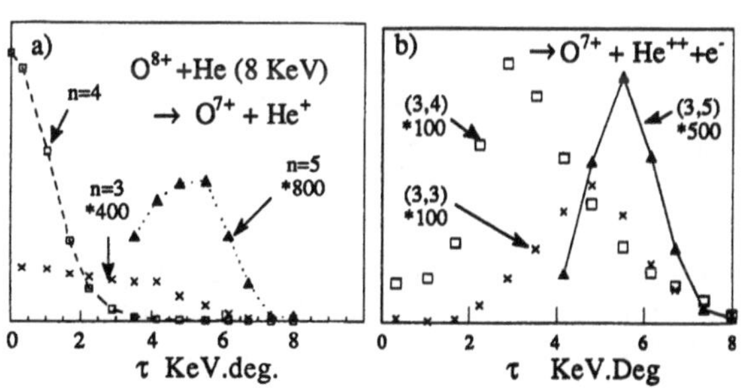

Figure 3: Relative cross sections $d\sigma/d\Omega$ for the various S.E.C. and D.E.C. processes a) and b) respectively as a function of the reduced scattering angle τ.

Taking the example of the O^{8+} He system at 8 keV collision energy, measurement[13,5] of energy gain and scattering angle of the O^{7+} ions (fig 3) allowed to check the population mechanisms. The n_1=4 channel is indeed directly populated at R_{n1}, population of the n_1-1=3 channel derives from a small "leak" at the adiabatic R_{n1-1} crossing whereas the n_1+1=5 channel is fed through promotion of the dominant n_1=4 channel. D.E.C. into (3,4) takes place in two separate steps along the A branch whereas D.E.C. into (3,3) derives from "virtual" transition to n=3 at $\mathcal{R}c$ (actual curve crossing schemes of fig 2a and 2b respectively), and finally D.E.C. into (3,5) comes from promotion of the n=4 electron of the (3,4) channel. Due to the inner location of the radii for electron promotion, the n=5 and (3,5) S.E.C. and D.E.C. channels should be favored at higher collision energy. On the other hand, more flux should pass the crossing for S.E.C. into n=4 diabatically, the n=3 and (3,3) channel should gain importance relative to the n=4 and (3,4) S.E.C. and D.E.C. channel. This qualitatively explains the electron spectroscopy data of Mack *et al*[18] taken at 80 keV collision energy.

Another interesting consequence of the *one electron transition hypothesis* is that D.E.C. channels should follow the same trends as S.E.C. channels. For instance assuming high l population and orientation[10] (captured electrons rotating in a preferred sense) for both S.E.C. processes, then D.E.C. is expected to populate $(nl,n'l')$ states such that l and l' add up to produce high total angular momentum L values. This trend has indeed been observed by high resolution electron spectroscopy[18,19]. Of course if none of the individual l, l' angular momenta is known the argument may be reversed to invoke both high individual l population and strong orientation propensity.

III Transfer Excitation processes

To complete this "optimistic" simplified description of electron capture processes, an important mechanism for *one electron capture* must be added namely the transfer excitation process (T.E.). In T.E. processes an electron is transferred (captured) from target to projectile simultaneously to the excitation of another electron, either from the target or from the projectile. Since two electrons are concerned, T.E. has often be interpreted as governed by the di-electronic interaction $1/r_{12}$ with r_{12} the inter-electron separation. We suggest here a different approach based on "independent" transfer and excitation.

Since it is not obvious that this process has exactly the same physical origin as in high energy atomic collisions where it has first been studied[20], it may be worth beginning with some experimental evidences from multiply charged ion-atom collisions at low keV energies. Using energy gain spectroscopy Kimura et al[21] found that T.E. is the dominant electron capture process in the N^{4+} He system. Coincident energy gain spectroscopy measurements[22] confirmed the importance of T.E.($\sigma \approx 20\ a_0^2$) in collision of $Ne^{7+}(1s^2\ 2s)$ ions on He target. For this system the angular differential cross sections unambiguously demonstrate that T.E. into $Ne^{6+}(1s^2\ 2p\ 3l)$ is directly populated at the crossing between the potential curves corresponding to the entrance channel and the T.E. channel (fig 5). Similar conclusion has been reached[23] for the F^{6+} He system. Additional evidence is

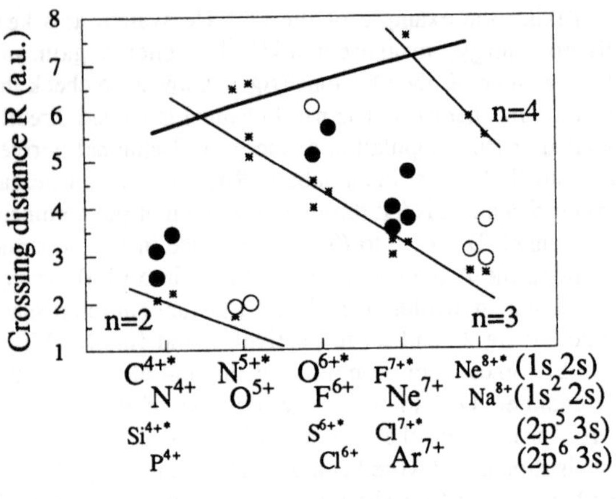

Figure 4 Location of the crossing radii for T.E. processes as a function of the ionic charge q of selected projectiles. O when they exist and ● when found effective. The * indicate the location of the crossings with S.E.C. channels, the full bold line is for R_{cbm} the internuclear distance for S.E.C. predicted by the coulombic barrier model. Thin lines which connect crossings associated in a given n are to guide the eye. No effective T.E. crossings ● were found for $R>R_{cbm}$

found for Helium-like ^3S metastable C^{4+*}(1s 2s)[24] and O^{6+*}(1s 2s)[25] ions colliding with He. T.E. into $O^{5+}(1s^2\,2l)$ with simultaneous He^+ 1s→2*l* *target excitation* has been observed[26] in $O^{6+}(1s^2)$+He collisions.The relatively small cross section was found in agreement with the He^+ 2*l*→1s photon emission cross section measured by Hoeksta et al[27].

Illustrating the location of the crossing radii associated with the observed T.E. processes (●) in fig 4), it is readily seen that the favored transitions are located at internuclear distances much smaller than those predicted by the C.B.M. The relevant crossings with the T.E. will be actually populated only if reached by the system.When no S.E.C. crossing is available close to R_{cbm}, the full incident flux may reach the crossing with the T.E. channel (ionic charge 4 and 6) whereas if a S.E.C. crossing is encountered after R_{cbm} it may behave rather adiabatically and hinder the T.E. crossing (ionic charge 5 and 8). An intermediate situation with a moderately diabatic crossing with the n=4 channel is found with the Ne^{7+} ion.

If we select the Ne^{7+} He system the curve crossing analogy is striking with the "one step virtual transition" discussed above.(fig 2b and 5). Transfer excitation (into n_l-1)

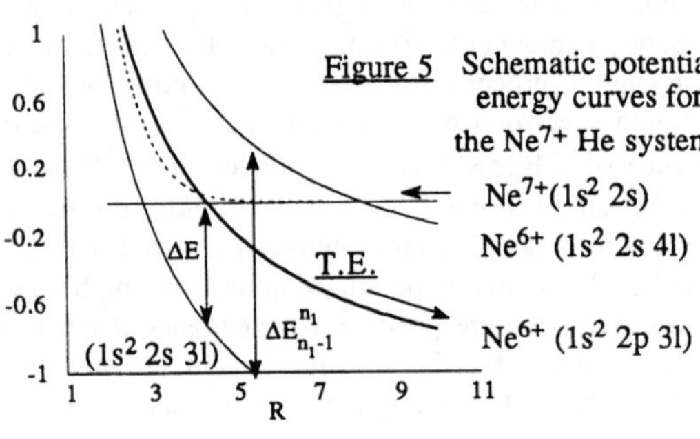

Figure 5 Schematic potential energy curves for the Ne^{7+} He system.

takes place between the Rn_l and Rn_l-1 crossing (on the B branch in fig 1). The crossing is located close to the very adiabatic "n_l-1" S.E.C. channel which, in a way, is part of the final (core*,n_l-1) state. Pushing the analogy one step beyond, one might consider that T.E. is likely to occur when the entrance channel is already virtually the capture channel, with a projection $P_{01}=\sin(\chi)^2\approx(2\mathcal{V}_{01}/\Delta E)^2$, ΔE is here at the same time the energy separation and the excitation energy. Following the formulation (2), the effective interaction $\mathcal{H}_{TE}$ for T.E.processes could be decomposed in the form $\mathcal{H}_{TE}\approx\mathcal{V}_{01}w/\Delta E$, where $\mathcal{V}_{01}$ is still the one electron interaction[2] of § I and w would be the one electron interaction responsible for the excitation ΔE of an electron by a charged particle (which here is the singly ionized target). In this approach T.E. still appears as a two electron process but due to two "simultaneous" one electron processes. ΔE appears as the key parameter, from which both the curve crossing scheme and the interaction could be estimated. In the diabatic limit the transition probability should scale as ΔE^{-2}. This partly explains why T.E. processes have been found important mainly when the projectile offers an easily excitable

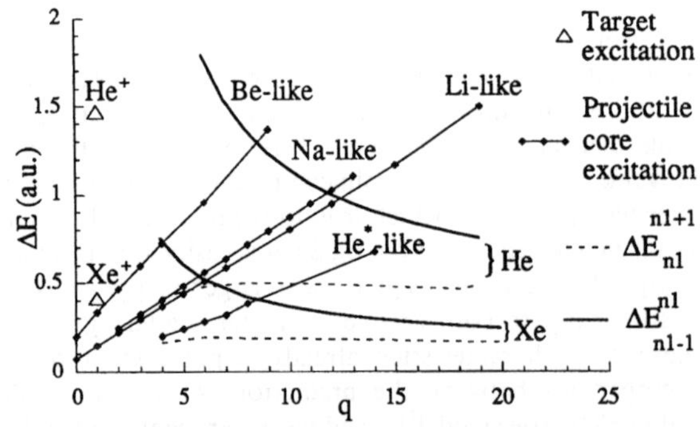

Figure 6. Some Li-like ($1s^2$ 2s→2*p*), Na-like (3s→3*p*) and Be-like ($2s^2$→2s2*p*) excitation energies[28] as well as some excitation energies of He-like ^3S metastable states (1s 2s→2*p*). Alkali-like or Ne-like metastable configurations could have been reported and probably found comparable. All these excitation energies increase rapidly with the ionic charge q of the projectile. Singly ionized rare gas target excitation levels have been reported (Δ on the left side).

energy level (fig 4). Note that the bi-electronic hypothesis probably leads to a strong ΔE dependence too.

Fig 6 shows that there should be no more "small" excitation energies available when the ionic charge q of the projectile increases (whatever the projectile core configuration). An intuitive limit could be set at $\Delta E=\Delta E^{nl}_{nl-1}$ where ΔE^{nl}_{nl-1} (— in fig 6) is the energy separation between the dominant S.E.C. n_l level and the one for capture into n_l-1. If $\Delta E>\Delta E^{nl}_{nl-1}$ then the T.E. channels associated with capture into n_l-1 lie above the n_l S.E.C. channel which then masks the n_l-1 channel, therefore T.E. would probably loose its (n_l-1) S.E.C "trigger". Assuming hydrogen-like capture levels and setting Ip as target ionization potential, the C.B.M. gives $nl=q[\,(1+2q^{1/2})/(Ip(2q^{1/2}+q))\,]^{1/2}$. For large q value

(3) $\Delta E_{n1-1}^{n1} \approx 2\, I_p^{3/2}\, q^{-1/4}$. From fig 6, T.E. is supposed to become of little importance for one electron capture if q is larger than 10 or 15 depending on the target ionization potential Ip and on the projectile electronic core configuration.

Transfer excitation can also participate in the capture of the second electron. The above upper q limit for T.E. would apply even more drastically to the capture of the second electron if a first one had not already been captured into n_1. Indeed this electron can easily be excited from n_1 to n_1+1 ($\Delta E \approx \Delta E_{n1-1}^{n1}$ (calculated for q-1) - - - - in fig 6). For very highly charged systems T.E. is likely to occur only associated with the capture of the second (third..) electron. Such processes where the first captured electron is further excited have already been observed[29,30,26] in $O^{6+} + He \rightarrow O^{4+}(1s^2\ 2p\ nl) + He^{++}$ with n=4,5..9. Indications of important T.E. processes can be found in the photon spectroscopy data of Martin *et al*[31]. Studying the multiple electron capture by Kr^{18+} on Ar and Kr target they observed lines identified as emitted by $Kr^{16+}(nl,n'l')$ with n<<n' and n'≈10,11..15. For collisions on Ar, the C.B.M. predicts S.E.C. into n_1=10 and D.E.C. into (n_2=8,n_1=10) whereas intervention of T.E. could explain population of (n_2-1 - n_1+1,2,3..)=(7 - 11,12,13..). Unfortunately the relative weight of the n'=11,12...15 configurations is not known.and n_2 is not resolved (it should have decayed first[31]). For Kr target T.E. would populate (8;12,13,14..). As another indication they found a capture cross section ratio for the targets Kr and Ar close to four whereas the C.B.M. (πR_{2cbm}^2) cross sections ratio should only scale as $(Ip/Ip')^2$ ≈1.3. An additional factor may arise from the ΔE_{n1}^{n1+1} ratio $\approx(Ip'/Ip)^{3/2}$ direct contribution to $\mathcal{H}_{TE}$ or ΔE^{-2} ratio $\approx(Ip/Ip')^3 \approx 1.5$ to the transition probability. It can be shown also, that the number of crossings available for T.E. increases with decreasing target ionization potential.

The very fact that these Rydberg lines have been observed brings up the important question of the multiply excited states decay. Until recently, the doubly excited states formed by electron capture were supposed to decay almost completely via electron emission. The argument raised by Hansen[32], is that the number of continua available for autoionization increases with projectile ionic charge q. There should always be a possibility for low energy electron emission (supposed to be very fast). Extensive calculations on $Ar^{8,9+}$ doubly excited configuration decay by Luc-Koenig and Bauche[33] showed that indeed autoionization prevails for nearly symmetric (n_2,n_1) $n_2 \approx n_1$ configurations, but that the fluorescence yield rapidly increases if the outermost electron is further excited to outer shells. Since n_2 has to be much larger than n_1 the states may be called "assymetric". At variance with one electron capture processes which favor the formation of nearly "symetric" configurations, T.E. would lead to the population of "assymetric" states. In this respect T.E. may explain the large average fluorescence yield observed[34] with the coincident energy gain technique for doubly and triply excited states populated in N^{7+}, O^{8+} and Ne^{8+} collisions on Ne, Ar,Kr and Xe. Since crossings for T.E. are always located ahead of their related S.E.C. channel, one may imagine situations where several closely located T.E. channels could remove a substantial part of the flux away from the "simple" S.E.C. crossing. Though triggered by a given S.E.C. process, T.E would hinder the corresponding crossing. With high resolution energy gain measurements Cederquist et al[35] could even find indications that radiative decay starts to dominate in Xe^{q+} + Xe collisions if q>25, but that autoionization still prevails for He target and Xe^{35+} projectile. With the naive picture of ΔE_{n1}^{n1+1} as the dominant parameter for T.E., eq.(3) suggest that for He target the situation equivalent to Xe^{25+} + Xe may only appear with $q \approx 25*(I_{pXe}/I_{pHe})^6 > 1000$. In the same spirit, alkali targets may show comparable effects for moderate projectile charge states or, for high q values, lead to production of very high Rydberg states close to the ionization limit. Because T.E. processes take place on the way in of the collision they could be important for collisions onto solid targets where the successive ionization potentials are particularly low.

Transfer excitation is a two electron process which seems to derive from "simultaneous" S.E.C. and one electron excitation. For high enough projectile charge states, the only "available excitation" comes from the first captured electron. Though responsible for the capture of only one electron at a time, T.E. is associated with multiple electron capture processes. Strong evidences for important contribution of "assymetrical" configurations are found in the decay of the characteristic multiply excited states[31,32,33,34,35]. Further work may clarify the role of transfer excitation in the formation of these asymetric states.

Conclusion

For the highly charged ion collisions investigated, it seems that the interaction responsible for single electron capture is so important that most electron capture processes can be explained associated with a single electron capture process. Considering, in addition, the possibility for one electron excitation or promotion, the present quasimolecular description explains the wide variety of experimental results. In some circumstances, when the direct electron capture is "waiting" for a proper level on the projectile, then two one electron processes may take place "simultaneously". When a first electron is captured into a high enough Rydberg state, transfer excitation could be important for multiple electron capture processes.

References:

1 H.Ryufuku, K. Sasaki, T. Watanabe *Phys.Rev.A* **21**, 745 (1980)

2 R.E. Olson et A. Salop, *Phys.Rev.A* **14**, 579 (1976)

3 Janev R.K. and Winter H, Physics Report **117** (1985)

4 Gordeev Yu S,Dijkkamp D, Drentje A G and de Heer FJ,*Phys.Rev.Lett* **23** 1842 (1983)

5 P. Roncin, MN. Gaboriaud, L. Guillemot, H. Laurent and M. Barat,*J.Phys.B : At.Mol.Opt. Phys.* **23**, 1215 (1990)

6. Barat M and Lichten W , *Phys.Rev.A* **6**, 211 (1972)

7 Taulbjerg K , *J.Phys.B : At.Mol.Phys.***19** L367 (1985)

8 Dijkkamp D, Gordeev Yu S, Bradzuk A, Drentje A G and de Heer FJ, *J.Phys.B : At.Mol.Phys..* **18** 737 (1985)

9 P. Roncin, C Adjoury, MN. Gaboriaud, L. Guillemot, M. Barat and N O Andersen *Phys. Rev. Lett.* **65** 3261 (1990)

10 N O Andersen, S E Nielsen, *Europhys. Lett* **1** 15 (1986)

11 Barany A., Astner G., Cederquist H., Danared H., Huldt S., Hvelplund P., Johnson A., Knudsen H., Liljeby L., Rensfelt K.G. *Nucl. Instrum.Meth.* **B 9** 397 (1985)

12 A. Niehaus, *J.Phys.B : At.Mol.Phys.* **19**, 2925 (1986)

13 H. Laurent, M. Barat, MN. Gaboriaud, L. Guillemot, P. Roncin , *J.Phys.B : At.Mol.Phys.* **20**, 6581 (1987)

14 Nikitin E E and Umanskii S Ya, *Springer Verlag* (1984)

15 M. Barat, P. Roncin, L. Guillemot, MN. Gaboriaud, and H Laurent *J.Phys.B : At.Mol.Opt. Phys.* **23**, 2811 (1990)

16 L. Guillemot P. Roncin, MN. Gaboriaud, H Laurent and M. Barat, *J.Phys.B: At.Mol.Phys.* **23**, 4293 (1990)

17 H. Danared, H. Anderson, G. Astner, A. Barany, P. Defrance, S. Rachafi, *J.Phys.B : At.Mol.Phys.* **20**, L165 (1987)

18 M. Mack, J.H. Nijland, P.v.d. Straten, A. Niehaus, and R. Morgenstern, *Phys.Rev.A* **39**, 3846 (1989)

19 A. Bordenave-Montesquieu, P. Benoit-Cattin, M. Boudjema, A. Gleizes and H.Bachau *J.Phys.B : At.Mol.Phys.* **20**, L695 (1987)

20 Tanis J A, Bernstein E M, Graham W G, Stockli M P, Clark M, Mc Farlan R H, Morgan T J, Berkner K H, Schlachter A S and Stearn J W *Phys. Rev. Lett.* **53** 2551 (1984)

21 M. Kimura, T. Iwai, Y. Kaneko, N. Kobayashi, A. Matsumoto, S. Ohtani, K. Okuno, S. Takagi, H. Tawara and S.Tsurubuchi, *J.Phys.B : At.Mol.Phys.* **15**, L851 (1982)

22 P. Roncin, MN. Gaboriaud, H. Laurent and M. Barat, *J.Phys.B : At.Mol.Opt. Phys.* **19**, L691 (1986)

23 MN. Gaboriaud, P. Roncin, L Guillemot, H. Laurent and M. Barat, to be published

24 L Guillemot, P. Roncin, MN. Gaboriaud, H. Laurent and M. Barat, S Bliman, M G Suraud, D Hitz, J J Bonnet, M Bonnefoy, A Chassevent and A Fleury *J.Phys.B : At.Mol.Opt. Phys.* **23** 3353 (1990)

25 J.E. Rubenson, J. Nordgren, A. Barany, M.G. Suraud, S. Bliman, D. Hitz, E.J. Knystautas, Proceeding of the MCI IVth (ed. Journal de Physique) **50**. C1-321 (1988)

26 Roncin P., Barat M., Gaboriaud M.N., Guillemot L. and Laurent H. *J.Phys.B: At. Mol. Phys.* **22** 509.(1989)

27 R Hoekstra, H Winter and F J de Heer, Nucl. Inst.Meth. **B23** 104 (1987)

28 S. Bashkin et J.O. Stoner, Atomic Energy levels and Gotrian diagrams, North Holl Amst. Oxford N.Y.(1975)

29 Stolterfoht N., Havener C.C., Phaneuf R.A., Swenson J.K., Shafroth S.M. and Meyer F.W. *Phys. Rev. Lett.* **58** 958 (1987)

30 Winter H., Mack M., Hoekstra R., Niehaus A. and de Heer F.J. *Phys. Rev. Lett.* **58** 957 (1987)

31 S Martin, A Denis, Y ouerdane, A Salmoun, A El Motassadeq, J Desesquelles M Druetta , D Church andT Lamy *Phys. Rev. Lett* **64** 22 2633 (1990)

32 J E Hansen *Journal de Physique* **50** Cl 603 (1989)

33 E. Luc-Koenig and J. Bauche, *J.Phys.B : At.Mol.Opt Phys.* **23**, 1763 (1990)

34 Roncin P, Guillemot L, Barat M., Gaboriaud M.N., and Laurent H to be published

35 H Cederquist, H andersson, G Astner, P Hvelplund and JOP Pedersen *Phys. Rev. Lett.* **62** 13 1465 (1989)

Supplement to Z. Phys. D - Atoms, Molecules and Clusters 21, S 99-104 (1991)

Atoms, Molecules
Zeitschrift für Physik D and Clusters

Radiative stabilization following transfer of two electrons to Xe^{q+} ($q \leq 35$) in slow collisions with He and Xe

H. Cederquist
Manne Siegbahn Institute of Physics, S-104 05 Stockholm, Sweden

Abstract

We present a derivation of an ECB-based relation between the branching ratio for autoionization F and the ratio between the cross sections for transfer ionization and single-electron capture. We apply this method to two different experimental results. The method yields semi-empirical upper limits for F of unity for double Rydberg levels formed in Xe^{q+}-He collisions for charge states up to q=29. The fact that direct measurements of $\sigma_{TI}/(\sigma_{TI}+\sigma_{DC})$ leads to values in the range $0.9<F<1.0$ for the He collisions gives strong support for using the model for the Xe^{q+}-Xe results, where no direct measurements of true double-capture cross sections were made. The latter results show a rather strong decrease of the upper limits for F to values significantly below unity for q>25. We point out that the apparent discrepancy between the results obtained for He and Xe targets at high q could be explained by population of higher angular momentum (l, l') states in the latter case.

I. Introduction

In this paper we will present experimental results on branching ratios for autoionization, F, for doubly excited states of $Xe^{(q-2)+}$ populated through two-electron transfer in Xe^{q+}-He collisions at $4q$ keV [1]. We will also review earlier results for semi-empirical upper limits of F for double Rydberg states formed in $4q$ keV Xe^{q+}-Xe collisions [2,3]. The latter results are obtained using the extended classical over barrier (ECB) model [4,5] and measured ratios between the cross sections for transfer ionization and single-electron capture, R [2,3]. In the present energy regime, where the collision velocity is much lower than the orbital velocities of active electrons ($v \ll 1$ a.u.), transfer ionization has been found to proceed through the formation of an intermediate doubly excited state $(nl, n'l')$ which autoionizes at large internuclear separations [6,7]:

$$Xe^{q+} + B \longrightarrow Xe^{(q-1)+}(nl) + B^{+} \longrightarrow$$

$$Xe^{(q-2)+}(nl, n'l') + B^{2+} \longrightarrow Xe^{(q-1)+} + B^{2+} + e^{-}. \quad (1)$$

Here (n, n') and (l, l') denote the principal and angular momentum quantum numbers of the two transferred electrons, respectively, while B represents the target species. As indicated by (1), the two-electron transfer process proceeds through two consecutive one-electron transitions. It has been shown through numerous experimental studies of projectile scattering angles that such two-step processes dominate for collision systems where the densities of projectile capture states are sufficiently high [8-10].

The electric dipole decay rate scales as q^4, while the Coulombic autoionization rate is independent of q. Since a typical allowed radiative decay rate is about four orders of magnitude smaller than a typical autoionization rate in a neutral atom, one would expect that the two decay branches will be of equal importance for projectile charge states in the vicinity of q=10 [11]. This prediction is, however, partly contradicted by calculations by Hansen [12], who found that autoionization dominates the decay of doubly excited Ar^{14+} states due to the possibility of low-energy electron emission when both electrons are highly excited. Radiative stabilization of $Xe^{(q-2)+}(nl, n'l')$ levels redistributes intensity from the transfer ionization process (1) to the true double-electron capture process:

$$Xe^{q+} + B \rightarrow Xe^{(q-2)+}(nl, n'l') + B^{2+}$$

$$\longrightarrow Xe^{(q-2)+} + B^{2+} + photons. \quad (2)$$

The branching ratio for autoionization is defined by

$$F = \omega_a/(\omega_a + \omega_r), \quad (3)$$

where ω_a and ω_r are autoionization and radiative decay rates, respectively. As pointed out, experimental information [6-10] suggest that the two first steps in (1) and (2) are identical and hence F can be written

$$F = \sigma_{TI}/(\sigma_{TI} + \sigma_{DC}), \quad (4)$$

where σ_{TI} and σ_{DC} are the cross sections for processes (1) and (2), respectively.

The final charge states of the projectile and target ions were determined in coincidence in the He experiment. This gives us the opportunity to present experimental results for F in the projectile charge state region between q=11 and q=29. Since the true double capture cross sections were measured to be very small in comparison to the cross sections for transfer ionization we arrive at F-values close to unity. This leads to the conclusion that the $(nl, n'l')$ configurations populated through two-electron transfer predominantly relax through electron emission. This is seemingly in contradiction with the results obtained with the Xe target for q>25, where the measured rather strong decrease in the ratio between the cross sections for transfer ionization and single-electron capture as a function of q was interpreted as due to an increasing tendency for radiative stablization of the $(nl, n'l')$ configurations. We will argue below that this apparent ambiguity can be resolved if the differences in angular momentum quantum levels populated in collisions with He and Xe targets are taken into account.

The experimental information on electron capture in slow collisions between highly charged ions and atoms is abundant for projectile charge states below q=20, whereas only a few experiments have been reported for larger q. Iwai *et al.* [13] and Tawara *et al.* [14] measured total cross sections for transfer of one electron to the projectile in I^{q+}-He ($q \leq 41$) and Kr^{q+}-He ($q \leq 25$) collisions, respectively, while Mann [15] measured these cross sections for 0.2q keV I^{q+}-He, H_2 collisions.

The two experiments under discussion here are only described very briefly (section II), since more thorough accounts have been given elsewhere [1-3]. We have relied on the validity of the extended classical over barrier (ECB) model at high q for the interpretation of the measured q-dependence of the ratio between the cross sections for transfer ionization and single-electron capture R. The experimental support for the ECB model at high q is presented in section III together with a model-dependent relation between R and F. In section III we also estimate the densities of final capture states, which have been used in order to derive semi-empirical upper limits for F (F_{max}) for the $(nl, n'l')$ states populated in collisions with Xe. We demonstrate the reliability of this method by applying it to the He data, for which F_{max} can be compared to F-values determined directly through (4). The experimental F-values for He are found to lie below F_{max} obtained through the ECB model, estimated capture-state densities, and the measurements of R, for all q investigated. Finally, we suggest an explanation for the higher tendency for radiative stabilization after transfer of two electrons from the Xe target than after two-electron transfer from the He target in the projectile charge-state regime above q=25.

II. Experimental techniques

Absolute single-electron capture, transfer ionization, and true double-electron capture cross sections were measured by means of coincidence registration of final projectile and target charge states for 4q keV Xe^{q+}-He collisions. Projectile charge states were measured with a hemispherical electrostatic energy analyzer, while the recoil ion charge states were determined by means of the time-of-flight technique. A position-sensitive detector was mounted in the focal plane of the energy analyzer, which had an angular acceptance of 8°. F can be determined directly in this experiment through the measurements of σ_{DC} and σ_{TI} (c.f. (4)).

Relative cross sections for different processes leading to reduction of the projectile charge by one atomic unit were determined for 4q keV Xe^{q+}-Xe collisions utilizing the translational energy-gain technique. The ratio $R = \sigma_{TI}/\sigma_{SC}$ (where σ_{SC} is the cross section for Xe^{q+}-$Xe \rightarrow Xe^{(q-1)+} + Xe^{+}$) was determined as a function of q by recording a large fraction of the post-collisional energy distribution of the $Xe^{(q-1)+}$ beam with a position-sensitive detector located in the focal plane behind a parabolical retardation lens and a hemispherical energy analyzer. The angular acceptance in this experiment was only 0.8°, but it was experimentally verified that this was sufficient for full transmission of projectiles from both transfer ionization and single-electron capture processes. This was done through measurements of R for different positions on the detector. For each q, R was found to be constant over a large region around the center of the detector and care was taken to position the peaks corresponding to transfer ionization and true single-electron capture within this region for every measurement [3]. A simple estimate of maximum projectile scattering angles gave values close to 0.04°, essentially independent of q. In contrast to the He experiment this experiment does not provide straightforward information on F and we will derive a relation between F and R from the ECB model in the following section.

III. The relation between F and R

The extended classical over barrier (ECB) model [4,5] relies on the basic assumption that target electrons are transferred to the projectile with unit probability when the maximum of the internuclear potential barrier coincides with the binding energy of the outermost (remaining) target electron. This assumption directly implies that multiple-electron transfer proceeds as a series of one-electron transitions at well-defined internuclear separations (r). Absolute cross sections for one-, two-, and m-electron removal from the target are then simply given by

$$\sigma_1 = \pi({r_1}^2 - {r_2}^2), \tag{5}$$

$$\sigma_2 = \pi({r_2}^2 - {r_3}^2), \tag{6}$$

and
$$\sigma_m = \pi({r_m}^2 - {r_{m+1}}^2), \quad (7)$$
where r_1, r_2, and r_m are the internuclear distances at which the first, second and m:th electrons are transferred, respectively. It was recently pointed out [16] that it is necessary to include transfer-excitation processes
$$Xe^{q+} + He \rightarrow Xe^{(q-1)+}(nl) + He^{+}(n_t = 1) \rightarrow$$
$$Xe^{(q-2)+}(nl, n'l') + He^{2+} \rightarrow Xe^{(q-1)}(nl) + He^{+}(n_t \geq 2), \quad (8)$$
for two-electron targets. This is because an infinite number of transfer-excitation (TE) channels cross with the DC/TI potential curve within a very narrow region of r inside r_2 [16]. The principal quantum number of the excited target energy level ranges between n_t=2 and $n_t = \infty$. The target excitation (te) process is included through the replacement of (5) and (6) by
$$\sigma_1 = \pi(r_1^2 - r_2^2 + r_{te}^2) \quad \text{and} \quad (9)$$
$$\sigma_2 = \pi(r_2^2 - r_{te}^2), \quad (10)$$
respectively, where r_{te} is given by
$$r_{te} = r_2/(1 + E_{ex} r_2/(q-3)). \quad (11)$$
$E_{ex} = E_{ex}(n_t)$ is the excitation energy of the target after the collision. In (9) and (10) we assume that impact parameters smaller than r_{te} contribute to the cross section for removing one electron from the target (σ_1) with unit probability through the transfer excitation process. In figure 1 we show comparisons between experimental and model cross sections for removing one (σ_1) and two (σ_2) electrons from the He target. The upper data points (circles) are σ_1, while the lower ones (squares) are σ_2. The full curves represent model cross sections for n_t=2, while the dashed curves show model values for $n_t = \infty$. The agreement between ECB-model predictions and the Xe^{q+}-He data set, which so far is the only one giving absolute values for both σ_1 and σ_2 in the q>20 charge-state regime, is rather satisfactory.

It has been implicitely assumed above that projectile capture states form quasi-continua. This leads to energy resonances with Stark-shifted target energy levels as soon as the top of the internuclear barrier is reached. Justifications for such reasoning are abundant in experiments performed at fairly moderate q, but strong support for classical over barrier models is indeed also supplied by slow collision experiments at q>20. The ECB model predicts energy inelasticities and thereby n quantum numbers of capture levels [3-5]. For the first and the second of the transferred electrons one gets
$$\frac{1}{n^2} = \frac{2}{q^2}(\frac{q-1}{r_1} + I_1) = \frac{2I_1}{q^2}(\frac{q-1}{2\sqrt{q}+1} + 1), \quad (12)$$
and
$$\frac{1}{n'^2} = \frac{2}{q^2}(\frac{q-3}{r_2} + I_2) = \frac{2I_2}{q^2}(\frac{q-3}{2\sqrt{2(q-1)}+2} + 1), \quad (13)$$

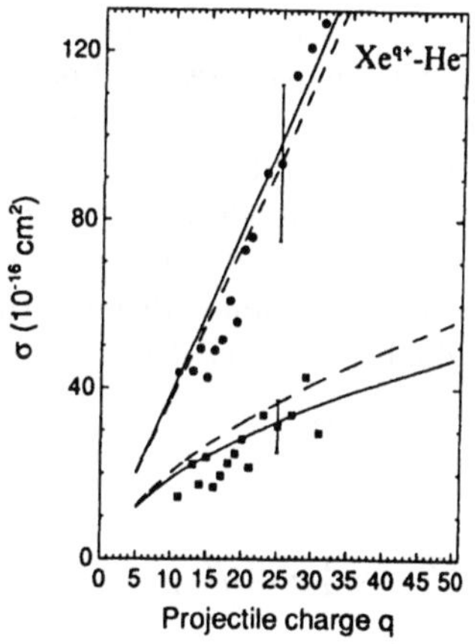

Fig. 1. A comparison between experimental and ECB-model cross sections for removal of one and two target electrons from He in $4q$ keV Xe^{q+}-He collisions. The upper data set consists of cross sections for one-electron removal from the target, while the lower data set shows cross sections for removal of two electrons (i.e. the sum of transfer ionization and true double-electron capture cross sections). The full and dashed lines indicate model cross sections for the r_{te} values obtained with n_t=2 and $n_t = \infty$, respectively.

where I_1 and I_2 are the first and second ionization potentials, respectively. For large q, equations (12) and (13) imply that $n \sim q^{3/4}$ and $n' \sim q^{3/4}$, respectively. The $n = n(q)$ behaviour has been confirmed by the measurements of translational energy gain for I^{q+}-He$\rightarrow I^{(q-1)+}$ collisions by Iwai *et al.* [13], showing that the basic classical over barrier assumption of electron transfer at the top of the barrier works also at high q. The essentially linear q-dependence of total electron removal cross sections predicted by classical over barrier models through $\sigma_{tot} = \pi r_1^2$ is confirmed by a wealth of experiments at moderate q [17,18] and for higher q by the Xe^{q+}-He experiment [1]. Classical over barrier models furthermore predict that σ_{tot} is proportional to I_1^{-2}. This has been investigated by Müller *et al.* [19] who measured $\sigma_{q,q-1}$ (the total cross section for capture of one electron by the projectile) for Xe^{10+}-$B \rightarrow Xe^{9+}$ processes as a function of the first ionization potential of the target B. They used eleven different targets of ionization potentials ranging between 3.89 and 24.58 eV, arriving at a best fit of $I_B^{-1.94}$. The assumption of a quasi-continuum of projectile capture state becomes better with increasing projectile charge (c.f. (12)), decreasing ionization potential, and increasing number of target electrons on the projectile for a given q [20]. For sufficiently high q, however, it has been shown that the number of electrons on the projectile becomes unimportant [20]. A high density of capture states is also necessary to assure that capture cross sections are virtually independent of the collision velocity in a large region of low velocities [21-23].

We now define $S = (\sigma_{TI} + \sigma_{DC})/\sigma_{SC}$, the ratio between the cross sections for removing two and one electrons from the target ($S = \sigma_2/\sigma_1$). Since $R = \sigma_{TI}/\sigma_{SC}$, the branching ratio for autoionization is simply $F = R/S$ according to (4). When two electrons are transferred consecutively to the projectile at the r-values given by the ECB model the branching ratio for autoionization of the resulting doubly excited $(nl, n'l')$ state can be written

$$F = R/S^{ECB}, \tag{14}$$

where

$$S^{ECB} = \sigma_2^{ECB}/\sigma_1^{ECB} = (r_2^2 - r_3^2)/(r_1^2 - r_2^2) \tag{15}$$

for multiple-electron targets and

$$S^{ECB} = \sigma_2^{ECB}/\sigma_1^{ECB} = (r_2^2 - r_{te}^2)/(r_1^2 - r_2^2 + r_{te}^2) \tag{16}$$

for two-electron targets. Measurement of $R = \sigma_{TI}/\sigma_{SC}$ thus gives semi-empirical results for F if the model assumptions are valid.

In a realistic case with finite densities of capture states the internuclear separations for electron transfer will be smaller than ECB model values since the energy resonance with the projectile capture level will occur inside the separation where the Stark shifted target level reaches the top of the internuclear barrier. Assuming that projectile capture levels, nl and $n'l'$, are evenly spaced in excitation energy within their n and n' manifolds, Andersson *et al.* [3] arrived at a maximum relative deviation between an nl-capture radius r_{nl} and r_1 of

$$\delta_1^{max} = \frac{(r_1 - r_{nl})^{max}}{r_1} = \frac{4I_1(\sqrt{q}+2)^2}{q(q-1)(2\sqrt{q}+1)}. \tag{17}$$

Similarly the maximum relative deviation between $n'l'$ and r_2 capture radii is

$$\delta_2^{max} = \frac{(r_2 - r_{n'l'})^{max}}{r_2} = \frac{2I_2(q+2\sqrt{2(q-1)}-1)^2}{q^2(q-3)(\sqrt{2(q-1)}+1)}. \tag{18}$$

This leads to lower and upper limits for S^{ECB} which are

$$S_{min}^{ECB} = (r_2^2(1-2\delta_2^{max}) - r_3^2)/(r_1^2 - r_2^2(1-2\delta_2^{max})) \tag{19}$$

and

$$S_{max}^{ECB} = (r_2^2 - r_3^2(1-2\delta_3^{max}))/(r_1^2(1-2\delta_1^{max}) - r_2^2), \tag{20}$$

for multiple-electron targets. Similar expressions are easily obtained for He through (16), (17), (18), and the relation between the relative uncertainties in r_{te} and r_2. ECB-dependent upper limits for F are then obtained through

$$F_{max} = R/S_{min}^{ECB}. \tag{21}$$

IV. Results and discussions

In figure 2, we have plotted our measured values of R together with model results for S^{ECB} (full line), S_{max}^{ECB} and S_{min}^{ECB} (dashed lines) as functions of q for the Xe^{q+}-He case. All measurements of R except the one for q=31 lie above S_{min}^{ECB}.

The full line in figure 3 shows F_{max} for He according to (21), while the data points are experimental branching ratios for autoionization (F) according to (4). The fact that all measured F-values (except the one at q=15 [1]) fall below F_{max} supports the use of semi-empirical

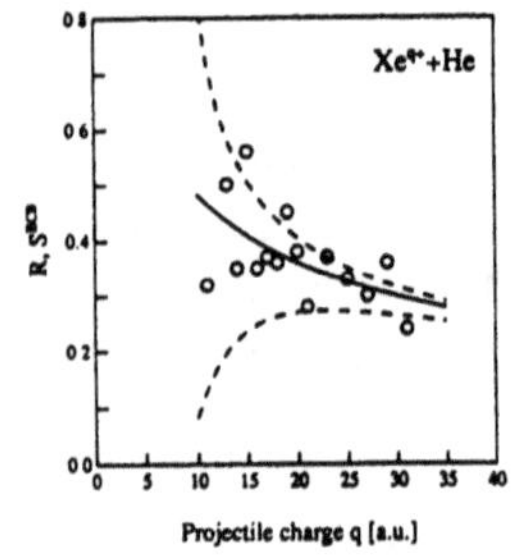

Fig. 2. The experimental ratio between the cross sections for transfer ionization and single-electron capture, R, as a function of q for $4q$ keV Xe^{q+}-He collisions. The model-dependent lower limit for experimental R-values consistent with full autoionization (F=1) S_{min}^{ECB} is represented by the lower dashed curve. The upper dashed curve is S_{max}^{ECB}, while S^{ECB} is indicated by a full line.

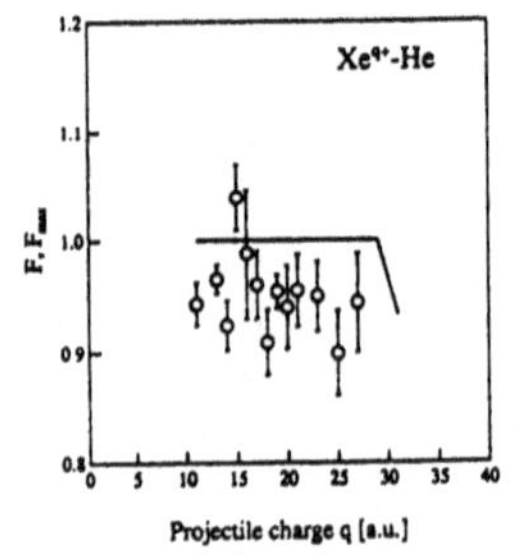

Fig. 3. Experimental branching ratios for autoionization of doubly excited projectiles $Xe^{(q-2)+}(nl, n'l')$ as a function of q for $4q$ keV Xe^{q+}-He collisions. The data points are obtained by means of measured values of $R = \sigma_{TI}/\sigma_{SC}$ and $S = (\sigma_{TI} + \sigma_{DC})/\sigma_{SC}$ through $F = R/S$. The full line shows semi-empirical upper limits for F, $F_{max} = R/S_{min}^{ECB}$ (c.f. text).

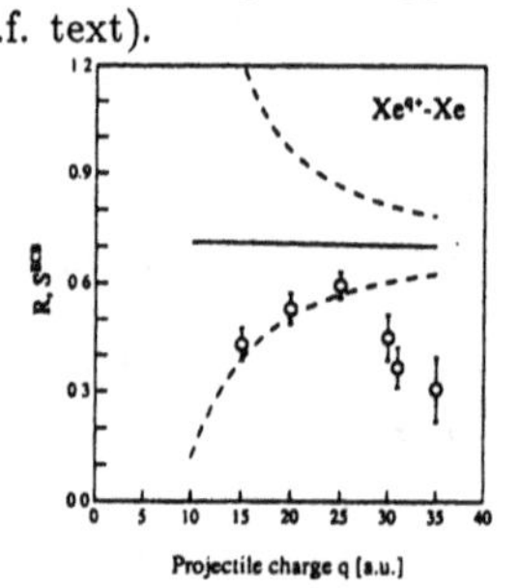

Fig. 4. Measured values of the ratio, R, between the cross sections for transfer ionization and single-electron capture in $4q$ keV Xe^{q+}-Xe collisions. The lower limit for experimental R values consistent with full autoionization (F=1), is taken to be S_{min}^{ECB}. S_{max}^{ECB} and S_{min}^{ECB} are indicated by dashed lines. The full line shows the model value for the ratio between the cross sections for removing one and two electrons from the Xe target.

upper limits for F in order to interprete the Xe^{q+}-Xe results. Figure 4 displays experimental R-values together with S_{max}^{ECB} and S_{min}^{ECB} for $4q$ keV Xe^{q+}-Xe collisions. Measurements of R between S_{min}^{ECB} and S_{max}^{ECB} are taken to be consistent with full autoionization (F=1), whereas measurements of R falling below S_{min}^{ECB} are interpreted as due to radiative stabilization of the $(nl, n'l')$ double Rydberg levels. Semi-empirical upper limits for the branching ratios for autoionization F_{max} obtained through (21) are shown for q=15, 25, 30, 31, and 35 in figure 5 [3].

It has been shown in many experiments at moderate q and low collision velocities ($v \ll 1$ a.u.) that the single most important factor that determines the cross section for population of a certain nl energy level is the position of its crossing with the incident channel [24,25]. It is thus reasonable to assume a uniform distribution of l states for the slow collisions ($30q$ eV/amu) under discussion here. The average values of l and l' are then proportional to n and n', respectively. The radiative decay rate is expected to decrease as $(n')^{-3}$ and as $\sim(l')^{-2}$ for a certain value of n' [26]. The autoionization rate, however, decays as n^{-3} and as l^{-5} for a certain n at

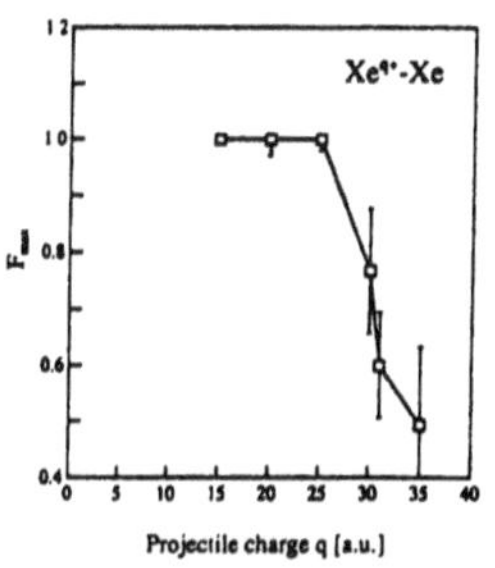

Fig. 5. Semi-empirical upper limits, F_{max}, for the branching ratios for autoionization, F, for doubly excited states created in $4q$ keV Xe^{q+}-Xe collisions ($F_{max} = R/S_{min}^{ECB}$). The error bars show one standard deviation.

sufficiently large l [27]. Thus, since (n, n') and thereby (l, l') will be higher after two-electron transfer from Xe (according to (12) and (13) and assuming that l and l' are proportional to n and n', respectively) it is understandable that F is higher for $Xe^{(q-2)+}(nl, n'l')$ levels formed in Xe^{q+}-He than in Xe^{q+}-Xe collisions when q>25. However, calculations by Hansen [12] show that there is a boost of autoionization decay rates when excited $Xe^{(q-1)+}$ energy levels are close to the doubly excited $Xe^{(q-2)+}(nl, n'l')$ levels, due to the possibility of low-energy electron emission. This effect counteracts the decrease in the autoionization rate due to the n^{-3} and l^{-5} scalings, since the average energy separation to the closest $Xe^{(q-1)+}$ level decreases with increasing excitation. That is, separations are smaller for $(nl, n'l')$ states produced in collisions with Xe. Still, the experimental results are that F falls off rapidly with q for q>25 for capture from Xe but not for capture from He.

We have presented experimental results for the branching ratios for autoionization for double Rydberg levels formed in $4q$ keV Xe^{q+}-He collisions, as well as semi-empirical upper limits for F for states populated in Xe^{q+}-He and Xe^{q+}-Xe collisions at $4q$ keV. We suggest that the observed differences between the decay modes of double Rydberg state formed in Xe^{q+}-Xe and Xe^{q+}-He collisions at q >25 are to be explained by capture to higher angular momentum states (l, l') in the former case.

References

1. H. Andersson, G. Astner, and H. Cederquist, J. Phys. B: At. Mol. Opt. Phys. **21**, L187 (1988).
2. H. Cederquist *et al.*, Phys. Rev. Lett. **62**, 1465 (1989).
3. H. Andersson *et al.*, Physica Scripta, **42**, 150 (1990).
4. A. Bárány *et al.*, Nucl. Instr. Meth. B **9**, 397 (1985).
5. P. Hvelplund *et al.*, Nucl. Instr. Meth. B **9**, 421 (1985).
6. A. Bordenave-Montesquieu *et al.*, J. Phys. B: At. Mol. Phys. **17**, L127 (1984).
7. P. Moretto-Capelle *et al.*, J. Phys. B: At. Mol. Opt. Phys. **22**, 271 (1989).
8. P. Rocin *et al.*, Europhys. Lett. **3**, 53 (1987).
9. H. Laurent *et al.*, J. Phys. B: At. Mol. Phys. **20**, 6581 (1987).
10. C. Biedermann *et al.*, Phys. Rev. A **41**, 5889 (1990).
11. R.K. Janev and L.P. Presnyakov, Phys. Rep. **70**, 1 (1981).
12. J.E. Hansen J. de Phys. (Paris) **50** C1-603 (1989).
13. T. Iwai *et al.*, J. Phys. B: At. Mol. Phys. **17**, L95 (1984).
14. H. Tawara *et al.*, J. Phys. B: At. Mol. Phys. **18**, 337 (1985).
15. R. Mann, Z. Phys. D **3**, 85 (1986).
16. H. Cederquist, Phys. Rev. A, submitted (1990).
17. A. Müller and A. Salzborn, Phys. Lett. **62A**, 391 (1977).
18. E.H. Nielsen *et al.*, J. Phys. B: At. Mol. Phys. **18**, 1789 (1985).
19. A. Müller, C. Achenbach, and E. Salzborn, Phys. Lett. **70A**, 410 (1979).
20. D.H. Crandall, R.A. Phaneuf, and F.W. Meyer, Phys. Rev. A **22**, 379 (1980).
21. E.H. Nielsen *et al.*, J. Phys. B: At Mol. Phys. **17**, L139 (1984).
22. Y. Nakai and T. Shirai, Physica Scripta **T28**, 77 (1989).
23. C. Biedermann *et al.*, Phys. Rev. A, accepted (1990).

24. D. Dijkkamp *et al.*, J. Phys. B: At. Mol. Phys., 4763 **18** (1985).

25. D. Vernhet *et al.*, J. Phys. B: At. Mol. Opt. Phys. **21**, 3949 (1988).

26. E.U. Condon and G.H. Shortly, *The theory of Atomic Spectra*, pp. 136, London: Cambridge University Press 1935.

27. W.E. Cooke and T.F. Gallagher, **19**, 2151 (1979).

Supplement to Z. Phys. D - Atoms, Molecules and Clusters 21, S 105-109 (1991)

Atoms, Molecules
Zeitschrift für Physik D and Clusters

Interaction of heavy ions with hot ionized matter

Dieter H. H. Hoffmann, K.-G. Dietrich, W. Laux
GSI Darmstadt, Postfach 110552, W-6100 Darmstadt, FRG

E. Boggasch, K. Mahrt-Olt, H. Wahl
Max-Planck-Institut für Quantenoptik, W-8046 Garching, FRG

A. A. Golubev, V. P. Dubenkov
Institute for Theoretical and Experimental Physics, 117259 Moscow, USSR

The energy loss of heavy ions in a hydrogen plasma has been measured in an energy range from 1.4 to 6 MeV/u. A z-pinch has been used as a plasma target with a maximum free electron density of 1.5×10^{19} cm^{-3}. Our data show a strong enhancement of the stopping power of the plasma compared to that of a cold gas with equal density. Charge state analysis of the ions also show a higher charge state of the ions in the plasma target, relative to the cold hydrogen gas targets. A plasma lens effect of the high power z-pinch discharge was observed in our experiments.

PACS: 52.40. Mj; 34.50.Bw; 52.80 s; 41.80. s

The production of safe and abundant energy from thermonuclear fusion processes was a major aim of intense research programs that have been launched since the 1950s in many countries. Both methods to confine the fuel during the burn phase, magnetic confinement fusion (MCF), and inertial confinement fusion (ICF), are now believed to be technically feasible. The ICF approach to fusion was addressed about two decades later than the MCF path, and today it is a serious competitor and offers a high potential to turn into an industrial source of energy.

Recent advances in laser fusion and the progress toward high gain laser fusion considerably raise the probability, that it will be a laser driver to ignite the first high gain fusion pellet under laboratory conditions. However, for a competitive commercial power reactor heavy ions offer a number of most favorable properties. The high efficiency (25%) and proven reliability of accelerators, as well as their ability to deliver high intensity beam pulses at a high repetition rate ($\geq$ 1 Hz) are main assets of ion driven fusion. Also the problem of coupling the beam energy to the target is advantageous for heavy ion beams. Due to the high nuclear charge (Z) heavy ions have a high deposition power ($\propto Z^2$) in matter, and the energy is deposited in the target volume rather than in a thin surface plasma layer. Even multi-TW pulses can be transported ballistically and can be focused to a small spot size.

The demands on a heavy ion driver are, however, very severe if we consider for a moment the key figures that are involved in ion driven fusion. It is expected that the energy delivered to the target in a 10 ns pulse has to be in the range of 5 to 10 MJ, which amounts to a pulsed power of 10^{15}W. The symmetry requirements of ICF pellets during the heating and compression process are difficult to meet with directly driven targets and make it appear likely that the indirect drive approach will be pursued. In this case it is necessary to heat a hohlraum to temperatures of approximately 300 eV to obtain a circulating flux of 10^{15} W/cm^2. Thus the ion beam kinetic energy has to be absorbed and converted into thermal radiation in a small amount of matter like 100 mg of high Z material (e.g. Au). Using these figures the order of magnitude for the deposition power P, which is the key parameter for heating matter with heavy ion beams, can be estimated to be 10^{16} W/g [1]. The advanced state of present day accelerator technology, and novel techniques and ideas to produce high intensity beam pulses [2] make it appear possible that this parameter regime can be reached.

There is, however, still a number of basic physics issues which need more experimental and theoretical investigation. One of these topics is a detailed study of interaction phenomena of intense heavy-ion beams with ionized matter, and especially the energy deposition profiles of heavy ion beams in a plasma environment.

The theoretical description of the energy loss of charged particles in cold, non ionized matter, is well established, well understood, and corroborated by a large number of experimental data during a number of decades. This is, however, not the case for hot ionized matter and existing theories have not yet been verified by experiments for a broad range of ion species. Moreover, the energy deposition of heavy ions is an important issue in the design of high gain fusion pellets for ion beam driven inertial confinement fusion (ICF).

When the ion velocity (v) is much larger than the thermal velocity of the free plasma electrons, an increase in stopping power is generally predicted [3,4], since the minimum energy transfer in a collision is no longer given by the average ionization potential ($\bar{I}$) but by the potential energy of the electrons in the Debye sphere. At high plasma temperatures, however, when the thermal electron velocity exceeds the projectile velocity, even a decrease of the stopping power combined with a large increase in range straggling is predicted [5].

Enhanced Stopping of Ions in Plasmas

In the energy regime considered in the experiments reported here (1.4 - 11 MeV/u) the well known Bethe formula without polarization correction and shell correction term describes the energy loss of non relativistic charged particles in a hydrogen plasma ($Z_2 = 1$) in the following form

$-(dE/dx) = (Z_{eff}^2 e\omega/v)^2 \times \ln\Lambda,$ (1)

where Z_{eff} is the effective charge of the ion while it traverses the target, e is the electronic charge and ω the plasma frequency, determined by the density of particles. For a cold target, where only bound electrons are present the argument of the Coulomb logarithm Λ is given by

$\Lambda_{bound} = 2mv^2 / \bar{I}$ (2)

This can be extended in a straightforward way to fully ionized material with free electrons using binary collision

Table I.: Comparison of Coulomb logarithm in cold and fully ionized hydrogen

theory. The Coulomb logarithm in this case reads:

$\Lambda_{free} = 0.764\, v / b_{min}\, \omega$ (3)

The minimum impact parameter b_{min} depends on the velocity of the ion. Since the electronic mass (m) is negligible compared to the heavy ion mass, the impact parameter is given by

$b_{min} = \max(e^2 Z_{eff} / mv^2 \;;\; \hbar / 2mv).$ (4)

For the conditions of our experiment at $E_0 = 1.4$ MeV/u ($v = 7.49\alpha c = 1.64\times10^9$ cm/s) b_{min} is always given by

$b_{min} = e^2 Z_{eff}/mv^2$ (5)

due to the high charge state Z_{eff} of the incident ions, and thus we have

$\Lambda_{free} = 0.764\, mv^3 / Z_{eff} e^2 \omega$ (6).

For comparison we list the Coulomb logarithm $\ln\Lambda_{bound}$ and $\ln\Lambda_{free}$ for cold and ionized hydrogen in columns 2 and 3 of table I. The third column displays the ratio ($\ln\Lambda_{free} / \ln\Lambda_{bound}$), which is a factor of about 2, thus the enhancement in stopping power of a hydrogen plasma as compared to cold hydrogen gas is also of the order of a factor 2, for all heavy ions.

Ion	$\ln\Lambda_{bound}$	$\ln\Lambda_{free}$	$\ln\Lambda_{free}/\ln\Lambda_{bound}$
Ca^{13+}	5.42	10.42	1.92
Ge^{18+}	5.42	10.09	1.86
Kr^{18+}	5.42	10.09	1.86
Pd^{26+}	5.42	9.72	1.79
Pb^{30+}	5.42	9.58	1.77
U^{33-}	5.42	9.48	1.75

Model Description of the Heavy Ion Effective Charge State

The charge state of heavy ions passing through matter is fluctuating rapidly, due to ionization and electron capture processes. A dynamic equilibrium of these processes finally determines the effective charge state Z_{eff} which is one key parameter in eq. 1, and it turns out to be a function of the target thickness (x), the ion velocity (v) and the temperature (T) of the target material ($Z_{eff} = Z_{eff}(x,v,T)$). For a complete evaluation of the energy loss equation (1), the x-dependence of Z_{eff} has to be analyzed. A detailed treatment of this problem is found e.g. in ref. [6]. For our purpose we developed a Monte-Carlo code to track the development of the heavy ion effective charge state, as the projectile traverses the plasma target, and used the results as an input for the energy loss calculation. In this code we treated ionization processes due to collisions of the heavy ion projectile with bound and free target electrons, and target nuclei in the framework of the binary encounter model of Gryzinski [8].

The dominant electron capture process in our experiment was recombination by capture of bound electrons, and in the case of the fully ionized plasma target, radiative electron capture (REC). Dielectronic recombination processes are also considered by the code according to ref. 6.

For highly charged ions as U^{33+} which we regard as our standard example, electron capture processes occur to states with high principal quantum numbers, which are difficult to treat. The Oppenheimer-Brinkmann-Kramers (OBK) approximation provides, however, a simple method to describe these processes. Since in this experiment we are dealing with a hydrogen target, we used the Nikolaev formula [9] for K-shell electron transfer to a projectile state with principal quantum number n and performed a summation over all possible n-states up to the numerical convergence limit. The resulting cross sections are known to overestimate experimental data [10], and we therefore applied the Chan-Eichler scaling factor to the OBK cross section [11].

An empirical scaling rule for the electron capture cross section into highly charged ions described by Schlachter and coworkers [12] represents our data equally well and due to its parametrization is especially suited to be used in a computer code.

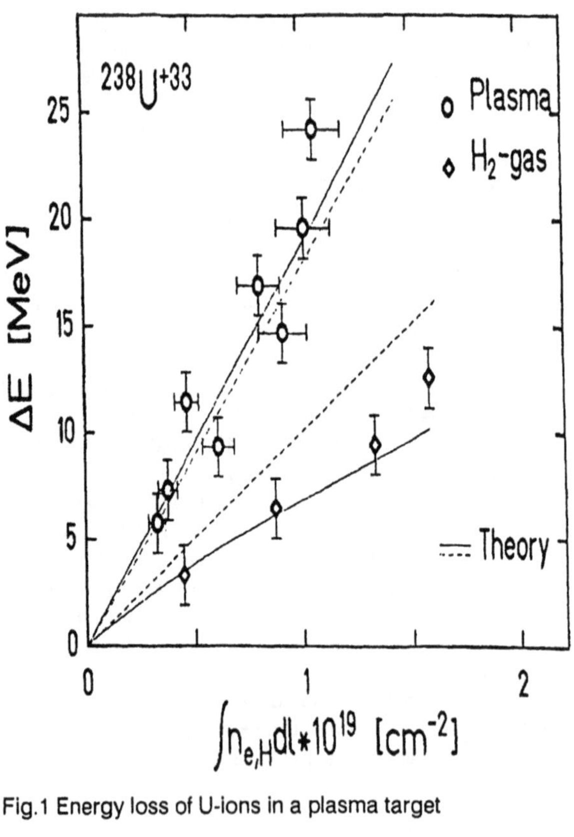

Fig.1 Energy loss of U-ions in a plasma target

Fig. 1 shows the experimental results for the energy loss of U ions in a low density plasma target. The calculation shown in this figure is the result of the Monte-Carlo code developed for this experiment and includes the three most important atomic cross sections for charge changing processes:

- Ionization by collisions with target ions and electrons [8]
- Recombination by capture of bound electrons, and [13]
- Radiative recombination of free electrons [14]

While the dashed lines represent the energy loss of Uranium ions, assuming a constant value of Z_{eff}, the full lines take into account the variation of the effective charge state, while the ion traverses the plasma target. It is an important feature of this experiment, that the ion charge state is initially above the equilibrium charge state in cold hydrogen gas. Therefore even in a target of low density ($n_e \approx 10^{17} cm^{-3}$), as it was available for the first part of the experiment, the charge state of the ion starts to develop in opposite directions, depending on whether the target is cold hydrogen gas or a fully ionized plasma.

The energy loss data for cold hydrogen gas are well represented by the full curve, which takes into account, that the ion charge state drops from 33+ to the equilibrium value of 27+. Thus the deviation from the straight dashed line assuming a constant charge state of 33+ is quite obvious.

In the plasma case however our code predicts an increase of only one charge state, thus the full line (charge state variation) is still very close to the dashed straight line (constant charge state) and the experimental uncertainties do not allow to distinguish between these two cases. We therefore extended our measurements into the regime of a high density hydrogen plasma.

Energy loss of heavy ions in a dense hydrogen plasma

The z-pinch plasma target used in our experiment provides a fully ionized hydrogen plasma with free electron densities exceeding $10^{19} cm^{-3}$. With this target we measured the energy loss of ^{208}Pb and ^{238}U ions at an energy of 1.4 MeV/u, and the charge state distribution of Ar ions (E_0 = 4.8 MeV/u) moving through the plasma.The z-pinch target was designed and constructed at the Fraunhofer-Institut für Lasertechnik in Aachen especially for the energy loss experiments [15]. The capacitor bank has a capacity of 4 μF and an inductance of 15nH. It is typically charged to 20-40 kV corresponding to an energy of 0.8-3.2 kJ stored in the bank. The discharge vessel consists of a cylindrical quartz tube with a length of 20 cm and an inner diameter of 10.4 cm, with copper electrodes on both ends of the tube. It is filled with hydrogen at typical pressures of 0.5 to 3.5 mbars. Spark gaps are used as fast high power switches. After triggering the spark gaps the discharge starts at the inner surface of the quartz tube forming a hollow plasma cylinder. Driven by the strong azimuthal magnetic field generated by the high axial current (typically 400kA) the plasma is compressed onto the axis, where in the pinch phase a very dense and hot plasma column with a diameter of about 10 mm is formed.

The time interval from ignition to the pinch phase is about 1.2 μs. It is followed by an expansion of the plasma where within 1 μs the plasma density decreases from above $10^{19} cm^{-3}$ to $10^{17} cm^{-3}$. Measuring the energy loss during this time interval allows us to obtain energy loss data for a wide density range in one experiment.

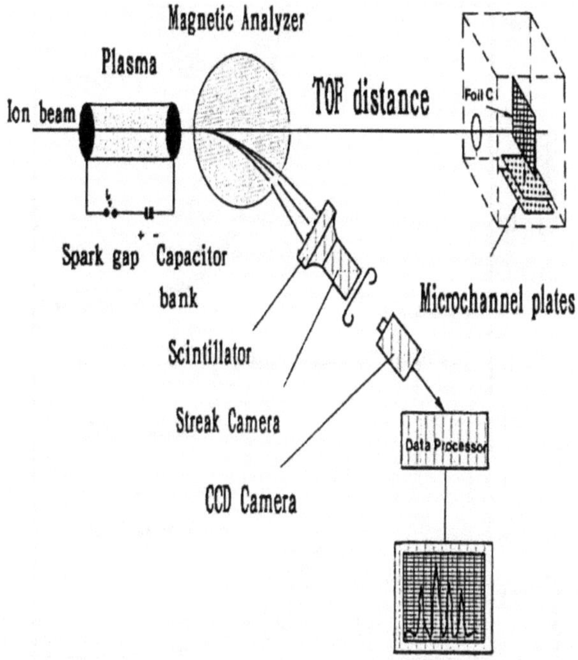

Fig.2: Experimental set-up

The z-pinch has been integrated to the beamline of the UNILAC accelerator with a differential pumping system. The experimental setup is shown in fig. 2. With the magnetic analyzer turned off, energy loss measurements with time-of-flight technique are performed. The magnet serves for charge state analysis of the beam passing through the plasma. The beam passes through the tube on the axis of symmetry through small (3 mm diameter) apertures in the electrodes.

In fig. 3 we present the results of the energy loss measurement as a function of time (upper part of fig. 3) together with the light output signal of the plasma (lower part of fig. 3). Up to 600 ns the data correspond to the energy loss (2 MeV) in the cold hydrogen gas target (length = 20cm, p = 2.5 mbar). At that point the pinch reaches the beam axis and the target density and the associated energy loss rises steeply, exceeding the range accessible to our time of flight spectroscopy.

The plasma light signal is measured through the exit aperture of the z-pinch cathode, thus the line of sight for this signal is along the beam plasma interaction zone at the center of the pinch column. The intensity of this signal is correlated to the free electron density of the plasma and therefore it shows the same features as the energy loss measurements, with a very steep rise of the intensity, when the pinch collapses on the axis. At this time the plasma density is above $5x10^{18} cm^{-3}$ and the ions are stopped completely inside the plasma column and a time of flight energy measurement is no longer possible. This explains the lack of energy loss data between 0.8μs and 1.2μs in the upper part of fig. 3. During the expansion phase of the pinch the energy loss is again within the limits accessible for time-of-flight spectroscopy and we observe an energy loss corresponding to 45% of the incident beam energy of 291 MeV.

The experiment consisted of two parts: Measurement of the energy loss of heavy ions and diagnostics of the plasma

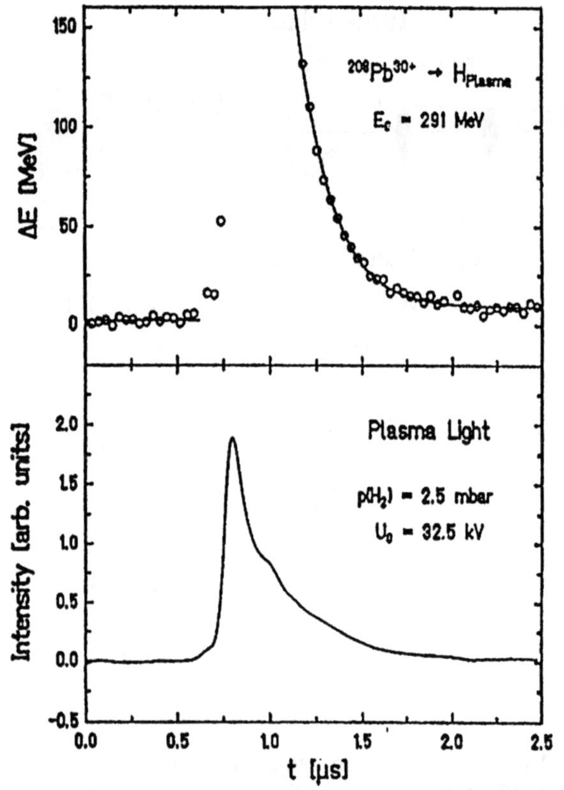

Fig. 3: Time dependence of the energy loss and the plasma light emission

density and temperature. The energy loss was determined with a time-of-flight method using a 430 cm time-of-flight distance. The time resolution achieved was 2 ns corresponding to an accuracy of the energy loss measurement of ±2 MeV for Pb ions. With this system energy losses up to 45% of the incident energy could be detected.

The plasma density diagnostics were performed with laser interferometry in Mach-Zehnder geometry along the z-pinch axis [16]. On-axis interferometry gives the most accurate information about the quantity $\int n_e dl$ e.g. the free electron density integrated along the interaction zone of ion beam and plasma. With this method area densities up to 5.2×10^{19} cm^{-2} could be diagnosed with an error of $\pm 1.5 \times 10^{18}$cm^{-2}. In the phase of higher plasma density this diagnostic technique fails because the laser light is absorbed in the plasma. The plasma temperature was determined by spectroscopy of the emitted light, analyzing the $H\alpha$ emission. In the pinch phase it is above 5 eV. During the following expansion phase temperature decreases but for several microseconds stays high enough to guarantee a degree of ionization of more than 99%. Thus the shown energy loss data were all obtained in a completely ionized plasma.

In fig. 4 the measured energy loss of $^{208}Pb^{30+}$ ions is displayed as a function of the free electron area density. The given error bars represent the full error comprising the error of the energy loss measurement the error of the interferometry and the uncertainty in matching the time scale of both. The energy loss for cold hydrogen gas is derived from tables [17]. The observed energy loss in the plasma is by a factor of three higher than the energy loss in cold gas.

We compare the data to our calculation. The ion charge state in the plasma increases because the dominant electron capture process in cold matter e.g. the capture of bound electrons is strongly suppressed in a highly ionized plasma. Our calculation predicts an increase of the Pb ion charge from 30+ to 38+ causing the energy loss to rise stronger than in the case of a constant charge state. Note that the theoretical curve has an ascending slope. The agreement of the calculation with the experimental data is very good, also for the ^{238}U data not shown here.

First experiments to measure the charge state distribution of an Ar beam passing through the plasma have been performed. The initial charge state is 10+ at an energy of 4.8 MeV/u. We observe that the beam is almost completely stripped into

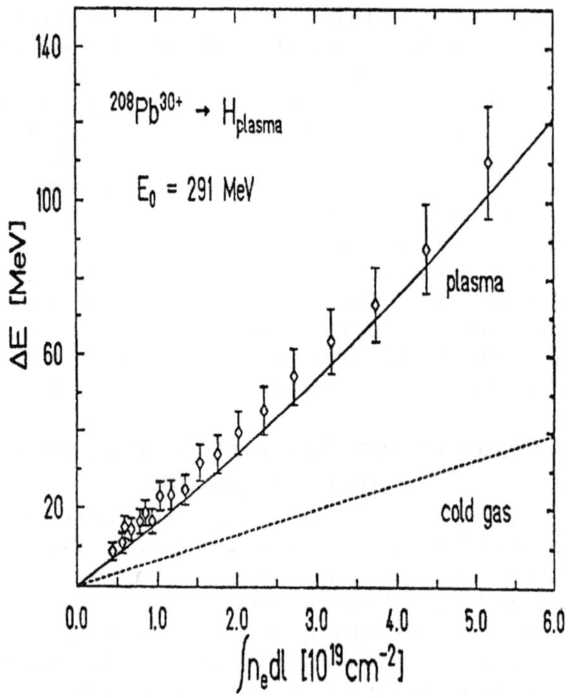

Fig. 4: Energy loss of Pb ions in a dense hydrogen plasma.

charge state 16+, a He-like configuration. Dielectronic recombination processes prevent the ions from being ionized into higher charge states and a very narrow charge state distribution centered at Ar^{16+} is the observed result.

Non Conventional Focusing Techniques

Our planned experiments to produce high energy density in matter require the highest possible specific deposition power of the ion beam. Non conventional techniques will be necessary for the final focus of an advanced ICF test facility, or even more for the envisioned power reactor. For our experimental program, focusing systems consist of electric and magnetic quadrupoles. Plasma lenses are presently under discussion as an attractive focusing device [18].

For this kind of lenses the plasma serves as an almost transparent current-carrying and space-charge-neutralizing medium. If a homogeneous current density distribution is obtained it will result in a linearly rising magnetic field. In contrary to quadrupoles these cylindrically symmetric "wire lenses" provide simultaneous focusing in both planes with very high field gradients of several hundred Tesla per meter. One possibility for producing the plasma is realized by a z-pinch discharge. Driven by the strong magnetic force during a pulsed high-current discharge an uniform axial plasma column is formed (fig. 1).

Already in the early phase of our energy loss experiments we were aware of plasma lens effects of our discharge device. [19]. With our Monte-Carlo code we were able to simulate the focusing properties of the discharge plasma (Fig. 3). The beam spot was observed experimentally with a streak camera to provide high time resolution. As a result we found that the z-pinch indeed acts as a lens. The beam spot was reduced from 1cm in diameter to about 2 mm. Measurements of the magnetic field inside the pinch with an inductive coil showed a field gradient of 25 T/m.
The reproducibility of the plasma parameters and the high reliability of the z-pinch device make it appear likely that a plasma lens device can be developed for high energy deposition experiments at the new accelerator facility of GSI.

References

[1] Murakami, M., Meyer-ter-Vehn, J., and Ramis, R., J. X-Ray Sc. and Techn. **2**,127(1990

[2] Rubbia, C., Nucl. Instr. and Meth. A278, 253(1989)

[3] Mehlhorn, T.A., J. Appl. Phys., **52**,6522(1981)

[4] Deutsch, C., Fromy, P., Garbet, X., Maynard, G., Fus. Technol. **13**,362(1988)

[5] Basko, M.M., Sov. J. Plasma Phys. **10** ,689(1984) [Fiz. Plazmy **10**,1195(1984)]

[6] Peter, Th.,and Meyer-ter-Vehn, Phys. Rev. *A*(1990) in print and Peter, Th., diploma thesis (1985), University of Munich, 1985

[7] Peter, Th., Arnold, R., Meyer-ter-Vehn, J., Phys.Rev. Letter **57**,1859(1986)

[8] Gryzinski, M., Phys. Rev. **A138**,305(1965)

[9] Nikolaev, V., Zh. Eksp. Theor. Fiz. **51**,1263(1966)

[10] Halpern, A.M., Law, J., Phys. Rev. **31**,4(1973)

[11] Chan, F.T., Eichler, J., Phys. Rev. Lett. **42**,58(1979)

[12] Schlachter, A.S., Stearn, J.W., Graham, W.G., Berkner, K.H.,Pyle, R.V., and Tanis, J.A., Phys. Rev. **A27**, 3372(1983)

[13] Oppenheimer, J.R., Phys. Rev. **31**,349(1928)

[14] Spitzer, Jr., L., Astrophys. J. **107**,7(1948)

[15] Noll, R., Kunze, H., Haas, C.R., Nucl. Instr. and Meth. A278, 85(1989)

[16] Kunze, H., 5th Int. Workshop on Atomic Physics for Ion Driven Fusion, Schliersee, 1990

[17] Northcliffe, L.C., and Schilling, R.F., Nucl. Data Tables A7, 223(1970)

[18] Heimrich, B., Nucl. Instr. and Meth. A278, 202(1989)

[19] Gardès, D., Bimbot, R., Della-Negra, S., Dumail, M., Kubica, B., Richard, A., Rivet, M.F., Servajean, A., Fleurier, C., Sanba, A., Deutsch, C., Maynard, G., Hoffmann, D.H.H., Weyrich, K., Wahl, H., Europhys. Lett., **8**(1988)701

Supplement to Z. Phys. D – Atoms, Molecules and Clusters 21, S 111–115 (1991)

Atoms, Molecules and Clusters
Zeitschrift für Physik D

The status of the electron beam ion sources

M. P. Stöckli
J. R. Macdonald Laboratory, Kansas State University, Manhattan, KS 66506-2604, USA

Abstract

More than twenty years after its invention, 14 Electron Beam Ion Sources are in operation worldwide. The substantial progress in operation and insight, achieved over the last few years, made the EBISes become reliable tools for the production of beams of very highly charged, low energy ions. For example, 8 EBISes produce bare argon on a standard basis. The successful production of hydrogen-like xenon presents the ions with the highest ionization energy, whereas the production of Th^{80+} presents the highest achieved charge state. Several synchrotrons are fed by EBIS-injectors, taking advantage of the EBIS batch mode production, which yields the highest charge states. A few EBISes are used for ion source development. However, most of the EBISes' efforts are directed to research the physics of highly charged ions. Some of those are used to study the electron-ion interaction inside the source. But normally, most EBISes deliver the ions for external experiments, which so far concentrate on the recombination of the highly charged ions with atoms, molecules and surfaces. The ions are typically produced at a potential of 1 to a few kilovolts per charge, but in most cases, the EBIS is mounted on a high voltage platform or is followed by an RFQ, and therefore can generate ion energies from a few hundred volts up to a few hundred kilovolts per charge. The delivered beams have a low emittance and a low energy spread, which is an advantage for high resolution experiments. This paper presents briefly all operational EBISes, their capabilities, their achievements, and their contribution to physics research.

I. Introduction

More than twenty years ago Donets proposed the concept for an Electron Beam Ion Source (EBIS) to produce highly charged ions [1]. Briefly, ions trapped inside an electron beam are stripped electron by electron. As shown in Figure 1, a strong magnetic solenoid is commonly used to confine and compress the electron beam, which increases the ionization rate. The highest charge states are obtained with the long confinement times achieved with a batch mode. In this mode, the electrostatic potential of the drift tubes is altered in a time sequence, first to provide for the injection of low charged ions, then to form a trap in which the ions are cooked and finally to expel the ions when the desired charge state is reached. However, the capture of electrons from the residual gas often limits the ultimate obtainable charge state, a limitation which is extended by the low residual pressures achieved in CRYogenic EBISes (CRYEBIS). Although the basic operation schema appears simple, there are complex interactions involved. In particular, the last "International Symposium on Electron Beam Ion Sources and their Applications" held at Brookhaven National Laboratory in November 1988 revealed a substantial progress in the operational performance and understanding of the characteristics of the various EBISes, such as the recognition and intentional application of ion-ion cooling which counteracts the electron beam induced ion heating, pioneered by the Livermore group [2,3] and Donets [4]. In addition to the symposium's proceedings [5], there are many other publications [6-8] which discuss the physics and technology of such sources, as well as the operation, performance, and achievements of the individual EBISes. In summary, there are quite a few fully operational EBISes, which are producing highly and very highly charged ions for physics research, or in some cases are used to study physics inside the source. Beams of bare argon became an operational standard and naturally beams of heavier, much higher charged ions can be easily produced. However, the construction and operation of an EBIS remains a challenging task and over the years several EBIS programs were shut down before their device was fully operational. Despite their misfortune, those groups have contributed substantially to the knowledge of our field and therefore helped some of us to succeed. But we do not have time to discuss the history of the development, as we want to concentrate on today's status of the operational EBISes, their contribution to physics research, and the new opportunities for the highly charged ion physicists.

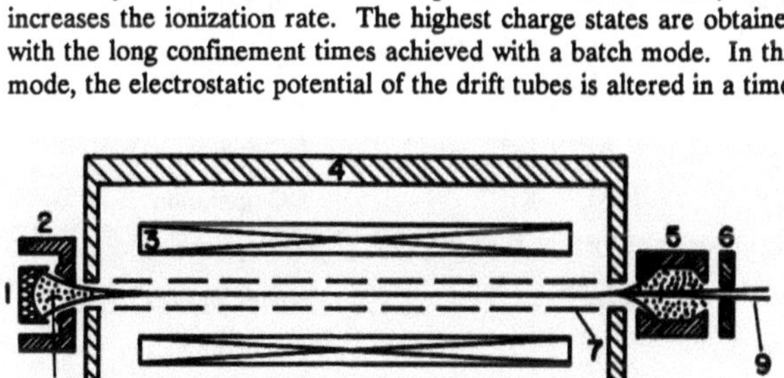

Fig. 1. Schematic of an Electron Beam Ion Source with the cathode (1) and the anode (2) of its electron gun, the solenoid (3) with its magnetic shield (4), the collector (5) with its repeller (6) and the drift tubes (7). The electron beam (8) is absorbed by the collector whereas the ion beam (9) is extracted by the repeller.

II. The Russian Challenge

The Synchrophasotron at the Joint Institute of Nuclear Research required a charge-to-mass ratio of at least 0.5 because of its insufficient vacuum and the acceptance of its injector LINAC and therefore required the injection of fully stripped ions for heavy ion operations. Starting in 1977 this challenge was met by KRION which served as an injector until 1984, when it was replaced with a higher yielding laser source [9]. KRION was upgraded to KRION-C and now serves as a test EBIS to study a variety of problems associated with the production of ultra-high charge states [10]. KRION-3 took on the challenge for high yields of very highly charged ions to replace or complement the present injector ion source [6]. At times KRION-3 exceeds 1000 amps/sqcm current density [11] and produces more than 20 million bare argon ions in 0.2 seconds [12], but futher development is required. However, it is KRION-2 which keeps producing the record high charge states of low energy ions such as bare Ar in 1977 [13] and bare Kr in 1986. Convincing proof for the production of bare krypton was delivered with the observation of the Kr K-alpha hypersatellite with increasing confinement time, when the extracted ions impacted a copper surface [9,14]. A ten- to twenty-fold increase of the yields of bare Ar and Kr ions was observed [12] when ion-ion cooling was applied [15]. Recently the production of hydrogen-like and bare xenon was announced [6,15]. These statements were backed at the 1989 Ion Source conference with the proof for the emission of the K-X-rays when the extracted xenon ions impacted on a surface [12]. Although that proof indicated only the production of hydrogen-like ions, the production of at least some fully ionized xenon is likely. This outstanding result was achieved with a 52 kV electron beam which bombarded the ions for 40 seconds [12]. Besides breaking the charge states' records, Krion-2 also provided over the years many novel data on atomic physics, such as ionization cross sections by electron impact for highly charged ions [12,16,17] and recombination of highly charged ions with gaseous targets [11,17] or with surfaces [6,9,11,14,15,17-19].

Two EBISes with classical 1 Tesla solenoids and room temperature drift tubes were constructed at the Institute of Nuclear Physics in Novosibirsk, USSR. The latest model IMI-2 is presently under startup in Novosibirsk. The first model IMI-1 was delivered to the Institute of Radium in Leningrad, where it produced O^{8+} and Ti^{12+} within the required repetition rate of 50 Hz. Presently IMI-1 is waiting to go on line as an injector on the synchrotron complex B-5.

III. The West-European Electron Beam Ion Sources

The injector LINAC of the 3 GeV synchrotron SATURNE at the Centre d'Etude Nucleaire in Saclay required a charge-to-mass ratio of 0.45. Heavy ion operations were initiated with an injector consisting of a CRYogenic EBIS (CRYEBIS-I) followed by an RFQ [20]. MIMAS, a low energy accumulator ring was added in 1988 to increase the yields. In 1987, after four years of successful operation, CRYEBIS-I was replaced with DIONE, a second generation EBIS. The reduced residual gas pressure, the higher electron beam energy and the tenfold increase in electron beam density yielded a dramatic improvement. Up to 10^9 highly charged ions such as bare carbon, nitrogen or neon, can be extracted after very short confinement times [21]. For example MIMAS can be stacked with up to 7 pulses of bare nitrogen within a quarter of a second [22]. DIONE can easily produce bare argon, but for injection it is preferable to use the higher yields of Ar^{16+} which can be produced in less than 50 milliseconds and stacked in MIMAS [21]. DIONE serves also as an ionizer to completely strip polarized ^6Li. Another breakthrough was achieved with the injection of Kr^{30+} which SATURNE accelerated up to 59 GeV to study high energy peripheral collisions [22,23]. Only recently was DIONE involved in its first atomic physics experiment, which investigated the angular momentum state of the electrons captured from the residual gas by the hydrogenlike argon ions at 7.5 MeV [24]. And very recently a newly installed atomic hydrogen target was used to refine this experiment [25].

The French CRYEBIS program in Orsay was terminated when CRYEBIS-II did not deliver the expected results, but interestingly enough a sister of CRYEBIS-II is working successfully in Stockholm: CRYSIS, the CRYogenic Stockholm Ion Source, the only EBIS which is operated with a closed cycle refrigerator to keep the superconducting solenoid and the drift tubes at cryogenic temperatures. It was outlayed as the primary injector for CRYRING, a synchrotron storage ring for very highly charged, heavy ions with an energy range of 0.3 to 24 MeV/amu [26]. So far CRYSIS produces with each cycle up to 100 million highly charged ions, such as bare Ne, Ar^{16+} or Xe^{35+}, and less of the higher charge states such as bare argon and neon-like xenon, which require longer confinement times [27]. The highly charged xenon ions were used to study charge changing processes in collisions with helium [28] and xenon. For the highest charge states, the latter experiment revealed evidence for radiative stabilization of the two electron transfer processes [29-31]. Recently CRYSIS, mounted on its 50 kV platform, was moved to its final location where it can serve as an injector to CRYRING. An external ion injector was added in the new setup for the production of non gaseous ions. This final arrangement is very versatile, because either CRYSIS boosted by an RFQ feeds CRYRING to study highly charged ions at intermediate energies or CRYSIS can feed one of four low energy atomic physics beam lines with up to 50 kV/q of highly charged ions. The first experiment using a new atomic beam line is in progress. This new facility, which welcomes guest researchers, provides a new opportunity for physicists interested in highly charged physics.

The Frankfurt EBIS is a small compact EBIS made with a 1 m long, 5 cm bore, 5 Tesla superconducting solenoid. It reaches high electron beam densities, and normally it is operated in a DC-mode. The production of Ne^{10+}, Ar^{16+}, Kr^{26+} and Xe^{36+} are impressive results considering the short confinement times obtained in a DC mode [32]. The DC mode is preferred because of its advantages for atomic physics experiments, for which novel, high resolution instruments were developed. However, as a team of the Institute for Applied Physics, their contributions concentrate on the invention of new EBIS techniques and equipment. For example, recently the effect of gas mixing to boost the yields through ion-ion cooling was investigated [33].

Table 1: Operational EBIS-Devices 1990:

LOCATION/LAB	KEY PERSONNEL	DEVICE	TRAP	PRODUCED IONS	REMARKS
Dubna/JINR	Pikin/Donets	KRION-2	LHe	Ar^{18+}, Kr^{36+}, Xe^{53+}	atomic physics
	Vadeev/Donets	KRION-3	LHe	Ar^{18+}	future injector
	Ovsyannikov/Donets	KRION-C	LHe	Ar^{18+}, U^{40+}	test EBIS
Leningrad/IR		IMI-1	warm	C^{6+}, O^{8+}, Ti^{12+}	future injector
Novosibirsk/INP	Abdul'manov	IMI-2	warm		under startup
Saclay/CEN	Faure	DIONE	LHe	Ar^{18+}, Kr^{30+}, I^{41+}	injector
Stockholm/MSI	Beebe/Liljeby	CRYSIS	LHe-R	Ar^{18+}, Xe^{44+}	injector & 4 beam lines
Frankfurt/IAP	Kleinod/Becker	Frankfurt-EBIS	LHe	Ar^{16+}, Kr^{26+}, Xe^{36+}	DC-operation
Nagoya/IPP	Tawara/Ohtani	NICE	LHe	Ne^{10+}, Kr^{25+}, I^{42+}	DC-operation
Tokyo/Metro U.	Okuno	mini-EBIS	LN2	O^{8+}, Ne^{9+}, Ar^{16+}	DC-operation
Ithaca/Cornell	Kostroun	CEBIS-I	warm	Ar^{16+}, Xe^{28+}	atomic physics
		CEBIS-II	LHe	Ar^{16+}	atomic physics
Livermore/LLNL	Marrs/Levine	EBIT-I	LHe		under upgrade
		EBIT-II	LHe	Ar^{18+}, Xe^{48+}, Th^{80+}	atomic physics
Manhattan/KSU	Stockli/Cocke	KSU-CRYEBIS	LHe	Ar^{18+}, Kr^{34+}, Xe^{44+}	7 atomic beam lines

The group is also working on the development of simplified and/or miniaturized EBISes using permanent magnets or no magnets at all, which naturally compromises the charge states because of the reduced electron beam densities [32].

IV. The Japanese Electron Beam Ion Sources

In 1977 the Institute of Plasma Physics at Nagoya University started to invite Japanese physicists to Nagoya to study atomic processes involving highly charged ions, considered to be very important in high temperature plasmas. The initial Proto-NICE EBIS (Naked Ion Collision Experiments) was followed by a cryogenic version NICE-1, which became operational in 1980. The source was operated in a DC-mode, which reduced the energy spread to less than 1 eV per charge [34]. The ion production included fully stripped ions up to neon, and very highly charged ions such as Kr^{25+} and I^{42+}. The few thousand ions per second obtained for the highest charge states were sufficient to measure the processes of interest, mainly electron capture in collisions with neutral atoms, which was studied in great detail, and included energy gain spectroscopy, taking advantage of the low energy spread. The results of this research were reported in 18 refereed publications over a 4 year period [35]. Although NICE-1 is still operational, there is limited activity since the NICE group was dissolved in 1984.

Okuno, a member of the former NICE group, constructed at the Tokyo Metropolitan University a mini-EBIS with a normal-conducting solenoid, cooled with liquid nitrogen. This compact, low-cost EBIS produces up to bare oxygen, up to Ne^{9+} and Ar^{16+} ions, which is excellent, considering the DC-operation and the limited electron beam density obtained with the 1 kG field [36,37]. Electron capture from helium was studied in the very low energy regime using ions with an energy between 1 eV and 2 kV [38].

V. Electron Beam Ion Sources in the U.S.

The first EBIS built at Cornell University, CEBIS-I, is the best "warm" EBIS ever built and accordingly is still in operation. Room temperature EBISes were typically limited by the low electron beam compression obtained with classical solenoids and the ion recombination in the high residual gas pressures. However, an integrated ion pump creates a very low internal pressure inside CEBIS-I, which permits very long confinement times and hence the production of ions like Ar^{16+} and Xe^{28+}. A detailed study of the trapped ions regarding heating and cooling was published [39] and studies of electron capture by extracted ions colliding with beams are underway. CEBIS-II is Cornell's cryogenic EBIS, which was recently upgraded and redesigned to fit into an economical, off-the-shelf 3 Tesla solenoid [40]. So far it has been operated with an electron beam energy up to 3.5 kV, sufficient to produce the desired Ar^{16+}. Despite the pulsed operation, a beam with an energy spread as little as 2 eV per charge was produced, very important for the energy gain and loss studies in progress [41]. A study of electron emission induced by the highly charged ions incident on a surface is also underway.

EBIT-I and EBIT-II, the Electron Beam Ion Traps at Livermore National Laboratory in Livermore, were designed to produce and study the highly charged ions *in situ*. A pair of Helmholtz coils (in lieu of the traditional long solenoid) generates the rather short 3 Tesla solenoidal field. The shortness reduces complexity and problems which can be encountered with the electron beam transmission [42]. Indeed the traps became operational after a very short time required for design, construction, installation and startup. The two traps, which were essentially identical, reach high current densities, high electron energies and have a very low background pressure, and therefore produce record charge states. Beryllium windows in the center between the two Helmholtz coils permit the observation of X-rays emitted in a radial direction. The emitted X-rays were used to demonstrate that the highly charged ions can be stored for hours, if proper cooling is applied [43]. The results on ion cooling are in good agreement with computer model calculations [44]. These novel tools for research on energetic electrons interacting with highly charged ions produced a large amount of novel data, such as electron impact excitation and radiative recombination [45,46], dielectronic recombination [47], X-ray excitation [48,49] and polarization [50], and precision wavelength measurements [51]. However, the strong magnetic field limits the research to processes which lead to X-ray emission. This limitation was recently overcome when the group started to extract the ions through a hole in the collector and analyzed ion beams of up to Ar^{18+}, Xe^{48+}, U^{71+}, and Th^{80+}, the record charge state for low energy ions [52]. Naturally the short trap length limits the ion yield, although it is more than adequate for most experiments, such

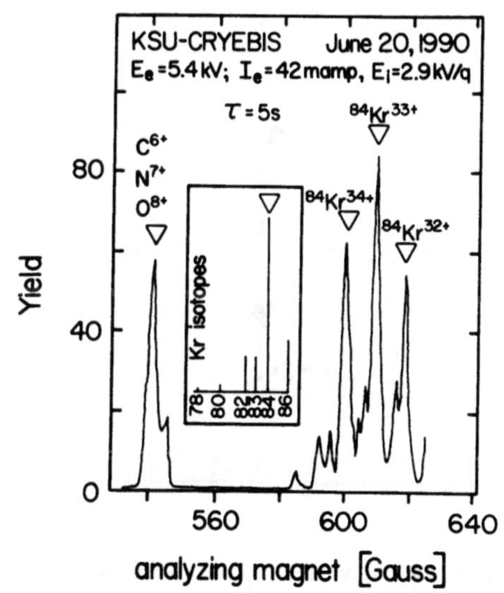

Fig. 2. Relative ion yields of highly charged krypton obtained with an electron beam energy below the K-ionization threshold. The insert shows the scaled isotope pattern. The triangles point to the dominant ioeotope 84 and shows Kr^{33+} to be the most abundant charge state.

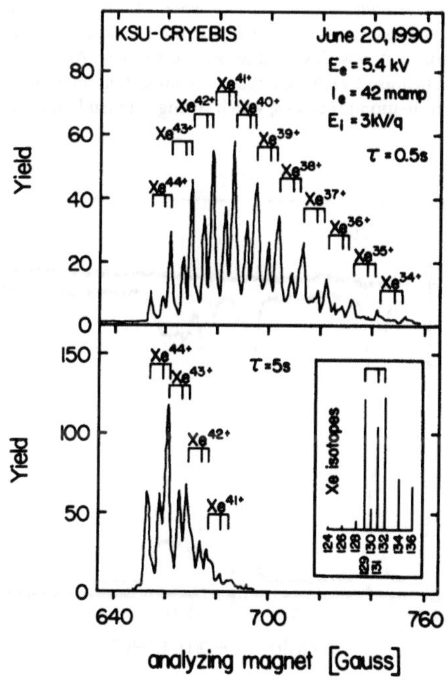

Fig. 3. Relative ion yields of highly charged xenon obtained with an electron beam energy below the L-ionization threshold. The insert shows the scaled isotope pattern. The fork symbols point to the three dominant isotopes. The figure demonstrates how the intensity of the highest charge states grow, when the confinement time is increased from 0.5 seconds (top) to 5 seconds (bottom).

as the studies of recombination on surfaces underway [52]. Presently EBIT-I is upgraded in a quest for higher charge states obtained with higher electron beam energies.

VI. The Latest Addition: The KSU-CRYEBIS Facility

The KSU-CRYEBIS was designed as a standard cryogenic EBIS, similar to CRYEBIS-II and DIONE and incorporated some ideas from EBIT [53,54]. After several years of construction and installation, it started to operate on Easter Saturday 1989. Within a few months the production of bare argon was confirmed and shown to be the most abundant argon charge state for a containment time of two seconds [55]. More recently Kr^{34+} and Xe^{44+} were produced as shown in Figures 2 and 3. All these results were obtained with an electron beam energy of approximately 5.5 kV, the present hardware limit, which will be upgraded in the future. However, novel physics is not restricted to the highest charge states. For example, it was shown that the EBIS is a perfect device for an accurate measurement of the dielectronic recombination probabilities in comparison to the ionization probabilities, because no solid angle or beam overlap has to be estimated [56]. The case of the dielectronic recombination of the helium-like ions versus the ionization of the lithium-like ions is amazingly simple to analyze, because it involves only two charge states, as long as the electron beam energy is below the K-ionization threshold. The equilibrium yield of the helium-like ions is determined by the ionization rate of the lithium-like ions and the loss rate of helium-like ions through radiative recombination and through capture from the residual gas. But, whenever the electron beam energy is in resonance with a double excited state of the lithium-like ion, the dielectronic recombination will dramatically increase the loss rate of the helium-like ions, as shown in Figure 4 [from ref. 57]. These measurements have been completed for argon [57] and neon by probing the equilibrium ion yields. Naturally this method does not reveal processes without a net change of charge, such as dielectronic recombination followed by autoionization of the double excited state. Therefore, these studies were complemented by observing the X-rays emitted in an axial direction through the collector and through a window in the 0 degree port of the analyzing magnet [57,58]. The small solid angle of the X-ray detector is partly compensated through the apparently high source density when observing on axis. This is another example in which an EBIS is used to study atomic physics *in situ*, although it was designed as an ion source. As an ion source it delivered beams of highly charged ions impacting on a surface to study the recombination processes, which strongly depend on the ion velocity [59].

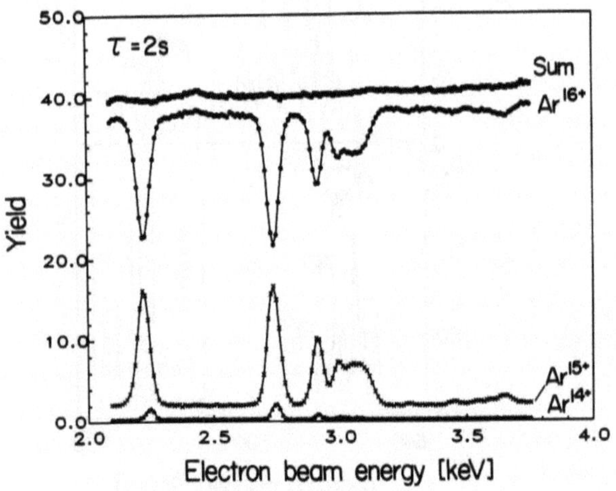

Fig. 4. Equilibrium ion yields (after confinement of 2 seconds) for Ar^{14+}, Ar^{15+}, and Ar^{16+} as a function of the electron beam energy. Whenever the electron beam energy is in resonance with a doubly excited state of Ar^{15+}, the Ar^{15+} yield will increase through dielectronic recombination of the Ar^{16+}. The top line shows the sum of the three charge states [from ref. 57].

The KSU-CRYEBIS was designed as a facility for atomic physics with highly charged, low energy ions. Therefore it is located on a high voltage platform which allows one to accelerate the ions up to 200 kV/charge. The extraction beam line with the necessary diagnostic devices and ion optical elements transports the beam to the switching magnet [60] which injects the beam into one of the seven beam lines. This setup, shown in Figure 5, allows for a very efficient switching between different experiments and will dramatically broaden the number of problems which can be studied in the same time period. The first experiment using the new beam lines started recently. Being a national user facility we strongly encourage the atomic physics community to make use of this novel atomic probe in our facility.

VII. Conclusions

As discussed, the presently operational Electron Beam Ion Sources produce very highly charged ions. The produced ion beams and *in situ* experiments continue to contribute substantially to physics research, especially to the atomic physics of highly charged ions. The activity is expected to increase, as new projects will come on line. A second mini-EBIS is under construction in Tokyo and an EBIT-like source is under construction in Oxford, England. A similar project is under development in Troitzk, USSR [61]. The upgraded EBIT-I will withstand higher electron beam energies and therefore produce even higher charged ions. The new projects underline the fact that Electron Beam Ion Sources have become reliable tools to study the physics of highly charged ions.

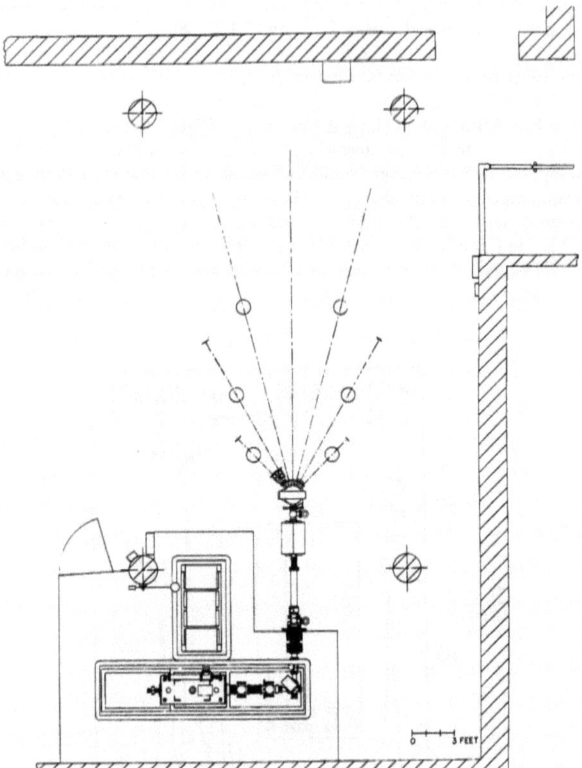

Fig. 5. Setup of the KSU-CRYEBIS facility. The CRYEBIS is located on a 200 kV T-shaped platform inside a protective cage. After being analyzed in a 90 degree magnet, the ions enter the extraction beam line through the accelerator columns. The switching magnet directs the beam into one of the seven indicated atomic beam lines. The circles show the tentative target locations. The shaded areas represent the building walls and support columns.

Acknowledgments

I am grateful to V.G. Abdul'manov, A. Bárány, R. Becker, J.P. Briand, C.L. Cocke, E.D. Donets, J. Faure, W.G. Graham, M. Kleinod, C. Koci, V. Kostroun, L. Liljeby, R. Marrs, K. Okuno, D. Richard, and D. Schneider who contributed to this paper through discussions, suggestions, comments, and proof reading. Especially appreciated is the help from V. Rieger with the production of the Krypton and Xenon charge state spectra. This work is supported by the Division of Chemical Sciences, Off. of Basic Energy Sciences, Office of Energy Research, U.S. Department of Energy.

References

1. Donets E.D.: USSR Inventor's Certificate No. 248860, March 16, 1967, Byull. OIPOTZ No. 23, 65 (1969).
2. Levine M.A., Marrs, R.E., Henderson J.R., Knapp D.A., Schneider M.B., Physica Scripta T22, 157 (1988).
3. Levine M.A., Schneider M.B.: in "Proceedings of the International Conference on ECR Ion Sources, East Lansing, 1987", National Superconducting Cyclotron Laboratory, Michigan State University Report MSUCP-47, 233 (1988).
4. Donets E.D., Shirkov G.D., Avtorskoe Svidetelstvo USSR, N1225420 (02.07.1984), Bul. OI, N44, 69 (1989).
5. Hershcovitch A., editor: "International Symposium on Electron Beam Ion Sources and Their Applications, Upton, NY, 1988", AIP Conf. Proc. No. 188, (American Institute of Physics, New York, 1989).
6. Donets E.D.: in "The Physics and Technology of Ion Sources" Brown I.G. edt., (John Wiley & Sons, New York) 245 (1989).
7. Schmieder R.W.: in "Physics of Highly-Ionized Atoms" Marrus R. edt., (Plenum Publishing Corp.) 321 (1989).
8. Arianer J., Cabrespine A., Goldstein C.: Nucl. Instrum. and Meth. 193, 401 (1982).
9. Donets E.D., Monchinsky V.A.: "Quatriemes Journees D'edudes Saturne La Physique avec Mimas, LaLonde-les Maures, 1986" (Laboratoire National Saturne, 91191 Gif-sur-Yvette Cedex, France), 310 (1987).
10. Ovsyannikov V.P.: Vth Int. Conf. on the Physics of Highly-Charged Ions, Giessen, Germany, 1990, Book of Abstracts, 5.13 (1990).
11. Donets E.D.: Nucl. Instrum. Methods B9, 522 (1985).
12. Donets E.D.: "Int. Conf. on Ion Sources, Berkeley, CA, (1989).
13. Donets E.D., Ovsyannikov V.P.: JINR Commun. P7-10438, Dubna, (1977).
14. Donets E.D., Kartashov S.V., Ovsyannikov V.P.: JINR Rapid Commun. 20-86, 27 (1986).
15. Donets E.D.: Rev. Sci. Instrum. 61, 225 (1990).
16. Donets E.D., Ovsyannikov V.P.: Sov. Phys. JETP 53, 466 (1981).
17. Donets E.D.: Physica Scripta T3, 11 (1983).
18. Donets E.D., Trifonov V.A., Tutin G.A., Eismont V.P.: JINR Commun. P7-83-573, Dubna, (1983).
19. Donets E.D., Dudnikov, V.G., Dunin V.B., Kartashov S.V.: this conference (1991).
20. Olivier M., Faure J., Laclare J.L., Lefebvre J.M., Leleux G., Ropert A., Tkatchenko A., Tkatchenko M., IEEE Trans. Nucl. Sci. NS-30, 1463 (1983).
21. Faure J., Antoine P., Ciret J.C., Degueurce L., Gros P., Courtois A., Gastineau B., Gobin R., Leaux P., Leroy P.A., Penicaud J.P. in reference [5], 102 (1989).
22. CERN Courier, September 1989, p. 17.
23. CERN Courier, November 1989, p. 37.
24. Briand J.P., Charles P., de Billy L., Essabaa S., Mchami A., Bliman S.: Bull. APS 35, 1167 (1990).
25. Briand J.P.: private communication (1990).
26. Herrlander C.J.: in ref [5], 379 (1989).
27. Liljeby L., Engstrom A.: in ref [5], 27 (1989).
28. Andersson H., Astner G., Cederquist H.: J. Phys B21, L187 (1988).
29. Cederquist H., Andersson H., Astner G., Hvelplund P., Pederson J.O.P.: Phys. Rev. Lett. 62, 1465 (1989).
30. Anderson H., Cederquist H., Astner G., Hvelplund P., Pedersen J.O.P., Physica Scripta 42, 150 (1990).
31. Cederquist H.: this conference (1991).
32. Kleinod M., Becker R., Klein H.: in ref [5], 15 (1989).
33. Becker R., Kleinod M.: this conference (1991).
34. S. Ohtani, Physics Scripta T3, 110 (1983).
35. "The collected Papers of NICE Project / IPP, Nagoya" Tawara H.: edt., IPP-AM-43, (Institute of Plasma Physics, Nagoya University, 1985).
36. Okuno K.: in ref. [5], 33 (1989).
37. Okuno K.: Jap. J. of Appl. Phys. 28, 1124 (1989).
38. Okuno K., Soejima K., Kaneko Y.: submitted for publication in Nucl. Instrum. and Meth. (1989).
39. Beebe E.N.: in ref. [5], 166 (1989).
40. Kostroun V.O.: in ref. [5], 65 (1989).
41. Kostroun V.O., private communication (1990).
42. Levine M.A., Marrs R.E., Bennett C.L., Henderson J.R., Knapp D.A., Schneider M.B.: in ref. [5], 82 (1989).
43. Schneider M.B., Levine M.A., Bennett C.L., Henderson J.R., Knapp D.A., Marrs R.E., :in ref. [5], 158 (1989).
44. Penetrante B.M., Levine M.A., Bardsley J.N.,: in ref. [5], 145 (1989).
45. Marrs R.E., Levine M.A., Knapp D.A., Henderson J.R.: Phys. Rev. Lett. 60, 715 (1988).
46. Marrs R.E., Levine M.A., Knapp D.A., Henderson J.R.: in "Electronic and Atomic Collisions" Gilbody H.B., Newell W.R., Read F.H., Smith A.C.H. eds., (Elsevier Science Publisher) 209 (1988).
47. Knapp D.A., Marrs R.E., Levine M.A., Bennett C.L., Chen M.H., Henderson J.R., Schneider M.B., Scofield J.H.: Phys. Rev. Lett. 62, 2104 (1989).
48. Marrs R.E., Beiersdorfer P., Bennett C., Chen M.H., Cowan T., Dietrich D., Henderson J.R., Knapp D.A., Osterheld A., Schneider M.A., Scofield J.H.: in ref. [5], 445 (1989).
49. Beiersdorfer P., Osterheld A.L., Chen M.H., Henderson J.R., Knapp D.A., Levine M.A., Marrs R.E., Reed K.J., Schneider M.B., Vogel D.A.: Phys. Rev. Lett. 65, 1995 (1990).
50. Henderson J.R., Beiersdorfer P., Bennett C.L., Chantrenne S., Knapp D.A., Marrs R.E., Schneider M.B., Wong K.L., Doschek G.A., Seely J.F., Brown C.M., LaVilla R.E., Dubau J., Levine M.A.: Phys. Rev. Lett. 65, 705 (1990).
51. Levine M.A., Marrs R.E., Bardsley J.N., Beiersdorfer P., Bennett C.L., Chen M.H., Cowan T., Dietrich D., Henderson J.R., Knapp D.A., Osterheld A., Penetrante B.M., Schneider M.A., Scofield J.H.: Nucl. Instrum. and Meth. B43, 431 (1989).
52. Schneider D., DeWitt D., Clark M.W., Schuch R., Cocke C.L., Schmieder R., Reed K.J., Chen M.H., Marrs R.E., Levine M., Fortner R.: Phys. Rev. A 42, 3889 (1990).
53. Stockli M.P., Arianer J., Cocke C.L., Richard P., Nucl. Instrum. and Meth. B40/41 1020 (1989).
54. Stockli M.P., Cocke C.L., Richard P.: in ref. [5] 114 (1989).
55. Stockli M.P., Cocke C.L., Richard P.: Rev. Sci. Instrum. 61, 242 (1990).
56. Ali R., Bhalla C.P, Cocke C.L., Stockli M.P.: Phys. Rev. Lett. 64, 633 (1990).
57. Ali R., Bhalla C.P, Cocke C.L., Schulz M., Stockli M.P.: to be published in Phys. Rev. A (1991).
58. Ali R., Bhalla C.P, Cocke C.L., Schulz M., Stockli M.P.: this conference (1991).
59. Schulz M., Cocke C.L., Stockli M.P., Hagmann S.: this conference (1991).
60. Courtesy of Western Michigan University.
61. e.g. Antsiferov P.S., Movshev V.G.: this conference (1991).

Supplement to Z. Phys. D - Atoms, Molecules and Clusters 21, S 117-121 (1991)

Atoms, Molecules
Zeitschrift für Physik D and Clusters

ECRIS: the electron cyclotron resonance ion sources - (status)

R. Geller
Département de Recherche Fondamentale, Centre d'Etudes Nucléaires de Grenoble, 85X, 38041 Grenoble Cédex, France

ABSTRACT

Ion beam characteristics are often decisive factors for experimental atomic physics and ECR sources for highly charged ions play then an important role [1]. Therefore we recall their basic principles and possibilities, propose practical scaling laws and give recent performances. But we also emphasize that the radiations issuing from the ECR plasma are valuable for atomic physics.

THE MULTIPLY CHARGED ION PRODUCTION

The probability of obtaining a multiply charged ion by a single electron impact collision is rather poor. (σ between 10^{-21} and 10^{-24} cm^2) Subsequently the only efficient way for obtaining an important yield of many-times ionized ions is by successive ionization. We are then led to increase τ the exposure time of the ions to a cloud of electrons of density n and velocity v. The parameter $n v \tau$ determines the achievable q. But the electron impact velocity v should exceed some 10^9 cm.s^{-1} in order to reach the ionization potential of the inner shell electrons i.e. some KeV. For this purpose the ECR is an ideal method whereas arcs don't give enough energy to the electrons [2]. Fig. 1 shows mean charges <q> for Argon ions versus $n\tau v$. They are obtained through calculations with the help of ionization cross sections and assume that only the confinement time τ limits the achievable charge. Recombination mechanisms are not considered. Note that the highly charged ions appear for $n\tau v > 10^{19}$ cm^{-2} and that q increases very slowly with $n\tau v$ (in a quasi logarithmic scale) which underlines the difficulty for producing highly charged ions. In short, one needs the highest possible electron density n with longest confinement time τ. The problem is : how can one obtain $[n\tau]$ maximum with energetic electrons and minimum neutral gas pressure to avoid recombinations [2]. Since in the usual ECR plasmas the electron density is generally $\lesssim 10^{12}$ cm^{-3}, with $v > 3\ 10^9$ cm s^{-1} one needs ionic lifetimes of $\gtrsim 10^{-2}$ s. Such ion lifetimes can only be obtained in special magnetic confinement systems. The so called B_{min} structures are the simplest of these systems. As for n, its upper limit is linked to the microwave penetration condition : $n \lesssim 1.2\ 10^{-8}\left(\frac{\omega_{ECR}}{2\pi}\right)^2$, which becomes particularly valid for increasing ω i.e. when the wavelength becomes small with respect to the plasma size.

BASIC INGREDIENTS AND MECHANISMS OF ECRIS

a) <u>The magnetic confinement principle</u>. The magnetic configuration plays a crucial role because it affects the particle lifetime τ, which in turn strongly influences the RF power necessary to couple to the plasma. The better the magnetic confinements, the higher are the efficiencies of ionization and of the production of highly charged ions. For these reasons, radial mirrors are added to axial mirrors and the structures are then called "B_{min} structures" because they exhibit a minimum magnetic field in the center (Fig. 2). Their confinement times, τ, generally exceed 10^{-3} seconds whereas for axial mirrors alone, τ is at least 50 times smaller. In general, one needs mirror ratios $B_{max}/B_{min} \approx 2$, and $B_{max}/B_{ECR} > 1.5$.
b) <u>The stochastic ECR heating</u>. The ECR heating mechanism of the electrons can be briefly summarized as follows. Consider an empty metallic box of undetermined shape, filled with microwave power (for instance f = 10 GHz, λ = 3 cm). If the dimensions of the box are large with respect to the wavelength λ, the box can be considered as a multimode cavity. If the box is now immersed in a minimum B structure where the magnetic field strength is between, for example, 0.2 and 0.5 T, then there must exist a magnetic surface with a spherical or "egg" shape where the field strength is B_{ECR} = 0.36 T (Fig. 2). On this surface, the gyrofrequency of the electrons is 10 GHz and thus equals the frequency of the microwaves. The ECR necessarily occurs because in a multimode cavity there is always a component E of the electrical field of the waves that is perpendicular to the magnetic field lines. Thus, the electrons gain essentially perpendicular momentum $\mu_\perp$. In the magnetic well, the electrons pass many times through the ECR surface. They can sometimes be decelerated (according to their phase). However, after many passages, they undergo a global heating called stochastic ECR heating and rapidely reach the keV energy range [11].

c) <u>The ECR mirror plugging</u>. As the electrons gain essentially perpendicular momentun $\mu_\perp$ with respect to B, a special ECR mirror plugging improves their confinement. This improvement is due to the invariance of $\mu_\perp$ in absence of collisions ($\mu_\perp = W_\perp/B$ = inv.) where the perpendicular energy $W_\perp = w'_{\perp ECR} + w_{\perp thermal}$. As the total energy of the electron is constant, ($W_\perp + w_{/\!/ thermal}$ = Cte) beyond the ECR surface where B increases, $W_\perp$ also augments and thus $w_{/\!/ thermal}$ necessarily drops. When $w_{/\!/} = 0$... the electrons being stopped, the force $F = -\mu$ grad B reflects them. Without $w_{\perp ECR}$, i.e. with only $w_{\perp thermal}$, the mirror is less effective because the cancellation of $w_{/\!/}$ becomes problematical for a larger population of electrons. Therefore the ordinary mirror effect for thermalized electrons is weaker than the ECR mirror plug (fig. 3). As for the ions, as usually they are confined in the electron cloud by a self created electrostatic well.

d) <u>The gaseous pressure problems</u>. Generally the sources have two stages in cascade. In the first, a cold plasma is ignited and then diffuses toward the second stage with the hot electrons in the minimum B confinement system. The neutral gas pressure in the second stage must be as low as possible in order to avoid charge exchange recombination losses. As long as the ECR surface does not intercept a solid obstacle, the confinement and out-gassing remain good. If not, the confinement is destroyed and the power is dissipated in the solid, which is vaporized or melted. This property is utilized for direct, in situ metal ion production. The metallic sample is on a movable piston and can approach the ECR surface (but metallic vapor can also be obtained with ovens or laser impact). In the second stage, the ion pumping (burn out) due to ECR and the effects of gas mixing and of wall recycling also play a positive role in the source behavior. Figure 2 shows the main components of a typical ECR ion source. For feeding RF power into the first and the second stage, microwaves are injected through a tight window.

e) <u>ECRIS pulses with and without ECR mirror plugging</u>. ECRIS are fitted for continuous wave operation (CW) and subsequent external beam chopping enables all kinds of beam shapings. But RF power-pulsing offers better opportunities. If the RF duration is long enough, ($\tau_{RF} > \tau > 10$ ms) better results can be achieved (due to less thermal gas release from the walls). However as predicted and shown by the Grenoble team [3][4], at the end of the RF pulse, there must exist a short and intense ion burst... This burst is made up by the axially released ions from the plasma bulk, which is no longer contained by the ECR plugging effect when the RF stops. By optimizing the tunings we could obtain peak intensities up to 10 times larger than the CW intensities and similar results were recently obtained at GANIL [5]. This is very important because these short pulses are suitable for synchrotron injection and the CERN "Lead Injector" will utilize this new technique (Fig. 4). Finally an even stronger ion pulse is expected if one succeeds in pulsing adequately the B_z component of the B_{min} structure. Such an attempt is now prepared at G.S.I. Darmstadt by U. Ratzinger et al.

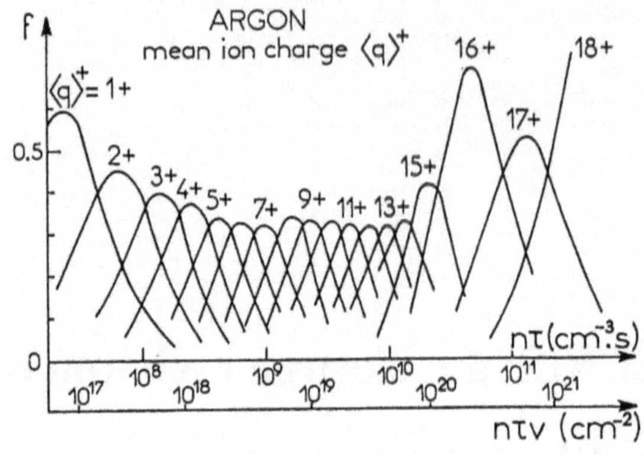

Figure 1 : A calculated charge distribution of ARGON mean charges $<q^+>$ (vs) $n\tau v$ for electrons of 10 keV.

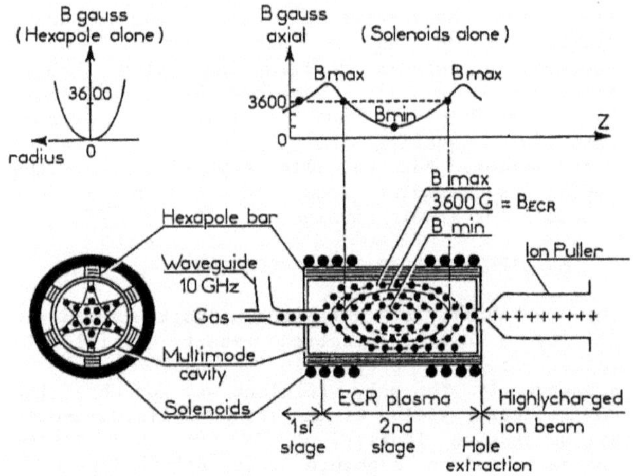

Figure 2 : B_{min} structure and ingredients for a 10 GHz ECRIS obtained with first -and second-stage cascading plasmas. B_{max} is the highest closed magnetic surface (not intercepting the walls) and B_{ECR} is the "egg" shaped surface where electrons are energized.

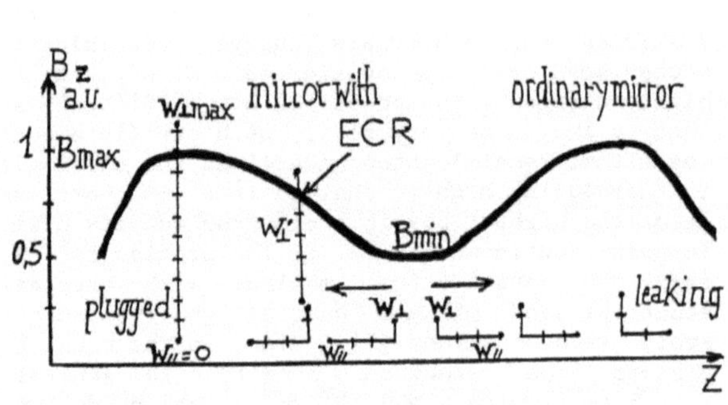

Figure 3 : Idealized ECR plug for electrons with mainly parallel thermal energy at start ($w_{/\!/}/w_\perp \simeq 3$).

THE RELATIVISTIC ELECTRONS AND THEIR ROLE IN THE PHYSICS OF ECR PLASMAS

Stochastic ECR yields in addition to the bulk of electrons inthe keV range, roughly ~ 1 % of relativistic electrons which are longliving because they are nearly collisionless. These hot electrons are capable in a single impact, to eject electrons from the K-shell of a heavy ion. The study of the electronic rearrangements inside this target through recombinations, radiations, Auger cascades, etc... leads to an original approach to these atomic processes. Therefore, not only the ion beams issuing from the source, but also the radiations issuing from this ECR plasma are usefull ingredients for atomic physics. Let us evaluate for instance the K shell ionization rates per cm^{-3} and second. $R \sim n^-_{rel}.n_T\ \sigma\ v^-$ where v^- is the relative impact velocity, n_T the target density (atoms or ions) and σ the single impact ionization cross section. For a krypton plasma, obtained in a standard ECRIS at 16 GHz, with plasma density $n{\sim}10^{12}$ (i.e. $n_T{=}n^+{=}n/{<}q{>}{\simeq}10^{11}cm^{-3}$), $n^-_{rel} \sim 10^{10}$ cm^3 and v^- and σ respectively equal to $8\ 10^9$ cm s^{-1} and 10^{-24} cm^2, we obtain a K shell ionization rate of $R \sim 8\ 10^6$ cm^{-3}/s. This big rate is provided by the impressive equivalent current density of relativistic electrons available in continuous regime : $i = qn^- v^-_{rel} \sim 6$ Amp/cm^2. Many radiative collisions can be generated by these electrons and should be observable. Thus, ECRIS working in continuous regime can deliver very important amounts of photons resulting from the inner rearrangements of the targets after K shell ionizations. For instance, in our group, spectroscopic measurements with SiLi detectors have been performed by Paul Briand on the 16 GHz MINIMAFIOS facility [6] with different gases. When looking on the axis of the ECRIS, the Kα and Kβ lines of krypton are easily monitored with different RF power levels and the FWHM are plotted. (fig. 5) We note that due to the big amount of available photons the Kα and Kβ lines are already detected at very low RF power levels. Thus, the relativistic electrons, as predicted by stochasticity, are generated at low RF power and as the plasma is not yet strongly ionized, the target particles are not only ions but also neutral atoms. Then, the FWHM increase suggests that different states of ions with one K and L vacancy are produced... Thus the study of these photons provides an original approach of inner shell processes occuring in ECR plasmas.

ECRIS SCALINGS : THE IMPORTANCE OF ω and B_{max}.

For improving ECRIS one needs scaling laws. Unfortunately the ECR plasma theory is very complex and handles parameters of which for most of them, we have not a grasp. Therefore we had to look for plasma rules resulting from somewhat simplified equations but involving only controlable characteristics [7]. As a result we found in summary the following relations where B is the average magnetic field, P the RF power, ω the incident RF frequency, q_{opt} the peak charge state of the ions spectrum giving the maximum ion current, I^q the value of this current and M the

Figure 4 : Short Bi^{28+}ion pulse when the ECR plug stops (MINIMAFIOS 18 GHz ; $P_{RF} \sim 3$ kW).

Figure 5 : Kα and Kβ lines FWHM vs RF power in krypton and krypton with oxygen (MINIMAFIOS 18 GHz).

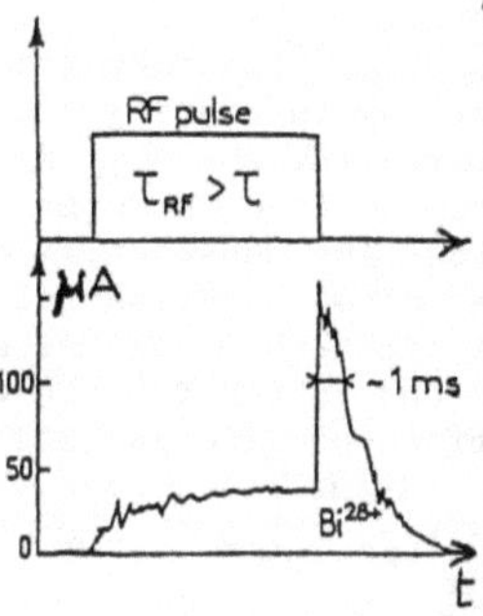

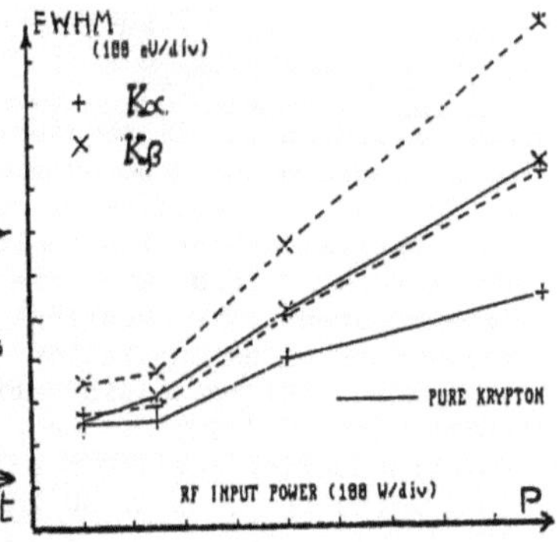

Figure 6 : Recent ECRIS prototypes
a) Superconductor ECRIS built at MSU, b) CAPRICE, c) NEOMAFIOS.

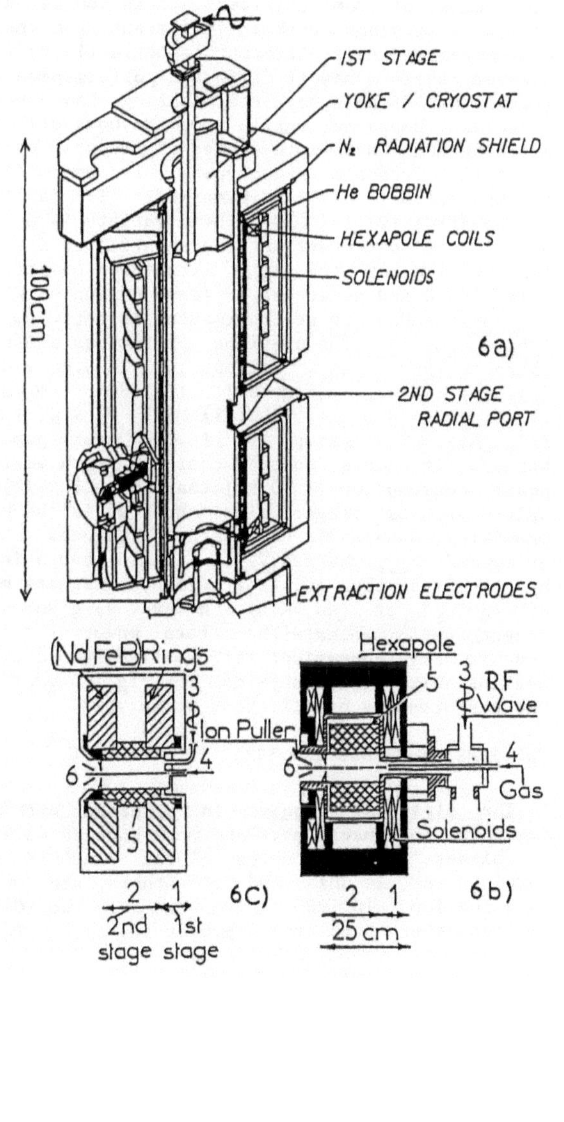

ion's mass : $n v \tau \propto B^{1.5}$; $\boxed{q_{opt} \propto \log B^{1.5}}$ (when only B is variable, $B \approx \frac{1}{2}(B_{ECR} + B_{max})$) $n v \tau \propto \omega^{3.5}$; $\boxed{q_{opt} \propto \log \omega^{3.5}}$ (when ω and B are changed together for ECR). In addition we found $q_{opt} \propto P^{1/3}$ (for RF power densities $\lesssim 1\ W/cm^3$) and $\boxed{I^q \propto \omega^2\ M^{-1}}$. These relations indicate possible but not dramatic improvements by increasing ω and B_{max}. Experiments done by increasing the B_{max}/B_{ECR} ratio and by varying ω up to 18.0 GHz confirm qualitatively these tendencies. The dependences of I^q vs M^{-1} and ω^2 were also experimentally investigated. The scalings, though inaccurate, allow further extrapolations and an ECRIS built by T. Antaya [8] in a joint effort of Michigan State University (MSU) and Grenoble involves superconducting coils for $B_{max} \leqslant 2.5$ T and RF generators for ω up to 30 GHz.

COMPARATIVE ECRIS PROTOTYPES FOR HIGHLY CHARGED IONS AND THEIR PERFORMANCES

More than 30 ECRIS work all over the world. The majority of them was built in small series at Grenoble. Of course, according to the needs and the means of the utilizers, these sources have various sizes and work at different frequencies and correspondingly different magnetic fields [9]. However, their somewhat different performances are always explainable by the scaling laws and no miracles happened until now. Anyhow all the existing ECRIS were more or less influenced by the following Grenoble prototypes : I) Supermafios, II) Cryomafios, III) Minimafios, IV) Caprice, V) Neomafios [2] [3] [10]. Their variety is mainly due to the different materials utilized for the construction of the B_{min} structure comprising solenoids S and hexapoles H. (made of copper wires c, superconductors sc, permanent magnets p, and iron yokes i). The previous prototypes are then characterized as follows : I) S.c. + H.c, II) S.sc + H.sc, (fig. 6a) III) S.c. + H.p, IV) S.c + H.p + i, (fig. 6b) V) S.p. + H.p (fig. 6c). The group IV) is now very popular because it allows compact sources, with smaller power consumption (< 50 kW) than III). But II) in spite of its larger size and cost is the most promising because it can reach the highest values of ω and B_{max} whereas V) is limited to 0.5 Tesla by the presently existing permanent magnets made of ($SmCo_5$) or (Nd,Fe,B). However V) gives very decent performances with a total power of < 5 kW and is very convenient for smaller laboratories dealing with atomic physics. The Fig 7 summarizes the performances of different prototypes.

THE PRESENT TRENDS FOR ECRIS IMPROVEMENTS

A) **<u>The classical trends</u>**. In agreement with the scalings, one uses more and more sources working at higher ECR freqencies. Today 7 ECRIS are working at 14 GHz and 5 others are under construction. In 1985 there were only two ECRIS. On the other hand, the compact sources (CAPRICE type) with important Iron yokes replace little by little the more power consuming MINIMAFIOS type (some of which are already working 10 years without discontinuity). The compact sources also promote the development of B_{min} structures allowing to reach $B_{max}/B_{ECR} > 2$. However for both, - frequency and mirror-ratio - rises, the MSU super-conducting ECRIS will be the most flexible facility, because it allows a complete study of the mirror dependance with variable solenoidal and hexapolar fields whereas for the other ECRIS the hexapoles are made of permanent magnets.. Another trend is the reduction of the first stage and the utilization of only one microwave generator for both stages.

Figure 7 : Typical ion currents I^q obtained with various ECRIS vs charge states q for different elements. *(left)* Gaseous elements ; *(right)* Solid-state elements : (A) NEOMAFIOS 8 GHz, (B) MINIMAFIOS 10 GHz, (C) CAPRICE 10 GHz, normal B mode, (D) CAPRICE 10 GHz $B_{max}/B_{ECR} > 2$, (D') CAPRICE 14 GHz $B_{max}/B_{ECR} > 2$, (E) MINIMAFIOS 16 GHz, (E*) MINIMAFIOS 18 GHz short pulses.

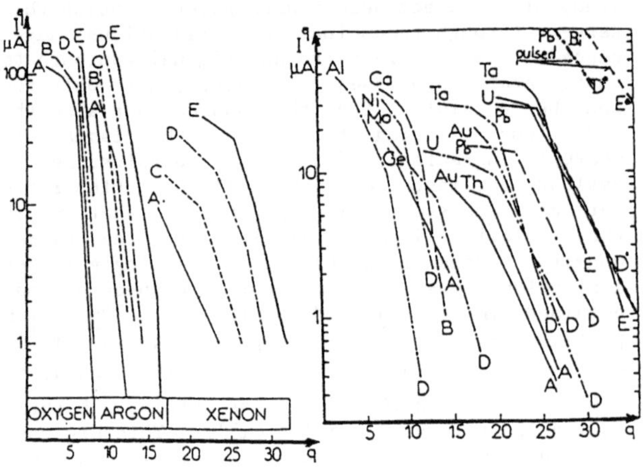

B) **<u>The empirical trends</u>**. Other improvements were obtained by less classical ingredients ; a) gas mixings with a great proportion of additional light gases ; b) different wall coatings and more recently c) additional filaments d) biased electrodes and eventually e) the possibility of harmonic and parametric resonance surfaces [10]. The observed effects are not questionable but the causes are still strongly debated and different explanations are proposed. Therefore, without excluding any particular explanation, I suggest for the time being a provisory but general rationale ; "During the improvement tests, one always tries to maximize [nτ] without departing from the neutrality condition $n^- = \langle q\rangle n^+$. But, in powerfull, low pressure plasmas, one should consider the possibility of electron starvation which can lead to turbulence or relaxations. Of course, an efficient first stage provides an excellent reserve tank of electrons. But the above mentioned ingredients also fight against electron starvation... Ingredient a) decreases the average $\langle q\rangle$ of the plasma and thus lowers the electron drain of the ambipolar diffusion leaving the plasma ; b) c) and d) clearly favor secondary or thermoionic electron emission from some external sources towards the ECR plasma bulk. Thus all of them contribute to the restoration of departing plasma neutrality and avoid collective plasma losses. May be all the above ingredients are just remedies against the deterioration of the plasma quiescence and not positive improvements".

C) The trends towards better comprehension. Until now the ECRIS developments were mainly based on plasma studies done before 1975 [2]. Today one needs more experiments on ECR plasmas in order to improve the theory. A big effort is now launched in our group by G. Melin [6] and a versatile ECR facility with modern diagnostics is under completion. It will surely renovate our thinkings.

CONCLUSION : ECRIS performance are still improving. The upper limits according to our scalings are not yet foreseeable. ECRIS opened new fields of research in particle science and play an important role in the economy of heavy ion physics. They enable considerable reductions in size and cost of the new accelerator projects or considerable energy upgrade of existing machines. The recent development of short pulsed beams with high intensities should improve the possibilities of synchrotrons. In addition the ECRIS plasma also enables some direct studies of Atomic physics.

REFERENCES

[1] Proc. of the Int. Conf. on Physics of multiply charged ions and Int. Workshop on ECRIS. Ed. S. BLIMAN, Journal de Physique 50, Coloque C1 (1989).

[2] R. Geller, Nucl. Sci. NS 23 (2) p. 904 (1976) ; Nucl. Sci. NS 26 (2), p 2120 (1979).

[3] G. Melin et al., Ref [1], p. 673.

[4] P. Briand et al., N.I.M., A 294, p. 673 (1990)

[5] M.P. Bourgarel, P. Sortais et al., 2nd Europ. Part. Accel. Conf., Nice (1990), in press.

[6] G. Melin et al., Rev. Sc. Instr., 61 No 1 II (1990), p. 236.

[7] R. Geller, ECRIS scaling concepts, Proc. 1989 IEEE Particle Accel. Conf. IEEE Catalog No 89CH 2669-0 (Chicago).

[8] T.Antaya, Ref [1], p. 707.

[9] C. Lyneis, T. Antaya, Ref [6] p. 221.

[10] B. Jacquot et al. NIM., A 269 1(1988).

[11] M.A. Liebermann, A. J. Lichtenberg, Plasma Phys. 15, p. 125 (1973)

APPENDIX

Tentative Assumptions for ECRIS scalings to improve and supplement as soon as possible

Until recently all ECR plasma calculations involved the quantitative knowledge of RF fields in the ECR zones, particle temperature, density gradients, velocity distribution, ES plasma potentials, and diffusion times and lengths. But these parameters cannot be evaluated. Therefore we try to use only directly measurable parameters such as magnetic fields B, microwave frequency ω, microwave power P, ion current I^q, ion charges q, and the experimental peak charge of the ion spectrum q_{opt}. Necessarily our assumptions are then oversimplified but the results are verifiable. We make the following assumptions.

1. In steady-state ECR plasmas, electrical neutrality is preserved, $n_e = \langle q \rangle n_i$, where $\langle q \rangle$ is the average ion charge inside the plasma. In addition, we assume that the equation of continuity is respected, i.e. the sum of particle creations and destructions is zero per time unit. At equilibrium one loses $\langle q \rangle$ times more electrons than ions, so one has to create $\langle q \rangle$ time more electrons.

2. The upper electron density is fixed by the cutoff frequency of wave penetration, $n \propto \omega_{co}^2$.

3. In order to maintain electron cyclotron resonance inside the plasma, the magnetic trap field B must vary proportionally to the frequency: $B \propto \omega$.

4. τ_i, the ion lifetime in the B_{min} trap, increases according to ambipolar diffusion laws: $\tau_i \propto B^{\sim 1.5}$, where $B \approx \frac{1}{2}(B_{min} + B_{max})$, i.e. $\tau_i \propto B_{max}^{1.5}$.

5. The energy lifetime θ of a medium is given by the ratio of stored energy to power input. The value of θ inside our plasma is limited by the electron losses. This is because, as a result of ECR heating, the energy is contained in the electrons. But we also know that for every q charged ions arriving at the wall, q plasma electrons follow in order to maintain the electrical neutrality inside the plasma and to neutralize the ion at the wall. Thus we assume that $\theta \propto \tau_i / \langle q \rangle$.

6. The electron energy W_e needed for reaching a given q is proportional to the ionization potential (IP), which increases roughly as q^2 for a given element: $W_e \propto q^2$.

7. In stochastic ECR heating the electron energy W_e increases more or less linearly with the RF power up to a certain limit and we assume that $\langle W_e \rangle$ is proportional to P. But beyond this limit one considers the nonlinear behavior, where $\langle W_e \rangle \propto P^{1/2}$.

8. For simplicity we write that $\langle q \rangle \approx q$ and $\frac{1}{2} m v^2 \approx \langle W_e \rangle$ and that the necessary RF power for achieving $\langle q \rangle$ depends on $P \propto n \langle W_e \rangle V \theta^{-1}$ in the linear case and $P^{1/2} \propto n \langle W_e \rangle V \theta^{-1}$ in the nonlinear case. (V = plasma volume).

Combining the above assumptions 2–6 and 8, one obtains $P \propto \omega^2 q_3 V B^{-1.5}$ and $P \propto \omega^{1/2} q^3 V$ when only P is varied (i.e. ω, B, V are fixed). Then in the linear case one obtains $q \propto P^{1/3}$ and $q \propto P^{1/6}$ in the nonlinear case.

If we consider that the parameters n, W, P, V, and ω are fixed but only B is varied, then according to assumption 4, $n v \tau \propto B^{1.5}$. As we have observed that in many cases q_{opt} varies roughly with $\log n v \tau$, then $q_{opt} \propto \log B^{1.5}$. The same rationale is employed when only ω and B are changed together according to assumptions 2 and 3. One then obtains $n v \tau \propto \omega^{3.5}$ and $q_{opt} \propto \log \omega^{3.5}$.

Independently we write that the extractable ion current is balanced by the plasma ion flow arriving on the meniscus, which gives

$$I^q \propto q n_i v_i \approx q \frac{n_e}{\langle q \rangle} \sqrt{\frac{W_i}{M}} \approx n_e \sqrt{\frac{W_e}{M}\left(\frac{m}{M}\right)}$$

and

$$I^q \propto \omega^2 M^{-1},$$

where we assume that the ion energy W_i is $\sim W_e(m/M)$. This implies that the elastic electron-ion collisions are too rare to allow substantial ion heating. Otherwise one would obtain $I^q \propto \omega^2 M^{-1/2}$.

Supplement to Z. Phys. D - Atoms, Molecules and Clusters 21, S 123-126 (1991)

Atoms, Molecules
Zeitschrift für Physik D and Clusters

Study of the interaction of very slow hollow atoms with a solid

J. P. Briand, L. de Billy, and P. Charles
Laboratoire de Physique Atomique et Nucléaire-Institut du Radium, Université P&M Curie, 4 Place Jussieu, F-75252 Paris Cedex 05, France

J. P. Desclaux, P. Briand, R. Geller, S. Bliman, and C. Ristori
DRF/SPhAt, CEN Grenoble, 85X, F-38041 Grenoble Cedex, France

Abstract

The behavior of very slow hollow atoms traveling inside a solid is presented. The radiative electron capture of the target conduction electrons by these ions has been observed owing to the infrared divergence cross section of this process. The very specific properties of the hollow atoms are used to study their interaction in the first mono-atomic layers below the surface.

I. Introduction

The interaction of singly ionized ions with clean metallic surfaces (1) has been extensively studied in the last two decades. In such cases the capture of a target electron by the ion, through Auger neutralization or resonant transfer processes, can be accurately characterized in well-defined experiments as well as by very precise calculations. More recently there has been a great interest for the very highly charged ions that are now produced by the EBIS or ECR (2) sources. In the complex interaction, where many electrons are captured, the main interest has been focused on the study of the species formed during the collision, at relatively large distances from the surface, and their decay.
These processes have mainly been studied through the observation of the ejected electrons, the absolute number of extracted particles γ, as well as their energy spectrum. Very lately some experimental investigations looking at the X rays emitted by the ions have been reported (3)(4)(5)(6)(7). One of the most difficult questions to answer when studying these interactions is whether or not the ejected particles, electrons, or photons, are emitted prior to reaching the surface. In the past few years, many experiments with relatively charged ions (q~9) at low velocities have been performed looking at the ejected electrons, which led to various answers depending on the nature of the ions and their velocity (8)(9)(10)(11).
In the case of the most highly charged ions, produced at low velocity with enough intensity to allow high resolution X-ray spectroscopic studies, e.g., with the 16 GHz ECR source of Grenoble (2), some clear answers (4)(5)(6) have been provided. The X rays observed with Ar^{17+} and Ar^{18+} ions impinging at normal incidence metallic targets, and presented in Refs (4)(5)(6), have been interpreted as mainly coming from inside the target. A detailed spectroscopic study of the characteristic X rays emitted by the ions led to the conclusion that they were formed in exotic states where most, or all, of the K and L electrons were excited in the M, N... shells (hollow atoms). It is the purpose of this talk to study the interaction of these very slow ions inside a solid by using the original properties of the hollow atoms. We shall then briefly overlook the model used to describe the formation of the hollow atoms and their specific properties, and then present some new and not yet published results on their interaction with a solid.

II. The hollow atoms (4)(5)(6)(7)

The hollow atoms have been observed by looking at the X rays emitted by Ar^{17+} ions, at a velocity of 0.6 a.u., impinging, at normal incidence, a silver metallic surface. These ions were produced by the 16 GHz ECR source built in Grenoble which delivers Ar^{17+} ion beams of a typical intensity of 10^9 particles per second (2). The emitted X rays were detected at approximately 90° with respect to the ion beam direction, using a crystal spectrometer, as shown in Fig.1. This spectrometer was made of a flat mosaic graphite crystal that reflected the X rays on a position sensitive detector of the backgammon type. The focusing properties of this crystal allowed to get all together a good energy resolution (few eV) and a very high transmission.

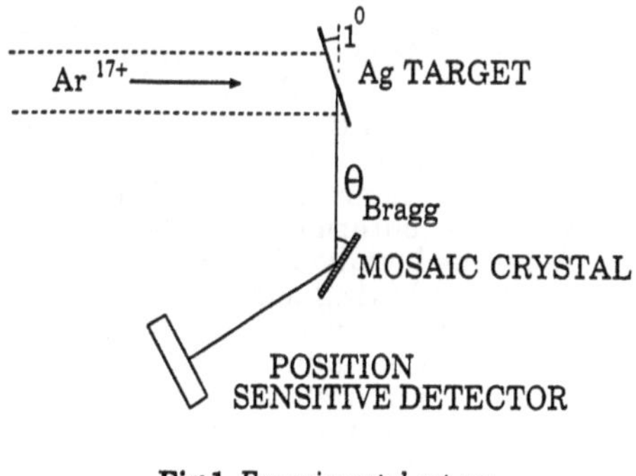

Fig.1. *Experimental set-up.*

While approaching, or crossing the surface of the target, the ion captures many electrons, in a very complex way, depending upon the velocity, the charge, and the projectile incidence angle. After a certain time, one of the electrons of the ion fills the K shell leading to the emission of characteristic K lines that were observed with the above described experimental set-up.
In the considered experimental situation, as described in Refs (3) and (4), many electrons of the solid are extracted by the ion at quite large distances from the surface (25-50 Å). As discussed in detail, these electrons are either ejected into the continuum of the ion, due to the energetics of the collision, or captured into high Rydberg quasi-stationary states. In the last case, the lifetimes of the states are much too long to allow the K shell to be filled prior to reaching the surface. The ions would then hit the surface with many electrons in quasi-stationary states of high principal quantum number and explode while hitting the surface. At distances close to the surface, at around 2-3 a.u., there is an approximate energy matching between the M, N... shells of the ion and, e.g., the N shell of the silver target, or K shell of a carbon deposit whose electrons can be transferred to the ion through the well known resonant neutralization process (Fig.2).

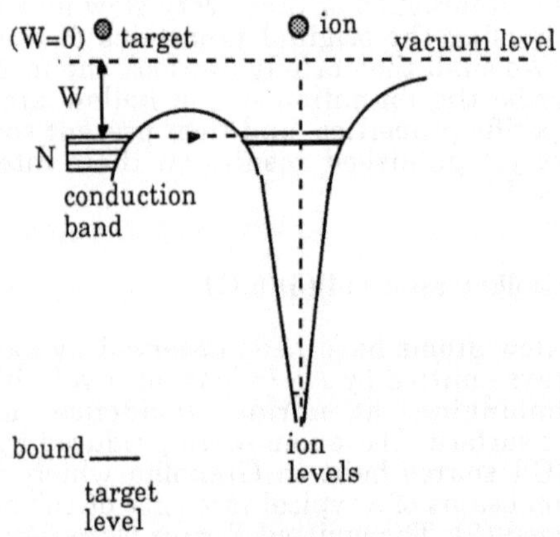

Fig.2. *Resonant transfer of a metal electron, e.g., in the conduction band, to an empty level of an ion (1). In the considered case, charge exchange may arise from target N-shells towards M- and N-shells of the ion.*

The charge exchange cross section for these very highly charged ions (~10^{-15} cm^2) is so large that a great number of electrons can be captured on a single mono-atomic layer of the target (N~3×10^{15} atoms cm^{-2}). The M, N shells of the ions are then very suddenly populated (few 10^{-16} s) when crossing the surface, leading to the formation of hollow atoms, where few, or none (6), K and L electrons are present, while the outermost shells are strongly populated. A detailed study of the properties of these super-excited states having such an extremely large population inversion can be found in Refs (4) and (6). These properties can be summarized as follows.
(i) *The reduction of the lifetimes.* The first property of the hollow atoms is the reduction of the lifetime of the states with an inner core ionized in the K shell, e.g., a heliumlike excited ion whose lifetime $\tau=10^{-14}$ s is very well known. When a heliumlike excited core [(1s)(2p) state] is surrounded by a large number of M, N... electrons (hollow atoms) the non-occupancy of the L shell means the presence of 7 L holes. There is then in these ions a competition for the filling of the single K vacancy and for the 7 L holes. The filling of the L holes (mainly via the LMM, LMN... Auger transitions) is much more probable than that of the K hole, owing mainly to the large number of available L vacancies. Therefore the total decay rate of these states is strongly enhanced with respect to the case of a free heliumlike ion [by a factor up to 30, see Ref. (4)]. The lifetime of the states at any stage of the cascade is then reduced at least by a factor up to 30. This also means that the K fluorescence yield is strongly reduced and at any step of the L shell filling the Kα transition probability is always very small (12). This explains why one observes (Fig.3) all the characteristic Kα lines (KL^n satellites) in presence of any number of L spectator electrons or why the K hole survives so long (e.g., the observation of the $K^1L^8M^8$ states where the K hole is only filled after 8 electrons have, stepwise, filled the L shell).
(ii) A detailed study of the dynamics of the relaxation of these ions (4) led to the conclusion that the lifetime of each of the 8 steps of the cascade filling the L shell is quite constant (~ $4\text{-}7 \times 10^{-16}$ s). The mean time needed for an electron to fill **one** L hole is then of the order of 5×10^{-16} s and roughly independent of the number of L holes (or electrons).
(iii) Fig.3. shows the characteristic Kα spectrum emitted during the relaxation of these ions. The eight lines displayed on this spectrum correspond to

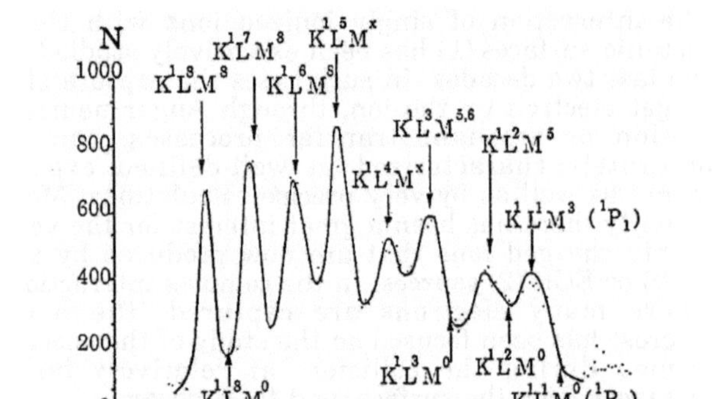

Fig.3. *Ar Kα spectrum observed during the interaction of Ar^{17+} ions with silver.*

the eight steps of the L-shell filling (each line being a satellite of the Kα transition in presence of any number n of L-spectator electron labeled KL^n). These time-ordered lines then constitute so to say the tops of an internal clock whose unit of time is of the order of 0.5 fs. They have been used for instance to study some fast processes such as the hypersatellite-satellite cascade with Ar^{18+} ions (7). Whereas the ion travels inside the solid at a roughly constant velocity, these lines may also be considered as some boundary marks along the ion trajectory of about ~ 5-10 Å spacing. The energy of the KL^n lines depending on the penetration depth of the ion into the solid, it is possible to study selectively the interaction of the ion along the first few tens of angstroms of its travel below the surface, a property used in the experiments presented in this talk.

III. The interaction of hollow atoms with a solid

The internal structure of an ion inside a dense medium is affected by the screening induced by the (free) electrons and sometimes the lattice ions. This electronic screening is known to lower all the electronic binding energies of the ion; lowering usually roughly constant for the inner, K,L... shells. The Kα line energy, for example, is then not appreciably changed by the ion environment. This allowed, in the considered situation, the study of the hollow atom intrinsic behavior, even inside the solid, by looking at these Kα lines (Fig.3). It wont be the case for the transitions such as Kβ or Kγ that involve the outermost electrons whose properties are more affected by the environment than the inner ones. We present in Figs 4 and 5 the Kβ,γ... spectra corresponding to the Kα one.
For experimental reasons this Kβ,γ spectrum that extends over a larger energy range than the Kα one has been recorded in two separated experiments. The spectrum in Fig.4 shows the last four lines of the temporal sequence filling stepwise the L shell (KL^8... KL^5 groups corresponding to the states with 5 to 8 L spectator electrons). The spectrum shown in Fig.5, which represents the first four lines of the

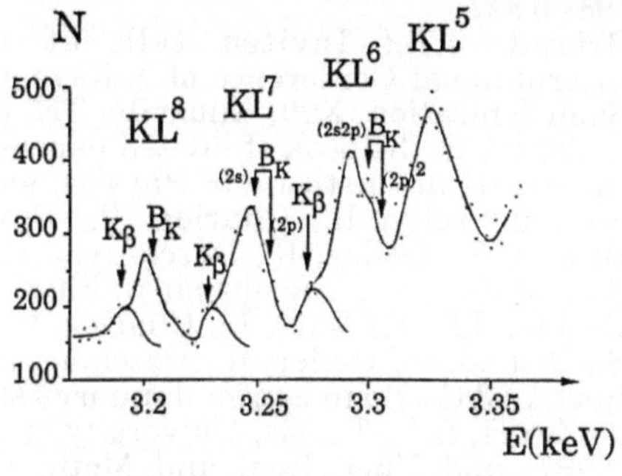

Fig.4. *Kβ,γ... lines of KL^8 to KL^5 states.*

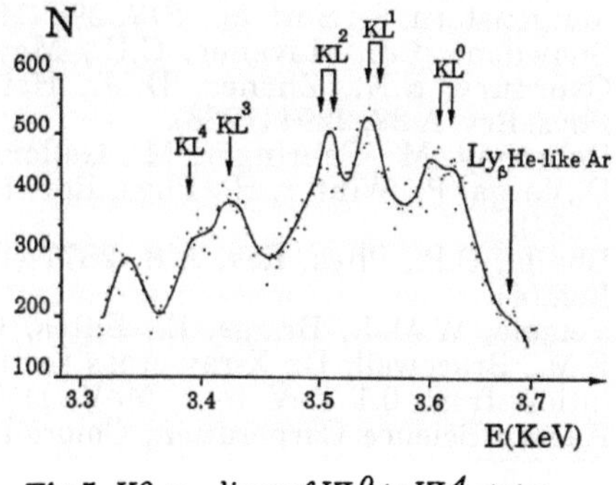

Fig.5. *Kβ,γ... lines of KL^0 to KL^4 states.*

cascade filling the L shell, looks much more complicated because of the overlap of many Kβ and Kγ lines. We shall only discuss in this talk the last four lines displayed in Fig.4 which can be unambiguously identified.
The electronic configuration of the considered KL^n states being well known one can predict the intensity of each of the Lyman β lines corresponding to the satellite Lyman α ones presented in Fig.3. The energy of the predicted Kβ lines is shown in Fig.4 (vertical arrows) and the intensity of the first of these lines (K^1L^8) is represented by the dashed curve. One can observe in Fig.4, associated to each of the Kβ lines, another broad line, more intense, having always an energy slightly smaller than the corresponding B_K binding energies.
In order to understand the origin of these rays, we must first concentrate on the study of the line accompanying the KL^8M^8 group and corresponding to a neutral argon atom ionized in the K shell. The energy of this broad line is exactly equal to that of the N→ K transition which does not exist in neutral argon, but due to its large intensity (and width) it cannot be assigned to such a transition (the intensity of a Kγ line is always smaller than that of the Kβ in contradiction with what is observed in Fig.4; should these lines be some true Kγ lines the ion would then be exceedingly negative : $\bar{q}>10$) Therefore one can be led to think that if the K shell is not filled by any of the bound electrons, this transition might result from the filling of the K shell by a free electron of the solid. In this hypothesis, one should observe the characteristic line following a radiative electron capture which must lie at an energy $h\nu=B_K+E_e$ (E_e energy of an electron at rest in the laboratory frame as seen in the frame of reference of the ion). A simple calculation, described in more details in Ref. (14), and taking into account the lowering of the binding energy B_K of the ion in the solid, predicts that this line must be equal to that of the N →K transition, in agreement with our experimental result. To support qualitatively this interpretation we have also calculated the REC rate for this line, whose intensity may be deduced from the photoexcitation cross section (13), through the microreversibility principle (radiative electron capture is by time reversal the same process than photoionization). The intensity of the REC line can easily be measured in comparing it with that of the Kβ line which can be normalized to the Kα one. It is then possible to determine the absolute REC rate Nσ, i.e., the number of events per ion (N being the number of target electrons per cm^2 calculated as described below). This number 2% is found to be in reasonable agreement with the number calculated (1.5%) from the photoionization cross sections [see Ref. (14) for further details). Similar calculations for the other KL^n states have also been performed and found in good agreement with the experimental data presented in Fig.4. The shape of a REC line reflects, in theory, the energy distribution of the free electrons in the solid, i.e., the structure of the conduction band. The width of the REC lines, of about 10-13 eV, observed in Fig.3, is in good agreement with this assumption. The exact shape of these lines is however, in the considered case, difficult to predict. The relative velocity of the electron to be captured and the ion, which determines the value E_e of the REC line, actually results from the composition of the velocity of the ion (well-defined in modulus and orientation) and that of the electrons of the metal, which are randomly oriented, and whose momentum distribution is well known. A much better analysis of this momentum distribution could then be obtained by using slower ions in order to avoid the addition of the the two velocities. In this case, the REC cross section scaling like E_e^{-1} would conse-

quently diverge, leading to more intense lines. At the lowest velocities, however, the ion velocity is much lower than that of the electrons in the conduction band, and the ion may be considered at rest. Therefore the "projectile" ion would form some quasi molecular compound with the lattice ions and the considered transition would be a co-operative transition in the transient molecule.
The interaction of the hollow atoms in a solid largely differs in two ways from what is usually observed. The radiative electron capture is usually mostly studied with the very energetic ions produced by high energy ion accelerators, such as LINAC, Van de Graaf, cyclotrons or synchrotrons. The ion velocity is then much larger than the velocity of the electrons in the solid. The electronic cloud appears for the ion as a frozen electron gas. In the case under study, the electrons have a velocity equivalent to that of the ions, and the electron gas may quickly relaxe when the ion is entering the solid. The electrons, as in the case of a sudden inner-ionization of an atom of the solid, are quasi-instantaneously attracted by the ion in order to screen it. The electron gas is, in these circumstances, active instead of being frozen as in the case of very energetic ions. The number of target electrons to be captured by the ions is then the number of electrons required for the neutralization. We have used this property for the calculation of the REC rate. This also leads to the interesting characteristics of the REC rates $N\sigma$ which must linearly increase with the charge of the ions (N: the number of available electrons to be radiatively captured being proportional to the number of electrons screening the ion, i.e. the charge of the ion). This prediction is easily verified by looking at the relative intensity of the REC lines in Fig.4 [the KL^n lines decrease linearly in intensity with n]. The REC rate is therefore independent of the nature of the target metal (e.g., the number of the electrons in the conduction band). This has been recently experimentally checked on various metals (14).
In the most useful cases, the ions that are ionized in their outermost shells are in their ground state and do not spontaneously radiate. In the case of hollow atoms that are core excited some characteristic X rays are emitted by the ions along their trajectory. The main property of these hollow atoms is that the energy of the emitted lines, as quoted above, depends on the penetration depth of the ion in the solid: the KL^1 line, emitted at the beginning of the L shell filling (close to the surface where the hollow atom is formed) has a different energy from that of the KL^8 lines (Fig.4) emitted at the end of the life of the hollow atom (deeper in the solid). One can expect in this situation to have different signatures of the ion interaction at various penetration depths (lines of various widths reflecting different band structures) in the first tens of mono-atomic layers below the surface, for example. A useful property in the study of surface substrats.

IV. Conclusion

In a more general way, the fact that most of the lines, composing the broad "Kβ" peak observed with a SiLi detector (3)(4)(5) identified as the radiative electron capture lines, proves once more that the decay of the hollow atoms occurs inside the solid. In more complex cases, with lighter elements at lower energies or at grazing incidence (15)(16)(17), the competition between the emission of the Kβ and Kγ lines and the radiative electron capture (both possible at such a low energy interaction) allows, in high resolution experiments, to answer the question whether or not the radiative decay takes place prior to reaching the surface.

References

(a) Unité associée au CNRS No771.

1. Hagstrum, H.D.: Phys. Rev. **96**, 336 (1954).
2. Geller, R.: Invited talk, this conference.
3. Donets, E.D., Nucl. Inst. and Meth.: B **9**, 522 (1985).
4. Briand, J.P., de Billy, L., Charles, P., Essabaa, S., Briand, P., Geller, R., Desclaux, J.P., Bliman, S., and Ristori, C.: Phys. Rev. Lett. **65**, 159 (1990).
5. Briand, J.P., Charles, P., Essabaa, S., Mayo, M., Bliman, S., Briand, P., Geller, R.: Proceedings of the XVth International Conference on the Physics of Electronic and Atomic Collisions, Brighton (UK) 1987, Abstracts of contributed papers, edited by J. Geddes, H.B. Gilbody, A.E. Kingston, C.J. Latimer and H.J.R. Walters (XV ICPEAC), Brighton, UK, 1987 p 827.
6. Briand, J.P.: Invited talk at the 16th International Conference of X-Rays and Inner-Shell Ionization, X90, Knoxville TN, July 1990, to appear in the book of invited papers edited by the American Institute of Physics, and Briand, J.P., de Billy, L., Charles, P., Essabaa, S., Briand, P., Geller, R., Desclaux, J.P., Bliman, S., and Ristori, C.: to appear in Phys. Rev. A.
7. Briand, J.P., de Billy, L., Charles, P., Essabaa, S., Briand, P., Geller, R., Desclaux, J.P.: Phys. Rev. A **43** (1991) to appear January 1st issue.
8. de Zwart, S.T.: Thesis, University of Groningen (1987) and Nucl. Inst. and Meth. B **23**, 239 (1987).
9. de Zwart, S.T., Drentje, A.G., Boers, A.L., Morgenstern, R.: Surf. Sci. **217**, 298 (1989).
10. Snowdon, K.J., Havener, C.C., Meyer, F.W., Overdury, S.H., Zhener, D.M., Heiland, W.: Phys. Rev. A **38**, 2294 (1988).
11. Delaunay, M., Fehringer, M., Geller, R., Hitz, D.,Varga, P., Winter, H.: Phys. Rev. B **35**, 4232 (1987).
12. Bhalla, C.P., Phys. Rev. A **8**, 2877 (1973), and Ref. (4).
13. Veigele, W.M.J., Briggs, E., Bates, L., Harry, E.M., Bracewell, B.: X-ray cross section compilation from 0.1 keV to 1 MeV, DNA 2433F, Kaman Science Corporation, Colorado Springs, C0 80907.
14. Briand, J.P., et al: to be published.
15. Donets, E.D., Dudnikov, V.G., Dunin, V.B., Kartashov, S.V.: Contributed paper, this conference.
16. Andriamonje, S., Lamy, T., Brenec, A., Simionovici, A., Andrä, J.: Contributed paper, this conference.
17. Schulz, M., Cocke, C.L., Stöckli, M., Hagmann, S.: Contributed paper, this conference.

Supplement to Z. Phys. D - Atoms, Molecules and Clusters 21, S 127 (1991)

Atoms, Molecules
Zeitschrift für Physik D and Clusters

New trends in ion-surface interactions

E. S. Parilis
Arifov Institute of Electronics, Tashkent 700143, USSR

A brief discussion of the present situation and new trends in the investigation of sputtering from solid surfaces under bombardment by slow multiply charged ions is given. The observation and further investigation of sputtering via "Coulomb explosion" require the use of highly charged ions of as low energy as possible and insulators or semiconductors with controlled low conductivity. The exploration of this phenomenon will provide a method for the emission of clusters and gives an opportunity to study collisions of highly charged ions with clusters in merged or crossed beams.

When a multiply charged ion approaches a solid surface several step-by-step ion neutralizing Auger transitions transfer a large positive charge to the surface atoms neighbouring the point of projectile impact. In case of a non-metallic target (an insulator or a semiconductor with large resistance) "Coulomb explosion" of the charged microscopic domain was predicted to result in an additional sputtering of the solid even below the threshold of the ordinary ion beam sputtering [1] .

The first experimental proof of the Coulomb explosion mechanism [2-4] was followed by a more detailed calculation [5] and has found a further verification, at least for an increase of secondary ion emission with increasing ion charge [6-9]. With respect to the total sputtering yield the experimental data were contradictory. In Ref. [6] this yield was observed to be independent of the ion charge state but in Ref. [7] it was found to be increasing with the ion charge state. The secondary ion emission in Ref. [6] was not as large as in Ref. [4] although the same experimental conditions were claimed in both measurements. A more sophisticated theoretical analysis [10] has shown that the Coulomb explosion sputtering is the result of a competition between the Auger cascade neutralizing the multiply charged ion and the neutralization of the positively charged surface domain by the conductive electrons.

1. To record finally and unambiguously the proper phenomenon of sputtering via "Coulomb explosion" it is necessary to use:

   (a) highly charged ions with total neutralization energy sufficient to cover the losses via Auger electron emission, Coulomb repulsion, bonds breaking and creating recoils inside the solid;

   (b) as low energy of the ions as possible, preferentially with 10 - 100 eV per one charge unit;

   (c) an insulator or a semiconductor with small conductivity. Controlled conductivity changes by changing temperature or doping would give the opportunity to record the dependence of Coulomb sputtering on target conductivity. For example, a drastic change of the sputtering yield in the high temperature superconductivity point is expected.

2. The experimental study of the emission of large clusters under surface bombardement by slow highly charged ions, which was never done, could give a direct evidence of the "Coulomb explosion" both by the relatively high ionization degree of the emitted clusters and by their fragmentation in flight.

3. Experiments with merged or crossed beams of large clusters and highly charged low energy ions would give the unique possibility to study Coulomb explosion of clusters which accumulate a large positive charge within 10^{-14} - 10^{-15} sec during an encounter with a highly charged ion via multiple Auger neutralization with cross-sections as large as 100 Å^2.

References

1. Parilis, E.S. Proc. IX International Conf. on Phenomena in Ionized Gases, Bucharest, 1969, p. 94, Atomic collisions in Solids Ed. by Palmer, North-Holl.Publ. Comp., 1970, p. 324
2. Arifov, T.U., Vasiljeva, E.K., Gruich, D.D., Kovalenko, S.F. and Morozov, S.N., Izv. AN SSSR,, Ser. fiz. 40, 2621 (1976)
3. Radjakov, Sh.S., Rakhimov, R.R. and Abdusalamov, D., Izv. AN SSSR, ser. fiz. 40, 2543 (1976)
4. Morozov, S.N., Gruich, D.D. and Arifov, U.A., Izv. AN SSSR, ser.fiz. 43, 612 (1979)
5. Bitensky, I.S., Murachmedov, M.N. and Parilis, E.S. Sov. Phys.Tech.Phys., 24, 618 (1979)
6. DeZwart, S.T., Fried, T., Boersma, D.O., Hoekstra, R., Drentje, A.G. and Boers, A.L., Surf. Sci., 177, L 939 (1986)
7. Eccles, A.J., van der Berg, J.A., Brown, A. and Vickerman, J.C., Appl.Phys.Lett. 49, 188 (1986)
8. Weathers, D.L. and Tombrello, T.A., Prior M.H. and Stokstad, R.G., Tribble, R.E. Nucl.Instr. and Methods in Physics Research B 42, 307 (1989)
9. Schmieder, R.W. and Bastasz, R.J., Nucl.Instr. and Meth. in Phys.Res. B43, 318 (1989)
10. Bitensky, I.S. and Parilis, E.S., Journ. de Physique 50, C2-227 (1989)

Supplement to Z. Phys. D - Atoms, Molecules and Clusters 21, S 129-133 (1991)

Atoms, Molecules
Zeitschrift für Physik D and Clusters

What ions experience on their way into a metal surface

H. Winter
Institut für Allgemeine Physik, Technische Universität Wien, Wiedner Hauptstraße 8-10, A-1040 Wien, Austria

Abstract
We discuss interaction of slow multicharged ions with metal surfaces, by considering recent experimental results for emission of electrons and soft x-ray photons as well as charge states of scattered particles. Simple concepts are developed to explain the main processes which take place before, during and after projectile impact on the surface.

I. Introduction

Bombardment of solid surfaces with fast heavy particles initiates a number of reactions which may also be of great practical interest, as elastic and inelastic scattering, sputtering, electron emission and desorption. As a rule, such processes require *close collisions* of the projectile with the surface atoms. However, if the projectiles carry sizeable potential energy due to their excitation or ionization, they may also strongly interact with a surface already *before* touching it. The resulting processes, in particular the socalled *potential electron emission*, become generally more important for higher ion charge states, i.e. for multicharged ions ("MCI"), cf. /1,2/. Powerful MCI sources /3/ and the now commonly available UHV technology have greatly facilitated the investigations of MCI-surface collisions, which are of interest e.g. for plasma-wall interaction in magnetic fusion devices. Furthermore, such collisions create fancy *multiply-excited atomic systems* and may also lead to excitation of *plasmons, "Coulomb explosion"* and *shock wave production* in the target /4/.

Effects caused by the penetration of fast MCI (projectile velocity v >> 1 a.u.) into solids have been discussed e.g. in refs. /5,6/. We confine ourselves to *lower impact velocities* ($v \leq 1$ *a.u.*), where the influence of the potential energy of projectiles may become much more important than that of the kinetic energy. We consider primarily the time interval during approach of an MCI until it actually hits the surface. By discussing various results from recent experimental studies we try to present a consistent picture of processes

II. The MCI-metal surface complex

In fig. 1a we show the situation of a MCI with given impact velocity v and charge state q far away from a metal surface, which may be characterized in the *jellium approximation* by its surface work function W_ϕ and Fermi energy E_F. At the Fermi edge the quasi-free metal electrons assume the Fermi velocity v_F. Coming closer to the surface, the MCI will eventually reach a distance $d_q \approx 3\,q$ (Å) /7/ where electrons from the Fermi edge may start to tunnel resonantly into empty states on the MCI (cf. fig. 1b). The sequence of further processes will now depend on q, v_F and the MCI velocity v. We will neglect finer details as for example the shift of MCI states with distance from the surface, acceleration toward the latter because of the projectile mirror charge and the influence of the MCI velocity component parallel to the surface /8,9/.

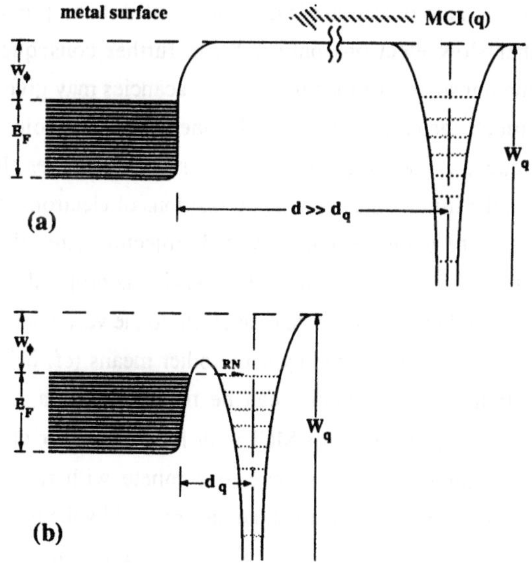

Fig. 1: Situation for a MCI approaching a metal surface.

As long as v stays below v_F (the latter being of the order of 10^6 m/s), a number of electrons can tunnel into the MCI before it actually arrives at the surface. Consequently, multiply-excited states are transiently populated on the MCI and then become subject to rapid autoionisation. The autoionisation rates depend critically on the particular electronic configurations (consider, for example, ref. /10/ for the case of Ar ions), but generally remain well below the rates for resonant transitions of electrons from the metal into the MCI. Therefore, a slowly approaching MCI, once having reached the distance d_q, will become and stay neutral until its impact on the surface. Consequently, the total number of electrons emitted by autoionisation processes during the available time of approach will markedly depend on v, as readily shown for moderate q-values /1/, and more recently also for higher charged ions, cf. fig. 2.

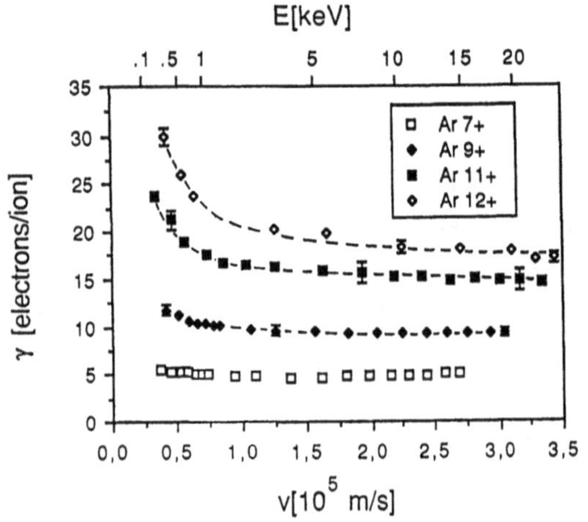

Fig. 2: Electron yield for multicharged Ar ions impinging on clean W (from ref. /11/).

Whereas autoionisation of the highly excited states produces rather slow electrons only /2,8/, in further consequence some transiently formed inner shell vacancies may give rise to much faster electrons. Such inner shell vacancies are produced if the original MCI carries an open inner shell, since the initial quasi-resonant transitions of electrons from the surface populate highly excited projectile states almost exclusively. Only after some electrons in the projectile have cascaded down or upper states adjacent to the vacancy levels have become populated by some other means (cf. below), the inner shell vacancies can be recombined via Auger electron transitions. For MCI with higher nuclear charge these Auger transitions have to compete with radiative transitions, the rates of which are determined by the relevant fluorescence yields which also depend on the actual projectile configurations, cf. /10/.

III. MCI - induced electrons from a metal surface

Considering first *electron emission*, useful informations on slow MCI-surface impact are offered by the *total electron yield* (usually denoted by γ), the *electron energy distribution* (to be denoted by "ED"), and the set of probabilities W_n for given numbers n of electrons emitted from single MCI impacts (viz. the *electron emission statistics*, "ES", cf. /12/). Total electron yields at low impact energy fall with increasing E, then level off and finally rise again because of the onset of *kinetic emission* (KE), which can only arise after the projectile has hit the surface. KE contributions may be subtracted from the total ("apparent") yields if the former can be assumed as independent of the ion charge state /13/. An unambiguous separation of contributions made before MCI impact (commonly named *potential emission* - "PE") and the KE contribution is feasible by analysing the electron emission statistics (ES) for MCI. Fig. 3 shows corresponding data for Ne^{q+} (q= 0,1,2,3) impact on clean gold. If the MCI becomes fully neutral before hitting the surface, the subsequent KE is produced by a neutral projectile and the ES of the preceding PE contribution will thus not be related to the ES for KE. The latter can be precisely determined with originally neutral projectiles. With this information at hand, the ES for PE is easily obtained by deconvolution of an ES measured with neutral particles out of the ES measured for a particular MCI. Fig. 3(c) shows, however, that this deconvolution procedure works only properly for Ne^{q+} impact energies of up to 6 keV (corresponding to projectile velocities $v_n \leq 2{,}4\times10^5$ m/s). At higher v, appearance of "negative probabilities" in the deconvoluted ES shows onset of an interrelation of PE- and KE contributions, i.e. at $v > v_n$ the MCI can no more become completely neutral before hitting the surface. Therefore, the determination of v_n delivers the necessary neutralisation time $t_n \geq d_n/v_n$. When comparing total electron yields measured for given projectile species and -velocities but different q, it turns out that a predicted proportionality of γ with total potential energy W_q of the MCI /2/ only holds up to certain q limits, above which the yields clearly increase slower with W_q than for the lower q /11,13/. This change is related to the already discussed onset of inner shell vacancy formation in the approaching projectile, as first clearly observed in /15/. The resulting emission of a fast electron requires a comparably large fraction of the total potential energy W_q, which at low q can be utilized to eject a number of rather slow electrons, such providing a considerably larger contribution to the total electron yield.

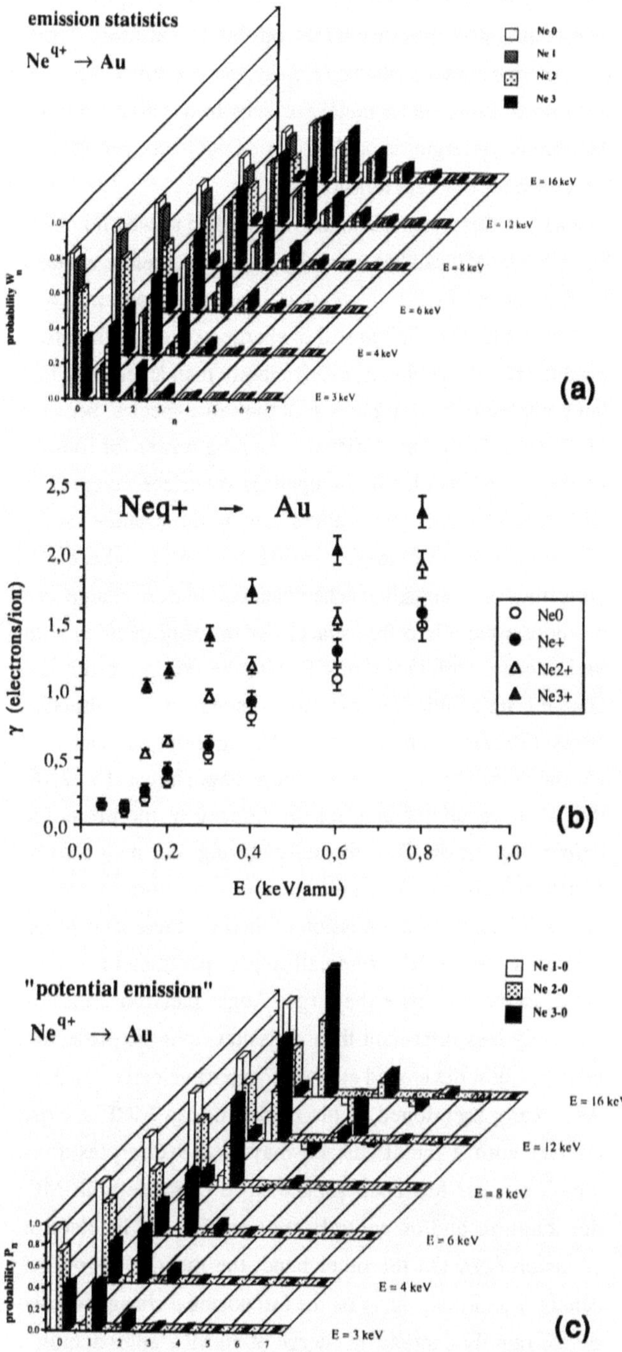

Fig. 3: (a) ES for impact of Ne^{q+} (at any given n, bars from left to right belong to q = 0,1,2,3, respectively) on clean gold /14/.
(b) Total yields derived from the ES shown in (a).
(c) ES for exclusive potential emission (at any given n, bars from left to right belong to q = 1,2,3, respectively), for different ion charge, as obtained by an unfolding procedure. Note the onset of "negative probabilities" above E = 6 keV.

In electron energy distributions measured for slow MCI impact on clean metal surfaces /1,2/ the bulk of ejected electrons appears at rather low energy. This is now well understood, because the autoionisation among the highly excited states is the principal source for emission of electrons, see also /8/. However, structures at higher electron energies have been found e.g. for impact of slow He^{2+} and metastable $He^{+}(2s)$ on Ni /16/ or for Ne^{2+} and Ar^{2+} on W /17/ and Cu /18/. These faster electrons result from autoionisation of transiently produced doubly excited neutral species via resonant capture of two electrons from the surface. It became gradually clear that MCI are not neutralised in a step-by-step manner as earlier assumed /1/, but rather via the already described multiple-resonance transitions with subsequent autoionisation, as first pointed out in /2/. Experiments with higher ion charge states delivered further information on these transient multiply excited states as the source of electron emission. The neutralisation is started by quasi-resonant transitions which populate outer shells almost exclusively. Lateron the resulting inner shell vacancies will decay via Auger transitions and give rise to emission of fast electrons. This has been studied for MCI with open inner shells, eg. N^{6+}, O^{7+} and Ar^{9+} /19,20,21/. As an example, in fig. 4 we present an ED measured for N^{6+} impact on clean W /20/. A comparably small amount of rather fast electrons can be observed in an energy region roughly coincident with the well established KLL-Auger electron energies as being measured for N atoms with one K-shell vacancy.

Positions, shapes and impact velocity dependences of such Auger electron peaks have recently been studied in more detail /15,22-24/. The Doppler-shifted electron energies observed at different electron take-off angles proved that at low MCI velocity the fast electrons are ejected from the projectiles still on their way *toward* the surface. Slight shifts of electron peaks from the expected values (see above) may be explained by the complete neutrality of the emitting projectile, whereas free atoms with one inner shell vacancy are singly ionized. In addition, a repulsive action on emitted electrons by the mirror charge of the incompletely shielded ion core near the surface has been postulated /22/.

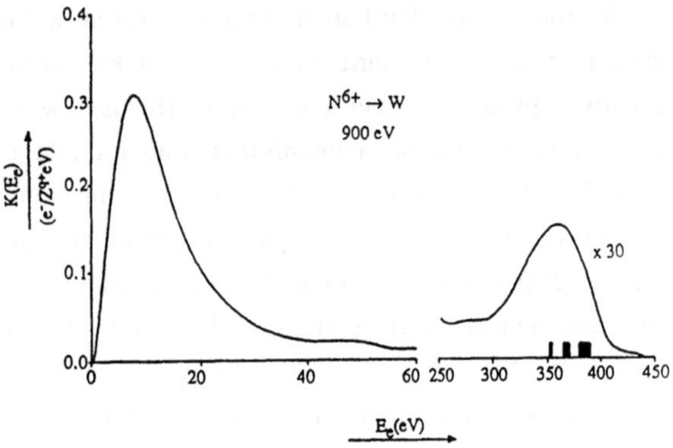

Fig. 4: ED for impact of 900 eV N^{6+} on clean W /20/. The peak around 350 eV shows KLL- and/or KLM-Auger electrons from neutralised projectiles with one K-shell vacancy.

For impact of H-like MCI (C^{5+}÷Ne^{9+}), together with KLL-Auger electrons also much slower electrons from LMM-, LMN-, etc. Auger transitions can be observed, and the latter can be found for lower q as well /23/. If both groups are produced, their relative intensities indicate that the L-shells are not exclusively filled via cascading from the initially populated highly excited states, but to a comparable extent also via resonant transitions from deeply bound target states. The yield of the high energy Auger electrons was shown to follow the inverse MCI impact velocity /24/. If termination of resonance neutralisation of the MCI upon surface impact is assumed, from this impact-velocity dependence a characteristic time for population of the L-shell may be deduced, since from the latter the observed KLL-Auger electron emission is fed. This characteristic time compares quite well with corresponding data derived by means of ES deconvolution, as has been explained earlier.

Inner shell-vacancy formation during MCI approach toward a metal surface has already been observed in scattering of 20 keV Ne-, Ar- and Kr MCI on clean W /25/. Whereas for charge states up to q=8 only singly and doubly charged species appeared in the scattering peaks, for q=9 also triply charged ions could be observed. This can be related to an inner shell vacancy which has been produced in the originally nine-fold charged projectile during its approach toward the surface and could survive scattering, after which the resulting Auger electron emission causes the scattered particle charge state to increase by one unit.

There are also measurements on MCI scattering from gas-covered metal surfaces /26/. Systematic investigations with preferrably clean metal surfaces in a wider range of MCI impact energies should be carried out to deliver further details on the relevant transitions.

The earlier discussed ES studies showed up to which impact velocity an MCI can become fully neutralized before actually hitting the surface. Above that limit the projectile will penetrate into the surface still in charged state and therefore produce KE somewhat more efficiently than an equally fast but completely neutral particle of the same species, because of the slightly higher electronic stopping power for ionized vs. neutral projectiles in the uppermost surface layers /27/. Only these uppermost layers are of importance, since electrons released deeper inside the solid, where the projectiles have already reached their equilibrium charge, can no more escape into free space. For faster projectiles with higher q the correlation of KE with the original projectile charge state should become more distinct. A new situation arises if during its approach the MCI develops an inner shell vacancy (see above). Several groups have shown /15,22,28/ that such vacancies can not only decay in the projectile before, during or after impact/scattering, but may also be transferred in close collisions into the target atoms, which in turn will decay under emission of their characteristic Auger electrons. The necessary small impact parameters of such close collisions make the target Auger electron emission generally less important than emission from the projectile, which is best developed at lowest impact velocity /15,22/.

As already mentioned in the beginning, an MCI in close contact with a metal surface may also excite plasmons, which process has been suggested to compete with MCI deexcitation and/or neutralisation ending up in electron emission /29/. On the other hand, the metal electron gas cannot react immediately on the fast potential changes caused by the rapidly decreasing charge of an MCI approaching a surface, i.e. a finite "response time" of the electron gas has to be expected. Indeed, from ED measured for keV He^{2+} impinging under grazing incidence on clean Cu(110), a finite response time of the order of 10^{-16} s has been derived /30/.

IV. MCI-induced soft x-ray photons

If for MCI with medium- to high Z inner shell vacancies develop as the result of resonance neutralisation near a surface, they may also be filled via radiative transitions. This results in soft x-ray emission, as first shown by Donets et al. for impact of Ar^{17+} (v ≈ 3x10^5 m/s) on Be and Cu /31/, and for Kr^{35+} (v ≈ 1,2x10^6 m/s) on Cu /32/.

The x-ray photons have been observed with solid state detectors, as also applied for another study on impact of Ar^{17+} and Ar^{18+} (v ≈ 1,1x10^6 ms) on Al /33/. In all cases the achievable detector energy resolution did not permit a precise analysis of the projectiles electronic configuration during x-ray emission. This could recently be drastically improved with a high resolution crystal spectrometer for observing Ar^{17+}-Ag collisions at an impact velocity of ca. 1,4x10^6 m/s /34/. In place of the formerly as "K_α" classified broad x-ray line now ten adjacent transitions could be resolved and assigned to radiative transitions into the K-shell vacancy from higher states being distributed in different configurations. The similar intensities of these lines have been explained by the decisive role of the L-shell population for the rates of both KLM- and LM≥M Auger transitions. Consideration of involved Auger transition rates shows, however, that the observed x-rays are emitted long after the projectile has penetrated into the surface, and that even the earlier population of the L- and M-shells has already taken place inside the solid. The situation in such collisions is rather complicated, since projectile vacancies carried into the surface may also be transferred to the target atoms. Further experiments of similar kind but at lower impact energies will provide an extended time interval for studying the principal MCI neutralisation processes.

Summary

In the last few years several studies have been devoted to the impact of slow multicharged ions on metal surfaces. They provided a better understanding of the rather intricate processes taking place during such collisions. More systematic work using all experimental techniques described in this paper in a wider range of ion charge states and impact energies may eventually lead to a detailed, quantitative description for these reactions, which also offer several rather promising applications.

Acknowledgments

I am grateful for the continuous cooperation with Drs. F. Aumayr, G. Lakits and P. Varga in the context of this work, which has been supported by Austrian Fonds zur Förderung der wissenschaftlichen Forschung (Projekt Nr. P6381PH) and Kommission zur Koordination der Kernfusionsforschung at the Austrian Academy of Sciences.

References

1. Hagstrum, H.D.: Phys.Rev. **96**,325;336(1954)
2. Arifov, U.A., Kishinevskii, L.M., Mukhamadiev, É.S., Parilis, É.S.: Sov.Phys.-Tech.Phys. **18**,118(1973)
   Arifov, U.A., Mukhamadiev, É.S., Parilis, É.S, Pasyuk, A.S.: Sov.Phys.Tech.Phys. **18**,240(1973)
3. Geller, R.: Rev.Sci.Instrum. **61**,659(1990)
   Donets, E.D.: Rev.Sci.Instrum. **61**,225(1990)
4. Bitenskii, I.S., Parilis, É.S.: J. de Phys. **50**,C2-227(1989)
5. Datz, S.: Physica Scr. T 3,79(1983)
6. Clouvas, A., et al.: Phys.Rev. B **39**,6316(1989)
   Koschar, P., et al.: Phys.Rev. A **40**,3632(1989)
7. Apell, P.: Nucl.Instrum.Meth.Phys.Res. B **23**,242(1987)
8. Andrä, H.J.: Nucl.Instrum.Meth.Phys.Res. B **43**,306(1989)
9. Varga, P.: Comments At.Mol.Phys. **23**,111(1989)
10. Bhalla, C.P.: Phys.Rev. A **8**,2877(1973)
11. Delaunay, M., Fehringer, M., Geller, R., Hitz, D., Varga, P., Winter, H.: Phys.Rev. B **35**,4232(1987)
12. Lakits, G., Aumayr, F., Winter, H.: Rev.Sci.Instrum. **60**,3151(1989)
13. Fehringer, M., Delaunay, M., Geller, R., Varga, P., Winter, H.: Nucl.Instrum.Meth. Phys.Res. B **23**,245(1987)
14. Lakits, G.: thesis, TU Wien (1989, unpublished)
15. de Zwart, S.T.: thesis, University of Groningen (1987, unpublished), and Nucl.Instrum.Meth.Phys.Res. B **23**,239(1987)
16. Hagstrum, H.D., Becker, G.E.: Phys.Rev. B **8**,107(1973)
17. Varga, P., Hofer, W., Winter, H.: Surface Sci. **117**,142(1982)
18. Wouters, P., Zeijlmans van Emmichoven, P.A., Niehaus, A.: Radiat.Eff.Def.Sol. **109**,111(1989)
19. Zehner, D.M., Overbury, S.H., Havener, C.C., Meyer, F.W., Heiland, W.: Surface Sci. **178**,359(1986)
20. Delaunay, M., Fehringer, M., Geller, R., Varga, P., Winter, H.: Europhys.Lett. **4**,377(1987)
21. Varga, P.: in Electronic and Atomic Collisions, ed. H.B. Gilbody et al., Elsevier, Amsterdam 1988, p. 793
22. de Zwart, S.T., Drentje, A.G., Boers, A.L., Morgenstern, R.: Surface Sci. **217**,298(1989)
   Folkerts, L., Morgenstern, R.: J. de Phys. **50**, C1-541(1989)
23. Folkerts, L., Morgenstern, R.: Europhys.Lett. (1990, in print)
24. Folkerts, L., Koopmans, P., Morgenstern, R.: KVI Groningen Annual Report 1989, p. 94
25. de Zwart, S.T., Fried, T., Jellen, U., Boers, A.L., Drentje, A.G.: J.Phys.B:At.Mol.Phys. **18**,L623(1985)
26. Hughes, R.H., Miller, R.D., Wattuhewa, X.M.Ye., Pederson, D.O.: Phys.Rev. B **40**,43(1989)
27. Lakits, G., Arnau, A., Winter, H.: Phys.Rev. B **42**,15(1990)
28. Snowdon, K.J., Havener, C.C., Meyer, F.W., Overbury, S.H., Zehner, D.M., Heiland, W.: Phys.Rev. A **38**,2294(1988)
29. Apell, P.: J.Phys.B:At.Mol.Opt.Phys. **21**,2665(1988)
30. Zeijlmans van Emmichoven, P.A., Wouters, P.A.A.F., Niehaus, A.: Surface Sci. **195**,115(1988), and
   Zeijlmans van Emmichoven, P.A., Niehaus, A.: Comments At.Mol.Phys. **24**,65(1990)
31. Donets, E.D.: Physica Scr. T 3,11(1983), and Nucl.Instrum.Meth.Phys.Res. B **9**,522(1985)
32. Donets, E.D., Kartashov, S.V., Ovsyannikov, V.P.: Proc. ICPIG XVII, Budapest 1985, contr. papers
33. Briand, J.P., et al.: Proc. XV ICPEAC, Brighton 1987, contr. papers, p. 827
34. Briand, J.P., et al.: Phys.Rev.Lett. **65**,159(1990)

Supplement to Z. Phys. D – Atoms, Molecules and Clusters 21, S 135–142 (1991)

Atoms, Molecules
Zeitschrift für Physik D and Clusters

X-ray- and Auger-spectra from multi-charged ions interacting with surfaces and solids

H. J. Andrä, A. Simionovici, T. Lamy, A. Brenac, G. Lamboley
LAGRIPPA, CEN Grenoble, BP 85X, F-38041 Grenoble Cedex, France

J. J. Bonnet, A. Fleury, M. Bonnefoy, M. Chassevent
Laboratoire de Physique des Collisions Atomiques et Moléculaires, CNAM, 292 rue St. Martin, F-75141 Paris Cedex 03, France

S. Andriamonje
CEN Bordeaux-Gradignan, Le Haut-Vigneau, F-33175 Gradignan Cedex, France

A. Pesnelle
Service de Physique des Atomes et des Surfaces, CEN Saclay, F-91190 Gif-sur-Yvette Cedex, France

Abstract

Low and high resolution X-ray spectra of the K_α satellites emitted from Multi Charged Ions (MCIs) in collisions with surfaces at variable vertical velocity allow for the first time to differentiate between X-rays emitted from ions inside the bulk and from ions experiencing a pure surface interaction. Another proof for the bulk interaction at high vertical velocity is the first observation of the two electron - one photon double K_α (or $K_{\alpha\alpha}$) emission with low energy MCIs. Auger-electron spectra emitted from hydrogenlike MCIs approaching a surface at very low velocities unveil new aspects of the atomic dynamics in front of the surface which may require the revision of previous models.

Introduction

When a Multi Charged Ion (MCI) of charge "q" approaches a conducting solid it starts to capture electrons via Resonance Neutralization (RN) into Rydberg levels at rather large distances $z_0 \approx 1.6 \langle r_{ns} \rangle = 1.2\ q/\phi_W \approx 59\ a_0$ for $q = 9$ and $\phi_W = 5$ eV, where ϕ_W is the work function in atomic units [1] and $n=q/\sqrt{2\ \phi_W}$. From z_0 towards the surface, it will then experience electron exchange, Auto Ionization (AI) cascading, and eventually Auger electron and X-ray emission dynamics yet to be explored. In the excellent work of de Zwart [2] on the electron spectra observed during the interaction of very slow Ar^{9+} ($1s^2\ 2s^2\ 2p^5$) with a metal surface it has been shown that these electrons are emitted in front of the surface. The electron spectra have been interpreted by assuming an electron distribution in the various electronic shells of the atom which is consistent with an AI-cascade from Rydberg levels where every electron emitted to the continuum is immediately replaced by a new resonantly transfered electron from the surface.

A more recent interpretation [3] of the spectra assumes an emission from neutral Ar with one 2p-electron excited to the 4s-state and thus sheds some doubt on the AI-cascade model.

During the electron exchange dynamics the ion is in addition exposed to an attractive force towards the surface due to the image charge. Very close to the surface ($z \leq 4a_0$) the repulsive Coulomb force between the MCI and all partially screened surface atoms will dominate over the image force and the MCI can be reflected. It will penetrate the solid only if its vertical velocity $v_v = v_0 \sin(\theta_{in})$ (or "vertical energy" $E_v = E_0\ \sin^2(\theta_{in})$) is sufficient to overcome this repulsion ; θ_{in} is the angle of incidence with respect to the surface plane and E_0 is the energy of the incident MCI. When receding, the MCI will continue to experience the electron dynamics mentioned. When penetrating, the electron exchange dynamics will become more violent in close encounters with target atoms such that capture into inner shells of the MCI from inner shells of the target atoms may become very important. Hints of such a bulk interaction stem from low resolution X-ray measurements of Ar^{17+} interacting with a Be target at 18 keV [4, 5] which have been interpreted as arising from an electron distribution at the moment of the X-ray emission [6] which is not compatible with the AI-cascade model.

In order to differentiate between these two regimes (i) in front of the surface and (ii) inside the bulk, we have carried out measurements of low and high resolution X-ray spectra as well as of Auger electron spectra from MCIs interacting with conducting surfaces and solids at systematically varied vertical velocities. To this end, the vertical velocity v_v has been varied for the X-ray measurements via the angle of incidence at a rather high energy E_0 (340 and 35 keV) of the incident MCI, while for the Auger electron spectroscopy the energy E_0, has been varied at fixed $\theta_{in} = 45°$ from 10 keV to as low as 54 eV (!). By systematically varying v_v one can vary the probability of reflection P_r of ions from the surface potential and simultaneously the probability $1\text{-}P_r$ of penetrating the solid. This holds true for ideally flat and clean monocrystalline surfaces as well as for moderately prepared polycrystalline surfaces in moderate vacuum, however with modified probabilities. For singly charged neon at normal incidence on atomically flat Si or Ta, e.g. close to total reflection is expected for $E_0 < 100$ eV while for $E_0 > 20$ keV complete penetration takes place. Both limits of energy are shifted to higher values with increasing effective charge of the ion in the region very close to the surface $z \leq 4\ a_0$.

Low resolution X-ray spectra

The goal of this first experiment is to decide whether the previously observed X-ray spectra in the interaction of Ar^{17+} with solid targets stem from an interaction (i) in front of the surface or (ii) inside the bulk.

Ar^{17+} from the NEW LAGRIPPA FACILITY [7] has been impinging on targets of Ta and doped Si at a vacuum of 10^{-5} Pa. The Si-target was a piece of 10×25 mm of n(P)-doped Si 111 single crystal wafer surface, cut and polished for the production of integrated circuits. The Ta target was a disk of 20 mm diameter of pure polycrystalline Ta, polished to optical mirror quality (λ/20).

The X-rays emitted in the interaction between the ions and these targets have been detected by a Si(Li) detector of low resolution (220 eV at 5.9 keV) but high efficiency. The detector which had a useful area of 70 mm^2 and a Be window of 25 μm has been placed at a distance of 30 mm from the interaction region and viewed the targets at 90° with respect to the ion beam axis. The energy calibration has been obtained from X-ray standard sources which yield an absolute calibration of the Ar^{17+} data around 3 keV. Electric ion beam currents of 80 pA of Ar^{17+} at 340 keV (well separated from the dominant neighbor $^{14}N^{6+}$ in the mass spectrum) have been provided on a 3 mm wide and 5 mm high spot in the center of the target chamber with a convergence of ±15 mrad in both directions.

The X-ray spectra obtained with Ar^{17+} impinging at 340 keV on the Ta target at two angles of incidence are presented in Fig.1. For demonstration purposes the He-like Ly_α line of Ar^{16+} has also been measured in Ar^{17+} - He collisions and is included in Fig.1. The spectra correspond to the spectrum observed earlier by Donets et al. [4,5], although the absolute energies are obviously different due to different experimental conditions. The spectra are composed of two dominant contributions which are compared with the K_α and K_β lines of singly ionized Ar (full vertical lines) and with the Lyman series of He-like Ar^{16+} (broken vertical lines). The left and dominant component is fitted with a gaussian and attributed to $1s\ 2s^m\ 2p^n \rightarrow 1s^2\ 2s^m\ 2p^{n-1}$ - K_α satellites with the maximum shifted with respect to the K_α line. After subtraction of this gaussian fit from the spectrum, a second spectral structure is left over at higher energies. This right component is assumed to be dominated by K_β satellites and includes K_γ satellites as principal perturbation at the high energy end. The centroids of these right components are indicated in Fig.1 and are located well between the K_β and Ly_β lines.

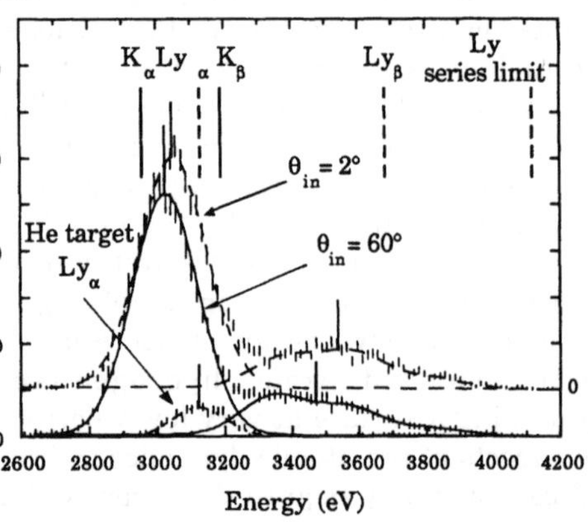

Fig.1 K X-ray spectrum of Ar^{17+} impinging at 340 keV and $\theta_{in} = 2°$ (full line) and 60° (broken line) on a Ta surface. The spectrum obtained with a He target (dot-dashed line) is also shown for calibration purposes.

The energetic positions of the Ar K_α satellite peaks in Fig.1 for the smallest (2°) and largest (60°) angle of incidence measured are well confined within the limits defined by the K_α and the Ly_α line positions. This observation strongly suggests, in agreement with the earlier interpretations, that the energetic position of the Ar K_α satellite peak is directly correlated to a different average population of the L-shells of the emitting ions [6], neglecting the minor influence of the population of the M-shell. This correlation can be used here to conclude that the average population of the L-shell of the emitting Ar ion decreases with decreasing angle of incidence, or equivalently with decreasing probability of penetration of the MCI into the bulk. The result at "vertical energy" $E_v = E_0 \sin^2(2°) = 335$ eV in comparison to the one at 60° (222 keV) provides thus a strong indication for an important contribution of an interaction in front of the surface which supposedly produces via AI-cascades less electrons in the L-shell than the bulk interaction. If this holds true for further refined experiments one can also conclude from this observation that all previous results with Ar^{17+} at "vertical energy" E_v well above 10 keV have nothing to do with an ion-surface interaction; they are on the contrary, clearly dominated by the interaction of the ion with the bulk.

High resolution X-ray spectra

In order to verify the interpretation of the above experiment a repetition with considerably improved spectral resolution had become necessary in order to resolve the K_α satellites and thus directly deduce the populations in the L-shell. Unfortunately, our equipment was only fit for measuring X-rays in the range of those from Ne^{9+} projectiles, which yield however, the same type of information as Ar^{17+}.

Ne^{9+} ions have been impinging in a vacuum of 10^{-5} Pa at energies E_0 = 140 and 35 keV on Ta and doped Si targets at systematically varied angles of incidence θ_{in}. The X-rays have been observed with a Bragg spectrometer [8] supplied with a flat RbAP-crystal (2d = 2.61 nm). The Soller slit angular divergence has been 0.11° to yield an experimental resolution of $E/\Delta E \approx 150$ around E = 900 eV. The photons have been detected by a flowing gas proportional counter with propane at a regulated pressure of 2 10^{-4} Pa, supplied with a Makrofoil window of 0.4 µm nominal thickness. The spectra have not been corrected for the variation of the crystal reflectivity with the X-ray wavelength.

In Fig.2a, the K X-ray spectrum in the wavelength range of 1.1 to 1.5 nm is shown for Ne^{9+} impinging at 35 keV on the highly polished polycrystalline Ta surface at θ_{in} = 50°. It exhibits the K_α satellite group KL^0 to KL^7 with a maximum of intensity corresponding to a completely filled L-shell, especially when taking into account the monotonic increase of the fluorescence yield from KL^0 (0.013) to KL^7 (0.1 approx.). The spectrum obtained from the same experiment at θ_{in} = 5° is presented in Fig.2b. It exhibits a striking modification of the relative intensities of the whole K_α satellite group KL^0 to KL^7 with a tremendous increase of the KL^7 emission corresponding to a single 2p-electron in the L-shell.

The spectrum in Fig.2a can be interpreted in terms of an emission during the direct and rapid filling of the L and M shells from inner shells of the target atoms when Ne^{9+} is penetrating the solid at the given experimental conditions and is slowing down inside the bulk. The faster the filling, the more K_α emission with a completely filled L-shell can take place, and the probability to emit with less electrons in the L-shell is greatly reduced.

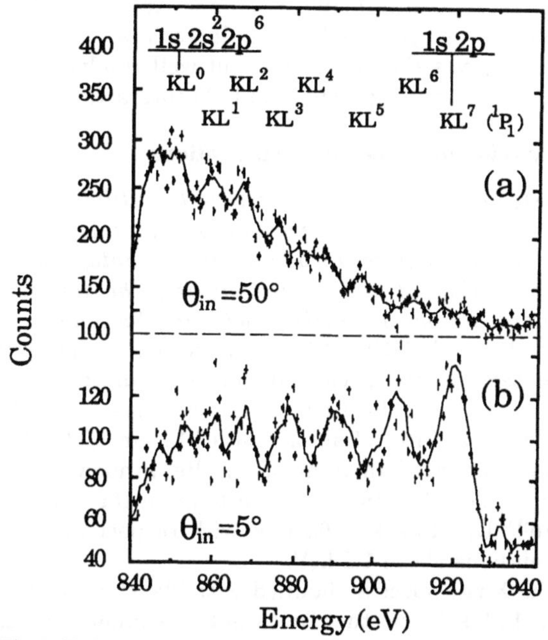

Fig.2 K X-ray spectrum in the wavelength range of 1.3 to 1.5 nm emitted from Ne^{9+} impinging at 35 keV and θ_{in} = 50° (a) and 5° (b) on a Ta surface. The full lines are 5 point smooths to guide the eye.

The spectrum in Fig.2b has to be interpreted as a statistical mixture of three different regimes. The MCI experiences at a "vertical energy" E_v of only 215 eV an extended ion-surface interaction during the approach up to very short distances from the first atomic surface layer. This interaction can be assumed to produce RN-AI cascades which yield a maximum intensity for the KL^7 peak and decreasing intensities for KL^6, KL^5, etc. Very close to the surface, the supposedly strong surface corrugation will be responsible for a large fraction of penetration and only a minor probability of total reflection. The reflected projectiles may have accumulated some additional L-shell population during the approach but continue to emit X-rays following RN-AI-cascades with slightly higher L-shell population than during the approach. The projectiles which penetrate the bulk after their ion surface pre-history accumulate very rapidly L-shell population to emit a spectrum very close to the one in Fig.2a. The superposition of the spectra of all three interaction regimes explains sufficiently well the spectra of Fig.2b. A more detailed discussion is presented in the contributed paper of J.J. Bonnet et al. [9] in these proceedings. One can conclude that a dominant KL^7 peak is the first unequivocal proof for the existence of a Ne^{9+}-surface interaction in front of the surface which produces a small L-shell population at the time of the X-ray emission. However, this argument has to be taken cautiously for the recently observed Ar^{17+} solid interaction [10]. For Ar^{16+}, the KL^7 fluorescence yield is 13 times larger than for Ne^{8+}, which gives rise to a minor but well resolved KL^7 peak even during a pure ion-bulk interaction.

Two electron - one photon transitions

Instead of immediately pursuing the ultimate goal of the observation of a pure ion-surface interaction, an attempt has been made to take advantage of the fast filling of the L-shell during the ion-bulk interaction when clearly penetrating the solid.

To this end, MCIs of Ne^{10+} have been impinging on targets of C, Ta and doped Si in the same conditions as the first experiment described. An additional filter of absorption of 10 μm Mylar has been mounted in front of the Be window of the Si(Li) detector, in order to attenuate the detected Ne^{9+}-Ly_α line at 1.02 keV with respect to X-ray emissions above 1.2 keV.

The X-ray spectra obtained with Ne^{10+} impinging at 140 keV on the Ta target are shown as an example in Fig.3. Despite the absorption function, the spectra are dominated by a strong peak at about 1.1 keV which stems from unresolved K_α, K_β, K_γ satellites (one K-hole) and hypersatellites (two K-holes) of neon. At higher energies, around 1.8 keV, a smaller peak appears well above the limit of 1.36 keV of the Ne^{9+} Lyman series, which obviously varies systematically in intensity as a function of the angle of incidence.

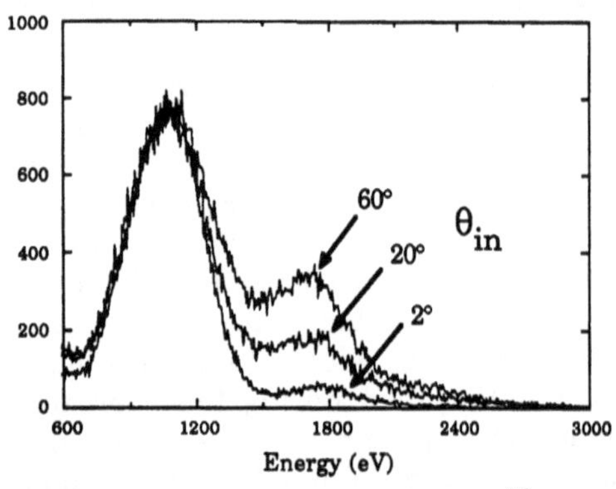

Fig.3 Si(Li) X-ray spectrum emitted from Ne^{10+} interacting at 140 keV and angles of incidence of θ_{in} = 2°, 20°, and 60° with the Ta target.

Its energy corresponds to two electron - one photon satellite transitions between $2s^2\,2p^6 \rightarrow 1s^2\,2s\,2p^5$ - double K_α (or $K_{\alpha\alpha}$) - and $2s\,2p \rightarrow 1s^2$ - double Ly_α (or $Ly_{\alpha\alpha}$) - transitions respectively, observed for the first time in low-energy (≈ keV/u) collisions. Further consistency to this interpretation is given by the fact that this peak does not occur with Ne^{9+} projectiles. Such two electron - one photon transitions are known to appear in energetic (≈ MeV/u) heavy ion - solid collisions [11-13], where two K-holes are created in the projectile via ionization while its L-shell is partially conserved. For these transitions to occur at low projectile energies, a bare nucleus has to approach the surface with sufficient vertical velocity that the two K-holes survive in front of the surface and that the bare nucleus penetrates the solid. Inside the solid, the L-shell has to be filled up very rapidly by direct capture from inner shells of the target atoms or by cascading from the M, N shells after direct capture into these shells. The rate of filling the L-shell has to be higher than the decay rate of a single "p"-electron in the L-shell either by radiative or by KLM-Auger transitions. The fact that a well defined peak is observed suggests that its average energy is determined by an average of 4 ± 1 electrons in the L-shell. The increase of the intensity of the peak as a function of the angle of incidence can thus roughly be explained by an increasing probability of penetrating the solid and a higher survival probability for the bare ion in front of the surface. A more detailed analysis will be given in a forthcoming paper but we can already state that the variation of the intensity with angle is smaller for the Si target than for the Ta one, and much smaller for the C target. This is clear evidence for a strong target dependence of the L-shell filling rate of the projectile.

Auger electron spectra

Although our high resolution X-ray measurements with Ne^{9+} at grazing incidence have given unequivocal proof for the existence of a Ne^{9+}-surface interaction in front of the surface, these experiments still suffered from a large fraction of ions penetrating into the solid due to surface imperfections. The only way to reduce this perturbation by an ion-bulk interaction is by decreasing the energy of the incoming ions to a level where penetration into the solid can be neglected. Furthermore, the small X-ray fluorescence yields of Ne strongly support the observation of the Auger (AI) electron emission instead, which may carry more detailed information on the AI-cascading.

These spectra have been measured at the end of the center beam line of the NEW LAGRIPPA FACILITY [7] with hydrogenic ions of C, N, O, F, and Ne delivered at 2 keV + plasma potential U_p per charge. In order to further decelerate these MCIs, they entered, through a decelerating lens system of 1 mm diameter entrance and exit apertures, into a metallic box which could be brought to a potential V_d as shown in Fig. 4.

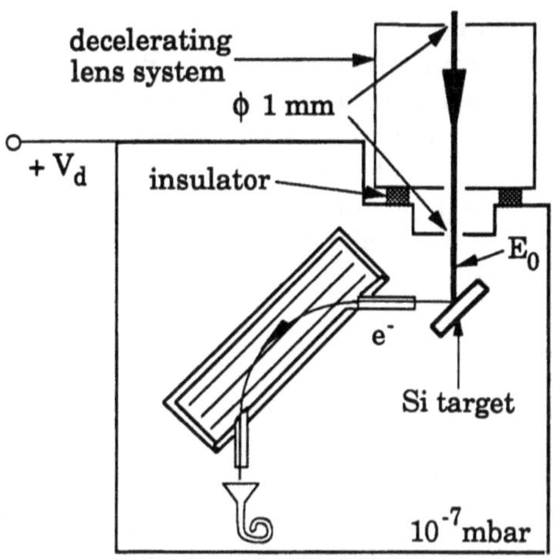

Fig.4 Experimental setup for Auger electron spectroscopy with decelerated ion beams.

The MCIs hit the doped Si monocrystalline target inside the box with an energy $E_0 = q\,(2\text{ keV} + U_p - V_d)$ at an angle of 45°. The Auger electrons have been analyzed with a small parallel plate electron spectrometer with a resolution of $E/\Delta E \approx 63$ and detected by a channeltron, both mounted inside the box. The ion beam current on the insulated target has been measured at any value of V_d for normalization purposes. With U_p ranging from 12.7 to 25.3 volts, MCI beams of $E_0 \geq 8$ q eV have been used with currents $i \geq 0.1$ nA on target for ions C^{5+} through Ne^{9+}.

The vacuum for all measurements has been limited to 10^{-5} Pa only and the Si surface has been used as delivered for the production of electronic integrated circuits. In order to prove that such experimental conditions are sufficient for this kind of investigation, the complete spectra from O^{7+} interacting at 10.54 keV and at 845 eV with this target are shown in Fig.5 without transmission corrections. Above a very rapidly decreasing background at low energies, a "low energy" spectral structure seems to show up in the lower spectrum which is clearly resolved in the upper spectrum and can be assigned to LMM transitions of oxygen. The electron intensity then passes through a minimum before steadily rising towards a "high energy" spectral structure which is evidently narrower in the upper spectrum compared to the lower one and can be assigned to KLL transitions of oxygen.

Between these "low energy" and "high energy" spectral structures, one observes in the lower spectrum additional structures which disappear completely in the upper one.

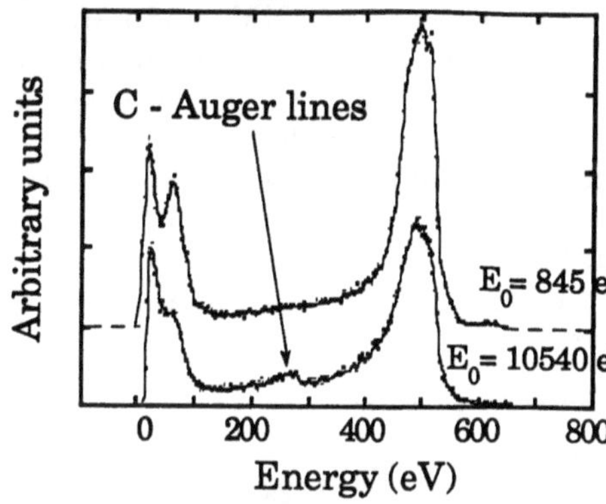

Fig.5 Auger electron spectrum emitted from O^{7+} interacting at $\theta_{in} = 45°$ and energies of 10.54 keV and 845 eV with a doped, monocrystalline Si-target.

They are attributed to carbon Auger transitions and thus indicate a nonnegligible surface contamination. This has been strongly reduced by baking the whole experimental setup at 100 °C for 24h but could not be eliminated completely. The upper spectrum shows, however, that such a surface contamination may still exist but does not show up in experiments at low incident energy $E_0 \leq 1$ keV. It may, of course, influence the target work function, the density of states of the conduction band, and thus the surface conductivity, but the variation of these quantities is considered to be negligible with respect to the transition energies measured here [15].

A spectrum similar to the one of Fig.5 has been observed for all the ions investigated at medium energies E_0. The most significant modifications have been recorded for Ne^{9+} when E_0 has been reduced to the lowest values possible. The "low energy" spectral structure from Ne^{9+} is therefore shown in Fig.6 for E_0 = 2830 eV and E_0 = 145 eV.

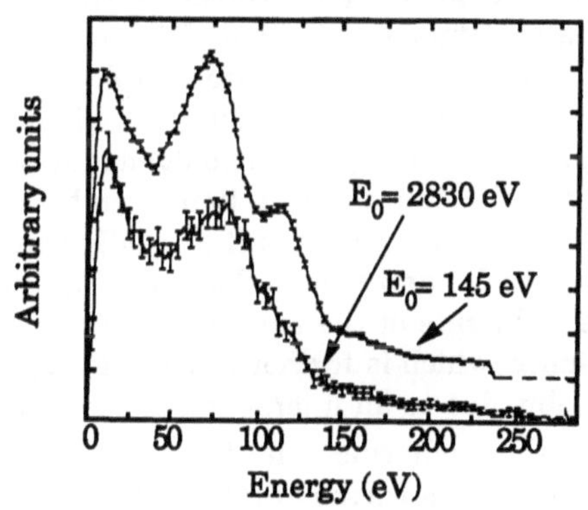

Fig.6 "Low energy" Auger electron spectrum emitted from Ne^{9+} interacting at energies of 2830 eV and 145 eV and θ_{in} = 45° with a doped, monocrystalline Si-target.

One clearly observes an increase of intensity of the overall structure at the lowest energy with respect to the background and most importantly a clear separation of the two peaks. These two dominant peaks can be assigned to LMM and LMN Auger transitions of neon which evidently increase when the MCI velocity toward the surface decreases. The modifications of the "high energy" spectral structure are shown in Fig.7 for three energies E_0. To witness, the limits of the 1s $2s^2 \rightarrow 1s^2$ and 1s $2s^2$ $2p^6 \rightarrow 1s^2$ 2s $2p^5$ KLL satellite transitions of neon ions with no M-shell electrons [15] are indicated in the figure.

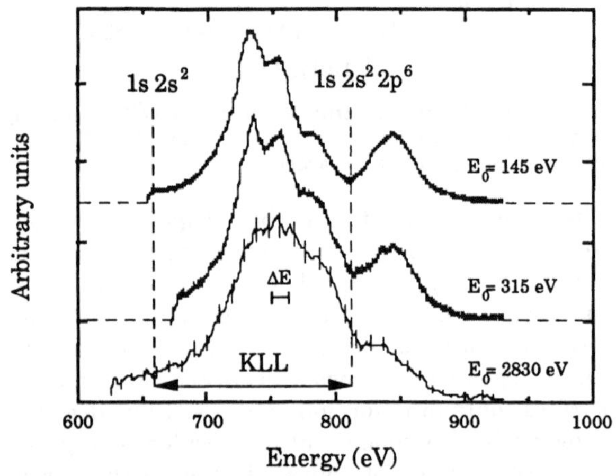

Fig.7"High energy" Auger electron spectrum emitted from Ne^{9+} interacting at 2830 eV, 315 eV, 145 eV and θ_{in} = 45° with a doped, monocrystalline Si-target.

Within these limits, one observes with decreasing E_0 a striking decline of the higher energy part while an enhancement of the two peaks in the center is clearly visible, accompanied by a steepening of the left slope of the spectrum. Simultaneously, the large peak above the KLL limit which is assigned to KLM transitions, grows significantly with decreasing E_0. We interpret this overall behavior of the spectrum with decreasing E_0 as a continuous reduction of the contributions from ions penetrating the solid while the interaction in front of the surface contributes more and more. Essentially, the two dominant peaks at E_0 = 145 eV within the KLL limits can be assigned to KLL transitions with initial configurations of one K , two L , and five M electrons. The shoulder on the right side of the main structure within the KLL limits, which is further reduced at E_0 = 97 eV is probably due to an interaction at the distance of closest approach when the ion is reflected. This interpretation is corroborated by the "low energy" LMM and LMN transitions in Fig.6 which feed the L-shell and therefore the two dominant peaks within the KLL limits.

At the lowest energy E_0, we thus observe several of the final steps of the autoionization cascade of the MCI-surface interaction in front of the surface, superimposed on a minor broad structure within the KLL limits due to the MCI-surface interaction at the distance of closest approach. At increasing E_0, these two contributions are reduced in favour of a more and more pronounced penetration which is accompanied by a more and more rapid filling of the L-shell which causes the broad structure between the KLL limits to increase considerably and to shift its maximum towards the higher KLL energy limit.

Conclusion

The systematic variation of the "vertical energy" of Ar^{17+} or Ne^{9+} approaching a metal surface combined with a low or high resolution X-ray spectrometer, respectively, has created experimental conditions which allow to identify for the first time X-rays emitted from ions purely interacting with the surface. All previous observations of X-rays emitted from MCIs interacting with solids [4, 5, 10] can now undoubtedly be attributed to an emission from the ions inside the bulk and as such are not related to the ion-surface interaction. Future experiments with lower energies E_0 have yet to show whether X-ray spectroscopy is a valuable tool for the investigation of the MCI-surface interaction.
The Auger electron spectroscopy in connection with decelerated MCIs is evidently very well suited for the investigation of the MCI-surface interaction as first demonstrated by S. de Zwart [2]. The systematic change of the electron spectrum with increasing deceleration of the incident ions allows to conclude that at the lowest energy E_0, the dominant Auger electron emission of an initial Ne^{9+} takes place in front of the surface. The analysis of the whole spectrum allows to deduce a distribution of electrons (DOE) [1] at the time of the electron emission with a maximum in n = 3 with negligible population in n = 5. This sheds some doubt on the interpretations of the Ar^{9+} data by de Zwart [2] and Andrä [1], where the maximum of the DOE is found in n = 5,6 or by Folkerts and Morgenstern [3] where the maximum of the DOE is in n = 3. From the present results, one would expect the maximum in n = 4, assuming that the difference in atomic structure between F-like Ar^{9+} and H-like Ne^{9+} has no significant influence. Therefore, a revision of the interpretations of the Ar^{9+} data seems to be necessary; for this purpose, experimental data from higher Z F-like ions is required in order to better resolve these spectral features which are as

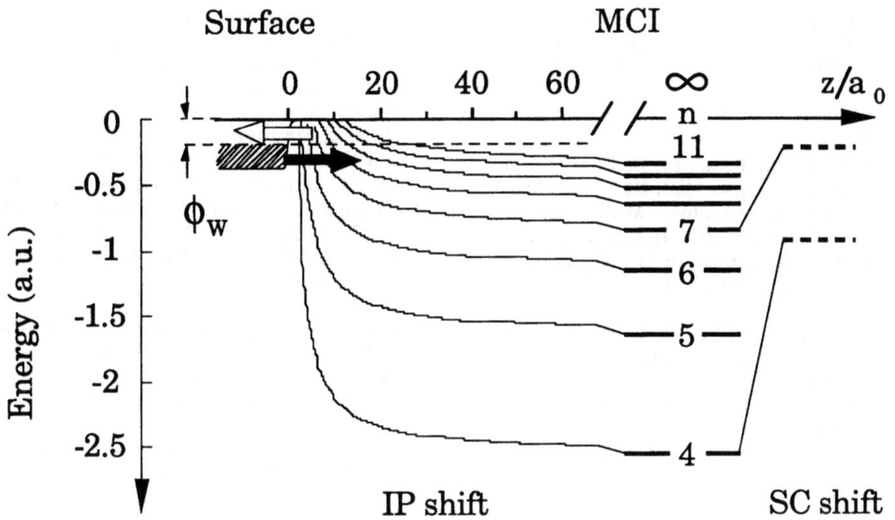

Fig.8 Energetic shift of Rydberg levels due to the image potential of a Ne^{9+} ion as a function of distance z from a conducting surface with respect to the conduction band with a work function $\phi_W \approx 5$ eV. Typical screening shift of the n = 4 and n = 7 levels due to neutralization in the n = 5 and n = 8 levels respectively is shown for the 1s ns Ne ion.

yet buried in a single structure.
The DOE deduced here, the much too low autoionization rates of Rydberg electrons [1], and the much too small integrated secondary electron yields measured elsewhere [16] indicate the necessity for a modification of the RN-AI-Cascading Model (CM) developed by Hagstrum [17], Arifov et al. [18], de Zwart [2], and Andrä [1]. For q = 9, the RN-population of n = 11 followed by AI-cascading down to the observed DOE would require more than 40 steps of cascading, producing over 40 secondary electrons to be compared to the 12 electrons per ion reported [16]. Mechanisms are therefore required which allow to populate lower n-shells directly and resonantly from the conduction band. To this end, we propose to add to the RN-AI-CM model, the shift of the energy levels as a function of z due to the Image Potential (IP) [19], and to add the upward-shift of the energy levels due to Screening (SC) as soon as two or more electrons populate these or neighboring levels (see Fig.8). The resonant capture (full arrow in Fig.8) of two or more electrons into the IP-shifted n = 11 at z_0, for instance, will SC-shift this level above the Fermi edge so that these electrons will be reionized and recaptured by the solid (empty arrow in Fig.8). These electrons had also SC-shifted the neighboring n = 10,9 levels such that the next electrons could already be captured into these lower n levels during the reionization of the first ones. They will be SC and IP-shifted upwards again during the further approach to the surface so that recapture into the solid becomes possible when they pass above the Fermi edge. These "cycles" continue to act during the approach towards the surface such that at $z \approx 15\ a_0$, rather low n levels ($4 \le n \le 6$) may directly and resonantly be populated from the conduction band. The complexity of this process precludes a straightforward quantum calculation so that a semi-classical model seems to be the most promising tool for its better understanding [20]. The whole dynamics is of course accompanied by AI-transitions which are responsible for the now reduced secondary electron yield. Since the AI-rates increase above $10^{14}\ s^{-1}$ for lower n-values, the final cascade can take place in front of the surface during the available time Δt before entering the solid, where $\Delta t = 15\ a_0/v_v = 2.1\ 10^{-14}$ s for Ne^{9+} at 145 eV.

Acknowledgement

This work has been performed at LAGRIPPA, a joint laboratory of the CNRS and of the CEA.

References

1) H.J. Andrä, "Electronic Interaction of Multi-Charged Ions with Metal Surfaces at Low Velocities", in Proceedings of the NATO Advanced Study Institute on "Atomic Physics of Highly Charged Ions", ed. R. Marrus, Plenum Press 1989, p. 377, see references therein; see also Nucl.Instr.Meth. **B43**, 306 (1989).
2) S.T. de Zwart, Thesis, Groningen (1987); ibid. Nucl.Instr.Meth. **B23**, 239 (1987).
3) L. Folkerts and R. Morgenstern, J.Phys.(Paris) **50**, C1-541 (1989).
4) E.D. Donets, Phys.Scr. T3, 11 (1983).
5) E.D. Donets, Nucl.Instr.Meth. B9, 522 (1985).

6) G. Zschornak, G. Musiol, and W. Wagner, invited talk at the Workshop on "Energiereiche atomare Stöße" in Oberstdorf/FRG, Jan. 1985; see also Phys.Scr. T3, 194 (1983).
7) T. Lamy, G. Lamboley, D. Hitz, and H.J. Andrä, Rev.Sci.Instr. **61**, 336 (1990).
8) A. Fleury, J. Debernardi, M. Bonnefoy, S. Bliman, J.J. Bonnet, and M. Chassevent, Nucl.Instr.Meth **B14**, 353 (1986).
9) J.J.Bonnet, A.Fleury, M.Bonnefoy, M.Chassevent, T.Lamy, A.Brenac, A.Simionovici, H.J.Andrä, and S.Andriamonje in Proceedings of the Vth Int. Conf. Phys. Highly Charged Ions, Sept. 1990, Giessen-FRG, ed. E.Salzborn, Z.Phys. D., (1991)
10) J.P. Briand, L. de Billy, P. Charles, S. Essabaa, P. Briand, R. Geller, J.P. Desclaux, S. Bliman, and C. Ristori, Phys.Rev.Lett. **65**, 159 (1990).
11) W. Wölfli, Ch. Stoller, G. Bonani, M. Suter, and M. Stöckli, Phys. Rev. Lett. **35** (1975) 656.
12) T. Åberg, K.A. Jamison, and P. Richard, Phys.Rev.Lett. **37**, 63 (1976).
13) J. Auerhammer, H. Genz, A. Kumar, and A. Richter, Phys.Rev. **A38**, 688 (1988), and references therein.
14) For a comparison with results obtained under UHV-conditions see the contributed paper by L. Folkerts and R. Morgenstern in these Proceedings.
15) D.L. Matthews, B.M. Johnson, and C.F. Moore, Atomic and Nucl.Data Tables **15**, 41 (1975).
16) M. Delaunay, M. Fehringer, R. Geller, D. Hitz, P. Varga, and H. Winter, Phys.Rev. **B35**, 4232 (1987).
17) H.D.Hagstrum, Phys.Rev.**96**, 325 (1954); ibid. **96**, 336 (1954); in "Electron and Ion Spectroscopy of Solids", edited by L.Fiermans, J.Vennik, and W.Dekeyser (Plenum, New York, 1978).
18) U.A.Arifov, L.M.Kishinevskii, E.S.Mukkamadiev, and E.S.Parilis, Zh.Tekh.Fiz. **43**, 181 (1973); Sov.Phys.-Tech.Phys. **18**, 118 (1973).
19) See also the contributed paper by U. Wille in these Proceedings.
20) N. Bardsley and B.M. Penetrante, private communication.

Supplement to Z. Phys. D - Atoms, Molecules and Clusters 21, S 143-146 (1991)

Atoms, Molecules
Zeitschrift für Physik D and Clusters

Dielectronic recombination measurements at EBIT

D. A. Knapp
High Temperature Physics Division, Lawrence Livermore National Laboratory, Livermore, CA 94550, USA

Abstract

The Electron Beam Ion Trap at the Lawrence Livermore National Laboratory has proved an ideal device for the study of interactions between electrons and highly-charged ions. I describe measurements of one such interaction, dielectronic recombination, in several ion species. The results are in marginal agreement with theoretical predictions.

Introduction

Measurements of electron-ion interactions for high-energy electrons with highly-charged ions have been difficult to perform using traditional plasma or accelerator sources. The Electron Beam Ion Trap (EBIT) at the Lawrence Livermore National Laboratory offers a superior environment for such experiments.

To understand the unique utility of EBIT, it is important to review briefly its operation. The EBIT trap has been described in detail elsewhere[1]. It consists of a high current density (4000 A/cm^2) electron beam passing through a series of drift tubes. A simplified schematic of the trap is shown in Fig. 1. Ions are trapped radially by the space-charge of the beam and axially by potentials applied to two end-cap drift tubes. The energy of the electrons in the trap is determined by the potential applied to the center drift tube. The ions are nearly at rest, with typical kinetic energies of 20 eV/q. As a result, there is essentially no Doppler shifting or broadening of the spectroscopic lines. The electron-ion interaction rates are relatively low, so essentially all the ions are in the ground state; equilibrium population of metastable states is not a concern. The electron-ion interaction energy distribution is quite narrow, typically about 50 eV FWHM. The interaction energy can be varied with a settling time of about 1 ms, limited only by the slew rate and settling time of the voltage regulator on the drift tubes. X rays are observed at 90° to the electron beam direction through ports in the central drift tube.

As the electrons in the beam move through the trap, they interact with the ions by impact excitation and ionization, radiative recombination, resonant excitation, and dielectronic recombination. In this paper, I will concentrate on measurements of dielectronic recombination. However, EBIT has also been used for a variety of other measurements on highly charged ions, including spectroscopy[2], polarization[3], ionization[4], and innershell ionization[5].

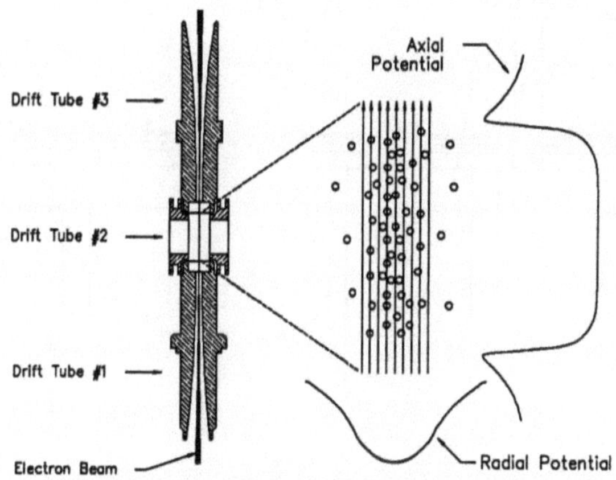

Fig. 1. Schematic view of EBIT operation. The ions are bound radially by the space-charge of the electron beam and axially by a set of drift tubes. The energy of the electron beam in the trap is controlled by the voltage applied to the drift tubes.

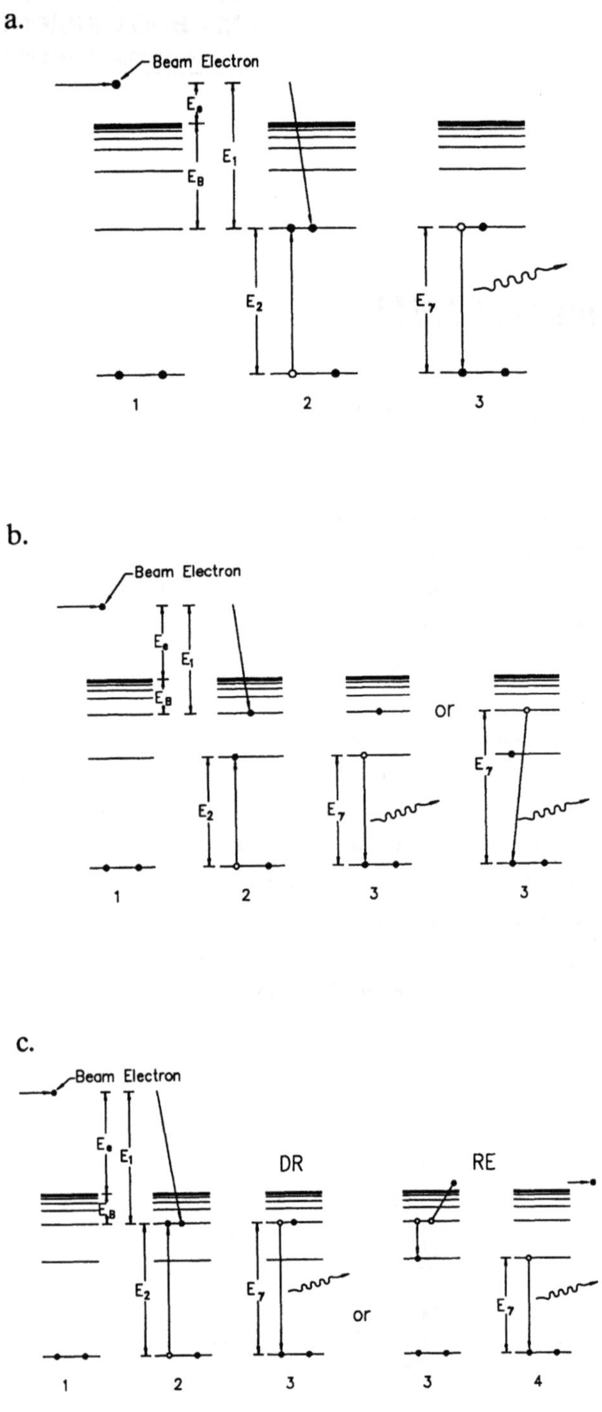

Fig. 2. Diagrams showing dielectronic recombination and resonant excitation of He-like ions. a.) KLL DR. A beam electron is captured into an ion, exciting a bound electron. The intermediate doubly excited state can then stabilize radiatively. b.) KLM DR. If the intermediate state has electrons in two states, two possible x ray energies can result. One K x ray always results from the DR of a He-like ion. c.) KMM DR and RE. For some resonances, the intermediate state can undergo Auger decay, resulting in resonant excitation.

Dielectronic Recombination Measurements

Dielectronic recombination (DR) is an important process in plasmas. The cross sections are resonant and very large, so DR affects the charge balance of the ions and has a significant effect on the emitted x ray spectrum. Schematic illustrations of DR are shown in Figure 2. A beam electron is captured to a resonant intermediate state by an inverse Auger process. The intermediate state can then stabilize either radiatively or by Auger emission. The former process is dielectronic recombination. If the latter process leaves the ion in an excited state, it subsequently decays radiatively, and the process is known as resonant excitation.

In EBIT, the intensity of DR x rays is normalized to that of radiative recombination to obtain cross sections. The DR cross sections are so large that the ionization balance in the trap changes quickly when the beam energy is on a DR resonance. Our initial measurements of DR[6] were performed by flipping the beam energy between an "ionization energy" far from any DR resonance, (and thus one at which a stable ionization balance could be generated), and "probe energies," at which the DR x rays could be observed. It proved difficult to switch energies quickly enough to prevent significant depletion of the ionization state of interest, so the duty factors for such measurements were only about 10%.

As a result, we have developed a new measurement technique for DR. The beam energy is ramped linearly between the ionization energy and a minimum energy below the DR resonances of interest, and then ramped back up, as shown in Fig. 3. The ramping time is about 10 ms in each direction, so the time spent on DR resonances is very short.

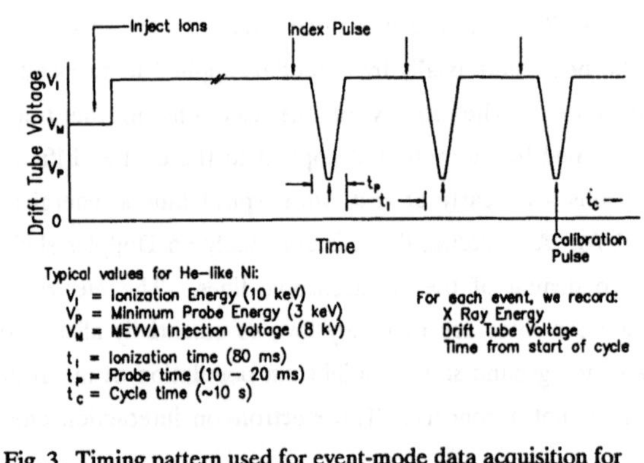

Fig. 3. Timing pattern used for event-mode data acquisition for dielectronic recombination measurements. Parameters shown are those used for He-like Ni^{26+}.

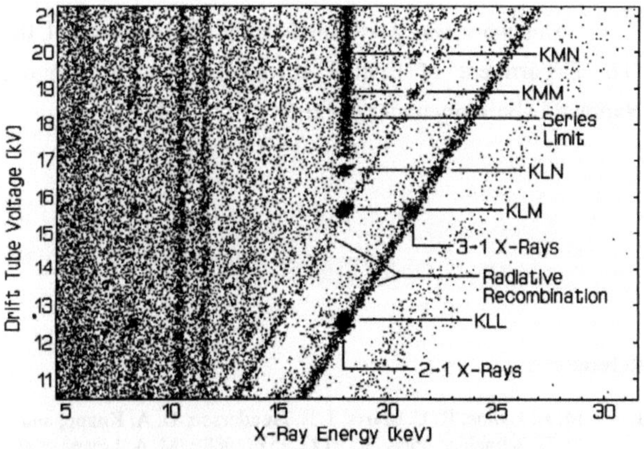

Fig. 4. Scatter plot of event data from Mo^{40+}. The physical processes giving rise to the events form distinct patterns. Cuts can be applied to select events from individual x ray energies or along the radiative recombination lines.

We record the time, beam energy, and x ray energy for each x ray detected, using a fast data-acquisition system to minimize the effect of deadtime.

The resulting data can be shown as a scatter plot of x ray energy vs. drift tube voltage. A typical scatter plot for Mo^{40+} is shown in Fig. 4. The x rays arising from various physical processes can easily be distinguished. Bound-state x ray transitions appear as vertical lines, while photons from radiative recombination appear as slanted lines, since their energy varies with the electron beam energy. The dielectronic recombination resonance appear as bright spots at the intersections between x ray and radiative recombination lines.

We generate excitation functions by selecting events at particular x ray energies or along radiative recombination lines. Typical DR excitation functions for He-like Ni^{26+} are shown in Fig. 5. The KMM and KMN resonances are apparent in the n=2-1 x ray excitation but not in the cut along the n=2 RR. This behavior indicates that the intermediate state decays primarily by resonant excitation.

We have measured DR for helium-like Ni^{26+}, helium-like Mo^{40+}, and neon-like Au^{69+}. Comparisons of the excitation functions for events observed during the down-going and up-going voltage ramps confirm that the effect of recombination on the ionization balance during the ramps is negligible.

Another advantage of this method becomes apparent in the normalization of the DR x ray intensity to radiative recombination (RR). The radiative recombination appears as a slowly varying background under the DR peaks in the RR excitation function. We take the ratio of the observed flux of RR x rays to the theoretical RR cross section[7] for electron beam energies not on DR resonance. This procedure corrects for the effects of variations in detector efficiency and ion-beam overlap. Surprisingly, the observed ratio varied little with electron beam energy, indicating that the ions are nearly completely trapped in the beam.

The results of our measurements of dielectronic recombination cross sections are compared to the results of a Dirac-Fock calculation[8] in Fig. 6. The calculation now includes an important effect not included in our previous paper, that of the angular distributions of polarized x ray lines.

The data for all three ions agree with the predictions within experimental uncertainties. However, the agreement for Ni^{26+} and Au^{69+} is marginal. The primary uncertainty in these results is the normalization to radiative recombination. It is worth noting that this normalization is much more difficult for Ne-like ions than for He-like ions, because there are many DR resonances on the RR lines, obscuring the RR flux. These results are subject to additional uncertainty in the ionization-state distribution. The x ray detector used did not have adequate resolution to allow an independent determination of the ionization balance from satellite or radiative recombination lines. Instead, the shape of the theoretical dielectronic recombination resonances was fitted to the observed resonances, determining the beam energy resolution and ionization balance simultaneously. We then compared the fit total cross section to the predicted total cross section. It is therefore difficult to unambiguously differentiate between a simple disagreement of amplitude and a disagreement of shape. Nonetheless, it is apparent that the general shapes of the dielectronic recombination resonances generally agree with calculations.

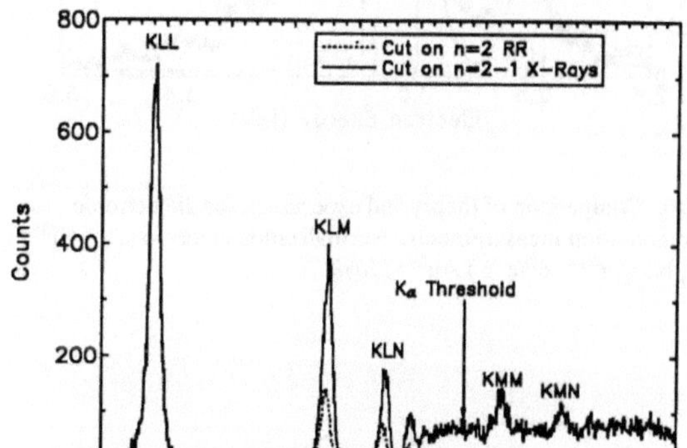

Fig. 5. Excitation functions of Ni^{26+}, taken along the n=1 radiative recombination line and the n=2-1 x ray line.

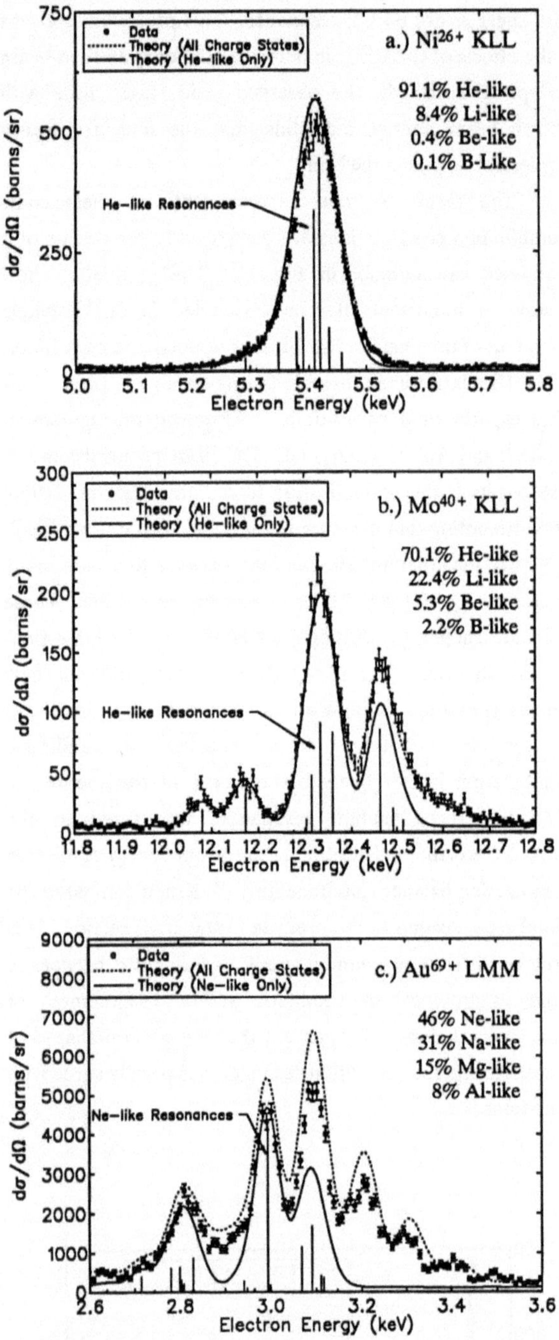

Fig. 6. Comparison of theory and experiment for dielectronic recombination measurements. Normalization errors are: a.) Ni^{26+}: 8%; b.) Mo^{40+}: 6%; c.) Au^{69+}: 20%.

This work was performed under the auspices of the U.S. Department of Energy at the Lawrence Livermore National Laboratory under Contract No. W-7405-Eng-48.

References:

1. M. A. Levine, R. E. Marrs, J. R. Henderson, D. A. Knapp, and M. B. Schneider, Phys. Scr. T22, 157 (1988); M. A. Levine *et al.*, Nucl. Inst. Meth. B43, 431 (1989).
2. C. L. Bennett *et al.*, Bull. Am. Phys. Soc. 35, 1160 (1990); P. Beiersdorfer, M. H. Chen, and R. E. Marrs, Phys. Rev. A 41, 3453 (1990).
3. J. R. Henderson *et al.*, Bull. Am. Phys. Soc. 35, 1159 (1990).
4. K. Wong *et. al.*, private communication; see also K. Wong *et. al.* in this volume.
5. D. A. Vogel *et al.*, private communication; see also D. A. Vogel *et al.* in this volume.
6. D. A. Knapp *et al.*, Phys. Rev. Lett. 62, 2104-2107 (1989).
7. J. Scofield, private communication.
8. M. H. Chen, Phys. Rev. A 33, 994 (1986); J. Scofield and M. H. Chen, Bull. Am. Phys. Soc. 35, 1153 (1990).

Supplement to Z. Phys. D - Atoms, Molecules and Clusters 21, S 147-148 (1991)

Atoms, Molecules
Zeitschrift für Physik D and Clusters

Spectroscopic data tables for highly ionized atoms

W. L. Wiese, J. R. Fuhr, W. C. Martin, A. Musgrove, and J. Sugar
National Institute of Standards and Technology, Gaithersburg, MD 20899, USA

Abstract

Two data centers at the National Institute of Standards and Technology (NIST) are engaged in the critical compilation of atomic spectroscopic data including those for highly ionized atoms--the Atomic Energy Levels Data Center and the Data Center on Atomic Transition Probabilities. Several major compilations have been recently completed, centered on the iron-group elements.

1. Introduction

The principal data characterizing an atomic or ionic spectral line are:

its wavelength;
its classification, i.e., spectroscopic notation in terms of quantum numbers;
the energies of the initial and final atomic states; and
its line strength (or transition probability or oscillator strength).

2. Spectroscopic Data Sources

Spectroscopic data for highly stripped ions have been produced with experimental and theoretical techniques that are quite different[1,2] from those involving lower ions or neutral atoms. With respect to wavelengths, measurements are, except for H- and He-like ions, much more accurate than calculations and are therefore the data of choice for critical compilations. The wavelength data are also utilized to derive energy levels from the analysis of spectra according to the well-known Ritz principle. The typical accuracy of wavelength measurements is about 1 part in 10^5 to 10^6, while the best theories produce accuracies of only 1 part in 10^3 for simple atomic structures such as alkali-like ions--involving 1 valence electron. The crucial problem for experiments is to generate the highly ionized atoms in sufficient numbers for accurate observations. To accomplish this goal, the largest high-power glass laser facilities as well as Tokamak plasmas at fusion research installations have been utilized as radiation sources.

Atomic transition probabilities for highly ionized atoms are mainly obtained from atomic structure calculations, which are mostly based on the relativistic (Dirac-Fock) multiconfiguration self-consistent field approach[2]. This work is typically done for isoelectronic ions so that scaling with nuclear charge may be utilized. Only few experimental contributions exist for transition probabilities of highly ionized atoms, and these are all due to the beam-foil spectroscopy lifetime technique. Recent lifetime results are estimated to be accurate to about 10-20 percent. Theory, on the other hand, does not provide any accuracy estimates.

3. Tables of Critically Evaluated Data

Several extensive spectroscopic compilations involving higher ions have been recently completed. Three major compilations, all published as supplements to the Journal of Physical and Chemical Reference Data are as follows:

<u>For Wavelengths</u>: Atomic and Ionic Spectrum Lines below 2000 Angstroms: Hydrogen through Krypton, by Kelly.[3]

<u>For Energy Levels</u>: Atomic Energy Levels of the Iron-Period Elements: Potassium through Nickel, by Sugar and Corliss[4].

<u>For Atomic Transition Probabilities</u>: Atomic Transition Probabilities--Scandium through Nickel, by J. R. Fuhr, G. A. Martin, and W. L. Wiese[5,6].

In Table 1, we list the references for all the most recent compilations[3-23] containing data for highly ionized atoms. We begin somewhat arbitrarily with ions of charges larger than 10 and thus start the table with the element sodium (nuclear charge Z = 11). However, we should mention that the elements lighter than sodium are well covered with all their ions[26]. As is clearly seen from the table, evaluated data start to become scarce for most elements heavier than nickel. Only for a few elements of particular interest to either fusion research,

TABLE 1.

References for Critical Tables on Wavelengths, Energy Levels and Transition Probabilities of Highly Ionized Atoms with Charge States >10. Only the most recent tables are listed.

Z*	Element Symbol	Data Tabulations on Wave-Lengths	Energy Levels	Transition Probabilities
11	Na	3	7	-
12	Mg	3	9	8
13	Al	3	10	8
14	Si	3,11	12	8
15	P	3,11	13	8
16	S	3,11	14	8
17	Cl	3,11	--	-
18	Ar	3,11	--	8
19	K	3,11	4	8
20	Ca	3,11	4	8
21	Sc	3,15	4,15	5
22	Ti	3,11,16	4,16	5,16
23	V	3,11	4	5
24	Cr	3,11	4	5
25	Mn	3,11	4	5
26	Fe	3,11,17	4,17	6,17
27	Co	3,11	4	6
28	Ni	3,11,18	4,18	6,18
29	Cu	3,11,19	19,20	--
30	Zn	3,11	--	--
31	Ga	3,11	--	--
32	Ge	3,11	--	--
33	As	3,11	--	--
34	Se	3,11	--	--
35	Br	3,11	--	--
36	Kr	3,11	21	--
37	Rb	11	--	--
38	Sr	11	--	--
39	Y	11	--	--
40	Zr	11	--	--
41	Nb	11	--	--
42	Mo	11,22	23	--
.	.	--	--	--
.	.	--	--	--
.	.	--	--	--

*Atomic number or nuclear charge.

X-ray laser development, or astrophysics has some material been critically compiled for such ions.

For highly ionized heavy atoms, the original literature must be searched to obtain spectroscopic data. The recent bibliographies published by NIST[24,25] should be of considerable assistance. For the most current references, the NIST databases may be consulted[26].

References

1. I. Martinson, Rept. Progr. Physics 52, 157 (1989).
2. Y.-K. Kim, in "Atomic Processes in Plasmas," AIP Conf. Proc. 206, 19 (Americ. Instit. Phys., New York, 1990).
3. R. L. Kelly, J. Phys. Chem. Ref. Data 16, Suppl. 1 (1987).
4. J. Sugar and C. Corliss, J. Phys. Chem. Ref. Data 14, Suppl. 2 (1985).
5. G. A. Martin, J. R. Fuhr, and W. L. Wiese, J. Phys. Chem. Ref. Data 17, Suppl. 3 (1988).
6. J. R. Fuhr, G. A. Martin and W. L. Wiese, J. Phys. Chem. Ref. Data 17, Suppl. 4 (1988).
7. W. C. Martin and R. Zalubas, J. Phys. Chem. Ref. Data 10, 153 (1981).
8. W. L. Wiese, M. W. Smith and B. M. Miles, atl. Std. Ref. Data Ser.-Natl. Bur. Stds. (U.S.) 22 (1969)
9. W. C. Martin and R. Zalubas, J. Phys. Chem. Ref. Data 9, 1 (1980).
10. W. C. Martin and R. Zalubas, J. Phys. Chem. Ref. Data 8, 817 (1979).
11. V. Kaufman and J. Sugar, J. Phys. Chem. Ref. Data 15, 321 (1986).
12. W. C. Martin and R. Zalubas, J. Phys. Chem. Ref. Data 12, 323 (1983).
13. W. C. Martin, R. Zalubas, and A. Musgrove, J. Phys. Chem. Ref. Data 14, 751 (1985).
14. W. C. Martin, R. Zalubas, and A. Musgrove, J. Phys. Chem. Ref. Data 19, 821 (1990).
15. V. Kaufman and J. Sugar, J. Phys. Chem. Ref. Data 17, 1679 (1988).
16. K. Mori, W. L. Wiese, T. Shirai, Y. Nakai, K. Ozawa, and T. Kato, At. Data Nucl. Data Tables 34, 79 (1986).
17. T. Shirai, Y. Funataka, K. Mori, J. Sugar, W. L. Wiese, and Y. Nakai, J. Phys. Chem. Ref. Data 19, 127-275 (1990).
18. T. Shirai, K. Mori, J. Sugar, W. L. Wiese, Y. Nakai, and K. Ozawa, At. Data Nucl. Data Tables 37, 235-332 (1987).
19. T. Shirai, T. Nakagaki, Y. Nakai, J. Sugar, K. Ishii, and K. Mori, J. Phys. Chem. Ref. Data, in press (1990).
20. J. Sugar, and A. Musgrove, J. Phys. Chem. Ref. Data, 19, 527 (1990).
21. J. Sugar, and A. Musgrove, J. Phys. Chem. Ref. Data, in preparation (1990).
22. T. Shirai, Y. Nakai, K. Ozawa, K. Ishii, J. Sugar, and K. Mori, J. Phys. Chem. Ref. Data 16, 327 (1987).
23. J. Sugar, and A. Musgrove, J. Phys. Chem. Ref. Data 17, 155 (1988).
24. Bibliography on Atomic Energy Levels and Spectra (L. Hagan, W. C. Martin, A. Musgrove, R. Zalubas), Natl. Bur. Stds. (U.S.), Spec. Publ. 363 (1972); Suppl. 1 (1977); Suppl. 2 (1980); Suppl. 3 (1985); Suppl. 4 (to be published).
25. Bibliography on Atomic Transition Probabilities (J. R. Fuhr, G. A. Martin, B. J. Miller) Natl. Bur. Stds. (U.S.), Spec. Publ. 505 (1978); Suppl. 1 (1980).
26. Inquiries to: Data Center on Atomic Energy Levels, or, Data Center on Atomic Transition Probabilities, National Institute of Standards and Technology, Gaithersburg, MD 20899, USA.

Supplement to Z. Phys. D - Atoms, Molecules and Clusters 21, S 149-150 (1991)

Atoms, Molecules
Zeitschrift für Physik D and Clusters

Satellite spectra for highly-charged H-like ions

U. I. Safronova
Institute of Spectroscopy, Academy of Sciences of the USSR, Troitsk, Moscow Region, 142092, USSR

Abstract

Satellite spectra of H-like ions have been calculated for Z=16-56. New channels of radiative transitions were added in contradictions to previous calculations. Comparison of two results are discussed.

1. Introduction

Usually, radiative channels 2lnl'-1snl", 2lnl'-1s2l" are take into account for data of H-like satellite spectra. The relative intensity of the satellite corresponding to transitions γ-γ' (γ=2lnl', γ'=1snl" or γ'=1s2l") for the case of dielectronic excitation is given by [1]

$$Q_d(\gamma-\gamma')=g_\gamma \frac{\Gamma(\gamma)\ A(\gamma-\gamma')}{A(\gamma)+\Gamma(\gamma)}\ ,\quad A(\gamma)=\sum_{\gamma'} A(\gamma-\gamma') \qquad (1)$$

where $A(\gamma-\gamma')$ and $\Gamma(\gamma)$ are probabilities of radiative transitions and autoionising rates [2]. For 2lnl' states we have only one channel of decay 2lnl'- 1skl" and many radiative channels. For autoionization states 2l3l' it is need to take into account three radiative channels:1s2l" 1s3l", 2l2l" and for 2l4l' there are five channels:1s2l", 1s3l" 1s4l", 2l2l" and 2l3l". The most important channels for both autoionization states are 1s2l" and 1snl" but for some levels it is need to take into account another channels.

For example, there are not direct dipole transitions from 2sndLSJ levels to levels of 1s2p and 1snp states, but there are direct transitions to 2s2pLSJ levels. In this article we consider the influence of these transitions on the values of Q_d.

II. Results of calculations

For calculation of atomic characteristics we used MZ code (see [1]) based on perturbation theory method. Relativistic and radiative effects are taken into account. The radiative transition probability and autoionosing rates are calculated in intermediate coupling scheme.

The results of calculations are given in the Tables 1-3.

Table 1. The relative intensity (Q_d in units $10^{13}s^{-1}$) with (a) and without (b) including 2lnl'-2l2l" transitions

		2s3d $^3D_3$-1snp $^3P_2$		2s4d $^3D_3$-1snp $^3P_2$	
Z		n=3	n=2	n=4	n=2
18	a	1.13	0.295	0.175	2.18-2
	b	1.16	0.300	0.280	3.49-2
26	a	0.396	0.102	0.0645	8.16-3
	b	0.686	0.180	0.391	4.92-2
30	a	0.308	0.0799	0.0304	3.86-3
	b	0.863	0.224	0.381	4.84-2
36	a	0.146	0.0381	0.0109	1.38-3
	b	0.983	0.246	0.356	4.54-2

The Table 1 shows the changing of Q_d with adding of 2snd $^3D_3$-2s2p $^3P_2$ transitions. One should say that we added not only this one transition because 2snd $^3D_3$ level is mixed with another levels (2s3d$^3D_3$ + 2p3p $^3D_3$, 2s4d $^3D_3$+2p4p $^3D_3$+2p4f$^3D_3$.$^1F_3$ $^3F_3$,$^3G_3$). This mixing of levels is rather big, espessially for 2s3d $^3D_3$, 2p3p $^3D_3$ levels since for Z-20 it takes place the changing of levels name.

We can see from the Table 1 that the including of new transitions gives decreasing of Q_d because we in this case (see Eq.1) we increased a value of $A(\gamma)$. One should say that such decreasing is more important for high Z, since for high Z the value of A_r(2snd $^3D_3$-2c2p$^3P_2$) is increased as aZ^4. In the result of this increasing of A_r we obtained new satellite spectra 3-2 and 4-2 line.

Table 2. New satellite lines for H-like ions (A_r , Γ , Q_d in units $10^{13}s^{-1}$)

Z	λ(A)	A(γ-γ')	Γ(γ)	Q_d(γ-γ')
		2s3d $^3D_3$ - 2s2p $^3P_2$		
24	11.652	2.69	0.102	0.230
26	9.8990	3.65	0.128	0.384
30	7.4009	6.13	0.163	0.792
36	5.1122	11.2	0.187	1.10
42	3.1390	21.5	0.197	1.28
48	2.8507	36.2	0.201	1.36
54	2.4432	57.7	0.204	1.40
		2s4d $^3D_3$ - 2s2p $^3P_2$		
24	8.7217	0.746	8.84-2	0.308
26	7.4082	0.989	9.36-2	0.400
30	5.5355	1.71	9.97-2	0.543
36	3.8206	3.47	0.104	0.658
42	2.7929	6.40	0.106	0.708
48	2.1289	10.9	0.107	0.732
54	1.6751	17.6	0.108	0.744

The Table 2 gives wavelengths and A(γ-γ'), Γ(γ), Q_d(γ-γ') for two lines with the same upper levels as there were in the Table 1, but with new ground level. The relative intensity of these lines increases very rapidly with increasing of Z. One should say that Q_d of these lines is not the biggest Q_d between lines of new satellite spectra. For example, Q_d(2p3p $^1D_2$-2s2p $^1P_1$)= 2.35; 2.88; 3.10; 2.89; 2.40; 1.48 for Z=18;20;22;24;26;30.

It is interesting that these values only in 3-10 times less than the maximum values of Q_d for 21nl'-1snl" transitions. The Table 3 gives results for two satellite lines with maximum Q_d. We included two numbers for Q_d: new data -a and old data - b (with and without including 21nl'-2121" transitions)

Table 3. New data for two important lines for H-like satellite spectra

Z	λ(A)	A(γ-γ')	Γ(γ)	Q_d- a	Q_d- b
		2s3d $^1D_2$ - 1s3p $^1P_1$			
20	3.0199	5.49	10.2	15.1	16.0
22	2.4928	8.85	11.0	19.5	20.6
24	2.0917	13.4	11.3	22.4	23.7
26	1.7795	18.7	11.0	23.1	24.3
42	0.66930	113	6.26	14.1	14.5
		2s4d $^1D_2$ - 1s4p $^1P_1$			
18	3.7328	4.39	4.74	9.65	9.88
20	3.0200	6.81	4.65	10.6	10.8
22	2.4924	9.67	4.38	10.4	10.5
24	2.0911	13.1	3.97	9.60	9.75
26	1.7788	17.4	3.65	8.80	8.93
42	0.66904	109	2.59	5.92	6.01

We can see from this Table that the influence of new transitions for Q_d is smaller than for the transitions from the Table 1. This is we can explain the mixing of these levels with levels 2pnp configuration since the level 2s3d $^1D_2$ changes his name two times: 2p3p $^1D_2$ for Z=18-30,$^3P_2$ for Z=31-54 and the level 2s4d $^1D_2$ is 2p4p $^3P_2$ for Z=31-54.

We recalculated Q_d for all satellite lines H-like spectra with taking into account 2131'-2121" and 2141'-2121" transitions. In result we obtained new data of Q_d for satellite spectra of np-1s line and all data for satellite spectra of nl-21'lines for n=3,4.

References

1. Vainshtein.L.A.,Safronova.U.I.:Atomic Data and Nuclear Data Tables. 21. 49 (1978), 25. 311 (1980)
2. Safronova.U.I., Urnov.A.M.: J.Phys. B 12. 3171 (1979)

Supplement to Z. Phys. D - Atoms, Molecules and Clusters 21, S 151-152 (1991)

Atoms, Molecules
Zeitschrift für Physik D and Clusters

Autoionization states 1s2l4l′-energy and autoionization rates

I. A. Ivanov and U. I. Safronova
Institute of Spectroscopy, Academy of Sciences of the USSR, Troitsk, Moscow Region, 142092, USSR

Abstract

The energies of the levels of the 1s2l4l' autoionization states in the range Z=6-54 have been calculated. The calculation was based on the 1/Z perturbation theory allowing for the relativistic effects in the framework of the Breit operator, higher relativistic effects and radiative corrections.

Theory

An expression for the energy matrix consists of five parts:

$$E(a,a')=E_N+E_R+E_S+E_L+E_D \qquad (1)$$

Where E_N - nonrelativistic energy, E_R and E_S the contributions of Breit operators, giving rise to the shift and splitting of the term respectively, the values E_N, E_R, E_S having been calculated in two orders of perturbation theory by 1/Z. E_L and E_D are the Lamb and Dirac corrections An energy matrix is diagonalised in the subspace of states with given parity and angular momentum. As a result of the diagonalisation we obtain the coefficients of intermediate coupling scheme . Using these functions the probabilities of autoionization decay are calculated.

Results and discussion

In table 1 a number of calculated transitions is listed (energies in units 10^4 cm^{-1}). The designations are:

I'458- $1s2p(^3P)4d\ ^4D_{7/2}$
I'278- $1s2p(^3P)4d\ ^2F_{7/2}$
I'478- $1s2p(^3P)4d\ ^4F_{7/2}$
S 212- $1s^22s\ ^2S_{1/2}$

Z	I'458 -S 212	I'278 -S 212	I'478 -S 212
26	6642.3413	6643.9102	6630.0137
28	7747.6895	7750.5200	7731.9375
30	8942.7295	8945.8330	8921.2695
32	10227.494	10230.888	10198.835
34	11565.545	11606.817	11603.108
36	13022.395	13074.846	13070.782
38	14570.460	14636.300	14631.834
40	16210.909	16292.603	16287.667
42	17945.004	18045.283	18039.801

Simaltaneously with the energies the coefficients of intermediate coupling scheme are calculated. The name is attributed to the level according to the largest intermediate coupling scheme coefficient. One can see from the table 1 that with Z is equal to 34 an intersection of the levels occurs. For obtaining the smooth Z-dependences of different physical values we deal with: the energies, the probabilities of the radiative transitions and autoionization decays we use the Wigner-Neumann theorem about the nonintersection of the levels having identical symmetry. The autoionization rates are calculated in the first order of perturbation theory with wavefunctions of intermediate coupling scheme. The formula for the amplitude of the autoionisation decay has the form: (see (1))

$$\Gamma(aLSJ)=\sum_{aa''L'S'}\sum C^J(a,L,S;a',L',S')\times \Gamma(a'L'S',a''L'S')\times C^J(a'',L',S';a,L,S) \quad (2)$$

Table 2 gives an example of an exibition of levels intersection when calculating the autoionization rates. The results of A_a are given in units 10^{13} s^{-1} , 2.27-2 implies 0.0227

Table 2

level	Z=33	Z=34
$1s2p(^3P)4d\ ^4D_{7/2}$	2.27-2	2.21-1
$1s2p(^3P)4d\ ^4F_{7/2}$	2.09-1	2.42-2

level	Z=42	Z=43
$1s2p(^3P)4p\ ^2S_{1/2}$	2.61-1	5.93-4
$1s2p(^1P)4p\ ^2P_{1/2}$	7.14-4	2.59-1

level	Z=21	Z=22
$1s2p(^3P)4d\ ^2D_{5/2}$	1.24-3	1.29-2
$1s2p(^3p)4D\ ^4D_{5/2}$	1.36-2	1.28-3

level	Z=16	Z=17
$1s2s(^1S)4p\ ^2P_{3/2}$	2.17	3.16-5
$1s2p(^3P)4s\ ^2P_{3/2}$	6.99-3	2.17

We can see that the changing of level's name will give smooth Z curves. These examples illustrate the considerations which we were ruled by, using the Wigner-Neumann theorem for the naming of the levels.

Table 3 gives autoionization rates (A_a) for even levels with J=1/2, units are the same as in table 2, the designations of the levels are:

1- $1s2s(^1S)4s\ ^2S$, 2- $1s2s(^3S)4s\ ^2S$
3- $1s2p(^1P)4p\ ^2S$, 4- $1s2p(^3P)4p\ ^2S$
5- $1s2p(^1P)4p\ ^2P$, 6- $1s2p(^3P)4p\ ^2P$
7- $1s2p(^3P)4p\ ^4P$, 8- $1s2p(^3P)4p\ ^4D$
9- $1s2s(^3S)4d\ ^4D$

Table 3

level	Z=20	Z=26	Z=42	Z=54
1	1.04	1.02	9.82-1	9.75-1
2	1.79	1.75	1.66	1.62
3	6.28-1	6.79-1	7.95-1	8.57-1
4	3.92-1	3.23-1	2.61-1	1.03-1
5	1.44-3	1.07-1	7.14-4	1.28-1
6	7.12-3	5.90-3	3.48-4	3.37-3
7	4.92-2	2.08-2	2.03-1	2.21-1
8	2.13-3	6.36-3	9.34-3	8.98-5
9	1.23-9	2.28-7	1.82-5	4.25-4

References

1. Aglitsky E.V, Safronova U.I : "The Spectroscopy of Autoionisation States of Atomic Systems" (in russian), Moscow,"Energoatomizdat",1985.
2. Vainshtein L.A., Safronova U.I.:Atomic Data and Nuclear Data Tables. 21. 49 (1978), 25. 311 (1980)
3. Ivanov I.A., Safronova U.I.:Optics and Spectroscopy,1988, 65,5

Supplement to Z. Phys. D - Atoms, Molecules and Clusters 21, S 153-154 (1991)

Atoms, Molecules and Clusters
Zeitschrift für Physik D

X-ray emission rates from MCDF wave functions including electron exchange corrections

I. Reiche and G. Zschornack
TU Dresden, Sektion Physik, WB AKP, Mommsenstraße 13, O-8027 Dresden, FRG

Abstract
A method to calculate X-ray emission rates for MCDF wave functions considering non-orthogonality of initial and final state orbitals is presented. The basic idea is to select those contributions to the emission rate which contain at most two overlap integrals between orbitals of different principal quantum numbers.

The agreement between calculations of X-ray emission rates, for instance of K_β-transitions, and experimental data can substancially be improved by considering in calculations the non-orthogonality of initial and final state orbitals /1/. In /1/ the necessary corrections were described for atomic wave functions each given by a single Slater determinant. Interaction matrix elements between angular momentum coupled multiconfiguration Hartree-Fock wave functions were derived by Hibbert et al. /2/ allowing up to two subshells in each configuration function to be described by orbitals which are not orthogonal to orbitals in the other configuration function of the matrix element.

In the present paper we give a way to calculate X-ray emission rates using atomic wave functions constructed from orbitals that are all non-orthogonal to orbitals of the other state with the same angular momentum quantum numbers. Instead of the restrictions on the number of non-orthogonal orbitals of /2/ we select a limited number of the largest contributions to the matrix element to reduce complications in calculation.

The atomic wavefunctions of the initial and final state Φ' and Φ are described in the MCDF-model /3/ by linear combinations of configuration state functions

$$\Phi(J,M)=\sum_{s=1}^{n_c} c_s\Phi_s(\gamma_s JM) \qquad (1)$$

which can be expanded in terms of unsymmetrized functions

$$\Phi_s=(N!/\prod_\mu N_\mu!)^{-1/2}\sum_q(-1)^{p(q)}V_s(q) \qquad . \quad (2)$$

Here N_μ is the number of electrons in subshell μ and q denotes a distribution of the electron coordinates with parity p(q), excluding redistribution of electrons within one subshell. The V_s are constructed from central-field Dirac orbitals by coupling the angular momenta in the jj-coupling scheme. Within one state the orbitals u_n form an orthonormal set

$$\langle u_m | u_n\rangle=\delta_{mn} \qquad . \quad (3)$$

But consider the matrix of overlap integrals between initial and final state orbitals $\{\langle u_m | u'_n\rangle\}$. In this matrix overlap integrals between orbitals of different angular momentum quantum numbers vanish as in (3) because of the structure of the one electron wave functions /1/. On the other hand, if all quantum numbers of the two orbitals are identical the overlap integral has a value close to one (instead of being equal to one as in (3)). In the third possible case of orbitals that differ only in the principal quantum number, the overlap integrals are non-zero because of electron relaxation. In the following the latter overlap integrals will be called 'non-diagonal' because they -assuming identical lists of orbitals in both states - appear outside the diagonal of the matrix.

According to the rules of time dependent pertubation theory the rate of spontaneous X-ray emission of multipole L for the transition from the state described by Φ' to the state Φ in atomic units e=m=ħ=1 is given by /4/

$$A=[J']^{-1}\sum_M\sum_{M'}2\pi|\langle\Phi|T^{(L)}_{M''}|\Phi'\rangle|^2 \qquad , \quad (4)$$

where $T^{(L)}$ is the tensor operator describing the interaction between atom and radiation field. $T^{(L)}$ may be written as a sum of operators acting on one electron coordinate

$$T^{(L)}=\sum_{r=1}^{N} t_r^{(L)} \qquad , \quad (5)$$

and applying the theory of symmetric one electron operators acting on antisymmetric wave functions gives /4,2/

$$A=2\pi[J',L]^{-1}\,|\sum_{s=1}^{n_c}\sum_{t=1}^{n'_c}c_s c'_t\langle\Phi_s\|T^{(L)}\|\Phi'_t\rangle|^2$$

$$\langle\Phi_s\|T^{(L)}\|\Phi'_t\rangle=(\prod_\mu N_\mu!N'_\mu!)^{1/2}(N-1)!^{-1}*$$

$$\sum_q\sum_{q'}(-1)^{p(q)-p(q')}\langle V_s\|t_N^{(L)}\|V'_t\rangle \qquad . \quad (6)$$

The integration in (6) yields - apart from coefficients of fractional parentage (cfp), a recoupling coefficient and a one electron matrix element, which will be discussed later - overlap integrals between one electron wave functions of initial and final states.

In the case of all non-diagonal overlap integrals vanishing (that means: assuming orthogonality between initial and final state orbitals) there are only contributions for configurations s,t whose orbital occupation numbers differ by one electron (the case of identical occupation schemes will not be discussed in this paper). Then all contributions from one pair of configurations are identical, the sum over the electron coordinate distributions can be replaced by a multiplicative and phase factor, giving /2/

$$\langle \Phi_s \| T^{(L)} \| \Phi'_t \rangle = (N_i N'_j)^{-1/2} (-1)^{\Delta p} \langle V_s \| t_N^{(L)} \| V'_t \rangle \quad .(7)$$

Here i and j are the subshells with different occupation in s and t, and

$$\Delta p = \sum_{\mu=j+1}^{j} N_\mu \quad \text{if } i<j, \quad \Delta p = \sum_{\mu=j+1}^{i} N'_\mu \quad \text{if } j<i \quad .(8)$$

$$\langle V_s \| t_N^{(L)} \| V'_t \rangle = (cfp)_i (cfp)'_j c_{recoup} * \langle u_i \| t_N^{(L)} \| u'_j \rangle \tag{9}$$

with the cfp

$$(cfp)_i (cfp)'_j = (j_i^{N_i} \alpha_i J_i \{| j_i^{N_i-1} \bar{\alpha}_i \bar{J}_i) * \\ * (j_j^{N'_j-1} \bar{\alpha}_j' \bar{J}_j' |\} j_j^{N'_j} \alpha'_j J'_j) \tag{10}$$

and the recoupling coefficient

$$c_{recoup} = \langle J_1 .. J_{i-1} (\bar{J}_i j_i) J_i .. J \,|\, J'_1 .. J'_{j-1} (\bar{J}'_j j_j) J'_j .. J' \rangle \ . \tag{11}$$

Starting from (6) the aim of the present paper is to show a treatment to calculate X-ray emission rates accounting for non-zero non-diagonal overlap integrals. In this case there may contributions for pairs of configurations that differ in the orbital allocation of more than one electron, and there may be several types of contributions for a given pair of configurations:

$$\text{(i)}\ k_0 (-1)^{\Delta p_0} \prod_\mu (\langle u_\mu | u'_\mu \rangle)^{N_\mu - \delta_{\mu i}} (cfp)_i (cfp)'_j * \\ * c_{recoup} * \langle u_i(r_N) \| t_N^{(L)} \| u'_j(r_N) \rangle \quad ,$$

$$\text{(ii)}\ k_1 (-1)^{\Delta p_1} \prod_\mu (\langle u_\mu | u'_\mu \rangle)^{N_\mu - \delta_{\mu o} - \delta_{\mu i}} \langle u_o | u'_p \rangle (cfp)_i * \\ (cfp)_o (cfp)'_j (cfp)'_p c_{recoup} \langle u_i(r_N) \| t_N^{(L)} \| u'_j(r_N) \rangle \ ,$$

$$\text{(iii)}\ k_2 (-1)^{\Delta p_2} \prod_\mu (\langle u_\mu | u'_\mu \rangle)^{N_\mu - \delta_{\mu q} - \delta_{\mu o} - \delta_{\mu i}} \langle u_q | u'_r \rangle * \\ * \langle u_o | u'_p \rangle (cfp)_i (cfp)_o (cfp)_q (cfp)'_j (cfp)'_p (cfp)'_r * \\ * c_{recoup} \langle u_i(r_N) \| t_N^{(L)} \| u_j(r_N) \rangle \quad , \ \dots \tag{12}$$

The orbitals i,j,o,p,q,r in (12) are given by the coordinate distributions in (6). The contribution (i) is the same as the (only) contribution in the case of mutually ortho-normal one electron wave functions, except that the 'diagonal' overlap integrals in (12) are different from one. To find the multiplicative factor k_1 one of the coordinates 1...N-1 is selected for the integration variable in the non-diagonal overlap integral, say (N-1), multiplying the matrix element by (N-1). For fixed orbitals i,j,o,p the number of distributions q of the remaining coordinates 1...N-2 (distribution q' being completely defined by q) is

$$(N-2)! \ / \ \{ \ (\prod_{\mu \neq i,j,o,p} (N_\mu !)) (N_i - 1 - \delta_{io})! * \\ * (N_j - \delta_{jp})! (N_o - 1)!^{1-\delta_{io}} \ (N_p - 1)!^{1-\delta_{jp}} \ \} \tag{13}$$

Inserting this into (6) one gets

$$k_1 = (N_i N_j N_o N_p)^{1/2} \tag{14}$$

where N_o (N_p) has to be replaced by N_i-1 (N_j-1) if i=o (j=p). In the same way k_2 is found to be

$$k_2 = (N_i N_j N_o N_p N_q N_r)^{1/2} \tag{15}$$

with replacements in analogy with that in (14) if more than one electron is decoupled from one shell. The phase factors Δp_1 and Δp_2 can be calculated directly from the parity difference between the electron coordinate distributions q and q'. All cfp and reduced one electron matrix elements are defined in the same way as in (9,10). To get the recoupling coefficients for the contributions (ii) and (iii) at first the recoupling is written down for the two configurations as in (11) but accounting for the decoupling of electrons from the subshells i,o(,q) in Φ_s and j,p(,r) in Φ'_t . Finally j_p is replaced by j_o (,j_r by j_q).

Contributions with more than two non-diagonal overlap integrals will not be considered here because the non-diagonal overlap integrals have been found to be less than 0.05 in all cases checked. Therefore , contributions with a large number of non-diagonal overlap integrals are important only if they contain one electron matrix elements much greater than that in the addends with less non-diagonal overlap integrals, or if the number of contributions gets very high. But the number of contributions is strongly limited by angular momentum selection rules, and to consider all possible one electron matrix elements it is sufficient to include into calculation the contributions with 0,1 and 2 non-diagonal overlap integrals. For atoms having only closed shells the correctness of the above conception has been shown by a comparison between calculations neglecting all contributions with more then two non-diagonal overlap integrals and calculations based on /1/ accounting for all contributions.

References

1. J.H. Scofield Phys.Rev. A9(1974) p.1041
2. A. Hibbert, C. Froese Fischer, M.R. Godefroid Comp.Phys.Commun. 51(1988) p.285
3. K.G. Dyall, I.P. Grant, C.T. Johnson, F.A. Parpia and E.P. Plummer Comput.Phys.Commun. 55(1989) p.425
4. I.P. Grant J.Phys. B7(1974) p.1485

Supplement to Z. Phys. D - Atoms, Molecules and Clusters 21, S 155-156 (1991)

Atoms, Molecules
Zeitschrift für Physik D and Clusters

Effects of final-state interactions on the L-MM Auger spectra of krypton

S. Fritzsche and G. Zschornack
Institut für Angewandte Kernphysik, Technische Universität Dresden, Mommsenstraße 13, O-8027 Dresden, FRG

Abstract

Relativistic Auger rate calculations in intermediate coupling are presented for the Kr L_3-MM spectra. Beyond the MCDF-model final-state interchannel interactions are taken into account.

I. Introduction

Extensive Auger transition rate calculations have been performed for the K-LL and K-MM spectra for medium and heavy elements. These investigations show the influence of

- the coupling scheme
- relativistic effects (spin-orbit interactions, transition energies) and
- initial- and final-ionic state configuration interactions (CI).

For almost all elements the pure LS- or jj-coupling scheme is inappropriate and intermediate coupling has to apply. This scheme is incorporated in the MCDF-model by configuration interactions.

However, computations for the K-LL spectra in light elements demonstrated that final-state interchannel interactions can also play an important role. This was shown in an MBPT-calculation by Kelly [1] as well as by Howat et al. [2] in the frame of resonant scattering.

In the krypton L-MM Auger spectra substantial relativistic as well as CI effects have to be considered. From experimental view krypton is suitable, because it can be studied by gas-phase electron spectrometry. Recently, Levin et al. [3] reported on a new high-resolution measurement of the Kr L-MM Auger spectra.

II. Theory

In the radiationless decay of inner-shell vacancy states it is usually assumed that the emission of electrons occurs independently from their excitation. In this two-step model the decay-rate for a nonradiative transition between initial-hole state ψ_i and continuum state $\psi_{\alpha\varepsilon}$ normalized per unit energy is given by

$$T_{\alpha i} = 2\pi \left| \langle \psi_{\alpha\varepsilon} | H - E | \psi_i \rangle \right|^2 \quad (1)$$

where H is the total Hamiltonian and E the total energy of the initially ionized atom. Atomic units are used throughout.
Auger transitions arise due to electron-electron interactions. In the MCDF-package of Grant et al. [4] (for the computation of the initial and final bound-state orbitals) as well as in our relativistic investigations we use the sum of Coulomb- and Breit-interaction in the two-electron operator. In the frozen-core approximation only these parts in the Hamiltonian are significant, thus the E term in Eq.(1) can be dropped.
Nevertheless, the MCDF-model regards only the CI of bound states in atomic systems. To include also interactions between a quasidiscrete Auger state and the final continua Aberg and Howat [2] treated the nonradiative decay in the broader context of resonant scattering.
In the time-independent scattering theory the wavefunctions involve the information about the scattering process in their asymptotic behavior. Therefore, in an Auger transition the wave functions have to satisfy the correct "wave boundary" conditions, which is not fulfilled by continuum wave functions generated in an central-field potential. To incorporate these effects of final-state intercontinuum interactions the

partially transition amplitude in Eq.(1) have to expanded by an additionally principle value integration [2]. In this deviation it is supposed that the final ionic state $\psi_{\alpha\varepsilon}$ is degenerated with (N-1) other continuum states $\{\psi_{\beta\tau}\}$ whereas the initial state should be rather isolated from states of the same symmetry.

III. Results and discussion

Although we made calculations for the whole L_{23}-MM spectra, here we discuss as an example the effects of final-state interactions only for the L_3-$M_{45}M_{45}$ spectrum.

Table 1: Effects of final-state interactions on theoretical relative intensities in the L_3-$M_{45}M_{45}$ spectrum of krypton

Final state	Relative intensity[a] MCDF[b] (S)	MCDF[c] (L)	MCDF-IC[d] (L)	Exp.[e]
$^3F_4$	21.6	21.7	23.7	17.3
$^3F_3$	12.7	12.8	15.6	33.5
$^3F_2$	11.8	12.8	15.1	
$^1D_2$	9.6	10.0	12.9	12.6
$^3P_1$	1.0	1.2	1.6	
$^3P_2$	0.4	0.6	0.8	9.1
$^3P_0$	6.9	7.0	6.8	
$^1G_4$	100.	100.	100.	100.
$^1S_0$	2.9	3.7	8.1	11.6

a MCDF-calculations with different numbers of final-state configurations.

b results with the smallest basis set involving only the required $[M_{45}M_{45}]$-configurations.

c large basis set calculation with all [MM]-vacancy configurations.

d results[c] including the interchannel continuum interaction

e ref. [3]

For this spectrum we have performed several MCDF calculations. The Auger energies were taken from different computations (ΔSCF-method). To study the influence of final ionic-state CI we start with the smallest basis set required to build the spectrum. In following calculations we enlarged the basis in a few steps incorporating finally at least all [MM]-vacancy states. The differences between the second and third column in Table 1 show the effects due to final ionic-state CI. But these give only slight corrections compared with the inclusion of inter-continuum interactions (column 4). For instance, these interchannel interactions shift the relative $^1S_0$ intensity by more than a factor of 2^0 and yield much better agreement with experiment [3] than a simpler MCDF-calculation. The total transition rate in the L_3-$M_{45}M_{45}$ group is scarcely altered by final ionic-state CI whereas continuum interactions also result in a redistribution among different groups of the L_3-MM spectra.

Recently, we studied final-state interactions for relativistic K-LL Auger transition rates in the neon-isoelectronic sequence and obtained that inter-continuum interactions result in a remarkable redistribution of individual transition rates even in high-energetic transitions.

IV. Conclusion

For the L-MM spectra we performed detailed and expensive calculations. Even though the MCDF-method is a powerful tool to obtain the wave functions in intermediate coupling we conclude that in a theoretical accurate Auger rate treatment also interactions of the continuum states have to be regarded.

References

1. Kelly, H.P.: Phys. Rev. **A11**, 556 (1975)
2. Howat, G., Aberg, T., Goscinski, O.: J.Phys. B11, 1575 (1978); Aberg, T., Howat, G.: in Handbuch der Physik, Vol. 31, W. Mehlhorn (ed.), (Springer, Berlin, 1982), p. 469
3. Levin, J.C., Sorensen, S.L., Crasemann, B., Chen, M.H., Brown, G.S.: Phys. Rev. **A33**, 968 (1986)
4. Grant, I.P., McKenzie, J.B., Norrington, P.H. , Mayers, M.F., Pyper,N.C.: CPC **21**, 207(1980); McKenzie et al.: CPC **21**, 233 (1980)

Supplement to Z. Phys. D - Atoms, Molecules and Clusters 21, S 157-158 (1991)

Atoms, Molecules
Zeitschrift für Physik D and Clusters

Soft x-ray spectroscopy of high-Z ions in a cool dense plasma

L. P. Presnyakov, A. P. Shevelko and D. B. Uskov
P. N. Lebedev Physical Institute, Academy of Sciences of the USSR

PACS: 52.40 Hf, 32.30 Rj

Abstract

Spectra of multiply-charged ions with one and two electrons are investigated when a laser-produced plasma interacts with a solid obstacle. The level population densities of the Mg ions, recombining in the region of dense cool plasma near the obstacle, are studied both experimentally and theoretically. Values of the gain coefficient are calculated for the case of carbon ions.

I. Introduction

The interaction of a laser-produced plasma with a solid obstacle is of interest for the ion spectroscopy and many current applications /1-5/. In the previous work /4/ we employed high-resolution x-ray spectroscopy to investigate the emission of the He-like and H-like magnesium ions and obtained the temperature, density, charge and exitation state distributions of plasma ions in the near-obstacle region. It was shown that the plasma near the magnesium target had the electron temperature of T_e= 500-700 eV and the electron density of N_e=10^{20}-10^{21} cm^{-3}, whereas near the aluminium obstacle T_e=70-100eV and N_e=2-3* 10^{19} cm^{-3}. Suitable diagnostics of the cool and dense near-obstacle plasma have been developed on the basis of the experimental and theoretical study of the upper members of the principal series and of the adjacent recombination continuum. In the present paper we give some results concerning the level population densities along the Lyman series and point out the conditions corresponding to comparatively large values of the gain coefficient.

II. Experiment and Calculation

The plasma was produced by making use of the second harmonic of a neodymium-glass laser focused onto the plane surface of a magnesium target. A flat aluminium obstacle was placed parallel to the target surface in the path of the laser-produced plasma burst /3-4/. The distance between the target and the obstacle was 1.2mm. The following techniques were used in the experiments: (i) investigation of the temporal-spatial emission parameters in the visible range with the use of a streak camera; (ii) recording an image of the plasma burst in the soft x-ray wavelength range λ<10Å, using a pinhole camera; (iii) recording an image of the plasma burst in the light of spectral lines of H-like and He-like magnesium ions (λ=6-10Å) using an x-ray crystal Hamos spectrograph /3-4/ with spectral resolution $\lambda/\Delta\lambda \sim 10^2$-$10^3$; and (iiii) recording a narrow spectral interval (λ=9.0-9.4Å) near the resonance line of MgXI (λ=9.168A) with high spectral resolution ($\lambda/\Delta\lambda \sim 10^4$)

using a crystal Johann spectrograph /4/. All the spectrograms were recorded in a single laser shot. The typical examples of the spectra recorded are published earlier /4/.

Calculations for the level population densities are carried out using a stationary model.

III. Results

Both experimental and theoretical data for the population of excited levels are given in Fig 1.

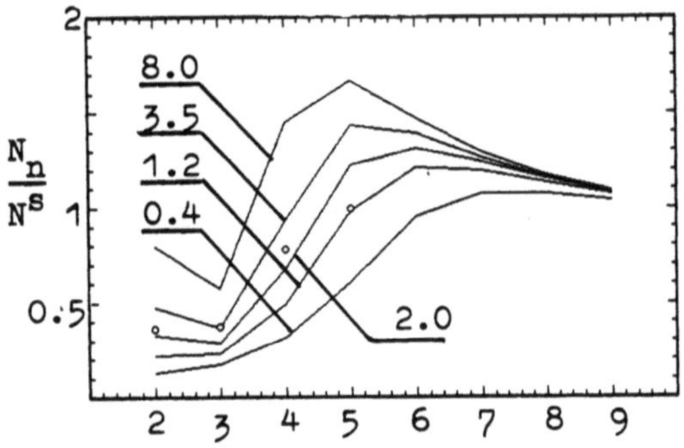

Fig.1. The magnitude of level population densities N_n of the $1sn^1P$ states of Mg XI divided by $N^s = \lim_{n\to\infty} N_n$ for T_e=70eV; o - results obtained from experimental data for line intensities along the principal series of the He-like magnesium ions and intensity of the photorecombination continuum at the edge; — theoretical calculations. The values of N_e is given on the curves in $10^{19}cm^{-3}$

Under the experimental conditions a population inversion is observed for the levels with n=4 and 5 relative to the levels with n=3. The calculated gain coefficient for the Mg XI ion is about $10^{-2}cm^{-1}$, close to that obtained in /5/. We have also calculated the magnitude of the gain coefficient for the case of CVI Balmer-α transition for some temperatures and densities which could be expected in experiment. The results are presented in Fig. 2.

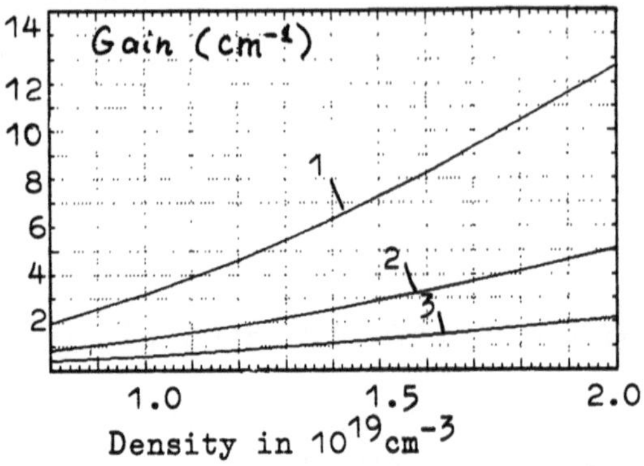

Fig. 2. The magnitude of the gain coefficient for 18.22nm transition in H-like carbon. 1 - Te=10eV; 2- Te=15eV; 3- Te=20eV.

IV. Conclusion

Plasma near the obstacle is rather cool and dense. However, it contains bare nuclei, H-like and He-like ions with a very small addition of Li-like ions. A population inversion observed near the obstacle shows a promising direction for the x-ray spectroscopy of highly charged ions.

Acknowledgments

One of us (D.U.) acknowledges support by the DFG for a stay at Bochum.

References

1. Bhagavatula V.A. and Yaakobi B.: Opt. Commun. 24, 311(1978)
2. Presnyakov L.P. and Shevelko A.P.: JETP Lett. 36, 44 (1982)
3. Mazing M.A., Pirogovsky P.Ya., Presnyakov L.P. and Shevelko A.P.: Phys. Rev. A32, 3695 (1985).
4. Beigman I.L., Pirogovsky P.Ya., Pressyakov L.P., Shevelko A.P and Uskov D.B.: J.Phys. B: At., Mol., Opt. Phys. 22, 2493 (1989)
5. Kodama R. and Mochizuki T.: Opt. Lett. 12, 990 (1987)

Supplement to Z. Phys. D - Atoms, Molecules and Clusters 21, S 159-160 (1991)

Atoms, Molecules
Zeitschrift für Physik D and Clusters

Theoretical analysis of iron X-ray spectra observed in solar active region

M. Cornille, J. Dubau, M. Loulergue
Observatoire de Meudon, URA 812, 92195 Meudon Cedex, France

F. Bely-Dubau, P. Faucher
Observatoire de la Côte d'Azur, BP 139, Nice Cedex, France.

Abstract

The theoretical analysis of X-ray spectra observed by the Flat Crystal Spectrometer of the Solar Maximum Mission in the 13Å-19Å wavelength range is presented. The emission lines correspond to Fe ions (Fe XVI, Fe XVII, Fe XVIII) and some O-ions (O VII,O VIII). The fitting of a synthetic spectrum to the observed one shows that it is difficult to represent the emissive region with one temperature model.

I. Introduction

A set of spectra obtained by the Flat Crystal Spectrometer of the Solar Maximum Mission is analysed. It concerns a non flaring active solar region and was observed on 1986 september 20. The emission corresponds to the 13Å - 19Å wavelength range which is of a great interest for astrophysical and laboratory plasmas.

The refined analysis of such soft X-Ray spectra presents many difficulties due to the multiplicity of emitted lines which correspond either to different elements or different ion stages of the same element. Such an analysis includes identification of the numerous lines and the building of a synthetic spectrum which allows to diagnose the plasma from a fitting with the observed one.

This paper is devoted to the analysis of the 13Å - 19Å wavelength range. In this range, a great number of Fe-lines which have to be identified and some O-lines are observed. The Fe-lines correspond to the important Fe XVII resonance lines and their satellites from Fe XVI. For temperatures greater than 2.5 $10^6 K$ some lines of Fe XVIII appear in the spectra. The O-lines correspond to some resonance lines of the H-like ion O VIII and their satellites from O VII.

II. Theory and Results

The atomic processes responsible for the resonance line intensities are mainly collisional excitation by electron impact including direct and cascade contributions, whereas satellite lines are due to dielectronic recombination and inner-shell excitation processes and the atomic parameters which have to be calculated are energy levels, collisional excitation rates, autoionisation and radiative probabilities.

Solar active regions correspond to low density hot plasmas which are optically thin at the X-ray energies of interest here. Then, in every ionisation stage, the excited states are underpopulated in regard to the ground state so that the excitation processes occur predominantly by collisions of ground state ions with electrons having a Maxwellian velocity distribution.

The atomic parameters have been calculated using the University College London computer package which was completed and adapted for X-spectrum studies. The different codes are Superstructure[1,2], Distorted Waves[3], JJOM[4], and AUTOLSJ[5], and the methods used to obtain the different parameters are described elsewhere[6,7].

For Fe XVII, energy levels and radiative probabilities were obtained using Superstructure by including the 15 following configurations : $1s^2 2s^2 2p^6$, $1s^2 2s^2 2p^5 3l$ and $1s^2 2s 2p^6 3l$ for l=0,1,2, $1s^2 2s^2 2p^5 4l$ and $1s^2 2s 2p^6 4l$ for l=0,1,2,3. Collisional excitation rates were calculated for transitions between the ground level $1s^2 2s^2 2p^6\ {}^1S_0$ and the 36 levels of the following configurations : $1s^2 2s^2 2p^5 3l$ and $1s^2 2s 2p^6 3l$ for l=0,1,2.

For Fe XVI, wavelengths, radiative and autoionisation probabilities were calculated for the satellite lines (≈ 500) emitted from the 175 levels corresponding to the configurations $1s^2 2s^2 2p^5 3l 3l'$ with l, l'=0,1,2 while the inner-shell excitation rates were calculated between the ground level $1s^2 2s^2 2p^6 3s\ {}^2S_{1/2}$ and the levels corresponding to the following configurations : $1s^2 2s^2 2p^6 3d$, $1s^2 2s^2 2p^5 3s^2$, $1s^2 2s^2 2p^5 3s 3d$ and $1s^2 2s^2 2p^5 3p^2$.

For Fe XVIII, wavelengths and radiative probabilities were calculated for transitions between the levels corresponding to the configurations $1s^2 2s^2 2p^5$, $1s^2 2s 2p^6$, $1s^2 2s^2 2p^4 3l$ and $1s^2 2s 2p^5 3l$ for l=0,1,2. The same configurations were considered for the collisional excitation rate calculation for transitions between the two levels of the ground configuration $1s^2 2s^2 2p^5$ and the levels corresponding to the other configurations.

At last, for the oxygen ions, wavelengths and radiative probabilities were calculated for O VIII for transitions be-

tween levels corresponding to the configurations nl with $n \leq 5$, while, for O VII, wavelengths, autoionisation and radiative probabilities were calculated for all the transitions $2l'n''l'' \rightarrow 1snl$ for n,n"=2,3,4.
The entire set of these parameters is then used to built a synthetic spectrum. Each line (transition from level i to level j) is caracterised by its emissivity ϵ_{ij} :

$$\epsilon_{ij} = N_i(X^{+m})\ A_{ij}\ E_{ij} \quad (erg\ cm^{-3}\ s^{-1})$$

where $N_i(X^{+m})$ is the number density of the emitting level i of the ion X^{+m}. A_{ij} and E_{ij} are respectively the radiative probability and the transition energy between the two levels. The quantity $N_i(X^{+m})$ is calculated by solving the time dependent statistical equilibrium equations. In the present case, the resolution was done from the following assumptions :

-state stationnarity : $dN_i/dt = 0$

-limitation of the levels to those having a quantum number $n \leq 3$.

-separation between calculation of level population and ionisation equilibrium.

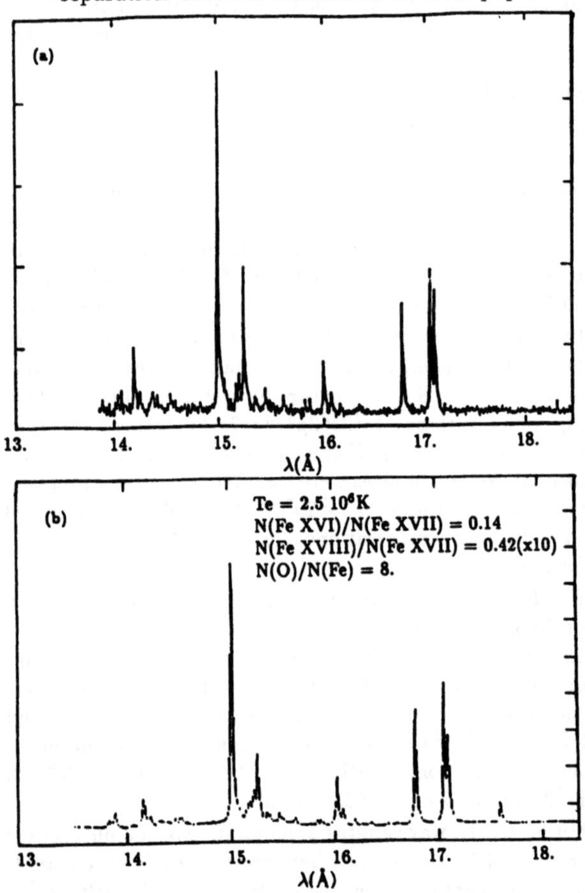

Figure 1. Solar active region spectrum.
(a) Spectrum observed by the FCS of SMM at 13:25:2.788 UT SEPT. 20 1986.
(b) Synthetic spectrum. Abundance ratios of Fe ions are taken from Arnaud and Rothenflug(8). Abundance ratio N(O)/N(Fe) is taken from Walker et al.(9).

Then, the synthetic spectrum is built taking into account for profiles of the emitted lines. These profiles are a combination of intrinsic width due to the Doppler motion of ions and instrumental effects due to the crystals.

A fitting between the synthetic spectrum and the observed one is obtained from a set of trial plasma parameters. The trial parameters are adjusted to obtain the best fit to the principal diagnostic lines. The parameter set thus determined includes electron temperature, equivalent ion temperature, abundance ratio between ions...
An illustration of such a fitting is given in figure 1. Fig.1a represents the spectrum observed by the spectrometer FCS of the satellite SMM at 13:25:2.788 UT Sept.20,1986 in the 14Å-18Å wavelength range. Fig 1b represents a synthetic spectrum which fits well enough the observed one. Such a fit is obtained for the plasma parameters noted on the figure.

III. Discussion

The principal trial parameter is the electron temperature which can be estimated from the intensity ratio of a pure dielectronic line to its parent line[10]. When the temperature is estimated,the abundance ratios between ions are then taken from the tables of Arnaud and Rothenflug[8]. Such a technique implies that the emissive region has only one temperature. The fit of the figure 1 which represents the best fit that we have obtained needs to take an abundance ratio N(Fe XVIII)/N(Fe XVII) = 0.42 which is 10 times greater than the value (0.042) given by Arnaud and Rothenflug at 2.5 $10^6 K$. Such a result may indicate that the whole emissive region is not at the same temperature, and thus, it can be necessary to consider a multi-temperature model from emission measure to represent correctly the analysed solar active region.

References

1 Eissner W., Jones M., Nussbaumer H., Comp. Phys. Commun., **8**, 270(1974).
2 Eissner W., Nussbaumer H., J. Phys. B, **2**, 1028(1969).
3 Eissner W., Seaton M.J., J. Phys. B, **5**, 218791972).
4 Saraph E., Comp. Phys. Commun.,**3**, 256(1972).
5 TFR group, Dubau J., Loulergue M., J. Phys.B, **15**, 1007(1981).
6 Bely-Dubau F., Dubau J., Faucher P., Gabriel A. H., Mon. Not. R. Astr. Soc., **198**, 239(1982a).
7 Bely-Dubau F., Dubau J., Faucher P., Gabriel A. H., Loulergue M., Steenman-Clark L., Volonté S., Antonucci E., Rapley C.G., Mon. Not. R. Astr. Soc., **201**, 1155(1982b).
8 Arnaud M., Rothenflug R., Astr. Ap. Sup.,**60**, 425(1985)
9 Walker A.B.C.Jr, Rugge H. R., Weiss K., Ap. J., **192**, 169(1974).
10 Gabriel A. H., Mon. Not. R. Astr. Soc., **160**, 999(1972).

Supplement to Z. Phys. D - Atoms, Molecules and Clusters 21, S 161-162 (1991)

Atoms, Molecules
Zeitschrift für Physik D and Clusters

Absolute radiometry technique for VUV and SXR radiation fluxes

S. V. Bobashev, N. N. Salashchenko*, L. A. Shmaenok
A. F. Ioffe Physico-Technical Institute, Leningrad
* Institute of Applied Physics, Gorki, USSR

In research on hot plasmas in addition to detailed spectroscopic measurements determination of soft-X-ray (SXR) radiation intensity distributions in a wide range of photon energies are urgently needed. In the case of an inertial plasma that created by a powerful driver (laser produced plasma, megaampere Z-pinch), such a distribution especially should be obtain in a single flash with a nanosecond (or subnanosecond) time resolution.

Our approach to solve this problem is analogous to that used in [1]: it is based on dividing the spectral range of interest into a number of narrow intervals by dispersion elements-multilayer mirrors (MLM), providing a moderate resolution ($\lambda/\Delta\lambda$~20-100), and detecting radiation in all intervals simultaneously by fast detectors: PIN diodes or microchannel plates. Owing to high reflectivities of MLM's at nongrazing angles (R>0.1) it is possible to use small- size samples of MLM, placing many spectral channels in one camera

This technique was applied recently in the work [2]: the 10-channel polychromator has been used for experiments with a distant Z-pinch plasma source at the ''ANGARA-5-1'' installation. The polichromator contained low-selective MLM's, prepared for specific spectral points in the photon energy region 0.26-0.93 keV.

Now, owing to improvements of the parameters of the elements and of the polychromator design, more extensive possibilities of its application may be considered. The new 10-channel polychromator (Fig.1) incorporates as previously the same separate channel blocks, each of them containing a MLM, a detector and a filter, but they are attached independently to a vertical flat (well polished) bar. So every block can be rotated in the vertical plane and adjusted practically to any given distance to a radiation source (L>25 cm). Thus experiments with small-scale laboratory X-Ray sources, among them laser produced plasmas being most widespread, are feasible . A possibility of a fine adjustment ($\Delta\theta$ <1') of the angle between the incident beam axis and the MLM surface has been reached. This geometrical precision corresponds to the accuracy of setting wavelength values for all the channels not worse than ($\Delta\lambda/\lambda \simeq 10^{-3}$).

A necessary (subnanosecond) time resolution for laser produced plasma experiments can be reached using microchannel plate detectors, which are to be mounted in the polichromator instead of PIN diodes. These 8mm-diameter chevron MCP's placed into the same universal bodies and connected with vacuum feedthroughs by 50 Ohm cables provide quite measurable signals up to several volts for short pulses. The main disadvantage of MCP's (based on the outer photoeffect) as compared with PIN diodes is lower reliability and stability of the absolute calibration, so a recalibration should be done more often.

In the new polychromator a considerable improvement of the channel spectral selectivity as well as an availability of an arbitrary choice of spectral points has been realized by using new types of MLM's. Two groups of MLM's have been prepared for the device. The first one for measurements with a rather high spectral resolution has been built mainly by pairs of elements, which appeared to be effective for specific spectral points (Cr, Ni, Co, Fe, V, Sc with C, Ni-Ti). Up to this time in our work all the measurements of MLM's parameters have been done with a grazing incidence monochromator supplied with an X-Ray tube. The polychromator channels are adjusted usually to L-lines of many elements and the K-lines of C, Al and Mg. The MLM's of the first group have been optimized for the mentioned photon energies in the range 0,1-1,5 keV.

The second group of MLM's includes samples of W-Si structures of different periods d=2.0-3.7 nm, providing for the polychromator the photon energy 0,3-1,7 keV.At the highest energies these MLM's compete successfully with those of the first group,being quite comparable at lower energies.The important advantage of these MLM's is the possibility of chosing in the polychromator any spectral points within the range 0,3-1,7keV.

As earlier the polychromator contains thin (submicron) filters,made of different metals (mainly by means of laser evaporation),with L-edges,located as close as possible to chosen spectral points.The filter thickness h < 0,4 micrometer appears to be quite sufficient for suppression of the visible and VUV light in the measurements of the soft X-ray radiation of plasma sources.

The polychromator may be considered as an absolute radiometric device due to the calibration of all the elements. We would like to emphasize that the technique of absolute calibration of detectors using the photoionization quantometer [3] in combination with different pulsed sources [4] has proved its utility in the work of the polychromator.

One may hope that future application of a synchrotron radiation beamline to calibrations of MLM's (direct measurements of R(l) instead of indirect determination via measurements R(Q) using X-Ray tube lines) will improve the accuracy and reliability of multichannel spectral selective radiometry in the range 0.1-2 keV.

REFERENCES

[1]. R.L. Blake, R.G. Hockaday, J.S. Grosso et.al.,Proc. SPIE, vol. 688 (1986), p. 146.
[2]. S.V. Bobashev, L.A. Shmaenok and V.P. Smirnov. Proceedings of 15-th International Conference on X-Ray and Inner-Shell Processes, 1990, to be published.
[3].S.V.Bobashev, L.A.Shmaenok.Rev. Sci. Instr. 52 (1980),16.
[4].Z.A.Al'bikov,V.P.Belik,S.V.Bobashev et.al., Plasma Diagnostics, vol.6,Energoatomizdat,Moscow,1989,p.48.

Supplement to Z. Phys. D - Atoms, Molecules and Clusters 21, S 163-164 (1991)

Atoms, Molecules
Zeitschrift für Physik D and Clusters

He^{2+} - H collisions: experimental charge exchange cross sections and their relevance for alpha particle diagnostics at JET

G. J. Frieling[1], R. Hoekstra[1,2], R. Morgenstern[2] and F. J. de Heer[1]

[1] FOM-Institute for Atomic and Molecular Physics, P. O. Box 41883, 1009 DB Amsterdam, The Netherlands

[2] Kernfysisch Versneller Instituut, Zernikelaan 25, 9747 AA Groningen, The Netherlands

Abstract

We have determined absolute cross sections for electron capture into the quasi-degenerate $4l$ states of He^+ produced in collisions of He^{2+} on atomic hydrogen. The cross sections have been deduced by deconvoluting spatial HeII($n = 4 \rightarrow n = 3$) emission profiles. The measurements have been performed in the impact energy range of 2 - 130 keV/amu. The results are compared with theory. In addition we discuss the relevance of these cross sections to α-particle diagnostics in fusion plasmas. At the JET tokamak the HeII($n = 4 \rightarrow n = 3$) emission can be used to determine related quantities like ion temperature, plasma rotation and local α-particle densities.

1. Charge exchange cross sections

In several publications (e.g. [1-4]) we have reported on our Photon Emission Spectroscopy experiments of electron capture into excited states in collisions of bare ions with atomic hydrogen, which are highly relevant for the diagnostics of fusion plasmas [5,6]. Here we shall confine ourselves to the process

$$He^{2+} + H(1s) \longrightarrow He^{+}(4l) + H^{+} \qquad (1)$$

followed by HeII($n = 4 \rightarrow n = 3$) photon emission at 468.6 nm.

To produce He^{2+} ion beams in the energy ranges of 2 - 13 and 25 - 130 keV/amu two accelerators have been used, the ECR ion source installed at the KVI (Groningen) and the Van de Graaff accelerator at AMOLF (Amsterdam) respectively. In the latter case He^{2+} arises from stripping the primary accellerated He^+ beam. In a collision chamber the ion beam crosses a partly dissociated hydrogen beam effusing from a radio frequency discharge source. The photon emission is observed with a monochromator (see e.g. [1-4]).

The $4l$ state selective capture cross sections have been determined from the HeII($n = 4 \rightarrow n = 3$) emission, measured along the ion beam axis, by means of deconvolution technique [2,3,4], which exploits the fact that the lifetimes of the $4l$ states are different.

The results for the HeII($n = 4 \rightarrow n = 3$) are given in fig. 1 together with only the theoretical results of Atomic Orbital (AO) calculations by Fritsch [7] and the distorted wave approximation

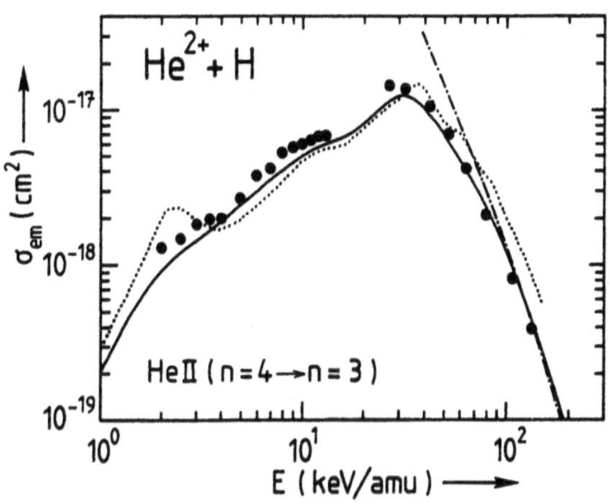

Fig. 1 HeII($n=4 \rightarrow n=3$) emission cross sections. Experimental errors: systematic 20%, relative the size of the points. Theory: ···· AO [7], — — — DWA [8] and ——— JET data base [11]

Table 1 Experimental and theoretical (AO [7], DWA [8] and JET data base [11]) total $n = 4$ (in 10^{-18} cm^2) and relative $4l$ cross sections (in %)

E(keV/amu)		$n = 4$	$4s$	$4p$	$4d$	$4f$
10	exp	18	18	33	31	17
	AO	13	14	39	23	24
	JET	15	11	38	30	21
50	exp	31	30	55	11	4
	AO	41	37	48	13	2
	DWA	64	25	45	24	6
	JET	29	32	49	14	5
100	exp	5	32	55	12	1
	AO	9	42	48	10	1
	DWA	5	31	46	19	3
	JET	5	36	49	12	3

(DWA) of Belkic [8]. Results of the unitary distorted wave approximation [9] and the classical trajectory Monte Carlo [10] calculations are not shown, since in their range of applicability the overall agreement is less good. Fig. 1 also shows the satisfactory agreement with the (preferred)

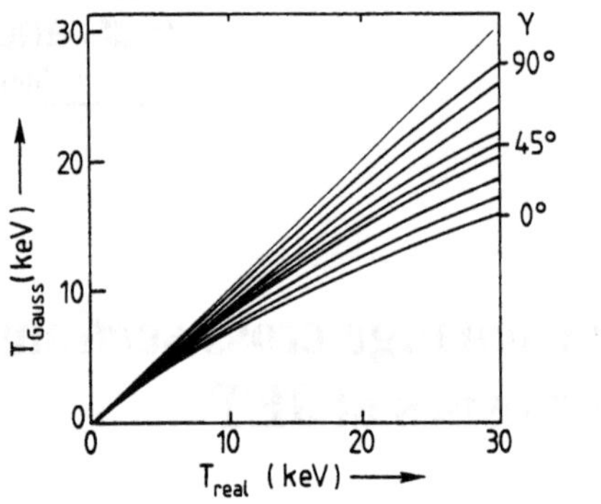

Fig. 2 Relation between the real ion temperature and the one deduced from Gaussian fits for different observation angles γ.

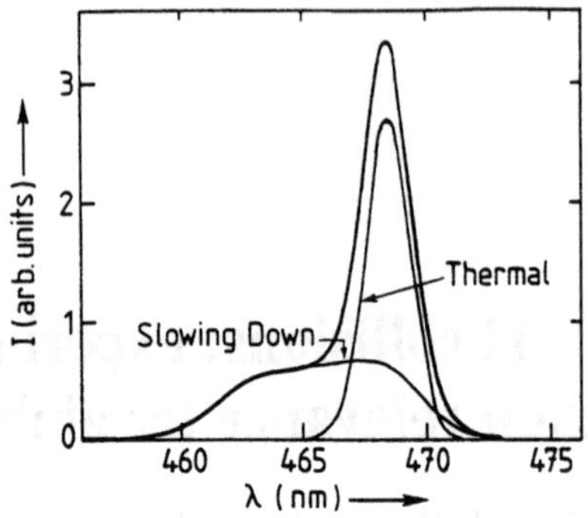

Fig. 3 The expected HeII($n = 4 \rightarrow n = 3$) emission spectrum in the tritium phase of JET.

the available theoretical and experimental data, including our lower energy results [2,3].Table 1 gives some of the deconvolution results for the $4l$ and the total $n = 4$ cross sections together with AO [7], DWA [8] and preferred JET data [11]. The good agreement between JET data base results and experiments indicates that it is a reliable set of cross sections which may be very beneficial for charge exchange recombination spectroscopy purposes.

2. Alpha particle diagnostics at JET

Observation of spectral emission from impurity ions in fusion plasmas following charge transfer from injected heating beams (40 - 80 keV/amu H or D) is a major tool to determine local ion temperatures, ion densities and the plasma rotation velocity [5]. First we consider the validity of deducing the ion temperature of thermalized alpha particles at JET, exploiting reaction (1). Generally the observed HeII($n = 4 \rightarrow n = 3$) emission spectrum which is Doppler broadened can be fitted by two Gaussians: a narrow one corresponding to charge exchange at relatively low temperature near the plasma boundary and a broad one due to charge exchange in the hot plasma core. However fitting with a Gaussian implicitly assumes a rate coefficient (product of the cross section and the relative velocity of the α-particles with respect to the neutral heating beam) with a constant emission cross section usually choosen at the velocity of the heating beam. For plasma quantities like temperature and rotation velocity this may give rise to large deviations from the case where the velocity dependence of the rate coefficient (the emisson cross sections) is correctly taken into account by integrating over the velocity distribution of the α-particles [6]. In order to get the relation between quantities determined with Gaussians and the true plasma conditions, computer calculations have been done which first create emission profiles as function of the plasma temperature and thereafter fit these profiles with Gaussians. For different observation angles γ (defined with respect to the heating beam) Fig. 2 shows the relation between the real ion temperature and the one deduced from a Gaussian fit.

Further studies [5,6], also considering Bremsstrahlung show that the "charge exchange recombination spectroscopy" is also promising for the diagnostics of slowing down alpha particles (α's produced by the nuclear fusion reaction) in the active phase of JET. Hence it may provide a new future tool for measuring e.g. the local source rate of alpha particles. Fig. 3 shows an expected HeII($n = 4 \rightarrow n = 3$) emission spectrum in the tritium phase of JET. In this figure it is seen that the slowing down and thermal component are clearly resolved.

Acknowledgment

We wish to thank A.N. Zinoviev, E. Smulders and W. Dickson for their substantial contributions to the high energy measurements. This work is part of the research program of the "Stichting voor Fundamenteel Onderzoek der Materie" (FOM), which is financially supported by the "Stichting voor Nederlands Wetenschappelijk Onderzoek" (NWO). This work is also part of the association agreement between FOM and EURATOM with financial support by the "Stichting voor Nederlands Wetenschappelijk Onderzoek" (NWO).

References

1. Hoekstra, R., Čirič, D., de Heer, F.J., Morgenstern, R.: Phys.Scr. **T28**, 81 (1989)
2. Hoekstra, R., de Heer, F.J., Morgenstern, R. *Electronic and Atomic Collisions* (edited by Dalgarno, A. et al. (American Institute of Physics, New York, 1990) 390
3. Hoekstra, R., Ph.D. Thesis, R.U.Groningen 1990
4. Hoekstra, R., these proceedings
5. Boileau, A., von Hellermann, M., Horton, L.D., Spence, J., Summers, H.P.: Plasma Phys. and Controlled Fusion **31** 779 (1989)
6. Frieling, G.J., Hoekstra, R., de Heer, F.J., Mandl, W., Boileau, A., Summers, H.P., von Hellermann, M.: Contr. Papers 17th EPS conf. on controlled fusion, Amsterdam 1664 (1990)
7. Fritsch, W., Phys.Rev. **A38** 2664 (1988) and private communication
8. Belkic, D., private communication
9. Ryufuku, H., Rep. JAERI-M-82-031 (1982)
10. Olson, R.E., private communication
11. Boileau, A., Frieling, G.J., von Hellermann, M., Mandl, W., Summers, H.P.: to be published

Supplement to Z. Phys. D - Atoms, Molecules and Clusters 21, S 165-166 (1991)

Atoms, Molecules
Zeitschrift für Physik D and Clusters

The laser ion source of multiply charged ions

V. B. Kutner
Joint Institute for Nuclear Research, Dubna, USSR

Yu. A. Bykovski, V. P. Gusev, Yu. P. Kozyrev, V. D. Peklenkov
Moscow Engineering Physics Institute, Moscow, USSR

Abstract

The effect of the transverse and logitudinal magnetic field on the charge spectrum of the laser plasma, and the quantitative characteristics of the plasma ion component have been investigated.

I. Introduction

The use of laser-produced plasmas as a multiply charged ion source for the accelerators has been successful and promising owing to the features of ion emission from the laser plasma, to the possibility of the universal application for all types of accelerators and to comparatively simple design of the ion source.
In the given paper the results of an investigation of the effect of the transverse and longitudinal magnetic field on the laser plasma parameters are presented.

II. Experiment

The investigation of the influence of magnetic fields on the laser plasma characteristics has been performed in the ion source bench tests[1]. The magnetic field B was varied from 0.01T to 1.5T. In the bench tests the charge separation and the analysis of extracted ions were performed at a pressure better than 10^{-5}mbar.
In the investigations two types of lasers were used: a CO_2-laser with a wavelength $\lambda = 10.5$ µm and a peak energy of up to 6mJ allowing to get a laser radiation power of $3\cdot10^{10}$W·cm^{-2} on the target surface, and the second one based on the commercial neodymium doped glass laser with $\lambda = 1.06$ µm at an energy of up to 70 mJ and with luminous flux $q = 6\cdot10^{9}$ W·cm^{-2}. Solid Li, LiF, C, Zr, Ta, etc. were used as target materials. The analysis of the characteristics of ion emission from laser-produced plasmas in a magnetic field was carried out using both the time-of-flight and mass-spectrometric methods.

III. Results

1. Three distinct domains of the interaction of the plasma with a magnetic field during its expansion in a transverse magnetic field may be ascertained, depending on the value of the plasma and magnetic field pressure:

$$\beta = (n_i m_i v_i^2/2)/(B^2/2\mu_0),$$

where n_i, m_i, v_i are the density, mass and ion velocity, respectively.
From the analysis of the characteristics of ion emission of the laser plasma expanding in a transverse magnetic field in the region of $\beta<1$ the following has been established [2-6]:
- the compression of the angle of ion divergence during expansion takes place. The main part of the plasma is expanding within the angle limits $\varphi\in(-20°$- $+20°)$ counted off the normal to the target surface. The angle $\varphi>0$ corresponds to the angle counted off in the bend direction of the ions by means of the magnetic field. The value of the angle (φ) decreases with increasing magnetic field and decreases with luminous flux (q). In this case the formation of one or two plasma plums drifting in a transverse magnetic field is possible, depending on the physical conditions. For instance, the spatial distribution of the plasma torch produced on the carbon target at $\beta\leq 0.003$ (the arrival time of the plasma front $\approx$1 µs) is shown in figure 1;

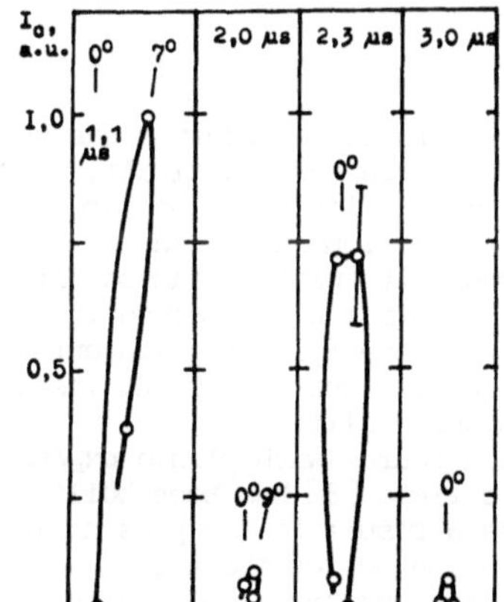

Fig.1. The ion current from the probe situated at distance L = 6.5 cm from the target surface at different time points after plasma production. Target: carbon, B = 0.5 T, $q\approx 4\cdot10^{9}$W·cm^{-2}.

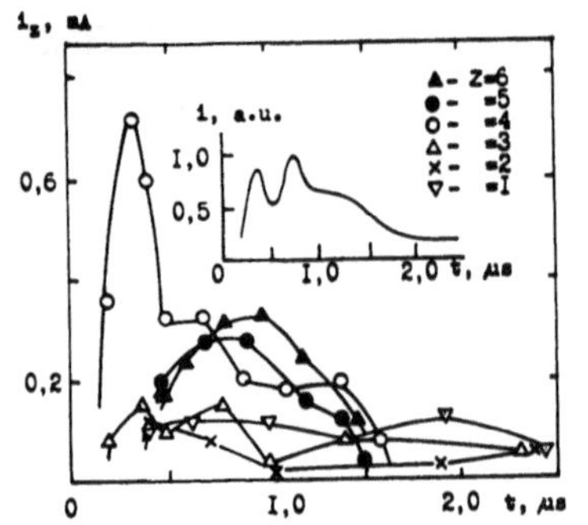

Fig.2. Variation of the carbon ion current for different Z in plasma plum passing through the transverse magnetic field, as a function of time after plasma production. L = 5 cm, B = 0.2 T, q ≈ $3\cdot10^{10}$ W·cm^{-2}. Upper fig.(a) is the total current of the ions with different Z.

- the magnetic field leads to a deformation of the energy spectra of the multiply charged ions.

Figure 2 shows the time variation of the carbon ion current distribution for different charges in the plasma blob, obtained by using the mass-spectrometric method.

The obtained results about the spatial distribution of the plasma plum charge and the energy distributions of the plasma ions have formed the basis for the construction of the laser heavy ion source with the transfer of the plasma from the target to the emission slit across the U200 cyclotron magnetic field [7,8].

2. The investigation of the influence of a longitudinal magnetic field (the normal to the target surface is parallel to the magnetic field lines) on the characteristics of ion emission from the laser plasma expanding at a distance of 13 cm have shown [8,9] the following:

- the production of a longitudinal magnetic field leads to the compression of the spatial angle of ion expansion. Under certain physical conditions the plasma takes the form of a cylinder whose axis is the normal to the target surface. The plasma column diameter decreases monotonically with increasing B (see figure 3).

IV. Discussion

Thus the investigation performed has shown that the spatial, charge and quantitative characteristics of the plasma ion component are entirely suitable for the development of an ion laser source for cyclotrons with plasma transportation from the target to the emission slit along the magnetic field, and with the following ion extraction from the lateral plasma surface during its streaming near the emission slit.

The laser ion source with plasma expansion along the magnetic field [8] has been built for the U-200 cyclotron as a result of the present investigation. In the given ion source the target was shifted toward one of the cyclotron poles at distance L=2 cm relative to the extraction slit. The target surface was placed perpendicular to magnetic lines. With the same laser parameters used in transverse plasma

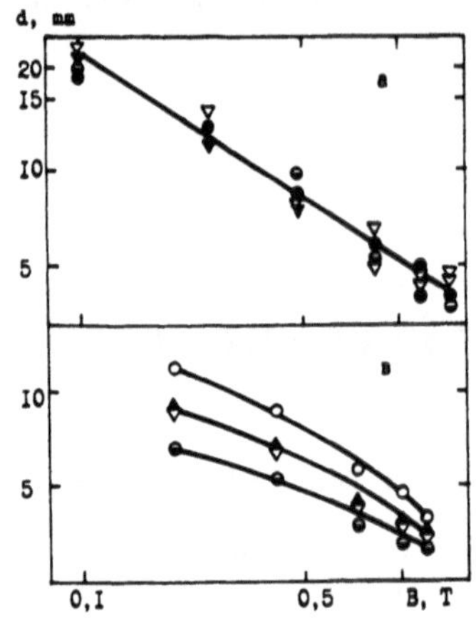

Fig.3. Dependence of the laser plasma diameter d as a function of the magnetic field B for different elements: ● - C; ∇ - Al; ♦ - Ti; ▼ - Cu; ● - Ta; o - Bi. q = $4\cdot10^{9}$ W·cm^{-2}, a - λ = 10.6 μm, b - λ = 1.06 μm. (The measuring error is smaller than 25%).

injection version the accelerated ion beams were obtained with charges analogous to the transverse magnetic field situation.

However, the quantity of the accelerated ions per laser pulse in the given version was smaller by a factor of approximately 2-3. On the other hand, taking into account $\tau \sim L$ in the first approximation the simple way of optimization exists for the given source by increasing L, for cyclotrons having a large enough clearance between the magnet poles.

The main advantage of the given ion source is the possibility of using comparatively low energy lasers for the production of ions. The maximum value of the average accelerated ion current in using the neodynium glass laser was equal to 100 nA for $^{7}Li^{2+}$ and to 50 nA for $^{12}C^{3+}$ at a laser frequency of 25 Hz.

V. Conclusion

The future development of the multiply charged ion laser source is associated, first of all, with a search for the ways of increasing the efficiency of ion extraction from the laser plasma and of enhancing the repetition rate of the ion source.

References

1. Pasjuk, A.S., Go-tsi-tsjan, Tretjakov, Yu.P. Preprint JINR, 1523, Dubna, 1964
2. Ananyin, O.B. et al.: Plasma Physics, 8. 1,92 (1982)
3. Ananyin, O.B. et al.: JTPhys. 53, 1, 94 (1983)
4. Ananyin, O.B. et al.: JTPhys. 52, 7, 1472 (1982)
5. Bykovski, Yu.A. et al.: JETP 96, 11, 1698 (1989)
6. Bykovski, Yu.A. et al.: Preprint JINR, P9-89-36, Dubna, 1989
7. Ananyin, O.B., et al.: JETP Lett. 9, 5, 261(1983)
8. Bykovski, Yu.A.et al.:JTPhys.58,7, 1291 (1988)
9. Bykovski, Yu.A. et al.: Plasma Physics 13, 10 1240 (1987)

Supplement to Z. Phys. D - Atoms, Molecules and Clusters 21, S 167-168 (1991)

Atoms, Molecules
Zeitschrift für Physik D and Clusters

Multicharged ion-atom collisions - application for Tokamak edge plasma physics

E. Unterreiter, F. Aumayr and H. Winter
Institut für Allgemeine Physik, Technische Universität Wien, Wiedner Hauptstraße 8-10, A-1040 Wien, Austria

Abstract
Studies on electron capture in slow ion - atom/molecule collisions are presented in view to diagnostics and modelling of the tokamak edge plasma region.

Introduction
Precise data on various atomic collision processes are of great importance for thermonuclear reactor development, especially in edge plasma diagnostics and for tests of impurity transport calculations. As in today´s fusion machines carbon is the principal impurity, in recent years we have become interested in carbon ion - atom collisions mainly for two purposes.

- Information on *balance processes* /1/ is needed for modelling of the edge plasma in tokamak discharges, and
- *Li activated charge exchange spectroscopy* (Li-CXS) /2/ involves electron capture from Li into multiply charged impurity ions for diagnostic purposes. Knowledge of time-varying radial density distributions of impurity ions in the plasma is of importance for understanding impurity transport into and out of the plasma center.

1. Absolute electron capture cross sections for C^+, C^{2+} - H_2, He collisions

To improve the data base for edge plasma-relevant balance processes /1/, absolute electron capture cross sections have been determined for the following reactions:

$$\begin{aligned} C^+ + H_2 &\rightarrow [C^0] + [H_2^+] \\ C^{2+} + H_2, He &\rightarrow [C^+] + [H_2^+], He^+ \end{aligned} \qquad (1)$$

C^+ and C^{2+} each have one low lying metastable state with excitation energies of 5.34 eV (C^+ ^4P) and 6.49 eV (C^{2+} $^3P^o$), which cause big differences at low impact energy for cross sections related to ground and metastable primary ion states.

For these measurements singly and doubly charged C ions have been obtained from a Nier type ion source, involving electron impact ionization of CH_4. The metastable fraction in the ion beam can be varied via the electron impact energy E_e and becomes zero for $E_e \leq 21$ eV for C^+ and $E_e \leq 70$ eV for C^{2+}, respectively. For $E_e \geq 150$ eV for both ion species constant metastable fractions are observed, which can be determined by an attenuation method as applied e.g. by Turner et al. 1968 /3/.

As attenuation gases H_2 for C^+ and He for C^{2+} were used, since in both systems the electron capture cross sections for the metastable ions are much larger than for the ground state ions. The such determined metastable fractions for $E_e \geq 150$ eV were 12% and 10% for C^+ and C^{2+}, respectively. Electron capture cross sections have been measured with pure ground state ion beams (E_e=21 and 70 eV, respectively) as well as "mixed" ion beams (with metastable fractions as given above), from which also cross sections for the metastable ions could be deduced, cf. fig.1.

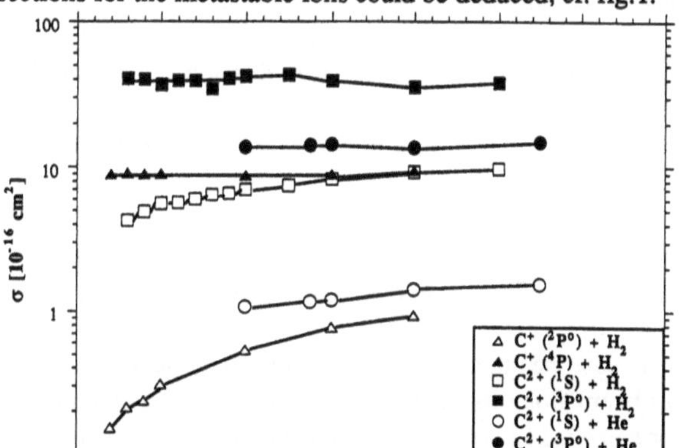

Fig.1: Absolute single electron capture cross sections for C^+ - H_2 and C^{2+} - H_2, He collisions. Open and full symbols apply to ground state and metastable primary ions, respectively.

Experimental details for determination of relative electron capture cross sections have been given by Aumayr and Winter 1985, 1986 /4/. Their methods have been slightly modified for particle counting, as described by Schweinzer and Winter /5/. The relative cross sections were calibrated to absolute data for $H^+ + H_2$,He $\rightarrow$ H + H_2^+,He^+ of Stier and Barnett 1956 /6/.

Total errors for the ground state primary ions are dominated by the applied calibration cross sections (20%). Because of the comparably small metastable fractions in the "mixed" beams, errors for the metastable primary ions are about 70%. More details and a critical comparison to results of many other groups will be given in a forthcoming paper.

As can be seen in fig.1, the importance of metastable ion admixtures increases with decreasing impact energy. In all collisions systems the related cross sections are at least a factor of 10 larger than for the corresponding ground state ions. Since in most earlier investigations the metastable fractions in the primary ion beams have not been well defined, the resulting cross sections cannot be regarded as reliable.

2. Li-CXS edge plasma diagnostics

A fast Li-diagnostic beam, if injected into a tokamak plasma discharge, can serve to study the temporal development of radial density distributions of multicharged impurity ions, since charge exchange (CX) processes as

$$Z^{q+} + Li(n\ell) \rightarrow Z^{(q-1)+*} + Li^{+} \qquad (2)$$

provide for emission of characteristic line radiation from the excited product ions /2/.

An experimental setup has been installed at the TEXTOR tokamak of KFA Jülich, FRG /7/. The Li-activated CX-radiation is detected spatially- and temporally resolved by means of a large aperture spectrometer at the end of a radiation transport system, permitting radial scanning across the plasma edge. Furthermore, LiI resonance radiation at 670.8 nm induced by collisions of the beam atoms with plasma particles is observed with a filter monochromator.

The intensity of characteristic radiation $S_\lambda(z)$ of a CX - excited impurity ion $Z^{(q-1)+*}$ is related to its absolute density $n_q(z)$ according to

$$S_\lambda(z) = k_\lambda \, v_{Li} \, n_q(z) \, \Sigma_i \, n_i \, \sigma_\lambda{}^i \qquad (3)$$

$\sigma_\lambda{}^i$ denotes the cross section for emission of characteristic photons with wavelength λ after state-selective CX production of a certain impurity ion state $Z^{(q-1)+}(n\ell)$ in collisions of the Z^{q+} ions with individual atomic Li states i. k_λ is an apparative parameter and v_{Li} the velocity of the neutral Li atoms.

In order to obtain absolute impurity ion densities from such measurements, a computer code has been developed to calculate the attenuation and fractional composition n_i of the Li beam along its penetration distance, involving either 1, 4 or 8 excited Li states. So far, in the model a clean plasma has been assumed with only electron- and proton-related collision processes taken into account. The time evolution of the Li beam composition, i.e. the density n_i of the various states can be described by a system of coupled linear differential equations, which are solved under the conditions that at the plasma boundary all Li atoms are in the Li(2s) ground state and the total number density of all Li states (atoms and ions) remains constant.

To check the reliability of our beam attenuation model, such calculated Li(2p) profiles have been compared with measured radial profiles of LiI resonance radiation (fig.2), showing generally a satisfactory agreement.

According to equ.(3), results of these calculations together with measured line emission profiles give access to radially resolved C^{q+} concentrations (q=4-6) in TEXTOR discharges /7/, as examples shown in fig 3.

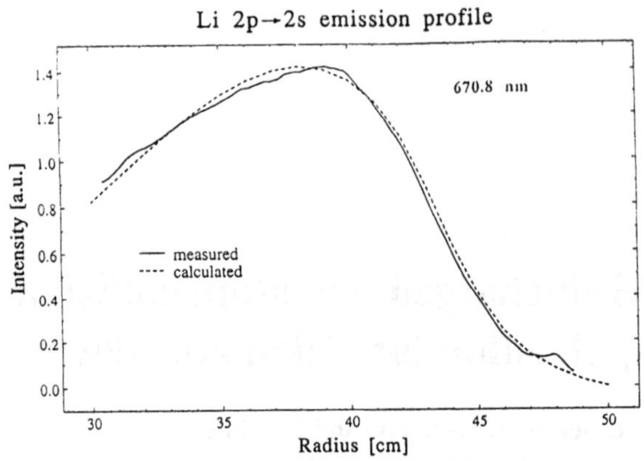

Fig.2.: Comparison of measured LiI radial emission profile (resonance line at 670.8 nm) with the modelled density of the Li(2p) component. Both profiles have been normalized to the same maximum value.

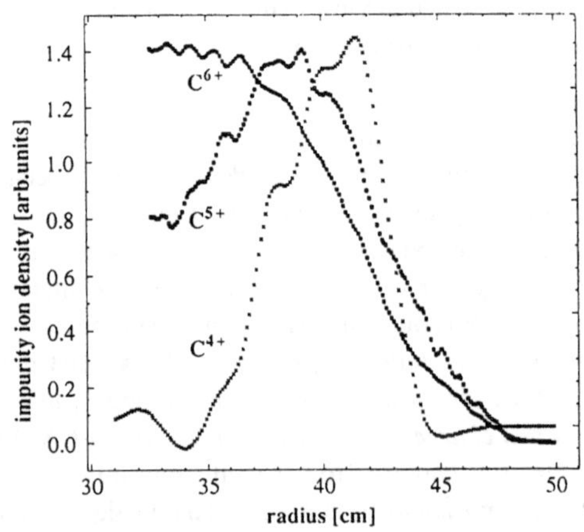

Fig.3.: Relative radial density profiles for C^{4+},C^{5+} and C^{6+} impurity ions in TEXTOR. After calibration, the maximum density values are 1.25 x $10^{11}cm^{-3}$ for C^{5+} and 2.5 x $10^{11}cm^{-3}$ for C^{6+}.

Acknowledgments

This work has been supported by Kommission zur Koordination der Kernfusionsforschung at the Austrian Academy of Sciences. Cooperation with IPP Jülich, FRG, is gratefully acknowledged.

References

/1/ R.K.Janev, R.A.Phaneuf, and H.T.Hunter, At.Data Nucl.Data Tables 40 (1988) 249

/2/ H.Winter, Comments At.Mol.Phys. 12 (1982) 165

/3/ B.R.Turner, J.A.Rutherford, and D.M.J.Compton, J.Chem.Phys. 48 (1968) 1602

/4/ F.Aumayr and H.Winter, Phys.Rev.A31 (1985) 67; F.Aumayr and H.Winter, Phys.Rev.A33 (1986) 846

/5/ J.Schweinzer and H.Winter, J.Phys.B (1990, in print)

/6/ P.M.Stier and C.F.Barnett, Phys. Rev. 103 (1956) 896

/7/ A.Pospieszczyk et al., J.Nucl.Mat. 162-164 (1989) 574; R.P.Schorn et al., (1990, submitted to Appl. Phys. B)

Supplement to Z. Phys. D - Atoms, Molecules and Clusters 21, S 169-170 (1991)

Atoms, Molecules
Zeitschrift für Physik D and Clusters

Heavy ion beam excitation of rare gases*

G. Ribitzki, A. Ulrich, B. Busch, W. Krötz, R. Miller, J. Wieser
TU München, Fakultät für Physik E12, W-8046 Garching, James-Franck-Straße, FRG

Abstract
The emission of light from rare gas targets at pressures of more than 100 Pa excited by a pulsed heavy ion beam has been studied. The absolute intensity of several spectral lines has been measured as a function of time at different target gas pressures. Population densities, excitation cross sections, and rate constants for collisional quenching were determined from the line intensities and the lifetimes of the excited states.

Introduction
The measurements were performed in a program to study lasers pumped by heavy ion beams[1]. Therefore it was of interest to measure the emission of light at pressures well above single collision conditions. Here we report on experiments with neon. The partial pressure of neon was kept constant and other rare gases were added in order to study the influence of the energy deposited in the target gas on the level population.

Experimental setup
An ion beam from the Munich Tandem van de Graaf accelerator was used for the experiments. The experimental setup is shown in Fig. 1. The ion beam entered the target cell through a titanium entrance foil (1.1 mg/cm^2) and was stopped in the target gas. Light emitted from a small volume on the beam axis was focused onto the entrance slit of a VUV monochromator with an Al(MgF_2) mirror. At the exit slit the light was detected with a photomultiplier in photon counting mode. Pulses of 90 MeV $^{32}S^{9+}$ ions with a width of 2 ns and containing about 5×10^5 projectiles were used for the excitation.

Experimental results
Data were taken for 3p-3s transitions at 585.25 nm (NeI), 337.828 nm (NeII), 331.975 nm (NeII) and 202.2 nm (NeIV). The neon pressure was 1000 Pa and argon or

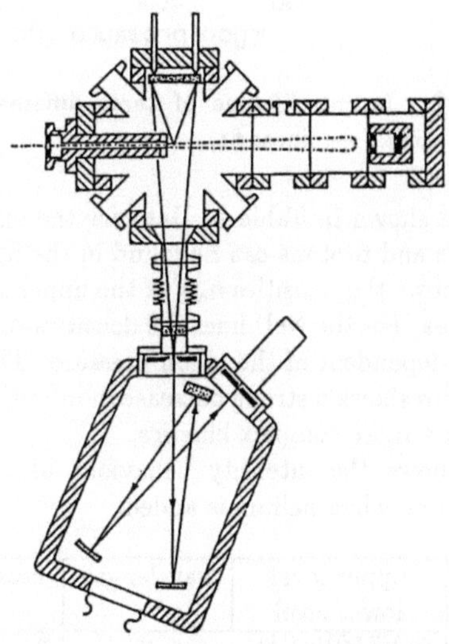

Figure 1: Schematic drawing of the experimental setup.

helium were added with partial pressures up to 7500 Pa. The intensity of the emitted light and its time dependence was measured at each pressure. The inverse lifetimes (τ^{-1}) of the transitions are shown versus pressure in Fig. 2. The linear increase of τ^{-1} with increasing argon pressure is interpreted as collisional quenching. The following quenching rate constants were determined from the data:

NeI $2p^5(^2P^0_{\frac{1}{2}})3p'[\frac{1}{2}]_0$ k=3.4×10^{-11}cm^{3}s^{-1}
NeII$2p^4(^3P)3p\,^2P^0_{\frac{3}{2}}$ k=2.4×10^{-10}cm^{3}s^{-1}
NeII$2p^4(^3P)3p'\,^2P^0_{\frac{1}{2}}$ k=8.0×10^{-10}cm^{3}s^{-1}

The intensity of the Ne lines decreased by adding the other gases. Assuming the quenching to be non radiative and using the tabulated Einstein A coefficients A_{ik}[2] one can determine the integral number of the excited target atoms per beam pulse and unit volume (n_{exc}) from the number of emitted photons and the collisional quenching rates. From these densities and the beam current effective excitation cross sections can be calculated which still may include cascade contributions from higher lying levels. The re-

*This work has been funded by the German Federal Minister of Research and Technology(BMFT) under contract number 06TM310I.

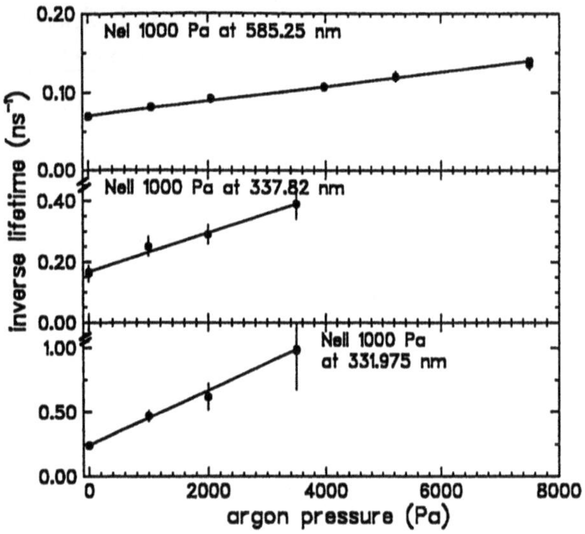

Figure 2: Inverse lifetime of three different neon lines. (•: data points, —: linear fit)

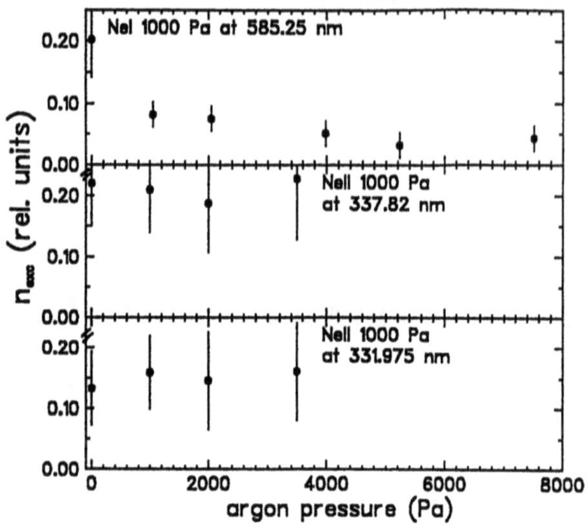

Figure 3: Excitation densities n_{exc} for three neon lines versus argon pressure.

sults are shown in Table 1. Data for the excitation with electrons and protons can be found in the literature [3,4]. Fig. 3 shows the densities n_{exc} of the upper level for three neon lines. For the NeII lines the densities n_{exc} were found to be independent of the argon pressure. The NeI line at 585.25 nm shows a strong decrease in intensity which may indicate a more complex kinetics.

Fig. 4 shows the intensity behaviour of the 202.2 nm (NeIV) line when helium is added.

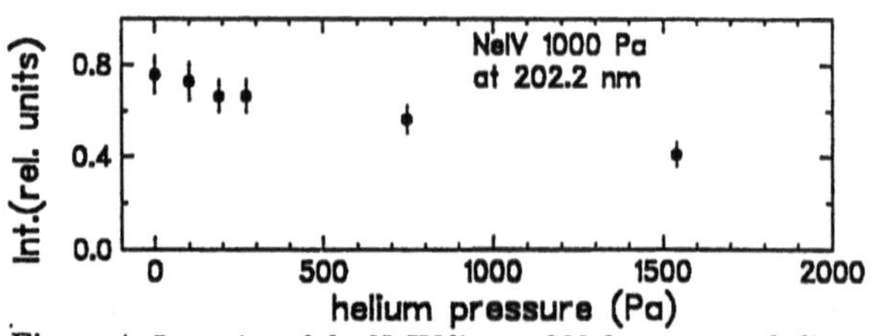

Figure 4: Intensity of the NeIV line at 202.2 nm versus helium pressure.

ion	upper level lower level	wavelength	cross section
NeI	$2p^5(^2P^0_{\frac{1}{2}})3p'[\frac{1}{2}]_0$ $2p^5(^2P^0_{\frac{1}{2}})3s'[\frac{1}{2}]_0$	585.25 nm	9×10^{-17}cm^2
NeII	$2p^4(^3P)3p\,^2P^0_{\frac{1}{2}}$ $2p^4 3s\,^2P^0_{\frac{1}{2}}$	337.828 nm	3×10^{-18}cm^2
NeII	$2p^4(^3P)3p'\,^2P^0_{\frac{1}{2}}$ $2p^4 3s'\,^2D_{\frac{3}{2}}$	331.975 nm	1×10^{-18}cm^2
NeIV	$2p^2(^1D)3p'\,^2D^0$ $2p^2 3s'\,^2D$	202.2 nm	4×10^{-19}cm^2

Table 1: Effective cross sections for the emission of several lines in 1000 Pa neon excited with 90 MeV $^{32}S^{9+}$ ions. Optical cross sections are given in the case of NeI and NeIV.

Conclusion

The excitation scheme for the NeII lines can be described by direct excitation and collisional and radiative deexcitation of the target atoms. Therefore we can deduce effective cross sections for the excitation of these levels. An increase in intensity of the neon lines with total pressure which might be expected from the increasing energy deposition in the target volume could not be observed. On the contrary the NeI and NeIV data show a decrease in intensity with increasing pressure.

References

[1] A. Ulrich et al.: in Proc. of 5th Int. Workshop on "Atomic Physics for Ion Driven Fusion", accepted for publication in "Laser and Particle Beams".

[2] W.L. Wiese, M.W. Smith, and B.M. Glennon: Atomic Transition Probabilities, NSRDS-NBS 4 (1966)

[3] M. Eckhardt, D. Hasselkamp, and K.-H. Schartner: Z. Physik A292, 337(1979)

[4] G. Keith and R.M.St. John: Phys. Rev. A6 240(1972)

Supplement to Z. Phys. D - Atoms, Molecules and Clusters 21, S 171-173 (1991)

Atoms, Molecules
Zeitschrift für Physik D and Clusters

Mean ionization state and composition of a dense high-Z plasma

A. Förster[1], T. Kahlbaum[2], and A. Rickert[3]

[1] Sektion Physik der Humboldt-Universität zu Berlin, Invalidenstraße 42, O-1040 Berlin, FRG
[2] Zentralinstitut für Elektronenphysik, Hausvogteiplatz 5-7, O-1086 Berlin, FRG
[3] Max-Planck-Institut für Quantenoptik, Ludwig-Prandtl-Straße 10, W-8046 Garching, FRG

Abstract

Ionization state and composition of dense copper plasma are calculated with the help of the chemical model and the average atom model, respectively, and compared with other published data.

I. Introduction

Mean ionization state and detailed composition as main characteristics of plasmas are of principle interest as well as required for the calculation of thermodynamic, transport, and optical properties. Plasmas produced by gas discharge, laser irradiation, or mechanical compression can reach high densities where strong interactions predominate. Therefore, a theoretical description of these dense plasmas has to account for nonideal effects such as continuum lowering and pressure ionization.

This is done, for instance, in the chemical model (section II) which was used to study the equation of state and plasma phase transitions [1 – 4], and in the average atom model (section III) recently applied to compute radiation extinction coefficients [5]. In this paper we explain both models, present numerical data for ionization state and composition of a dense copper plasma, and compare them with other results known from literature (section IV).

II. The Chemical Model

The chemical model [1] treats different bound states of nuclei and electrons as individual species each of them distinguished by its total charge, energy spectrum, mass, and effective radius. Consequently, we consider a plasma consisting of heavy particles (atoms and ions) and free electrons with number densities $n_z = N_z/V$ (charge number $z = 0, \ldots, Z_{nucl}$) and n_e, respectively. The equilibrium properties of such a system are provided by the Helmholtz free energy F which is chosen as the basic thermodynamic potential.

Following earlier work [2 – 4] F includes the ideal parts for heavy particles and electrons according to Boltzmann and Fermi-Dirac statistics, respectively, and contributions due to Coulomb interaction and short-range repulsion. The Coulomb contributions are taken into account by means of Padé approximants to interpolate between the quantum-corrected Debye law and expressions for the classical multi-component ion lattice screened by a highly degenerated electron fluid. The short-range contribution is represented by an exact expression for an arbitrary mixture of hard spheres with different radii [6]. For the sake of comparison with the average atom model we use hydrogen-like orbital radii of the outermost bound electrons in the ground states of the atoms and ions. Additionally, the presence of spatially extended particles reduces the volume in the ideal part for the electrons (Pauli exclusion principle). The partition functions of the heavy particles are assumed to obey the Planck-Brillouin-Larkin convention [1]. The energy spectra (unperturbed levels and statistical weights) are taken from [7], for $z > 2$ only the ground states are available [8].

The detailed plasma composition $\{n_z\}$ can be obtained from the coupled set of Z_{nucl} Saha equations $\mu_z = \mu_{z+1} + \mu_e$ corresponding to the ionization-recombination processes $i_z \longleftrightarrow i_{z+1} + e$ under the constraint of electroneutrality $n_e = \sum_z z n_z$. The chemical potentials $\{\mu_z\}$ and μ_e are defined as usual, e.g. $\mu_e = \partial F/\partial N_e$. Finally, we express the plasma composition in terms of the relative fractions $\alpha_z = n_z/n$ where $n = \sum_z n_z$ is the total density of heavy particles, and introduce the mean ionization state $\bar{z} = \sum_z z\alpha_z = n_e/n$.

III. The Average Atom Model

The average atom model [9] describes the plasma as composed of heavy particles in a mean ionization and excitation state and of free electrons with densities $n = n_{\bar{z}}$ and $n_e = \bar{z} n_{\bar{z}}$, respectively. This is equivalent to the replacement of the different z-fold charged atomic configurations by one $\bar{z}$-fold charged mean configuration (averaged atom). The subshells (n, l) of this mean configuration are occupied with fractional numbers P_{nl} of electrons according to Fermi-Dirac statistics and have radii r_{nl} which are taken hydrogen-like with screened nuclear charges. The occupation number P_{nl} depends on the one-electron energies $\{E_{nl}\}$ and vice versa each E_{nl} is determined by the whole set $\{P_{nl}\}$ what demands a selfconsistent solution to obtain $\bar{z} = Z_{nucl} - \sum_{nl} P_{nl}$.

Among the various approximation schemes for the explicit calculation of the energies we chose the screened hydrogenic model with ℓ-splitting [10] appropriate to the undisturbed level structure of isolated high-Z atoms. The shifts of the energy levels in the plasma due to Coulomb interaction are calculated within the framework of the ion-sphere model [11]. At high densities r_{nl} could be larger than the ion-sphere radius leading to an overlap between subshells of neighbouring configurations. To avoid this violation of the Pauli principle the corresponding degeneracy $2(2l+1)$ is smoothly reduced to zero [12]. The chemical model treats the same effect via the hard-core contribution to the free energy.

In the literature [11] the use of the binomial distribution is suggested to estimate the fraction of a configuration in a given ionization and excitation state from the mean configuration. Since in the present context only the plasma composition with respect to z is of interest, we can use the simple formula

$$\alpha_z = \binom{Z_{nucl}}{z}\left(\frac{\bar{z}}{Z_{nucl}}\right)^z\left(1-\frac{\bar{z}}{Z_{nucl}}\right)^{Z_{nucl}-z}$$

IV. Results and Discussion

Numerical results for the chemical model are presented in Figs. 1 and 2, and for the average atom model in Figs. 3 and 4. The detailed plasma composition is in general quite complex as can be seen for one temperature in Fig. 1. Nevertheless, two or three species always dominate. At low densities (up to $n \approx 10^{19}\ cm^{-3}$ at $T = 10^5\ K$) the plasma behaves ideal, at medium densities (up to $10^{22}\ cm^{-3}$) its composition is essentially determined by Coulomb interaction, and at high densities in addition the short-range repulsion becomes important.

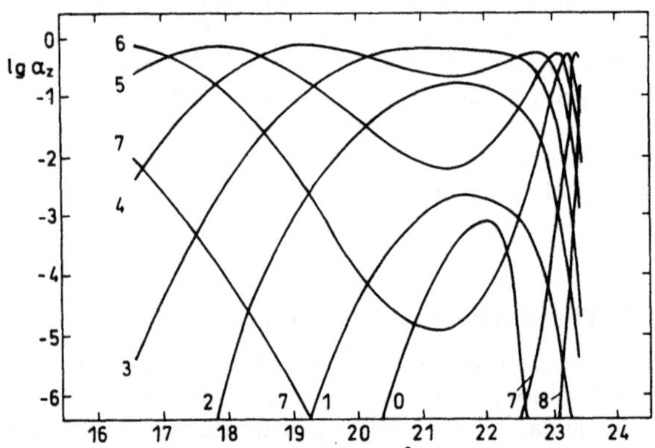

Fig. 1. The relative fractions $\{\alpha_z\}$ versus the density n in the chemical model for a temperature $T = 10^5\ K$. The curves are labeled by the charge numbers z.

As a direct consequence of the Coulomb interaction the mean ionization state (Figs. 2 and 3) has a minimum which leads to the formation of the "valley of low-ionized states" [2]. With increasing temperature the onset of interaction and, thus, the minimum of $\bar{z}$ is shifted to higher densities. The subsequent vanishing of the low-ionized states due to the combined influence of Coulomb forces and high-density effects is known as pressure ionization. The chemical model yields more or less pronounced steps in $\bar{z}$ (Fig. 2) if the ratio of the temperature to the relevant ionization energies does not exceed a certain value.

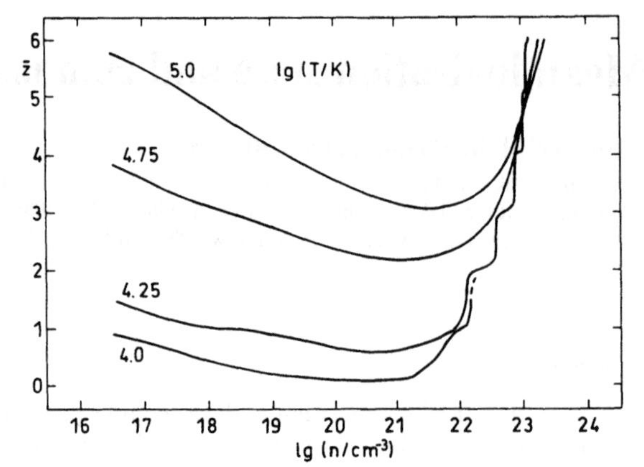

Fig. 2. The mean ionization state $\bar{z}$ versus the density n in the chemical model for four temperatures.

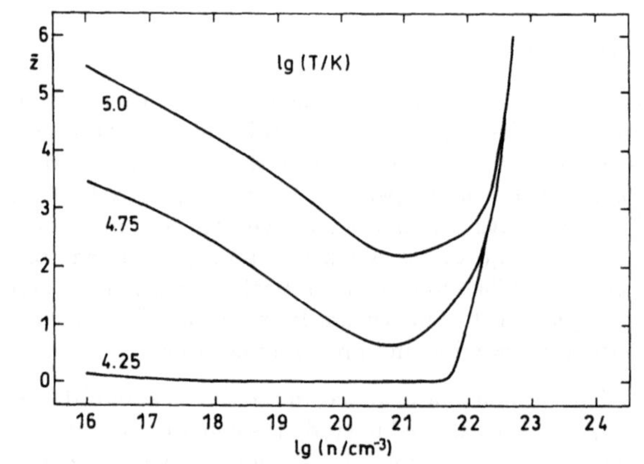

Fig. 3. The mean ionization state $\bar{z}$ versus the density n in the average atom model for three temperatures.

Fig. 4 shows another possibility to exemplify temperature ionization and pressure ionization. The steps in $\bar{z}$ occur at temperatures where the stable Ne- or He-like configuration is reached at $\bar{z} = 19$ and $\bar{z} = 27$, respectively. For the density $n = 10^{22}\ cm^{-3}$ pressure ionization is indicated by a crossing of the $\bar{z}$-curves.

In order to check the assumption of a binomial distribution (section III) of the relative fractions we have calculated $\{\alpha_z\}$ from the mean ionization state. $\bar{z}$ was taken from the chemical model to allow for a direct comparison between this estimate (Fig. 5) and the actual composition (Fig. 1). At first glance both distributions exhibit some common characteristics: the sequences of the species for a given density are nearly the same. However, the relative fractions of the non-dominating species are overestimated, i.e. the binomial distribution generally appears to be too flat. Specific features of the actual distribution originating from individual properties of the species, e.g. different ionization energies, cannot be reproduced.

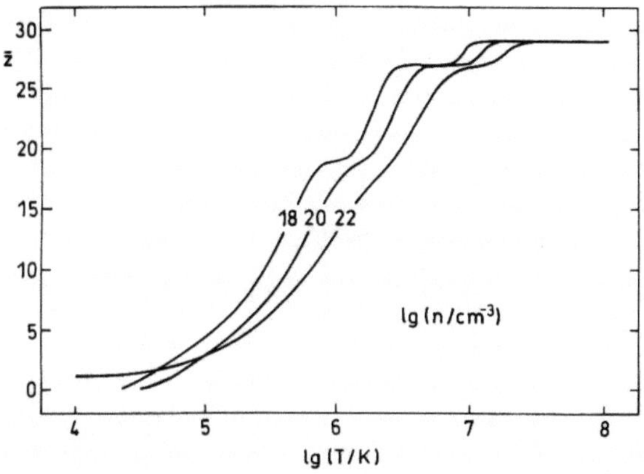

Fig. 4. The mean ionization state $\bar{z}$ versus the temperature T in the average atom model for three heavy particle densities.

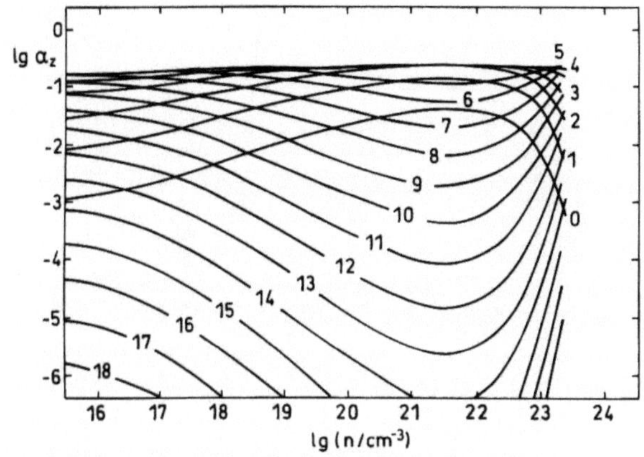

Fig. 5. Binomial distribution of the relative fractions $\{\alpha_z\}$ versus the density n calculated from $\bar{z}$ in the chemical model for a temperature $T = 10^5$ K (see Fig. 1). The curves are labeled by z.

As can be seen in Fig. 6, chemical model and average atom model give qualitatively similar results for the mean ionization state. The lower values of $\bar{z}$ in the average atom model can be explained by too high ionization energies provided by the screened hydrogenic model with ℓ-splitting for the first ionization stages. Additionally, we include in our comparison with the chemical model two other sets of tabulated data for copper: the SESAME tables from the T-4 Group [13] and the IVTAN tables from the Fortov group (in [4]). The SESAME tables are widely used although details of their physical background are not yet published to our knowledge. Except for the region where pressure ionization takes place the coincidence may be regarded as excellent. The minimum at $n = 10^{23}$ cm^{-3} has no physical significance and may be due to a mismatch between different theories entering SESAME. The IVTAN tables follow from a semiempirical approach based on the chemical model and adjusted to experimental data. Here the agreement at low temperatures is still satisfactory, especially with respect to the overall shape of the valley. The deviation is about 30% in the worst case.

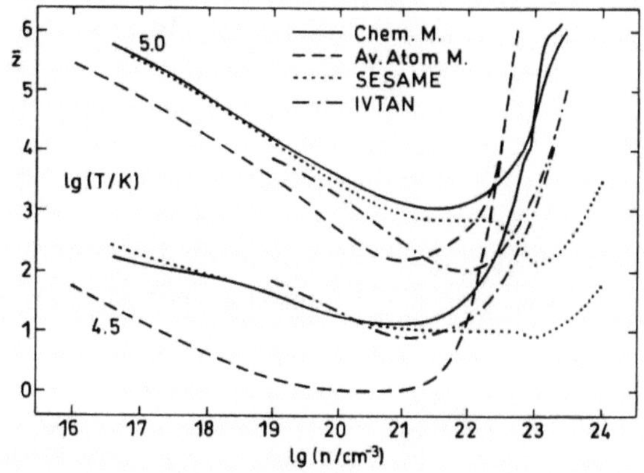

Fig. 6. The mean ionization state $\bar{z}$ versus the density n for two temperatures: comparison between the chemical model (full lines), the average atom model (dashed lines), the SESAME data (dotted lines), and the IVTAN data (dashed-dotted lines).

V. Conclusions

We have shown that the composition of a copper plasma in thermodynamic equilibrium can be well described within the chemical model whereas the binomial distribution is only a rather crude approximation. A reduction of the quantitative discrepancies in the mean ionization state seems to be possible by improving the atomic physics underlying the average atom model. Pressure ionization in copper most probably occurs in the density range of 10^{22} – 10^{23} cm^{-3}.

References

1. Kraeft, W.-D., Kremp, D., Ebeling, W., Röpke, G.: Quantum Statistics of Charged Particle Systems. Berlin: Akademie-Verlag 1986
2. Förster, A., Kahlbaum, T. in: Proc. 2nd Int. School on Quantum Statistics (QUAST II, Flessenow 1989), ed. D. Blaschke, Rostock: Universität Rostock 1990 (to appear)
Kahlbaum, T., Förster, A.: Laser Part. Beams **8**, no. 4 (1990)
3. Förster, A., Kahlbaum, T., Ebeling, W.: High Press. Res., to appear (1991)
4. Ebeling, W., Förster, A., Fortov, V.E., Gryaznov, V.K., Polishchuk, A.Ya.: Thermophysical Properties of Hot Dense Plasmas. Leipzig: Teubner 1990
5. Rickert, A.: Report MPQ-148. Garching: Max-Planck-Institut für Quantenoptik 1989
Rickert, A., Meyer-ter-Vehn, J.: Laser Part. Beams **8**, no. 4 (1990)
6. Mansoori, G.A., Carnahan, N.F., Starling, K.E., Leland Jr., T.W.: J. Chem. Phys. **54**, 1523 (1971)
7. Moore, Ch.E.: Atomic Energy Levels. NBS Circular No. 467, Vol. II, pp. 111 - 123. Washington: National Bureau of Standards 1952
8. Radcig, A.A., Smirnov, B.M.: Parameters of Atoms and Atomic Ions - Handbook. (in Russ.), 2nd Edn., pp. 108,114. Moscow: Energoatomizdat 1986
9. Pomraning, G.C.: The Equations of Radiation Hydrodynamics. Oxford: Pergamon 1973
10. Perrot, F.: Phys. Scr. **39**, 332 (1989)
11. Huebner, W.F. in: Physics of the Sun. Vol. I, ed. P.A. Sturrock, Dordrecht: Reidel 1986
12. Zimmerman, G.P., More, R.M.: J. Quant. Spectrosc. Radiat. Transfer **23**, 517 (1980)
13. T-4 Group: Report LALP-83-4. Los Alamos: Los Alamos National Laboratory 1983
T-4 Group: Report LA-10160-MS. ed. K.S. Holian; SESAME #3330, pp. 1-6. Los Alamos: Los Alamos National Laboratory 1984

Supplement to Z. Phys. D - Atoms, Molecules and Clusters 21, S 175-177 (1991)

Atoms, Molecules
Zeitschrift für Physik D and Clusters

Electron and positron impact experiments on metallic foils at the Giessen positron source TEPOS

W. Faust, C. Hahn, M. Rückert, H. Schneider, A. Singe and I. Tobehn*
Abteilung Angewandte Kernphysik, Strahlenzentrum der Universität Giessen, Leihgesterner Weg 217, W-6300 Giessen, FRG

Abstract

The sources TEPOS for slow positrons were installed at the Giessen 65 MeV electron linear accelerator. Ratios of K- and L-shell ionization cross-sections by electron and positron impact were measured on Au-/Ag-foils as function of the incident projectile energy. With energies near threshold a departure is observed of the ratio $\sigma(e^-)/\sigma(e^+)$ from unity (increasing with decreasing energy).

I. Introduction

The sources for low-energy positrons TEPOS I, with a maximum output of about 10^8 positrons/s, and TEPOS II, with a shorter beam-line, were installed at the Giessen pulsed 65 MeV electron linear accelerator (LINAC) of the Strahlenzentrum [1, 2]. TEPOS I delivers 80% of the maximum positron intensity at an aperture of 20 mm. For positron energies of $E(e^+)$= 600 eV, $\Delta E/E$ is about 16%.

II. Experimental Set-up

The facilities TEPOS I and II to produce slow positrons are described in detail elsewhere [1,2]. The slow positrons emerging from the tungsten moderator are accelerated by applying a positive potential ranging between 50 V and 2 kV. A solenoidal beam guide with maximum magnetic fields of 0.02 T transports them into the experimental hall. To improve the beam quality (smaller beam diameter, lower energy spread) we are testing various remoderation arrangements in transmission mode (1000 Å W-foil) [2]. We have installed an additional electrostatic positron extraction system consisting of different lenses and a spherical deflecting condenser. This allows to separate positrons of different energies. The slow positrons (e^+_{in}) penetrate into the W-foil, where a fraction of them obtains thermal energies. Some of those reach the surface without annihilation and are subsequently reemitted (e^+_{remod}) with energies corresponding to the work-function. Another fraction of the positrons (e^+_{in}) passes through the W-foil without thermalization (e^+_{trans}). Eventually, all positrons traversing the W-foil are postaccelerated.

With a former remoderating set-up in transmission and a subsequent magnetic time-of-flight arrangement consisting of a 3 m long vacuum tube [2] the remoderated positrons (e^+_{remod}) could be separated from the traversing positrons (e^+_{trans}) because of their different energies. The ratio was determined of the area of the peak belonging to e^+_{remod} in the measured time-of-flight spectra to the mean electron current of the LINAC as a function of the energy of e^+_{in}. The maximum relative yield of e^+_{remod} was attained at an energy of e^+_{in} of about 5 keV.

The electrostatically confined remoderated positrons will be well-focussed, and thus provide smaller beam diameters with lower energy spread. Further, a decisive advantage is the possibility to perform experiments outside disturbing magnetic fields. Thus, it may be possible, in particular, to investigate in atomic physics processes as scattering of positrons or the production of excited positronium.

III. Inner-Shell Ionization

The theory predicts differences in electron and positron impact ionization near threshold. Therefore, measurements of cross section ratios of K- and L-shell ionization by positron and electron in-

* In part excerpt from D 26.

teraction on metallic foils were performed. A vacuum chamber (fig. 1) with an adequate target holder for postaccelerating the positrons has been installed at the end of the beam-line TEPOS I. The emitted K- and L-X-rays were registered with a Si(Li)-detector [3].

The double ratio of cross sections

$$\frac{\sigma_K(e^-)/\sigma_L(e^-)}{\sigma_K(e^+)/\sigma_L(e^+)}$$

for K-/L-shell ionization was investigated on Ag- and Ag/Cu-foils [3-6]. An increasing departure of this ratio from unity with decreasing projectile energy is observed (fig. 2).

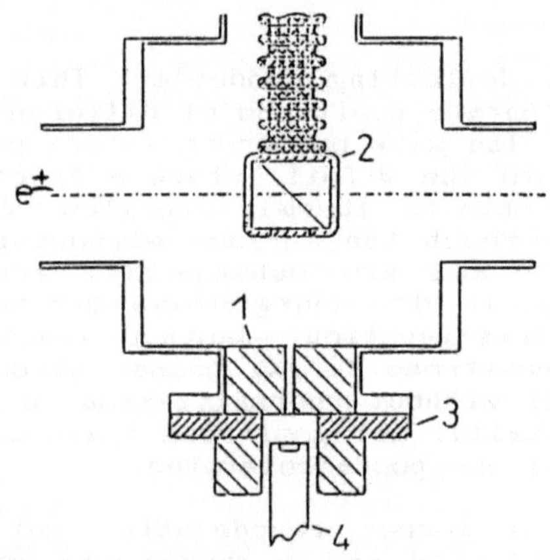

Fig 1: Experimental set-up. 1: Pb-collimator; 2: Al-target-holder; 3: Al-flange with a 50 μm Hostaphan window; 4: Si(Li)-detector.

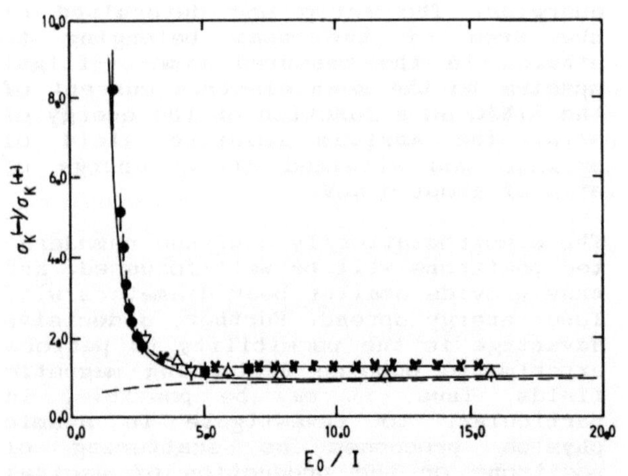

Fig. 2: Ratio of K-shell ionization cross sections as a function of the projectile energy E_0 (for e^- and e^+)/I (I= K-shell binding energy). Our results [3]: ● = for Ag; ■ = for Cu. Other results: △(Ag): by Ito et al [4], ▽ (Cu): by Schultz et al [5]; ——— : calculations of Hippler [6], — — —: PWBA, Hippler et al [7], ——×—— : Theory of Hock [8].

In the investigated energy range the coulomb interaction between projectile and target nucleus for the K-shell is dominant.

To analyse the processes for L-shell ionization [9] we are performing various experiments with Au/Ag-targets of different thicknesses [2]. A first measurement of a corresponding X-ray spectrum is shown in fig. 3.

Similar to the investigations of the K-shell ionization the double ratio of cross sections

$$\frac{\sigma^-}{\sigma^+} = \frac{[Au(L_\alpha)/Ag(\Sigma L)]_{e^-}}{[Au(L_\alpha)/Ag(\Sigma L)]_{e^+}}$$

was measured. The silver L-shell values could be taken as reference since the ratio

$$\frac{[Ag(\Sigma L)]_{e^-}}{[Ag(\Sigma L)]_{e^+}}$$

is practically unity in the investigated energy range. Our measurements confirm the theoretical results of Hippler [6], who considered the acceleration or dece-

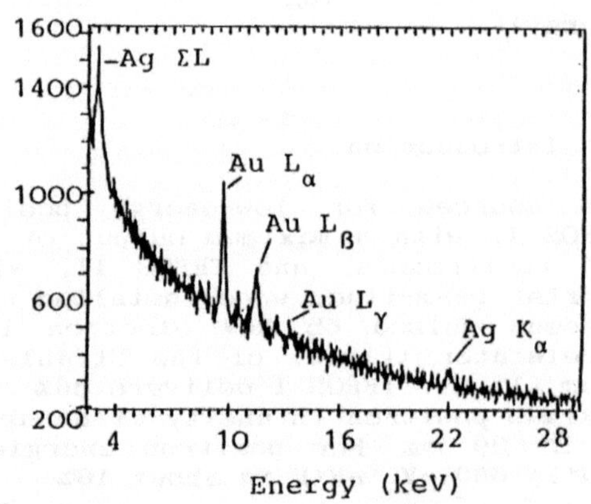

Fig. 3: X-ray spectrum by positron impact on a 1450 Å Au/ 3500 Å Ag target [E(e⁺) = 50 keV].

leration of the incident electrons or positrons in the field of the nuclei. In future experiments thinner foils, if necessary, will be used to reduce systematic errors.

Acknowledgements

We thank the Deutsche Forschungsgemeinschaft, Bonn-Bad Godesberg (Germany) for generous financial support.

References

1. F. Ebel, W. Faust, C. Hahn, S. Langer, M. Rückert, H. Schneider, A. Singe and I. Tobehn; Nucl. Instr. a. Meth. A272, 626 (1988)
2. F. Ebel, W. Faust, C. Hahn, M. Rückert, H. Schneider, A. Singe and I. Tobehn; Nucl. Instr. a. Meth. B50, 328 (1990)
3. F. Ebel, W. Faust, C. Hahn, M. Rückert, H. Schneider, A. Singe and I. Tobehn Phys. Lett. A140, 114 (1989)
4. S. Ito, S. Shimizu, T. Kawaratani, and K. Kubota; Phys. Rev. A22, 407 (1980)
5. P.J. Schultz and J.L. Campbell; Phys. Lett. 112A, 316 (1985)
6. R. Hippler; Phys. Lett. A144, 81 (1990) and private communications
7. R. Hippler, W. Jitschin Z. Physik A307, 287 (1982)
8. G. Hock Proc. XIth ICPEAC, Kyoto, 1979 (North Holland) p. 974
9. A.N. Lennard, P.J. Schultz, G.R. Massoumi and L.R. Logan; Phys. Rev. Lett. 61, 2428 (1988)

Supplement to Z. Phys. D - Atoms, Molecules and Clusters 21, S 179-180 (1991)

Atoms, Molecules
Zeitschrift für Physik D and Clusters

Highly-charged ions in a penning trap: mass measurements, etc.

R. Jertz, G. Bollen, H.-J. Kluge, L. Schweikhard, H. Stolzenberg
Institut f. Physik, Universität Mainz, Postfach 3980, W-6500 Mainz, FRG

I. Bergström, C. Carlberg, R. Schuch
Manne Siegbahn Institute for Physics, Frescativägen 24, S-10405 Stockholm, Sweden

Abstract

The use of a Penning trap will start a new generation of precision experiments on highly charged ions. The long storage time of the ions in combination with a controlled confinement in a very small volume will enable accuracies in mass determination better than $\delta m/m = 10^{-8}$.

I Introduction

A Penning trap facility is currently under construction at Mainz, which will be installed at the electron beam ion source (EBIS) facility CRYSIS [1] at the Manne Siegbahn Institute at Stockholm. CRYSIS is designed to produce naked Ar- or Xe- ions up to heavier ions of charge +70. Singly or moderatly charged ions are ionized successively by an intensive electron beam of 50 keV. These ions will be injected into the Penning trap facility. The experiment is aiming at the development of new precision methods for determining masses, atomic binding energies and decomposition of large atomic molecules. The first tests will be performed at Mainz by mass measurements of moderatly charged ions to increase resolving power and accuracy.

II Mass measurement

Figure 1 shows the principal layout of a Penning trap. In a static magnetic field B, directed along the z - axis, ions of charge q and mass m perform a cyclotron motion of frequency

$$\nu_c = (q/2\pi m) \cdot B. \qquad (1)$$

Due to the hyperbolical shape of the electrodes an electrostatic quadrupole field can be created to confine the ions in axial direction. The ion motion is not a pure cyclotron

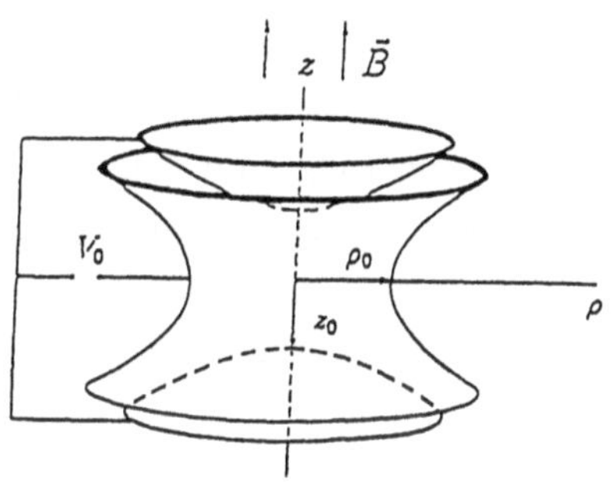

Figure 1: The Penning trap.

motion, but a superposition of a slow drift around the trap axis, called magnetron motion with frequency ν_-, a modified cyclotron motion with frequency $\nu_+ = \nu_c - \nu_-$ and an axial motion with frequency ν_z [2].

The cyclotron frequency and hence the q/m - ratio is detected by a time-of-flight method [3]. By exciting the ions with an additional electric quadrupole rf-field with frequency ν_c the radial energy of the ion motion increases. The ions are ejected and drift to an ion detector with which the time of flight from the trap is measured. In the inhomogeneous region of the magnetic field their radial energy is converted into axial energy. Hence, ions in resonance reach the detector faster than out of resonance.

In fig. 2 a cyclotron resonance is shown obtained by the Penning trap mass spectrometer at CERN. A resolving power up to $2.3 \cdot 10^6$ and an accuracy of $\delta m/m = 2 \cdot 10^{-7}$

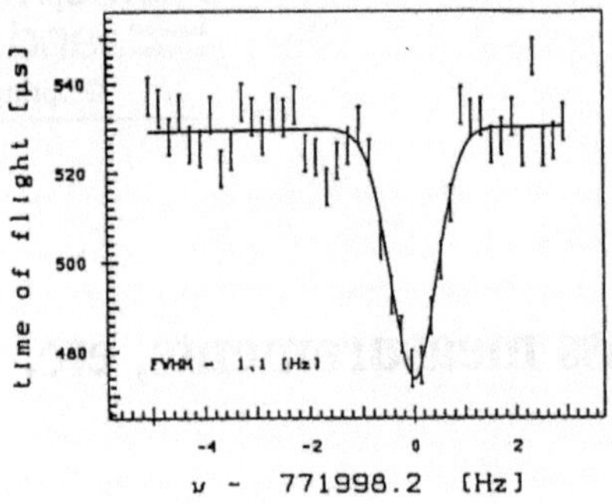

Figure 2: Cyclotron resonance.

have been achieved for short-lived Cs isotopes in the CERN experiment [4].

III Experimental setup

Figure 3 shows schematically the setup of the new trap facility. The apparatus consists of a superconducting magnet, the Penning trap in the homogeneous region of the magnet, a transfer section and finally a drift tube with the detector. Diaphragms allow to controll precisely the injection of the ion bunch into the trap. The trap which is identical to the one used in the CERN experiment [5] is highly compensated for trap imperfections in order to allow highly accurate mass determinations.

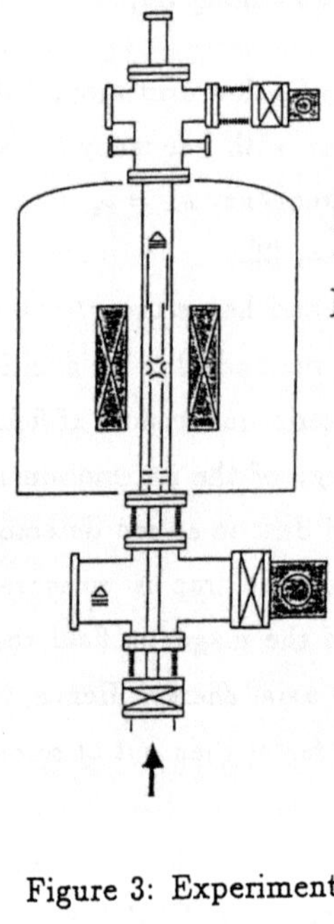

Figure 3: Experimental setup.

IV Discussion

The width of the cyclotron frequency and hence the resolving power depend on the coherent interaction time T of the ions with the rf-field. For the example of fig. 2 a width of $\Delta\nu_c$ (FWHM) = 1.1 Hz was obtained for T = 0.9 s. An obvious way to increase the resolving power is to increase the charge state q of the ions, since the frequency linearly increases with q. Also relative frequency shifts due to certain trap imperfections will decrease, since $\delta\nu_c/\nu_c \sim 1/q$. Hence, an accuracy of $\delta m/m = 10^{-8}$ and better seems feasible and one of 10^{-7} will be obtained already with T = 100 ms. The latter might be important in case of short-lived species.

Another application of highly charged ions concerns precision mass spectrometry of ^{28}Si proposed to serve for a new definition of the kilogram [6]. Since $^{12}C^{3+}$ and $^{28}Si^{7+}$ are isobaric, doublet mass spectrometry might result in an accuracy of $\delta m/m = 10^{-9}$ for the $^{12}C/^{28}Si$ mass ratio.

Finally the unique situation of highly-charged ions stored in an ion trap in combination with the possibility to determine the charge states and to prepare an ensemble in a single charge state will allow to study the interaction of these ions with atoms, molecules and photons with unpreceded accuracy and sensitivity.

References

[1] L. Liljeby and A. Engstrom, Int. Symp. on Electron Beam Ion Sources and their Appl., Am. Inst. of Ph., 256 (1988).

[2] L. S. Brown and G. Gabrielse, Rev. Mod. Phys., 58 (1986) 233.

[3] G. Gräff, H. Kalinowsky and J. Traut, Z. Phys., A297 (1980) 35.

[4] H. Stolzenberg et al., Phys. Rev. Lett. (submitted).

[5] St. Becker et al., Int. J. Mass Spectrom. Ion Proc. (in press).

[6] P. Seyfried, PTB-Mitteilungen, 99 (1989) 336.

Supplement to Z. Phys. D - Atoms, Molecules and Clusters 21, S 181-182 (1991)

Atoms, Molecules
Zeitschrift für Physik D and Clusters

Parity violation in two-electron systems

G. von Oppen
Institut für Strahlungs- und Kernphysik, Technische Universität Berlin, Hardenbergstraße 36, W-1000 Berlin 12, FRG

Abstract:

An experiment for detecting parity violation in He-like ions with Z ≈6 by investigating 2^3S_1 - 2^1S_0 transitions is proposed. The proposal is based on the near degeneracy of the $1s2s\ ^1S_0$ and the $1s2p\ ^3P_0$ state.

I. Introduction

Recently I discussed a possibility for detecting parity nonconservation (PNC) in helium [1] by investigating level crossings of $m = 0$ Stark substates with opposite reflection parity. Though enhanced by the degeneracy of opposite parity states, the expected PNC effects are, nevertheless, relatively small since the weak-interaction matrix elements of the $1s3d\ ^1D$ and ^3D substates considered in [1] are non-zero only due to interconfiguration mixing. In this contribution I draw attention to the near degeneracy of the opposite-parity states $1s2s\ ^1S_0$ and $1s2p\ ^3P_0$ of He-like ions with Z ≈ 6 for which the weak-interaction matrix element [2]

$$\langle H_{PNC}\rangle \approx i\frac{\sqrt{3/2}}{32\pi}\cdot Z^4_{eff}\cdot N\cdot\kappa_1\cdot\alpha\cdot G_F \text{ a.u.}$$

is large ($Z_{eff} \approx Z - 1$ effective charge, N neutron number, $\kappa_1 \approx -0{,}5$ electron-neutron coupling parameter, G_F Fermi's coupling constant, a.u. atomic units). On the other hand it is disadvantageous that in this case the levels cannot be crossed by applying external fields. Still, PNC effects may become detectable by studying the highly forbidden E1 transitions between the metastable 2^3S_1 and 2^1S_0 levels using the field configuration considered in [1] with an electric field F_z and a magnetic field B_y. Some related ideas for detecting PNC in two-electron systems have been discussed already in 1974 by V. G. Gorshkov and L. N. Labzovsky [3].

II. $2^3S_1 - 2^1S_0$ transitions

As long as PNC effects are neglected and no external fields are applied, the 2^1S_0 and 2^3S_1 states are extremely pure singlet and triplet states, respectively, since there are no singly excited triplet states with $J = 0$ and positive parity, and no singly excited singlet states with $J = 1$ and positive parity. Therefore, all multipole transitions between these levels are highly forbidden. However, triplet components are admixed to the 2^1S_0 state by the PNC perturbation H_{PNC} and by the interaction of the atom with an electric field F_z in z direction (quantization axis). H_{PNC} provides a small 2^3P_0 admixture with the amplitude

$$\epsilon_{PNC} = \langle H_{PNC}\rangle/\Delta E_Z,$$

where the energy separation $\Delta E_Z = E(2^1S_0) - E(2^3P_0)$ changes from negative to positive values near Z = 6 [4]: $\Delta E_5 = -1221\ cm^{-1}$, $\Delta E_6 = -139\ cm^{-1}$, $\Delta E_7 = +995\ cm^{-1}$. For Z = 6, $|\epsilon_{PNC}| \approx 0.6\cdot 10^{-11}$ is largest *. Accordingly, the quantitative data mentioned below refer to C^{4+}. The Stark interaction causes a $|2^3P_1, 0^+\rangle$ admixture. Second-order perturbation theory yields for its amplitude

$$\epsilon_F = \langle 2^1S_0|ez|2^1P_1,0^+\rangle\cdot F_z\cdot \langle 2^1P_1,0^+|H_{SO}|2^3P_1,0^+\rangle\cdot [E(2^1S_0)-E(2^1P_1)]^{-1}\cdot[E(2^1S_0)-E(2^3P_1)]^{-1} \approx 1.1\cdot 10^{-5}$$

if $F_z = 10$ kV/cm. Here the Stark substates are designated by $|m|$ and R_y-reflection parity [1] and the usual phase convention $(R_y\cdot T|\ \rangle = |\ \rangle)$ is assumed, ez is the electric dipole operator and H_{SO} the spin-orbit interaction for which the matrix elements are given in [5]. The magnetic field B_y (in y direction) also admixes triplet components to the 2^1S_0 state, however, only in third order. More important

*According to [4], there are more zeros of $\Delta E(2^1S_0 - 2^3P_0)$ at higher Z (Z ≈ 64 and 91) where ϵ_{PNC} is possibly larger than 10^{-6}. These near degeneracies have been considered elsewhere[7].

is the Zeeman splitting of the 2^3S_1 level resulting from the interaction with B_y. The state vectors of the three substates are $\sqrt{1/2}\{|2^3S_1,0^-\rangle \pm i|2^3S_1,1^-\rangle\} \equiv |\pm 1\rangle_y$ and $|2^3S_1,1^+\rangle \equiv |0\rangle_y$. Owing to the triplet admixtures to the 2^1S_0 state, E1 transitions between the 2^3S_1 states and the 2^1S_0 state can be induced. Due to the splitting of the 2^3S_1 -level, the light beam ($\lambda = 229$ nm) can be tuned to a selected Zeeman component, e.g. the $|+1\rangle_y$ - $^1S_0$ transition. For circularly polarized light propagating in x direction, one then calculates a transition rate

$$A \propto |\epsilon_F\langle 2^3P_1,0^+|y|2^3S_1,1^-\rangle \pm\epsilon_{PNC}\langle 2^3P_0,0^-|z|2^3S_1,0^-\rangle|^2$$

which depends on the sign of the circular polarization and the field configuration if ϵ_F and $\epsilon_{PNC} \neq 0$. The asymmetry expected for this transition rate is due to PNC and is of the order $|\epsilon_{PNC}/\epsilon_F|$.

III. Outline for the realization of an PNC experiment

In this short note a possible PNC experiment based on these ideas can only briefly be sketched. Whether it becomes feasible depends on the future development of storage rings for highly ionized ions. $^{12}C^{4+}$ ions of a sufficiently intense beam can be excited to the metastable $1s2s$ levels by electron impact. Under stationary conditions, the 2^3S_1 level is much more populated than the 2^1S_0 level owing to the radiative decay rates [6] $\Gamma(2^3S_1) = 48.6$ s^{-1} and $\Gamma(2^1S_0) = 3.31 \cdot 10^5$ s^{-1} differing by several orders of magnitude. Using an intense laser beam, transitions can be induced to the 2^1S_0 level, where Doppler tuning may be helpful. By using an applied magnetic field, the motional electric field and collinear laser excitation, a convenient geometry of the experimental set-up is achieved. For an efficient detection of the induced transitions, the 2^1S_0 atoms can be excited by a second laser beam ($\lambda = 353$ nm) to the 2^1P_1 level and the subsequently emitted X rays at $\lambda(2^1P_1$ - $1^1S_0) = 4.03$ nm can be observed.

IV. Concluding remarks

The experimental technique sketched above, in principle, allows a detection of parity violation in two electron systems. However, the feasibility of the experiment depends on the asymmetry and the counting rate achievable for the 4.03 nm photons. For $F_z \approx 10$ kV/cm, the asymmetry is of the order 10^{-6}. It can only be measured if at least 10^{12} photons are detected. The following consideration may suffice to demonstrate the difficulties of such an experiment. Under the most favourable conditions, one detects one 4.03 nm photon per C^{4+} ion excited to the metastable 2^3S_1 level. This assumption implies (i) that the mean rate of the laser-induced transitions to the 2^1S_0 level is larger than the spontaneous decay rate of the 2^3S_1 level, (ii) that the 2^1S_0 - 2^1P_1 transitions are saturated and (iii) that the detection efficiency for the 4.03 nm photons is close to one. With this assumption the prospects for a successful experiment depend essentially on an efficient generation of metastable 2^3S_1 ions. Let us consider a storage ring with, say, 10^8 stored C^{4+} ions interacting with an electron beam with density $n_e \approx 10^8$ cm^{-3} and length $l \approx 1$ m. In this case only one ion per circulation period is excited to the 2^3S_1 level if a total cross section $\sigma \approx 10^{-18}$ cm^2 for the excitation of triplet states is assumed, that is there are 10^{12} excitation processes only in about 10 days. At present these estimates are not very promising. However, ion storage rings are still new and, therefore, the prospects for a PNC experiment may become better in the future. A more promising situation would arise, in particular, with a more effective excitation technique. One may think, e.g., of filling the storage ring with H_2 gas at sufficiently low pressure. The weakly bound electrons would act as quasi-free particles in collisions with the energetic ions. Furthermore, even if a parity violating asymmetry cannot be detected, the measurement of the 2^3S_1 - 2^1S_0 transition frequency by the experimental technique outlined above would be of great interest for the investigation of relativistic and QED effects in two-electron systems.

Acknowledgements: I am grateful to Y. Kriescher and A. Wolf for valuable discussions.

References

1. v. Oppen, G.: Europhys. Lett. **11** 25 (1990)
2. Khriplovich, I.B.: Parity Nonconservation in Atomic Phenomena (in Russian), 2nd edition (Nauka, Moskow) 1988
3. Gorshkov, V.G., Labzovsky, L.N.: Sov. Phys. -JETP **42** 581 (1976) and JETP Lett. **19** 394 (1974)
4. Drake, G.W.: Can. J. Phys. **66** 586 (1988)
5. Drake, G.W.: Phys. Rev. A **19** 1387 (1979)
6. Drake, G.W.: Phys. Rev. A **3** 908 (1971) and Phys. Rev. **180** 25 (1969)
7. Schäfer, A. et al.: Phys. Rev. A **40** 7362 (1989)

Supplement to Z. Phys. D - Atoms, Molecules and Clusters 21, S 183-184 (1991)

Atoms, Molecules
Zeitschrift für Physik D and Clusters

Coupled channel analysis of electron-positron pair production in relativistic heavy ion collisions*

K. Momberger, N. Grün, and W. Scheid
Institut für Theoretische Physik, Justus-Liebig-Universität, Giessen, FRG

Abstract

We performed coupled-channel calculations for the nonperturbative description of electron-positron pair production in relativistic heavy ion collisions. For the system $U^{92+}+U^{92+}$ at E_{lab}=10 GeV/nucleon and impact parameter b=386 fm we obtained probabilities for pair production with inner-shell capture that exceed the results of first order perturbation theory by two orders of magnitude.

I. Introduction

Today there is interest in electron-positron production by relativistic heavy ions. In relativistic heavy ion colliders the pair production plays the role of a background effect that might lead to an impairment of the ion beams since the cross sections for pair production with capture are very large. From the theoretical side the pair creation process can serve as a testing ground for nonperturbative methods in QED, because the coupling constant $Z\alpha$ is nearly one and a failure of first order theories can be expected.
In this contribution we study nonperturbative effects on the electron-positron creation in relativistic heavy ion collisions. We carry out coupled channel calculations for a collision of U^{92+} on U^{92+} at an energy of E_{lab}=10 GeV/nucleon and compare the results with those of first order perturbation theory.

II. Coupled channel formalism

In order to describe nonperturbatively electron-positron pair production we solve the time-dependent Dirac equation for the electron wave function by the coupled channel formalism [1,2]. Expansion of the electron wave function in terms of target-centered hydrogenic Coulomb-Dirac functions results in the coupled channel equations:

$$i\,\dot{a}_{ji} = \sum_k \langle\phi_j\,|V(r,t)|\,\phi_k\rangle \exp(i(E_j-E_k)t)a_{ki}. \qquad (1)$$

For the positive and negative continuum a discretization by means of time-dependent relativistic wave-packets has been employed. The perturbation operator V describes the interaction between the electron and the projectile. We neglect the recoil of the target, the deflection of the projectile and the electron-electron interaction. Then the operator V is the Lorentz contracted electromagnetic potential of the projectile moving with contant velocity in the target system.

After solving the coupled-channel equations (1) by starting from each initially occupied state of the negative continuum, the time-evolution of the vacuum state is described with a finite Slater-determinant [3]. From this time-dependent Slater-determinant the probabilities for the production of free electron-positron pairs and for pair production with atomic inner-shell capture of the electron are extracted by projection on the corresponding final states.

III. Results

The formalism was applied to the system of U^{92+} colliding with U^{92+} at E_{lab}=10 GeV/nucleon at an impact parameter b=386 fm. The atomic basis set consists of 5 bound states ($1s_{1/2}$, $2s_{1/2}$, $2p_{1/2}$, $2p_{3/2}$, $3s_{1/2}$) and wave-packets at 20 energies for both continua (E_j=1.3,...,7.3; -1.3,...,-6.1; ΔE=0.6) for each of the three angular momenta $s_{1/2}$, $p_{1/2}$, $p_{3/2}$. The negative continuum is then described by 72 magnetic substates. The restriction in the angular momentum summation is the most important shortcoming of our present results, but calculations with much larger basis sets are not practically feasible. Therefore, our results should not first be compared with experimental data, but may mainly serve as a demonstration of a nonperturbative calculation in comparison with the one in first order perturbation theory.

In Fig.1 we show results for the differential probability to observe a positron of energy E_{e+} as a function of E_{e+}. The histogram-like curves show the total probability and its decomposition into contributions from pair creation of free pairs and with simultaneous capture. Also the total probability for the production of a single pair is indicated. The continuous lines below show the results of the first order perturbation theory. The

* Work supported by BMFT (06 GI 709) and GSI (Darmstadt)

production of free pairs is increased by one order of magnitude and the capture from vacuum by roughly two orders of magnitude. The production of a single electron-positron pair during the collision is dominant to about 90%, i.e. the enhancement can not be explained by the production of many pairs.

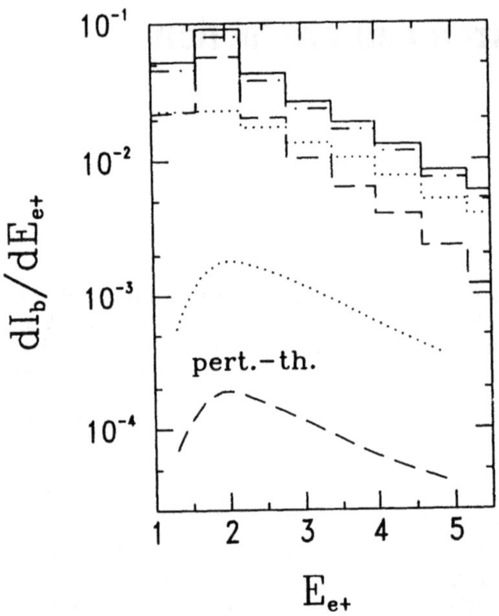

Fig.1: Differential probability for observation of a positron of energy E_{e+} in the collision U^{92+}+ U^{92+} at E_{lab}=10 GeV/nucleon and impact parameter b=386 fm. The histogram-like curves show the coupled-channel results, the continuous curves below are the results of the first order perturbation theory. The solid curve gives the total positron spectrum and the dotted-dashed curve the contribution from the production of a single electron-positron pair during the collision. The dotted curves show the probabilities for the production of free pairs, the dashed lines those for the production of pairs with simultaneous inner-shell capture. The energy E_{e+} is measured in units of mc^2=0.511 MeV.

IV. Conclusions

We obtained a larger enhancement of the results calculated with the coupled channel formalism in comparison with those of the first order perturbation theory. By successively switching off the different types of coupling matrix elements, we found that the enhancement effect for the pair production with inner shell capture can be ascribed to its major part to the conservation of unitarity for all times during the collision. Unitarity becomes most important for times at the point of closest distance of the ions, when the transition amplitudes are strongly peaked and may reach maximum values nearby one. In this case the conditions for the validity of perturbation theory are no more fulfilled. To a minor part the enhancement is due to the couplings of higher order, for which the continuum-continuum couplings play the essential role.

References

1. Mehler, G., Soff, G., Rumrich, K., Greiner, W.: Z.Phys. D13 193 (1989)
2. Momberger, K., Grün, N., Scheid, W., Becker, U.: J.Phys. B23,2293S (1990)
3. Reading, J.F.: Phys. Rev. A8, 3262 (1973)

Supplement to Z. Phys. D - Atoms, Molecules and Clusters 21, S 185-186 (1991)

Atoms, Molecules and Clusters
Zeitschrift für Physik D

A Monte Carlo model of complex spectra for opacity calculations**

M. Klapisch*
Lawrence Berkeley Laboratory

P. Duffy, W. H. Goldstein
Lawrence Livermore Laboratory

* On leave from Hebrew University, Jerusalem, Israel

** Work performed under the auspices of the U.S. Department of Energy by LLNL under contract W-7405-ENG-48

Abstract

We are developing a Monte Carlo method for calculating opacities of complex spectra. It should be faster than atomic structure codes and is more accurate than the UTA method. We use the idea that wavelength-averaged opacities depend on the overall properties, but not the details, of the spectrum; our spectra have the same statistical properties as real ones but the strength and energy of each line is random. In preliminary tests we can get Rosseland mean opacities within 20% of actual values.

Introduction

Current approaches to calculating opacities of complex spectra are not always satisfactory. In particular, when transition arrays are partly or completely resolved, "exact" atomic structure calculations can take too long to be practical, and the Unresolved Transition Array (UTA) method gives very poor opacities. We are therefore developing a new, Monte Carlo method for calculating opacities in these cases which gives much better results than UTA method and promises to be much faster than atomic structure calculations.

The key idea is that opacities calculated over small wavelength ranges do not depend on the exact wavelengths or strengths of individual lines. If certain *statistical* properties of the spectrum are correct, calculated opacities will be about right.

All the atomic physics in our model is of a statistical nature; the strength and energy of each line is random. Our model reproduces 4 statistical properties of each transition array which can be calculated exactly and quickly using UTA theory. The model makes spectra in which individual line strengths and energies are random but these statistical properties are correct; their Rosseland mean opacities are roughly the same as those of the real spectra.

The **Overview** section describes the approach in more detail. In **Results to Date**, we show tests of our model on two simple transition arrays of iron, where "exact" atomic structure calculations are also possible for comparison. Next, we discuss **Sensitivity Tests**, where we show which statistical properties of the spectra the Rosseland mean is most sensitive to. Finally, we discuss **Future Work** which may improve the results.

Overview of Our Method

In our approach, transition arrays are modelled one at a time. Each array is characterized by 4 statistical properties:

1) the number of lines
2) the total strength of all lines, Σs_i
3) the mean energy (strength-weighted) of all lines, $\Sigma e_i s_i$
4) the variance of the strength-weighted line energies, $[\Sigma e^2{}_i s_i - (\Sigma e_i s_i)^2]^{1/2}$.

(Here s_i and e_i are strength and energy of the ith line.) The last three of these are the first 3 moments of spectrum of the array, as defined in UTA theory. In addition, we assume that the line strengths for each array obey a chi-squared distribution (i.e., there are many more weak lines than strong ones).

The above numbers (4 per array) are only inputs the model needs. These can be calculated exactly and quickly from UTA theory[1].

Spectra are simulated as follows. Lines are generated one at a time, each being characterized by a strength and an energy (h nu). The strength is a random number picked such that all the strengths together obey the proper chi-squared distribution. The energy is also random, picked so that the combined energies obey a gaussian distribution whose mean and width are given by inputs 3 and 4 above. This results in a spectrum where, if the distribution of intensity as a function of energy is called I(e), the 0th, 1^{tst}, and 2nd moments of I(e) are correct. We next convolve each line with an assumed shape (a Doppler-broadened gaussian). This completes the calculation of the bound-bound opacity. Next we add free-free opacity - inverse bremstrahlung and Compton scattering - and finally, we calculate the Rosseland mean opacity.

Results To Date

So far, we have tested our method only on simple individual transition arrays where "exact" atomic structure calculations are possible as a check. We show results here for the $3d^3$ - $3d^2 4p$ array in iron 5^+ and the $3d^4$ - $3d^3 4p$ array in iron 4^+. For each one, we show Rosseland mean opacities calculated by the atomic structure code RELAC, by our Monte Carlo model, and by an Unresolved Transition Array (UTA) model. For each model, Table 1 lists opacities calculated over various energy ranges: 4, 6 and 8 time the array width (inout 4 above) centered on the middle of the array.

Rosseland Mean Opacities (cm^2/gm)

$3d^3$-$3d^2 4p$ Transition Array in Iron 5^+

Energy Range	RELAC	Monte Carlo	UTA
4 sigmas	$6.12 \cdot 10^{-3}$	$6.3 \cdot 10^{-3}$	$2.18 \cdot 10^5$
6 sigmas	$4.30 \cdot 10^{-3}$	$2.0 \cdot 10^{-3}$	$4.27 \cdot 10^3$

$3d^4$-$3d^3 4p$ Transition Array in Iron 4^+

Energy Range	RELAC	Monte Carlo	UTA
4 sigmas	0.1619	0.12	$4.77 \cdot 10^4$
6 sigmas	$4.42 \cdot 10^{-2}$	$0.96 \cdot 10^{-2}$	$9.49 \cdot 10^3$
8 sigmas	$2.90 \cdot 10^{-2}$	$4.45 \cdot 10^{-3}$	$5.50 \cdot 10^2$

Table 1: Rosseland mean opacities for single transition arrays in iron 4^+ and 5^+ calculated by the atomic structure code RELAC, our Monte Carlo model, and UTA theory. We show opacities averaged over several energy ranges to show how our model works both in the center and the edges of the arrays. The 1-sigma array widths are 2.127 and 2.098 eV for iron 4+ and iron 5+, respectively. Each Monte Carlo calculation took about 90 seconds on a Macintosh personal computer.

Our results are are quite good for the smaller energy range and are all much better than the UTA results, which are useful only as upper limits. However, our results are too low over the wider energy range in both arrays. The reason is that, in the real arrays, the weaker lines are spread over a much wider energy range than the strong ones. Even though these lines contain negligible strength, they significantly increase the Rosseland mean opacity, which is heavily weighted by places where the opacity is low, because they eliminate gaps in the spectrum. In our current model, all the lines have the same energy distribution; thus we have too few lines and too little opacity away from the centers of the arrays. We have experimented with an *ad hoc* modification to the model which gives the weak lines a much broader energy distribution, and the results are indeed dramatically better. As discussed under **Future Work**, we plan to incorporate a better implementation of this into the model, which improve our opacities significantly.

Sensitivity Tests

We used our model to investigate the sensitivity of the Rosseland mean opacity to various statistical properties of transition arrays. This tells us which of these properties must be modeled to get good opacities, how accurately they must be known, and gives an indication of the uncertainty in the results. We found essentially that the Rosseland mean opacity is determined largely the gaps in the spectrum, not by the strengths of the lines. (This is due to the $1/K_{nu}$ factor in the definition of the Rosseland mean.) Our results are summarized briefly here.

These tests were done with the above-mentioned *ad hoc* modification to the model in which the weak lines have a broader energy range than the strong ones. (As described above, this is an important property of real arrays which we had not modelled.) Specifically, the weakest 1081 lines, which comprise about 1% of the total strength, were spread over a 1-sigma energy range of 8.33 eV, the same as in the real spectrum. This gave us a Rosseland mean for iron 4^+ within 15% of the correct value. We then varied statistical properties of the array one at a time and measured the effect of this on the Rosseland mean. We found that varying the fraction of total strength given to the weak lines (without varying the number of these lines) has remarkably little effect on the Rosseland mean. Changing this fraction by more than 3 orders of magnitude has no statistically significant effect on the opacity; however, removing the weak lines entirely reduces the opacity by a factor of 10! This is because varying the strength of the weak lines has almost no effect on the gaps in the spectrum, whereas removing them opens up many gaps and drastically reduces the opacity. (In fact, the weak lines contribute most of the opacity of this array, even though they comprise only 1% of the total strength.) For the same reason, increasing the number of weak lines - while holding their total strength fixed - does increase the opacity, because more lines means fewer gaps in the spectrum. Similarly, the opacity increases as the energy range of the weak lines is increased, because this gives them less tendency to overlap.

Future Work

Our first priority is to change the model as dicussed above so that the weaker lines are distributed over a wider energy range than the strong lines. A simple linear variation of energy range with strength should suffice. We suspect that the energy range for the weakest lines is approximately the sum of the widths of the upper and lower configurations forming the array. The range of the strongest lines can then be found by forcing the overall width of the spectrum to be as dictated by UTA theory (as the current model does).

(1) J. Bauche, C. Bauche-Arnaut, and M. Klapisch Advances in Atomic and Molcular Physics, vol. **23**, p. 131 (1988)

Supplement to Z. Phys. D - Atoms, Molecules and Clusters 21, S 187-188 (1991)

Atoms, Molecules
Zeitschrift für Physik D and Clusters

Relativistic calculations of inelastic scattering of electrons by highly-charged ions

D. L. Moores
Department of Physics and Astronomy, University College London, London, UK

M. S. Pindzola
Department of Physics, Auburn University, Auburn, Alabama, USA

Abstract

In this paper we present some further results of applying the theory described by Pindzola et al.(1) to excitation and ionization processes in highly charged ions.

Electron Impact Ionization of U^{91+}

In a previous work(2) electron impact ionization cross sections for the 1s state of U^{91+} were calculated and the results were found to lie well below the measurements by Claytor et al.(3) at an incident electron energy of 222 keV. The results presented in that paper did not include exchange when the full relativistic interaction was used. Results including exchange have now been obtained and are shown in Table 1. It is seen that the inclusion of exchange leads to cross section significantly larger than those obtained without it. This is due to the "collapse" of the interference term which occurs when retardation and magnetic terms are included. However, the increase is still insufficiently large to bring the calculation into agreement with experiment: the cross section at 222 kev increasing from 1.09 to 1.64 barns, still well below the measured value of 3.9 barns.

Also shown in Table 1 are calculations for ionization out of the 2p states of U^{91+}: it is found that the effects of retardation and magnetic interactions are quite small.

Table 1

Electron Impact Ionization Cross Section for U^{91+} (Barns)

X	COULOMB ONLY NE	E	FULL INTERACTION NE	E
1s state	(I = 4861.2 a.u.)			
1.125	0.33	0.35	0.30	0.52
1.25	0.58	0.60	0.53	0.89
1.5	0.94	0.92	0.89	1.40
1.678	1.12	1.07	1.09	1.64
2.0	1.35	1.24	1.34	1.91
$2p_{1/2}$ state	(I = 1257.4 a.u.)			
1.125	4.97	6.21	4.62	6.40
1.25	8.70	10.30	8.01	10.56
1.5	13.57	14.86	12.84	15.71
2.0	18.18	18.20	17.24	19.38
$2p_{3/2}$ state	(I = 1089.6 a.u.)			
1.125	6.88	8.58	6.57	8.98
1.25	11.94	14.10	11.37	14.78
1.5	18.53	20.23	17.84	21.55
2.0	24.51	24.48	23.82	26.43

Electron impact excitation of Fe^{25+} and U^{91+}

We have also calculated cross sections for excitation of 1s→2s and 1s→2p transitions in the hydrogenic ions Fe^{25+} and U^{91+}. The results are shown in Table 2. As might be expected, retardation and magnetic effects

Table 2

Electron Impact Excitation Cross Section (Barns)

X	COULOMB ONLY	FULL INTERACTION
Fe^{25+}	$1s \rightarrow 2s$	($\Delta\varepsilon$ = 255.6 a.u.)
1.125	87.5	90.6
1.25	79.7	82.5
1.5	67.8	70.2
2.0	52.7	54.4
Fe^{25+}	$1s \rightarrow 2p_{1/2}$	($\Delta\varepsilon$ = 255.6 a.u.)
1.125	118.	116.
1.25	112.	110.
1.5	104.	102.
2.0	93.	91.
Fe^{25+}	$1s \rightarrow 2p_{3/2}$	($\Delta\varepsilon$ = 256.4 a.u.)
1.125	229.	227.
1.25	218.	216.
1.5	203.	201.
2.0	183.	181.
U^{91+}	$1s \rightarrow 2s$	($\Delta\varepsilon$ = 3603.8 a.u.)
1.125	1.04	1.61
1.25	0.96	1.45
1.5	0.83	1.22
2.0	0.68	0.94
U^{91+}	$1s \rightarrow 2p_{1/2}$	($\Delta\varepsilon$ = 3603.8 a.u.)
1.125	0.76	1.07
1.25	0.69	0.95
1.5	0.61	0.84
2.0	0.53	0.68
U^{91+}	$1s \rightarrow 2p_{3/2}$	($\Delta\varepsilon$ = 3771.6 a.u.)
1.125	0.90	1.05
1.25	0.86	0.97
1.5	0.83	0.93
2.0	0.80	0.86

X: incident energy/threshold energy
NE: no exchange terms
E: exchange terms included
DE: threshold for excitation
I: ionization potential

are small for Fe^{25+}, and the $1s\rightarrow2p_{1/2}$ and $1s\rightarrow2p_{3/2}$ cross sections are roughly in the non-relativistic ratio of 1:2. The effects are significant however in U^{91+}, and when the full interaction is included the $1s\rightarrow2p_{1/2}$ and $1s\rightarrow2p_{3/2}$ differ by only a few percent near threshold.

References

1. M.S. Pindzola, N.R. Badnell, D.L. Moores and D.C. Griffin, in these proceedings.

2. D.L. Moores and M.S. Pindzola, Phys. Rev. A41, 3603 (1990).

3. N. Claytor, B. Feinberg, H. Gould, C.E. Bemis, Jr., J. Gomez del Campo, C.A. Ludemann and C.R. Vane, Phys. Rev. Lett. 61, 2081 (1988).

Supplement to Z. Phys. D - Atoms, Molecules and Clusters 21, S 189-191 (1991)

Atoms, Molecules
Zeitschrift für Physik D and Clusters

Resonances in the electron-impact single and multiple ionization of ions

G. Hofmann, A. Müller, B. Weißbecker, M. Stenke, K. Tinschert, E. Salzborn
Institut für Kernphysik, Strahlenzentrum der Justus-Liebig-Universität, W-6300 Giessen, FRG

Abstract

Cross sections for electron-impact single ionization of Li-like ions B^{2+}, C^{3+}, N^{4+}, O^{5+} *and* F^{6+} *and for n-fold ionization of* Li^+ *(n=1),* Na^+ *(n=1),* K^+ *(n=1,2,3),* Rb^+ *(n=1,2,3),* Cs^+ *(n=1,2,3,4) and* Ba^+ *(n=1,2,3,4) have been measured with extremely high precision (lowest limit of statistical uncertainties as low as 0.01%). Detailed studies of indirect ionization mechanisms proceeding via multiply-excited intermediate states could be carried out.*

Introduction

A fast energy-scanning technique which had been used before [1] in dielectronic-recombination experiments was introduced [2] to the measurement of electron-impact ionization cross sections of ions where it provides a powerful tool to study indirect ionization mechanisms. In particular, with the new technique it became possible to identify complex reaction paths and intermediate multiply-excited states involved in such mechanisms [2,3,4].

Beside removal of a target electron by direct ionization, a net increase of the target ion charge state can result from excitation-autoionization (EA), resonant-excitation-double-autoionization (REDA) and resonant-excitation-auto-double-ionization (READI) processes [5]. All of these processes involve an inner-shell excitation and subsequent autoionization(s), REDA and READI proceed via intermediate compound states of the parent ion and the projectile electron and thus are closely related to the process of dielectronic recombination.

At the level of precision reached in the present experiments, EA, REDA and READI can be clearly seen in the single ionization of Li-like ions [3,6]. In addition, we could make quantitative measurements of resonant and non-resonant two-electron (K-shell) excitations in the net ionization of He-like Li^+ ions [4]. Double excitation of both K-shell electrons at a time involves small cross sections. For further studies of such processes we started an experimental program with heavier ions (also with rare-gas configurations) assuming that double excitation from filled p orbitals in L-, M-, N-,... shells is much more likely than for the two K-shell electrons in Li^+ ions. Thus we performed experiments on single and multiple ionization of the singly charged alkali ions and used the energy-scanning technique to detect fine details in the cross sections.

Results and Discussion

For the measurement of cross sections of processes

$$(1) \qquad e + A^{q+} \rightarrow \ldots \rightarrow A^{(q+n)+} + (n+1)e$$

we detect the product ions $A^{(q+n)+}$ long after the electron collision so that practically all possible relaxation processes happen before the final charge-state analysis. Thus we observe a net n-fold ionization which may involve several intermediate steps after the primary electron-ion collision.

Figure 1 shows measured cross sections for the net single ionization of O^{5+} ions. The zero on the cross section axis is strongly suppressed in order to make the observed features visible. The energy range covers the onset of EA processes $1s^22s \rightarrow 1s2s^2$ and reaches up to the threshold of $1s^2 \rightarrow 1s2s4\ell$ excitations. Above 550 eV the cross section increases by

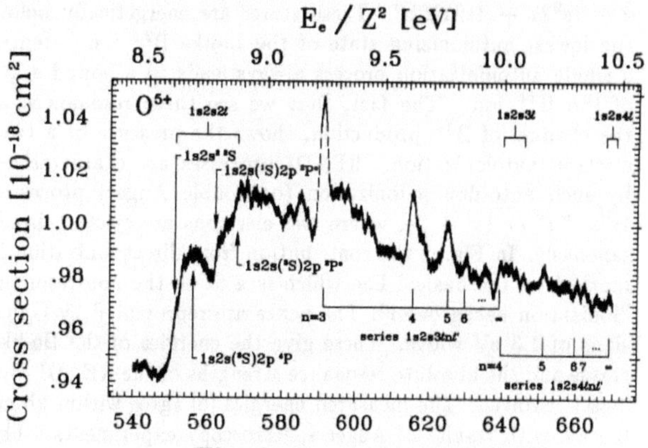

Fig. 1 Cross sections for electron-impact ionization of O^{5+} ions obtained by the energy-scanning technique. Assignments of intermediate excited states to the cross section features are given.

roughly 8% due to the contributions of indirect ionization mechanisms. In the experimental data a number of clear steps can be recognized which are due to excitation of particular levels in the $1s2s2\ell$ configurations. The most interesting feature observed here is the relatively strong contribution from a REDA Rydberg series. Be-like states are populated by the dielectronic capture $e + 1s^2 2s \rightarrow 1s2s3\ell n\ell'$ of the projectile electron into the target ion with simultaneous $\Delta n=2$ core excitation of the ion. The Be-like states decay predominantly by successive emission of two electrons $1s2s3\ell n\ell' \rightarrow 1s2s2\ell'' + e \rightarrow 1s^2 + 2e$ and thus contribute to the production of He-like ions, i.e. net single ionization of the Li-like parent ions. There are even indications for another Rydberg series involving $\Delta n=3$ excitations of the target ion. Positions of energy levels are indicated in the figure.

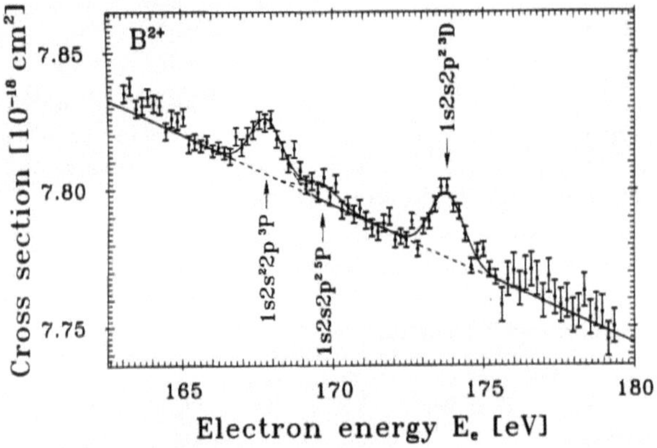

Fig. 2 Details of the cross section for electron-impact ionization of B^{2+} ions obtained by the energy-scanning technique. Direct ionization is indicated by the dashed line. READI peaks were fitted by Gaussian distributions (solid line) sitting on top of the direct ionization curve. The peaks correspond to Be-like states (in B^+) $1s2s^2 2p\ ^3P$, $1s2s2p^2\ ^5P$ and $1s2s2p^2\ ^3D$ which contribute to the observed net single ionization via simultaneous two-electron emission.

Also in the Li sequence we could clearly identify READI processes. Figure 2 shows the cross section of B^{2+} ions in an energy range where READI processes are the only possible resonant ionization mechanisms. The resonances observed at 168 eV and 174 eV can be associated with the Be-like states $1s2s^2 2p\ ^3P$ and $1s2s2p^2\ ^3D$ populated by dielectronic capture $e + 1s^2 2s \rightarrow 1s2s2\ell 2\ell'$. These states are energetically below the lowest autoionizing state of the Li-like B^{2+} ion. Hence, a single autoionization process always leads to a bound state of the B^{2+} ion. The fact, that we see these resonances in the channel of B^{3+} production, shows the presence of a two-electron autoionization. READI processes are characterized by such auto-double-ionization (or double Auger) processes $1s2s2\ell 2\ell' \rightarrow 1s^2 + 2e$, where two electrons are ejected simultaneously. In Fig. 2 the contribution from direct ionization is marked by the dashed line which is a fit to the non-resonant "ionization background". The peaks are represented by Gaussians of 1.3 eV width. These give the energies of the Be-like states and the absolute resonance strengths of the READI processes involved. The measured energies [6] agree within about 0.5 eV with results of Auger spectroscopy experiments. The peak areas can be compared with calculations [7] and in particular serve to test theoretical decay rates for double Auger processes.

By pushing the precision of our experimental method to the unprecedented level of 0.01% statistical uncertainty, we were able to see and to measure [4] ionization contributions from direct two-electron excitation of Li^+ ions $1s^2 \rightarrow 2\ell 2\ell' \rightarrow 1s + e$ and even from the trielectronic capture $e + 1s^2 \rightarrow 2\ell 2\ell' n\ell''$ followed by successive or simultaneous emission of two electrons $2\ell 2\ell' n\ell'' \rightarrow 1s + 2e$. The results are given in Fig. 3 where cross section differences $\sigma_{1,2} - \sigma^d_{1,2}$ are shown, i.e. direct ionization ($\sigma^d_{1,2}$) was subtracted from the originally observed net single ionization ($\sigma_{1,2}$). The two dips in the cross section are due to destructive interference of the resonant (trielectronic capture) channels with direct ionization channels. The step in the cross section above 146 eV is due to direct two-electron excitations and additional trielectronic capture contributions.

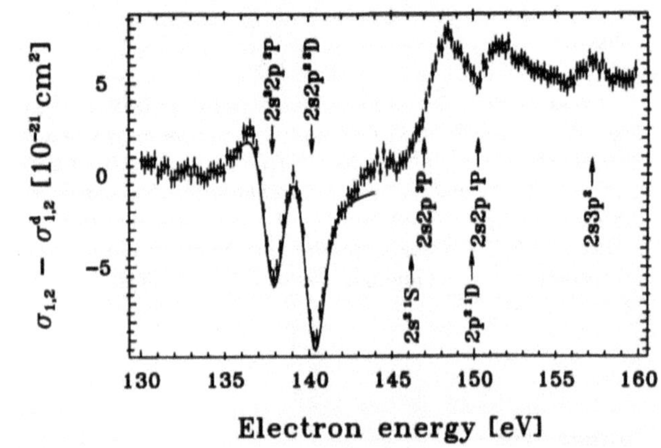

Fig. 3 Details of the Li^+ electron-impact ionization cross section which was measured with 0.01% statistics [4]. The points were obtained by subtracting the direct ionization cross section $\sigma^d_{1,2}$ from the total measured single ionization cross section $\sigma_{1,2}$. A number of energy levels are marked by arrows. The corresponding cross section structures indicate complex ionization channels like resonant-double-exitation-auto-double-ionization.

Simultaneous excitation of two electrons is much more likely when there are more equivalent electrons in a less tightly bound shell. We could see for instance resonant or non-resonant two-electron excitations $(e+)2p^6 \rightarrow 2p^4 3\ell 3\ell'(n\ell'')$ in the ionization of Na^+ ions with cross sections of a magnitude of about 100 times bigger than for Li^+. Unfortunately, double excitation of two p-electrons in ns^2np^6 configurations is nearly always masked by single excitation of one s-electron since both excitation processes require about the same projectile electron energies for the ions studied here and both excitation processes lead to autoionizing states.

One of the important experimental results concerning resonances is the observation of branching into different decay channels [2] from one intermediate highly- excited state produced by dielectronic capture of the projectile electron into the target ion. The same initial compound states can be observed in net single, double, triple or quadruple ionization at identical energies.

An example is given in Fig. 4 which shows features in cross sections of Ba^+ ions [8]. Here, the background from direct multiple ionization was determined from the total cross section measured below the resonance features. A straight line was fitted to these data and then subtracted from the total experimental cross section. What remains is a contribution involving the 3d subshell. There are two dominant peaks in both cross sections which are due to dielectronic capture plus subsequent emission of four or five electrons, respectively. Additional to the resonances there are contributions from 3d excitation with subsequent emission of three or four electrons, respectively, and probably also 3d ionization with subsequent emission of two or three electrons, respectively, as evidenced by an increased "background" level at energies above the resonances. All these processes can produce the net three-fold and net four-fold ionization observed in the experiments.

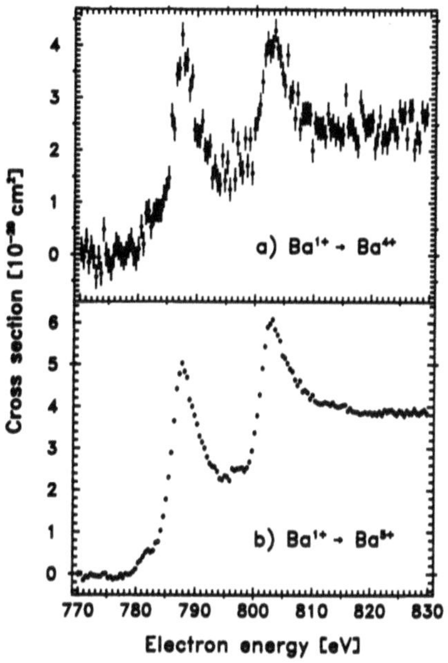

Fig. 4 Comparison of energy-scan measurements [8] of net triple (a) and net quadruple (b) ionization of Ba^+ ions. Direct contributions to the cross section were represented by straight lines and subtracted from the measured total cross section in order to isolate the resonance features displayed here.

It is particularly interesting to note the different decay probabilities for the intermediate resonant 3d-excited states. Branching into net three-fold ionization yields a resonant cross section contribution of about $4{\cdot}10^{-20}$ cm² (for the peak at 787 eV) and branching into net four-fold ionization yields a cross section of about $5{\cdot}10^{-20}$ cm². This means that the branching ratio for the emission of five electrons from the resonant intermediate excited state exceeds that for the emission for four electrons by a factor of roughly 1.3. When there is a vacancy in a deep lying shell the cascading Auger processes lead to a certain probability distribution of the final charge states. In the case of a 3d vacancy in an intermediate Ba atom the maximum of the probability distribution may be around a charge state q=5. Since the non-resonant contributions to the production of Ba^{q+} from Ba^+ ions are much higher for lower q it becomes increasingly difficult to find the low q resonance contributions in experiments measuring the total cross section. This is also the reason why the statistical uncertainty of the data displayed in Fig. 4 is higher for three-fold compared to four-fold ionization: the "background" from non-resonant multiple ionization is a factor 7 higher in $\sigma_{1,4}$ than in $\sigma_{1,5}$.

Conclusion

The measurement of electron impact ionization cross sections with high precision and good energy resolution offers attractive opportunities to study indirect ionization mechanisms. Careful observation of fine details in the cross sections allows to measure energy levels of multiply- excited states and to get information on the decay probabilities of such states.

We gratefully acknowledge support by Deutsche Forschungsgemeinschaft (DFG), Bonn-Bad Godesberg, and Max-Planck-Institut für Plasmaphysik (IPP), Garching.

References

1. Müller, A., Belić, D.S., DePaola, B.D., Djurić, N., Dunn, G.H., Mueller, D.W., Timmer, C., Phys. Rev. A **36**, 599 (1987)
2. Müller, A., Tinschert, K., Hofmann, G., Salzborn, E., Dunn, G.H., Phys.Rev. Lett. **61**, 70 (1988)
3. Müller, A., Hofmann, G., Tinschert, K., Salzborn, E., Phys.Rev. Lett. **61**, 1352 (1988)
4. Müller, A., Hofmann, G., Weißbecker, B., Stenke, M., Tinschert, K., Wagner, M., Salzborn, E., Phys.Rev.Lett. **63**, 758 (1989)
5. Müller, A., in PHYSICS OF ION IMPACT PHENOMENA, ed. by Mathur, D., Springer, Berlin, Heidelberg, New York, 1991, in preparation
6. Hofmann, G., Müller, A., Tinschert, K., Salzborn, E., Z.Phys.D - Atoms, Molecules and Clusters **16**, 113 (1990)
7. Pindzola, M.S., Griffin, D.C., Phys.Rev.A **36**, 2628 (1987)
8. Weißbecker, B., Müller, A., Hofmann, G., Stenke, M., Tinschert, K., Salzborn, E., to be published

Supplement to Z. Phys. D - Atoms, Molecules and Clusters 21, S 193-195 (1991)

Atoms, Molecules
Zeitschrift für Physik D and Clusters

Inner-shell ionization of lithium-like chromium ions

D. Vogel, P. Beiersdorfer, R. Marrs, K. Wong, R. Zasadzinski
High Temperature Physics Division, University of California, Lawrence Livermore National Laboratory, Livermore, CA 94550, USA

Abstract

We have used high-resolution x-ray spectroscopy to investigate inner-shell ionization of Cr^{21+} ions by electron impact using the Electron Beam Ion Trap at Lawrence Livermore Laboratory. Our measurements indicate that inner-shell ionization enhances the intensity of the radiative transition $1s2s\ ^3S_1 \rightarrow 1s^2\ ^1S_0$.

I. Introduction

Helium-like ions occur in many types of plasma. Examples include solar flares and tokamaks, as well as plasmas produced by intense laser pulses. Since the ions in a plasma repeatedly undergo excitation by colliding electrons, they continually emit x-rays, most of which escape the plasma. These impact excitation x-rays provide important clues to the composition, charge balance, density and temperature of the plasma. The helium-like x-ray spectrum is of particular interest because satellite lines from neighboring lithium-like and beryllium-like charge states mingle with the He-like lines. Their relative intensities therefore provide a convenient indication of the charge balance. Processes such as dielectronic recombination and inner-shell ionization complicate the picture, however, because they populate excited levels in adjacent charge states.

In hot ionizing plasmas, inner-shell ionization can have a profound effect on the intensity of the transition $1s2s\ ^3S_1 \rightarrow 1s^2\ ^1S_0$. It is precisely the x-rays resulting from this transition that allow us to isolate the inner-shell process. The Electron Beam Ion Trap (EBIT) has already been used to investigate both impact excitation and dielectronic recombination [1,2]; now we have studied ionization interactions as well (Wong et al. [3] have investigated outer-shell ionization of Ba^{53+} ions).

II. Theory

Figure 1 shows the energy levels involved (He-like and Li-like), as well as their radiative decays to the ground state. The resonance line w, an electric dipole (E1) transition, is the strongest. Next in intensity is y, also an E1 transition, but weaker than w because it is an intercombination line. The other intercombination line, x, is a magnetic quadrupole transition. Line z, a magnetic dipole transition, is a forbidden line, but in EBIT, even z is strong because the plasma density is low. Neither the $^3P_0$ nor the $^1S_0$ level decays radiatively to the ground state since $j=0$ to $j=0$ transitions are strictly forbidden (the latter state does decay through two-photon emission, while nuclear hyperfine splitting enables the $1s2p\ ^3P_0 \rightarrow 1s^2$ transition for the ^{53}Cr nucleus, whose abundance is about 9.5%). The satellite lines q, r, and t, come from doubly excited Li-like ions. Such doubly-excited ions can be produced by dielectronic recombination; these three levels also have significant cross sections for direct (inner-shell) excitation from the Li-like ground state.

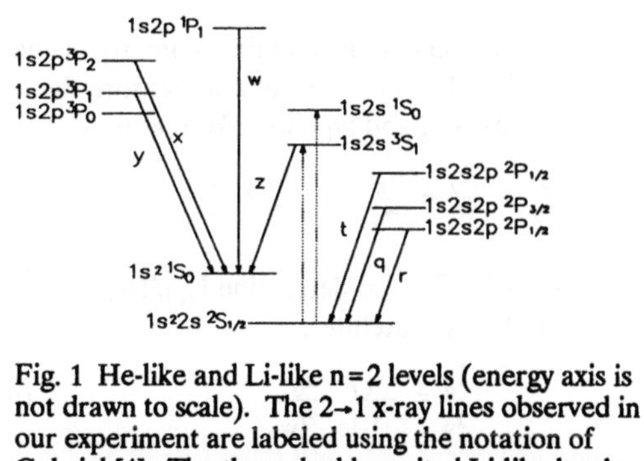

Fig. 1 He-like and Li-like n=2 levels (energy axis is not drawn to scale). The 2→1 x-ray lines observed in our experiment are labeled using the notation of Gabriel [4]. The three doubly excited Li-like levels shown are the only ones with appreciable cross sections for direct excitation. The vertical dotted lines indicate inner-shell ionization.

Line z can be enhanced strongly by inner-shell ionization. Inner-shell ionization of a Li-like ion removes one of the two K-shell electrons, e.g.,:

$$Cr^{21+} + e^- + \rightarrow (Cr^{22+})^* + 2e^- \qquad (1)$$

leaving the resulting He-like ion in either the 1s2s $^3S_1$ or 1s2s $^1S_0$ level. The threshold for this process is 7.4 keV, much higher than the (outer-shell) ionization potential of Cr^{21+}, and close to the ionization potential of Cr^{22+}. A possibility also exists (if the electron energy is slightly greater than 7.4 keV) that the ion ends up in one of the other He-like levels, a process known as shake-up. The $^3S_1$ level, being the lowest, is also fed by cascades. Cascades from the $^3P_0$ and $^3P_2$ levels, as well as states with n>2 provide about 80 percent of z's intensity.

At the low electron and ion densities present in EBIT, collisional de-excitation is rare and metastables decay before they can undergo additional excitations. Hence, line intensities can be expressed in terms of the ground state density and excitation cross sections [5]. The photon emission rate for line w is given by:

$$R_w = j_e N_{He} \sigma_w \qquad (2)$$

where j_e is the electron current density, N_{He} is the number of ground-state He-like ions in the beam, and σ_w is the cross section for electron impact excitation. For z, we must add terms for cascades and inner-shell ionization:

$$R_z = N_{He} j_e (\sigma_z + \sum_i \sigma_i B_{iz}) + N_{Li} j_e \sigma_{ii} \qquad (3)$$

The second term is a sum over all the states i that can feed z, where σ_i is the state's excitation cross section and B_{iz} is the branching ratio. The last term gives the inner-shell contribution; it is the only one that depends on N_{Li}, the Li-like ion density. The observed emission intensity is proportional to the photon rates:

$$I_i = R_i \cdot G \qquad (4)$$

where the constant G depends on the geometry of the detector and the EBIT. Since we cannot measure the ion number densities N_{He} and N_{Li}, we take the intensity ratio:

$$\frac{I_z}{I_w} = \frac{R_z}{R_w} = \frac{\sigma_z'}{\sigma_w} + \frac{N_{Li}}{N_{He}} \frac{\sigma_{ii}}{\sigma_w} \qquad (5)$$

where $\sigma_z' = \sigma_z + \Sigma \sigma_i B_{iz}$. To determine N_{Li}/N_{He}, we can use one of the Li-like satellites:

$$\frac{I_q}{I_w} = \frac{N_{Li}}{N_{He}} \frac{\sigma_q}{\sigma_w} \qquad (6)$$

where I_q and σ_q correspond to line q. Combining (5) and (6) gives a linear relationship among I_z, I_q, and I_w:

$$I_z = \frac{\sigma_z'}{\sigma_w} I_w + \frac{\sigma_{ii}}{\sigma_q} I_q \qquad (7)$$

Thus, by measuring the line intensities at different charge states, one can determine σ_{ii}/σ_q.

III. Experimental Procedure

We used the Livermore Electron Beam Ion Trap (EBIT), described by Levine et al. [6], to generate highly charged chromium ions. By adjusting the beam accelerating voltage, one can probe electron-ion interactions at energies from 4-25 keV. At energies above the inner-shell threshold, Cr^{22+} ions predominate overwhelmingly. To increase the fraction of Cr^{21+} ions, we alternate the electron beam energy between the KLL dielectronic resonance, at about 4.2 kV, and the observation energy (9-23 kV). Since the DR cross section is large, many Cr^{22+} ions recombine into the Li-like ground state. The DR resonance has a gaussian shape, so slightly changing the lower level voltage reduces the Cr^{21+}/Cr^{22+} ratio. Typical dwell times are 5 ms for observation and 10 ms at the KLL resonance. Figure 2 shows the relative charge balance, indicated by the ratio I_q/I_w, as a function of beam energy near the DR resonance.

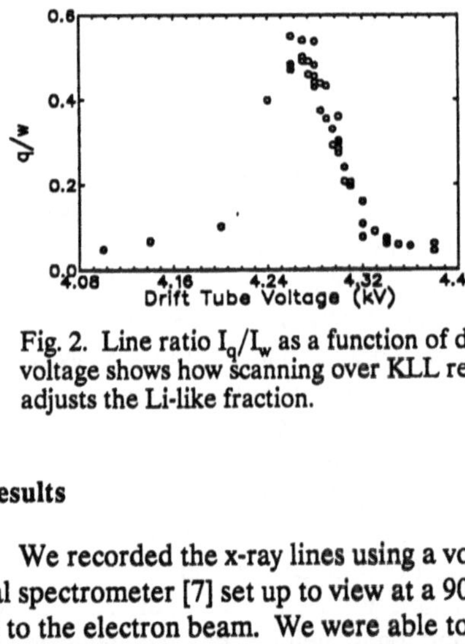

Fig. 2. Line ratio I_q/I_w as a function of drift tube voltage shows how scanning over KLL resonance adjusts the Li-like fraction.

IV. Results

We recorded the x-ray lines using a von Hámos crystal spectrometer [7] set up to view at a 90-degree angle to the electron beam. We were able to observe inner-shell ionization at six different interaction energies, ranging from 8.8 keV to 23.5 keV. Figure 3 shows the effect of inner-shell ionization in the 20.3-keV spectrum. Compared to line w, satellite lines q and r are larger because more Cr^{21+} is present; at the same time, z has been noticeably enhanced.

When the charge balance is varied, the line ratios show the expected linear relationship. Figure 4 is a plot of the line ratio I_z/I_w versus I_q/I_w for 8.8 keV and for 20.3 keV, along with a linear fit to each data set, derived using a nonlinear least squares procedure. We find the same relationship between I_z/I_w and I_r/I_w. Table 1 summarizes the results for all six interaction energies. Note that the fits to I_q and I_r give approximately the same intercept in each case. Although the slopes are proportional to σ_q and σ_r respectively, the intercept should equal σ_z' / σ_w in both cases.

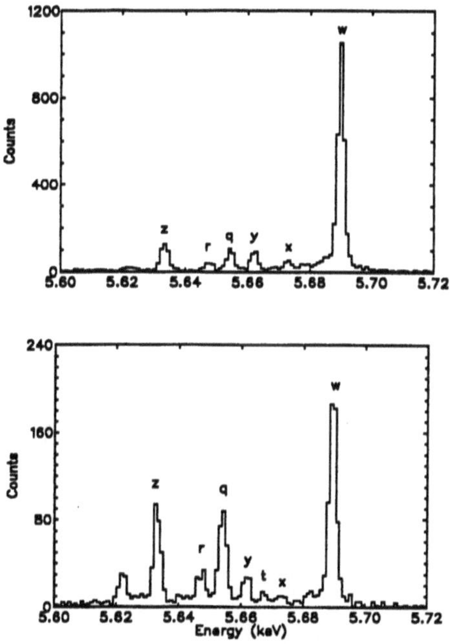

Fig. 3 High-resolution x-ray spectra at beam energy of 20.3 keV ($\lambda/\Delta\lambda$ = 1800). (a) Cr^{21+} fraction is small so satellite lines are small and line z is little enhanced. (b) With more Cr^{21+} present, q and r are larger; corresponding enhancement of z from inner-shell ionization is evident.

The linear correlation between I_z/I_w and I_q/I_w in each case makes it evident that inner-shell ionization has a significant effect on z. Additional evidence comes from the increase in slope with energy (up to 20.3 keV) since σ_{ii} is predicted to increase correspondingly. At the same time, the possibility of shake-up cannot be excluded. We also checked for correlations between lines y and x and the Li-like satellites. Both showed a slight positive slope at each energy, although in the case of x, the value was too small to be conclusive. Line y shows a definite linear correlation, but the slopes do not vary monotonically with energy as with z. The enhancement of y may be caused by overlap with line t since these two were not completely resolved by the spectrometer.

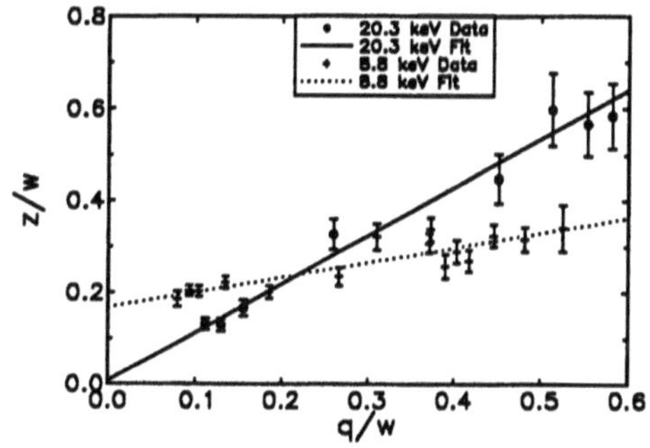

Fig. 4.. I_z/I_w versus I_q/I_w for 8.8 keV and 20.3 keV. Linear relationships show line enhancement due to inner-shell ionization. Slopes are equal to σ_{ii}/σ_q.

Table I. Linear fit parameters for intensity ratio I_z/I_w versus I_q/I_w and I_q/I_w for different energies (in keV)

	z/w vs. q/w		z/w vs. r/w	
Energy	Slope	Intercept	Slope	Intercept
8.8	0.32 ± .03	0.17 ± .01	0.92 ± .11	0.17 ± .01
11.8	0.56 ± .06	0.12 ± .02	1.78 ± .23	0.11 ± .02
14.8	0.72 ± .07	0.06 ± .02	2.47 ± .34	0.05 ± .03
17.8	0.91 ± .08	0.02 ± .02	2.99 ± .37	0.01 ± .03
20.3	1.10 ± .07	0.01 ± .01	2.85 ± .27	0.00 ± .02
23.3	1.13 ± .08	0.03 ± .01	2.90 ± .28	0.03 ± .02

We would like to thank E. Magee and D. Nelson for expert technical support throughout the experiment. This work was performed under the auspices of the U. S. Department of Energy by Lawrence Livermore National Laboratory under contract No. W-7405-ENG-48.

REFERENCES

1. R.E. Marrs, M.A. Levine, D.A. Knapp, and J.R. Henderson, *Phys. Rev. Lett.* **60**, 1715 (1988).
2. D.A. Knapp, R.E. Marrs, M.A. Levine, C.L. Bennett, M.H. Chen, J.R. Henderson, M.B. Schneider, and J.H. Scofield, *Phys. Rev. Lett.* **62**, 2104 (1989).
3. K. Wong, P. Beiersdorfer, D. Vogel, R. Marrs, and M. Levine (in this volume).
4. A. H. Gabriel, *Mon. Not. R. Astron. Soc.* **160**, 99-119 (1972).
5. J. Dubau and S. Volonte, *Rep. Prog. Phys.* **43**, 199 (1980).
6. M.A. Levine, R.E. Marrs, J.R. Henderson, D.A. Knapp, and M.B. Schneider, *Physica Scripta* **T22**, 157 (1988).
7. P. Beiersdorfer, R. Marrs, J. Henderson, D. Knapp, M. Levine, D. Platt, M. Schneider, D. Vogel, and K. Wong, *Rev. Sci. Instrum.* **61** (1990).

Supplement to Z. Phys. D - Atoms, Molecules and Clusters 21, S 197-199 (1991)

Atoms, Molecules
Zeitschrift für Physik D and Clusters

Electron impact ionization cross section measurement of lithium-like Ba

K. L. Wong, P. Beiersdorfer, D. Vogel, R. Marrs, M. Levine
High Temperature Physics Division, University of California, Lawrence Livermore National Laboratory, Livermore, CA 94550, USA

Abstract

The electron impact ionization cross section of Li-like Ba^{53+} has been measured at 22 keV using an electron beam ion trap. The measurement makes use of a novel technique which is discussed in detail. We find a value that is somewhat higher than predicted.

I. Introduction

Accurate knowledge of ionization, excitation, and recombination cross sections of highly charged ions along an isoelectronic sequence are needed to check the scaling laws of theoretical calculations. In this regime relativistic effects become important. Until recently cross section measurements for ionization have been limited to medium charge states using crossed beams (U^{16+}) [1], theta pinches (Ti^{8+}) [2], and tokamaks (Fe^{21+}) [3]. However, with the advent of the Livermore electron beam ion trap (EBIT) very high charge states can now be studied up to U^{82+}.

In the following we describe a measurement of the ionization cross section of Li-like Ba^{53+}, i.e. of the process $1s^2 2s + e^- \rightarrow 1s^2 + 2e^-$. This is the first measurement based on a new technique that relies on x-ray observations and represents the highest charge state ionization measurement to date. Barium was chosen in the study because it is readily available as it evaporates from the electron gun. The technique will be applied in future measurements to study the ionization cross section of several ions along the Li-like isoelectronic sequence.

II. Description of Technique

EBIT was designed as an x-ray source [4,5]; therefore, we have devised a technique for measuring the electron impact ionization cross section of Li-like ions based on studying the x-rays that emanate from EBIT at 90° with respect to the electron beam. The measurement scheme uses the fact that if the electron beam energy is set to the KLL dielectronic resonance energy for He-like ions, dielectronic (DR) and radiative (RR) recombination of He-like ions is to a good approximation balanced by ionization of Li-like ions. The scheme is illustrated by the diagram in Figure 1. In this approximation, where we have neglected charge exchange recombination (CX), the density ratio of the He and Li-like charge states is approximately equal to the ratio of the cross sections for ionization and the sum of DR and RR. As a result, an x-ray emitted following DR is a good indicator that an ionization event has occured. Taking into account recombination due to RR, we can therefore determine an ionization cross section by counting the n=2 to n=1 x-rays.

The general equation for the density n_{He} of ions in the He-like charge state in steady-state is given by:

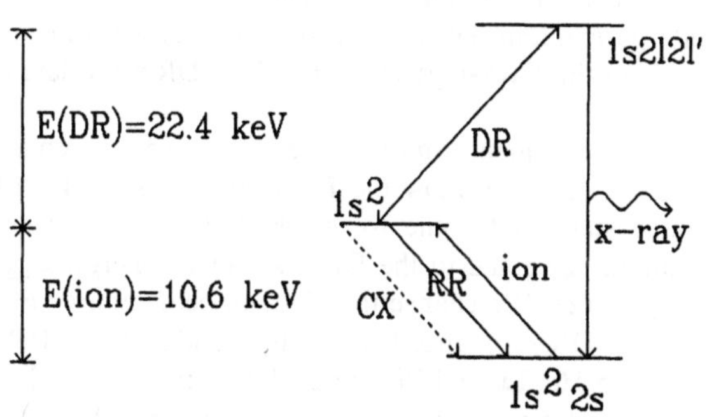

Figure 1: Measurement scheme for Li-like Ba. By counting the x-rays given off in the DR on He-like ions the ionization cross section can be measured, provided RR and CX are accounted for.

$$dn_{He}/dt = n_e v_e[\sigma_{ion}{}^{Li} n_{Li} - \sigma_{DR}{}^{He} n_{He} - \sigma_{RR}{}^{He} n_{He}] - n_0 v_0 \sigma_{CX}{}^{He} n_{He} = 0 \quad (1)$$

where n_e, v_e is the electron density and velocity, $\sigma_{ion}{}^{Li}$ the ionization cross section of Li-like ions, $\sigma_{DR}{}^{He}$ the cross section of dielectronic recombination onto He-like ions, $\sigma_{RR}{}^{He}$ the cross section of radiative recombination onto He-like ions, $\sigma_{CX}{}^{He}$ the cross section of charge exchange with He-like ions, n_q the ion charge density of charge state q, and n_0, v_0 the density and velocity of neutral background gas. Solving for $\sigma_{ion}{}^{Li}$:

$$\sigma_{ion}{}^{Li} = (n_{He}/n_{Li})[\sigma_{DR}{}^{He} + \sigma_{RR}{}^{He} + (n_0 v_0/n_e v_e)\sigma_{CX}{}^{He}]. \quad (2)$$

The x-ray intensities at 90° produced by DR onto He-like ions and RR onto Li-like ions can be expressed as:

$$I_{DR}{}^{He}(90°) = n_e v_e \sigma_{DR}{}^{He}(90°) n_{He} \quad (3)$$

$$I_{RR}{}^{Li}(90°) = n_e v_e \sigma_{RR}{}^{Li}(90°) n_{Li} \quad (4)$$

Substituting these into Eq. (2), we get:

$$\sigma_{ion}{}^{Li} = [I_{DR}{}^{He}(90°)/I_{RR}{}^{Li}(90°)]\sigma_{RR}{}^{Li}(90°) [\sigma_{DR}{}^{He}/\sigma_{DR}{}^{He}(90°) + \sigma_{RR}{}^{He}/\sigma_{DR}{}^{He}(90°) + (n_0 v_0/n_e v_e)\sigma_{CX}{}^{He}/\sigma_{DR}{}^{He}(90°)]. \quad (5)$$

We note that the term $\sigma_{DR}{}^{He}/\sigma_{DR}{}^{He}(90°)$ in Eq. (5) accounts for the possibility that the DR x-ray intensity may be different at 90° from the 4π average. This factor arises because the x-rays are polarized and are emitted anisotropically [6] due to the fact that the atoms are excited by a unidirectional electron beam.

III. Experimental Procedure

The RR and DR x-rays are measured with a solid-state germanium detector. Typical results are shown in Fig. 2 for the RR x-rays and in Fig. 3 for DR x-ray data for Ba.

The sequence in which the spectra were taken is as follows: (1) We allow the Ba to ionize to the He-like charge state and come to steady-state at the electron beam energy equal to the DR resonanace energy, $E_{beam} = E_{DR}$. (2) We change the electron beam energy away from the DR resonance $E_{beam} \neq E_{DR}$ and measure RR x-rays onto He, Li, and Be-like Ba for 20 ms, Fig. 2. Note that this time is short compared to the theoretical ionization time of 200 ms for changing the charge balance. From this we determine the relative abundance of He and Li-like Ba ions.

(3) Finally, we switch the beam energy back to the DR resonance $E_{beam} = E_{DR}$ and measure the DR x-rays, Fig 3. This period takes 40 ms which is long compared to theoretical DR time of 14 ms.

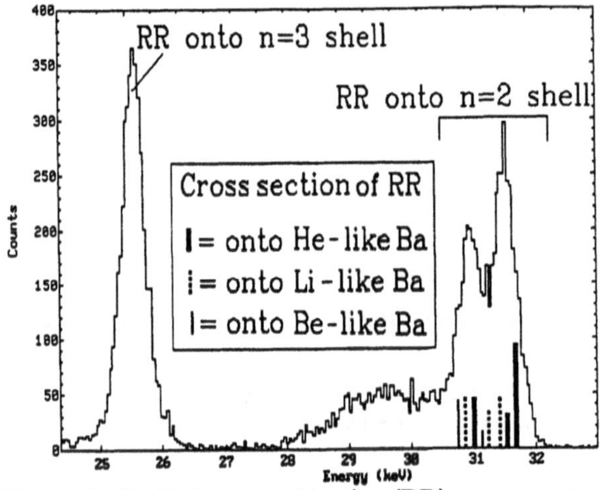

Figure 2: Radiative recombination (RR) x-ray spectrum, E_{beam} = 21.2 keV. The RR x-rays onto the n = 2 shell are fitted to determine a charge balance. The solid and dashed lines represent the relative theoretical cross sections of RR onto the different charge states.

The RR x-ray signal, $I_{RR}(90°)$, shown in Fig. 2 contains contributions from He, Li, and Be-like ions. The relative He, Li and Be-like ion densities are determined from a fit of the RR data with known RR cross sections and photoionization energies. A 100-eV detector resolution (gaussian width) is assumed in the fitting procedure.

Because $\sigma_{DR} > \sigma_{RR}$, the DR x-ray signal in Fig. 3 is more intense than the RR signal in Fig. 2. $I_{DR}(90°)$,

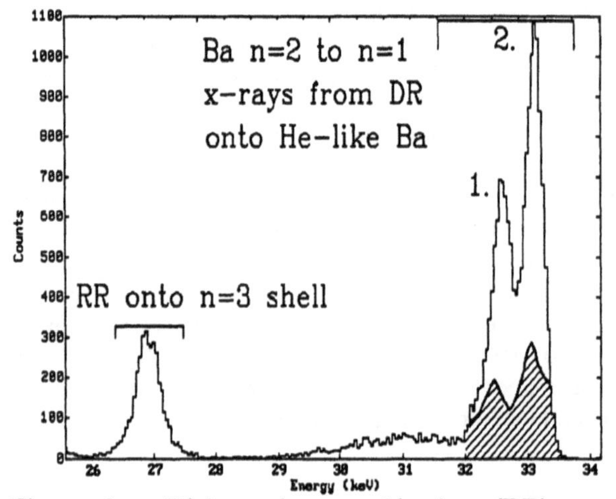

Figure 3: Dielectronic recombination (DR) x-ray spectrum, E_{beam} = 22.4 keV. The n = 2 to n = 1 feature contains a contribution from x rays due to RR onto the n = 2 shell of Ba (hatched area). Peak 1 is due to transition $[(1s2p)_1 2p_{3/2}]_{5/2} \rightarrow 1s^2 2p_{3/2}$; Peak 2 is due to transtion $[(1s2s)_0 2p_{3/2}]_{3/2} \rightarrow 1s^2 2s$.

however, has contributions from x rays due to RR onto He, Li, and Be-like ions due to the fact that at $E_{beam} = E_{DR}$ the DR x rays and RR x rays appear in the same energy region. Because E_{beam} (Fig. 2) is approximately equal to E_{beam} (Fig. 3), RR is approximately the same in both spectra. Thus, we can determine $I_{DR}^{He}(90°)$ by subtracting the $I_{RR}(90°)$ signal in Fig. 2 from the $I_{DR}(90°)$ signal in Fig. 3, $I_{DR}^{He}(90°) = I_{DR}(90°) - I_{RR}(90°)$. We checked the validity of this approach by comparing the x-ray signal of RR onto the n=3 shell, which is seen in both the RR and DR x-ray spectra, and verified that they are equal.

Substituting the experimental quantities into the theory expression in Eq. (5) we obtain:

$$\sigma_{ion}^{Li} = [I_{DR}(90°)-I_{RR}(90°)/I_{RR}^{Li}(90°)]\sigma_{RR}^{Li}(90°) [\sigma_{DR}^{He}/\sigma_{DR}^{He}(90°) + \sigma_{RR}^{He}/\sigma_{DR}^{He}(90°) + (v_0n_0/v_en_e)\sigma_{CX}^{He}/\sigma_{DR}^{He}(90°)]. \quad (6)$$

Thus, we can determine σ_{ion}^{Li} from a measurement of $I_{DR}(90°)$ by normalizing to $\sigma_{RR}^{Li}(90°)$ and σ_{RR}^{He}. The result is affected by CX, which we treat as a pertubation.

IV. Results

The results for Li-like Ba are given in Table I for three energies corresponding to the energies of the dielectronic resonances in He-like Ba. We use theoretical calculations for $\sigma_{RR}^{Li}(90°) = 6.6 \times 10^{-23}$ cm^2, $\sigma_{RR}^{He} = 9.2 \times 10^{-23}$ cm^2 [7], $\sigma_{CX}^{He} = 2.3 \times 10^{-13}$ cm^2 [8], $\sigma_{DR}^{He}/\sigma_{DR}^{He}(90°) = 1.0, 1.0, 1.23$ for the three DR resonance energies in order of increasing energy [9], and v_e, n_e, and v_0 are determined by the operating parameters of EBIT. The only unknown in the equation is the neutral particle density n_0, which is related to the pressure of the background gas. This unknown determines the CX contribution and is discussed in the following paragraph. The results with and without CX are compared to theoretical values calculated by Reed using the distorted wave code of Zhang [10] and Moores [11] and to the semiempirical Lotz formula [12]. We measure an ionization cross section 1.8x theory without CX and 2.5x theory with CX.

In order to determine the effect of CX, we have measured the x-ray signal at energies away from the DR resonance. In this regime ionization is approximately balanced by CX and RR. Knowing RR we can infer CX. Since the CX contribution does not change as a function of E_{beam}, we can use this to correct the DR data.

Table I. Electron impact ionization cross section of Li-like Ba in units of 10^{-21} cm^2. * denotes sum of adjacent resonances. Energy in units of keV.

	Experiment		Theory		
Energy	no CX	with CX	Zhang	Moores	Lotz
21.83	0.24 ± 0.16	0.36 ± 0.28	0.143	0.141	0.142
21.97	0.25 ± 0.17	0.38 ± 0.29	0.143	0.141	0.142
22.44*	0.25 ± 0.13	0.36 ± 0.24	0.143	0.141	0.142

Because of the low resolution of the Ge detector there is a large uncertainty in the determination of the charge balance from the the RR data. The error bars are, therefore, calculated assuming maximum possible variations in the charge states n_{He}, n_{Li}, and n_{Be}. For example, the lower limit is calculated by assuming that n_{Li} = 100%, even though this assumption is unphysical. Measurements are in progress that determine the charge balance with improved accuracy.

Acknowledgments

We thank D. Nelson, E. Magee, and A. Hinz for their technical support. This work was performed under the auspices of the U. S. Department of Energy by Lawrence Livermore National Laboratory under contract No. W-7405-ENG-48.

References

1. D. C. Gregory et al., Phys. Rev. A41, 106 (1990)
2. R. U. Datla and J. R. Roberts, Phys. Rev. A **28**, 2201 (1983)
3. J. S. Wang et al., Phys. Rev. A **38**, 4761 (1988)
4. M. A. Levine et al., Phys. Scr. **T22**, 157 (1988)
5. R. E. Marrs et al., Phys. Rev. Lett. **60**, 1715 (1988)
6. J. R. Henderson et al., Phys. Rev. Lett. **65**, 705 (1990).
7. J. H. Scofield, (private communication)
8. A. Muller and E. Salzborn, Phys. Lett. **62A**, 391 (1977)
9. M. H. Chen, (private communication)
10. H. L. Zhang and D. L. Sampson, (to be published)
11. D. Moores, (private communication)
12. W. Lotz, Z. Phys. **216**, 241 (1968)

Supplement to Z. Phys. D - Atoms, Molecules and Clusters 21, S 201-202 (1991)

Atoms, Molecules
Zeitschrift für Physik D and Clusters

A new technique for excitation studies in electron-ion collisions

R. W. O'Neill, I. G. Hughes and I. D. Williams

The Queen's University of Belfast, Belfast, BT7 1NN, Northern Ireland, UK

Abstract

A new technique based on an electron energy–loss technique is being developed to measure absolute total cross sections for ion excitation by electron impact. Novel aspects of the instrument include collection of the electrons in the backward direction, and the use of curved trochoidal plates to minimise distortion of the inelastically scattered beam during analysis.

Introduction

Atomic data for excitation rates of positive ions by electron impact are required for quantitative diagnosis of the properties of both laboratory and astrophysical plasmas. With such data electron densities and temperatures can be deduced from measured ratios of emission line intensities. Theoretical rate calculations are heavily relied upon in the absence of sufficient experimental measurements; however, agreement between experiment and theory is poor in the region near excitation threshold in even the most simple systems [1].

Recently an effort has been made to adapt the electron energy–loss technique for use in electron–ion scattering studies. This offers some attractive advantages over the previously used technique of optical spectroscopy; eg cascade corrections are unnecessary, optically forbidden and intersystem transitions may be directly probed and measurements may be extended to highly charged ions where UV transitions would prohibit optical spectroscopy. In the early 80's Chutjian and co–workers carried out angular differential measurements in a crossed–beams configuration, for both optically allowed and forbidden transitions [2,3], detecting scattered electrons by means of an electrostatic hemispherical analyser. More recently absolute total excitation cross sections have been presented by the groups at JILA (C^+ [4], Si^{3+} [5]) and JPL (Zn^+ [6]). Both groups have adopted a merged–beams approach, with analysis of the scattered electrons being carried out by planar trochoidal analysers – acting as forward velocity selectors.

The Instrument

The QUB crossed–beams instrument is shown schematically in figure 1.

The electron gun, analyser and detector are situated in a uniform magnetic field of typically 40–50 Gauss. This magnetic field serves to constrain the motion of the electrons, even after scattering from the ion beam has occurred. In this way electrons scattered through all angles may be detected simultaneously. The electron beam exiting the gun is deflected by an $\underset{\sim}{E} \times \underset{\sim}{B}$ trochoidal field before intersecting the ion beam at 90°. Typical sizes and intensities in the interaction region are ~ 1 mm, 0.5 μA for the electron beam, ~ 2 mm, 1–5 μA for the ion beam. Beam profiles and overlap integral are obtained by scanning the beams with perpendicular narrow slits.

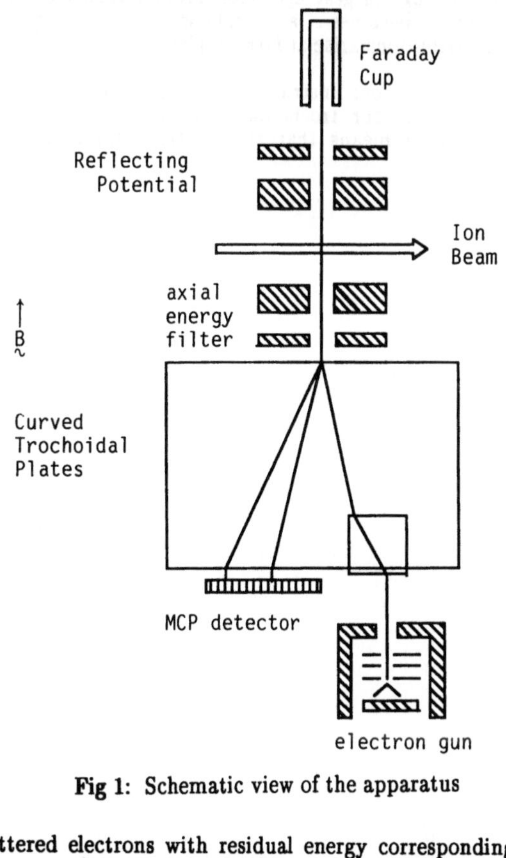

Fig 1: Schematic view of the apparatus

Scattered electrons with residual energy corresponding to an energy–loss associated with an excitation event are reflected backwards, deflected by the trochoidal analyzer and accelerated onto a position sensitive detector. This consists of three microchannel plates in chevron formation, with a resistive anode. Data is stored in a PC–type computer, which also provides control and data analysis for the overlap slits. The interaction and analysis region is maintained at a base pressure of ~ 10^{-10} mbar, with the ion beam being dumped into a Faraday cup in a differentially pumped chamber. Synchronized double–beam modulation is used to eliminate spurious signals from ion and electron interaction with the background gas, while those electrons elastically scattered at high angles (large cyclotron radii) from the ion beam are rejected by aperture stops.

Novel aspects of the technique are:

1) Analysis and detection of the scattered electrons occur in the backward direction. In this way the problem of completely separating primary and scattered beams is made simpler by allowing all of the primary beam to proceed in the forward direction, collecting in a Faraday cup well away from the mcp detector. Most importantly, scattered electrons through all of 4π sr are collected. This is extremely important for some forbidden and intersystem transitions, where theory shows the scattering to be very isotropic, eg 1s–2s in He^+ [7].

2) With parallel plate trochoidal analyzers beam distortion will occur due to the voltage differential across the finite beam. This is due to acceleration of those electrons that are closer to the positive plate and retardation of those nearer the negative plate. Since those inelastically scattered electrons at the extremities of the beam are those due to $\sim 90^0$ scattering (largest cyclotron radius) then clearly this effect becomes a problem when significant contributions arise at large angles.

   Hence for our instrument we have designed curved plates for our trochoidal analyzer. The cylindrical geometry means that the electric field is no longer constant in the gap between the plates. This,combined with the different path lengths experienced by the beam extremities in the curved plates, greatly reduces the beam distortion.

It is proposed to apply this method to excitation of multiply charged ions.

References

1 For review see Phaneuf R A: Atomic Processes in Electron–Ion and Ion–Ion Collisions (Plenum, New York), 117 (1986).

2 Chutjian A and Newell W R: Phys Rev A 26, 2271 (1982).

3 Williams I D, Chutjian A and Mawhorter R J: J Phys B: At Mol Phys 19, 2189 (1986).

4 Wahlin E, Timmer C, Forand L, Swenson D, De Paola B, Phaneuf R, Belic D, Rinn K, Muller A and Dunn G: Proc XVI Int Conf on Electronic and Atomic Collisions, New York, Abstracts, p 368 (1989).

5 Dunn G, Z Phys D, current issue.

6 Smith S J, Man K F, Mawhorter R J, Williams I D and Chutjian A: XII Int Conf on Atomic Physics, Ann Arbor, Mi, Conf Proceedings (1991), to be published.

7 Berrington K A, Kingston A E and Salvini S A: J Phys B: At Mol Phys 16, 2399 (1983).

Supplement to Z. Phys. D - Atoms, Molecules and Clusters 21, S 203-204 (1991)

Atoms, Molecules
Zeitschrift für Physik D and Clusters

Excitation of multiply charged ions by electron impact

B. A. Huber, Ch. Ristori, H. J. Andrä[1], P. A. Hervieux, M. Maurel and C. Guet
SPhAt/PIM, Centre d'Etudes Nucleaires de Grenoble, F-38041 Grenoble Cedex, France

[1] LAGRIPPA, Centre d'Etudes Nucleaires de Grenoble, F-38041 Grenoble Cedex, France

Abstract

A new crossed beam apparatus has been developed for measuring differential cross sections for the excitation of multiply charged ions by electron impact. First results have been obtained for sodium-like Argon (Ar^{7+}); the 3s - 3p transition with an excitation energy of ~ 17.5 eV has been studied at an electron energy of 100 eV. The preliminary experimental values are compared with theoretical results obtained within a Born-approximation.

I. Introduction

In nearly all types of electrical discharges the population of excited ion states is prevailingly determined by electron - ion collisions. Therefore, the electron impact excitation of ions not only represents an important fundamental atomic process but is also responsible for strong radiation losses of hot plasmas, particularly if the excitation of multiply charged ions is considered.

Nevertheless, the number of experimental investigations, which are necessary for testing existing theoretical results, is rather limited. In systems like C^{3+} and N^{4+} total excitation cross sections were measured in a crossed - beam experiment by analysing the emitted photon flux [1-2]. The excitation of Ba^{46+} ions has been studied within an electron beam ion trap [3]. For other singly charged species like Mg^+, Zn^+ and Cd^+ angular differential excitation cross sections have been obtained [4].

In the following we will briefly describe the new experimental setup, which has been used to study the 3s - 3p excitation of sodium - like argon. These are the first angular differential excitation cross sections which are reported for multiply charged ions.

II. Experiment

A beam of multiply charged ions intersects an electron beam at an angle of 90 degrees. The electrons are analysed with respect to their kinetic energy at scattering angles between ~ 10° and ~ 27°. Inelastically scattered electrons are separated from the elastic scattering signal due to the excitation process.

The beam of multiply charged ions is delivered by the NEW LAGRIPPA Facility with a maximum acceleration voltage of 20 kV. For an ion beam of Ar^{7+} (E_{ion} = 140 keV) the electrical current amounts to (6 - 8) μA with a beam diamater in the interaction region of about 3 mm.

At an electron energy of 100 eV the electron beam is characterized by a beam diameter of 2 mm and an electron current of ~ 4 μA. The halfwidth of the energy distribution is less than 500 meV.

The electrostatic spectrometer consists of 2 cylindrical units with an overall energetic resolution $E/\Delta E$ of 50. In order to allow for good vacuum conditions and to minimize the influence of background electrons it is shielded by a transparent double grid system.

The base pressure within the interaction region is in the order of $3\cdot 10^{-9}$ Pa increasing to a value of $(1-2)\cdot 10^{-8}$ Pa with both beams in operation. The effect of the earth magnetic field is reduced by a double-layer μ-metal shielding and a Helmholtz-coil device limiting the residual magnetic induction to values below 10 mG.

The elastic and the inelastic electron signals are measured at the same angle and the same ion and electron beam conditions. The differential excitation cross section is derived from the measured ratio of both signals. This avoids to determine accurately the formfactor of the 2 intersecting beams.

III. Results and discussion

The differential cross section for the electron impact excitation of the 3s - 3p resonance transition in Ar^{7+},

$e_f^- + Ar^{7+}(3s) \longrightarrow e_s^- + Ar^{7+*}(3p) - 17{,}5\ eV,$

has been measured at an electron energy of 100 eV and for angles between 17° and 27°.

Figure 1 shows the energy distribution of electrons measured at a scattering angle of 22°. Electrons which

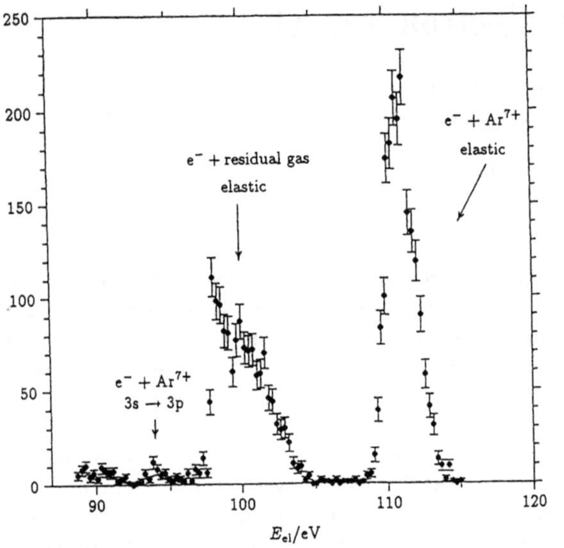

Fig. 1: Energy distribution of 100 eV - electrons scattered from the residual gas and an Ar^{7+} ion beam. (Scattering angle: 22°; for explanation see text)

are scattered elastically from the residual gas form the peak at E_{el} = 100 eV. Its lineshape is broadened and asymmetric. Due to the missing entrance slit of the spectrometer electrons are recorded which are scattered from the residual gas all along the electron beam. The structure at energies around 90 eV is caused by inelestic scattering with the background gas including excitation and ionisation processes.

The two signals at ~112 eV and ~ 94 eV, which are due to the electron ion interaction, are shifted in kinetic energy because of the finite beam velocities. Note, that this kinematical shift can be used to position the investigated excitation peak in a favourable background region by tuning the ion beam extraction voltage. The energetic halfwidth of both peaks is smaller as compared to those arising from the residual gas as the interaction volume of both beams is better defined and localized.

The ratio of both peaks has been corrected for the different transmission and detection efficiencies. Normalizing to the calculated differential cross sections for the elastic e⁻- ion scattering we obtain preliminary values for the differential excitation cross section which are given in figure 2. In the present case the results are restricted to scattering angles larger than 17° as the background signal arising from the electron gun increases strongly in the regime of small angles.

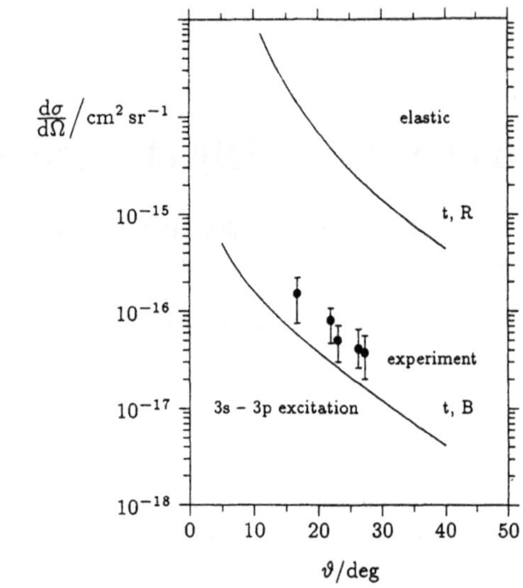

Fig. 2: Angular differential cross sections for the 3s - 3p excitation of Ar^{7+} at an electron energy of 100 eV. Full curves: theory, t,R: elastic Rutherford, t,B: inelastic Born; dots: experiment.

The experimental values are compared with theoretical results, obtained within the framework of a Born-approximation. The applicability of this approximation is verified by comparing the corresponding total cross sections with those obtained with a Distorted Wave Born calculation [5]. The experimental values tend to be larger than those of the Born calculation. However, Coulomb-Born calculations which are presently performed are expected to increase the theoretical cross section values to some extent.

Acknowledgments

These measurements have been performed at LAGRIPPA - a Joint laboratory of the CEA and of the CNRS. The authors are indebted to A.Brenac, G.Lamboley and T.Lamy for providing the ion beam and to J.Crançon, P.Perrin, R.Rocco and F. Zadworny for contributing in the early stage of this experiment.

References

1. Taylor P.O., Gregory D., Dunn G.H., Phaneuf R.H., Crandall D.H.: Phys.Rev.Lett. **39**, 1256 (1977)
2. Gregory D., Dunn G.H., Phaneuf R.H., Crandall D.H.: Phys.Rev.**A20**, 410 (1979)
3. Marrs R.E., Levine M.A., Knapp D.A.,Henderson J.R.: Phys.Rev.Lett. **60**, 1715 (1988)
4. Williams I.D., Chutjian A., Mawhorter R.J.: J.Phys.B: At.Mol.Phys. **19**, 2189 (1986)
5. Desclaux J.P.: private communication

Supplement to Z. Phys. D - Atoms, Molecules and Clusters 21, S 205-206 (1991)

Atoms, Molecules
Zeitschrift für Physik D and Clusters

Recombination of free electrons with U^{28+} ions

S. Schennach, A. Müller, M. Wagner, J. Haselbauer, O. Uwira, W. Spies
Institut für Kernphysik,
Strahlenzentrum der Justus-Liebig-Universität,
W-6300 Giessen, FRG

E. Jennewein, R. Becker, M. Kleinod, U. Pröbstel
Institut für Angewandte Physik, J.-W.-Goethe Universität,
W-6000 Frankfurt, FRG

N. Angert, J. Klabunde, P. H. Mokler, P. Spädtke, B. Wolf
Gesellschaft für Schwerionenforschung, W-6100 Darmstadt, FRG

Abstract

This paper reports a measurement of radiative recombination rates for U^{28+} ions and free electrons at energies from 0 to 2 eV. The results appear to contradict known theories.

Introduction

A free electron and an ion A^{q+} can recombine by radiative recombination (RR)

$$e + A^{q+} \rightarrow A^{(q-1)+} + h\nu \quad (1)$$

where the excess energy is carried away by a photon.

Observations of RR in collisions of free electrons with ions have been very scarce so far and there is only one published quantitative experimental result on RR rates by Andersen et al. [1] for C^{6+} ions.

At several laboratories, however, experiments have been set up to study recombination processes in colliding beams of electrons and ions. Much of the present interest in this field is related to the construction and operation of heavy ion storage rings. In particular, ion beam cooling in a cold electron beam raises the question of ion losses due to recombination processes. Experimental studies are needed in order to understand the loss mechanisms and in order to provide diagnostic tools for the determination of the degree of ion beam cooling.

Additional to raising such questions, the assembly of an electron beam cooler in a heavy ion storage ring itself provides unique possibilities to study electron-ion collisions [2] and to obtain experimental data which are required for the understanding of hot plasmas.

Experiment and discussion

In a collaboration of groups from the universities of Giessen and Frankfurt and from GSI an electron-ion merged beams experiment has been set up [3,4] at the heavy ion linear accelerator UNILAC of GSI. The arrangement is schematically shown in Fig.1.

Inside a strong magnetic field (up to 1T) produced by a superconducting solenoid a dense electron beam is formed [5]. The maximum density is $n = 10^{10}$ cm^{-3} which is reached for an electron energy of 8 keV with a current density of 10 Acm^{-2} drawn from a 3 mm diameter cathode and resulting in a total electron current of 660 mA. Elliptical windings on top of the solenoid provide transverse magnetic field components which serve to merge and demerge the electron and the ion beam. The merging section is about 50 cm long.

The electron target is incorporated in a beam line at the UNILAC accelerator. Behind a charge purifying magnet the ion beam is merged with the electron beam. With a second magnet the recombined ions are separated from the parent ion beam which is collected in a Faraday cup. The recombined ions are detected by a position-sensitive multi-channelplate detector. Six rotatable magnetic steerers, three in front of the electron target and three behind, serve to compensate the deflection of the ion beam by the transverse magnetic field components in the target.

The use of a superconducting solenoid for guiding the electron beam has the advantage that large surfaces around the electron beam vacuum are on liquid helium temperature thus providing efficient cryopumping. When the electron beam is switched on we do not measure a pressure increase outside the electron target even with a sensitivity of $0.1 \cdot 10^{-10}$ mbar.

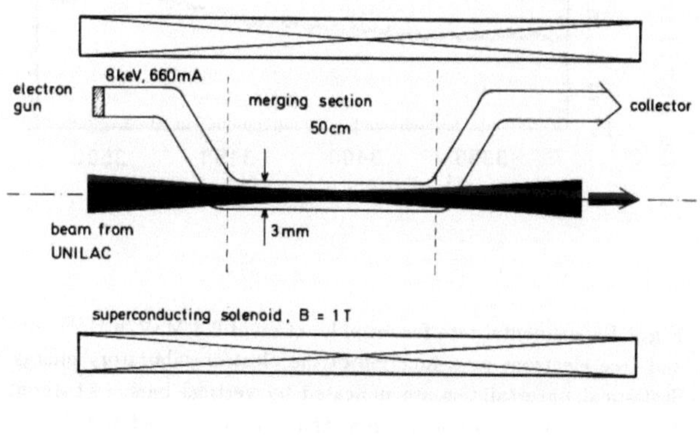

Fig.1 Schematic diagram of a merged beams experiment [3] designed for the UNILAC heavy ion accelerator of GSI, Darmstadt. The electron gun is inside the homogeneous field of a superconducting solenoid. Bending the electron trajectories onto the ion beam axis is accomplished by additional elliptical windings on top of the solenoid.

In a first experiment a beam of 6.3 MeV/u U^{28+} ions was merged with an electron beam whose laboratory energy was varied between about 3300 and 3550 eV. Within this energy range the parallel velocities of both beams can be matched so that the center of mass energy is only determined by the velocity spreads in both beams. The cross section σ_{RR} for RR [6,7], which diverges for the energy approaching zero, results in a finite rate r given by the convolution of the total velocity spread with σ_{RR}. For vanishing velocity spread in parallel direction and a transverse velocity spread of typically 0.2 eV one can calculate the rate of RR for U^{92+} ions to be about $4 \cdot 10^{-8}$ cm^3s^{-1} [8]. According to Hahn and Rule [9] the cross section for RR of an ion with nuclear charge Z and N electrons attached to it can be calculated in reasonable approximation by using an effective charge $Z_{eff} = Z - N/2$ in the conventional rate formulas for completely stripped ions [6,7]. For U^{28+} this approximation results in about $r = 2 \cdot 10^{-8}$ cm^3s^{-1} (for $kT_{\|} = 0$ and $kT_{\perp} = 0.2$ eV).

Fig. 2 shows our rate measurement for U^{28+} ions as a function of electron laboratory energy. At about 3500 eV the parallel velocities of electrons and ions are matched and there a pronounced maximum in the total recombination rate for $e + U^{28+} \rightarrow U^{27+}$ is observed. No background was subtracted. The energy dependence in Fig. 2 suggests that there is a background contribution of $0.5 \cdot 10^{-7}$ cm^3s^{-1} The peak rate after subtraction of this background amounts to more than $1 \cdot 10^{-7}$ cm^3s^{-1} and thus exceeds the expected rate by more than a factor 5. Storage of such ions as U^{28+} in a storage ring may become impossible because of excessive recombination losses of ions during the cooling process. Unfortunately, we do not have direct information on the velocity distribution f(**v**) in this particular experiment. From the width of the peak in Fig. 2 (which is 5.3 eV in the laboratory frame) an upper limit of $kT_{\|} = 1 \cdot 10^{-3}$ eV can be inferred. The calculations of electron trajectories in the present arrangement give a maximum transverse energy spread of the order of 0.5 eV. The lower limit in any case should be given by the temperature of the cathode which is about 1100 K (corresponding to roughly 0.1 eV). Because of the uncertainty of the energy spread in the U^{28+} experiment the recombination rate that has to be expected on the basis of the known theory may be another factor of 2 higher at the lower energy spread of 0.1 eV which still leaves a factor 2.5 discrepancy.

Another explanation of the unexpected high recombination rate could be in the presence of dielectronic recombination resonances on top of the radiative-recombination peak. At the electron densities around $4 \cdot 10^8$ cm^{-3} in this special experiment it may even be possible to have substantial contributions from three-body recombination [10] provided the electron temperature is well below 0.2 eV.

It has also been speculated that the strong magnetic field in the U^{28+} experiment ($B \approx 0.3$ T) together with the high charge state of the ions may lead to trapping [11] of an electron along a magnetic field line in the Coulomb field of an ion even at distances of 10^{-4} cm and thus the probability for RR might be enhanced.

Further experimental studies will be necessary to answer the new questions brought up by our experiment on RR of U^{28+} ions.

The experiments at the UNILAC accelerator reported here are supported by Bundesministerium für Forschung und Technologie BMFT.

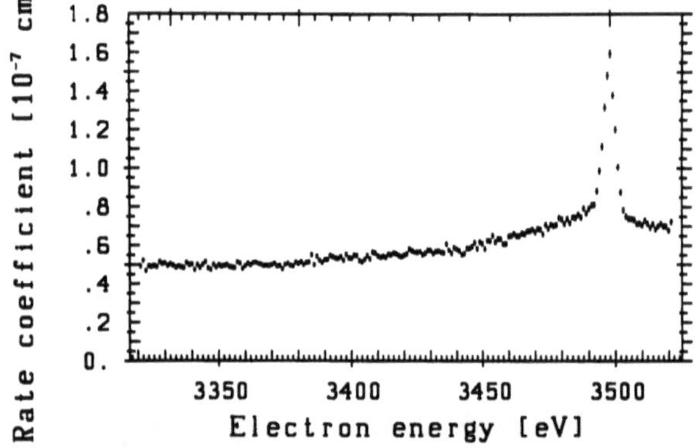

Fig.2 Experimental rate for recombination of 6.3 MeV/u U^{28+} ions and free electrons as a function of the electron laboratory energy. Statistical uncertainties are indicated by vertical bars. At about 3500 eV the velocities of electrons and ions is matched and the observed rate only depends on the energy (or velocity) spreads in both beams. Background is not subtracted.

References

1. L.H.Andersen, J.Bolko and P.I.Kvistgard, Phys.Rev.Lett. 64, 729 (1990)
2. G.Kilgus, J.Berger, P.Blatt, M.Grieser, D.Habs, B.Hochadel, E.Jaeschke, D.Krämer, R.Neumann, G.Neureither, W.Ott, D.Schwalm, M.Steck, R.Stokstad, E.Szmola, A.Wolf, R.Schuch, A.Müller, and M.Wagner, Phys.Rev.Lett. 64, 737 (1990)
3. A.Müller, S.Schennach, W.Spies, N.Djurić, R.Becker, E.Jennewein, M.Kleinod, U.Pröbstel, D.Brill, I.Müller, N.Angert, H.Emig, W.Enders, J.Klabunde, P.H.Mokler, K.Poppensieker, P.Spädtke, B.Wolf, GSI Scientific Report (1988) GSI-89-1, ISSN 0174-0814, p. 165
4. R.Becker, E.Jennewein, M.Kleinod, U.Pröbstel, K.Valadkhani, A.Müller, W.Spies, S.Schennach, N.Angert, F.Bosch, I.Hofmann, J.Klabunde, P.Mokler, H.Schulte, P.Spädtke, and B.Wolf, in: EPAC 88, Proceedings of the European Particle Accelerator Conference, Rome, 1988 (Edited by S.Tazzari), p. 607, World Scientific Publishers, Singapore (1989)
5. U.Pröbstel, R.Becker, E.Jennewein, M.Kleinod, and K.Valadkhani, GSI Scientific Report (1986), GSI-87-1, ISSN 0174-0814, p. 271
6. M.Stobbe, Ann.Physik 7, 661 (1930)
7. H.Bethe and E.Salpeter, Quantum Mechanics of One- and Two-Electron Systems, in: HANDBUCH DER PHYSIK, Vol. 35, Springer Verlag, Berlin, Heidelberg, New York (1957)
8. D.Liesen and H.F.Beyer, Preprint GSI-ESR/86-04 (1986)
9. Y.Hahn and D.Rule, J.Phys. B 10, 2689 (1977)
10. H.F.Beyer, D.Liesen and O.Guzman, Particle Accelerators 24, 163 (1989)
11. S.R.Goldman and I.Hofmann, Preprint LA-UR-89-2844, Los Alamos National Laboratory

Supplement to Z. Phys. D - Atoms, Molecules and Clusters 21, S 207-208 (1991)

Atoms, Molecules
Zeitschrift für Physik D and Clusters

X-rays from electron bombardment of heliumlike argon

R. Ali, C. P. Bhalla, C. L. Cocke, M. Schulz, and M. Stöckli
J. R. Macdonald Laboratory, Physics Department, Kansas State University, Manhattan, KS 66506-2604, USA

Abstract

We have observed x rays from the electron bombardment of heliumlike argon in the KSU EBIS ion source. The yield of x rays versus electron energy displays strong dielectronic recombination resonances and excitation thresholds. By normalizing to a single resonance strength, cross sections for dielectronic recombination and for 1s excitation are obtained from the data for different radiative decay channels. The results are in good agreement with theory.

I. Introduction

Over the past seven years several measurements of dielectronic recombination (DR) cross sections have been reported [1-8], subjecting previously untested theoretical calculations to experimental scrutiny. The theory has done well in general, and the availability of experimental results from many different sources is now becoming common. We recently reported one such result [8], namely the measurement of dielectronic recombination cross sections on ground state heliumlike argon from charge state equilibria established in the KSU EBIS ion source. These measurements were normalized to theoretical cross sections for ionization of lithiumlike argon. They provided resonance strengths which are free from radiation angular distribution problems, but provided no information on the relative strengths of different radiative decay branches of the resonances. They also do not provide information on excitation cross sections. In this paper we report measurements of the x ray spectra from these recombination and excitation processes which complement the information reported earlier.

II. Experiment

Argon target ions were produced in the KSU EBIS by bombarding argon with a 2.4 keV electron beam for typically 1.5 seconds. This is sufficient to drive the charge state equilibrium of argon ions held in the trap to greater than 95% Ar^{16+}, with less than 1% of the ion inventory in charge states below 15+. The electron beam energy was then swept rapidly over a range between 2 and 4 keV. Ar K x rays from the source interior were observed with a Si[Li] detector at 0 degrees relative to the electron beam direction. The data were taken in event mode, forming a two dimensional spectrum of x ray energy versus electron beam energy. From this two dimensional spectrum we extracted yields of x rays versus electron energy for different decay channels. For the sweep times used it was found that the contamination of the Ar^{16+} target with Ar^{15+} ions, from recombination on the DR resonances, could be kept below 10%. The yields were placed on an absolute cross section scale by normalizing to the KLL DR resonance, which had previously been found to be in agreement with the theoretical resonance strength.

III. Results

Fig. 1 shows cross sections for production of K_α, K_β, and K_γ radiation versus electron energy as deduced from projections from the two dimensional spectra. The solid curves are theoretical calculations done using procedures described in Ref. [9], folded into the experimental electron energy resolution. Since we observe the x rays at zero degrees, the comparison between theory and experiment is made for differential rather than total cross sections. The agreement is good. A disagreement between theory and experiment reported by us earlier for the KLN resonance has been traced to an experimental effect, namely a detector saturation, and is not observed here.

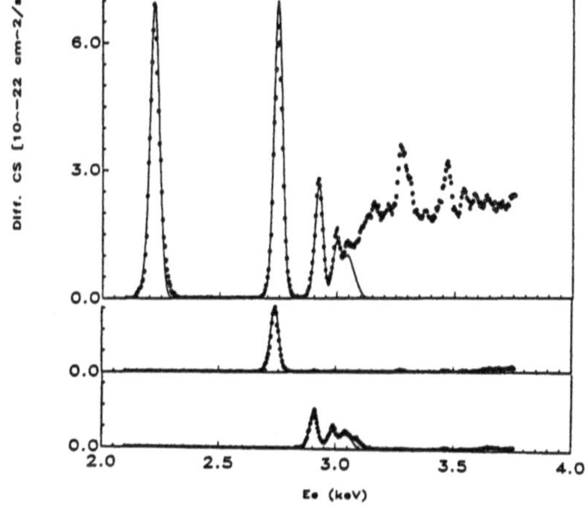

Fig. 1 Differential cross sections at zero degrees for K_α, K_β, and K_γ x rays for electrons on Ar^{16+}, respectively. The solid lines are theoretical calculations for KLn resonances.

Fig. 2 shows the total x ray production cross section and a comparison with theory which includes, in addition to the KLn series of DR resonances, the KMn series as well. Also shown is a comparison with calculated differential cross sections for n=1 to n=2 excitation of heliumlike argon [10]. Both theory and experiment show a rise in the x ray yield as the series of KLX DR resonances evolve into the excitation threshold. This rise is attributed to the fact that the excitation has substantial contributions from metastable states, ($^3P_1$, $^3P_2$, $^3S_1$) which, on the EBIS time scale, are "prompt" x ray emitters. These

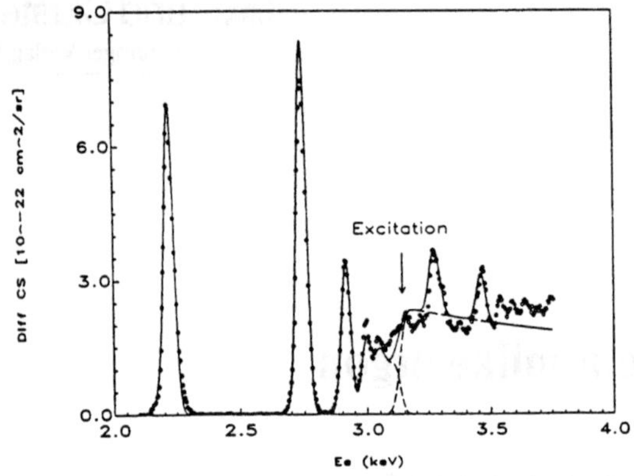

Fig. 2 Differential cross sections at zero degrees for total x ray production showing comparison with calculations for doubly excited resonances and excitation as well as the KLn DR resonances.

metastable cores can contribute to the KLn series only for n large enough that their Auger widths become comparable to their very small radiative widths and thus the rise appears near the threshold. Cores of this type have not yet been included in the DR calculations. The agreement between experiment and theory confirms that the calculations for both DR and excitation are on solid footing for these collisions.

Acknowledgment

This work was supported by the Division of Chemical Sciences, Office of Basic Energy Sciences, Office of Energy Research, U.S. Department of Energy.

References

1. J.B.A. Mitchell, *et al.*, Phys. Rev. Lett. 50, 335 (1983).
2. D.S. Belic, *et al.*, Phys. Rev. Lett. 50, 339 (1983).
3. P.F. Dittner, *et al.*, Phys. Rev. Lett. 51, 31 (1983); Phys. Rev. A 38, 2762 (1988).
4. A. Mueller, *et al.*, Phys. Rev. Lett. 56, 127 (1986).
5. L.H. Andersen, Phys. Rev. Lett. 62, 2656 (1989).
6. G. Kilgus, *et al.*, Phys. Rev. Lett. 64, 737 (1990).
7. D.A. Knapp, *et al.*, Phys. Rev. Lett. 62, 2104 (1989).
8. R. Ali, *et al.*, Phys. Rev. Lett. 64, 633 (1990).
9. C.P. Bhalla and T.W. Tunnell, J. Quant. Spectr. Radia. Transfer 32, 1431 (1984).
10. H.L. Zhang and D.H. Sampson, private communication (1990).

Supplement to Z. Phys. D - Atoms, Molecules and Clusters 21, S 209-211 (1991)

Atoms, Molecules
Zeitschrift für Physik D and Clusters

Dielectronic recombination into Rydberg levels of lithiumlike titanium

P. Beiersdorfer, S. Chantrenne, M. H. Chen, R. E. Marrs, D. A. Vogel, K. L. Wong, and R. Zasadzinski
High Temperature Physics Division, University of California, Lawrence Livermore National Laboratory, Livermore, CA 94550, USA

Abstract

To characterize line formation near threshold for electron-impact excitation, dielectronic capture into high-n levels in lithiumlike Ti^{19+} ions has been studied with high-resolution x-ray spectroscopy on the Livermore electron beam ion trap. The process is shown to result in satellite lines which are virtually indistinguishable from their dipole-allowed heliumlike parent lines. High-n satellite lines are absent from dipole-forbidden heliumlike lines.

I. Introduction

Dielectronic capture by heliumlike ions into levels with high principal quantum number n,

$$1s^2 + e^- \rightarrow 1s2lnl' \quad , \tag{1}$$

results in x rays

$$1s2lnl' \rightarrow 1s^2nl' + h\nu_1 \tag{2}$$

with energy very close to that of the heliumlike parent line

$$1s2l \rightarrow 1s^2 + h\nu_2 \quad . \tag{3}$$

In plasma, dielectronic capture and electron-impact excitation processes occur concurrently, and x rays emitted in the radiative stabilization blend with the parent line and form satellite lines on the low-energy side, which typically cannot be resolved [1,2].

We have studied the process of high-n dielectronic capture by heliumlike Ti^{20+} ions with the Livermore EBIT [3], which probes ions with a near-monoenergetic electron beam. If the beam energy is set to a value just below the threshold for direct excitation, dielectronic capture into high-n levels ensues, and x rays resulting from subsequent radiative stabilization can be observed in the absence of their parent lines.

The measurements show that only the two dipole-allowed transitions w and y, which proceed from the upper levels $1s2p\ ^1P_1$ and $1s2p\ ^3P_1$, respectively, are associated with high-n satellites lines. The magnetic quadrupole transition x and the magnetic dipole transition z, which proceed from the upper levels $1s2p\ ^3P_2$ and $1s2s\ ^3S_1$, respectively, are not accompanied by high-n satellites. This behavior is consistent with the relative magnitude of the radiative decay rates of the heliumlike transitions and of the Auger rates of the high-n doubly excited levels.

II. Experiment

The $n = 2$ to $n = 1$ transitions of Ti^{20+} are measured using the EBIT high-resolution von Hámos spectrometer [4], which observes x rays in a plane perpendicular to the direction of the electron beam. The spectrometer employs a 50-mm×25-mm×0.25-mm LiF crystal bent to a radius of curvature of 30 cm. The crystal is cut parallel to the (200) plane with a 2d-spacing of 4.027 Å. Consequently, the heliumlike transitions are observed at a near 45° Bragg angle. At this angle, the crystal acts as a polarizer, and detected x rays are polarized parallel to the beam direction [5]. A helical delay line proportional counter [6] with a sensitive area of 3×11 cm^2 is employed to record the spectra. The resolving power of the spectrometer is about $\lambda/\Delta\lambda = 1000$, which is sufficient to resolve all directly excited heliumlike lines.

K x-ray spectra of heliumlike Ti^{20+} obtained at three different beam energies are shown in Fig. 1. The spectrum in (a) is obtained at a beam energy of 4986 eV, which is well above the direct electron-impact excitation threshold for all four heliumlike transitions (cf. Table I), and all four heliumlike lines are seen. The spectrum in (b) is obtained at 4685 eV, which is 49 eV below the threshold for x. Indeed, x is not seen in the spectrum. On the other hand, z and lines that appear to be w and y continue to be present. z has disappeared in (c), which was recorded 53 eV below its threshold; apparent lines w and y, however, persist in the spectrum. The latter lines are formed by dielectronic capture into Rydberg levels and subsequent radiative stabilization. Because of a close match of their transition energies, they are virtually indistinguishable from their parent transitions. The spectra also show the two principal lithiumlike lines q and r, i.e., the transitions $1s2s2p\ ^2P_{3/2} \rightarrow 1s^22s\ ^2S_{1/2}$ and $1s2s2p\ ^2P_{1/2} \rightarrow 1s^22s\ ^2S_{1/2}$. Below threshold for direct excitation, apparent lines q and r are formed dielectronically like the apparent heliumlike lines.

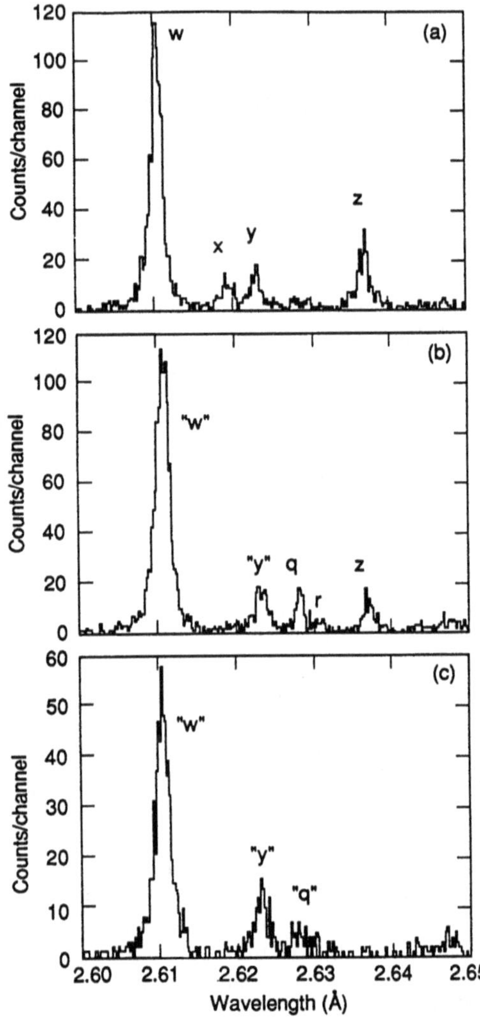

Fig. 1. K x-ray spectra of Ti^{20+} obtained at beam energies (a) 4985 eV, (b) 4685 eV, and (c) 4649 eV. Lines labeled in quotation marks are excited by dielectronic capture of beam electrons into high-n levels.

TABLE I. Radiative rates (Ref. 7) and threshold energies for direct electron-impact excitation (Ref. 8) of w, x, y, and z in Ti^{20+}.

Line	A_r (s^{-1})	E_{th} (eV)
w	2.41×10^{14}	4749.6
x	1.66×10^{9}	4733.8
y	1.06×10^{13}	4726.9
z	3.76×10^{7}	4702.0

The relative line intensities of the heliumlike lines and their associated high-n satellite lines are shown in Fig. 2 as a function of beam energy in the range $4.6 \leq E_{beam} \leq 5.1$ keV. All intensities are normalized to w. The figure demonstrates that x and z vanish about 50 eV below threshold; no apparent lines due to high-n satellites form below threshold that extend their excitation functions. Indeed, the onset of x and z, as the threshold is crossed, is consistent with a 50-eV FWHM spread in the energy distribution of the electron beam measured earlier with a K-edge filter by Levine *et al.* [9] This is not true for y nor, implicitly, for w. Their excitation functions do not vanish below threshold, but smoothly match those of their apparent lines, as indicated by Fig. 2(c).

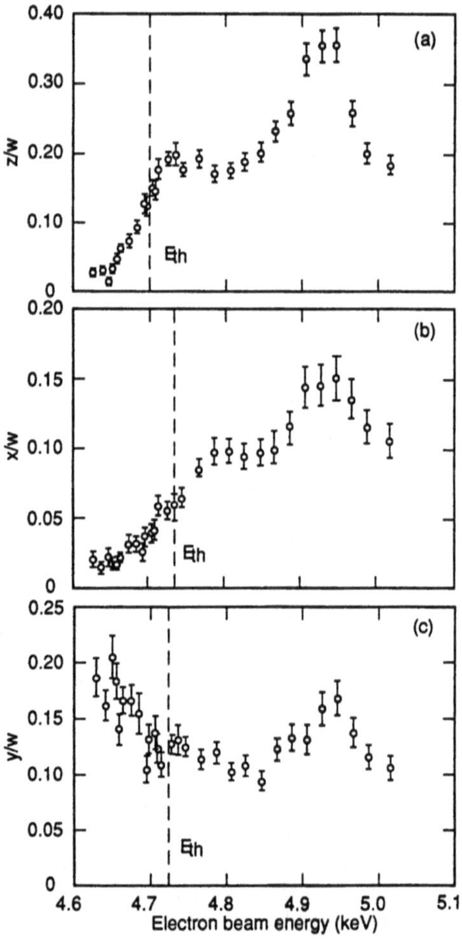

Fig. 2. Relative line intensities of the forbidden and intercombination lines in Ti^{20+}. E_{th} denotes the threshold energy for electron-impact excitation. The enhancement at 4.95 keV is due to resonance excitation.

The high-n spectator electron shifts the transition energies of apparent lines slightly from those of their heliumlike parent lines. We have measured the shift in case of apparent line w, and results are plotted in Fig. 3. All shifts are toward lower energy. In a plasma, where high-n satellites blend with their parent line, this casuses a redshift and broadening of w. As a result, the presence of high-n satellites can affect determinations of ion temperature from the line width and of plasma streaming from the line position, as demonstrated by Bitter *et al.* [2] In addition, their presence limits the accuracy with which transition energies can be measured [10].

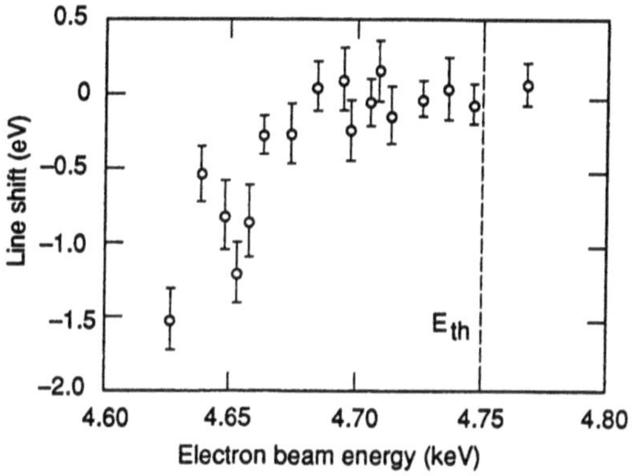

Fig. 3. Shift of apparent line w below threshold for electron-impact excitation (E_{th}). Measurements are relative to the position of w above threshold.

To complete the characterization of the behavior of the heliumlike lines near threshold, our measurements extend to the KMM dielectronic resonance centered at E_{Beam} = 4.95 keV. Here resonant capture of a beam electron substantially enhances emission of the heliumlike lines via the process

$$1s^2 + e^- \rightarrow 1s3l3l' \rightarrow 1s2l + e^- \quad . \tag{4}$$

III. Discussion

The marked differences in the threshold behavior of the four heliumlike lines can be explained with a look at their respective radiative rates and the Auger rates of the associated high-n configurations.

The satellite intensity is proportional to the so-called line factor [11],

$$F_2 = \frac{g_s}{g_i} \frac{A_a^{si} A_r^{sf}}{\Sigma_j A_a^{sj} + \Sigma_{f'} A_r^{sf'}} \quad . \tag{5}$$

A_a^{si} is the autoionization rate for decay of upper level $|s\rangle$ to the ground state $|i\rangle$ of the recombing ion; A_r^{sf} is the rate of radiative decay of upper level $|s\rangle$ to lower level $|f\rangle$; g_s and g_i are the statistical weights of the autoionizing level and of the ground state of the recombining ion, respectively. The sum over f' extends over all levels lower than $|s\rangle$; the sum over $|j\rangle$ extends over all levels that are populated by autoionization of $|s\rangle$. A_r^{sf} can be approximated by the radiative rate of the associated parent line (cf. Table I), while $A_a^{si} \approx 10^{+13}\ \mathrm{s}^{-1}$ for n=4 and drops as n^{-3}. Hence, line factors of high-n satellites to the heliumlike electric dipole lines are as large as $F_2 \approx 10^{+13}\ \mathrm{s}^{-1}$; those to the magnetic dipole and quadrupole lines have lines factors of only $F_2 \leq 10^{10}\ \mathrm{s}^{-1}$. Consequently, dielectronic capture into Rydberg levels is a strong process only for dipole-allowed parent transitions with large radiative rates, and their intensity matches smoothly with the intensity of their high-n satellites, as the threshold for direct excitation is crossed. By contrast, dipole-forbidden lines are devoid of matching high-n satellites as a result of the competition between autoionization and radiative stabilization of high-n, doubly excited levels, and turn on in a step-like fashion.

The differences in threshold behavior of dipole-forbidden and dipole-allowed transitions suggests a means for identifying contributions from different transitions in blends. In heliumlike ions that have a finite nuclear moment the $1s2p\ ^3P_0$ level decays to the ground state via the hyperfine interaction and blends with y. Its radiative decay rate is about $10^{+10}\ \mathrm{s}^{-1}$ for heliumlike ions of the transition metals [12]. Like x the line is, therefore, unaffected by high-n satellite transitions and should turn on in a step-like fashion as the direct excitation threshold is crossed. This step should be a distinct feature in the excitation function of y. Experiments involving V^{21+} or Mn^{23+} are planned to confirm these expectations. The dominant isotope of Ti has a vanishing nuclear moment, and no such step is seen in Fig. 2(c).

Acknowledgments

We thank D. Nelson, E. Magee, and A. Hinz for their technical support. This work was performed under the auspices of the U. S. Department of Energy by Lawrence Livermore National Laboratory under contract No. W-7405-ENG-48.

References

1. F. Bely-Dubau, A. H. Gabriel, and S. Volonté, Mon. Not. R. Astron. Soc. **189** (1979) 801; F. Bely-Dubau *et al.*, Phys. Rev. A **26** (1982) 3459.
2. M. Bitter *et al.*, Phys. Rev. A **32** (1985) 3011.
3. M. A. Levine, R. E. Marrs, J. R. Henderson, D. A. Knapp, and M. B. Schneider, Phys. Scr. **T22**, 157 (1988); R. E. Marrs, M. A. Levine, D. A. Knapp, and J. R. Henderson, Phys. Rev. Lett. **60**, 1715 (1988).
4. P. Beiersdorfer *et al.*, Rev. Sci. Instrum. **61**, (1990).
5. J. R. Henderson *et al.*, Phys. Rev. Lett. **65**, 705 (1990).
6. E. Källne *et al.*, Rev. Sci. Instrum. **58** (1987) 1077.
7. C. D. Lin, W. R. Johnson, and A. Dalgarno, Phys. Rev. A **15** (1977) 154.
8. G. W. F. Drake, Can. J. Phys. **66** (1988) 586.
9. M. A. Levine *et al.*, Nucl. Instrum. Methods **B43** (1989) 431.
10. P. Beiersdorfer, M. Bitter, S. von Goeler, and K. W. Hill, Phys. Rev. A **40** (1989) 150.
11. A. H. Gabriel, Mon. Not. R. Astron. Soc. **160** (1972) 99; J. Dubau and S. Voloné, Rep. Prog. Phys. **43** (1980) 199.
12. P. J. Mohr, in *Beam-Foil Spectroscopy*, edited by I. A. Sellin and D. J. Pegg (Plenum Press, New York, 1976), Vol. I, p. 97.

Supplement to Z. Phys. D - Atoms, Molecules and Clusters 21, S 213-214 (1991)

Atoms, Molecules
Zeitschrift für Physik D and Clusters

Measurements of resonant transfer and excitation cross-sections for U^{90+} projectiles in H_2

W. G. Graham[1], K. H. Berkner[2], E. M. Bernstein[3], M. W. Clark[4], B. Feinberg[2], M. A. McMahan[2], T. J. Morgan[5], W. Rathbun[3], A. S. Schlachter[2], and J. A. Tanis[3]

[1] Physics Department, Queen's University, Belfast, BT7 1NN, Northern Ireland, UK
[2] Lawrence Berkeley Laboratory, Berkeley, California 94720, USA
[3] Physics Department, Western Michigan University, Kalamazoo, Michigan 49008, USA
[4] Lawrence Livermore Laboratory, Livermore, California 94550, USA
[5] Physics Department, Wesleyan University, Middletown, Connecticut 06457, USA

Abstract
Resonant transfer and excitation (RTE), resulting from simultaneous electron capture and K–shell excitation in a single collision, has been measured for 97 to 150 MeV/u U^{90+} ions in molecular hydrogen. This measurement, the first for a very heavy projectile, provides a test of relativistic dielectronic–recombination (DR) theory.

I Introduction

Resonant transfer and excitation [1] (RTE) is one manifestation of the electron–electron interaction in ion–atom collisions, and involves the capture of a target electron simultaneously with the excitation of a projectile in a single collision, leading to the formation of an intermediate doubly–excited state of the projectile. The decay of this doubly–excited state through either x–ray [2,3] or Auger–electron [4] emission has been used to identify and measure RTE in collisions of multiply–charged ions with light gaseous targets through the observation of resonances in the energy dependence of electron capture cross sections associated with x–ray or Auger electron emission. Additionally, RTE has been observed in the energy dependence of the total single–electron–capture cross section in $Ca^{q+} + H_2$ collisions.[5] RTE, which involves the capture of an electron bound to a target atom, has been shown [1–3] to be analogous to dielectronic recombination [6] (DR) for which the captured electron is free. Since RTE cross sections can be calculated from DR cross sections by averaging over the momentum distribution of the target electrons,[7] measurements of RTE provide a test of DR theory. For relatively low Z projectiles ($Z \leq 32$) theoretical calculations [8,9] of RTE cross sections have shown excellent agreement with experiment. Recently, theoretical studies of DR and RTE have been extended to highly–stripped U ions incident on various targets. [10,11] The inclusion of relativistic effects in the calculations is found to have several effects; namely the calculated Auger transition rates can change significantly, the lowest energy transition splits into three peaks, and all the peaks shift to higher energies. The RTE cross sections calculated by Chen [10] for U^{90+} on H_2 are shown in Fig 1. The six peaks correspond to the population of various intermediate states. For example the first three peaks arise from, with increasing energy, the formation of the $1s2s^2$ + $1s2s2p_{1/2}$ + $1s2p^2{}_{1/2}$, $1s2s2p_{3/2}$ + $1s2p_{1/2}2p_{3/2}$ and $1s2p^2{}_{3/2}$ intermediate states. The cross sections calculated by both Chen [10] and Pindzola and Badnell [11] are in excellent agreement with each other. In the present experiment RTE for U^{90+} in H_2 has been measured by studying the energy dependence of the total single–electron–capture cross section. RTE resonances are expected [10,11] in the energy range 110 to 180 MeV/u (26 GeV to 43 GeV), based on the calculated Auger energies of U^{90+}.

II Experiment

The experimental measurements were carried out at the Lawrence Berkeley Laboratory using the Bevalac Facility. A well defined U^{90+} beam was incident on a 2.41–m–long gas cell.[12] Differential pumping and fast acting beam–line valves at the entrance and exit of the gas cell, which opened only when the 160ms beam pulse was present, allowed the gas cell to be operated at gas pressures of up to 6 Torr without significant increase in the pressure in the beam line on either side of the gas cell. The pressure in the gas cell was measured using a capacitance manometer.

The charge state components emerging from the gas cell were dispersed with a magnetic spectrometer onto a position–sensitive scintillation counter, with sufficient separation between the U^{90+} and U^{89+} components, to allow them to be counted separately. At each energy the fraction of U^{89+} in the beam was measured at three different gas pressures, including zero, and generally a minimum of 1000 total counts for the U^{89+} component was obtained. The maximum observed U^{89+} fraction was 0.02. Since single–collision conditions prevailed, as confirmed by the linear target–thickness dependence of the U^{89+} fraction, the electron–capture cross section was obtained from the slope of the U^{89+} fraction versus target–thickness.

The relative uncertainties in the measured electron–capture cross sections were taken to be the standard deviation of the total number of counts at that particular energy with an uncertainty in determining a background correction added in quadrature. Other possible sources of relative uncertainty in the cross section determination were small relative to the statistical and background uncertainties. Sources of absolute uncertainty in the electron–capture cross section measurement arising from the measurement of the beam intensity and target thickness, were estimated to be ± 12%. The energy of the beam could be determined from the Bevalac magnetic field and beam radius and the energy loss in the stripping foil used to prepare the U^{90+} charge state. The absolute uncertainty in the U^{90+} beam energy is estimated to be ± 0.5 MeV/u.

III Results and Discussion

The measured total single–electron–capture cross sections show two prominent peaks and some indication of a third peak superimposed on a monotonically decreasing background. The monotonically decreasing background, on which these peaks lie, is attributed to contributions from radiative and nonradiative electron capture. A smooth function can be fitted to the radiative and nonradiative electron capture background using the experimental data points below 109.9 MeV/u and from 138 to 145 MeV/u, where the RTE cross section is expected to be zero [10,11]. The RTE cross section for U^{90+} in H_2 can thus be found by subtracting this background from the measured total electron–capture cross section. The RTE cross section, determined in this manner, is shown in Fig 1.

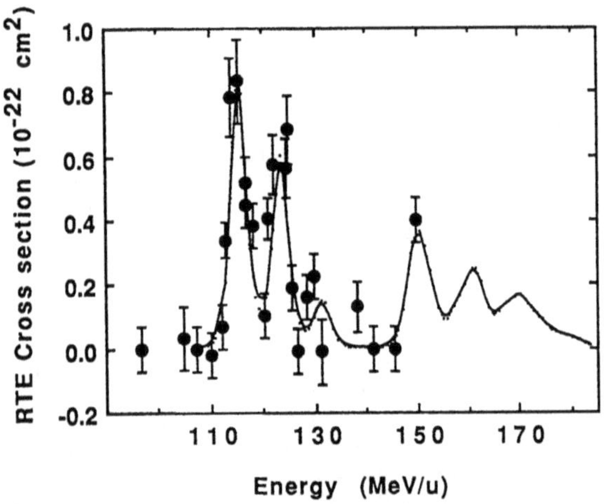

Figure 1 Experimental and theoretical cross sections for U^{90+} in H_2. •, experimentally determined RTE cross section (total single–electron–capture cross section minus empirical fit). Solid curve, theoretical calculations of RTE cross section. [10] Errors bars show the absolute uncertainty in the experimentally determined cross sections.

The absolute uncertainty in the measured RTE cross section includes uncertainties in the total electron–capture cross section and from fitting the electron–capture background. It can be seen that over the energy range covered the experimental results are in excellent agreement with Chen's theoretical calculations both in the magnitude of the cross sections and the positions of the maxima. It is emphasized that the magnitudes and energies of both the measured and calculated cross sections are absolute and have not been normalized to one another.

IV Conclusion

Resonant transfer and excitation (RTE) has been measured in collisions of U^{90+} with H_2 thereby extending significantly the projectile range for which RTE has been investigated. The measurements are found to be in excellent agreement with calculations based on relativistic dielectronic–recombination (DR) theory.

V Acknowledgements

This work was supported by the Science and Engineering Research Council of Great Britain, the US Department of Energy, Office of Basic Energy Sciences, Division of Chemical Sciences and Director, Office of Energy Research, Office of High Energy and Nuclear Physics, Nuclear Physics Division of the US Department of Energy under contract No DE–AC03–76SF00098.

The authors wish to thank the operations staff at the Bevalac, the SAS collaboration at the Bevalac for making the gas cell available for our use, Drs Nigel Badnell and Harvey Gould for their advice on aspects of this work and Dr Mau Chen for providing the results of his theoretical calculations prior to publication.

References

1 see J A Tanis, Nucl Instrum and Methods, A262, 52 (1987) and references therein.
2 J A Tanis, E M Bernstein, W G Graham, M Clark, S M Shafroth, B M Johnson, K W Jones and M Meron, Phys Rev Lett 49, 1325 (1982).
3 J A Tanis, E M Bernstein, W G Graham, M P Stockli, M W Clark, R H McFarland, T J Morgan, K H Berkner, A S Schlachter and J W Stearns, Phys Rev Lett 53, 2551 (1984).
4 A Itoh, T J M Zouros, D Schneider, U Stettner, W Zeitz and N Stolterfolt, J Phys B 18, 4581 (1985).
5 W G Graham, E M Bernstein, M W Clark, J A Tanis, K H Berkner, P Gohil, R J McDonald, A S Schlachter, J W Stearns, R H McFarland, T J Morgan and A Muller, Phys Rev A33, 3591 (1986).
6 H S W Massey and D R Bates, Rep Prog Phys 9, 62 (1942)
7 D Brandt, Phys Rev A 27, 1314 (1983).
8 Y Hahn and K L La Gattutta, Phys Rep 166, 196 (1988).
9 N R Badnell, Phys Rev A, 40, 3579 (1989).
10 M H Chen, Phys Rev A 41, 4102, 1990 and Private Communication.
11 M S Pindzola and N R Badnell submitted to Phys Rev A.
12 H–P Hulskotter, B Feinberg, A Belkacem, W E Meyerhof, J Alonso, L Blumenfeld, E Dillard, M J Rhoades–Brown, B S Rude, J Schweppe, D W Spooner, K Street, P Thieberger and H E Wegner (to be submitted to Phys Rev A).

Supplement to Z. Phys. D - Atoms, Molecules and Clusters 21, S 215-216 (1991)

Atoms, Molecule:
Zeitschrift für Physik D and Cluster:

Relativistic Auger rates for dielectronic recombination with highly-charged few-electron ions

P.Zimmerer, N. Grün, and W. Scheid
Institut für Theoretische Physik der Justus-Liebig-Universität, Giessen, FRG

Abstract

Dielectronic recombination (DR) cross sections are calculated for high-Z, two-electron ions using a distorted wave theory. We show DR cross sections for KLL dielectronic recombination of hydrogen-like U ions and the corresponding Auger rates. The Auger rates are calculated with correlated relativistic wave functions and include the Breit interaction. The dependence of the Auger rates on the nuclear charge Z is also investigated.

I. Introduction

Direct measurements of dielectronic recombination (DR) cross sections for few-electron ions with high Z values are planned at the new storage ring ESR at GSI in Darmstadt. It is well known that the DR cross section, averaged over a resonance, is essentially determined by the probabilities for radiationless capture and excitation, which are related to the Auger transition probabilities by the principle of detailed balance. For few-electron ions with high charge numbers Z recent calculations showed, that relativistic effects give large corrections to the nonrelativistic Auger rates in the range $Z > 54$ [1, 2].

II. DR cross section

The DR cross section of high-Z, few-electron ions can be calculated using the distorted wave theory of Y. Hahn [3]:

$$\sigma_{DR}(i \to d \to f) = \frac{2\pi}{p_c^2} V_a(i \to d)\omega_d \delta_d(E_i) . \qquad (1)$$

Here, p_c is the momentum of the incoming electron, and E_i the energy of the initial state. V_a is the radiationless capture and excitation probability, and ω_d the fluorescence yield of the intermediate state d. $\delta_d(E_i)$ is a Lorentz curve, which determines the shape of the cross section in the vicinity of the resonance.

III. Relativistic Auger rates

In order to study relativistic effects on the DR cross section, we calculate Auger rates, connected to V_a by the principle of detailed balance, with correlated relativistic wave functions and the Breit interaction between the electrons [4]:

$$V_{12}^{Br} = -\sum_{i,j=1}^{3} \alpha_{1i}\alpha_{2j} \left[\delta_{ij} \frac{\cos \omega R}{R} + \frac{\partial}{\partial R_i}\frac{\partial}{\partial R_j} \frac{\cos \omega R - 1}{\omega^2 R} \right] \qquad (2$$

with

$$\vec{R} = (R_1, R_2, R_3) = \vec{r}_2 - \vec{r}_1 . \qquad (3$$

Here, α_{1i} and α_{2i} are the ith Cartesian components of the c matrices of the Dirac equation acting on the spinor componen of the electrons 1 and 2, and R_i is the ith Cartesian component o the relative vector between the two electrons. The first term in V_{12}^{B} is the retarded interaction between two Dirac currents, the secon term is the retardation correction to the charge–charge interaction

IV. Results and Discussion

We have carried out nonrelativistic and relativistic calculations o KLL–Auger rates of two–electron systems. Fig. 1 shows the influ ence of the use of relativistic wave functions and different forms o the Breit interaction on the KLL–Auger rates for U ions. We fin that also nonrelativisticly large rates are strongly increased.

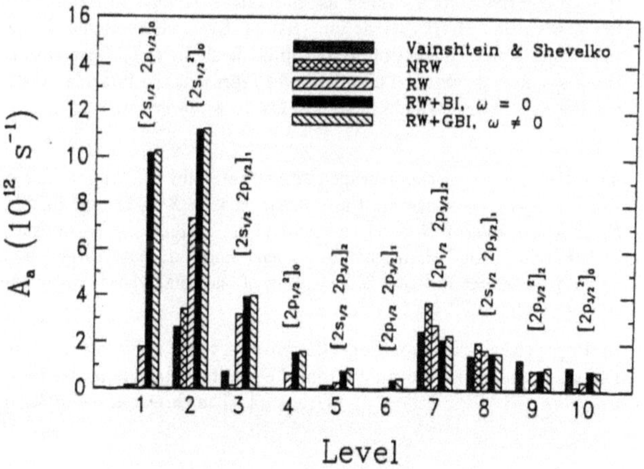

Figure 1: Auger rates for U^{90+} for the ten doubly occupied L-she states. The rates are shown by vertical bars. From left to right the represent calculations with $1/Z$–expansion theory of Vainshtein ar Shevelko [5], nonrelativistic wave functions (NRW), with relativi tic wave functions only (RW), with relativistic wave functions ar the Breit interaction (RW+BI,$\omega = 0$), and with relativistic wa functions and the generalized Breit interaction (RW+GBI, $\omega \neq 0$

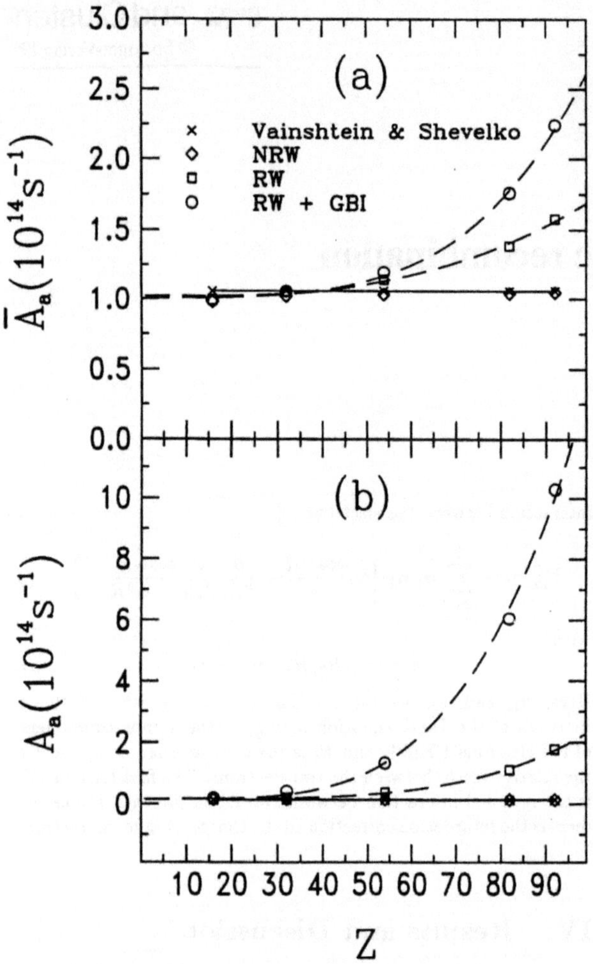

Figure 2: KLL-Auger rates for two-electron ions as a function of the nuclear charge Z. The symbols are crosses: $1/Z$-expansion of Vainshtein and Shevelko [5], open diamonds: use of nonrelativistic wave functions (NRW), open squares: use of relativistic wave functions only (RW), open circles: use of relativistic wave functions and the generalized Breit interaction (RW+GBI). (a) Averaged Auger rate of all levels of the doubly occupied L-shell. (b) Auger rate of the $[2s_{1/2}2p_{1/2}]_0$ state. The full curves represent the nonrelativistic scaling law. The dashed curves are fits to a power law.

The dependence of the averaged Auger rate and the Auger rate of the $[2s_{1/2}2p_{1/2}]_0$ state on the nuclear charge Z is shown in Fig. 3. In the range of $Z > 54$ the relativistic corrections are already of the same order of magnitude as the nonrelativistic rates. The calculated values indicate a Z-scaling of the relativistic correction in the range of Z^3 up to Z^5.

In Fig. 3 the corresponding DR cross sections are plotted. The radiative widths are obtained from the general relativistic structure program GRASP of Dyall and Grant [6]. The three curves differ in

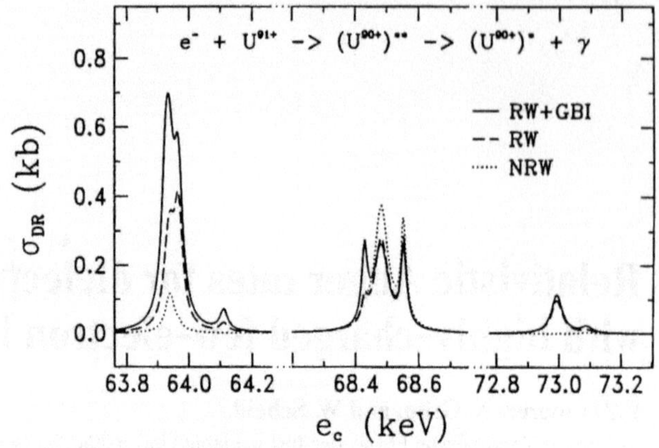

Figure 3: Total cross section for capture of a free electron via DR by U^{91+} as a function of the energy of the electron. The three curves are calculated with different Auger rates. The full curve is obtained from a calculation with relativistic wave functions and the Breit interaction, the dashed curve with relativistic wave functions only, and the dotted curve with nonrelativistic wave functions. The resonance energies are the same in these calculations.

the Auger rates used. Most of the resonances are strongly increased due to the use of the corrections to the nonrelativistic Auger rates discussed above.

V. Conclusions

The use of relativistic wave functions and the inclusion of the Breit interaction in the Auger transition matrix element leads to a drastic increase of the KLL-Auger rates and the DR cross sections of ions with high Z values. This holds not only for the rate of selected levels, but also for the averaged rate. For high nuclear charges Z the relativistic corrections become comparable to the nonrelativistic rates. From the results of our calculations we deduce a scaling law with Z^3 up to Z^5 for the relativistic corrections.

References

1. Chen M.H.: Phys. Rev. A **41**, 4102 (1990)
2. Zimmerer P., Grün N. and Scheid W.: Phys. Lett. A **148**, 457 (1990).
3. Hahn Y.: Adv. At. Mol. Phys. **21**, 123 (1985).
4. Mann J.B. and Johnson W.R.: Phys. Rev. A **4**, 41 (1971).
5. Vainshtein L.A. and Shevelko V.P.: unpublished.
6. Dyall K.G., Grant I.P., Johnson C.T., Parpia F.A., and Plummer E.P.: Comput. Phys. Commun. **55**, 425 (1989).

Supplement to Z. Phys. D - Atoms, Molecules and Clusters 21, S 217-218 (1991)

Atoms, Molecules
Zeitschrift für Physik D and Clusters

Channel-coupling effects in electron-atom collisions

S. Bougouffa

Phys. and Maths. Dept., Omar Al-Mukhtar University, El-Bieda, Libya

Abstract

We are concerned with the following interesting point: The existence of sharp peaks of inelastic cross section with increasing magnitude of coupling strength in e-atom collisions. A simple interpretation of this effect will be discussed.

I. Introduction

In the present work we consider the case of e-H collisions in transition from the state $nl \rightarrow nl\pm1$, and we analyse lhe "resonance" which originate from the channel coupling effect i.e. the enhancement or attenuation of the inelastic cross section according to the magnitude of the coupling strength between the two states. This kind of "exotic resonance" must not be confused with the more conventional (In energy) in the sense of the Breit-Wigner relation for example. We summarize the numerical results obtained so far and the simple interpretations of this effect will be discussed in the last paragraph.

II. The two-channel case

Using the same notations and conventions as in [1],the system of coupled differential equations (CDE) to be considered here is :

$$\left[\frac{d^2}{dr^2} + f_i(r)\right]F_i(r) = \sum_{j\neq i} B_{ij}(r)F_j(r) \qquad (1)$$

where $i,j=1,..,n$ and the r.h.s. of this system are unknown and must be chosen by use of appropriate simplifications of the real physical situation of the scattering process. We apply the theory on the separation of a system of CDE in order to obtain further simplifications [2].

In keeping the same notations and using the technique developed so far [1],[2] we note that in the two state approximation, the main task amounts to the search for solution of the two following separated differential equations :

$$\left[\frac{d^2}{dr^2}+K^2+V(r)-\frac{l(l+1)}{r^2}\pm\frac{1}{2}[(\Delta k^2)^2+4B^2]^{1/2}\right]Y^{\pm}=0 \qquad (2)$$

with $K^2=\frac{1}{2}(k_0^2+k_1^2)$; k_0^2, k_1^2 being the energy of these states, $\Delta k^2=k_1^2-k_0^2$ is the energy gap and $B(r)$ is the coupling function between these states. This system leads to the determination of two eigen phase shift $\delta_l^{\pm}$ from which the inelastic cross section Q_l can be inferred :

$$Q_1 = C.\sin^2(\delta_1^+ - \delta_1^-) \qquad (3)$$

C is a quantity depending on k_0^2, Δk^2, l and χ the mixing parameter of the theory [1]. A method to determine this parameter has already been discussed in previous papers and from this , C can be derived analytically. However as for the moment we are concerned with the positions of the extrema and not with their absolute magnitude, we simply set C=1.

In the schematic model [3] we have $B(r) = B_0/r^2$ being the coupling strength. (the case $B(r)$=constant has already been given in [2] for various l and Δk^2). The results are summarised in the following fig-I-. In this case, the occurence of the "resonance" are quite explicit.

III. The three-channel case

In order to analyse the comportment of the inelastic cross section according to the magnitude of the coupling strength, we continue to take the simple example of e-H collisions in transition $nl \rightarrow nl\pm1$, which lead to a system of 3 CDE . With our notations we have here :

$$f_0=k_0^2-\frac{l(l+1)}{r^2}, \quad f_1=k_1^2-\frac{l(l-1)}{r^2}, \quad f_2=k_1^2-\frac{(l+1)(l+2)}{r^2}, \quad B_{ij}=C_{ij}/r^2 \qquad (4)$$

More precisely, if we consider the coupling between the states ns,np, and neglect the interactions with all other states [3], we obtain :

$$C_{00}=C_{11}=C_{22}=C_{12}=0$$

and $$C_{01}=\beta(l)^{1/2}, \quad C_{02}=-\beta(l+1)^{1/2} \qquad (5)$$

where $$\beta=\frac{2s_1(np,ns)}{[3(2l+1)]^{1/2}}$$

where s_1^2 is the line strength [3].

Upon introducing the T matrix [3] defined by

$$T = 1 - S \qquad (6)$$

one can express the collision strength for $ns \rightarrow np$ transition as :

$$\Omega(np,ns) = \sum_l (2l+1)\left[|T_{01}|^2 + |T_{02}|^2\right] \qquad (7)$$

where the indices o,1,2 refer to the respective channels.

In the case of exact resonance ($\Delta k^2=0$) we display in figures-II- the exact results for the collision strength versus l for several values of the line strength s^2(s^2 is a measure of strength of coupling).

For the case of not exact resonance, the numerical calculations are actually in progress and we shall present it later on.

IV. Conclusion

These results leads us to the following remarks :

- The enhancement and attenuation of the inelastic cross section in a two state model is really a general effect and may affect the magnitude of the conventional resonances.
- It can be understood in considering the raising and lowering of the potential barrier $\left[\pm\frac{1}{2}[(\Delta k^2)^2 + 4B^2]^{1/2}\right]$ which depends on three factors; the function B(r), the energy gap Δk^2 and quantum number l.
- In the two-channel case the eigen phase shift have been determined numerically. However there are cases where they can also be obtained analytically specially for exact resonances (Δk^2=0) leading to a theoretical prediction of the extrema. We are in fact developing our investigation along this line using a JWKB approach and shall present the results later on.
- In fig-II,a- the collision strength for ns→np transition with exact resonance is given for several values of the line strength s^2. It has been shown that for exact resonance collisions, as one increases the strength of coupling (from weak coupling), the strength collision first increases rapidly, then reaches a certain saturation stage, and finally behaves in an oscillatory manner [4].
- This can be illustrated more clearly in fig-II,b- where $\Omega_1/(2l+1)$ is plotted against s^2 for l=2,..,5. The saturation effect is more pronounced for low values of l. Thus the curve for l=2 passes through a maximum around s^2=9.3, while the l=3 collision strength does not reach the highest value until s^2=23.4, and the collision strength for higher l require even stronger coupling for complete saturation [5]. This saturation effect is also responsible for keeping the collision strength below the conservation limit (see fig. II,a)

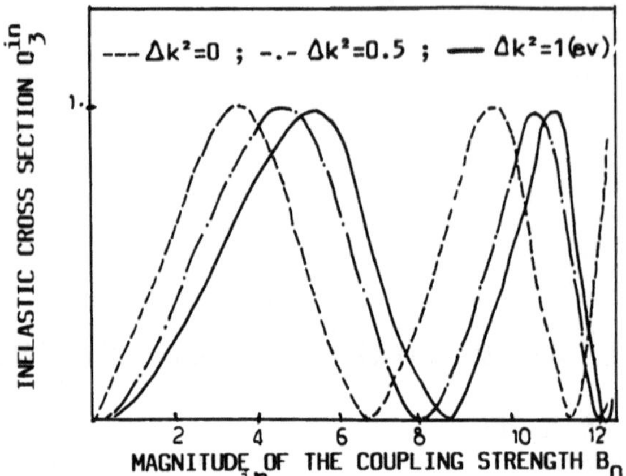

Fig.I,b. Q_1^{in} in terms of B_0 for different Δk^2, with V=0, l=3 and k_0^2=13.6ev

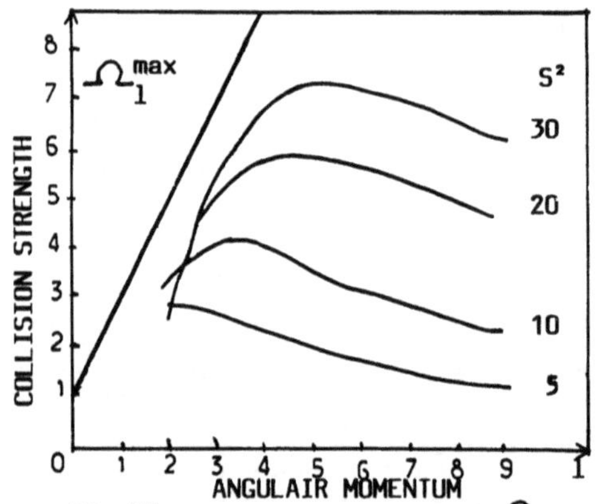

Fig.II,a. Collision strength Ω_1 in the exact resonance for several values of s^2

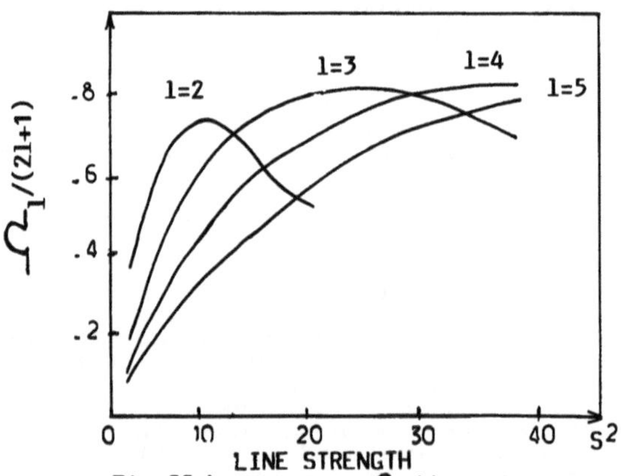

Fig.II,b. Values of $\Omega_1/(2l+1)$ for different l and Δk=0 in terms of s^2.

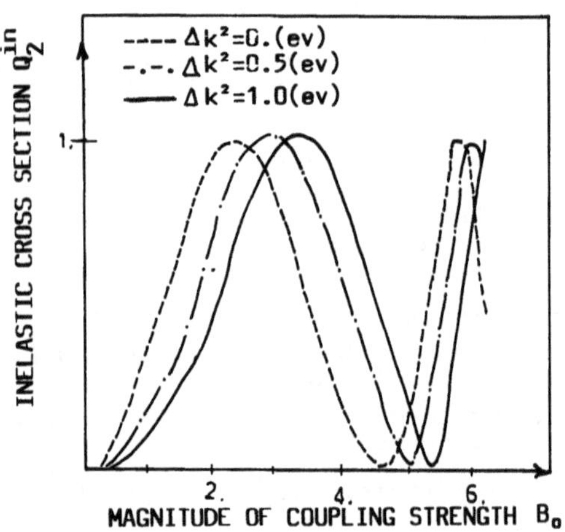

Fig.I,a. Inelastic cross section Q_1^{in} for several values of the magnetude of the coupling strength B_0 for different Δk^2, with V=0., l=2 and k_0^2= 13,6 (ev) .

References

[1] Cao X.C.: J. Phys. A Math. Gen. 25 617 (1988)
[2] Cao X.C., Bougouffa S.: J. Physique C1 Suppt N°3 49, 251(1988)
[3] Seaton M.J. Proc.Phys.Soc.(London) 77 175(1961)
[4] Mott N.F. and Massey H.S.W. "The theory of atomic collisions" Oxford(1965)
[5] Bougouffa S., Thèse de Doctorat Univ. de Constantine, (1990).

Supplement to Z. Phys. D - Atoms, Molecules and Clusters 21, S 219-220 (1991)

Atoms, Molecules
Zeitschrift für Physik D and Clusters

Doubly excited states in lithiumlike Si XII, S XIV and Ar XVI

S. Khardi, M. C. Buchet-Poulizac, J. P. Buchet, A. Denis, J. Desesquelles, S. Martin and Y. Ouerdane
Laboratoire de Spectrometrie Ionique et Moleculaire, Université Claude-Bernard Lyon-1
43, Bd du 11 Novembre 1918, 69622 Villeurbanne Cedex, France

Abstract

The 3d-4f transitions between doubly excited states have been studied in Si XII, S XIV, Ar XVI. Moreover three 3d - 5f and 4f-5g lines have been identified in Si XII..

I. Introduction

The non-autoionising multiply excited levels in atoms and ions are being intensively investigated at present. Most results concern 2-2 transitions which have been studied in a large number of lithiumlike ions [1] up to Si XII [2]. New results have been recently reported on higher n levels for several light elements in the three electron sequence [3]. We report here the first measurements of 3d - 4f transitions between doubly excited Si XII**, S XIV** and Ar XVI** and 3d-5f and 4f-5g transitions in Si XII**.

II. Experiment

The beam-foil experiment was performed at SARA, Grenoble, using silicon, sulfur and argon beams at 2-4.5 MeV/u energies. The foils successively used were sufficiently thick to obtain the highest ionisation stage corresponding here to Li and He like ions with a good intensity. Thus berylliumlike transitions were minimized and spectra were made clearer and easier to analyse. The EUV photons emitted by the excited fast ions were analysed by a McPherson 2.2 m grazing incidence spectrometer viewing the ion beam at 90° using a gold coated 600 g/mm concave grating blazed in the 15 nm region. The Doppler broadening depending on the observation solid angle was minimized by reducing by a factor 3 the available part of the grating inserting a mask just in front of it. The detector was a multichannel plates assembly (MCP) the first plate of which was magnesium fluoride coated to obtain a better efficiency. The MCP was set tangential to the Rowland circle to improve the resolution. Partial spectra have successively been recorded for given positions of the detector along the Rowland circle from 7 to 30 nm. The wavelength calibration obtained by fitting to known He- and Li- like lines has been found to be better than 0.003 nm.

III. Results

A partial spectrum recorded with a silicon beam at 2.1 MeV/u between 12.2 and 13.0 nm is displayed in Figure 1.

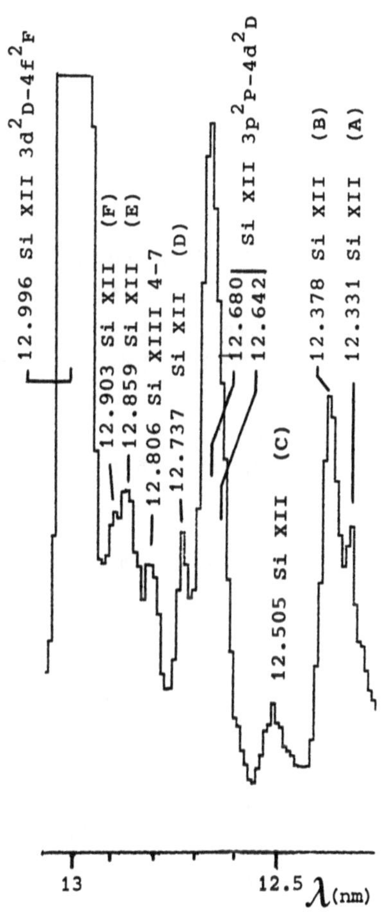

Fig. 1. Partial spectrum of foil-excited silicon (2.1 MeV/u). Transitions between doubly A-F, the meaning of which is given in Table 1.

Table 1. 1s2s3d - 1s2s4f and 1s2p3d - 1s2p4f transitions in Si XII**, S XIV** and Ar XVI** (nm).

Transition	Element	λ exp.	λ fit	I
$1s2p3d^4F°$	Si XII	12.331(3)	12.295	173
$1s2p4f^4G$	S XIV	9.121(3)	9.088	325
(A)	Ar XVI	7.030^b(5)	6.991	-
$1s2s3d^4D$	Si XII	12.378(3)	12.367	344
$1s2s4f^4F°$	S XIV	9.152(3)	9.144	645
(B)	Ar XVI	7.030(5)	7.035	-
$1s2p3d^2D°$	Si XII	12.505(3)	12.499	66
$-1s2p4f^2F$	S XIV	9.264(3)	9.233	215
(C)	Ar XVI	m	7.099	-
$1s2s3d^2D$	Si XII	12.737(5)	12.673	-
$-1s2s4f^2D°$	S XIV	-	9.324	-
(D)	Ar XVI	-	7.142	-
$1s2p3d^4D°$	Si XII	12.859(3)	12.835	78
$-1s2p4f^4F$	S XIV	9.466^b(5)	9.451	187
(E)	Ar XVI	7.248^b(5)	7.248	74
$1s2p3d^4P°$	Si XII	12.903(3)	12.880	144
$-1s2p4f^4D$	S XIV	9.466^b(5)	9.473	-
(F)	Ar XVI	7.248^b(5)	7.259	-

b : blend, m : masked by Ar XVI (3p - 4d)

Table 2. 1s2s3d - 1s2s5f and 1s2p3d - 1s2p5f transitions in Si XII** (nm)

Transitions	λ exp.	λ fit	I
$1s2p3d^4F°$ $-1s2p5f^4G$ (A)	8.579 (3)	8.477	50
$1s2s3d^2D$ $-1s2s5f^2F°$ (B)	8.579 (3)	8.586	50
$1s2p3d^4D°$ $-1s2p5f^4F$ (E)	8.777(5)	8.797	54

Table 3. 1s2p4f4G - $1s2p5g^4H$ and $1s2s4f^4F$ - $1s2s5g^4G$ transitions in Si XII**, SXIV** and Ar XVI** (nm)

	Si XII	SXIV	Ar XVI
λ exp.	27.930	20.544	15.755
λ th.	28.006	20.582	15.761

The linewidth (FWHM), mostly instrumental in origin, is 0.034 nm. The strongest features are due to singly excited Si XII 3d-4f unresolved doublet at 12.996 nm and Si XII 3p-4d resolved doublet at 12.642 and 12.680 nm. The heliumlike Si XIII (4-7) appears as a very weak feature between them at 12.808 nm. The other observed peaks have been labelled A-F following the notation of Blanke et al. for light elements [3]. They have been identified to 3d-4f transitions in doubly excited lithiumlike silicon. Charge state assigments have been confirmed by the line intensity variations with the ion beam energy. Going from 2.1 MeV/u to 3.6 MeV/u the heliumlike line as the doubly excited lithiumlike lines grow faster than the singly excited lithiumlike lines. Similar spectra have been obtained with sulfur and argon beams. Doubly excited line identification has been made by extrapolating previous data along the isoelectronic sequence. Results are shown in Table 1 for the 3-4 doubly excited transitions of Si XII, S XIV and Ar XVI where our experimental values obtained using a gaussian fit program are compared to the values obtained by extrapolation from a parabolic fit to known wavenumbers in B III** - Ne VIII** and MgX**. Relative intensities are given in the last column.

Results found for 3d-5f doubly excited transitions in Si XII** using the same technics are displayed in Table 2. These transitions have not been studied in sulfur and argon the corresponding lines are indeed expected far away in the EUV, outside of the scanned wavelength range.

This work has been tentatively extended to transitions between the 4-5 states. Proposed assignments and results are given in Table 3 where experimental data are tentatively compared to theoretical results obtained using the MCDF program of Desclaux [5]. The comparison with extrapolated values is not possible as no experimental data at all have been yet published for these sequences.

REFERENCES

1. Mannervik S. : Physica Scripta 40, 28 (1989)
2. Buchet J.P., Buchet-Poulizac M.C. Denis A., Désesquelles J., Druetta M, Martin S., Ouerdane Y. : J. de Phys. 49 (1988) C1-III.
3. Baudinet-Robinet Y, Dumont P.D., Garnir H.P. : Physica Scripta 33 (1986) 73.
   Dumont P.D., Garnir H.P.,Baudinet-Robinet Y, Chung K.T.: Phys.Rev. A 32 (1985) 229.
   Blanke J.H., Fricke B., Heinemann D., Sepp W.D., Thies B., Göbenli T., Heckemann P.H., Möller G., Träbert E. : Z.Phys. D 13 (1989) 1
4. Buchet-Poulizac M.C.: communication privée for Ne VIII**
5. Desclaux J.P. : Compt.Phys.Commun. 9 (1975) 31

Supplement to Z. Phys. D - Atoms, Molecules and Clusters 21, S 221-223 (1991)

Atoms, Molecules
Zeitschrift für Physik D and Clusters

Calculations of decay rates for triply-excited states in N^{2+}

N. Vaeck* and J. E. Hansen

Zeeman Laboratory, University of Amsterdam, Plantage Muidergracht 4, NL-1018 TV Amsterdam, The Netherlands

* Research Assistant of the Belgian National Fund for Scientific Research (FNRS)

Abstract

We report here calculations of energy level values and autoionization decay rates for all the terms belonging to the $1s^2 3\ell 3\ell' 3\ell''$ configurations in N^{2+}. Comparison are made with previous calculations for $1s^2 3\ell 3\ell'$ terms in N^{3+}. The properties of the triply-excited states and in particular the fast autoionization decay rates found in the present calculations can perhaps help in understanding the mechanism responsible for neutralisation of ions at surfaces.

I. Introduction

In experiments with neutralisation of ions at surfaces it has been observed that the ions are already neutralised at impact [1]. Since it is believed that the electrons picked up from the surface are initially in high n states, a mechanism must exist which brings the electrons to the ground state in the short time available before impact (10^{-13} - 10^{-14} sec). Calculations for multi-ionized ions have shown that the autoionization probabilities and radiative decay rates for doubly-excited states with large n values (d-e), at least for the ions studied so far [2,3,4], appear to be too long for these processes to be effective in bringing one or both electrons down towards the ground state in the available time. In other words, doubly-excited states can survive long enough to make the pick-up of additional electrons possible. The systematics of radiative and non-radiative decay probabilities of triply-excited states (t-e) are largely unknown and we have started calculations for such systems in order to study these quantities.

We have first considered the $1s^2 3\ell 3\ell' 3\ell''$ configurations in N^{2+}, since a number of charge-exchange experiments have been made for this ion [5] in which states belonging to configurations of the type $1s^2 3\ell 3\ell'$ have been positively identified. The $1s^2 3\ell 3\ell' 3\ell''$ states can decay by autoionization to $1s^2 2\ell 3\ell$ states which are bound. This means that this process is competing with the radiative decay of the $1s^2 3\ell 3\ell'$ states and if fast enough this type of process can be important in the neutralisation.

II. Calculational method

The calculations were carried out using the suite of programs written by Cowan [6]. As for the previous calculations on the $1s^2 3\ell 3\ell'$ terms in N^{3+} [7], the basis set includes all the configurations belonging to the Layzer complex $3\ell 3\ell' 3\ell''$ [8] but in addition all the configurations of the type $3\ell 3\ell' 4\ell''$. The lowest of these lie in the same range of energy as the highest $3\ell 3\ell' 3\ell''$ states. Non-relativistic HF basis functions were used. We have used a perturbation approach described by Cowan [6] to determine the autoionization rates (A^a). The limits $1s^2 2\ell 2\ell'$ and $1s^2 2\ell 3\ell'$ lie below the $1s^2 3\ell 3\ell' 3\ell''$ terms, but in a first-order theory there is no electrostatic interaction between $1s^2 3\ell 3\ell' 3\ell''$ and any $1s^2 2\ell n\ell' \epsilon\ell''$ ($n \neq 3$) continuum and the $1s^2 3\ell 3\ell' 3\ell''$ states therefore can not decay to the $1s^2 2\ell 2\ell'$ limits. In principle, the autoionization rates depend on the specific energy of the free electron. We have used the arithmetic average energy for the odd and even configurations to determine the free electron energy since the large mixing makes it difficult to identify particular terms with particular configurations. We assume that the variation in rates with energy is small for the rather high free-electron energies involved, as already found for the $1s^2 3\ell 3\ell'$ terms [7]. In order to check the validity of the average approach [3,4] for this type of configuration, we have also calculated the average autoionization rates for the $1s^2 3\ell 3\ell' 3\ell''$ configurations individually and evaluated the radiative decay rates (A^r) to all the $1s^2 3\ell 3\ell' 2\ell''$ terms.

III. Energies and eigenvectors

In Fig.1, we present an energy level diagram for the $1s^2 3\ell 3\ell' 3\ell''$ states of N^{2+}. The energies are shown as LS term values. Percentages indicate the main contribution to the eigenvector of the lowest J value of each term and the low percentage values reflect the break down of the mono-configurational picture due to the strong mixing. A number of factors increase the mixing in the $3\ell 3\ell' 3\ell''$ eigenvectors compared to the mixing for the $1s^2 3\ell 3\ell'$ terms:
i) the high density of terms of the same symmetry and the

strong interaction between states of same configurations but with different parent terms,
ii) the strong interaction with the $1s^2 3\ell 3\ell' 4\ell''$ states, especially for the higher members of the $3\ell 3\ell' 3\ell''$ configurations which overlap the former states in energy
iii) and finally, a few "accidental" LSJ interactions due to the near degeneracy of states of different LS symmetry but with the same J value.

It is also of interest to study the structure of the $3\ell 3\ell' 3\ell''$ configurations for their own sake. For $3\ell 3\ell'$ configurations it is known that the single-configuration picture is of limited validity and that these states have, at least in the He isoelectronic sequence, similarities with the vibrational-rotational structure of tri-atomic molecules [9]. For the triply-excited states virtually nothing is known. Strong configuration interaction effects are expected since the configurations constitute a Layzer complex but very little is known about this particular type of Layzer complex since in "normal" atoms at least one of the 3s and 3p shells is filled before the 3d shell is filling up. We have for example found very strong mixing for $3d^3$ states which normally are quite pure. For example, the eigenvector of the $1s^2 3d^3\ ^2D_1$ state is (only contributions of more than 10% are given):

$$
\begin{aligned}
3d^3\ ^2D_1 = \; & 0.72\ 3d^3\ ^2D_1 && - 0.41\ 3p^2(^1S)3d \\
+ \; & 0.22\ 3d^3\ ^2D_3 && + 0.20\ 3p^2(^3P)3d \\
- \; & 0.18\ 3d3p(^3D)4p && - 0.17\ 3d3p(^3D)4f \\
+ \; & 0.16\ 3s^2(^1S)3d && - 0.16\ 3d3p(^3F)4p \\
- \; & 0.15\ 3d3p(^3P)4d && + 0.15\ 3d3p(^3P)4f \\
+ \; & 0.11\ 3d3p(^1P)4f && + 0.10\ 3p^2(^3P)4d
\end{aligned}
$$

The configuration interaction perturbs also the usual ordering of the states in the $3d^3$ manifold which is ^4F, ^4P, ^2G, ^2P, $^2D_3$, ^2H, ^2F and $^2D_1$ in V III, for example, and becomes ^2P, ^4F, ^2H, ^2G, ^4P, $^2D_3$, ^2F and $^2D_1$ in N III.

IV. Decay rates

We have calculated level autoionization decay rates for all the $1s^2 3\ell 3\ell' 3\ell''$ terms. We have neglected the interaction between the $1s^2 2\ell 3\ell'$ limits but this approximation is not expected to affect the total decay values markedly. All the results have been converted to LS term values. The detailed results will be published elsewhere.

The comparison between the results for the $1s^2 3\ell 3\ell' 3\ell''$ terms in N^{2+} and for the $1s^2 3\ell 3\ell'$ terms of N^{3+} shows that the autoionization rates of the triply-excited states do not exhibit a strong variation between terms in contrast to the autoionization rates for the doubly-excited states which has a strong term dependence. The ratio between the largest ($3d^2\ ^1G$) and the smallest ($3d^2\ ^1S$) non-radiative decay in $3\ell 3\ell'$ in N^{3+} is about $7.5 \cdot 10^3$ but the equivalent ratio is only 7.7 in N^{2+} (between $3d^2(^3P)3p\ ^4D^o$ and $3d^3\ ^2H$). In view of this result, we have checked whether single configuration calculations of average values give a correct picture of the behaviour of the autoionization decay rates of the $3\ell 3\ell' 3\ell''$ triply-excited configurations. At the same time we have calculated the average radiative decays to the $1s^2 3\ell 3\ell' 2\ell''$ states which are known to be weakly dependent on the particular term for the doubly-excited states [7].

The average decay rate assuming a statistical population distribution of the $1s^2 3\ell 3\ell' 3\ell''$ states of N^{2+}:

$$A^{stat} = \Sigma_i\, g_i\, A_i / \Sigma_i\, g_i$$

are given in Table 1.

Fig.1: Triply-excited energy levels in N III

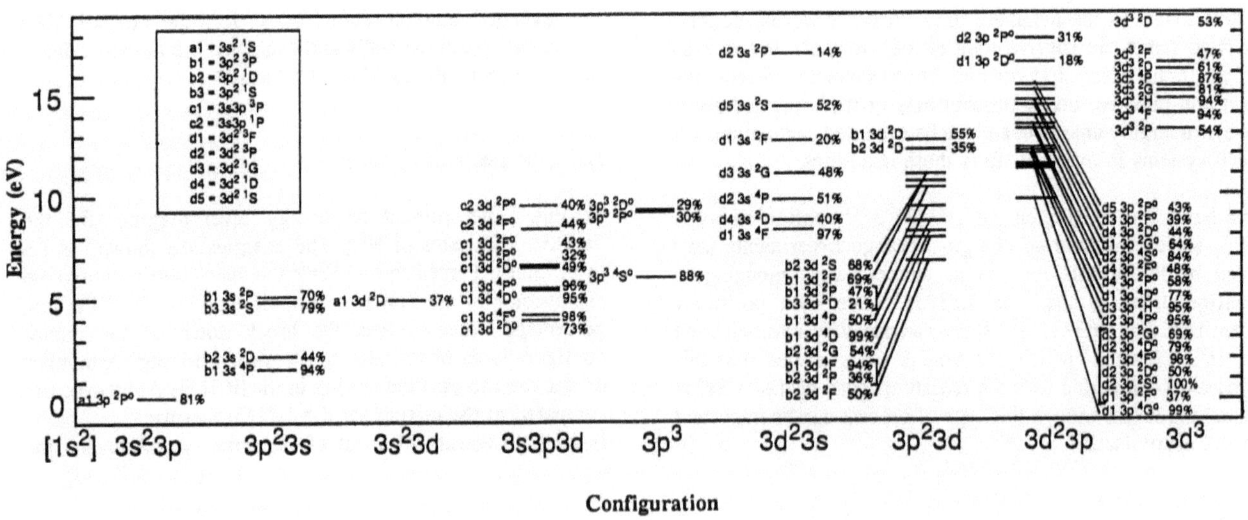

Table 1

Average autoionization decay rates assuming a statistical population distribution for the $1s^23\ell3\ell3\ell''$ terms of N^{2+}. The average autoionization rates and radiative decay rates obtained from single configuration calculations, labelled "average calc.", are given also.

A_a^{stat} (term calc.)	$4.87 \cdot 10^{14}$
A_a^{stat} (average calc.)	$5.10 \cdot 10^{14}$
A_r^{stat} (average calc.)	$5.69 \cdot 10^{10}$

From Table 1, we can see that the radiative decay of the triply-excited states does not play an important role in the total decay. The non-radiative decay rate obtained from single-configuration calculation is in good agreement with the results of the explicit term calculation even in this case which exhibit so strong configuration mixing. This fact suggests the use of this simple method to investigate the decay properties of the triply-excited configurations.

V. Conclusions

The average lifetime assuming a statistical population distribution is defined by:

$$\tau_{stat} = \Sigma_i\, g_i\, \tau_i / \Sigma_i\, g_i$$

For the $1s^23\ell3\ell'$ doubly-excited states $\tau_{stat} = 1.0 \cdot 10^{-13}$ is long enough to allow the pick-up of a third electron if these configurations can be formed in a neutralisation process. For the triply-excited states $\tau_{stat} = 2.7 \cdot 10^{-15}$ (neglecting the radiative decay). This is much faster than the radiative decay of the doubly-excited states with which it is in competition. On the basis of these calculations, it is thus possible to propose that pick-up of additional electrons in highly excited configurations can be important in the neutralisation of ions at surfaces. However, it is interesting to note that the ratio between the A^{stat} (t-e) and A^{stat} (d-e) is much smaller (2.3). The inverse of this number would be the ratio between the average lifetimes of the two complexes if a mechanism exists that continuously lead to an equalization of level distribution. This means that the above conclusion is rather dependent on the assumption of a statistical population distribution and the calculated ratio will be less extreme if, for example, the long lived doubly-excited $3d^2\ ^1S$ state is weakly populated as is found, for example, in ion-atom collisions [10].

References

1. Hagstrum, H.D.: Phys. Rev. **96**, 325 (1954)
2. Andrä, H.J.: Atomic Physics of Highly Charged Ions, Editor: R. Marrus. New York: Plenum Press (1989)
3. Hansen, J.E.: J. Phys. (Paris) **50**, C1-603 (1989)
4. Luc-Koenig, E., Bauche, J.: J. Phys. B **23**, 1763 (1990)
5. Mack, M.: Thesis, Groningen (1987)
6. Cowan, R.D.: The Theory of Atomic Structure and Spectra. Berkeley: University of Calif. Press (1981) Ch. 7, 14 and 18
7. Vaeck, N., Hansen, J.E.: J. Phys. **B 22**, 3137 (1989)
8. Layzer, D.: Ann. Phys., NY **8** 271 (1959)
9. Herrick, D.R.: Electronic and Atomic Collisions, p.487 Editors: J. Eichler, I.V. Hertel and N. Stolterfoht. Amsterdam: North-Holland (1984)
10. Posthumus, J.H., Morgenstern, R. : J. Phys. B **23**, 2293 (1990)

Supplement to Z. Phys. D - Atoms, Molecules and Clusters 21, S 225-226 (1991)

Atoms, Molecules
Zeitschrift für Physik D and Clusters

Electron spectroscopy of doubly-excited helium-like ions

Y. Kanai, H. A. Sakaue*, S. Ohtani**, K. Wakiya*, H. Suzuki*, T. Takayanagi*, T. Kambara, A. Danjo#, M. Yoshino##, and Y. Awaya

The Institute of Physical and Chemical Research (RIKEN), Wako, Saitama 351.01, Japan

Abstract

Ejected electrons from helium-like doubly-excited carbon and boron ions, produced by C^{6+} and B^{5+} with He collisions, were measured by the zero-degree electron spectroscopy technique. Electrons from configurations 2lnl' were observed in a wide energy range.

I. Introduction

Recently multi-electron transfer processes in slow highly-charged ion-atom collisions have been studied [e.g.1]. Those studies are being done for the following two purposes; (1) to study the atomic structure of highly charged ions, (2) to investigate the mechanism of the multi-electron transfer processes in highly-charged ion-atom collisions. In the double electron transfer into bare nucleus projectiles by collisions with singlet target atoms such as He, only the singlet states of the doubly excited configurations can be produced unless the collision has changed an electron spin[2]. When the doubly excited states produced in such collisions are the (2lnl') configurations, a possible final state after ejecting one electron should be the 1s state of H-like ions. Therefore, in these collision systems, the electron spectra should be simple enough to be easily analyzed. Here we report spectroscopic studies of ejected electrons from autoionising states of doubly-excited He-like ions.

II. Experiment

We measured ejected electrons produced by the following double-electron transfer collisions;

$$67\ \text{keV}\quad C^{6+} + He(1s^2) \longrightarrow C^{4+}(2lnl')^1L + He^{2+},\quad C^{4+}(2lnl')^1L \longrightarrow C^{5+}(1s) + e^- \tag{1}$$

$$50\ \text{keV}\quad B^{5+} + He(1s^2) \longrightarrow B^{3+}(2lnl')^1L + He^{2+},\quad B^{3+}(2lnl')^1L \longrightarrow B^{4+}(1s) + e^- . \tag{2}$$

Projectile ions were produced by the ECRIS which was constructed as an ion source for the RIKEN AVF cyclotron[3]. Ejected electrons were measured by the technique of the zero-degree electron spectroscopy. Our analyzer consists of a 45° parallel-plate electrostatic deflector and an electrostatic analyzer of simulated hemispherical type with a mean radius of 104 mm; the former was used to steer the electrons out of the ion beam and the latter determines the electron energy resolution. The precise experimental procedure will be described elsewhere[4].

III. Results and Discussion

Spectra observed for collisions of $^{13}C^{6+}$ with He are shown in figure 1. There are two features of the spectrum shown in figure 1:(1) no peaks of the 2l2l' configurations(around 270 eV) and (2) peaks of 2lnl' configurations in a wide range of n(≥ 3) are observed. These features can be explained in terms of a "reaction window". From an estimate of the reaction window by using the Niehaus' formula[5] shown in figure 1 as a broken line, the crossing radius R_c to produce the 2l2l' states in the double-electron transfer collisions is estimated to be outside the reaction window under our experimental conditions.

Spectra observed for collisions of $^{13}B^{5+}$ with He are shown in figure 2. In case of B^{5+} with He, R_c for 2l2l' states is inside the reaction window shown in figure 2 as a broken line.

We measured the ejected electrons from C^{4+}(2l3l')with high resolution. Measured values are listed in Table 1 together with the theoretical ones[6-8]. The determination of the energy for 2p3d $^1D^o$ has large uncertainty($\pm$ 0.5 eV) because the ejected electrons from this state do not show a clear peak. In Table 1, the term representations(K, T, and A) with the correlated

Permanent address
* Department of Physics, Sophia University, Chiyoda-ku, Tokyo 102, Japan
** Institute for Laser Science, University of Electro-Communications, Chofu, Tokyo 182, Japan
# Department of Physics, Niigata University, Niigata 950-21, Japan
Laboratory of Physics, Shibaura Institute of Technology, Omiya, Saitama 330, Japan

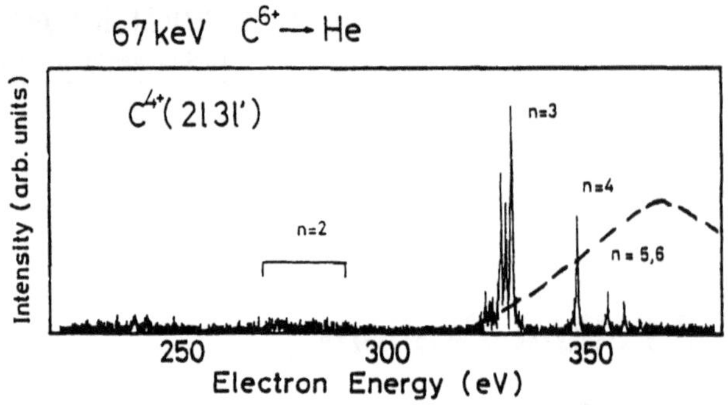

Fig. 1. Ejected electron spectra from $C^{4+}(2lnl')$ $n \geq 3$ configurations obtained under the single collision condition. Broken line indicates the "reaction window".

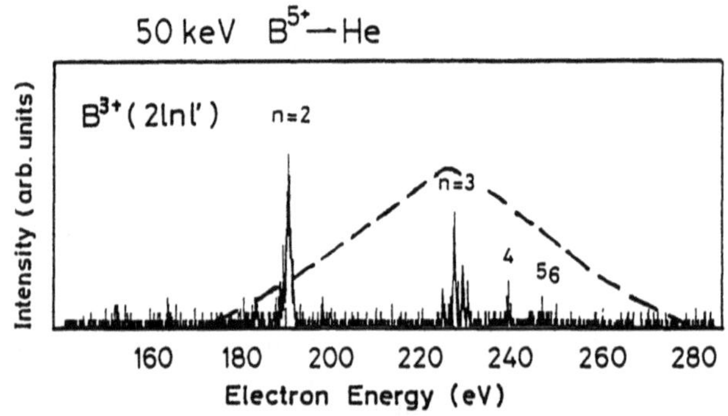

Fig. 2. Ejected electron spectra from $B^{4+}(2lnl')$ $n \geq 2$ configurations obtained under the single collision condition. Broken line indicates the "reaction window".

classification scheme are shown as well as the single-particle notations. Theoretical values noted as Lipsky are extrapolated values from those of Lipsky et al[6] by using a method described by Lin [9]. Observed peak positions in the energy spectra are shifted due to the post collision interaction(PCI) effect. Energy values of the PCI shift using the theoretical life time[10] are also listed in Table 1. The sum of the experimental values and the PCI shift values must be compared with the theoretical values. Experimental values of ejected electron energies from the 2l3l' states(^{1}S, ^{1}P) are in good agreement with theoretical values of Ho[7] after the PCI correction. Theoretical values by the MZ method of Safronova[8] are in good agreement with our results(^{1}S, ^{1}P, ^{1}D and ^{1}F).

Table 1 Energies of electrons ejected from $C^{4+}(2l3l')$. The sum of the observed values and the PCI values must be compared with theoretical values.

Observed values (eV)	PCI (eV)	$^{2S+1}L^{\pi}$	$(K,T)^A$	Theory Lipsky [6,9]	Safronova [8]	Ho [7]
323.7 ± 0.2	0.0	$^1P^o$	$(1,0)^-$	323.91	323.82	323.88
324.7 ± 0.2	0.2	$^1S^e$	$(1,0)^+$	325.16	324.86	324.96
325.5 ± 0.2	0.0	$^1P^e$	$(0,1)^-$	325.71	325.64	
327.0 ± 0.5	0.0	$^1D^o$	$(0,1)^0$	327.56	327.83	
327.5 ± 0.2	0.2	$^1D^e$	$(1,0)^+$	328.17	327.88	
328.7 ± 0.2	0.1	$^1P^o$	$(0,1)^+$	329.41	328.78	328.90
329.9 ± 0.2	0.0	$^1D^e$	$(0,1)^0$	330.34	330.06	
330.2 ± 0.2	0.0	$^1F^o$	$(1,0)^0$	330.71	330.16	
331.2 ± 0.2	0.0	$^1P^o$	$(-1,0)^0$	331.70	331.36	331.36

References

1. Bliman, S. ed.: Proceedings of the International conference on the physics of multiply charged ions and International workshop on E.C.R. ion sources, J. de Physique **50-C1** (1989).
2. Mack, M.: Nucl. Instrum. Methods **B23**, 74 (1987).
3. Hatanaka, K., Nonaka, H.: J. de Physique, **50-C1**, 827(1989).
4. Sakaue, H. A., Kanai, Y., Ohta, K., Kushima, M., Inaba, T., Ohtani, S., Wakiya, K., Suzuki, H., Takayanagi, T., Kambara, T., Danjo, A., Yoshino, M., Yohko, A.: J. Phys. **B23**, L401(1990).
5. Niehaus, A.: J. Phys. **B19**, 2925(1986).
6. Lipsky, L., Anania, R., Conneely, M. J.: At. Data Nucl. Data Tables **20**, 127(1977).
7. Ho, Y. K.: Phys. Rev. **A23**, 2137(1981).
8. Safronova, U. I.:, private communication (1990).
9. Lin, C. D.: Phys. Rev. **A39**, 4355(1989).
10. Safronova, U. I.: Physca Scripta, **T26**, 59 (1989).

Supplement to Z. Phys. D – Atoms, Molecules and Clusters 21, S 227–229 (1991)

Atoms, Molecules
Zeitschrift für Physik D and Clusters

Doubly excited states of three-electron systems ($Z = 2$ to 10) below the (1s2l)SLπ threshold

R. Bruch[1], H. Wang[1], D. Schneider[2], L. Lipsky[3] and M. J. Conneely[3,*]

[1] Atomic and Molecular Collision Laboratory, Department of Physics, University of Nevada, Reno, Reno, Nevada 89557, USA
[2] High Temperature Physics Division, Lawrence Livermore Laboratory, P. O. Box 808, Livermore, California 94550, USA
[3] Department of Computer Science, U-155, University of Connecticut, Storrs, Connecticut 06268, USA

Abstract

A systematic experimental and theoretical study of doubly excited 1s2lnl' states of three-electron ions has been performed. Theoretically the truncated diagonalization method (TDM) has been used to generate both nonrelativistic and relativistic energy values, wave functions and quantum defects for 42 different Rydberg series converging to four different thresholds, i.e. $(1s2s)^3S^e$, $^1S^e$ and $(1s2p)^3P^o$, $^1P^o$. This comprehensive data base has been used to classify recent high resolution Auger electron spectra arising from Li-like states. Specifically we have analyzed He^-, C^{3+}, O^{5+} and Ne^{7+} Auger electron data produced by double electron capture in slow ion-atom collisions. Numerous new lines associated with (1s2lnl') Rydberg levels have been identified.

I. Introduction

The development and availability of intense high charge state ion beams produced for instance by electron cyclotron resonance (ECR) ion sources [1] has stimulated a renaissance of fundamental research in the field of core-excited ionic states, which can be populated in slow multiple electron transfer processes [2]. The excitation and decay mechanisms of such multiply excited states provide important information on many body effects in ions. In particular the mutual interplay between electron correlation and relativistic interactions can be studied along isoelectronic series. In this investigation, we present an experimental and theoretical study on Li-like 1s2lnl' double excited states for Z=2 to 10. Experimentally we have studied the structure and collision dynamics of such core-excited ions by using the method of zero degree electron spectroscopy [3]. Theoretically we have performed comprehensive calculations of doubly excited states of three-electron systems, by systematically calculating the energies, wave functions and quantum defects.

II. Description of theoretical method

The method of Ahmed and Lipsky [4] is used to calculate the energies of doubly excited states of three-electron systems. The energies are obtained by diagonalizing the Hamiltonian

$$H_0 = \sum_{i=1}^{3}\left(-\frac{1}{2}\nabla_i^2 - \frac{Z}{r_i}\right) + \sum_{i<y}\frac{1}{r_{ij}} \qquad (1)$$

using the antisymmetric basis functions$\{\psi_o(n_1l_1, n_2l_2, n_3l_3)$ $LS\pi\}$ constructed from hydrogenic functions. In our basis we have included all configuration of this form where (for i=1,2,3) $0 \le l_i \le l_{max}=5$ and $n_i \le n_{max}=20$, with the following restrictions. Only if two of the three n's are less than or equal to three can the third one be as large as n_{max}. Otherwise, the maximum value is six. For any configuration, at most one of the n's can equal 1. This procedure yieds different numbers of configurations depending on the particular symmetry $LS\pi$ under consideration. For example, for $^4S^e$ symmetry this procedure gives 388 configurations, while for $^2D^e$ the number of configurations (NC) is 924 (see Table 1). Each configuration, in turn, produces anywhere from 0 to 10, totally

Table 1. Symmetries and series for doubly excited states of three-electron systems below the (1s2l)LSπ thresholds

LS	NC	NB	NS	Series Description {1}	{2}	{3}	{4}
$^2S^e$	621	1009	4	ns	ns	np	np
$^4S^e$	388	388	2	ns		np	
$^2P^e$	524	1257	2			np	np
$^2P^o$	827	1978	6	np	np	ns	ns
						nd	nd
$^4P^e$	524	694	1			np	
$^4P^o$	737	891	3	np		ns	
						nd	
$^2D^e$	924	2362	6	nd	nd	np	np
						nf	nf
$^2D^o$	604	1733	2			nd	nd
$^4D^e$	750	1064	3	nd		np	
						nf	
$^4D^o$	600	886	1			nd	
$^2F^e$	577	1701	2			nf	nf
$^2F^o$	803	2266	6	nf	nf	nd	nd
						ng	ng
$^4F^e$	577	883	1			nf	
$^4F^o$	723	1042	3	nf		nd	
						ng	

$\{1\} := (1s2s)^3S^e$ $\{2\} := (1s2s)^1S^e$ $\{3\} := (1s2p)^3P^o$ $\{4\} := (1s2p)^1P^o$

antisymmetric basis functions. For instance {1s,2s,2p} produces 2 orthogonal, independent antisymmetric functions. The computer automatically generates all the basis functions for a given configuration. The number of configurations for a given symmetry ($LS\pi$), the number of basis functions (NB-size of the matrix), and the number of Rydberg-like series (NS) of energy levels converging on the singly excited states of the residual ion are listed in Table 1. Theoretically we distinguish between 42 different series and 14 symmetries.

We have used the effective quantum number n^* of our calculated energy levels to identify the different series of 1s2lnl' Rydberg states which converge to the $(1s2s)^3S^e$, $(1s2s)^1S^e$, $(1s2p)^3P^o$ and $(1s2p)^1P^o$ thresholds. The effective quantum number is defined by the relation

$$E_n = E_T - \frac{1}{2}\left(\frac{Z-2}{n^*}\right)^2 \qquad (2)$$

where E_n is the calculated energy level (a.u.) and E_T is the threshold energy. We assume that the outermost electron "sees" a different threshold due to polarization than a free electron at infinity, and call these values the "Effective Threshold Energies". As an example we have listed the C^{4+} effective threshold energies in comparision with our two-electron TDM calculations, with and without relativistic terms.

Table 2. Threshold energies (a.u) for the four C^{4+} states $(1s2l)LS\pi$

Configuration and symmetry	Non-Relativistic		Relativistic	
	TDM	Effective	TDM	Effective
$(1s2s)^3S^e$	21.409999	21.40106	21.421292	21.41220
$(1s2s)^1S^e$	21.211496	21.18772	21.222739	21.19890
$(1s2p)^3P^o$	21.207470	21.19421	21.217334	21.20413
$(1s2p)^1P^o$	21.058636	21.03334	21.068347	21.04316

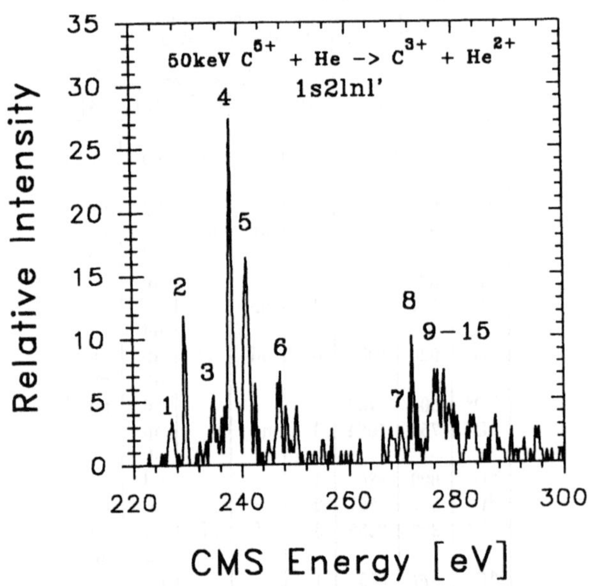

Fig. 1. 1s2lnl' Auger electron spectrum of C^{3+} produced by 50keV C^{5+}+He collisions. Electron energies are given in the projectile frame.

III. Experimental method

The experiments were performed using hydrogenic He^+, C^{5+}, O^{7+} and Ne^{9+} ion beams produced by the ECR ion source at the Lawrence Berkeley Laboratory. The experimental set-up consists of a target chamber, the differentially pumped scattering cell and two electron spectrometers in tandem, where the first parallel-plate analyzer separates the forward emitted electrons from the ion beam and the second high-resolution electrostatic analyzer is used to accumulate zero-degree Auger spectra. A more detailed description of the experimental procedure is given by Schneider et al. [3].

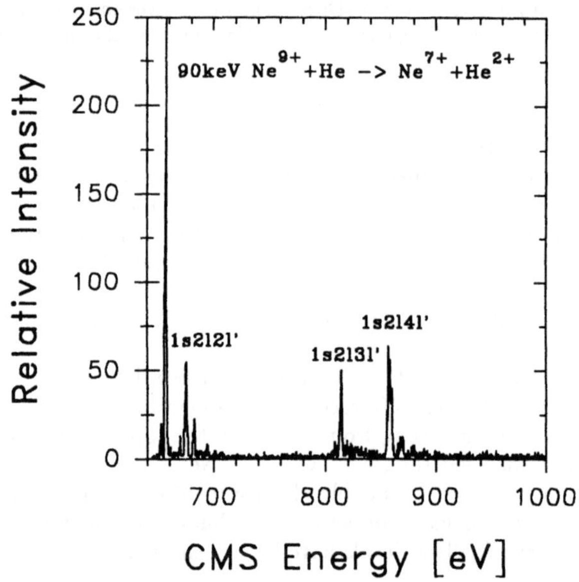

Fig. 2. Ne^{7+} 1s2lnl'(n=2 to 4) Auger electron spectrum following 90keV Ne^{9+}+He→Ne^{7+}(1s2lnl') +He^{2+}double electron capture. Electron energies are given in the projectile frame.

IV. Results and discussion

Experimentally we present here as prototype cases, C^{3+} and Ne^{7+} Auger electron spectra created by electron capture mechanisms. In Fig. 1 we have displayed a characteristic K-Auger spectrum following C^{5+}+He collisions. The most prominent peaks (1 to 15) originate from C^{3+} 1s2lnl' Rydberg levels created by double electron capture. The line assignment (see Table 3) is based on our TDM results and previous energy values calculated with the saddle-point variational method [5,6]. It is apparent that the lowest TDM energies are slightly higher than the coresponding saddle-point and experimental values. On the other hand the truncated diagonalization method is unique in generating a large set of Rydberg states simultaneously, which can be used to identify unknown Auger line structures, for example the features labled 6 to 15 in Fig. 1. We note that these new assignments disagree with calculations of Mann [7] using the Grant code.

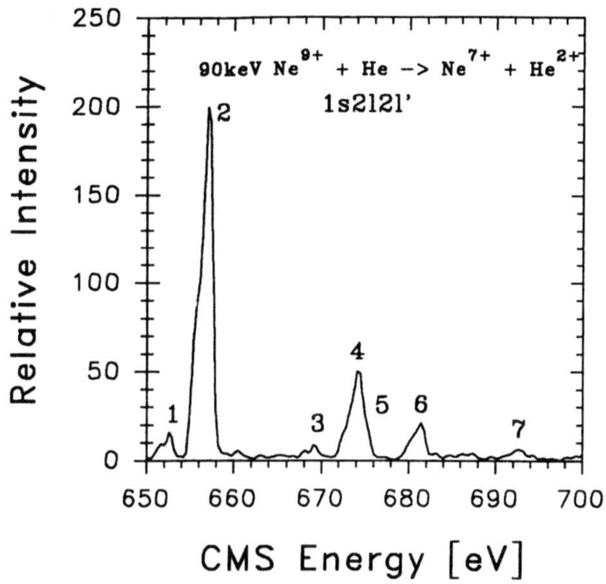

Fig. 3. High resolution 1s2l2l' Auger electron spectrum of Ne^{7+} produced by 90keV Ne^{9+}+He collisions. Electron energies are given in the projectile frame.
1: $(1s2s^2)^2S$ 2: $(1s2s2p)^4P^o$ 3: $[1s(2s2p)^3P]^2P^o$
4: $(1s2p^2)^4P$ 5: $[1s(2s2p)^1P]^2P^o$ 6: $(1s2p^2)^2D$
7: $(1s2p^2)^2S$

Table 3. Experimental and theoretical Auger transition energies for ($1s2ln l'$) states in C^{3+}

Peak	Experiment Energy(eV)	Theory	Classification This work
1	227.1^{d} 227.1^{e} 226.5^{f}	227.17^{a} 227.67^{b}	{1}2s, {2}3s $^2S(1)$
2	229.64^{*d} 229.6^{e} 229.5^{f}	229.64^{c} 230.31^{b}	{1}2p $^4P^o(1)$
3	234.6^{d} 234.4^{e} 235.5^{f}	235.55^{a} 236.36^{b}	{1}2p $^2P^o(1)$
4	238.2^{d} 238.5^{f}	238.50^{a} 239.66^{b} 239.57^{b}	{2}2p $^2P^o(2)$ {3}2p $^4P^e(1)$
5	242.6^{d}	242.08^{a} 243.01^{b}	{3}2p $^2D^e(1)$
6	248.3^{d}	249.10^{b}	{3}2p, {4}3p $^2S^e(2)$
7	271.5^{d}	271.45^{b}	{1}3s $^2S(3)$
8	272.1^{d}	272.47^{b}	{1}3p $^2P^o(3)$
9	275.6^{d}	275.34^{b}	{2}3s $^2S^e(4)$
10	276.2^{d}	276.16^{b}	{3}3s $^4P^o(2)$
11	276.8^{d}	277.07^{b}	{3}3s $^2P^o(4)$
12	277.8^{d}	278.54^{b}	{2}3p $^2P^o(5)$
13	278.9^{d}	278.93^{b} 279.71^{b}	{3}3p $^4P^e(2)$ {3}3p $^2D^e(2)$
14	279.7^{d}	280.51^{b} 280.64^{b} 280.75^{b}	{3}3d $^4P^o(4)$ {4}3s $^2P^o(6)$ {3}3p $^2S^e(5)$
15	283.0^{d}	282.57^{b} 283.57^{b}	{4}3p $^2D^e(11)$ {4}3p $^2S^e(6)$

[a] Saddle-point technique including mass correction Dawin term and retardation term, Bruch et al.[6]
[b] Three electron truncated diagonalization method(TDM) with relativistic mass correction and Dawin term, this work
[c] Configuration-interaction wave function with relativistic correction, Bruch et al.[6]
[d] Zero degree electron spectroscopy, this work
[e] Mann[7]
[f] Hofmann et al.[8]
* Energy calibration point

In this work we present also the first high resolution measurements of K-Auger electron spectra associated with the (1s2lnl')→($1s^2\varepsilon l$") Auger transitions in Ne^{7+}. According to the TDM calculation the most prominent groups of Auger lines observed refer to n=2 to 4 Rydberg states produced by double electron capture in
$Ne^{9+}(1s) + He(1s^2) \rightarrow Ne^{7+}(1s2lnl') + He^{2+}$ collisions. Finally, a high resolution spectrum in the vicinity of the $Ne^{7+}(1s2s2p)^4P^o$ state is indicated in Fig. 3. This spectrum is dominated by a few distinct lines originating from Li-like ($1s2s^2$), (1s2s2p) and ($1s2p^2$) configurations. The preferential population of the metastable autoionizing $(1s2s2p)^4P^o$ state may be caused by radiative cascade feeding from higher lying 1s2lnl' Rydberg levels [9].

V. Conclusion

The theoretical procedure described in this paper yields consistent sequences of energy levels and quantum defects for each Rydberg-like series. These data have been used to identify our high resolution Auger electron spectra. In particular we have found that the lowest members for each series are slightly too high, when compared to experimental data. Therefore we will repeat these calculations using generalized Laguerre functions to describe the core penetration of the two outer electrons more accurately. However, Laguerre functions are expected to give increasingly poorer results for the higher Rydberg levels.

References

[*] Permanent address: Department of Mathematical Physics, University College, Galway, Ireland

[1] C. M. Lyneis, Operating Experience with the LBL-ECR Source, LBL-23909 (1987)

[2] E. M. Mack, PhD. Thesis, University of Utrecht (1987)

[3] D. Schneider, N. Stolterfoht, G. Schiwietz, T. Schneider, W. Zeitz, R. Bruch and K. T. Chung, Nucl. Instr. Meth. **B24/25**, 173 (1987)

[4] M. Ahmed and L. Lipsky, Phys. Rev. **A12**, 1176 (1975)

[5] K. T. Chung and R. Bruch, Phys. Rev. **A28**, 1418 (1983)

[6] R. Bruch, K. T. Chung, W. L.Luken, and J. C. Culberson, Phys. Rev. **A31**, 310(1985)

[7] R. Mann, Phys. Rev. **A35**, 4988 (1987)

[8] G. Hofmann, A. Müller, K. Tinschert, and E. Salzborn, Z. Phys. **D16**, 113 (1990)

[9] D. Schneider, R. Bruch, W. Butscher W. H. E. Schwartz, Phys. Rev. **A24**, 1223 (1981)

Supplement to Z. Phys. D - Atoms, Molecules and Clusters 21, S 231-232 (1991)

Atoms, Molecules
Zeitschrift für Physik D and Clusters

Enhanced radiative Auger emission from lithiumlike $_{20}Ca^{17+}$

E. M. Bernstein[1], M. W. Clark[1], J. A. Tanis[1], W. G. Graham[2], T. J. Morgan[3], M. P. Stöckli[4], K. H. Berkner[5], A. S. Schlachter[5], and J. W. Stearns[5]

[1] Western Michigan University, Kalamazoo, MI 49008, USA
[2] Queen's University, Belfast BT7 1NN, Northern Ireland, UK
[3] Wesleyan University, Middletown, CT 06457, USA
[4] Kansas State University, Manhattan, KS 66506, USA
[5] Lawrence Berkeley Laboratory, Berkeley, CA 94720, USA

Radiative Auger emission (RAE) from lithiumlike $_{20}Ca^{17+}$ projectiles excited in collisions with He has been measured. The intensity of RAE photons relative to K_α x-ray emission is enhanced by a factor of 10 - 17 compared with theoretical calculations for ions with few electron vacancies. The enhancement of RAE for Ca^{17+} is consistent with the results reported previously for lithiumlike $_{16}S^{13+}$ and $_{23}V^{20+}$ and indicates a systematic dependence on Z. Both the enhancement and the relative RAE transition rate increase with Z.

PACS: 34.50.Fa, 32.80.Hd

1 Introduction

In the radiative Auger effect (RAE) [1], [2] an electron transition to a lower energy state is accompanied by simultaneous emission of a photon and the ejection of a second electron into the continuum. For example, K-LL RAE occurs when an L- shell electron fills a 1s vacancy while a second L-shell electron is ejected from the ion along with photon emission. Fig. 1 shows a schematic energy level diagram illustrating a K-LL RAE transition in lithiumlike $_{20}Ca^{17+}$. Since the total energy difference between the discrete initial and final atomic states is shared between the emitted electron and photon, the RAE photon spectrum is nonmonoenergetic.

RAE has been observed following K- shell vacancy production created by photons, electrons, protons, and alpha particles in atomic targets ($8 \leq Z \leq 22$). In these experiments the measured intensities of RAE for K-LL transitions relative to $K\alpha$ emission are less than 1% for atoms with $Z \geq 16$ in agreement with theoretical calculations [3],[4].

A large enhancement of the probability of RAE transitions compared to characteristic x-ray emission was recently reported for lithiumlike [5] $_{16}S^{13+}$ and lithiumlike [6] $_{23}V^{20+}$ ions which were excited in collisions with a He gas target. The measured ratios of K-Ln ($n \geq 2$) RAE emission compared to $K\alpha$ emission were about 5 and 20 times larger, respectively, than the expected values based on theoretical calculations [3],[4] and previous experimental results for ions (atoms) which had relatively few electron vacancies [1]. It was suggested [5], [6] that the large relative RAE transition rate was related to the small number of electrons in the lithiumlike ions.

The present paper reports RAE transition rates for 2.5 to 8.75 MeV/u lithiumlike $_{20}Ca^{17+}$ projectiles excited in collisions with a He target.

2 Experimental Technique

Since the RAE process involves the emission of an electron, those Ca^{17+} ions which are excited and decay by RAE will change their charge to 18+. Thus, the RAE photon spectra for Ca^{17+} were obtained by measuring x rays coincident with single electron loss in the Ca^{17+} + He system. These experiments were conducted at the Lawrence Berkeley Laboratory using the SuperHILAC accelerator. Relative uncertainties in the cross sections obtained are estimated to be ± 4% for the total projectile x- ray production and ± (8-10)% for the RAE photons. Systematic uncertainties in target thickness and x-ray detector efficiency and solid angle lead to an additional uncertainty of ±25 % in the absolute cross sections.

3 Results and Discussion

The relative RAE transition rate can be obtained by comparing the RAE photon intensity to the intensity of characteristic x rays from the same ionic species. The K-LL RAE photons observed in the spectrum coincident with electron loss (exiting Ca^{18+} ions) originate in projectiles which undergo K-shell excitation with no change in charge (Ca^{17+} ions).

The cross section for production of $K_{\alpha\beta}$ x rays from projectiles which do not change charge, $\sigma^q_{K\alpha\beta}$, which was not measured directly, can be deduced from the total $K_{\alpha\beta}$ x-ray production cross sections by subtracting the x-ray yields coincident with projectiles which either captured or lost an electron. The x-ray yield coincident with other charge states is negligible in the present case. About (10 - 15) % of the total characteristic x-ray yield is associated with projectiles which are excited in the collision and change charge.

Fig. 2 shows the energy dependence of the cross section for producing RAE photons, σ_{RAE}, along with $\sigma^q_{K\alpha\beta}$. In Fig. 2(a) the curve for $\sigma^q_{K\alpha\beta}$ is drawn through the data points.

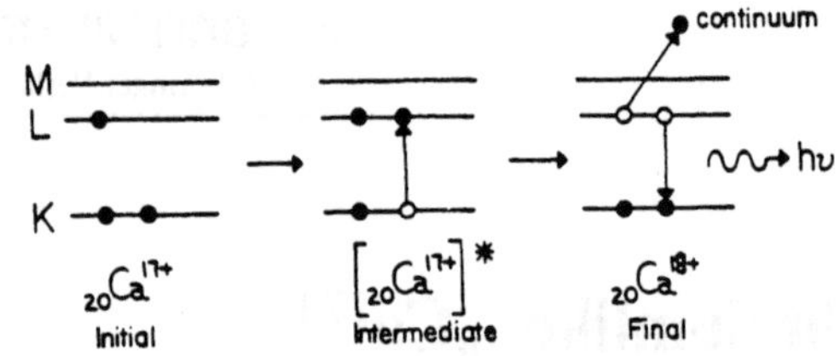

Fig. 1: Schematic energy level diagram illustrating a typical K-shell excitation event in lithiumlike $_{20}Ca^{17+}$ with subsequent decay by a K-LL radiative Auger transition.

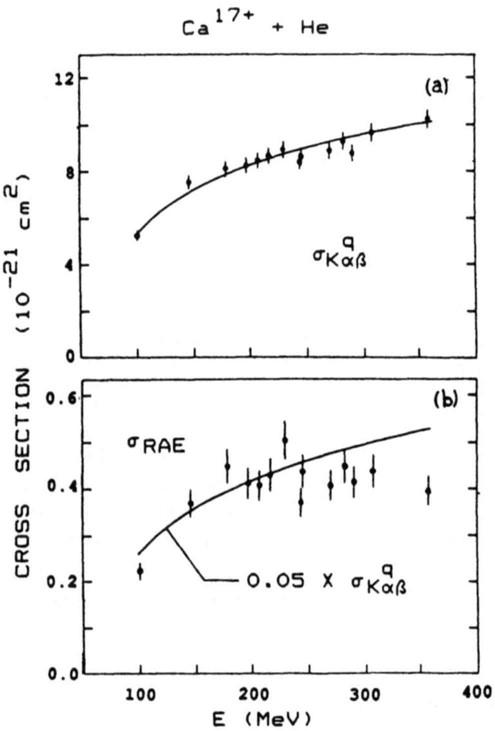

Fig. 2: Energy dependences of the following x-ray cross sections for $_{20}Ca^{17+}$ + He: (a) $\sigma^{q}_{K\alpha\beta}$ for $K_{\alpha\beta}$ x rays associated with projectiles which do not change charge (see the text). The line is drawn through the data. (b) σ_{RAE} for RAE photons coincident with electron loss. The curve was obtained by taking 5% of the curve for $\sigma^{q}_{K\alpha\beta}$ shown in (a). Relative uncertainties are indicated in both graphs.

The curve in Fig. 2(b) was calculated by multiplying the curve through $\sigma^{q}_{K\alpha\beta}$ by 0.05. Although it is evident from the results shown in Fig. 2 that the ratio of the intensity of RAE photons to characteristic K x rays associated with no charge change is not constant over the entire energy range, it is useful to determine the average value of this ratio in order to compare with theoretical calculations. The measured average value of the ratio of RAE to K_α x rays for Ca^{17+} is $(6.0 \pm 0.5)\%$. This result is a factor of 17 larger than the value calculated by Aberg [3] for few vacancy Ca ions and a factor of 10 larger than the value calculated by Scofield [4] for few-vacancy ions.

4 Conclusions

Strong enhancement of K-shell radiative Auger emission relative to K_α emission has been observed for lithiumlike Ca^{17+} ions excited in collisions with He target atoms. The enhancement factor of 10 to 17 is intermediate between the values of 5 and 20 obtained for S and V. The present results for Ca combined with the previous measurements for S and V provide evidence for a systematic enhancement of the relative radiative Auger rates for lithiumlike ions excited in ion-atom collisions. Both the relative probability of RAE emission and the mágnitude of the enhancement increase with Z for the ions studied to date. Measurements of RAE emission for lithiumlike $_{26}Fe^{23+}$ have been made and are currently being analyzed.

Acknowledgments

This work was supported in part by the U.S. Department of Energy, Office of Basic Energy Sciences, Division of Chemical Sciences and the Science and Engineering Research Council (UK).

References

[1] T. Åberg and J. Utriainen, Phys. Rev. Lett. 22, 1346 (1969).

[2] T. Åberg, Atomic Inner Shell Processes I. Ionization and Transition Probabilities, edited by B. Crasemann, (Academic, New York) p. 353 (1975).

[3] T. Åberg, Phys. Rev. A4, 1735 (1971).

[4] J. H. Scofield, Phys. Rev. A9, 1041 (1974).

[5] E. M. Bernstein, M. W. Clark, C. S. Oglesby, J. A. Tanis, W. G. Graham, R. H. McFarland, T. J. Morgan, B. M. Johnson, and K. W. Jones Phys. Rev. A41, 2594(1990).

[6] E. M. Bernstein, M. W. Clark, K. H. Berkner, W. G. Graham, R. H. McFarland, T. J. Morgan, A. S. Schlachter, J. W. Stearns, M. P. Stöckli and J. A. Tanis, J. Phys. B21, L509 (1988).

Supplement to Z. Phys. D - Atoms, Molecules and Clusters 21, S 233-234 (1991)

Atoms, Molecules
Zeitschrift für Physik D and Clusters

Two-photon decay of the 1s2s $^1S_0$ state in $_{36}Kr^{34+}$ produced by resonant transfer and excitation

Th. Stöhlker[1], Ch. Kozhuharov, A. E. Livingston[2], P. H. Mokler[1], J. Ullrich
GSI Darmstadt, FRG

B. Fricke
Gesamthochschule Kassel, FRG

The doubly excited 2s2p $^1P_1$ level of Kr^{34+} populated via resonant transfer and excitation (RTE) feeds selectively the metastable 1s2s $^1S_0$ state which can only decay via simultaneous emission of two photons to the ground state $1s^2$ $^1S_0$. X-ray/X-ray coincidence measurements in heavy ion-atom collisions enable the direct measurement of the spectral distribution of the two-photon decay in He-like ions. In addition, we observe strong photon cascades induced by radiative electron capture.

PACS: 32.30.Rj, 30.70.+e, 34.50.Fa

In two-electrons ions the primary decay mode for the 2^1S_0 state for all Z is the two-photon (2E1) decay to the 1^1S_0 ground state. The two photons are emitted simultaneously and the sum of their energies is equal to the 2 $^1S_0$ - 1 $^1S_0$ energy separation. As in the case of the 2E1 decay in H-like ions the resulting photon spectrum shows a continuous energy distribution, symmetric around the energy midpoint [1]. The theoretical spectral distribution for low Z helium-like ions [2] shows a broadening of the distribution with increasing Z as a result of decreasing importance of electron-electron interaction, whereas the hydrogen-like distributions [3] demonstrate a narrowing of the spectrum at higher Z due to the increasing importance of relativistic contributions. For the lowest Z the distribution for hydrogenic ions is wider than the corresponding one for He-like species.

The few experimental studies of the two photon decay have been concentrated on measuring the 2 $^1S_0$ lifetime ([4], reviewed in Ref. [5]) utilizing the beam foil technique. No information about the spectral distribution was given in these experiments. In this paper a method to study the spectral distribution is presented [6] [7]: the metastable 2 $^1S_0$ level is fed by the decay of the 2s2p $^1P_1$ state populated via the KLL-RTE resonance in collisions between hydrogenic projectiles and light gas atoms (H_2).

Within the impulse approximation RTE is considered as the time reversed Auger process: The energy gained by the capture of one quasi free target electron is used to excite one electron of the incident projectile forming a doubly excited intermediate state which may stabilize radiatively. For the KLL-resonances (the Auger-notation is used, i.e. excitation from K to the L shell by simultaneous capture into the L shell) of H-like projectiles this process leads to doubly excited states with two electrons in the L-shell and two K-shell vacancies. Therefore, the fingerprint for this process is the simultaneous emission of two K-photons. The resonance condition for the KLL-resonance is given by the adiabaticity parameter $\eta = 1/2$ ($\eta=(v/v_k)^2$, where v is the projectile velocity and v_k the velocity of the K-shell electron) [6] [7]. Only one of the KLL-resonances the 2s2p $^1P_1$ level feeds the metastable 1s2s $^1S_0$ state via emission of one K_α hypersatellite. The following 2E1 decay can then be observed via the x-ray/x-ray coincidence method (each detector measures the energy of one of the two photons) as already demonstrated in $Ge^{31+} \rightarrow H_2$ collisions [6] [7].

Measuring x-ray/x-ray coincidences for projectiles having captured one electron in the collision, we studied the two-photon decay of He-like Kr^{34+} ions in collisions between hydrogenic Kr^{35+} projectiles and H_2 gas atoms at the KLL-resonance, i.e. at 16.7 MeV/u. A charge state analysed beam was focused onto a differential pumped gas target. Additionally the ions were charge state analysed by a dipol magnet behind the target area. The projectiles associated with one captured electron were then registered by a fast scintillator counter and measured in coincidence with the x-rays from the target area. The target area was viewed by two Si(Li) detectors mounted perpendicular to the beam axis with an effective solid angle of 1% of 4π each.

The resulting cluster plot of coincident two photon events is shown in Fig. 1. A broad peak at (13 keV,13 keV) is found due to radiative stabilisation of the doubly excited intermediate states. Additionally, three prominent continuous ridges can be seen at x-ray energies corresponding to (K_α,E_1), (E_2,K_α), and (E_1,E_2) with $E_1 + E_2 = K_\alpha$. These structures are caused by the detection of the K_α-photon of the $2s2p^1P_1$ - $1s2s^1S_0$ transition in coincidence with one photon of the 2E1-decay, (E_2,K_α) or (K_α,E_1), and by the detection of both photons of the 2E1-decay (E_1,E_2).

Using the energy condition for the sum energy of the two-

[1]GSI Darmstadt and University of Gießen
[2]Notre Dame University, IN, USA

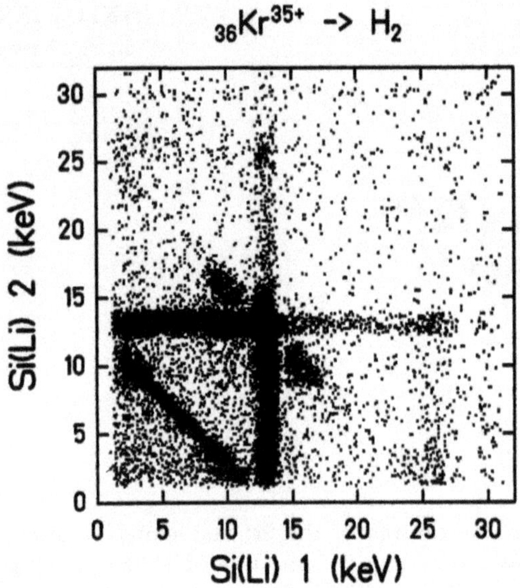

Fig. 1: x-ray/x-ray cluster plot for $Kr^{35+} \rightarrow H_2$ collisions measured in coincidence with projectiles having captured one electron.

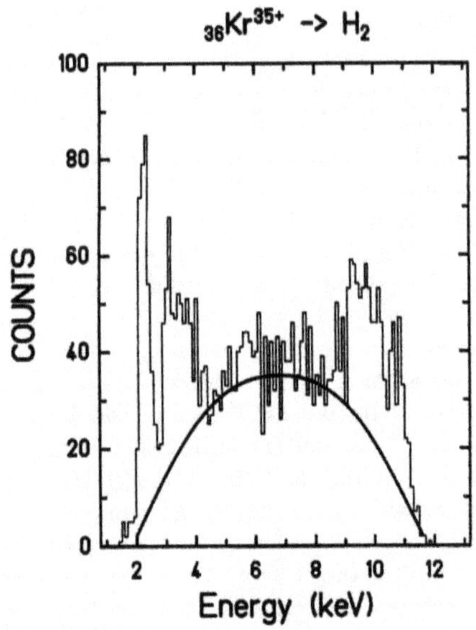

Fig. 2: Spectral distribution of the 2E1 decay of Kr^{34+} measured by one of the detectors.

photon decay the resulting spectral distribution measured by one detector is plotted in Fig. 2. In addition the theoretical predicted spectral distribution [8] convoluted with the response function for x-ray detection is shown in Fig. 2. The expected line shape is superimposed by broad structures on the high energy side and by Kr contributions like L radiation in the low energy regime. Both features can be explained by radiative electron capture (REC). Due to the resonance condition for KLL-RTE the energy of the L-REC photon has an energy equivalent to the K_α transition energy. Therefore REC into higher projectile shells and a following electron transition into the L-shell leads also to a simultaneouse emission of two photons with the sum energy equivalent to a K_α x-ray (e.g. M-REC/L_α coincidence). The additional diagonal ridge which is indicated partially in Fig. 1 at a sum energy of $2\times K_\alpha$ can also be explained by the REC process. Here the electron is captured radiatively into higher projectile shells followed by a direct transition into the K-shell (e.g. M-REC/K_β coincidence). Consequently, the diagonal has a cut off at an energy corresponding to capture into Rydberg states at the series limit.

Due to the additional REC contributions it is difficult to extract information on the spectral distribution for the 2E1 decay of our Z system in a straight forward manner. The REC contributions have first to be modelled very precisely and then subtracted from the measured distribution. This work is in progress now. At high Z systems this situation is more fortunate: Here the REC contributions are better resolved and can be separated more easily from the 2E1 spectral distribution.

References

[1] R. Marrus and P.J. Mohr, Adv. At. Mol. Phys. 14 (1978) 181

[2] G.W.F. Drake, G.A. Victor, and A. Dalgarno, Phys. Rev. 180 (1969) 25

[3] W.R. Johnson, Phys. Rev. Lett. 29 (1972) 1123

[4] R.W. Dunford, H.G. Berry, K.O. Groeneveld, M. Hass, E. Balke, M.C.A. Raphaelian, A.E Livingston, L.J. Curtis, Phys. Rev. A38 (1988) 5423

[5] G.W.F. Drake, Phys. Rev. A34 (1986) 2871

[6] P.H. Mokler, Th. Stöhlker, Ch. Kozhuharov, J. Ullrich S. Reusch, Z. Stachura, A. Warczak, A. Müller, R. Schuch, E.A. Livingston, M. Schulz, Y. Awaya, T. Kambara, invited paper to X-90 (Conference on X-Ray and Inner Shell Processes, Knoxville, 1990)

[7] P.H. Mokler, S. Reusch, A. Warczak, Z. Stachura, T. Kambara, R. Schuch , and M. Schulz, submitted to Phys. Rev. Lett.

[8] G.W.F. Drake, private communication 1988

Supplement to Z. Phys. D - Atoms, Molecules and Clusters 21, S 235-236 (1991)

Atoms, Molecules
Zeitschrift für Physik D and Clusters

Ionization of atoms by fully stripped ions in the Glauber approximation

A. C. Roy

Department of Physics, University of Kalyani, Kalyani - 741 235, West Bengal, India

Abstract

We report a method of evaluation of cross sections for single and double ionization of atoms by fully stripped ions. The method is based on the Glauber approximation. In the case of a hydrogenic target the probability amplitude turns out to be a simple two-dimensional integral which can be evaluated numerically with convenience.

I. Introduction

Since 1969 the Glauber approximation (GA) has been a useful theoretical tool for studying a wide variety of atomic collision processes [1]. It was first applied to the ionization of H and He by electron impact in 1974 [2]. Later on, Golden and McGuire [3,4,5] applied the GA with success to proton-impact ionization of atoms. These applications were, however, limited to single ionization only.

This paper reports an application of GA to calculate both single and double ionizations of atoms by bare ions. In Section II we present the method of evaluation of ionization cross sections in the case of a neutral hydrogenic target. Whereas, McGuire and co-workers adopt the partial wave decomposition of the scattering amplitude, we avoid the use of the partial waves since the method of McGuire and co-workers requires inordinately more computer time where many partial waves are involved. Our procedure leads to a two-dimensional integral for the probability amplitude. Section III contains the conclusions.

II. Theory

The single and double ionization cross sections are evaluated using [6] respectively,

$$\sigma_1 = 2\int d\boldsymbol{b}\; P(\boldsymbol{b})\,[1 - P(\boldsymbol{b})] \tag{1}$$

and

$$\sigma_2 = \int d\boldsymbol{b}\; [P(\boldsymbol{b})]^2 \tag{2}$$

where the probability amplitude $P(\boldsymbol{b})$ is calculated from the following expression :

$$P(\boldsymbol{b}) = |a(\boldsymbol{b})|^2 \tag{3}$$

where

$$a(\boldsymbol{b}) = \int e^{(i\boldsymbol{q}.\boldsymbol{b})}\, f(\boldsymbol{q})\, d\boldsymbol{q} \tag{4}$$

In the present formulation $f(\boldsymbol{q})$ represents the Glauber amplitude for the ionization of a neutral hydrogenic target by a bare ion of velocity v and charge Z_p and is given by [2] :

$$f(\boldsymbol{q}) = -(ik/2\pi)\int d\boldsymbol{b}\; d\boldsymbol{r}\;\; \chi_c^{-*}(\boldsymbol{k}_2, \boldsymbol{r})\,[1 - (|\boldsymbol{b} - \boldsymbol{s}|/b)^{2i\eta}]$$

$$\times\; u_0(\boldsymbol{r})\, e^{(i\boldsymbol{q}.\boldsymbol{b})}, \tag{5}$$

where $\boldsymbol{q} = \boldsymbol{k} - \boldsymbol{k}_1$ and $\eta = -Z_p/v$. Here $\boldsymbol{k}, \boldsymbol{k}_1$ and $\boldsymbol{k}_2$ are the momenta of the incident, scattered and ejected particles, respectively. $\boldsymbol{b}$ and $\boldsymbol{s}$ are the respective projections of the position vectors of the incident particle and the bound electron onto the plane perpendicular to the direction of the Glauber path integration. In Eq. (5), **q**, $\boldsymbol{b}$ and s are co-planer, $u_0(\boldsymbol{r})$ represents the wave function of the ground state of atomic H and is given by

$$u_0(\boldsymbol{r}) = \lambda^{3/2}\; \pi^{-1/2}\; e^{-\lambda r} \tag{6}$$

while $\chi_c^{-*}(\boldsymbol{k}_2, \boldsymbol{r})$ denotes the wave function of the ejected electron and is taken to be the Coulomb wave function

$$\chi_c^{-*}(\boldsymbol{k}_2, \boldsymbol{r}) = (2\pi)^{-3/2}\, e^{\gamma\pi/2}\,\Gamma(1-i\gamma)\, \exp(-i\boldsymbol{k}_2.\boldsymbol{r})\; {}_1F_1(i\gamma, 1, i(k_2 r + \boldsymbol{k}_2.\boldsymbol{r})) \tag{7}$$

with $\lambda = 1$ and $\gamma = Z_N/k_2$, Z_N represents the effective nuclear charge seen by the ejected electron.

Roy *et al.* [7] reduced the five-dimensional integral in Eq. (5) to a two-dimensional integral. Using this two-dimensional expression of the Glauber amplitude we have been able to reduce the integral in Eq. (4) to the following form :

$$a(\boldsymbol{b}) = A b^{-2i\eta-\varepsilon_2} \int d^2p_\rho\, \exp(-i\boldsymbol{b}.\mathbf{p}_\rho)\; p_\rho^{-2-2i\eta+\varepsilon_1}$$

$$\times\, [C_1\, (k_2^2+\lambda^2+p_\rho^2+2\mathbf{p}_\rho.\boldsymbol{k}_{2\rho})^{i\gamma+2}\, (\lambda^2-k_2^2-2i\lambda k_2+p_\rho^2)^{-i\gamma}$$

$$+\, C_2 (k_2^2+\lambda^2+p_\rho^2+2\mathbf{p}_\rho.\boldsymbol{k}_{2\rho})^{i\gamma-1}\, (\lambda^2-k_2^2-2i\lambda k_2+p_\rho^2)^{-i\gamma-1}] \tag{8}$$

$$A = \frac{4\pi^2 C\, D_1 D_2 \Gamma(i\eta+\varepsilon_2/2)\, 4^{i\eta+\varepsilon_2/2}}{\Gamma(1-i\eta-\varepsilon_2/2)}$$

$$C = -2^{-5/2}\, ik\, \pi^{-3} e^{\gamma\pi/2}\, \Gamma(1-i\gamma)$$

$$D_1 = \frac{\Gamma(1/2)\Gamma(2-2i\eta-\varepsilon_2)}{\Gamma(i\eta+\varepsilon_2/2)\Gamma(3/2-i\eta-\varepsilon_2/2)}$$

$$D_2 = \frac{\Gamma(1/2)\Gamma(2+2i\eta-\varepsilon_1)}{\Gamma(-i\eta+\varepsilon_1/2)\Gamma(3/2+i\eta-\varepsilon_1/2)}$$

$$C_1 = 2\lambda(i\gamma-1), \qquad C_2 = -2i\gamma(\lambda - ik_2).$$

Using (8), (3), (2), and (1), both σ_1 and σ_2 can be evaluated numerically by means of suitable Gaussian quadratures.

In helium there are observations of both single and double ionizations. The present procedure can easily be extended to He using hydrogenic wave function for the target.

III. Conclusion

We have presented a method of evaluation of cross sections for single and double ionization of atoms by bare ions. It is based on the Glauber approximation. The present method which avoids the use of partial waves has an advantage over the conventional partial wave technique which requires inordinately more computer time where many partial waves are involved. In the case of a hydrogenic target the probability amplitude turns out to be a simple two-dimensional integral which can be evaluated numerically with ease. Since theoretical calculations of double ionization at the present time are more difficult than single ionization the present simple method should be useful for obtaining estimates of double ionization cross sections.

References

1. See, for example, Chan, F. T., Lieber, M., Foster, G. and Williamson, Jr., W. : Advances in Electronics and Electron Physics, edited by Marton, L. and Marton, C., Vol. 49, p. 133, New York : Academic 1979.
2. Golden, J. E. and McGuire, J. H. : Phys. Rev. Lett. **32**, 1218 (1974).
3. Golden, J. E. and McGuire, J. H. : Phys. Rev. A **12**, 80 (1975).
4. Golden, J. E. and McGuire, J. H. : Phys. Rev. A **15**, 499 (1977).
5. McGuire, J. H. : Phys. Rev. A **26**, 143 (1982).
6. See, for example, McGuire, J. H., Müller, A., Schuch, B., Groh, W., and Salzborn, E. : Phys. Rev. A **35**, 2479 (1987).
7. Roy, A. C., Das, A. K., and Sil, N. C. : Phys. Rev. A **23**, 1662 (1981).

Supplement to Z. Phys. D - Atoms, Molecules and Clusters 21, S 237-238 (1991)

Atoms, Molecules
Zeitschrift für Physik D and Clusters

Distortion approximation for inner-shell ionization cross sections in ion-atom collisions using relativistic pseudostates

T. Mukoyama and C.-D. Lin*
Insitute for Chemical Research, Kyoto University, Uji, Kyoto, 611 Japan

* Department of Physics, Kansas State University, Manhattan, Kansas, 66506, USA

Abstract

Inner-shell ionization cross sections in ion-atom collisions are calculated in the distortion approximation using relativistic pseudostates. The obtained results are compared with the experimental data.

I. Introduction

In theoretical calculations for inner-shell ionization cross sections by heavy charged particles, the distortion effect of the target electrons due to presence of the projectile has usually been taken into account through the so-called binding-energy correction or the united-atom approach within the first-order Born approximation. Recently, we have shown [1] that we can go beyond the simple first Born approximation by using pseudostates, i.e. discretized states with positive energy eigenvalues, instead of final continuum states and calculated the K-shell ionization cross sections in ion-atom collisions in the distortion approximation [2]. It is well known that in the low-energy region where the distortion effect of the target atom is important, the electronic relativistic effect plays also an important role [3]. In the present work, we have developed the distortion approximation using relativistic pseudostates.

II. Theory

The electron wave function is expanded in terms of Slater-type orbitals with non-integer principal quantum numbers and the atomic Dirac Hamiltonian is diagonalized. The obtained atomic orbitals and pseudostates are used to calculate the ionization probabilities as a function of impact parameter. The atomic potential is generated by the Dirac-Fock-Slater method [4]. We have already shown [5] that the finite-basis-set solutions of the atomic Dirac equation by this procedure are quite stable and can yield good approximate energy eigenvalues and eigenfunctions for bound states. It is also tested that the pseudostate wave functions can well reproduce the behaviour of the continuum wave function for small radial distances in the low-energy region.

III. Results

In Fig. 1, the calculated K-shell ionization probabilities for 1-MeV protons on silver are shown as a function of impact parameter and compared with the experimental results of Andersen et al. [6]. The solid curve indicates the results in the distortion approximation with relativistic pseudostates, the dotted curve represents the values in the first Born appro-

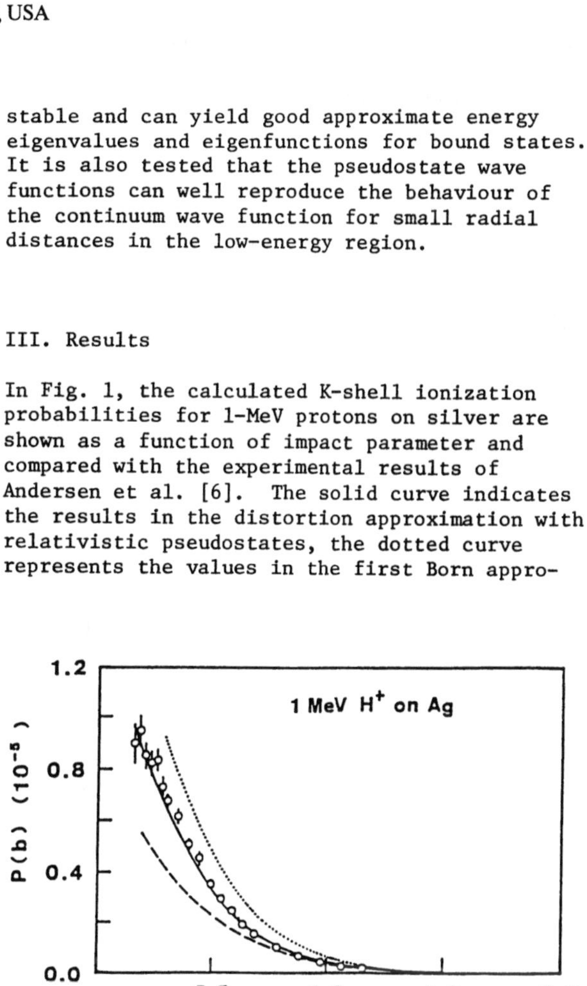

Fig. 1. K-shell ionization probabilities for 1-MeV protons on silver as a function of impact parameter. The solid curve indicates the distortion approximation using relativistic pseudostates, the dotted curve the first Born approximation with relativistic pseudostates, the dashed curve the distortion approximation with nonrelativistic pseudostates. The experimental data are taken from Andersen et al. (Ref. 6).

ximation with the same pseudostates. For comparison the results in the distortion approximation with nonrelativistic pseudostates, based on the model of Ref. 1, are shown in the figure. It is clear that the present calculation in the distortion approximation with relativistic pseudostates is in good agreement with the experimental results. On the other hand, the Born approximation overestimates the ionization probabilities for small impact parameters, while the nonrelativistic calculation underpredicts them.

IV. Conclusion

We have calculated the inner-shell ionization cross sections in ion-atom collisions in the distortion approximation using relativistic pseudostates. The obtained results are in good agreement with the experimental data.

References

1. Mukoyama, T., Lin, C.D.: Phys. Rev. A 40 (1989) 6886.
2. Bates, D.R.: In: Atomic and Molecular Processes. Bates, D.R. (ed.), p.549. New York: Academic 1962
3. Mukoyama, T., Sarkadi, L.: Phys. Rev. A 28 (1983) 1303.
4. Liberman, A., Cromer, D.T., Waber, J.T.: Comput. Phys. Commun. 2 (1971) 107.
5. Mukoyama, T., Lin, C.D.: Phys. Rev. A 35 (]987) 4942.
6. Andersen, J.U., Laegsgaard, E., Lund, M.: Nucl. Instr. and Meth. 192 (1982) 79.

Supplement to Z. Phys. D - Atoms, Molecules and Clusters 21, S 239-240 (1991)

Atoms, Molecules
Zeitschrift für Physik D and Clusters

Two-center effects in electron emission from the impact of fast, multiply-charged projectiles on helium

J. O. P. Pedersen, P. Hvelplund, and A. G. Petersen
Institute of Physics, University of Aarhus, DK-8000 Aarhus C, Denmark

P. D. Fainstein
Centro Atómico Bariloche and Institute Balseiro, RA-8400 Bariloche, Argentina

ABSTRACT

The double-differential cross sections (DDCS) for ejection of electrons from helium by bare ions of H, He, C, and O of 1.00 and 1.84 MeV amu^{-1} energy have been studied experimentally and theoretically. Two-centre effects resulting from the simultaneous projectile and residual target fields are found to be important in the DDCS.

I. INTRODUCTION

The situation, where electrons ejected from ion-atom collisions are subject to the simultaneous attraction of both the projectile and target, has been termed two-centre electron emission (TCEE). Effects of TCEE can be found by studying the projectile charge (q) dependence of DDCS. The first Born approximation (FBA), which treats the projectile-electron interaction as a pertubation, predicts a q^2 scaling, but if the projectile field is important also in the description of the final states of the electron, deviations from the q^2 dependence can be expected.

Effects of TCEE in collisions between heavy ions and helium have been studied experimentally both at low energies (≲120 keV amu^{-1}), with He^{2+} projectiles [1,2], and at high energies (5 MeV amu^{-1}), with C^{6+}, O^{8+}, and Ne^{10+} projectiles [3]. Results also exist for high energies and high charge states with 6 MeV amu^{-1} U^{38+} and Th^{38+} [4] and 25 MeV amu^{-1} Mo^{40+} [5]. It is desirable to use high impact energies when investigating the q^2 scaling because the FBA is known [6] to give good results for proton impact at high energies.

In all experiments, deviations from a q^2 dependence were observed. The results for higher charge states were compared with classical-trajectory Monte-Carlo calculations [4] and continuum-distorted-wave, eikonal-initial-state (CDW-EIS) calculations [7], and from these comparisons it was suggested that the deviations were actually due to TCEE effects.

Previous measurements, using high charge states and high impact energies, however, covered mainly emitted electrons with velocities smaller than the projectile velocity, and the high velocity structures in the DDCS could therefore not be observed.

II. EXPERIMENT

The experiments were performed at the Aarhus Tandem van de Graaff accelerator. We have used a electron spectrometer which could record spectra of electron energies at different observation angles between 20° and 160° with respect to the incoming projectile beam. Electron energies between 2 and 2500 eV were measured, thus including both the Electron-Capture-to-the-Continuum (ECC) region, where the eletrons have the same velocity as the projectile, and the binary-encounter peak.

III. MODEL

In the present CDW-EIS calculations, it is assumed that one electron remains frozen during the collision, while the active electron moves in an effective potential. The model accounts for the two Coulomb centers by distorting the initial wave function by an eikonal phase and by representing the final state as a product of Coulombic continu-

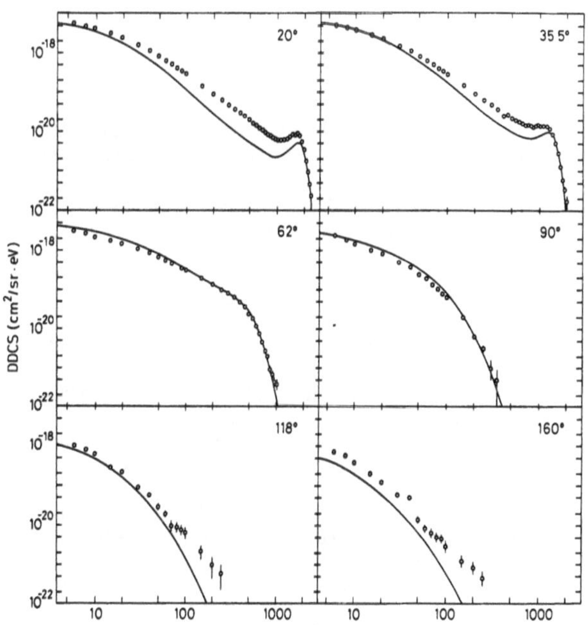

Fig. 1. DDCS for 1 MeV amu^{-1} O^{8+} impact on He as a function of electron energy for different emission angles. The points are experimental data and the solid lines CDW-EIS calculations

um waves, centred both at the target and projectile [7].

IV. RESULTS AND DISCUSSION

A full description of the experimental technique, the model calculations and a detailled comparison of the experimental and theoretical results has been given by Pedersen et al [8,9]. As an example of the results obtained, we present in Fig. 1 the DDCS for a single energy and projectile charge state, namely 1 MeV amu^{-1} O^{8+}. The peak at high electron energies in the spectra for 20^0 and 35.5^0 emission is the binary-encounter peak. The hump around 500 eV for 62^0 emission is a reminiscence of the peak. The general agreement between experiment and the CDW-EIS calculation is seen to be rather good.

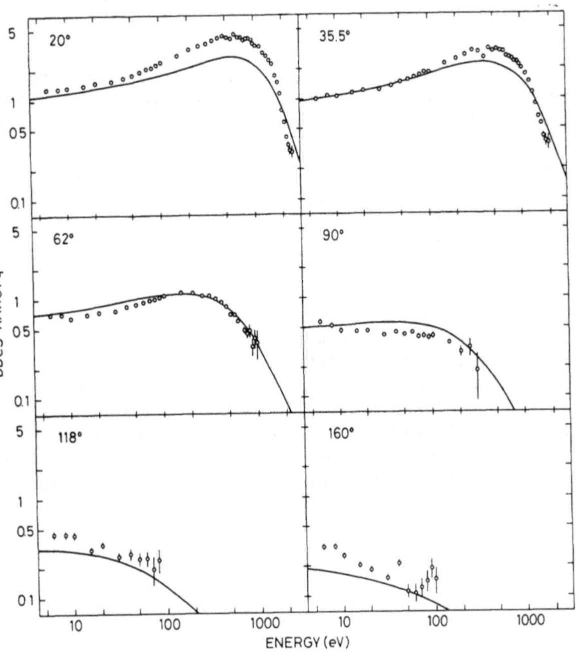

Fig. 2. Ratio between the DDCS for the 1 MeV amu^{-1} O^{8+}+He (from Fig. 1) and p^++He systems, divided by 64. Experimental ratios are shown as points and CDW-EIS calculations as solid lines.

In order to show deviations from the q^2 dependence more clearly, we present in Fig. 2 the DDCS ratios between those for O^{8+} and p^+ impact. The ratios have been divided by q^2 (=64). The FBA would give a constant ratio of one for all ejection energies and angles, and is seen to give good results for small electron energies, below about 100 eV. For large electron energies, however, the emission is reduced compared to the FBA at large emission angles, and the deviation is seen to increase both with electron energy and emission angle. For forward emission angles, the emission is enhanced at small energies, and the ratio reaches a maximum at an electron energy corresponding to the ECC peak energy (545 eV). For higher energies, the ratio decreases again and reaches a value of one at the energy corresponding to the binary-encounter peak, where the DDCS is again well predicted by the FBA. At energies above the binary peak, the ratio further decreases.

The CDW-EIS calculations are seen to reproduce the general behaviour of the experimental data, and the agreement with the experimental ratios is much better that the FBA. The energy intervals, where the experimental and model ratios are different, coincide with the differences in the absolute DDCS in Fig. 1.

V. CONCLUSIONS

The general behaviour of the ratio may be visualised in a classical picture: The enhancement at small emission angles and the reduction at large angles result from a deflection of the emitted electron due to the strong attraction by the projectile. In the forward direction, the influence of the projectile is strongest in the ECC region, where the electrons have a velocity comparable to the projectile. At even higher energies, the electrons feel the potential due to both Coulomb centers, and the ratio decreases again. These effects are included in the CDW-EIS model which gives a good description of the experimental data.

JOPP would like to acknowledge support from the Carlsberg Foundation, and PDF has been partially supported by the International Atomic Energy Agency, Research Contract No 5365/R1/RB.

REFERENCES

1) Bernardi, C.G., Suárez, S., Fainstein, P.D., Garibotti, C.R., Meckbach, W., and Focke, P., Phys.Rev. A 40, 6863 (1988).
2) Irby, V.D., Gay, T.J., Edwards, J.Wm., Hale, E.B., McKenzie, M.L., and Olson, R.E., Phys. Rev. A 37, 3612 (1988).
3) Schiwietz, G., Platten, H., Schneider, D., Schneider, T., Zeitz, W., Musiol, K., Kowallik, R., and Stolterfoht, N., Hahn-Meitner-Institut Report No. HMI-B447, Berlin (West), unpublished (1987).
4) Schneider, D., DeWitt, D., Schlachter, A.S., Olson, R.E., Graham, W.G., Mowat, J.R., DuBois, R.D., Loyd, D.H., Montemayor, V., and Schiwietz, G., Phys.Rev. A 40, 2971 (1989).
5) Stolterfoht, N., Schneider, D., Tanis, J., Altevogt, H., Salin, A., Fainstein, P.D., Rivarola, R., Grandin, J.P., Scheurer, J.N., Andriamonje, S., Bertault, D., and Chemin, J.F., Europhys.Lett. 4, 899 (1987).
6) Manson, S.T., Toburen, L.H., Madison, D.H., and Stolterfoht, N., Phys.Rev. A 12, 60 (1975).
7) Fainstein, P.D., Ponce, V.H., and Rivarola, R.D., J.Phys.B 21, 287 (1988).
8) Pedersen, J.O.P., Hvelplund, P., Petersen, A.G., and Fainstein, P.D., J.Phys.B L, in press, (1990).
9) Pedersen, J.O.P., Hvelplund, P., Petersen, A.G, and Fainstein, P.D., preprint, to be submitted to J.Phys.B (1990).

Supplement to Z. Phys. D - Atoms, Molecules and Clusters 21, S 241-242 (1991)

Atoms, Molecules
Zeitschrift für Physik D and Clusters

Two-electron transitions in high energy ion–helium collisions

A. L. Ford and J. F. Reading
Physics Department, Texas A&M University, College Station, TX 77843, USA

Abstract

Recent calculations and experiments on two-electron transitions in helium are reviewed. For double ionization and excitation ionization there is a large difference (about a factor of two) between the cross sections produced by protons and antiprotons. This difference is due to interference between first and second Born amplitudes, and correlation is required for first Born to be nonzero. The double ionization to single ionization ratio and the proton/antiproton difference is much larger for probabilities at small impact parameter than for the cross sections. The situation is qualitatively different for double excitation, where experimental and theoretical evidence points to a very small interference term and hence little positive projectile/negative projectile difference.

Results and Discussion

We have previously published coupled-states calculations that include the effects of electron correlation in the double ionization of helium by bare ions with MeV/u energy [1]. Correlation is essential to describe the difference in double ionization produced by protons versus antiprotons, and our calculations reproduce reasonally well the experimentally observed difference in the total cross sections. These calculations required the invention and implementation of two new theoretical methods. One is our forced-impulse method (FIM) for including dynamical electron correlation during the collision without the unacceptable computational expense of direct expansion in two-electron pseudostates. In the FIM the electron wavefunctions evolve independently (IPM) for short time steps; at the start of each of these time evolution steps the wavefunction is projected onto fully correlated states. The other new method, required for calculating double ionization, is projection of our continuum pseudostates onto single and double ionization channel states, constructed to satisfy the appropriate boundary conditions. This is necessary for solving the interpolation problem, of separating between the two overlapping single and double ionization continua.

An analysis of the impact parameter dependent probabilities $P(b)$ for single and double ionization that go into the calculation of the above cross sections has just been published [2]. We find that the double to single ionization ratio and the proton-antiproton difference is much larger in $P(b)$ at small impact parameter b than observed previously in the cross sections. The mechanisms for producing double ionization and that give rise to the proton-antiproton difference are most effective at small impact parameters. Experiments that probe small b would therefore be of great interest, particularly for antiprotons or bare projectile ions of larger charge, such as He^{2+} or C^{6+}. One such experiment is the double ionization cross section differential in the projectile scattering angle, $d\sigma/d\Omega$, and such measurements have in fact been done using proton projectiles [3,4]. However, our straight-line trajectory, classical path calculations cannot be converted directly to a cross section differential in projectile scattering angle using the classical relation between impact parameter and scattering angle for Coulomb scattering. The difficulty is that scattering of the projectile by close collisions with the target electrons, which results in the electrons being ejected with large kinetic energy, appears to have an important effect on $d\sigma/d\Omega$ for both single and double ionization [5,6]. It may be nonetheless that at very large scattering angles, well beyond the angle to which a proton can be scattered classically by a free electron at rest, the scattering of the projectile is dominated by the Coulomb interaction with the helium nucleus. If this is indeed the case, then the He^{2+}/He^{+} charge-state fraction observed in coincidence with these large projectile scattering angles (greater than 3 mrad) should agree with our calculated ratio of double to single ionization probabilities at small impact parameters. There is indeed fair agreement between our calculations and experiments. At $E = 1.0$ MeV the experimental ratio [3] is $1.3 \pm 0.2\%$; our calculation gives 0.9%. At $E = 0.30$ MeV the experimental result is $2.4 \pm 0.2\%$; our calculation gives 1.8% The reason for our calculations being somewhat low is probably the truncation of our basis to s and p states only. Similar measurements using antiprotons would be of considerable interest. Our calculations predict much larger values for this ratio for antiproton projectiles: 3.3% at $E = 1.0$

MeV and 5.5% at $E = 0.3$ MeV. Also of interest would be similar measurements for projectiles of larger charge; we predict a double to single ratio at small b (large projectile scattering angle) of 74% for C^{6+} at 0.64 MeV/u. Another possible experiment to probe the same effect would be a collision between neutral helium and a partially stripped ion, such as nitrogen, oxygen or neon. Double ionization of the helium in coincidence with electron removal from the projectile ion would select out small impact parameters, of the order of the orbital radius for the ion. Our calculations suggest substantial enhancement of the double to single ratio compared with that measured without the simultaneous removal of a projectile electron, if the projectile ion is stripped down to orbitals with radius much smaller than the helium Bohr radius.

Double ionization is a two-electron transition and therefore requires a minimum of two interactions, one to excite each electron. Two different interactions are available: the projectile-electron interaction and the electron-electron interaction (correlation). Our cross sections calculated for specific projectile charge Z_p can be analyzed by fitting them to an expansion in Z_p. The leading Z_p^2 term is first Born, first-order in the projectile interaction, and therefore must involve the correlation interaction at least to first order. The Z_p^4 term (second Born) comes from an amplitude second order in the projectile interaction. In this amplitude the projectile can interact once with each electron and no correlation interaction is required, or the projectile can interact twice with the same electron but then correlation at least to first order is required to get nonzero double ionization. If the correlation interaction is weak then the former contribution dominates. The Z_p^3 term comes from interference between these first and second Born amplitudes, and it is this term that is responsible for the proton/antiproton double ionization difference. It is also important to note that whereas dipole transitions ($s \rightarrow p$ etc.) dominate for single ionization, a second Born amplitude of the correct symmetry to interfere with the first Born requires one projectile-induced nondipole transition in either the first or second Born amplitude. The relative importance of the Z_p^2 versus Z_p^4 term changes with projectile energy and charge; different collisions probe different double ionization mechanisms.

Very recent measurements [7] of the two-electron process of excitation ionization in helium show the same sort of $Z_p = +1$ compared to $Z_p = -1$ (electrons) differences as seen in double ionization. On the other hand, recent measurements [8,9] of the two-electron transition to doubly excited resonance states of helium $(2s2p\,{}^1P + (2p)^2\,{}^1D)$ show very little $Z_p = +1$ versus $Z_p = -1$ difference. Preliminary calculations we have performed in the FIM agree qualitatively with the latter experiments, as do also coupled-channels calculations of Fritsch and Lin [10]. Our double excitation calculations show a very small Z_p^3 interference term at the experimental collision energy compared to the situation for double ionization, hence the absence of an effect due to the sign of the projectile charge. Model calculations by Pedersen [11] suggest that the lack of interference is due to different impact parameter dependence of the first and second Born amplitudes. He finds that the first Born amplitude is dominated by large impact parameters, whereas the second Born is dominated by small impact parameters. More theoretical work is needed before quantitative comparison can be made between collision calculations and the experiments, that measure the yield of ionized and autoionized electrons. The resonance states in question are embedded in the single ionization continuum, so there is a similar interpretation problem as that due to the overlapping single and double ionization continua.

Acknowledgments

This work was supported by NSF Grant PHY-8707383.

References

1. Reading J.F., Ford A.L.: J. Phys. B **20**, 3747 (1987); Ford A.L., Reading J. F.: J. Phys. B **21** L685 (1988)
2. Ford A.L., Reading J.F.: J. Phys. B **23**, 2567 (1990)
3. Giese J.P., Horsdal E.: Phys. Rev. Lett. **60**, 2018 (1988)
4. Kamber E.Y., Cocke C.L., Cheng S., Varghese S.L.: Phys. Rev. Lett. **60**, 2026 (1988)
5. Reading J.F., Ford A.L., Fang X.: Phys. Rev. Lett. **62**, 245 (1989)
6. Olson R.E., Ullrich J., Dörner R, Schmidt-Böcking H.: Phys. Rev. **A40**, 2843 (1989)
7. Pedersen J.O.P., Folkmann F.: J. Phys. B **23**, 441 (1990)
8. Giese J.P., Schulz M., Swenson J.K., Schöne H., Benhenni M., Varghese S.L., Vane C.R., Dittner P.F., Shafroth S.M., Datz S.: Phys. Rev. **A42**, 1231 (1990)
9. Pedersen J.O.P., Hvelplund P.: Phys. Rev. Lett. **62**, 2373 (1989)
10. Fritsch W., Lin C.D.: Phys. Rev. **41**, 4776 (1990)
11. Pedersen J.O.P.: Phys. Scr. **42**, 180 (1990)

Supplement to Z. Phys. D - Atoms, Molecules and Clusters 21, S 243-244 (1991)

Atoms, Molecules
Zeitschrift für Physik D and Clusters

Electron capture in slow collisions of multicharged ions with hydrogen atoms using merged beams

C. C. Havener, M. P. Nesnidal, M. R. Porter and R. A. Phaneuf
Physics Division, Oak Ridge National Laboratory, P. O. Box 2008, Oak Ridge, TN 37831-6372, USA

Abstract

Absolute total electron-capture cross-section measurements are reported for collisions of O^{3+} and O^{4+} with atomic hydrogen in the energy range 1-1000 eV/amu using merged beams. The data are compared with available coupled-states theoretical calculations.

I. Introduction

Electron capture by multiply charged ions from neutral atoms at near-thermal collision energies plays an important role in many plasma environments, and presents a challenge to our understanding of atomic collision processes. The process for capture from atomic hydrogen is denoted by:

$$X^{+q} + H \rightarrow X^{+q-1} + H^{+}.$$

Approximations which are useful in describing this process at higher collision energies become inappropriate as the time available for interaction becomes large compared to the atomic orbital periods. Reactant trajectories and the electronic structure of the temporary quasi-molecule formed in the collision must be accurately described and taken into account [1]. The attractive polarization potential in the initial state, proportional to q^2/r^4, where q is the ionic charge and r is the internuclear separation, enables collisions to sample values of r smaller than the impact parameter. Depending on the electronic structure of the system, this effect may increase the cross section and give rise to resonances at near-thermal energies [2]. Experimental measurements are needed in this low-energy region to verify these phenomena and to benchmark detailed theoretical calculations.

II. Experiment

The merged-beams approach which has been developed at ORNL to address this problem has been described in detail [3]. A beam of pure ground-state H atoms is created by intra-cavity laser photodetachment of a 6-9 keV beam of H^-. This neutral beam is merged electrostatically with a 25-75 keV beam of multiply charged ions from the ORNL-ECR ion source, and the beams remain merged in a region of ultrahigh vacuum ($<2\times10^{-8}$ Pa) for a common path of 80 cm. Two-dimensional intensity distributions of each beam are measured using rotating-wire and moving-slit scanners at four positions along the merged path. A dipole magnet demerges the beams, and directs the multicharged ions to a Faraday cup where their intensity is measured. The intensity of the primary neutral H beam is measured by secondary-electron emission from a stainless steel surface, which is calibrated *in situ*. The demerging magnet also deflects the H^+ reaction products into an electrostatic analyzer which removes them from the magnet dispersion plane and directs them to a channel electron multiplier where they are counted. A phased fast two-beam chopping scheme separates H^+ produced by beam-beam collisions from background H^+ produced by stripping on residual gas. Each cross-section measurement is independently absolute, since all experimental parameters are measured.

One advantage of the fast merged beams technique for the study of low-energy collisions is the compression of the product angular distribution effected by the transformation from the center-of-mass frame, where the collision occurs, to the laboratory frame, where the products are detected. This compression increases as the relative collision velocity decreases, and is especially important for very low-energy electron capture by multiply charged ions due to the exoergicity of the reaction and the Coulomb repulsion of the products. In the present technique, two electrostatic lenses form a "telescope" to increase the angular acceptance for the collision products. Careful ion trajectory modelling [3] indicates a mean angular acceptance of 1.8 degrees over the merge path, which corresponds to an acceptance of at least 90 degrees in the center-of-mass frame for the measurements reported here at 1 eV/amu. Such modelling, together with the ability to reproduce a given center-of-mass velocity for different laboratory conditions (using H or D, faster or slower than the ion beam), gives confidence that a significant fraction of the reaction products are not being lost.

III. Results and Discussion

The present electron-capture cross-section measurements for O^{3+} + H(D) collisions are compared in Figure 1 to previous measurements [4] and theoretical calculations [5],[6] for this system. The merged-beams data cover an energy range from 0.9 to 1500 eV/amu, demonstrating the wide dynamic range accessible to this technique. The comparison of theory and experiment in this case is complicated by the expected presence of a significant metastable fraction in the O^{3+} beam from the ECR ion source. The quantum-mechanical coupled-states

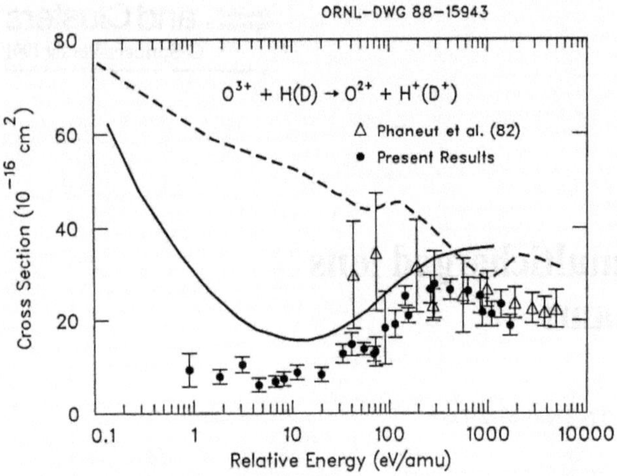

Figure 1. Comparison of present merged-beams measurements with other experimental results [4] and with theoretical calculations of Bienstock et al [5] (dashed curve) and of Gargaud et al [6] (solid curve).

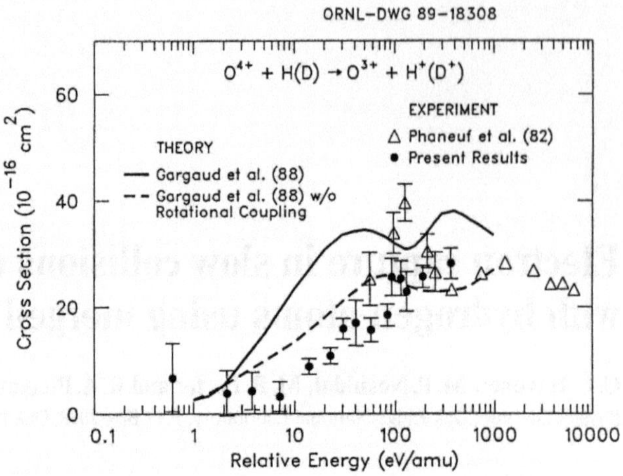

Figure 2. Comparison of present merged-beams measurements with other experimental results [4] and with theoretical calculations of Gargaud and McCarroll [8].

calculations of Bienstock et al. differ from those of Gargaud et al. by a factor of two or more over much of the energy range indicated. While the present data favor the calculations of Gargaud et al., there is still a significant discrepancy which becomes more serious at the lower energies. Their calculations indicate that capture goes predominantly into the 3s state at energies above 20 eV/amu, with capture to 3p producing the rise at decreasing energies below 10 eV/amu. The present data suggest that the 3p contribution may have been over-estimated in these calculations. In fact, a cross-section curve which adds only 1/3 of the calculated 3p contribution to the calculated 3s contribution describes the measurements well, both in magnitude and energy dependence. Recent state-selective measurements based on photon spectroscopy also suggest a smaller 3p contribution than predicted for this system [7].

Figure 2 compares present merged-beams results for O^{4+} + H(D) collisions with previous measurements [4] and theoretical calculations [8]. Experiment and theory are in better agreement when the effects of rotational coupling are excluded from the calculation. Again, however, the presence of a significant metastable fraction in the reactant O^{4+} beam complicates the comparison.

IV. Improvements

Based on differential-cross-section calculations for the O^{5+} + H system, Andersson et al [9] have suggested that the angular acceptance of the merged-beams apparatus [3] may have been insufficient in that case to collect all products in the energy range 2-10 eV/amu. To resolve this question, a number of modifications have been made to the apparatus to increase the angular acceptance and to improve the signal-to-noise ratio. These changes should permit the technique to be extended to lower energies and to more highly charged ions, where the strength of the attractive polarization potential is enhanced, and resonance effects may be observable in the total cross section.

V. Acknowledgments

This research was supported by the Division of Chemical Sciences, U.S. Department of Energy, under contract No. DE-AC05-84OR21400 with Martin Marietta Energy Systems, Inc. M. P. Nesnidal was sponsored by the Great Lakes Colleges Association/Associated Colleges of the Midwest. M. R. Porter was sponsored by Science Alliance, a state of Tennessee Center of Excellence, through the University of Tennessee.

References

1. Kimura, M. and Lane, N.F.: Adv. Atom. Molec. Phys. **26**, 79-159 (1990).
2. Rittby, M., Elander, N., Brandås, E. and Bárány, A.: J. Phys. B **17**, L677 (1984).
3. Havener, C.C., Krause, H.F., Schulz, P.A. and Phaneuf, R.A.: Phys. Rev. A **39**, 1725 (1989).
4. Phaneuf, R.A., Alvarez, I., Meyer, F.W. and Crandall, D.H.: Phys. Rev. A **26**, 1892 (1982).
5. Bienstock, S., Heil, T.G. and Dalgarno, A.: Phys. Rev. A **27**, 2741 (1983).
6. Gargaud, M., McCarroll, R. and Opradolce, L.: Astron. Astrophys. **208**, 251 (1989).
7. R. Hoeckstra, KVI Groningen (private communication, this conference).
8. Gargaud, M. and McCarroll, R.: J. Phys. B **21**, 513 (1988).
9. Andersson, L.R., Gargaud, M. and McCarroll, R.: XVI ICPEAC, Abstracts of Contributed Papers, New York (1989), p. 566.

Supplement to Z. Phys. D - Atoms, Molecules and Clusters 21, S 245-246 (1991)

Atoms, Molecules
Zeitschrift für Physik D and Clusters

Partial charge exchange and excitation cross sections for C^{3+} + H collisions

L. F. Errea, B. Herrero, L. Méndez, O. Mó and A. Riera
Departamento de Química, Universidad Autónoma de Madrid, Cantoblanco, 28049 Madrid, Spain

Abstract

C^{3+} + H collisions are analyzed in the low-intermediate energy range where a Common Translation Factor (CTF) modified molecular approach is adequate. Following a previous conclusion [1] we present partial charge exchange (C.E.) and excitation cross sections for relevant channels.

Method and molecular data

We have expanded, within the impact parameter formalism, the electronic wavefunction in terms of 22 molecular states multiplied by the CTF of Errea et al. [2].

An effective two-electron hamiltonian was chosen as in [1]. Molecular variational states were obtained using a full configuration interaction (CI) treatment in the atomic basis of (49) s, p and d contracted Gaussian Type Orbitals, including angular correlation effects.

Figure 1 shows the energy correlation diagram (C.D.) for the triplet states involved in the dynamical treatment. An analogous singlet C.D. enlarged with sigma states dissociating into $C^{2+}(1s^2 2p^2; {}^1S, {}^1D) + H^+$ has been considered.

Evaluation of the dynamical couplings [3,4] and integration of the coupled differential equations [5] were performed as in [1].

Cross sections

Fig. 2 presents our calculated total C.E. cross sections, compared with the experimental data [6,7].

We also show our results for the total excitation and for the contribution of triplet subsystem to the C.E. sections, together with the corresponding values of Opradolce et al. [8] and the measurements of Ciric et al. [7]. Overall agreement is found even for the triplet subsystem.

At low ($v < 0.2$ a.u.) velocities C.E. yields almost exclusively $C^{2+}(2s3s; {}^3S)$ as a consecuence of the radial coupling between this state and the entrance channel. Besides the lorentzian peak around the pseudocrossing ($R_0 = 11.4$ a.u.), this coupling presents in the inner R_0-region where energy curves are very close a tail due to small delocalization effects that, though very small, accounts for effective transitions. This can be checked looking at the estimated Landau-Zener section that gives values that are much smaller than the

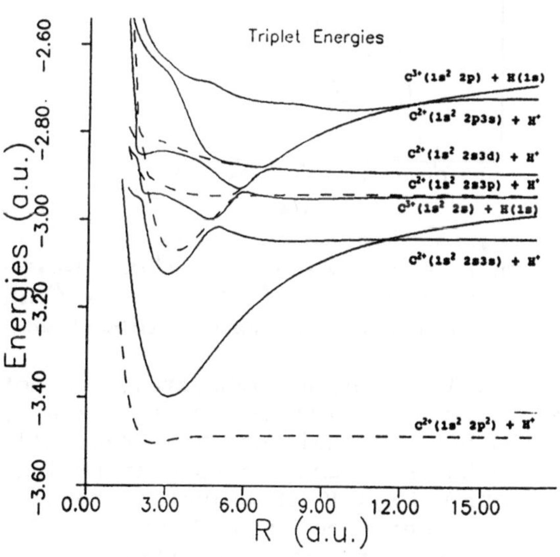

Figure 1. Correlation diagram for triplet states. —— Sigma states. - - - Pi states.

calculated except for very low velocities (v = 0.04 a.u.).

At higher impact energies (v > 0.2 a.u.) all other radial and rotational couplings become effective and transitions to other C.E. or excitations channels occur at small R values.

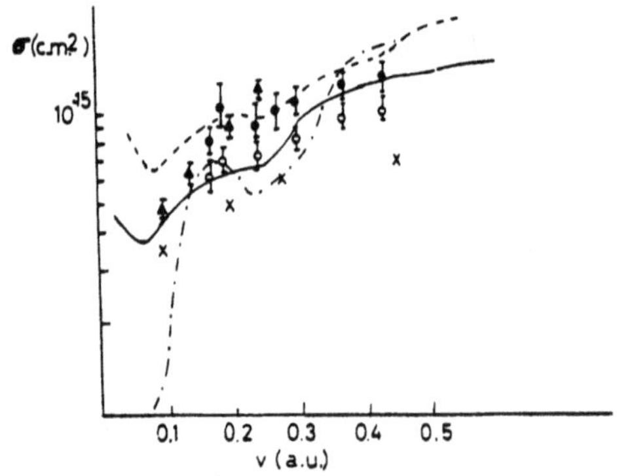

Figure 2. Total cross sections. - - - Our results; ▲ Mc Cullough et al[6]; ● Ciric et al.[7]; C.E. for triplets: —— This work; o Ciric et al.[7]; x Opradolce et al.[8];- · - This work total excitation (multiplied by 10).

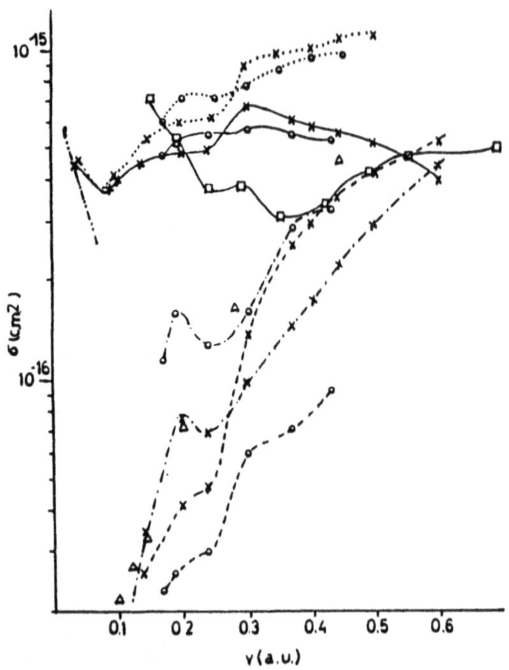

Figure 3. Triplet cross sections. x Our results; o Ciric et al [7]; □ Mc. Cullough et al.[6]; ... $C^{2+}(1s^2,2s3s+2s3p+2s3d)$; — $C^{2+}(1s^2,2s3s)$; - - - $C^{2+}(1s^2, 2s3p)$; — · — $C^{2+}(1s^2, 2s3d)$; ▲ Opradolce et al. $C^{2+}(1s^2,2s3p)$; —·· Estimated Landau-Zener.

In the singlet subsystem, at not too large velocities (v<0.2 a.u.), collisional transition preferentially occurs to the channel dissociating into $C^{2+}(1s^2 2p^2)$ (^1S)+ H^+ that avoids crossing with the entrance channel at R = 5.5 a.u. At higher energies other CE and excitation channels are reached by a similar mechanism to that for triplets.

Fig. 3 and 4 show our calculated C.E. partial cross sections for the triplet and singlet subsystems respectively, compared to the experimental data [6,7].

Agreement with experimental data of Ciric et al. is good for the C.E. sections to the manifold $C^{2+}(1s^2\ 2snl,\ n=3)$ although partial discrepances appear for particular submagnetic levels.

Acknowledgments

This research was partially supported by the DGICYT proyect PB 87-0112.

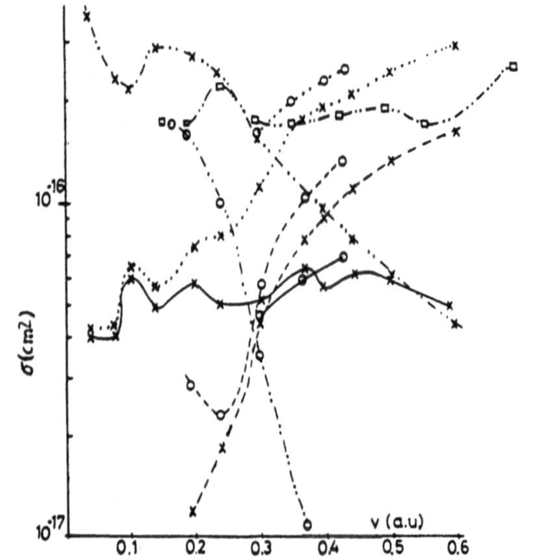

Figure 4. Singlet cross sections. x Our results; o Ciric et al [7]; □ Mc Cullough et al.[6]; ... $(1s^2,2s3s+2s3p+2s3d)$; —·· $C^{2+}(1s^2, 2p^2)$; — $C^{2+}(1s^2, 2s3s)$; - - - $C^{2+}(1s^2, 2s3p)$.

Bibliography

[1] Errea L.F., Méndez L. O. Mó and Riera A., Phys. Scrip. T28 67 (1989).

[2] Errea L.F., Méndez L. and Riera A., J.Phys. B 15 101 (1982).

[3] Macias A. and Riera A., Phys. Rep. 80 229 (1982).

[4] Mó O. and Riera A., J. Mol. Struct. (Theochem) 164 135 (1988).

[5] Gaussorgues C., Piacentini R.D. and Salin A., Comput. Phys. Comm., 10 224 (1975).

[6] Mc Cullogh R.W., Wilkie F.G. and Gilbody H.B., J. Phys. B17 1373 (1984). see also Gilbody H.B., Nucl. Instr. Meth. Phys. Res. B23 29 (1987).

[7] Ciric D., Brazuck A., Dijkkamp D., de Heer F.J. and Winter H., J. Phys. B18 3629 (1985).

[8] Opradolce L., Benmeuraiem L., Mc Carroll R. and Piacentini R.D., J. Phys. B21 503 (1988).

Supplement to Z. Phys. D - Atoms, Molecules and Clusters 21, S 247-248 (1991)

Atoms, Molecules
Zeitschrift für Physik D and Clusters

Diabatic description of electron capture by highly-charged ions from hydrogen

L. R. Andersson
Manne Siegbahn Institute of Physics, S-104 05 Stockholm, Sweden
R. McCarroll
Laboratoire de Dynamique Moleculaire et Atomique, Université Pierre et Marie Curie, T12-E5, F-75252 Paris Cedex 05, France

Abstract

A diabatic formulation of electron capture by highly charged ions from atomic hydrogen at low energies is presented. This takes account of Stark mixing and polarization effects with a minimal size of basis set. The method is applied to the C^{4+}/H system in the energy range 88–589 eV/amu. The results agree with earlier calculations based on an adiabatic formulation but do not resolve discrepancies with experiment.

I. Introduction

State selective electron capture by highly charged ions from atomic hydrogen targets is most simply conceptualized as a non-adiabatic transition in the vicinity of an avoided crossing of adiabatic potential energy curves. However, although calculations using an adiabatic state basis are successful in explaining the main features of the collision cross sections, practical applications are often beset with numerical difficulties. Amongst these are problems linked to the existence of very long range (diabatic) crossings where the non adiabatic coupling is highly peaked. Not only can the integration of the coupled differential equations (either in a semi-classical or quantum mechanical formulation) be quite time consuming, but non physical spurious effects can easily arise. In principle, the problem can be overcome by introducing diabatic states which make it much easier to exclude weakly coupled channels from the basis set.

However, if any real gain is to be achieved, some direct way of generating the diabatic states is essential. The simplest diabatic representation is that based on two-centre atomic wave functions, but given the importance of Stark mixing of the excited states of the final channel and strong polarization effects in the entry channel, it is unsatisfactory since a large number of states are effectively coupled even at low energies. In this work we propose a variant of the model potential approach to generate diabatic molecular orbitals of a kind which reduces the basis set to minimal size. This approach should also be particularly advantageous for the description of orientation and alignment effects, where very long range interactions are involved.

The method is applied to the C^{4+}/H system, which has been studied experimentally recently by Hoekstra *et al.* [1]. Their results show significant differences from the theoretical results of Gargaud *et al.* [2] at energies of around 100 eV/amu. Our objective is two-fold: firstly to test the validity of our diabatic representation, secondly to provide an independent check of the theory in the energy range where the disagreement with experiment is appreciable.

II. Theory

The diabatic states are obtained by a partial diagonalization of the total one-electron Hamiltonian. The states which correlate to the same centre in the asymptotic region are diagonalized in a set of basis functions centred on the ion for the final channels and on the hydrogen atom for the initial channel. The core of the highly charged ion is described by a model potential. The non-diagonal elements of the electronic Hamiltonian as well as the radial and rotational couplings are then calculated between these diabatic states. By diagonalizing the diabatic Hamiltonian matrix we recover the adiabatic states. In order to avoid the build up of orbitals localized on the wrong center, the basis set on the highly charged projectile is made orthogonal to the hydrogen orbital by a projection operation. For a number of systems with a highly charged projectile and a hydrogen target this procedure has proved useful down to about 6 a.u. For smaller distances the basis functions on the hydrogen nucleus start to build orbitals on the projectile ion. It should in principle be possible to overcome this difficulty by an extension of the projection procedure.

Because of the dependence of the radial and rotational couplings on the origin of electron coordinates, allowance must be made for translation effects. This has been achieved by the adoption of appropriate reaction coordinates as in Gargaud *et al.* [2]. In a semi-classical formulation of the collision dynamics, the use of reaction coordinates is equivalent to the introducion of a common translation factor. It should be remarked that, contrary to the case of an adiabatic representation,

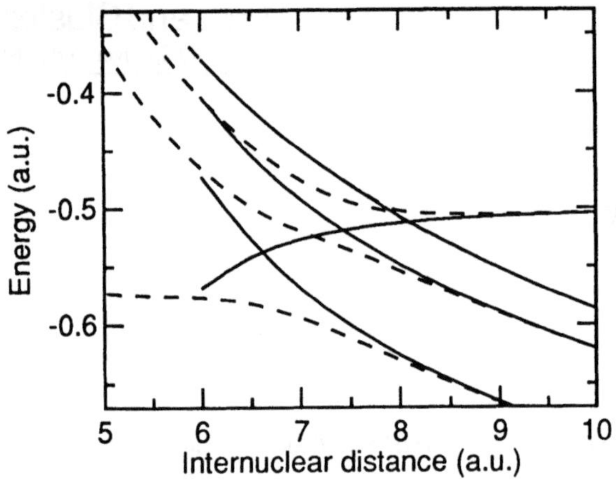

Figure 1: The diabatic (full curve) and adiabatic (broken curve) states of the C^{4+}/H system. From bottom and upwards the curves correlate to 3s,3p and 3d on C^{3+} and 1s on H, respectively. On this scale the adiabatic curves correlating to H(1s) and C^{3+}(3d) cannot be distinguished from the diabatic curve correlating to C^{3+}(3d).

there is no conceptual difficulty in introducing state dependent translation factors of the plane wave type.

The impact parameter equations are solved with a straight line trajectory using the diabatic states down to 6–8 a.u., where we switch to the adiabatic representation. There is no problem of probability conservation since there is an exact transformation connecting the diabatic and adiabatic states. The results are not significantly sensitive to the switching radius. The state-selective cross sections are obtained in the usual way by integrating the transition probabilities over the impact parameter.

III. Results

In figure 1 we show the diabatic (full line) and adiabatic (broken line) states for the C^{4+}/H system. The adiabatic states are identical to the ones calculated by Gargaud *et al.* [2] The diabatic curves are given down to 6 a.u. only. The procedure of calculating these breaks down at smaller distances, where the adiabatic states must be used in the dynamical calculation.

We have calculated the cross sections in the energy, region 88–589 eV/amu. The experiments by Hoekstra *et al.* [1] show a dip in the cross section around 100 eV/amu, while the calculations by Gargaud *et al.* predict a single maximum. The present calculations serve as a test of the diabatic states and as an independent check of the earlier calculations. The adiabatic states were used inside 7.5 a.u. We have checked that this distance is not critical to the results. In table 1 we compare the state-selective cross sections at 88–589 eV/amu with the ones reported by Gargaud *et al.* There are small deviations, but the general trend supports the earlier calculations.

Energy (eV/amu)	Cross sections (10^{-16} cm^2) 3s	3p	3d	Total
88	1.2^a	26.1^a	2.0^a	29.3^a
	1.6^b	28.7^b	1.7^b	32.0^b
124	2.5^a	30.9^a	2.3^a	35.7^a
295	5.2^a	27.7^a	6.0^a	38.9^a
	5.9^b	29.0^b	6.1^b	41.0^b
589	7.2^a	22.8^a	8.2^a	38.2^a
	8.9^b	21.1^b	9.3^b	39.4^b

Table 1: Cross sections for one-electron charge transfer in C^{4+}/H collisions. Present results (a), Gargaud *et al.* [2] (b).

The dip observed experimentally [1] is not reproduced.

IV. Conclusion

Calculations of state selective one-electron capture cross sections in C^{4+}/H collisions using a diabatic formulation support earlier findings from a fully quantum mechanical treatment using an adiabatic basis. However, disagreement with experiment is still unresolved. In particular, there is no theoretical evidence of more than one maximum in the cross section for capture to $C^{3+}(3p)$. We can therefore offer no speculative explanation as to why the experimental cross section exhibits an upturn for energies less than 100 eV/amu.

References

1. Hoekstra R., Beijers J.P.M., Schlatmann A.R., Morgenstern R., de Heer F.J.: Phys. Rev. **A41.** 4800 (1990)
2. Gargaud M., McCarroll R., Valiron P.: J. Phys. **B20.** 1555 (1987)

Supplement to Z. Phys. D - Atoms, Molecules and Clusters 21, S 249-251 (1991)

Atoms, Molecules and Clusters
Zeitschrift für Physik D

Neutralization efficiency of H^- beams based on measurements of one- and two-electron removal from H^- in $H^- + Ar^{q+}$ ($q \leq 8$) collisions

F. Melchert, W. Debus, S. Krüdener, R. Schulze, R. E. Olson* and E. Salzborn
Institut für Kernphysik, Strahlenzentrum der Justus-Liebig-Universität, W-6300 Giessen, FRG

* Physics Department, University of Missouri Rolla, Rolla, MO 65401, USA

Abstract
Employing the Giessen ion–ion crossed–beams facility, absolute cross sections have been measured for single– and double–electron removal from H^- in energetic collisions with Ar^{q+} ($q \leq 8$) ions. The data is compared to CTMC calculations and is discussed with respect to conversion efficencies of H^- into H^0 beams in plasma neutralizers proposed for efficient neutral beam heating of next generation fusion plasmas.

I. Introduction

The injection of multi–megawatt neutral $H^0(D^0)$ beams is a proven technique for auxiliary heating of magnetically confined fusion plasmas. The method presently applied for energies up to 60 keV/u is based on the neutralization of intense H^+ ion beams in a gas target. However, future fusion devices call for injection energies at hundreds of keV where the neutralization efficiency of H^+ ions is very small (few %).

For the latter requirements, the neutralization of an accelerated H^- ion beam offers a much higher efficiency of about 60% [1]. It has been proposed [2,3] that the neutralization efficiency can be further increased up to about 85% by the use of a plasma neutralizer consisting of free electrons and ions. Furthermore, a plasma neutralizer based on multiply–charged ions is expected [4] to offer the advantage of a significantly reduced optimum target thickness due to the larger electron removal cross sections.

However, modelling and design studies for plasma neutralizers have suffered from the lack of experimental cross sections for the single–electron removal

$$H^- + X^{q+} \rightarrow H^0 + \ldots \quad (\sigma_{-0}) \qquad (1)$$

and the double–electron removal reactions

$$H^- + X^{q+} \rightarrow H^+ + \ldots \quad (\sigma_{-+}) \qquad (2)$$

The latter reaction is especially important since it directly degrades the efficiency of a plasma neutralizer.

II. Experiment

We have employed the Giessen ion–ion crossed–beams facility to measure absolute cross sections for reactions (1) and (2) in $H^- + Ar^{q+}$ ($q \leq 8$) collisions. The experimental arrangement [5] and the signal–recovery technique [6] used for the present measurements have been described in detail previously. Briefly, two well– collimated (< 2 mm diameter) and charge–analyzed ion beams of adjustable energies (H^- beam: (8–300) keV, (4–70) nA; Ar^{q+} beam from a 5 GHz ECR ion source: (20–80) keV, (2–200) nA) are arranged to intersect at an angle of $\theta = 45°$ in an ultrahigh vacuum of about $2 \cdot 10^{-11}$ mbar with both beams "on". The collision products (H^0, H^+) formed in the H^- beam are separated immediately after the interaction region by electrostatic deflection and counted individually by a channeltron–based single–particle detector.

The most serious experimental problem to overcome has been provided by the high rate of background events, which result from stripping collisions of H^- ions with residual gas particles. Typically, the signal of 4 to 2000 cts/s was masked by a background which was a factor 200 to 50 higher, respectively. For signal recovery a beam–modulation technique [6] was employed. Typical measurement times were up to 4 hours for one cross section.

III. Results

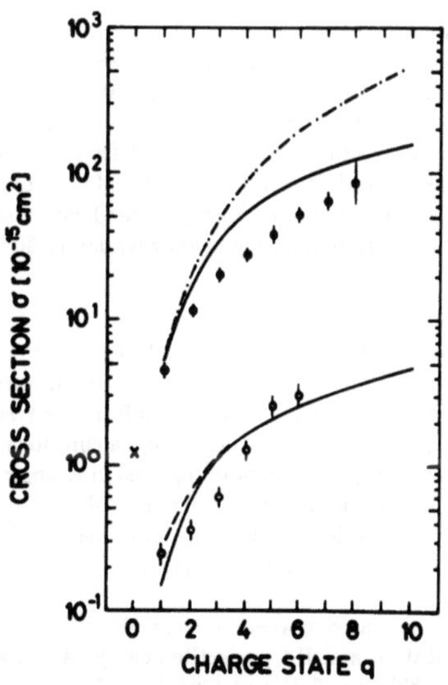

Fig. 1. Experimental single (•) and double (o) electron removal cross sections for $H^- + Ar^{q+}$ ($q \leq 8$) collisions at $E_{c.m.} = 50$ keV. An experimental data point (×) for single– electron removal in H^- + Ar^0 collisions [9] is shown for comparison. The solid lines are CTMC calculations for the respective processes. The dot–dashed line is the Bethe–Born calculation of ref. [8]. The dashed line is the electron removal cross section for neutral H^0 [10].

In Fig. 1 the experimental cross sections for single (σ_{-0})- and double (σ_{-+})-electron removal in $H^- + Ar^{q+}$ ($q \leq 8$) collisions at a center-of-mass energy $E_{c.m.} = 50$ keV are displayed. Also shown are three-body classical trajectory Monte Carlo (CTMC) calculations [7] as well as a Bethe–Born calculation [8] for reaction (1). Both the CTMC and the Bethe-Born results reproduce the q=1 experimental data. However, it can be seen that the q^2 scaling predicted by the first-order Born approximation for the single-electron removal becomes increasingly invalid as the charge state q of the ion increases. The experimental cross sections σ_{-0} scale, within error bars, with a $q^{1.3}$ dependence on the ionic charge. The CTMC calculations also overestimate to some extent the σ_{-0}-measurements whereas they are in qualitative agreement with the σ_{-+}-data.

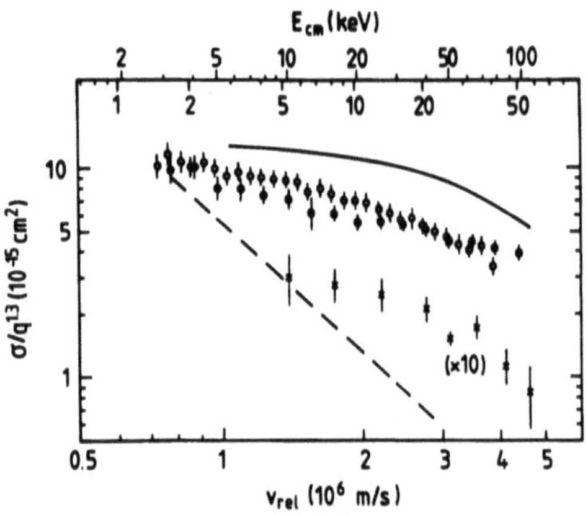

Fig. 2. Scaled experimental single-electron removal cross sections for $H^- + Ar^{4+}$ (•) and $H^- + H^+$ (∘, [11]). Scaled double-electron removal cross sections (multiplied by a factor 10) for $H^- + Ar^{3+}$ are shown by crosses. The dashed line indicates a $1/v^2$ dependence for $H^- + Ar^{q+}$ collisions. The solid line is the CTMC result [7] for the $H^- + Ar^{4+}$ system. The upper x-axis scales show the center-of-mass energies for the system $H^- + Ar^{q+}$ (above) and $H^- + H^+$ (below), respectively.

Fig. 2 displays the velocity dependence of the $H^- + Ar^{4+}$ single-electron removal cross section along with that for $H^- + H^+$. When scaled by $q^{1.3}$, the results are nearly identical. The velocity dependence of the cross sections is far removed from the $\sim 1/v^2$ scaling predicted by asymptotic theories. The CTMC results [7] are also shown in Fig. 2. The calculated energy dependence is similar to that observed, the absolute magnitude is approximately 50% too large.

Taking into account the present data, both the maximum neutralization efficiency η_{max} and the corresponding optimum target thickness π for a plasma neutralizer can be calculated from the respective equations given in Ref. [4]. For the modelling calculations, we have assumed a fully ionized plasma containing ions in a single charge state q which are at rest in the laboratory frame. All electron attachment processes have been neglected. The cross sections for electron impact ionization have been taken from Ref. [12-14].

Fig. 3 shows calculated maximum neutralization efficiencies η_{max} versus charge state q and H^- projectile energy. At 50 keV/u, η_{max}-values between 82% (q=5,6) and 89% (q=2,3) are realizable. For q=1 we obtain η_{max}=86%. It is obvious that a large ratio of cross sections σ_{-0}/σ_{-+} for ions in a given charge state results in a high value for η_{max}. In order to calculate neutralization efficiencies at higher energies we have extrapolated the energy dependence of the single-electron removal cross section of $H^- + Ar^{4+}$ shown in Fig. 2. Furthermore, it was assumed that all cross sections σ_{-0} and σ_{-+} show the same energy dependence, regardless of q. It can be seen from Fig. 3 that the neutralization efficiencies obtained decrease by only about 3% in the energy range from 50 keV/u up to 1 MeV/u.

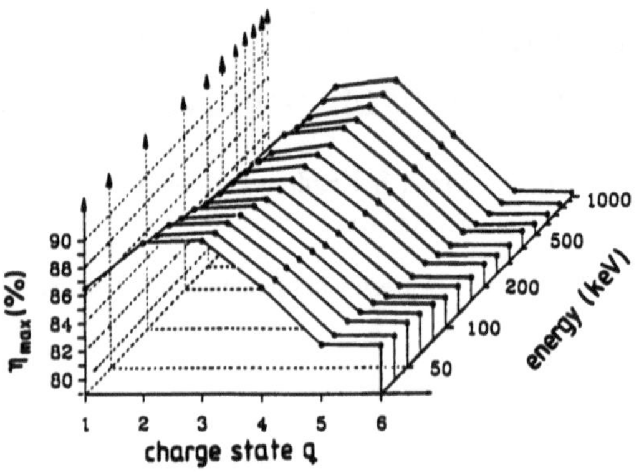

Fig. 3. Calculated maximum neutralization efficiencies η_{max} vs charge stage q and H^- energy. The solid lines are drawn to quide the eye.

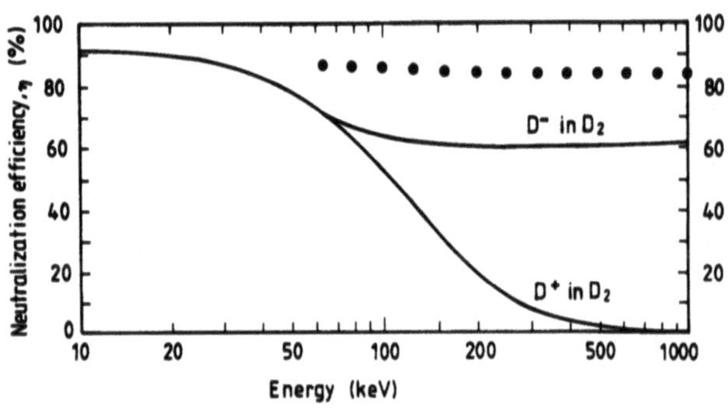

Fig. 4. Maximum neutralization efficiencies vs beam energy. • D^- in an Ar^{4+} plasma target (present results). The solid lines refer to D^- and D^+ beams, respectively, in a D_2 target [1].

Fig. 4 shows calculated η_{max}-values for D^- ions in an Ar^{4+} plasma target along with corresponding efficiencies [1] for D^- and D^+ ions, respectively, in a D_2 gas target. The significant efficiency enhancement of a plasma target for the neutralization of energetic D^--beams can be readily seen.

Our measurements, however, do not bear out the expectation [4] that plasma neutralizers can be made significantly more compact by employment of multiply-charged ions compared to singly-charged ions. For example, our modelling calculations predict a reduction of 3 in the optimum target ion line density for 1 MeV H^- ions compared to 15 for using a q=5 ion rather than a q=1 ion. Since the measurements indicate that the species of ions does not relevantly change the efficiency of H^- neutralization, a plasma neutralizer based on hydrogen appears to be most advantageous. However, also in this case modelling and design studies suffer from the lack of data since the cross section for double ionization σ_{-+} for $H^- + H^+$ collisions has not yet been measured.

This work has been funded by BMFT under contract no. 06 GI 307.

References

[1] Berkner, K. H., Pyle, R. V. and Stearns, J. W.: Nucl. Fusion **15**, 249 (1975)

[2] Dimov, G. I., Ivanov, A. A. and Roslyakov, R. G.: Sov. Phys. Techn. Phys. **22**, 1091 (1976)

[3] Ivanov, A. A. and Roslyakov, G. V.: Sov. Phys. Techn. Phys. **25**, 1346 (1980)

[4] Schlachter, A. S., Leung, K. N., Stearns, J. W. and Olson, R. E., in: *Production and Neutralization of Negative Ions and Beams*, ed. J. G. Alessi, Brookhaven 1986, AIP Conference Proceedings No. 158, p. 631

[5] Rinn, K., Melchert, F. and Salzborn, E.: J. Phys. B **18**, 3783 (1985)

[6] Rinn, K., Melchert, F., Rink, K. and Salzborn, E.: J. Phys. B **19**, 3717 (1986)

[7] Melchert, F., Debus, W., Liehr, M., Olson, R. E. and Salzborn, E.: Europhys. Lett. **9**, 433 (1989)

[8] Kim, Y. K. and Inokuti, M.: Phys. Rev. A **3**, 665 (1971)

[9] Anderson, C. J., Girnius, R. J., Howald, A. M. and Anderson, L. W.: Phys. Rev. A **22**, 822 (1980)

[10] Olson, R. E., Berkner, K. H., Graham, W. G., Pyle, R. V., Schlachter, A. S. and Stearns, J. W.: Phys. Rev. Lett. **41**, 163 (1978)

[11] Krüdener, S., Melchert, F., Schön, W. and Salzborn, E, to be published

[12] Peart, B., Walton, D. S. and Dolder, K. T.: J. Phys. B **3**, 1346 (1970)

[13] Defrance, P., Claeys, W. and Brouillard, F.: J. Phys. B **15**, 3509 (1982)

[14] Shah, M. B., Elliot, D. S. and Gilbody, H. B., J. Phys. B **20**, 3501 (1987)

Supplement to Z. Phys. D - Atoms, Molecules and Clusters 21, S 253-254 (1991)

Atoms, Molecules
Zeitschrift für Physik D and Clusters

Single electron capture of Ne^+ in He in the energy range 0.5 to 5 keV

H. Martínez, I. Alvarez, C. Cisneros and J. de Urquijo
Instituto de Física, UNAM, P.O. Box 139-B, Cuernavaca Mor., 62191 México

ABSTRACT

Absolute differential and total cross sections have been measured for the single electron capture of Ne^+ in He at impact energies between 0.5 and 5 keV. Oscillations in the differential cross sections are interpreted as the envelope of the fast oscillations due to the interference of the scattering amplitudes. Total cross sections are in agreement with previous experiments.

I. INTRODUCTION

Differential scattering cross sections in atom-atom or ion-atom collisions are of continous interest. They provide information about the interaction potential between the collision partners. In the system under study all the differential cross sections show well defined oscillations that are interpreted as the interference of scattering amplitudes. The Landau-Zener model is appropriate to compare the total cross sections in the energy range of study and to obtain some potential parameters[1].

II Experiment

A Ne^+ beam was formed in a Colutron-type ion source, accelerated to the desired energy and collimated before entering the scattering region. The target cell was located at the centre of a rotatable, computer-controlled vacuum chamber housing an electrostatic analyser, with two funnel-type channel electron multipliers. The Ne^0 formed by electron capture passed straight the analyser through a 1 cm diameter hole and impinged on a CEM so that the neutral counting rate could be measured. Separation of charged particles ocurred inside the analyser which was set to detect the Ne^+ species with a second CEM. Total beam current was measured by a retractable Faraday cup, hence the cross sections are absolute. The intensity I represents the number of Ne^+ incident per unit area per sec and was monitored by the Faraday cup. The number of Ne^0 formed by electron capture, $N(\theta)$ contained in the solid angle $d\Omega$ at angle θ was determined as function of θ, as the detector assembly was rotated about the He cell; the number of He atoms per unit volume was determined by the gas target density. With these measurements, the quantity

$$\frac{d\sigma_{10}}{d\Omega} = \frac{N(\theta)}{I}$$

was calculated. (Fig. 1) Total cross sections were evaluated by the integration over the solid angle.

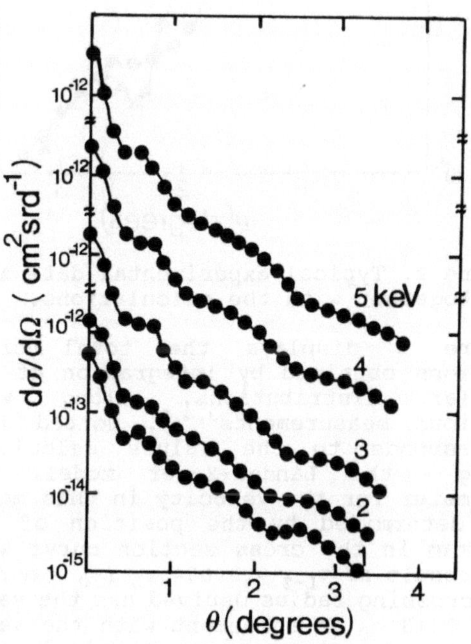

Fig. 1. Differential cross section data.

III Results and discussion

All differential cross sections show a well defined structure, which corresponds to the envelope of the fast oscillations and a semiclassical scattering treatment outside the small angle region can be applied. If we consider the structure as the interference of two incident angular momenta, the differential cross section for two branches becomes[2]

$$\sigma = |(\sigma_{cl})_1^{1/2} + (\sigma_{cl})_2^{1/2} \exp[i(\beta_2 - \beta_1)]|^2$$

In first aproximation and in order to compare with the experimental data, we have taken $(\sigma_{cl})_1 = (\sigma_{cl})_2$ and adjusted the other values to them. as shown in Fig. 2.

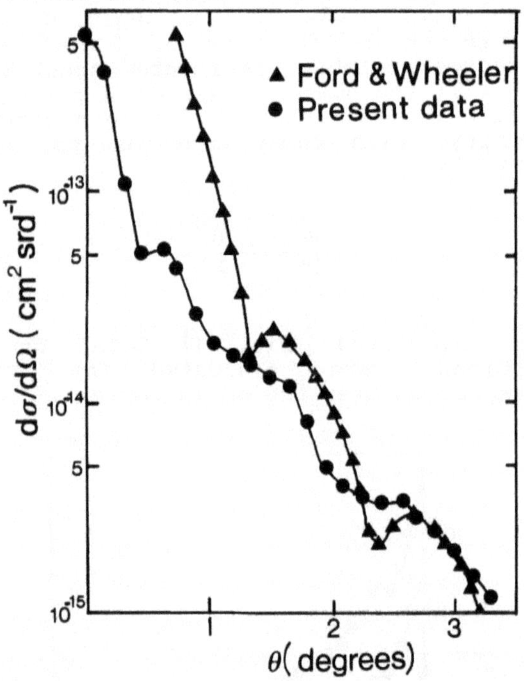

Figure 2. Typical experimental data at 1 keV together with the calculations.

Figure 3 displays the total cross sections obtained by integration of the angular distributions, along with previous measurements[3,4,5]. Dotted line corresponds to the values calculated using the Landau-Zener model. The parameter for the velocity in this model was determined by the position of the maximum in the cross section curve with the result of $v_{L-Z} = 0.034\ v_0$ ($v_0 = e^2/\hbar$). The crossing radius derived has the value $R_c = 3.13\ a_0$ in agreement with the value reported for the inverse reaction[6].

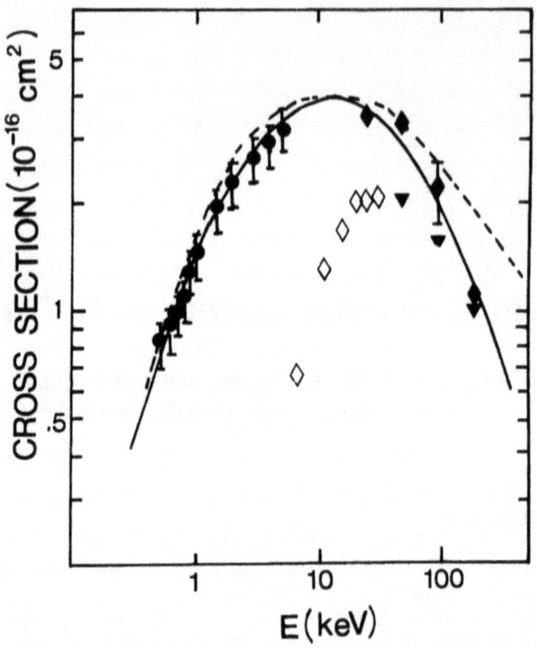

Fig. 3. Total cross sections. ● Present data, ♦ from Ref. 5, ▼ from Ref. 4, ◊ from Ref. 3.

IV Conclusion

We have measured the angular distribution of the single electron capture of Ne^+ in He. The oscillations have been treated through a semiclassical formalism. The total cross sections evaluated from the experimental results adjust well with previous measurements. The Landau-Zener model reproduces the trend of the data and the calculated crossing radius is the same as that previously reported for the inverse reaction.

References

1. R. B. Bernstein in <u>*Atom-Molecule Collision Theory*</u> (Plenum Press, New York, 1979) p. 449.
2. K. W. Ford and J. A. Wheeler, Annals of Physics, **7**, 259 (1959).
3. J. P. Flaks, Sov. Phys. Tech. Phys. **31**, 263 (1961).
4. A. B. Wittkower and H. B. Gilbody, Phys. Soc. **90**, 353 (1967).
5. H. Winter, E. Bloemen and F. J. de Heer, J.Phys. B,**10**, 1599 (1977).
6. B. Zygelman and A. Dalgarno, Phys. Rev. **A33**, 3853 (1986).

This work was supported in part by DGAPA grant number IN-01-48-89 and by CONACyT.

Supplement to Z. Phys. D - Atoms, Molecules and Clusters 21, S 255-256 (1991)

Atoms, Molecules
Zeitschrift für Physik D and Clusters

Core-changing versus core-conserving charge transfer in Ne^{6+}-He collisions

J. P. M. Beijers[1], R. Hoekstra[1,2], A. R. Schlatmann[1], R. Morgenstern[1], and F. J. de Heer[2]

[1] Kernfysisch Versneller Instituut, Zernikelaan 25, 9747 AA Groningen, The Netherlands
[2] FOM-Institute for Atomic and Molecular Physics, P. O. Box 41883, 1009 DB Amsterdam, The Netherlands

Abstract

We have measured VUV spectra of photons emitted by Ne^{5+*} ions subsequent to single-electron capture in Ne^{6+}- He collisions at impact energies of 1.2 and 0.11 keV/amu. Capture into the NeVI ($1s^22s^23d$) state has been identified as the dominant reaction channel at both energies. At 0.11 keV/amu capture into core-excited singlet states contributes 35% to the total cross section.

1. Introduction

Single-electron capture (SEC) is almost always the dominant charge transfer process in collisions between highly-charged ions and neutral gas atoms. However, more-electron transitions can still play an important role in specific collision systems and it is important to understand the relative contribution of these processes to the total charge transfer cross section of the system under study.

In this paper we focus on the single electron capture process

$$Ne^{6+} + He \longrightarrow Ne^{5+*} + He^{+} \qquad (1)$$

because there seems to be some confusion in the recent literature regarding the dominant charge transfer channel in this system. Andersson *et al.* [1] have measured energy-gain spectra of reaction (1) at impact energies $\lesssim$ 0.1 keV/amu. From Landau-Zener calculations they conclude that the main contribution to the capture cross section comes from core-changing collisions, i.e. two-electron processes where one of the target electrons is captured into a $3l$ state and simultaneously the core of the projectile ion is excited such that a transition $1s^22s^2 \rightarrow 1s^22s2p3l$ results. We will designate these transitions as core-exciting (CE) transitions. In contrast, Hansen and Andersson [2] who performed close-coupling calculations at a collision energy of 0.1 keV/amu found that single-electron capture into the $3d$ state with the core remaining passive, a so-called PC transition, is by far the dominant capture channel.

In order to resolve this issue we have used photon-emission spectroscopy [3] to study charge transfer in the Ne^{6+} - He system as both the CE states and PC states decay via photon emission.

2. Results and discussion

Figs.1 and 2 show the VUV spectra observed at impact energies of 1.2 and 0.11 keV/amu, respectively. Except lines from the $3s$, $3p$, and $3d$ PC states the spectra show also contributions from $2s2p(^{1,3}P^o)3l$ CE states. The strongest emission line in both spectra, however, is the one at 12.26 nm from the NeVI($1s^22s^23d$) state, thus confirming Hansen and Andersson's interpretation [2]. The strong emission line at 15.0 nm in the 0.11 keV/amu spectrum originates from two-electron capture into NeV($1s^22s^22p3d$) .

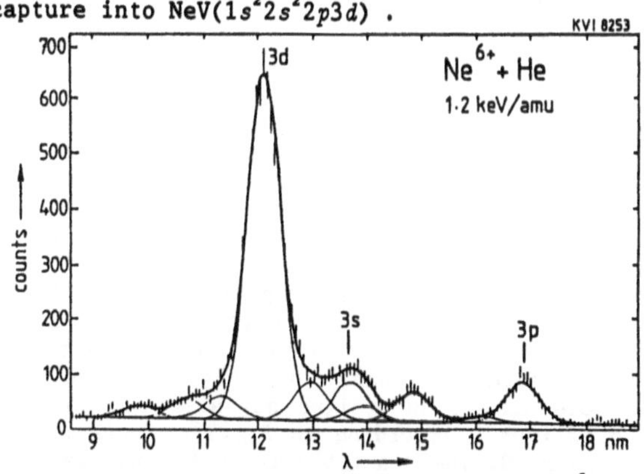

Fig. 1. Photon emission spectrum for Ne^{6+} - He collisions at an impact energy of 1.2 keV/amu.

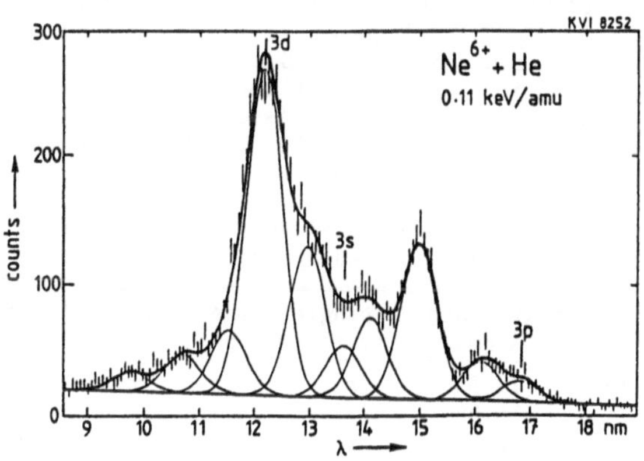

Fig. 2. As fig. 1 but for an energy of 0.11 keV/amu.

The quantitative analysis of these spectra is complicated by the fact that the primary ion beam contains not only ground-state $Ne^{6+}(1s^2 2s^2)\ ^1S$ ions but also metastable $Ne^{6+}(1s^2 2s2p)\ ^3P^o$ ions. We estimate this metastable fraction to be 0.3 ± 0.1, see ref.[4]. Thus the emission lines from the $2s2p(^3P^o)3l$ states either originate from a CE process involving a ground state primary ion or from a PC transition by a metastable primary ion. On the other hand, emission lines from the $2s2p(^1P^o)3l$ states all come from CE processes. In order to estimate the relative importance of these PC and CE states we have plotted their binding energies relative to the reaction window as calculated with the dynamic version of the classical overbarrier model [5]. This is shown in fig.3 for an impact energy of 0.11 keV/amu. This figure explains why out of the $1s^2 2s^2 3l$ multiplet the $3d$ state is most strongly populated by SEC in Ne^{6+} - He collisions.

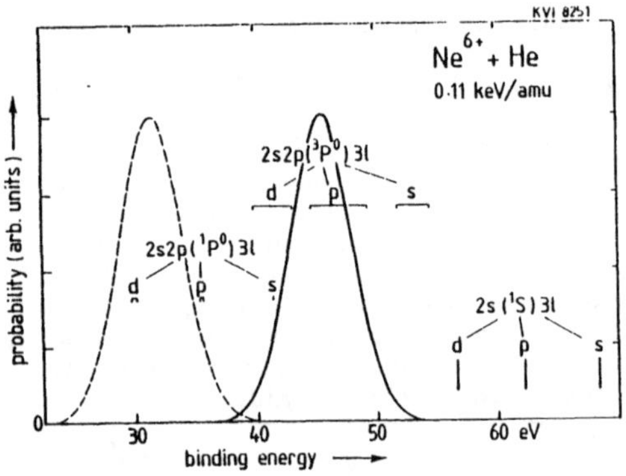

Fig. 3. Reaction windows for SEC in collisions of ground state (solid curve) and metastable (dashed curve) Ne^{6+} on He at an energy of 0.11 keV/ amu. The reaction window for metastables is shifted to lower binding energies according to its excitation energy because *total* binding energies are plotted.

Although the CE states, especially the triplet states, are much more favourably situated relative to the reaction window their coupling with the initial state is probably weak [2], since not only the core has to be excited but also the spin has to be flipped. According to fig.3 we have used 10 gaussians, defined in table 1, to fit the measured spectra. Only the heights have been varied, the positions and widths being determined by the experimental apparatus. Figs.1 and 2 show that this collection of peaks fits the spectra very well. The contribution of CE states to the capture cross section at an impact energy of 1.2 keV/amu is rather small (< 15%), whereas at 0.11 keV/amu CE states, in particular the $2s2p(^1P^o)3s\ ^1P$ state, do contribute significantly.

Table 1. Upper levels and emission lines used to deconvolute the measured spectra.

ground state core		singlet core		triplet core	
state	λ(nm)	state	λ(nm)	state	λ(nm)
$3s$	13.85	$3s$	13.1	$3s$	16.0
$3p$	16.88	$3s$	14.1	$3p$	10.9
$3d$	12.26	$3p$	10.0	$3p'$	11.4
$2p3d$	15.0				

This pronounced energy dependance indicates that the contribution of the metastable beam component to the observed CE peaks is rather small.

Finally, from the areas of the different peaks we have determined the relevant capture cross sections. The branching ratios for the PC states are unity, for the CE states they are not that well known. However, all decay channels are observed. We estimate the total cross section for one-electron capture from the populations of PC and SC singlet states and attribute the triplet CE peaks totally to the metastable beam component. The resulting cross sections are shown in table 2. Details of the present analysis will be published in a forthcoming paper.

Table 2. Relevant state selective and total electron capture cross sections for $Ne^{6+}(1s^2 2s^2)$ ions colliding with He in 10^{-16} cm^2.

E(keV/amu)	σ(3d)	σ(PC)	σ(CE)	σ_t	ref
1.2	16.2	20.4	3.4	23.8	
0.11	7.3	8.5	4.5	13.0	
Error(%)	25	30	30	25	
1.1	10.4	14.9			[4]
0.1				10	[1]
0.1				11.7	[2]
0.066				11	[6]

3. Conclusions

We have studied single-electron capture in Ne^{6+} - He collisions at impact energies of 1.2 and 0.11 keV/amu. By observing the VUV photons emitted by the decaying Ne^{5+*} product ions we identified the $NeVI(1s^2 2s^2 3d)$ state as the strongest capture channel at both impact energies. Capture into CE singlet states is not important at an impact energy of 1.2 keV/amu, but becomes much more important at 0.11 keV/amu.

Acknowledgment

We wish to thank J. Eilander and J. Sijbring for their excellent technical support. This work is part of the research program of the "Stichting voor Fundamenteel Onderzoek der Materie" (FOM), which is financially supported by the "Stichting voor Nederlands Wetenschappelijk Onderzoek" (NWO).

References

1. Andersson,L.R., Pedersen,J.O.P., Bàràny,A., Bangsgaard,J.P., Hvelplund,P.: J.Phys.**B22**, 1603 (1989)
2. Hansen,J.P., Andersson,L.R.: J.Phys.**B22**, L285 (1989)
3. Hoekstra,R., Beijers,J.P.M., Schlatmann,A.R., Morgenstern,R., de Heer,F.J.: Phys.Rev.**A41**, 4800 (1990)
4. Dijkkamp,D., Gordeev,Yu.S., Brazuk,A., Drentje, A.G., de Heer,F.J.: J.Phys.**B18**, 737 (1985)
5. Niehaus,A.:J.Phys.B19, 2925 (1986)
6. Schmeissner,C., Cocke,C.L., Mann,R., Meyerhof, W.: Phys.Rev.A30, 1661 (1984)

Supplement to Z. Phys. D - Atoms, Molecules and Clusters 21, S 257-258 (1991)

Atoms, Molecules
Zeitschrift für Physik D and Clusters

Multiply-charged ions and cluster ions produced from frozen CO_2 and H_2O targets in energetic, heavy ion collisions

H. Tawara[1], T. Tonuma[2], H. Kumagai[2], T. Matsuo[3] and H. Shibata[4]

[1] National Institute of Fusion Science, Nagoya 464-01, Japan
[2] Institute of Physical and Chemical Research (RIKEN), Wako-shi, 351-01, Japan
[3] Department of Pathology, Tokyo Medical and Dental University, Yushima, Tokyo 113, Japan
[4] Research Center for Nuclear Science and Technology, Tokai-mura, Ubaraki 319-11, Japan

Abstract
Some frozen gases (CO_2 and H_2O) are used to produce multiply charged recoil ions as well as cluster ions under high energy, heavy ion impact. The ions produced in the frozen gas targets are found to be quite different from those in gas phase targets.

I. Introduction
The production of multiply charged, recoil ions is known to be largely enhanced when highly charged projectile ions collide with gas target atoms [1]. The recoil ions produced, in particular very high charge ions, get relatively large recoil energy (>10 eV) during collisions. This recoil energy could be obstacles in some applications such as effective ion trapping in small electro-magnetic confining spaces. We have started to study if cold gas targets could provide multiply charged recoil ions with very low kinetic energy. And also we have observed various cluster ions from the frozen gas targets. Up to now the cluster ions are mostly generated through cooled expanding gas nozzle [2]. Thus, not much information on the frozen gas targets is available presently.

II. Experiment
1.5 MeV/amu Ar^{q+} (q=4-14) ions from RIKEN linear accelerator are sent into gas targets frozen on Al foils which are cooled down to the liquid nitrogen temperatures. The cooled foils, thin enough to pass projectile ions, are inclined 45° with respect to the incident ion beam, meanwhile the secondary ions from the frozen surface are extracted normal to the foils and their mass/charge is analyzed through a double focusing sector magnet. The pressure at the chamber wall is measured to be 3-5 $*10^{-7}$ Torr when the target gas is introduced, while the background pressure is about $1*10^{-7}$ Torr.

III. Results and discussion
1) Multiply charged ions
A typical mass/charge spectrum of ions from the frozen CO_2 target in low mass region is shown in Fig.1(a), together with that from gas target in Fig.1(b). Significant difference in spectra in both target phases is clearly noticed : the production of multiply charged carbon recoils are strongly suppressed over that of multiply charged oxygen ions in the frozen target, compared with that in gas target [3]. Though no clear reason could be assigned for this difference, the kinetic energy of recoil ions might be a key issue ; namely oxygen ions are known to get more kinetic energy than carbon ions during Coulomb explosion of multiply charged molecular ions produced in energetic collisions [3] and thus oxygen ions can escape the surfaces more easily than carbon ions. Also their chemical activities inside the target might influence such a behavior.

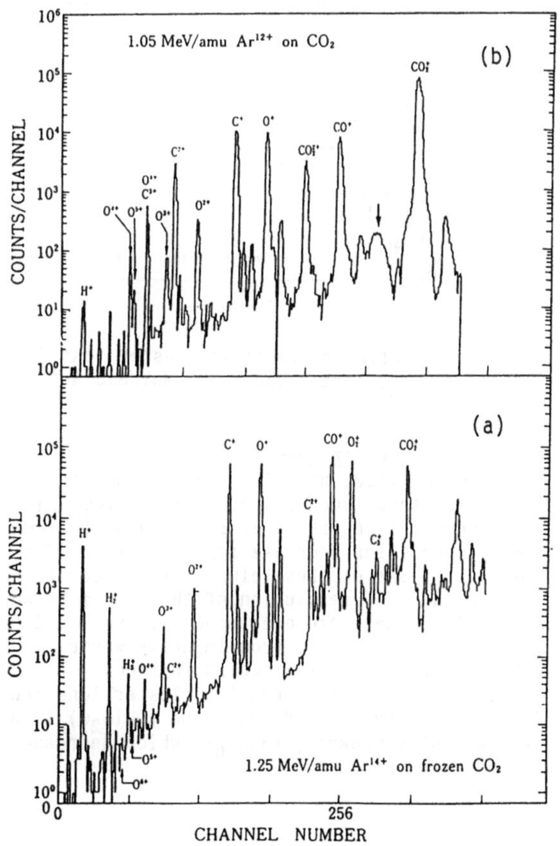

Fig.1 Comparison of multiply charged ions produced from frozen target(a) and gas phase target(b)

2) Positive cluster ions

Positive cluster ions from frozen CO_2 target are found to consist mainly of $[O_2(CO_2)_n]^+$, $[O(CO_2)_n]^+$ and $[CO_2(CO_2)_n]^+$, followed by slightly weak $[CO(CO_2)_n]^+$ cluster ions. In $[O_2(CO_2)_n]^+$ cluster ions, O_2 are formed through recombination of O atom, one of the dissociation products of CO_2, with similar O atom nearby during migration to the surfaces and become a core when cluster ions are being formed. This is a sharp contrast to the cluster ion distributions in gas nozzle targets [4] where the dominant constituent is $[CO_2(CO_2)_n]^+$ ion. Another difference is the fact that with increasing n the intensities of large cluster ions decrease relatively smoothly in the frozen targets, whereas in gas nozzle targets they are known to sharply increase at "magic numbers" corresponding to particular n [2,4].

Meanwhile, the composition of cluster ions from the frozen H_2O target (see Fig.2) is quite simple : mostly a series of $[H_3O(H_2O)_n]^+$, similar to that from gas nozzle target [4], with H_m^+ (m=1,2,3) and weak cluster ions composing H and O. The main cluster ions might have a structure with the core of H_3O but not H because H_3O^+ are far more intense, compared with H_2O^+ ions which are the dominant products in gas phase targets.

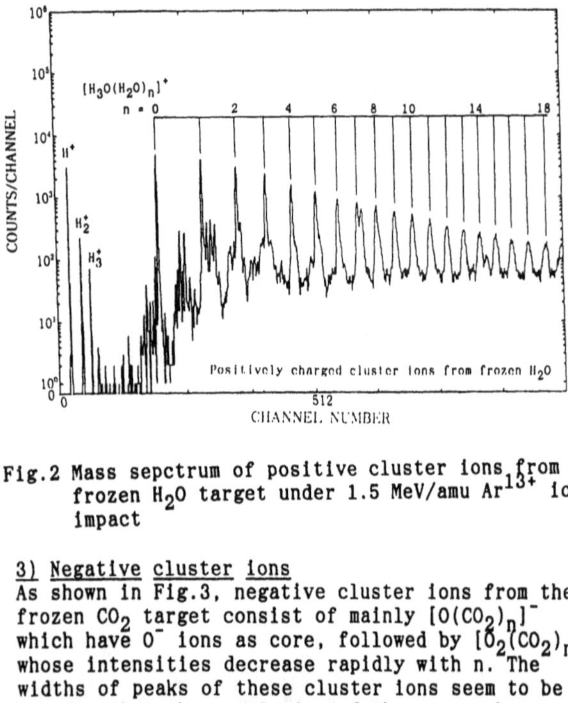

Fig.2 Mass sepctrum of positive cluster ions from frozen H_2O target under 1.5 MeV/amu Ar^{13+} ion impact

3) Negative cluster ions

As shown in Fig.3, negative cluster ions from the frozen CO_2 target consist of mainly $[O(CO_2)_n]^-$ which have O^- ions as core, followed by $[O_2(CO_2)_n]^-$ whose intensities decrease rapidly with n. The widths of peaks of these cluster ions seem to be broader than the resolution of the system because $[OH(CO_2)_n]^-$ peaks are overlapped. These H and OH are found to originate from impurity water molecules frozen on the cooled surfaces.

On the hand, the main negative cluster ions from frozen H_2O target are found to be $[OH(H_2O)_n]^-$ and with slightly weaker $[O(H_2O)_n]^-$ which decrease rapidly with n.

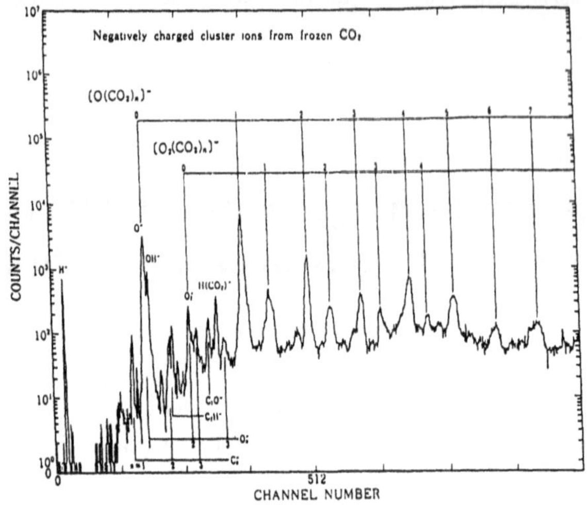

Fig.3 Mass spectrum of negative cluster ions from frozen CO_2 target under 1.5 MeV/amu Ar^{13+} ion impact

IV. Concluding remarks

We have shown that the frozen gas targets are efficient producers of multiply charged ions as well as of cluster ions under energetic, highly charged ion impacts and that spectra of both multiply charged recoil ions and positive as well as negative cluster ions from the frozen gas targets are significantly different from those observed in the cooled gas nozzle targets. This is probably related to the density of target and the collisions followed in the targets. Dissociation products, ions or atoms, can become a core, forming various (sometimes unexpected) cluster ions.

Some of the observed molecule and cluster ions in the present work have never been seen in laboratories yet but only in space. These ions could be used to get collision data for modelling and analysis of astrochemistry and other fields.

As pointed out already, the observed spectra of cluster ions are found to be significantly influenced by the presence of the frozen impurity species such as H_2O in particular. Some of them can be seen in Fig.3 as a form of H^- or OH^- for negatively charged ions and $[H_m(C_n)]^+$ and $[H_m(O_n)]^+$ (m=1-3) for positively charged ions.

References

1. T.Tonuma, H.Shibata, S.H.Be, H.Kumagai, M.Kase, T.Kambara, I.Kohno and H.Tawara, Phys. Rev. A33 (1986) 3047
2. D.Kreisle, O.Echt, M.Knapp and E.Recknagel, Surf. Sci. 156 (1985) 321
3. T.Matsuo, T.Tonuma, M.Kase, T.Kambara, H.Kumagai, I.Kohno and H.Tawara, Chem. Phys. 121 (1988) 93
4. O.Echt, K.Sattler and E.Recknagel, Phys. Letters 90A (1982) 185
5. O.Echt, D.Kreisle, M.Knapp and E.Recknagel, Chem. Phys. Letters 108 (1984) 401

Supplement to Z. Phys. D - Atoms, Molecules and Clusters 21, S 259-260 (1991)

Atoms, Molecules
Zeitschrift für Physik D and Clusters

Correlations between charged particles emitted in ion-molecule collisions

R. D. DuBois*, R. Herrmann*, J. Feng, R. Dörner*, J. Euler* and K. Ullmann***
Institute für Kernphysik, Universität Frankfurt/M., August Euler Str. 6, W-6000 Frankfurt/M., FRG

* permanent address: Pacific Northwest Laboratory, P. O. Box 999, Richland, WA 99352, USA
** permanent address: Physics Dept., University of Lanzhou, Lanzhou, Gansu, PRC

Abstract:

A new experimental apparatus for studying correlations between charged particles emitted in ion-atom collisions is described. The apparatus consists of two channelplate detectors, each equipped with position sensitive anodes. Time correlated events in both detectors provide information about correlated charged particle emission. Initial test studies of dissociative double ionization for H_2 and N_2 targets are reported here.

I. Introduction:

From numerous experimental and theoretical studies during the past decade it is well established that multiple ionization of target atoms, particularly for highly charged ion impact, is a common event. This results in the emission of two, or more, electrons. For molecular targets, the emission of two or more electrons can also result in the creation of energetic ionic fragments due to coulomb explosion of the resultant molecular ion. In all cases, investigating the energy and angular distributions of the emitted particles can be used to study multi-electron transition events.

In the case where only two particles are emitted, identifying and studying these events can be done by looking for correlated signals recorded by two detectors that can be positioned relative to each other. Unfortunately this typically means extremely small signal rates because of solid angle limitations for each detector. Thus an obvious way to improve the signal rate is to increase the detection solid angle while still maintaining a reasonable energy and angular resolution.

II. Experiment:

At the University of Frankfurt we have constructed an apparatus which is capable of detecting charged particles emitted into a solid angle of nearly 1.5π for each of two detectors. By using two-dimensional position sensitive anodes, information about the angular distributions of the emitted particles can be preserved. The apparatus, as used for test studies involving dissociative ionization of H_2 and N_2, is shown schematically in Fig. 1.

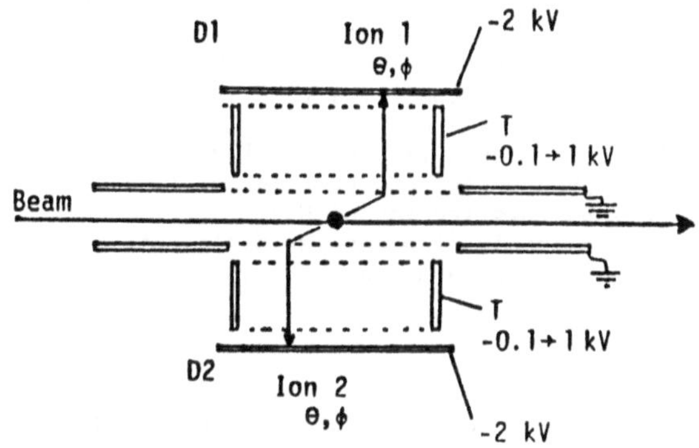

Fig. 1. Apparatus (top view) showing the two-dimensional time-of-flight detectors, D1 and D2.

The detectors (D1 and D2) are located at plus and minus 90° with respect to the beam direction. A point source target, approximated by a gas jet, is located between the two detectors. Due to physical restrictions, the detectors could not be positioned closer to the beam path. Thus an ion transport system (T) was used to increase the solid angle with respect to what could be achieved due to physical limitations. By applying a negative voltage to the transport system, low energy ions emerging from the target region through the first grid could be swept towards the detector. Thus particle emission between roughly 15 and 165° in both the theta and phi directions could be detected. Hence the probability of detecting correlated events is rather large, provided that the particles are emitted into opposite hemispheres.

Data were collected in the following fashion. Low intensity, collimated beams of H^+ and He^+ ions passed through the target region, thus inducing target ionization. The projectile beam was then charge state analyzed and one of the components was deflected onto a movable surface barrier detector. Amplified and discriminated signals from this detector served as stop pulses for two time-to-amplitude converters with the start signals being the target ions detected by either detector 1 or 2. Thus time-of-flight spectra were recorded for each detector.

In principle, energy information is obtainable from these time-of-flight spectra. However locating the detectors as close as possible to the beam axis in order to maximize the detection solid angle meant sacrificing time-of-flight energy information. It is possible however to identify energetic (eV) fragment atomic ions resulting from dissociative ionization events from low energy (meV) molecular ions resulting from simple ionization of the molecule. This is demonstrated in Fig. 2 for 0.5 MeV H^+ and He^+ impact on H_2. The sharp peak on the right is presumed to be photons or electrons.

Time-of-flight spectra were collected in List Mode fashion for both detectors and stored in a computer. By replaying the experiment, time correlated events between the two spectra could be investigated. An example of this is shown in Fig. 3 where events detected by detector D2 were searched with the condition that they were time correlated with the energetic atomic ions (H^+ ions) detected by detector D1. These events are directly proportional to the dissociative double ionization cross section of H_2.

In principle, the angular distributions of these correlated events are available due to the two-dimensional detection used, but for the examples shown limited statistics prohibited investigating this feature.

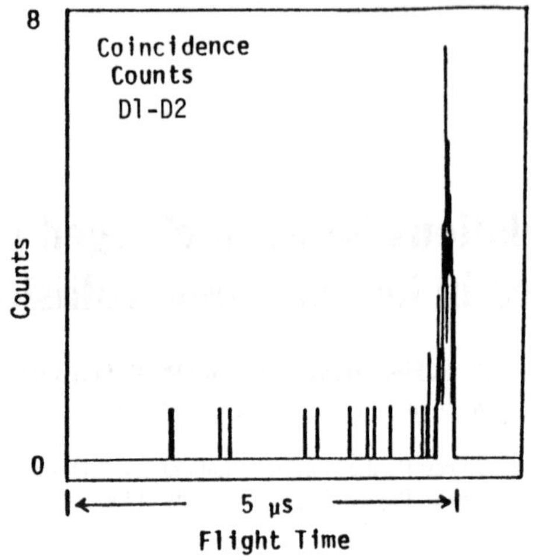

Fig. 3. Detector 2 signal coincident with D1 H^+ events.

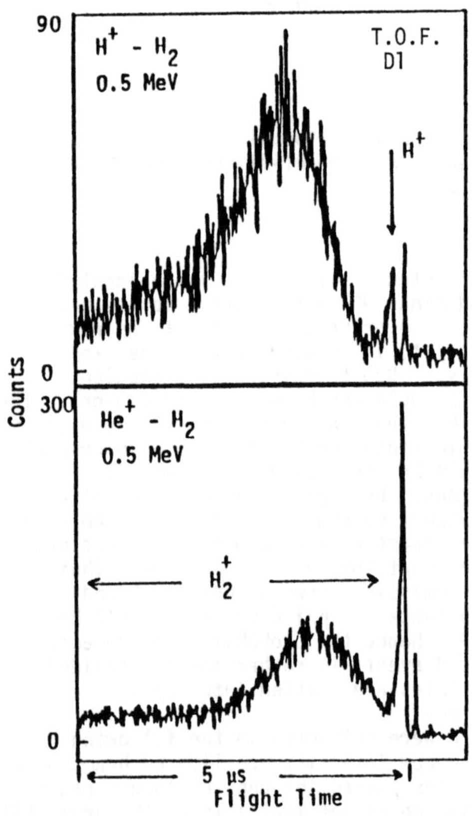

Fig. 2. Time-of-flight spectra for target ions produced by 0.5 MeV H^+ and He^+ impact on H_2.

III. Conclusions:

The capabilities of investigating correlated charged particle emission have been demonstrated. Using dissociative ionization of molecular hydrogen and nitrogen as a test example, it was found that dissociative ionization with respect to simple ionization of the molecule is too small to be studied using the present device. However, for He^+ impact adequate intensities were obtained. Therefore, experimental studies, including investigating the angular distributions, using highly charged ion impact appear to be feasible.

Experimental difficulties that need to be overcome or improved upon include a better approximation to a point source. The simple hypodermic needle used does not provide an adequate target/background gas ratio. The test data also showed that weak electric fields were present in the target region; these fields influence the very low energy molecular ion emission which are useful for normalization purposes. Lastly, the field configurations of the ion transport systems need to be altered to eliminate possibilities of secondary electrons created along the flight path between the target region and the detector from reaching the other detector.

Acknowledgments:

This work was funded in part by the Office of Health and Environmental Research (OHER) under contract DE-AC06-76RL01830, the Fulbright Commision and the Bundesminister für Forschung und Technologie grant 060F/73 (741336476).

Supplement to Z. Phys. D - Atoms, Molecules and Clusters 21, S 261-262 (1991)

Atoms, Molecules
Zeitschrift für Physik D and Clusters

Dissociation of D_2 molecules in collisions with multiply charged ions

H. Lebius, H. Köhler, K. Wiesemann
Institut für Experimentalphysik AG II, Ruhr-Universität, W-4630 Bochum, FRG

B. A. Huber
Service Physique Atomique/PIM, Centre d'Etudes Nucleaires, F-38041 Grenoble - Cedex, France

Abstract

Dissociative electron capture has been studied in low-energetic collisions of multiply charged ions with simple diatomic molecules (H_2, D_2 ; $E \sim (1*q)$ keV). The different possible capture reactions are separated with the aid of time-of-flight spectra, which are obtained by the coincident measurement of projectile and recoil ions. In particular, we discuss the relative importance of dissociative single electron capture, its dependence on the projectile scattering angle as well as the type of the involved molecular transition.

I. Introduction

The different electron capture reactions which may occur in collisions of multiply charged ions with deuterium are summarized as follows:

$$A^{q+} + D_2 \rightarrow A^{(q-1)+*} + D_2^+ \,, \qquad (a)$$

$$\rightarrow A^{(q-1)+*} + D^+ + D \,, \qquad (b)$$

$$\rightarrow A^{(q-1)+} + 2D^+ + e^- \,, \qquad (c)$$

$$\rightarrow A^{(q-2)+*} + 2D^+ \,, \qquad (d)$$

where q represents the charge state of the incoming projectile. It is obvious, that double electron capture (d) as well as the transfer ionisation process (c) cause a 'Coulomb-explosion' of the molecular ion, as in both cases two electrons are lost by the target. On the other hand pure single electron capture may take place with (b) and without (a) dissociation of the target ion. The relative importance of both processes may depend on the projectile charge state, and parameters like the impact parameter as well as the 'orientation' of the molecular axis.

In reaction (b) fast deuterium atoms are produced the energy of which depends on the type of molecular transition, i.e. the internuclear distance of the molecule during the electron capture process. As neutral particles can penetrate a magnetic confinement, this process may contribute to the cooling of a magnetically confined plasma.

The dissociation of deuterium molecules induced by double electron capture processes has been studied by Giese et al [1]. In this contribution we consider the importance of 'pure' dissociative single electron capture and its dependence on the scattering angle of the projectile.

II. Experiment

A beam of multiply charged ions A^{q+}, which is delivered by the LAGRIPPA facility in Grenoble or by an ECR ion source at the university of Bochum with an energy of $\sim(1*q)$ keV, is intersected by an effusive beam of deuterium gas. After the collision the projectiles enter a retarding field analyser, the voltage of which can be set to allow for a transmission of $A^{(q-1)+}$ and $A^{(q-2)+}$ ions at the same time or for $A^{(q-2)+}$ ions only.

The recoil ions are extracted by a weak electrostatic field (20 V/cm) parallel or perpendicular with respect to the ion beam. They pass a drift region before being accelerated to the cathode of a channeltron. The recoil extraction efficiency depends on the kinetic energy of the recoil ions. In particular, dissociation products, which gain energy during the dissociation process, are detected at low extraction fields only, if the dissociation occurs in the direction towards the detection system (or in the opposite direction). If the molecular rotation during the capture process is neglected, this means, that the intermolecular axis must be oriented in the same way.

Both particles, projectile and recoil ions, are measured in coincidence. The resulting time-of-flight spectra yield information on the mass, charge state and the kinetic energy of the recoil ion and, if measured for different retarding fields, on the charge state of the projectile.

III. Results and discussion

In the following we describe results which we have obtained in the collision system Ar^{4+}/D_2 at collision energies of ~ 4100 eV. In figure 1 the corresponding time - of - flight spectrum is shown for a projectile scattering angle of 0.4 °. The upper part shows the

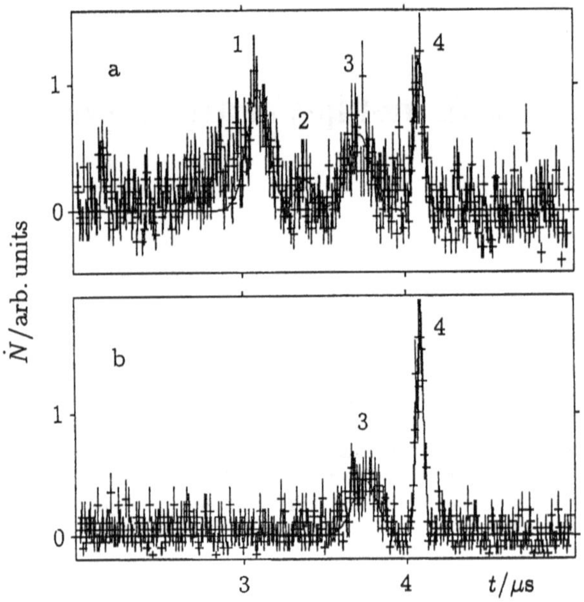

Fig. 1: Time - of - flight spectra obtained in Ar^{4+} / D_2 collisions. a) coincidences with Ar^{3+} and Ar^{2+} ions; b) coincidences with Ar^{2+}. Kinetic energy: 4100 eV; scattering angle: 0,4 °.

spectrum obtained when all secondary ions can pass the retarding field. Hence contributions from all reactions mentioned above are included. In the lower part only signals due to pure double elctron capture are recorded.

Whereas peak 1 corresponds to coincidences between Ar^{3+} and D_2^+ ions, according to reaction (a), peaks 2 and 3 are due to D^+ fragments which dissociate in two directions, antiparallel and parallel to the extraction field, respectively. Peak 4 is caused by coincidences between projetiles and photons which are emitted after the capture process.

In the case of double electron capture two deuterons are formed in the Coulomb - explosion, however, as to be seen in figure 1 b, only that one is registered, which arrives at first at the channeltron; i.e. that one which dissociates towards the detection unit. The same argumentation holds for the transfer excitation process. If a singly charged molecular ion dissociates (in reaction b), we obtain a D^+ fragment as well as a neutral deuterium atom, which is not measured. If the neutral atom escapes in forward direction, the corresponding D^+ ion, which starts in the opposite direction and which arrives later at the channeltron, contributes to peak 2 in figure 1 a. Therefore, this peak represents a monitor for dissociative single electron capture.

Peak 3 in figure 1 a may include 3 contributions from dissociative single electron capture, transfer ionisation and double capture as well. An analysis of the different peak heights shows the transfer ionisation process to be of minor importance in the present system. This is in agreement with earlier TES-measurements [2] and scaling law considerations [3].

The relative importance of reactions (a), (b) and (d) is illustrated in figure 2, which shows the corresponding ratios for different scattering angles. Comparing pure single electron capture with and without dissociation, one finds the latter one to be dominant in forward direction corresponding to large impact parameters. The dissociative single electron capture process becomes more important at larger scattering angles, which is correlated with a closer approach of both colliding particles. In this angular region contributions of reaction (b) are even larger than those of double

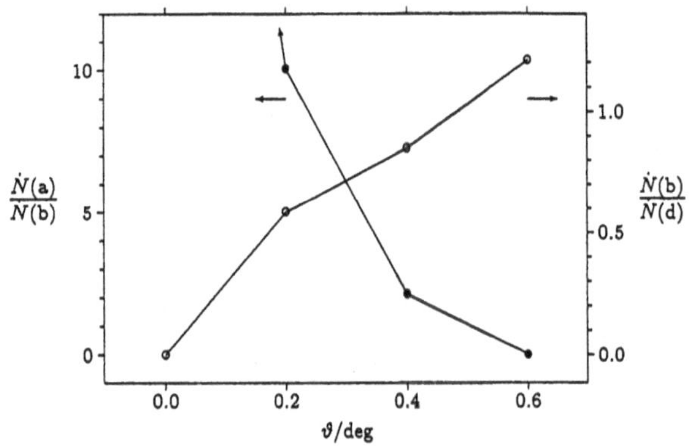

Fig. 2: Variation of counting rate ratios with the projectile scattering angle. Full symbols $\dot{N}$ (a) / $\dot{N}$ (b), left scale; open symbols $\dot{N}$ (b) / $\dot{N}$ (d), right scale.

electron capture. The dependence on parameters like charge state and orientation is not fully analysed at present; it will be discussed in a further paper.

Regarding the kinetic energy of the dissociation products, an analysis in the O^{5+}/He collision system at a collision energy of 4100 eV reveals a value of 8,1 eV which is close to the expected one for a Franck–Condon transition.

Acknowledgments

These experiments have been partly performed at LAGRIPPA - a joint laboratory of the CEA and of the CNRS. The authors are indebted to A. Brenac, G. Lamboley and T. Lamy for providing the ion beam. The financial support by the Deutsche Forschungsgemeinschaft is gratefully acknowledged.

References

1. Giese J.P., Cocke C.L., Waggoner W.T., Pederson J.O.K., Kamber E.Y., Tunnel L.N.: Phys. Rev. **A 38**, 4494 (1988)
2. Huber B.A., Kahlert H.J.: J.Phys.B: At.Mol.Phys. **13**, L159 (1980)
3. Müller A., Groh W., Salzborn E.: Phys.Rev.Lett. **51**, 107 (1983)

Supplement to Z. Phys. D - Atoms, Molecules and Clusters 21, S 263-265 (1991)

Atoms, Molecules
Zeitschrift für Physik D and Clusters

Satellite structure of chromium K-alpha x-rays, induced by 1.0 and 1.4 MeV/u heavy ions

F. Folkmann
Institute of Physics, University of Aarhus, DK-8000 Århus C, Denmark

Abstract

Cr Kα satellite lines are measured with high resolution from four Cr containing solid targets excited by 1.0 and 1.4 MeV/u ^1H, ^4He, ^{12}C, ^{16}O and ^{35}Cl ions. Differences are seen for heavy ions in the intensity pattern, which can be approximated by a binomial distribution, and are correlated to valency 3 or 6 of Cr (6 toxic).

1. Introduction

The purpose of this experiment is to test whether measurement of the satellite structure of characteristic x rays induced by heavy ions, which is known to depend on the number of nearby valence electrons[1], can be employed for useful determination of more detailed and interesting information on the chemical state of an element.

As an element of interest is chosen chromium, as it is much more toxic in the valency 6 state than in the valency 3 state, by roughly 3 order of magnitude. Cr is known to be dangerous to health and may cause irritant effects on the skin and respiratory passages. We find it therefore interesting to develop a method to measure not only the occurence of Cr in a sample, but also the valence state.

2. Experimental setup

To measure the Kα x rays of Cr with well separated satellite peaks corresponding to the number n of L vacancies present at the time when the K vacancy is filled by an L electron, we use diffraction in a cylindrically curved crystal and a position sensitive detector in a fixed setup [2].

Cr containing solid targets were excited by the following beams from the EN-tandem van de Graaff accelerator in Aarhus : $^1H^+$, $^4He^{++}$ $^{12}C^{4+}$, $^{16}O^{4+}$ and $^{35}Cl^{6-8+}$ at 1.0 and 1.4 MeV/u. To measure K alpha x rays from Cr, we used the same type of setup as described in [2] with a LiF(220) crystal of 2d=0.2848 nm curved to radius R=440 mm and the position sensitive flowmode proportional counter centered at $\theta_0=52.23^0$. This implies a distance from the center of the crystal to the detector of $R \cdot \sin\theta_0=348$ mm. The target-crystal distance was 160 mm.

The linearity of the position sensitive detector has been improved significantly, as shown in fig. 1. The amplifications in

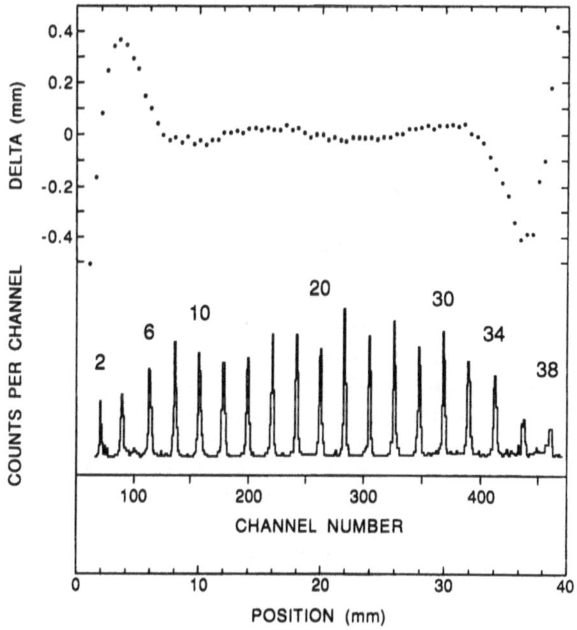

Fig. 1. Detector response measured for a ^{55}Fe source displaced by 2mm from 2 to 38 mm. POSITION depends by a linear fit on CHANNEL, and the deviation DELTA is shown on the top, so the real distance DIST = POSITION + DELTA.

the two branches were carefully adjusted to be equal and with zero offset using a pulser coupled parallel over a capacitor at the input to the preamplifier. For peaks in the central 20 mm of the detector, the mean deviation was previously [2] 172 µm, whereas it is now 18 µm in the same region, and even only 22 µm over a central optimum length of 27 mm, measured by 54 equidistant positions of a ^{55}Fe source [3]. The wings at the ends in fig. 1 are due to the backgammon readout pattern having the same length of 40 mm as the opening of the detector, and may be improved by a longer pattern. With this known length scale of the detector, wavelengths or energies are measured relative to one known x ray, e.g. proton excited V Kβ at 5.4273 keV using the Bragg relation [2]. This procedure is illustrated at the top of fig. 2, where it can be checked by Pr $L\beta_1$ at 5.489 keV, Sm $L\alpha_1$ at 5.636 keV and Nd $L\beta_1$ at 5.722 keV. Cr $K\alpha_2$ at 5.4055 keV and Cr $K\alpha_1$ at 5.4147 keV, are not resolved in these measurements.

3. Results and discussion

In addition to the 1.4 MeV proton excited spectra at the top of fig. 2 showing diagram lines, impact with heavier projectiles exhibits satellite peaks for Cr Kα in increasing degree for heavier, or higher nuclearly charged, projectiles at the same velocity. In figs. 3 and 4 are shown the measurements for fixed projectile and velocity incident on four different samples : green is Cr_2O_3, red is $K_2Cr_2O_7$, black is CrO_3 in formvar, and steel is stainless steel 304L with 18% Cr.

The satellite peaks are labelled by the number n of simultaneous L vacancies, and the intensity distribution among the lines was found by fitting lines of the same width to the spectrum. The satellite lines are seen to have nearly the same energy difference, measured near 30 eV, in accordance with a classical estimate [4] of (Z-4.15)·1.66 eV, giving 33 eV for Cr.

The extracted intensities I(n) can be fitted by a binomial distribution

$$I(n) = \binom{8}{n} \cdot p^{n} \cdot (1-p)^{8-n}$$

where p is the probability of ionization of the L shell. However, it turns out that the diagram line with n=0 comes generally stronger than from such a description, especially for steel samples bombarded with 1.4 MeV/u ^{16}O and ^{35}Cl, see fig. 4.

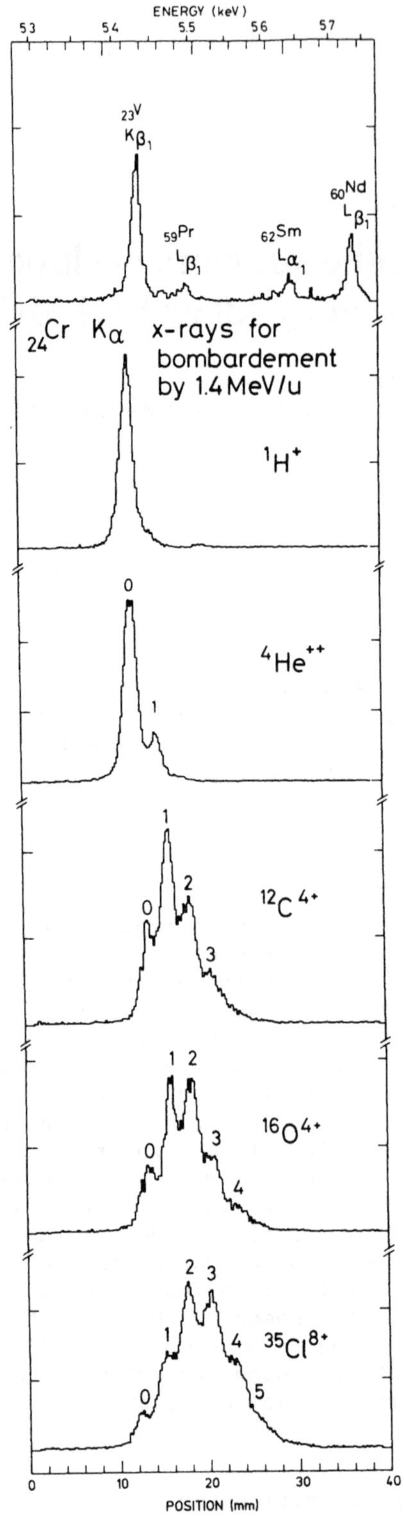

Fig. 2. Measured POSITION spectra, with energy scale on top, for solid targets bombarded with 1.4 MeV/u protons(top), alphas, carbon, oxygen and chlorine ions. The Cr was red $K_2Cr_2O_7$. For heavier ions a higher population of satellite lines, labelled by L vacancies, is seen.

There seems to be another mechanism populating this state, so we have chosen to exclude this line from the binomial fit, giving p=.227, .219, .218, .222 for fig.4; .181, .177, .176, .186 for fig. 3; and for n≥0 .009, .030, .177, .219, .317 in fig.2.

Based on our measurements with different projectiles we favour 1.4 MeV/u ^{16}O for seeing differences between Cr compounds. Although the primary ionization of Cr by heavy ion bombardment is the same, the number of valence electrons in the vicinity influences the measured distribution of satellite lines due to additional L-vacancies, probably because of capture prior to filling of the K vacancy [1]. A small difference in the relative intensity distribution of the Cr Kα satellite lines is found between the 4 analyzed compounds. In our data there is a tendency to be able to group the toxic valency 6 compounds having lower p values than the other (valency 3) compounds, which have only little or no toxicity. However, a more careful study involving also other solid effects should be undertaken.

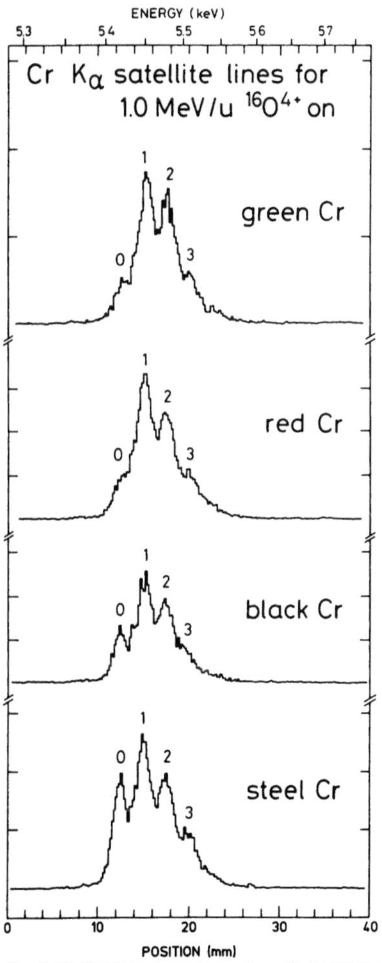

Fig. 3. POSITION spectra for 1.0 MeV/u $^{16}O^{4+}$ on four Cr targets where the red and black have valency 6 of Cr.

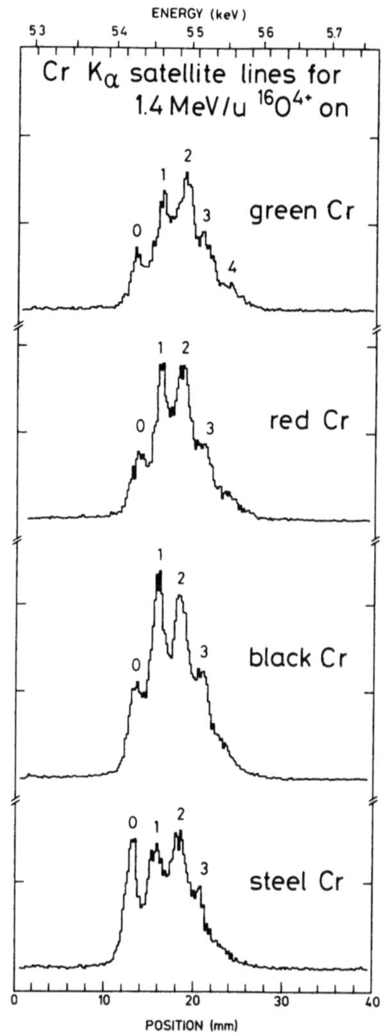

Fig. 4. POSITION spectra for 1.4 MeV/u $^{16}O^{4+}$ on four Cr targets where the red and black have valency 6 of Cr. Note the differences, e.g. the high intensity of the n=0 peak for the steel sample.

References

1. Watson, R.L., Leeper, A.K., Sonobe, B.I., Chiao, T., Jensen, F.E.: Phys.Rev.A15, 914(1977)
2. Folkmann, F., Frederiksen, F.: Nucl.Instr. Meth.B49, 126(1990)
3. Folkmann, F., Frederiksen, F., Loft Nielsen, H.: J.Physique(Paris)48, C9-99(1987)
4. Mokler, P.H., Folkmann, F.: "Structure and Collisions of Ions and Atoms", Sellin, I. A.(ed.), p.201. Berlin: Springer(1978)

Supplement to Z. Phys. D – Atoms, Molecules and Clusters 21, S 267-268 (1991)

Atoms, Molecules and Clusters
Zeitschrift für Physik D

A critical study of the molecular coulombic barrier model (MCBM) for multiple electron capture

L. Guillemot, P. Roncin, M. N. Gaboriaud, H. Laurent[1] and M. Barat

Laboratoire des Collisions Atomiques et Moléculaire[2], Université Paris-Sud bât. 351, F-91405 Orsay Cedex, France.

Experiments were performed at LAGRIPPA[3] CENG BP 85X, F-38041 Grenoble Cedex, France.

PACS: 34 70 – Charge transfer

Abstract: Multiple electron capture by highly charged ions in collision with rare gas targets is experimentally investigated. Differential cross sections are measured for specific processes in which both the number of electrons lost by the target and captured by the projectile are determined. This set of data is compared with the predictions of the Molecular Coulombic Barrier Model . The good predictive quality of this model is confirmed but deficiencies are pinpointed which are attributed to the neglect of "simultaneous" electron transfers as well as to the intrinsec static character of the model.

I Introduction

In multiply charged ions-atom collisions, at medium and low velocity, electron capture are by far the dominant inelastic processes. In this respect single capture has been the subject of many experimental and theoretical investigations and can now be considered to be fairly well understood. But the specificity of those ions is their large coulombic charge which makes them very greedy of electrons, so much that on multi-electron targets, processes where several electrons are transfered to the ionic projectile have important cross sections[1],[2]. They are generally described by the following type of reaction: $A^{q+} + B \rightarrow A^{(q-s)+} + B^{r+} + (r-s)\, e^-$, now on *referred to as a (s,r) process.* Such processes with many active electrons are quite exotic in the field of atomic physics and reach a high degree of complexity which still let us with only very partial understanding. This paper reports on a set of experimental data used to critically analyzed the Molecular Coulombic Barrier Model (MCBM). Although rather crude the MCBM offers an attractive framework for the description of multiple electron captures and it is known to fairly well account for the main physical properties of those exotic processes.

II Experimental study

This experimental study was carried out using a coincident multiparametric spectrometer which has been extensively described [3]. Details on the measurements made for the multi-electron captures can be found in [4]. We will just shortly recall here the very basics of our measurements. The scattered ionic projectiles are electrostatically analyzed and detected onto a bidimensional position sensitive detector. We thus determined there charge state $A^{(q-s)+}$, their energy (energy gain of the reaction) and the scattering angle in a vertical plane. A time of flight measurement of the ionized targets permits a *simultaneous* determinations of the recoil ion charge states B^{r+}.

Interesting information is brought by the angular information associated with the simultaneous determination of both partner charge states. It has been found that the angular scattering is strongly dependant on the the number **r** of electron removed from the target. On the contrary the angular scattering is very similar for processes of same r but different number **s** of electron found stabilized on the projectile. One can thus conclude that the main signature of the collision mechanisms lies in the degree of ionization of the target and on the other hand the s distributions should better be understood as resulting mainly from deexcitation of multiply excited states where the photon emission competes with electron ejection by autoionization.

III What about the theoretical description

We focused here on the MCBM presented by Niehaus in 1986[5] which is, after others[1],[6], the most sophisticated extension to multiple electron capture of the classical Over Barrier Model. The MCBM remains a classical and statical model in which the active electrons are considered to be independent particles. In its own special way the MCBM splits the collision trajectory in two phases: the 'way in' during which the target electrons are successively and systematically "molecularized" and the 'way out' during which the molecularized electrons can be either captured by the projectile or recaptured by the target with certain probabilities. This description brings a large set of different theoretical processes called "strings" each defining the number and the order (in the 'way out') of the captured electrons.

Since we found that the MCBM can not properly account for the different decays experimentally observed, the comparison between the MCBM predictions and the experimental data have been made in summing over the different decay schemes associated to a given number r of electrons removed from the target.

Our multiparametric measurements allow to perform this comparison for three different physical parameters : the relative cross sections, the angular differential cross sections and the energy gains.

- Turning to the relative cross sections σ_r of the processes where only the number r is defined it has been found, for all investigated systems, that the MCBM gives too much importance to the multiple captures (r>1) compared to single capture which is systematically measured to be more dominant.
- From the comparison of the angular profiles two main features emerge (i) the MCBM overestimates the contribution at large scattering angles (ii) the location of the model distribution maxima is often at too large angle even if some exception can be found, as for example in the Ar^{9+}+Ar system, the MCBM prediction is very satisfactory for capture of five electrons.
- It is noteworthy that the energy information is reduced to the mean values of the experimental distributions. The very large number of involved states and the limited energy resolution of the analyzer preclude from any more detailed identification. As it is well known the "Coulombic Barrier" model fairly well predicts

[1] I.P.N.O. – Bat. 100, BP No. 1, 91406 Orsay, France

[2] Unité associée CNRS No. 281

for multiple electron captures the MCBM tends to underestimate the energy gains, the discrepancies increasing with increasing number of captured electrons.

To further develop the analysis it is interesting to detail the model calculations in constructing partial cross sections associated to a given range of impact parameter trajectories. The string selections associated to given number r of captured electrons are splitted up in (i) the outer contribution given by the string describing the capture of the r outer target electrons (ii) the inner contributions by all the relevant strings requiring smaller impact parameters. The overall comparison ends up with the following paradoxical conclusion. The angular distributions clearly show an overestimation of the inner contributions. Better agreement is always reached when considering only the outer string (even if this outer string generally leads to too large mean scattering angles). In the same time these outer contributions of the MCBM predicts too small energy gains, that naturally leads to an improvement of the predictions when taking into account all the inner contributions.

IV Discussion

We now discuss the above mentioned conclusions in the framework of a quasi-molecular description of the collision in dealing with two enlightening examples: triple capture in the two isocharged systems O^{8+}+Ar and Ar^{8+}+Ar. They have been chosen because in the former case the MCBM properly reproduces the angular DCS and not in the latter case where the calculation gives too large scattering angles. In such complex processes a quasi-molecular description can only be achieved at the expense of a drastic simplification. The triple capture is reduced to a four molecular state problem with the entrance, one single, one double and one triple capture channel. We built some 'semi empirical' potential curves in taking for the three latter channels purely coulombic r(q-r)/R curves asymptotically adjusted on the mean values of the experimental energy gain distributions (fig 1). It appears that two very different crossing topologies are encountered. In the O^{8+}+Ar case triple capture channel is expected to be populated via three successive transitions at three well separated crossings, this basically should make the MCBM description relevant. Such is not the case for the Ar^{8+}+Ar system where a sort of triple crossing point between the three capture channels emerges and makes the transitions to double and triple capture channels likely to occur close to this internuclear distance.

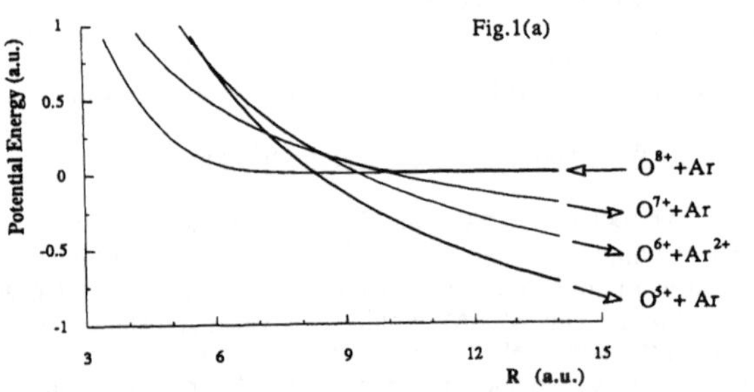

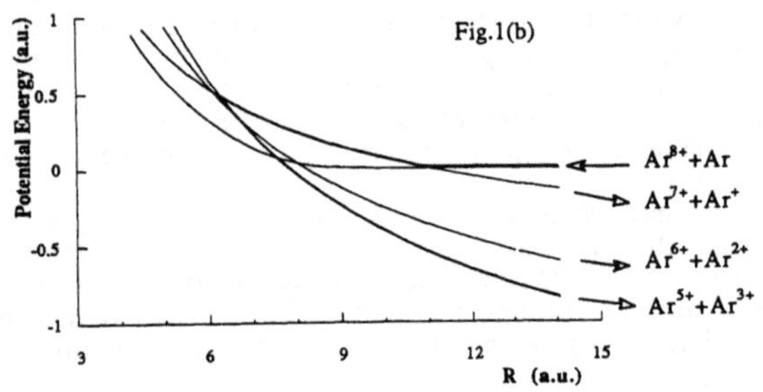

Fig 1 : Semi empirical purely coulombic potential energy curves for triple electron capture in (a) the O^{8+}+Ar system, (b) the Ar^{8+}+Ar system

V Conclusion

The present analysis points out one of the main conceptual weakness of the MCBM which is the absence of any simultaneous electron transitions. From the study of several double capture processes[7],[8] it is known that 'simultaneous' capture are commonly encountered when the first electron has not been captured in the main single capture channel, *on the way in*. It can thus be understood that indeed the lack of any simultaneous electron transitions in the MCBM should explain the overestimation of the scattering angle as well as the underestimation of the energy gains. It must be emphasized that any improvement of the model along the line of this analysis should lead to a reformulation of the MCBM in which several channels are open also in the 'way in' instead of systematic 'molecularization'.

References

[1] L., Astner G., Barany A., Cederquist H., Danared H., Huldt S., Hvelplund P., Johnson A., Knudsen H., Rensfelt K.G.1986 *Phys.Scr.* **33** 310
[2] Danared H., Anderson H., Astner G., Barany A., Defrance P., Rachafi S.1987 *J.Phys.B : At.Mol.Phys.* **20** L165
[3] Barat M., Gaboriaud M.N., Guillemot L., Roncin P., Laurent H.and Andriamonje S.1987 *J.Phys.B : At.Mol.Phys.* **20** 5771
[4] Guillemot L., Roncin P., Gaboriaud M.N., Laurent H.and Barat M.1990 *J.Phys.B : At.Mol.and Opt. Phys.* **23** 4293-4312
[5] Niehaus A.1986 *J.Phys.B : At.Mol.Phys.* **19** 2925
[6] Barany A., Astner G., Cederquist H., Danared H., Huldt S., Hvelplund P., Johnson A., Knudsen H., Liljeby L., Rensfelt K.G.1985 *Nucl. Instrum.Meth.* **B 9** 397
[7] Laurent H., Barat M., Gaboriaud M.N., Guillemot L.and Roncin P. 1987 *J. Phys.B: At.Mol.Phys.* **20** 6581
[8] Roncin P., Barat M., Gaboriaud M.N., Guillemot L. and Laurent H. 1989 *J.Phys.B: At. Mol. Phys.* **22** 509.

Supplement to Z. Phys. D – Atoms, Molecules and Clusters 21, S 269–270 (1991)

Atoms, Molecules
Zeitschrift für Physik D and Clusters

State-selective capture in collisions between ions and alkali-metal atoms

J. Pascale[1], R. E. Olson[2], and C. O. Reinhold[2]

[1] Service de physique des Atomes et des Surfaces, 91190 Gif-sur-Yvette Cedex, France
[2] Department of Physics, University of Missouri-Rolla, Rolla, MO 65401, USA

Abstract

We have calculated final n, l, m-distributions for the electron capture in Na^+ + Na(28d) collisions and N^{5+}, Ar^{8+} + Cs(6s) collisions using a three-body classical trajectory Monte-Carlo (CTMC) method. The collision velocity range is $0.5 \lesssim x \lesssim 1.7$. Here, $x = v_p/v_e$, where v_p is the ion projectile velocity in the laboratory frame and v_e is the initial orbital velocity of the electron bound to the alkali-metal core. The results are discussed along with experimental data.

I. Introduction

Electron capture between multiply-charged ions (MCI) and neutral atoms may occur with high efficiency in the low to intermediate energy range, and the electron is captured to highly excited level [1]. If the target atom is excited initially in a high Rydberg state, the efficiency of the process may be so large that it would be possible to capture two electrons in two successive collisions of the MCI with the Rydberg atom. Thus, di-excited MCI would be produced. However, in such an experiment, it is very important to know the n, l, m- distributions of the captured electron after the first collision. We have determined these distributions in the case of Na^+ + Na(28d) collisions and MCI + Cs(6s) collisions (MCI = N^{5+}, Ar^{8+}) , for which experimental data for the n-distributions are available [2, 3].

II. Theoretical approach

For these collisions involving Rydberg sates, an approach based on quantum or semiclassical methods is practically intractable in view of the large number of channels that are coupled, which must also include generally the ionization and excitation channels. Perturbative methods are also inappropriate because of the strength of the Coulomb interactions. In both types of collisions presently studied, the electron is captured in a high Rydberg state, and therefore the electron capture is expected to be well described by a classical model. However, the two types of collisions differ much in the initial state of the target atom. Then, the cross sections may be sensitive to the choice of the interaction between the electron and the ionic cores. For the present calculations we have used the CTMC method [4]. It is based on solving the Hamilton's equation of motion for the three body system (active electron, target-core, ionic projectile) for a given set of initial conditions for the target and the projectile. The method of Reinhold and Falcon [5], which avoids the integration of the Kepler's equations as usually done in other CTMC calculations, has been employed. A pure Coulomb interaction and a more realistic model potential have been used for the electron core interactions. The model potential has the correct behavior at small and large distances, and reproduces to a good accuracy the energy levels of the electron-core system. The final n, l, m-distribution of the captured electron is determined from the calculated binding energy of the electron relative to the ionic projectile, and from its classical angular momentum and its projection along the initial velocity of the projectile.

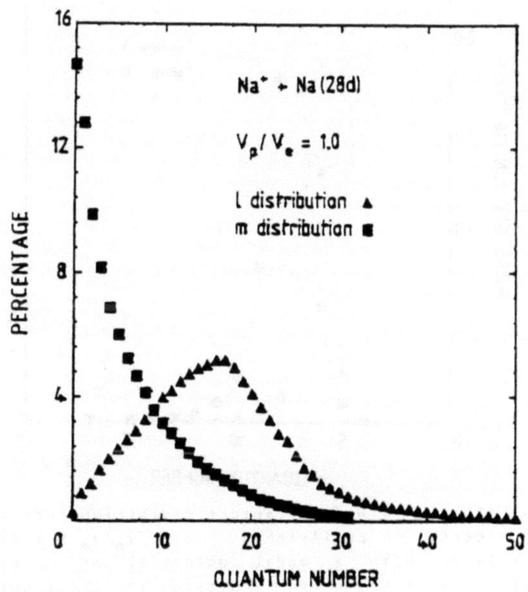

Fig. 1. The calculated percentage of final m and l values for the electron capture in Na^+ + Na(28d) collisions, at v_p/v_e = 1.0.

III. Results and discussion

The details and the full results of our CTMC calculations are reported elsewhere [6].

$Na^+ + Na(28d)$ collisions

The calculations have been performed with up to 10^6 trajectories, for x = 1.0, and 1.658. The total cross sections for the electron capture have not been found to depend very sensitively on the choice of the interactions. Our results confirm the disagreement which was previously found, for x = 1.0, between low scale CTMC calculations [7] and experimental data [2] for the position of the peak in the n-distribution. The relative experimental n-distribution was determined from a field ionization technique, with the basic assumption of a preferential population of final states with m = 0 and 1. The present calculations of the n, l, m-distributions show the importance of the electron capture into states with m>1 (see Fig. 1) for both values of x. Therefore, the analysis of the experimental data by the field ionization technique should be revided. Indeed, MacAdam et al [8] have come recently to the same conclusion. Large final l-values are also obtained (see Fig. 1), which are explained by the large impact parameters which contribute mainly to the electron capture.

MCI + Cs(6s) collisions

The calculations have been performed for x = 0.894 and 0.529, and with 2 10^4 and 1.2 10^4 trajectories for N^{5+} and Ar^{8+}, respectively. The target core effects are found to be quite significant for both the magnitude and the position of the peak of the n-distribution (see Fig. 2). Our results are in good agreement with recent measurements [3] if a

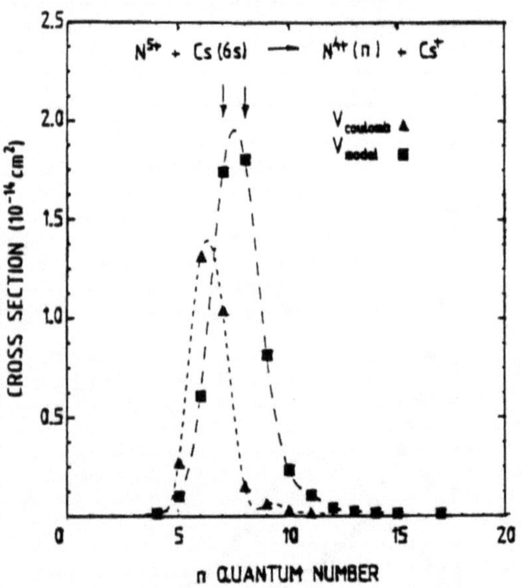

Fig. 2. The electron capture n-distributions for N^{5+} + Cs(6s) collisions, at v_p/v_e = 0.894, calculated with a model potential or a pure Coulomb interaction as indicated in the figure. The position of the peak is compared to the experimental findings of Martin et al [3] (arrows)

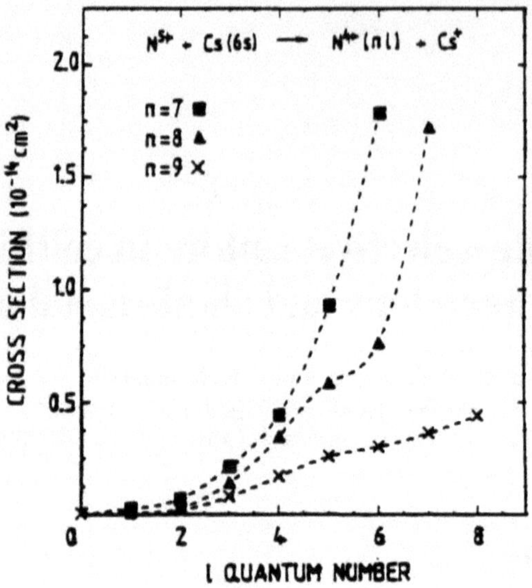

Fig. 3. The calculated n, l-distributions for the electron capture in N^{5+} + Cs(6s) collisions, at v_p/v_e = 0.894 and for n = 7, 8, 9.

realistic model potential is used for the electron-Cs^+ interaction (see Fig. 2). It is to be noted that the experimental data where obtained unambiguously from the relative intensities of the observed lines. Our calculations show also a predominance of the electron capture to states with large l-values (see Fig. 3).

IV. Conclusion

For both types of collisions presently studied, our CTMC calculations have found m-distributions of the final states which are peaked for m = 0, but broad. This result indicates that diabatic ionization processes should not be neglected in the analysis of experimental data by means of a field ionization technique. Our results also show that final states with large l-values are easily obtained in the electron capture process, particularly in the case of MCI + Cs(6s) collisions where a predominance of electron capture to nearly circular states is predicted.

References

1. Olson, R.E. : J. Phys, B13, 843 (1980)
2. Rolfes, R.G. and MacAdam, K.B. : J. Phys. B15, 459 (1982)
3. Martin, S., Salmoun, A., Do Cao, G., Bouchama, T., Andra, J. and Druetta, M. : Journal de Physique 50, C1-357 (1989)
4. Abrines, R. and Percival, I.C. : Proc. Phys. Soc. 88, 861 (1966)
5. Reinhold, C.O. and Falcon, C.A. : Phys. Rev. A33, 3859 (1986)
6. Pascale, J., Olson, R.E., and Reinhold, C.O. : to appear in Phys. Rev. A (1990)
7. Becker, R.C. and MacKellar, A.D. : J. Phys. B17, 3923 (1984)
8. MacAdam, K.B., Gray, L.G. and Rolfes, R.G. : Bull. Am. Phys. Soc. 35, 1167 (1990)

Supplement to Z. Phys. D - Atoms, Molecules and Clusters 21, S 271-272 (1991)

Atoms, Molecules
Zeitschrift für Physik D and Clusters

Electron capture by metastable Ar^{8+} ($2p^5$ 3s) ions in He

H. Lebius
Institut für Experimentalphysik AG II, Ruhr-Universität Bochum, W-4630 Bochum, FRG

B. A. Huber
SPhAt / PIM, Centre d'Etudes Nucleaires de Grenoble, F-38041 Grenoble - Cedex, France

Abstract

Single electron capture in low-energetic Ar^{8+} / He collisions has been studied, using projectiles in the ($2p^6\ ^1S_0$) ground state as well as in the excited ($2p^5\ ^3P_{2,0}$) metastable states. In both cases capture occurs dominantly into the n = 4 levels with slightly different l - distributions. In the case of the metastable species the relative fractions of the final states which decay by autoionising or by photon emission are discussed and compared with theoretical predictions.

I. Introduction

Single electron capture by Ar^{8+} in He produces singly excited states according to the equation:

$$Ar^{8+}(2p^6\ ^1S_0) + He \rightarrow Ar^{7+*}(2p^6nl\ ^2L) + He^+ , \quad (1)$$

which decay by photon emission. If the reaction is performed with metastable projectiles - $Ar^{8+}(2p^5 3s\ ^3P_{2,0})$ - states are populated, which are energetically degenerate with the ionization continuum of Ar^{7+}:

$$Ar^{8+*}(2p^5 3s\ ^3P_{2,0}) + He \rightarrow Ar^{7+*}(2p^5 3s\, nl\ ^{2,4}L) + He^+. \quad (2)$$

These states may autoionise (doublet states) or may be deexcited by photon emission (quartet states). The metastability of the quartet states is to be expected only if the LS - coupling scheme is applicable.

The collision system Ar^{8+} / He has been studied experimentally by several groups applying different techniques. Among the state - selective studies the spectroscopy of electrons [1-3] is sensitive to the autoionising states formed in reaction (2), whereas the observation of the emitted light [4] yielded information on the nl - distribution of states, which deexcite by photon emission (limited to reaction (1)). Furthermore, theoretical results [4,5] exist concerning the final state population of the 'ground state reaction' (1).

In the present paper we describe the results which we have obtained by applying the translational energy spectroscopy (TES) technique with a state prepared ion beam. These measurements complete the results of the above mentioned studies and allow a more detailed discussion of the decay of the states populated in reaction (2).

II. Experiment

A beam of multiply charged ions is delivered by the LAGRIPPA facility at a collision energy of about $(1 \cdot q)$ keV. In order to narrow the energetic width of the ion beam, it passes an energy monochromator with an overall resolution of ~ 2000 before entering the collision region. By this means the halfwidth of the energy distribution is limited to values below $(0{,}5 \cdot q)$eV.

Secondary ions are analysed at different scattering angles with respect to their charge state as well as to their kinetic energy with the aid of a second energy analyser. From an exact measurement of the kinetic energies of the projectiles the excitation energy and hence the final state population is determined. A detailed description of the translational energy spectrometer and the TES - technique is given elsewhere [6].

In order to increase the metastable fraction of the ion beam an electron capture reaction has been used in the beamline of LAGRIPPA between the two mass / charge analysing magnets. Whereas the first magnet was set to transmit primary Ar ions in charge state 9, the second one could be passed only by secondary Ar^{8+} ions, formed in a gas target in between. The corresponding capture reaction is described by the following equation:

$$Ar^{9+}(2p^5) + X \rightarrow Ar^{8+*}\left(2p^5 \begin{Bmatrix} 4l \\ 5l \end{Bmatrix}\right) + X^+ , \quad (3)$$

with $X = N_2$, He and air.

The excited secondary ions decay with high probability into metastable Ar^{8+} states, which are monitored by means of reaction (2). Analysing the autoionisation signal we find the metastable fraction f of the prepared ion beam to be increased by a factor of ~ 15 with respect to the unprepared ion beam if air is used as a target gas (estimated values for f: 49 % and 3 %, respectively). A similar effect is obtained with nitrogen; in the case of He the efficiency for producing metastable states is found to be smaller by a factor of ~ 3.

III. Results and discussion

At a collision energy of 8100 eV we have measured the translational energy spectra of Ar^{7+} using ion beams with low and with high metastable fractions. In the first case the spectrum is assumed to be produced only by ground state ions; the change of the spectrum found in the second case is attributed to the influence of the metastable ions. Taking into account the estimated values for the metastable fractions it is possible to derive a translational energy spectrum due to the metastable species only. Both spectra are given in figure 1.

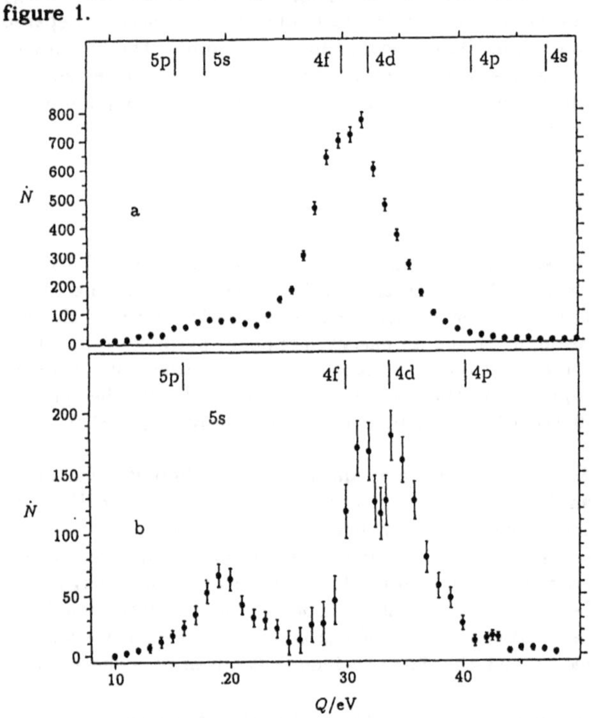

Fig. 1: Energy gain spectra of Ar^{7+} ions produced in Ar^{8+}/ He collisions. a) $Ar^{8+}(2p^6\,^1S_0)$, b) $Ar^{8+}(2p^5 3s\,^3P_{2,0})$. Collision energy: 8100 eV; scattering angle: 0°.

The capture by ground state ions occurs dominantly into the 4d and 4f levels and - to a smaller amount - into 5s and 4p. The measured relative populations obtained by a deconvolution procedure are in good agreement with theoretical and other TES-results as to be seen in Table 1. Some deviations occur with respect to the optical results at these low energies.

Considering metastable projectiles, we find single electron capture to create the $(2p^5 3s\, nl)$ configuration with the same nl - levels as before. However, the relative importance of the sublevels is slightly shifted towards higher energies (4d : 44%, 4f : 41 %, 5s : 16 %). In addition the peak positions (compared in figure 1 with theoretical values [7]) are shifted due to the different screening of a 3s and a 2p electron.

Table 1: Relative population of excited Ar^{7+} states in $Ar^{8+}(2p^6\,^1S)$ / He collisions given in %.

reference	4s	4p	4d	4f	5s
present exp	-	4	56	34	6
present MCLZ	-	3	57	29	6
VUV-spectr. [4]	9	31	41	29	
theory [4]	-	7	55	38	
theory [5]	-	11	61	28	
TES [9]	-	2	- 91	-	9

The spectrum (b) represents only that part of reaction (2), which is metastable with respect to autoionisation. The other part is seen in the energy gain spectrum of Ar^{8+}, which is less well resolved due to the broadening by the emitted electron. However, by an analysis of both spectra the relative fraction of final states which autoionise can be estimated. We found that 54 % of the reaction products decay by photon emission whereas 46 % autoionise. The peak positions in fig. 1b suggest that the measured non-autoionising signal is due to the quartet states.

If doublet and quartet states are populated according to their statistical weights and if the LS - coupling holds rigorously, an autoionising fraction of 33 % should be expected. Recent calculations [8] regarding the decay of the $(2p^5 3s\, 4l)$ - configuration yield an averaged value of ~ 80 %.

Acknowledgments

These expriments have been performed at LAGRIPPA - a Joint Laboratory of the CEA and of the CNRS. The authors are indebted to A.Brenac, G.Lamboley and T.Lamy for providing normal and state - prepared ion beams. We thank Dr. Luc - Koenig, P. Roncin and M. Barat for communicating data before publication.

References

1. Bordenave - Montesquieu A., Benoit - Cattin P., Boudjema M., Gleizes A., Dousson S.: Nucl. Instr. Meth. Phys. Res. **B 23**, 94 (1987)
2. Boudjema M., Benoit-Cattin P., Bordenave-Montesquieu A., Gleizes A.: J. Phys. B: At. Mol. Opt. Phys. **21**, 1603 (1988)
3. Hutton R., Prior M. H., Chantrenne S., Mau Hsiung Chen, Schneider D.: Phys. Rev **A 39**, 4902 (1989)
4. Druetta M., Martin S., Bouchama T., Harel C., Jouin H.: Phys. Rev. A **36**, 3071 (1987)
5. McCarroll R. in 'Recent studies in Atomic and Molecular processes', ed. Kingston A. E., Plenum Press New York and London (1987), p.113
6. Huber B.A.: Comments At. Mol. Phys. **21**, 15 (1987)
7. Desclaux J.P., private communication
8. Luc - Koenig E., Bauche J.: J. Phys. B.: At. Mol. Opt. Phys. **23**, 1763 (1990)
9. Barat M., Roncin P., private communication

Supplement to Z. Phys. D - Atoms, Molecules and Clusters 21, S 273–274 (1991)

Atoms, Molecules
Zeitschrift für Physik D and Clusters

Population of excited states by charge exchange in multiply-charged ion–atom collisions

V. V. Afrosimov
Ioffe Physical-Technical Institute, USSR Academy of Sciences, Leningrad 194021, USSR

Abstract
Using the "collisional spectroscopy" technique, absolute cross-sections for different (n,ℓ)-excited state populations in electron capture by multiply charged ions from hydrogen atoms have been measured. Comparison of experimental and calculated data for He^{2+} – H and Ar^{6+} – H shows that the population of (n,ℓ) states can be easily determined using scaling [1] and the multi-level Landau-Zener model.

I. Introduction

The multiply charged ion - hydrogen atom collision, A^{z+} + H, is close to a system "electron in the field of two Coulomb centers" and therefore it is convenient for making the choice of an adequate collision model from comparison between theoretical and experimental data. Charge exchange is the main process for this system at low energies, it has the largest cross section ($10^{-15} - 10^{-14}$ cm^2) and it leads primarily to the population of highly excited states forming $A^{(z-1)+}$ ions. The decay of these states is accompanied by emission of characteristic lines which are used for impurity diagnostics in high-temperature hydrogen plasmas used in CTR. This diagnostics is based on the injection of a beam of hydrogen atoms into the plasma and on the analysis of the burst of radiation associated with the injection. It is important for this diagnostics to choose a simple way to estimate the state population for different ions and for a wide range of velocities. The experimental data obtained in this work are used to proof some of such ways.

II. Results

1) n-state population

It follows from numerous available data on the charge exchange in $A^{z+} + H(n_0)$ collisions that the states of $A^{(z-1)+}$ ions with $n_m = n_0 z^{3/4}$ are mainly populated (see for example [2]). The Bohr velocity for the n_m level is $v_{em} = z^{1/4}/n_0$. An analysis of existing theoretical calculations of n-state population probabilities P(n) by charge exchange in A^{z+} – H systems [1] showed that the ratio F=P(n)/P(m) for excited H-like ions $A^{(z-1)+}$ has the same dependence on the parameter $t = n/n_m$ for different z values. Furthermore, the F(t) function depends, equally for different z, on another parameter, $S = v/v_{em} = vn_0/z^{1/4}$, where v is the relative velocity of the colliding particles.

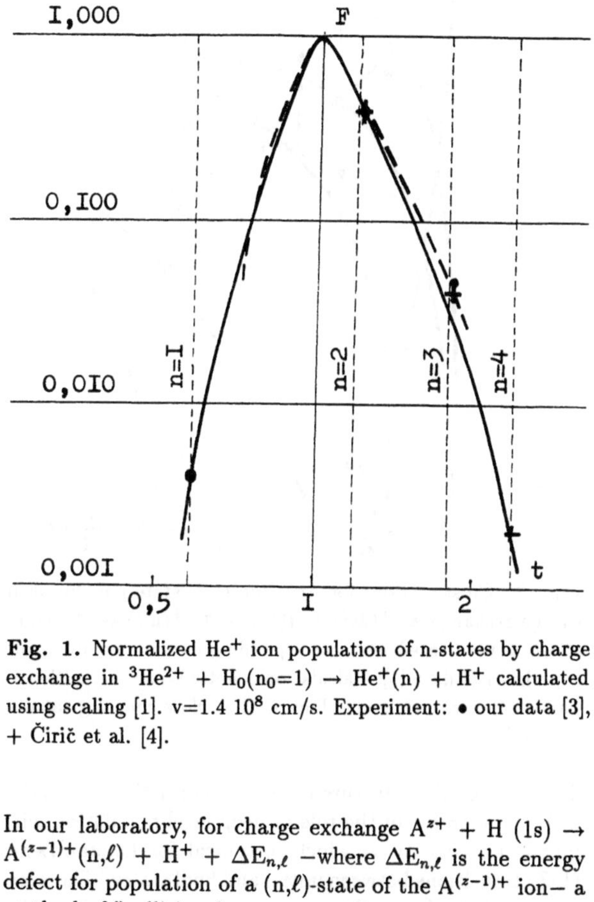

Fig. 1. Normalized He^+ ion population of n-states by charge exchange in $^3He^{2+} + H_0(n_0{=}1) \rightarrow He^+(n) + H^+$ calculated using scaling [1]. v=1.4 10^8 cm/s. Experiment: • our data [3], + Čirič et al. [4].

In our laboratory, for charge exchange A^{z+} + H (1s) $\rightarrow$ $A^{(z-1)+}(n,\ell) + H^+ + \Delta E_{n,\ell}$ –where $\Delta E_{n,\ell}$ is the energy defect for population of a (n,ℓ)-state of the $A^{(z-1)+}$ ion– a method of "collisional spectroscopy" was used to measure the $A^{(z-1)+}$ ion state population. The method consists in detection of all $A^{(z-1)+}$ ions with particular $\Delta E_{n,\ell}$ by precise analysis of their kinetic energy in an electrostatic analyzer with the resolution $\approx 6 \times 10^3$. Fig. 1, showing the function F (t) illustrates the applicability of the scaling proposed by Zinoviev and Korotkov [1].

It should be noted that the existence of such scaling seems natural because of similar change of parameters of the system Ar^{z+} + H with the charge z of one of the Coulomb centers and also because of the small role of the Ar^{z+} electron core in charge exchange which occurs at large internuclear distances.

2) (n,ℓ) state population

A more detailed way to choose the simplest and at the same time reliable model for estimation of the state population cross section is the comparison of experimental and calculated $\sigma_{n,\ell}$ data for states which differ not only in principal numbers n, but also in orbital ones, ℓ.

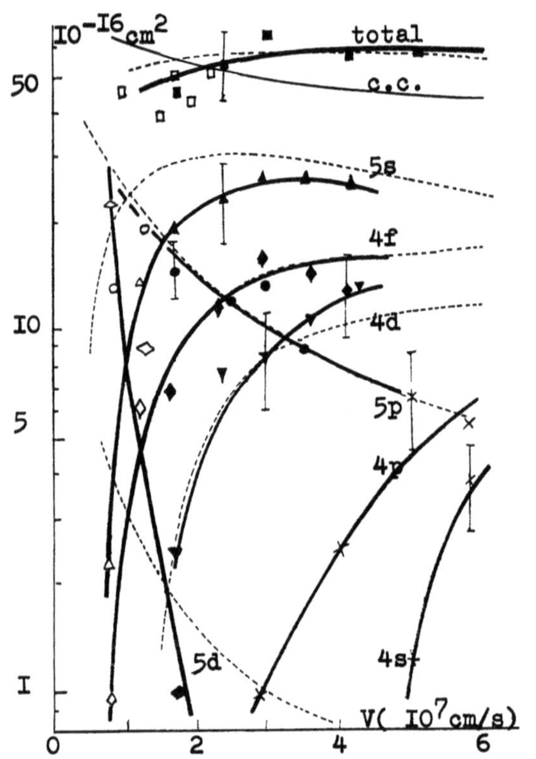

Fig. 2. Cross sections for Ar^{5+} ion (n,ℓ)-state population in charge exchange $Ar^{6+}(2s^2)$ + H(1s) → Ar^{5+}(n,ℓ) + H^+. (n,ℓ)-states are indicated near corresponding curves. Points – our experimental data [5]; Calculations: dotted lines – multi-level Landau-Zener model; solid lines C.C. – close coupling calculations [7].

The characteristic feature of the $\sigma_{n,\ell}(v)$ dependence (Fig. 2) is the increase of the role of deep level population with increasing velocity v, which is typical for Landau-Zener (L-Z) transitions between initial and final states. Therefore theroretical estimates have been performed using the multi-level L-Z model. The initial term was taken as $U_0(R) = -I + \alpha z^2/(2R^4)$, where I and α are the ionization potential and the polarizability of the hydrogen atom, respectively. The final term is $U_k(R) = -E_k + (z-1)/R$, where E_k is the binding energy of the electron in a k-state of the $A^{(z-1)+}$ ion. The locations of pseudocrossings R_{ck} can be derived from the equality $U_0(R_{ck}) = U_k(R_{ck})$. Matrix element values were calculated in accordance with [6]. The cross sections for each (n,ℓ)-state population were calculated using integration over all impact parameters taking into account transitions at all pseudocrossings which the system of colliding particles passes on the incoming and outgoing trajectory.

Fig. 2 testifies the good applicability of the relatively simple multi-level L-Z model to all the states which have their pseudocrossings with the initial term at internuclear distances:

$$R > 1/I \times (\sqrt{2I} - 1 + \sqrt{2(z-1) + 2I}) \text{ a. u.},$$

i.e. in the region where the k-state energy level of the Ar^{5+} ion and the 1s-level of the hydrogen atom are separated by the potential barrier.

III. Conclusion

At present there exist sufficiently simple and reliable means for both experimental determination and calculation of the state population by charge exchange of multi-charged ions. The spectroscopy of radiation accompanying plasma ion charge exchange A^{z+} + H, (CHES-diagnostics) proposed by us more than 10 years ago [8] still remains practically the only way for studying composition, concentration and spatial distribution of fully stripped ions in the plasma. One should note that quantitative CHES-measurements require taking into account beam H-atom excitation and mixing of states of the $A^{(z-1)+}$ ions by the influence of collisions and fields in the plasma. Fortunately, these corrections are developed now also quite well.

REFERENCES

1. A.N.Zinoviev, A.A.Korotkov. Pis'ma Zh. Eksp. Teor. Fiz. **50**, 276, 1989.
2. R.K.Janev, H.Winter. Physics Reports **117**, 378, 1985.
3. M.N.Panov et.al. Phys. electr. a. atom. coll., Leningrad, III, 25, 1987.
4. P.Čirič et.al. J.Phys B **18**, 4745, 1985.
5. V.V.Afrosimov et.al. Pis'ma Zh. Eksp. Teor. Fiz. **46**, 89, 1987.
6. R.E.Olson, A.Salop, Phys.Rev. A **14**, 579 (1976)
7. O.Y.Larsen, K.Taulbjerg. J.Phys. B **17**, 4523, 1984.
8. V.V.Afrosimov et.al. Pis'ma Zh. Tekhn. Fiz. **3**, 77, 1977.

Supplement to Z. Phys. D - Atoms, Molecules and Clusters 21, S 275-276 (1991)

Atoms, Molecules and Clusters
Zeitschrift für Physik D

Subshell selective electron capture in collisions of X^{5+} (X = C, N, O, Ar) with H_2 at 50 keV

A. Salmoun[1], **A. El Motassadeq**[2], **and M. Druetta**[3]

[1] Departement de Physique Université, Cadi Ayyad Boulevard, Moulay Abdellah B.P.S. 15, Marrakech (Maroc)
[2] Laboratoire de Spectrometrie Ionique et Moleculaire (U.A. CNRS 171), Campus de la Doua, F-69622 Villeurbanne Cedex, France
[3] Laboratoire Traitement du Signal et Instrumentation (U.A. CNRS 842), Université de Saint Etienne, 23, rue du Dr. P. Michelon, F-42023 Saint Etienne Cedex, France

Abstract :

In order to compare selective capture by ions of the same charge from H_2, we have studied charge exchange collisions of X^{5+} (X=C,N,O,Ar) with H_2 by employing VUV spectroscopy. Partial cross sections $\sigma_{n,\ell}$ are measured. The total single-electron capture finally leading to the ground state (excluding the TE process) is found to be the same for all ions.

I. Introduction:

Theoreticaly, a regularity must be felt in the behaviour of the total cross section of single electron capture for ions of identical charge. We have studied by U.V spectroscopy the charge exchange collisions of X^{5+} (X= C,N,O,Ar) with H_2 target. These measurements lead to the selective electron capture and the total cross section may be deduced .

II. Experiment:

The experimental set-up has already been described [1, 2]. The ion beam is produced by an E.C.R ion source. The light emitted during the collision is observed at 25° to the beam axis with a 3 m grazing incidence(82°) spectrometer equipped with 300 or 600 lines/mm gratings blazed at 55.2 or 27.6 nm respectively. The photon detection is insured by MgF_2 coated microchannel plates (MCP). The gas pressure was kept at 5×10^{-5} mbar. The calibration of the detection sensitivity has been obtained by using a set of known line emission cross sections [3]

III. Results and discussion:

Spectra have been recorded in the 10 -100 nm spectral range. The observed lines correspond to 2-3, 2-4 and 3-4 transitions. A typical spectrum of O^{5+} + H_2 collisions is displayed in figure 1.

This study concerns the single electron capture into the ground state of each ion. Capture into metastable statesis also observed (1s2s ^3S N^{5+}) as well as single capture and excitation of a core electron (transfer excitation TE ; $1s2p^2n\ell$ OV, $1s2pn\ell$ NV, $3s3p^2n\ell$ ArV). We have measured 20% as fraction of TE to the total single charge exchange for the O^{5+} + H_2 system. This result and other work in our laboratory [4] show that the TE may be important.

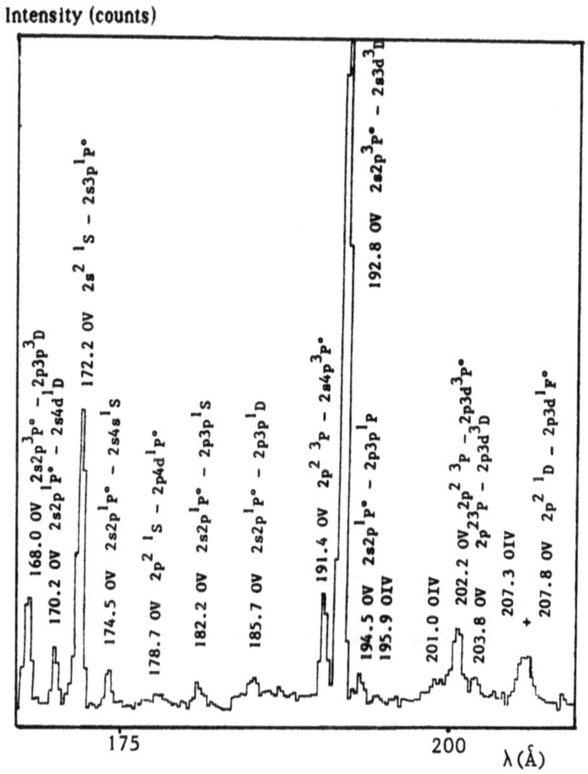

Figure 1. Partial spectrum obtained during the collision of O^{5+}(50 kev) with H_2

This process deserves a study to search for the involved parameters (energy, choice of target and projectile).

Transitions in X^{3+} ions have also been observed but with H_2 as gas target they are mainly due to double collision processes.

Partial cross sections ($\sigma_{n\ell}$) are deduced from line emission cross sections. For CV cross sections of Lβ and Lγ have been used [5]. The total single electron capture into the ground state (excluding the TE process) is found to be the same for all ions owing to the error bars. Our measurements compare favorably with the $\sigma_{q,q-1}$ results [6,7,8]. The results show also that triplet to singlet capture cross section for CV , OV and ArV is not found statistical as measured for other ions [9,10]. The Lennon et al study [11] on C^{3+} + He collision as function of the energy shows that the $^3P°$ state is less populated than the $^1P°$ state for energies < 20 kev and will be more populated over 20kev. This problem deserves a study in function of the energy and for differents projectiles.

Table 1. Excitation cross sections for X^{5+} + H_2 collisions $\sigma_{n\ell}$ ($x10^{-17}$ cm^2 ± 30 %)

σ(nℓ)	CV	NV	OV	ArV
4s	10	6	16	20
4p	33	9	8	100
4d	30	13	15	21
4f	20	19	19	36
Σσ(4ℓ)	93	47	58	177
3s	35	30	81	
3p	55	80	29	0
3d	61	70	54	80
Σσ(3ℓ)	151	180	164	80
*$\Sigma_{nl}\sigma(n\ell)$	244	227	222	257
other result	235[a]	250[a]	230[b]	342[c]

[a] Ref [5] [c] Ref [6] [b] Ref [7]

*single electron capture with core conservation.

The Results in Table.1 show capture into n=3 and n=4 levels, in agreement with Ryufuku model predictions [12], with different ℓ distributions. A qualitative agreement with Niehaus model predictions [13] is also observed for all the studied systems. But the multichannel Landau Zener model [14] do not predict correctly the population of the excited levels of OV, CV and NV. Contrarily to the ArV ion for which the experimental measures agree with the Landau Zener model predictions. n=3 levels are more populated than the n=4 levels for OV, CV and NV but for the ArV the n=3 are less populated than n=4 levels.

Conclusion :

We have measured subshell selective one electron capture in collision of X^{5+}(X=C,N,O,Ar) with H_2. Total excitation cross-sections have been deduced and are found the same for all ions at 50 keV.

References:

[1] S.Martin, M.Druetta and J.Désequelles, Nucl.Instr.and Meth. B14 (1986)254.

[2] M.Druetta, S.Martin and J.Désequelles Nucl.Instr. and Meth B23(1987)268.

[3] T.Bouchama and M.Druetta Nucl.Instr. and Meth. B40 (1987)1252.

[4] A.Salmoun Thesis , Lyon I University (1990).

[5] J.J.Bonnet private communication.

[6] D.H.Crandall, M.L.Mallory and D.C.Kocher Phys.Rev. A.15 (1977)61.

[7] D.H.Crandall, R.A.Phaneuf, and N.Meyer Phys.Rev. A.22 (1980)379.

[8] R.A.Phaneuf, I.Alvarez, F.W.Meyer, and P.M.Crandall Phys.Rev. A.26(1982)1892.

[9] M.Druetta, M.Mayo, P.H.Cotte, S.Martin, D.Dousson, D.Hitz and K.Tran Cong, Phys.lett. A108(1985)338.

[10] T.Bouchama Thesis, Lyon I University (1988).

[11] M.Lennon, R.W.McCullough and H.B.Gilbody J.Phys.B: At.Mol. Phys.16 (1983)2191.

[12] H.Ryufuku, Phys.Rev.A21(1980)745.

[13] A.Niehaus, J.Phys.B19(1986)2929.

[14] L.Landau, Phys. Z.Sowjetunion 2. 46 (1932).

Supplement to Z. Phys. D - Atoms, Molecules and Clusters 21, S 277-278 (1991)

Atoms, Molecules
Zeitschrift für Physik D and Clusters

Photon recoil ion-projectile coincidences in multicapture processes of Kr^{18+} + Kr collisions: „doubly excited Rydberg states with n = 10, ..., 15 and l = 9, ..., 14"

S. Martin[1], Y. Ouerdane[1,2], A. Denis[1], and M. Carré[1]

[1] Laboratoire de Spectrométrie Ionique et moléculaire, (URA CNRS No. 171), Université de Lyon 1, Campus de la Doua, F-69622 Villeurbanne Cedex, France

[2] Laboratoire de traitement du signal et instrumentation (URA CNRS No. 842), Université de Saint Etienne, 42 rue Docteur Paul Michelon, F-42023 Saint Etienne Cedex, France

Abstract

With the object of studying the collision of Kr^{18+} with Kr at low energy, we performed for the first time experiments using a coincidence technique between outgoing projectiles and recoil ions, photons and recoil ions or photons and outgoing projectiles. The photons were issued from transitions between Rydberg states with n=10,....,15 and l=9,...,14 populated in multicapture processes. We could then obtain the charge state distributions of recoil ions corresponding either to a given charge state of outgoing projectiles or to a Rydberg transition.

I. Introduction

Studies of multiexcited states produced by multicapture processes in collisions between multicharged ions and atoms are of considerable interest at present (1). The Classical Barrier Model (C.B.M.), which is an independent particle model, cannot explain all the experimental results, for example the Rydberg states populations resulting from the double capture process (2). In this case, the correlation effect during the collision between multicharged ions and equivalent electrons of the target plays an important role.

In a recent publication (3), we have reported measurements concerning the emission cross sections of Rydberg states with n=15 ->20 and l=14 ->19 populated by multielectron capture (Δq = 2,..., 6) in low energy collisions of Kr^{18+} with Kr. We present in this publication results obtained in experiments performed to try to answer many questions regarding the processes which populate these levels.

II. Experiment

The experiments were conducted at Lagrippa in Grenoble. Kr^{18+} ions were produced by an ECR source at an energy of 360keV. Our experimental arrangement is schematically shown in fig. 1. The interaction region is delimited by a short length cell where the Kr target pressure is low (about 5.10^{-6} mbar)

For projectile-recoil ion coincidence measurements, a charge state of the outgoing beam is selected with the electrostatic analyser ,the CEM provides a start pulse to the Time-Amplitude

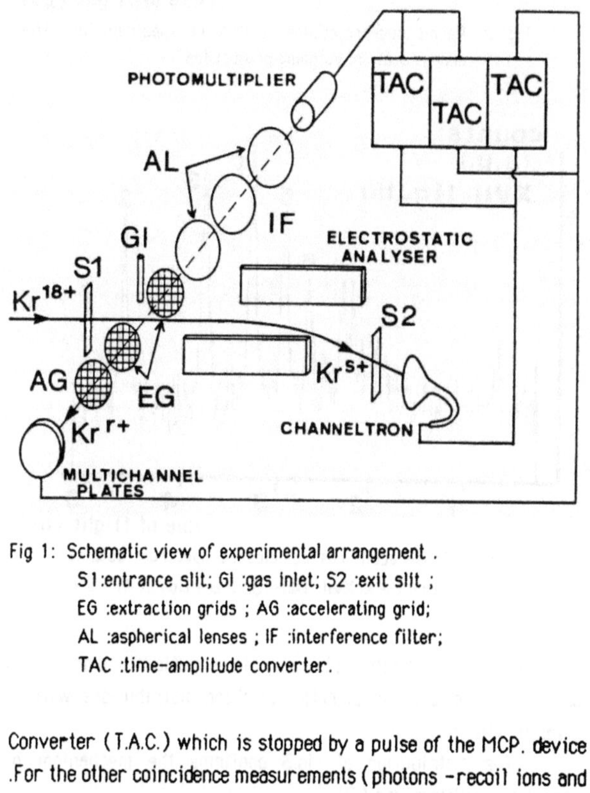

Fig 1: Schematic view of experimental arrangement . S1:entrance slit; GI :gas inlet; S2 :exit slit ; EG :extraction grids ; AG :accelerating grid; AL :aspherical lenses ; IF :interference filter; TAC :time-amplitude converter.

Converter (T.A.C.) which is stopped by a pulse of the MCP. device .For the other coincidence measurements (photons -recoil ions and photons-projectiles), the start pulse is provided by the PM and the stop by the MCP or CEM respectively.

A typical example of recoil ion charge state distribution is given in fig. 2 where Kr^{18+} projectile stabilizes two electrons in a single collision.

Fig.3 shows recoil ions distribution obtained in an experiment of photons-recoil ions coincidence .

III. Results and discussion.

The theoretical recoil ion charge distribution (4) obtained with the ECBM model (5) is slightly different from the experimental distribution. Kr^{5+}, Kr^{6+} and Kr^{7+} peaks of fig.2 are well reproduced and are attributed to the uncorrelated capture. Kr^{2+}, Kr^{3+} and Kr^{4+} peaks of fig.2 are not reproduced by this

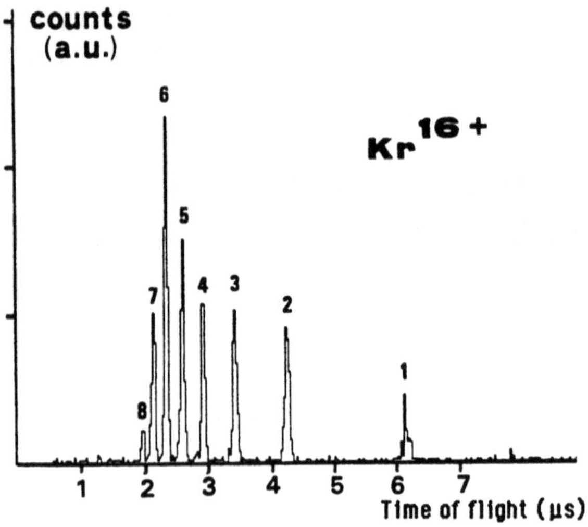

Fig. 2 : Recoil ions-projectiles coincidence spectrum for 16+ charge state of outgoing projectiles.

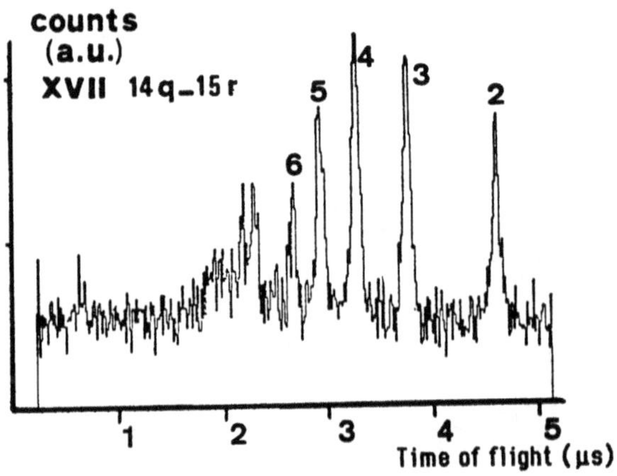

Fig.3: Photons-recoil ions coincidence spectrum obtained by selecting the KrXVII 14q-15r transition at 479.5nm.

model and must be attributed to a correlated muticapture process. A detailed theoretical interpretation of these distributions will be given in a future paper.

The distribution of Fig.3 confirms the mechanism of correlated multicapture (2).

Fig.4 displays results of coincidence measurements between photons corresponding to selected Rydberg transitions and outgoing projectiles with different charge states.

The charge state distribution of the projectile obtained with Kr XVII (14q-15r) transition shows that most of the photons are emitted after the stabilization of two electrons which follows the multicapture.

The observation of coincidence between photons and the Kr^{17+} outgoing projectiles is explained by the fact that the stabilization of two electrons can populate the $1s^2\ 2s^2\ 2p^6\ 3s^2$

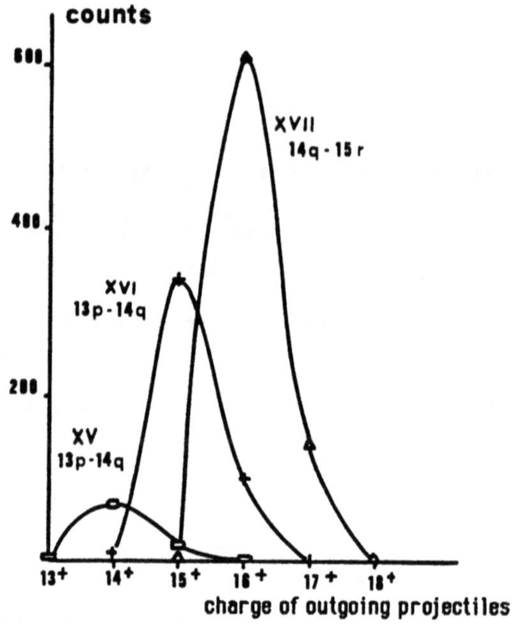

Fig.4 : Photon-projectile coincidence results obtained in the same experimental conditions.

$3p^6\ 4s\ 15r$ state which is metastable with respect to autoionisation (6). The Kr XVII ion populated in this state can autoionise in order to give Kr XVIII (17+) only after radiative decay of Rydberg electron along the Yrast sequence (15r--> 14q- -> 13p--> 12o....).

The same interpretation can be given for the distributions corresponding to Kr XVI (13p-14q) and KrXV (13p-14q) by considering respectively the metastable cores $1s^2\ 2s^2\ 2p^6\ 3s^2\ 3p^6\ 3d\ 4s$ and $1s^2\ 2s^2\ 2p^6\ 3s^2\ 3p^6\ 3d^2\ 4s$.

This work has been performed at LAGRIPPA - joint laboratory of the CEA and the CNRS. The authors would like to thank T. Lamy, A. Brénac and G. Lamboley for their assistance in running the ECR ion source.

REFERENCES

1. S.Martin, A.Salmoun, Y.Ouerdane, M.Druetta, J.Désesquelles and A.Denis: Phys. Rev. Lett. **62**, 2112 (1989).
2. F.W.Meyer, D.C.Griffin, C.C.Havener, M.S.Hug, R.A.Phaneuf, J.W.Swenan and N.Stulterfoht: Phys. Rev. Lett **60**,018 (1988).
3. S.Martin, A.Denis, Y.Ouerdane, A.Salmoun, A.El Motassadeq, J.Désesquelles, M.Druetta, D.Church and T.Lamy : Phys. Rev. Lett. **22**, 2633 (1990).
4. L.Liljeby, G.Astner, A.Barany, H.Cederquist, H.Danered, S.Huldt, P.Huelplund, A.Johnson, H.Kuadson and K.G.Rensfelt: Physica Scripta **33**, 310 (1986).
5. A.Niehaus: J.Phys. B At.Mol.Phys. **19**, 2925 (1986).
6. S.Martin, G. Do Cao, A.Salmoun, T.Bouchama, A.Denis, J.Andrä, J.Désesquelles and M.Druetta: Phys.Lett. **A 133**, 239 (1988).

Supplement to Z. Phys. D - Atoms, Molecules and Clusters 21, S279-281 (1991)

Atoms, Molecules and Clusters
Zeitschrift für Physik D

Electron emission in collisions of the bare ions He^{2+} and Li^{3+} with molecular hydrogen

R. Schulze, M. Liehr and E. Salzborn
Institut für Kernphysik, Strahlenzentrum der Justus-Liebig-Universität, W-6300 Giessen, FRG

Abstract

Electrons emitted in collisions of the bare ions He^{2+} and Li^{3+} with molecular hydrogen are detected and measured with a cylindrical mirror analyser (CMA). The projectile energy range covered extended from 5 keV/amu to 10.7 keV/amu for He^{2+} and from 2.6 keV/amu to 4.3 keV/amu for Li^{3+}. The electrons observed are found to originate from autoionization of doubly-excited states $He^{0**}(nln'l')$ and $Li^{+**}(nln'l')$ with n=2, n'≥2 populated by double-electron capture in a single collision.

I. Introduction

The investigation of electron spectra accompanying charge-changing collisions involving bare ions yields information about both electron capture and electron emission mechanisms in two-electron systems. The choice of the particular collision systems discussed here was prompted by the following reasons :

(i) Double-electron capture by bare ions leads to doubly-excited two-electron systems, where the interaction of the electrons can be examined in the unperturbed Coulomb-field of the nucleus.

(ii) H_2 is a two-electron target. As a consequence, only singlet-states will be formed by double capture.

(iii) The ECBM [1] predicts double capture to mainly occur to the above-mentioned states leaving only $He^+(1s)$ and $Li^{2+}(1s)$, respectively, as final states accessible via autoionization, which makes the electron spectra obtained easily to be interpreted.

II. Experiment

The bare ions He^{2+} and Li^{3+} used in this experiment are produced by means of the Giessen 5 GHz ECR-ion source.

A well-collimated beam (1mm×1mm) of the respective ion is made to impinge on an effusive H_2 gas jet. Since the target gas pressure in the interaction region is not accessible to direct measurement, the target thickness has to be inferred from the pressure in the H_2 storage volume. The target thickness thus estimated amounted to approx. 10^{13} cm^{-2}. Shortly before entering the interaction region, the ion beam is deflected electrostatically in order to clean it from particles that have undergone charge-changing collisions in the residual gas. Typical beam intensities were 30 nA electric current He^{2+} and 3 nA electric current Li^{3+}.

The CMA utilized to energy-analyze the emitted electrons subtends a solid angle of approx. 1.3% of 4π under an angle $\Theta = 38° \pm 2°$ with respect to the beam axis. The calculated transmission amounts to 74% and the energy resolution $\Delta E/E$ was measured to be approx. 0.65%. The energy-analyzed electrons are detected and counted with a channeltron based single-particle detector.

III. Results

A typical electron spectrum of the collision system $He^{2+} + H_2$ at a projectile energy of 10.7 keV/amu is displayed in fig. 1. The spectrum is dominated by a structure consisting of a number of lines between approx. 33 eV and 41 eV of electron energy corresponding to the Auger-decay of doubly-excited states $He^{0**}(2ln'l')$, n'=2,3 to the final state $He^+(1s)$ in reactions of the type

$$He^{2+} + H_2 \rightarrow He^{0**} + \ldots$$
$$\quad\quad\quad\quad\quad \hookrightarrow He^+(1s) + e^- + \Delta E \qquad (1)$$

With the help of theoretical data by Oza [2], it is possible to identify some of the electron lines observed. The lines at 33 eV and 37.4 eV can be ascribed to the decay of states $(2s^2)^1S$ and $(2p^2)^1S$, respectively, while the line at approx. 35 eV originates from a superposition of the decay of $(2s2p)^1P^o$- and $(2p^2)^1D$-states, which cannot be resolved experimentally. The broad structure above approx. 38 eV can be explained by transitions from $He^{0**}(2ln'l')$, $n' \geq 3$, to $He^+(1s)$.

In addition, a peak at 33.7 eV is observed. The emergence of this line can only be understood as resulting from the decay of $(2s2p)^3P^o$- state. Since H_2 is a two-electron target, the population of this state in a single collision would constitute a violation of Wigner's spin-conservation rule. However, when varying the target gas pressure, the $^3P^o$- line is found to increase quadratically, so that its population has to be ascribed to two subsequent one-electron capture processes via the formation of a long-lived, metastable $He^+(2s)$.

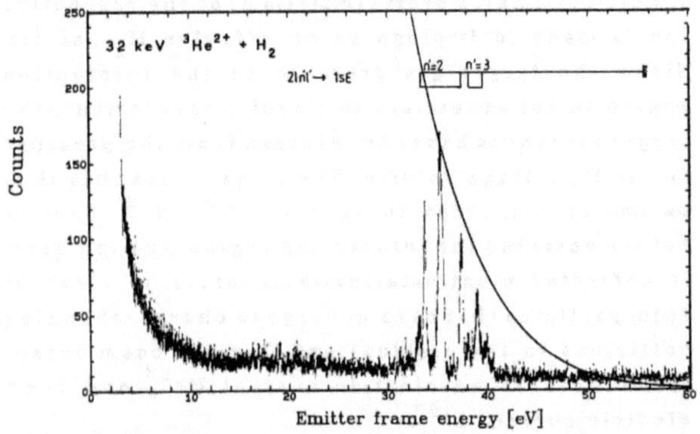

Fig 1 : Electron spectrum of the collision system $^3He^{2+} + H_2$ at a projectile energy of 10.7 keV/amu. The solid line represents the ECBM capture-window.

Nevertheless, the $^3P^o$-line is useful to calibrate the CMA, since its transition energy is known precisely, and it is well seperated from other structures in the spectrum. Within the statistical error margin, all autoionization lines resulting from the decay of (2l2l')-singlet-states do not exhibit any dependence on the target-gas pressure. The investigation of electron spectra from collisions of Li^{3+} with H_2 yield similar results. Here, also, only lines produced by formation of states $Li^+(nln'l')$, n=2, n' = 2,3 are observed.

IV. Discussion

In both collision systems, doubly-excited states are found to be populated in accordance with the ECBM. The line positions observed show an overall good agreement with theoretical predictions, although the electron energies of some transitions are somewhat smaller than predicted by theory [2,3]. This effect arises from post-collision-interaction (PCI) [4] acting on short-lived states ($\tau \approx 10^{-15}$ s).

The PCI induces a line-shift towards lower energies as well as an asymmetric line shape, both depending on the projectile ion velocity, the charge state of the recoil ion, and the term width of the doubly-excited state [5].

In case both the binding energy an the lifetime of a doubly-excited state is known, the shape and position of the corresponding autoionization line can be calculated for a given projectile velocity and collision system [6]. So, when an appropriate fit-proedure is applied, partly overlapped lines can be resolved.

With this method, it is possible e.g. to quantitatively determine the contributions from autoionization lines belonging to different values of total angular momentum L [7](see fig. 2).

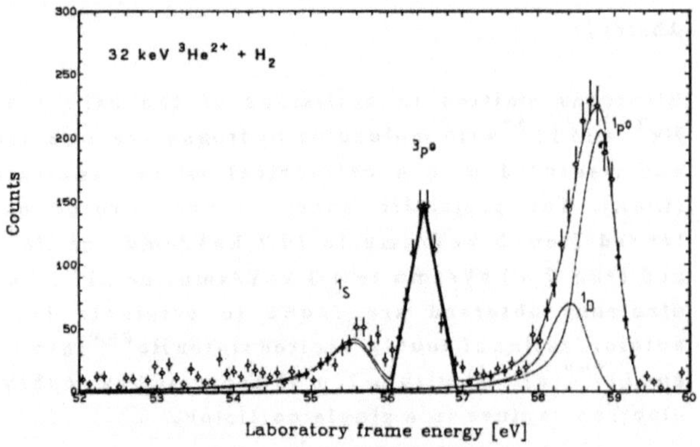

Fig. 2 Comparision of the electron spectrum of the collision system $^3He^{2+} + H_2$ at 10.7 keV/amu with a fit obtained using the data of Oza [2].

In the collision system $He^{2+} + H_2$, the relative population of the terms $(2s2p)^1P^o$ and $(2p^2)^1D$ shows a strong projectile velocity dependence.

V. Conclusion

Energy-differential measurements of electron spectra arising from collisions of the bare ions He^{2+} and Li^{3+} with molecular hydrogen were performed at various projectile energies. The electrons observed are found to originate from aotoionization of doubly-excited states He^{0**} and Li^{+**}, respectively, produced by double-electron capture in a single collision.

Double-capture occurs dominantly to states $(nln'l')^1L$ with n = n' = 2 and l,l' = 2,3 thus confirming the predictions of the ECBM. In addition, to fully account for the position and shape of the autoionization lines observed, it is necessary to include PCI-effects.

References

[1] A. Niehaus; J.Phys. **B19** (1986) 2925

[2] D.H. Oza; Phys Rev **A33** (1986) 824

[3] A.K. Bhatia, A. Temkin; Phys Rev **A11** (1975) 2018

[4] R.B. Barker, H.W. Berry; Phys Rev **151** (1966) 14

[5] P. van der Straten, R. Morgenstern; J.Phys. **B19** (1986) 1361

[6] A.Z. Ostrovskij, Yu.N. Sebyakin; Sov.Phys. JETP **46** (1977) 852

[7] A. Bordenave-Montesquieu, P. Benoit-Cattin, M. Boudjema, A. Gleizes, H. Bachau; J.Phys. **B20** (1987) L695

Supplement to Z. Phys. D - Atoms, Molecules and Clusters 21, S 283-284 (1991)

Atoms, Molecules
Zeitschrift für Physik D and Clusters

Collision velocity dependence of double capture processes in N^{7+} and Ne^{8+} + He systems (v = 0.18 to 0.61 au)

P. Moretto-Capelle, P. Benoit-Cattin, A. Bordenave-Montesquieu and A. Gleizes
Centre de Physique Atomique, URA CNRS 770, Université Paul Sabatier, 118 Route de Narbonne, F-31062 Toulouse Cedex, France

Abstract

Double capture processes depend on the collision velocity in a complicated way which is different for N^{7+} and Ne^{8+}+ He collisional systems. Some examples of energy dependence for differential and total cross sections are given.

I. Introduction

From an experimental point of view, the transfer of two electrons at low energy in a collision between multicharged ions and helium is far from being well known. We want to emphasize here the trends of the double capture (DC) with the collision velocity v (v = 0.18- 0.61 au) as observed by electron spectroscopy. N^{7+} and Ne^{8+} ions have been chosen because they are currently investigated by theoreticians [1]. For lower ion charges, the energy dependence in N^{5+}, O^{6+}+ H_2 systems, has already been studied (at 50°) by the group of Groningen mainly to compare with the prediction of the extended classical barrier model [2]. In the present work the electron spectra have been measured at several angles (in the range 10° to 160°), for each energy, in order to determine the true velocity dependence of the total cross sections. Doubly differential, singly differential (s.d.c.s.) and total cross sections are given in arbitrary units which agree with each other within the same system. They can be put on an absolute scale using the cross sections already given in previous works [3,4].

II. Ne^{8+}+ He system

An example of velocity dependence of the double capture in Ne^{8+}+ He is shown in Fig. 1 with electron spectra observed at 10°. The following reaction is illustrated:

$$Ne^{8+}(1s^2) + He \rightarrow Ne^{6+}(1s^2 3l nl')\,^1L + He^{++}$$
$$\hookrightarrow Ne^{7+}(1s^2 2l) + e^-$$

The total cross section $\Sigma\, \sigma_{T}(3,n)=\Sigma \int d\Omega\; d\sigma(3,n)/d\Omega$

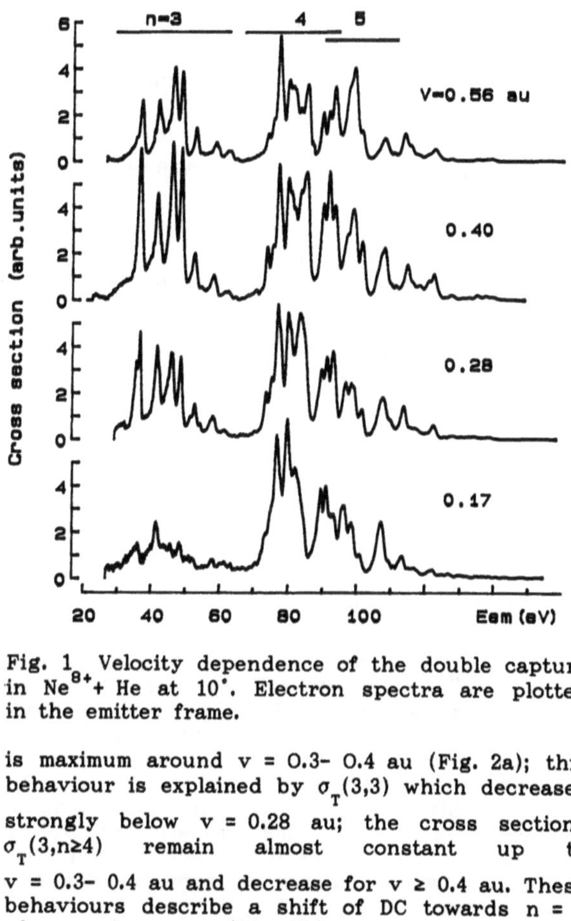

Fig. 1 Velocity dependence of the double capture in Ne^{8+}+ He at 10°. Electron spectra are plotted in the emitter frame.

is maximum around v = 0.3- 0.4 au (Fig. 2a); this behaviour is explained by $\sigma_T(3,3)$ which decreases strongly below v = 0.28 au; the cross sections $\sigma_T(3,n\geq4)$ remain almost constant up to v = 0.3- 0.4 au and decrease for v ≥ 0.4 au. These behaviours describe a shift of DC towards n = 4 when v decreases (Fig. 2b); at v = 0.18 au even the contribution of upper n > 4 terms becomes higher than the n = 3 one. A good agreement with [5] is noted for $\sigma_T(3,4)/\sigma_T(3,3)$.

A simple variation against velocity is no longer found for the s.d.c.s. $d\sigma(3,n)/d\Omega$ (Fig. 3). This means that no conclusion concerning the v dependence as well as the absolute value of σ_T can be drawn from a measurement at one angle. An intricated velocity dependence of lines related to individual states was also found; such a detailed discussion lies outside the scope of the present paper.

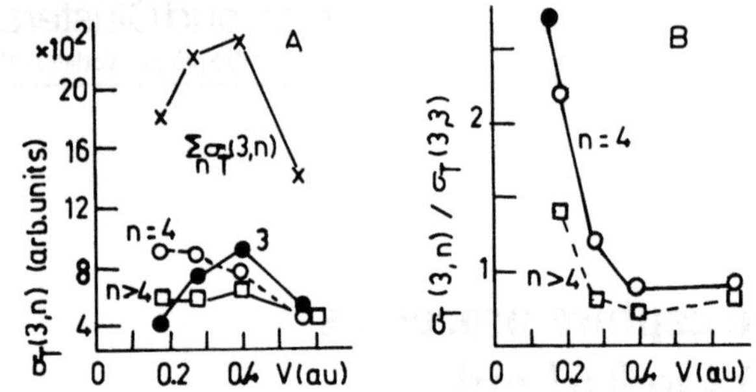

Fig. 2 Velocity dependence of total cross sections in Ne^{8+}+ He. A: absolute values (note the multiplying factor). B: relative values with respect to n = 3; black dot: result from [5].

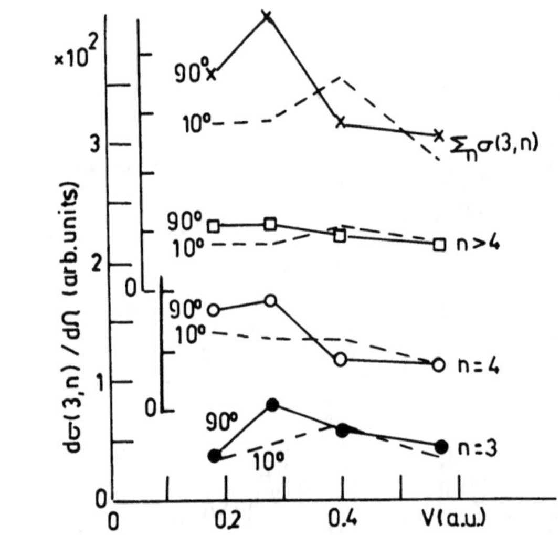

Fig. 3 Velocity dependence of singly differential cross sections in Ne^{8+}+ He. Note the multiplying factor and the zero shifts on the ordinate scale.

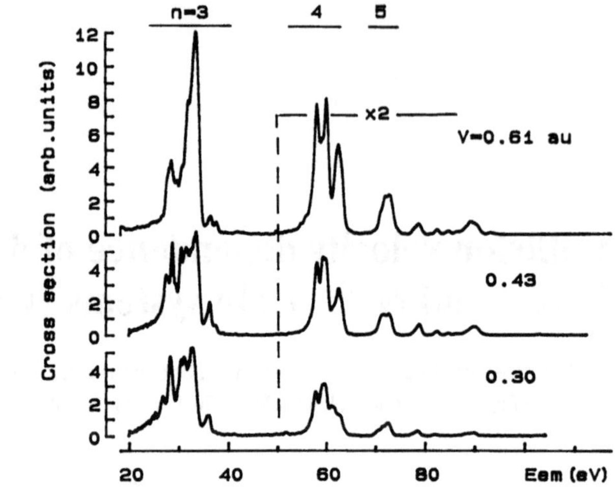

Fig. 4 Velocity dependence of the double capture in $^{15}N^{7+}$+ He at 10°.

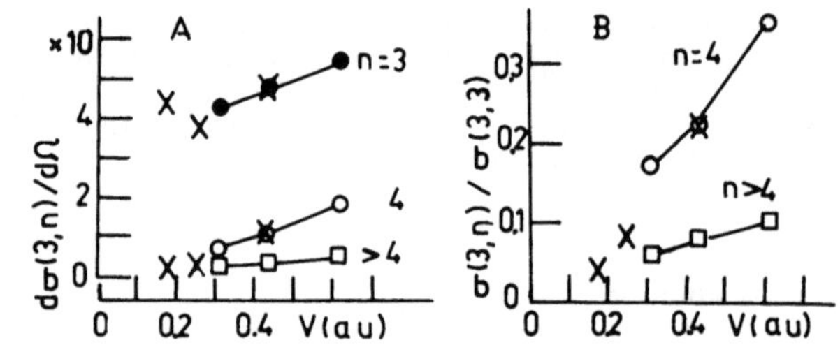

Fig. 5 Velocity dependence of s.d.c.s. in N^{7+}+ He at 10°. A: absolute values (arb.units); (×): theor. values [1] (see text). B: relative values with respect to n = 3; (×): theor. values [1].

III. $^{15}N^{7+}$+ He system

For N^{7+}+ He the DC process follows the same scheme as for Ne^{8+}+ He. The velocity dependence is illustrated at 10° in Fig. 4; as the spectra have been measured at few angles, accurate total cross sections cannot be given here. Compared to Ne^{8+}, the velocity dependence of the s.d.c.s. is seen to be very different (Fig. 5a). The (3,3) s.d.c.s. increases with v in the investigated range whatever the angle is; as it represents about 70% of DC, a similar increase is expected for the total cross section $\Sigma_n \sigma_T(3,n)$. The (3,3) theoretical total cross section [1], normalized to our value at v = 0.43 au / 90° (assuming an isotropic angle dependence), shows the same trend down to v = 0.25 au; a more valuable comparison would imply the measurement of total cross sections.

Since the population of the n > 3 terms increases at 10° more quickly with v than the n = 3 one, the *relative* weight of the (3,3) states decreases when v increases. The ratios $\sigma(3,n)/\sigma(3,3)$, measured at 10° (Fig. 5b), indeed show a shift of DC towards n = 4 and to some extent towards higher n > 4 terms of the series when v increases; the reverse has been observed for neon (Fig. 2b). A good agreement is again noted with [1] for n = 4 (Fig. 5b), within the isotropy hypothesis.

This work has been performed at Lagrippa, a common laboratory of the CNRS and of the CEA.

References

1. Harel, C., Jouin, H.: Europhysics Letters 11, 121 (1990) and this conference
2. Mack, M., Niehaus, A.: Nuclear Instr. Meth. B 23, 116 (1987)
3. Bordenave-Montesquieu, A., Boudjema, M., Benoit-Cattin, P., Gleizes, A., Moretto-Capelle, P.: J. de Phys., Colloque C1, 50, 305 (1989)
4. Bordenave-Montesquieu, A., Benoit-Cattin, P., Boudjema, M., Gleizes, A.: 15th International Conference on the Physics of Electronic and Atomic Collisions 1987, Brighton, Book of Invited Papers, pp.643-653, Amsterdam: North Holland 1988
5. Barat, M., Gaboriaud, M.N., Guillemot, L., Roncin, P., Laurent, H., Andriamonje, S.: J.Phys.B 20, 5771 (1987)

Supplement to Z. Phys. D - Atoms, Molecules and Clusters 21, S 285-286 (1991)

Atoms, Molecules
Zeitschrift für Physik D and Clusters

Double electron capture into highly charged ions: correlated or independent?

J. H. Posthumus, P. Lukey and R. Morgenstern

Kernfysisch Versneller Instituut, Rijksuniversiteit Groningen, Zernikelaan 25, 9747 AA Groningen, The Netherlands

Abstract

We propose an independent-electron model for describing quantitatively the population of doubly excited ($n\ell n\ell'$) states formed by two-electron capture into multiply charged ions. Population probabilities calculated by this model reproduce experimentally determined ones if one assumes that the classical electronic angular momentum is preserved, in magnitude and also in orientation. Correlation during the collision need not be invoked for the interpretation of the spectra.

I Introduction

Although correlation has a strong influence on the formation of stationary states, the question remains whether it is also important during the process of two- or many-electron transfer and if it directs the system into certain states [1,2] -e.g. states with such a mixing of configurations that they resemble to a large extent a triatomic molecule performing a rotational or vibrational motion. In order to obtain a quantitative basis for this discussion we have developed a model which predicts population probabilities for the various doubly excited states from an independent transfer of two electrons from the target to the projectile.

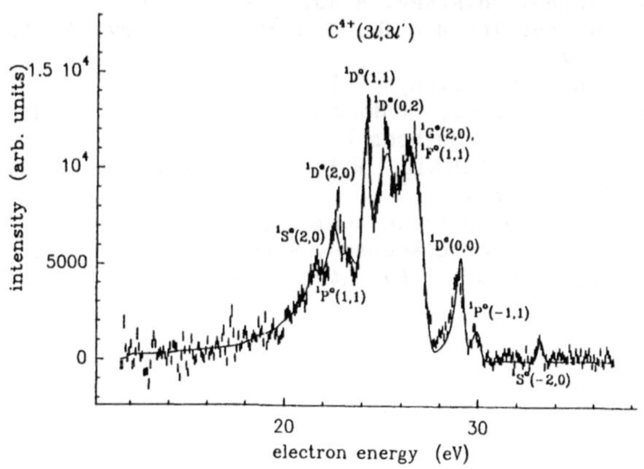

Fig.1. energy spectrum of autoionization electrons from 60 keV collisions of C^{6+} on H_2. The solid line represents a fit in which only the populations of the various states are taken as fit parameters.

II Experiment

We have measured energy spectra of autoionization electrons resulting from collisions of bare and He-like ions on the two-electron targets He and molecular hydrogen at energies below 10 keV/amu. Electrons ejected from doubly excited ($3\ell3\ell'$) projectile states were analyzed in a cylindrical mirror analyzer [2] at an emission angle of 50° with respect to the projectile beam direction. Fig.1 shows a spectrum resulting from collisions of 60 keV $^{13}C^{6+}$ on molecular hydrogen. A fit of these spectra was made, in which energetic positions and lifetimes of the states involved were taken from theory [3,4] and in which only the peak intensities were used as fit parameters. Peak shapes are strongly influenced by post collision interaction and this is taken into account in the calculations. The possibility of obtaining good fits in this way confirms the accuracy of the theoretical energies and lifetimes. In order to obtain state populations from peak intensities one has in principle to know the angular distributions $I(\vartheta)$ of ejected electrons. However we have performed calculations which indicate that $I(\vartheta)$ is mainly governed by the second order Legendre polynomial P_2. Our detection angle of 50^0 is sufficiently close to the magic angle of $\vartheta=54.7^0$ for which P_2 is zero, that the intensities found in the fit reflect the population probabilities to a large extent.

III Model calculations

In a first step we calculate amplitudes for population of a certain configuration ($n\ell m,n'\ell'm'$) for each impact parameter b as

$$a_{n\ell m,n'\ell'm'}(b)= \left[p(E_{n\ell,n'\ell'})\cdot p_\ell(b)\cdot p_{m_\ell}\cdot p_{\ell'}(b)\cdot p_{m_{\ell'}}\right]^{1/2} \quad (1)$$

Here E is the hypothetical energy of a state with the pure configuration ($n\ell,n'\ell'$), calculated as a weighted mean of energies of all states containing the configuration ($n\ell,n'\ell'$), the weights being taken as the square of the appropriate configuration mixing coefficients:

$$E_{n\ell,n'\ell'}= \sum_{KT}(c^{KT}_{n\ell n'\ell'})^2 \; E^{KT} \quad (2)$$

The approximate correlation quantum numbers K and T [5] are used to identify the eigenstates. The probability p(E) is calculated using the extended classical barrier (ECB) model [6]. The $p_\ell(b)$ are calculated according to the extension of the ECB model given by Burgdörfer *et al* [7] which, roughly said, attributes to a captured electron the classical angular momentum $b \cdot v$, with b the impact parameter and v the projectile velocity, together with a quantum mechanical uncertainty $\Delta\ell$. The $\Delta\ell$ as calculated by Burgdörfer *et al.* by classical considerations is for our system around 2, of the same size as $b \cdot v$. To reproduce the measurements we need better conservation of angular momentum. For the results shown in this paper we used in fact a value of $\Delta\ell=1$. The $p_{m\ell}$ finally are determined, by assuming that the orientation of the angular momentum is also preserved. With the z-axis defined by $\vec{r} \times \vec{v}$, the substates with $m=\ell$ would only be populated ($\vec{r}$ is the vector pointing from the target to the projectile). This restriction is however too strong. The experimentally observed population of D° demands for some population of substates with $m=\ell-1$. In our model we have taken therefore

$$p_{m_\ell} = 2/3\ \delta(\ell,m_\ell) + 1/3\ \delta(\ell-1,m_\ell) \qquad (3)$$

In a second step we project the constructed state vector onto the new basis of correlated eigenstates characterized by angular momentum quantum numbers L,M,S and correlation quantum numbers K,T. Like Jain *et al* [8] we then find probabilities P^{KT}_{LMS} for populating these states

$$P^{KT}_{LMS}(b) = \left[\sum_i c^{KT}_i \sum C^{\ell\ell' L}_{mm'M}\ a_{n\ell m,n'\ell' m'}(b)\right]^2 \qquad (4)$$

with the $C^{\ell\ell' L}_{mm'M}$ being appropriate Clebsch Gordan coefficients, the c^{KT}_i again the configuration mixing coefficients for the various correlated states taken from theory [9], and the index i running over all configurations. Finally we integrate over b up to a maximum value b^{max} given by the ECB model as the largest impact parameter for which two electron capture can occur. At a particular impact parameter our model predicts primarily the population of one configuration. The interference terms in (4) can therefore be omitted in determining the populations of the stationary states. Probabilities calculated in this way for the various states are shown in fig.2 in comparison with the probabilities obtained from the analysis of the experimental spectrum.

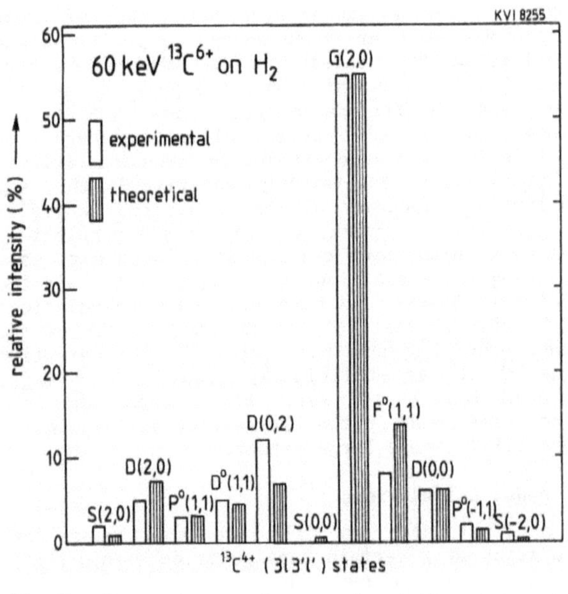

Fig.2. Comparison of experimentally determined peak intensities with those calculated from an independent electron model

IV Conclusion

The agreement between experimentally obtained population probabilities and the calculated ones leads us to two conclusions: (i) unlike the explanation given in [2], for an interpretation of the spectrum presented here one need not necessarily invoke correlated electron capture; (ii) the electrons are with the highest probability captured into $m=\ell$ sublevels with respect to a z-axis perpendicular to the scattering plane. The second conclusion is very much in line with the predictions of Hansen *et al* [10] based on a much more elaborate theory.

References

1. N.Stolterfoht et al, Phys.Rev.Lett **57**, 74 (1986)
2. M.Mack, J.H.Nijland, p.v.d.Straten, A.Niehaus, R.Morgenstern, Phys.Rev. A **39**, 3846 (1989)
3. N.Abu-Salbi and J.Callaway, Phys.Rev. **A24** 2372 (1981)
4. Y.K.Ho, Phys.Rev. A **35**, 2035 (1987)
5. D.R.Herrick and O.Sinanoğlu, Phys.Rev. A **11**, 97 (1975)
6. A.Niehaus, J.Phys.B **19**, 2925 (1986)
7. J.Burgdörfer, R.Morgenstern, A.Niehaus, Nucl. Instr. and Methods B **23**, 120 (1987)
8. A.Jain, C.D.Lin, W.Fritsch, Phys.Rev. A **39**, 1741 (1989)
9. J.Hansen, private communication
10. J.P.Hansen, L.Kocbach, A.Dubois and S.E.Nielsen Phys.Rev.Lett. **64**, 2491 (1990)

Supplement to Z. Phys. D - Atoms, Molecules and Clusters 21, S 287-288 (1991)

Atoms, Molecules
Zeitschrift für Physik D and Clusters

Coincidence measurements between electrons from projectile autoionization and target ions in specific charge states

J. H. Posthumus, P. Lukey and R. Morgenstern
Kernfysisch Versneller Instituut, Rijksuniversiteit Groningen, Zernikelaan 25, 9747 AA Groningen, The Netherlands

Abstract

For Ar^{9+} collisions on Ar we have - by means of a coincidence method - identified the number of primarily transferred electrons leading eventually to the emission of projectile Auger electrons at various energies. Auger cascades are found to play an important role: most of the high energy electrons are produced via secondary Auger processes in the last step of a cascade.

I. Introduction

As long as electron capture by multiply charged ions is investigated by using two-electron targets like He or molecular hydrogen, electron emission from the projectile can clearly be ascribed to two-electron capture into doubly excited projectile states. The situation is different for the case of many-electron targets like e.g. Ar. A large variety of processes including transfer of several electrons or two-electron capture together with additional target ionization can occur. In order to investigate such many-electron processes we have installed a setup in which autoionization electrons are detected in coincidence with charge state selected target ions.

II. Experiment

In our apparatus, described in more detail elsewhere [1], the ejected electrons are analyzed in a cylindrical mirror analyzer. The target ions produced during the collisions are extracted in backward direction (180° with respect to the projectile beam direction) and are energy analyzed in a double hemispherical analyzer (fig. 1). In order to separate ions originating from the scattering center from those produced at other places along the beam direction the scattering center is kept at a unique potential. The charge state of the energy analyzed ions is measured via a Time Of Flight (TOF) method.

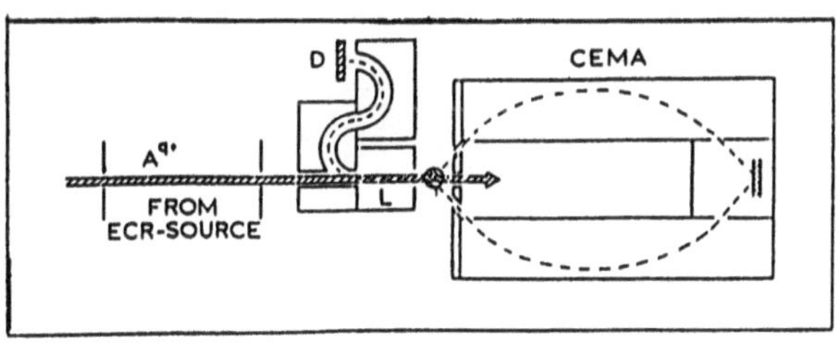

Fig.1. Experimental setup to measure coincidences between ejected electrons and corresponding target ions.

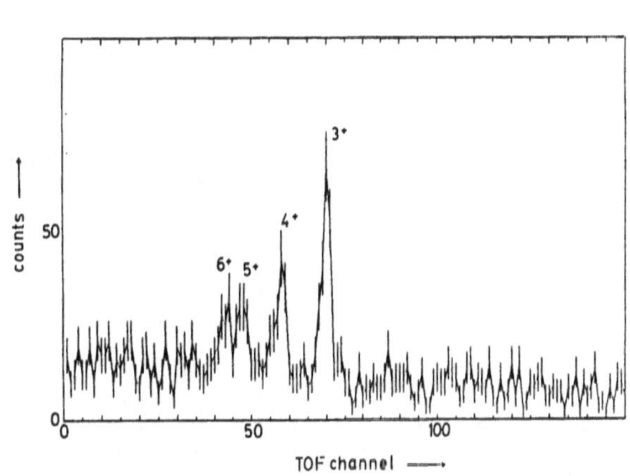

Fig.2. Time-of-Flight spectrum of Ar target ions with charge q arising from Ar^{9+} on Ar collisions, measured in coincidence of projectile emitted autoionization electrons at 183 eV.

III. Results

Fig.2. shows a typical TOF spectrum of Ar^{q+} ions. The corresponding electrons are clearly due to decay of doubly excited $Ar^{7+}(2p^5 3\ell 3\ell')$ projectile ions and therefore one would in the first place suspect that a two-electron capture process is leading to these electrons. From this point of view it is amazing to see that no coincident Ar^{2+} target ions are observed. This situation is also met for all the other high energy electron peaks. This can be seen in fig.3, which shows an electron energy spectrum together with a bar-graph above several peaks which indicates the charge state distribution of target ions corresponding to the electrons. Electrons around 103 eV from decay of $Ar^{7+}(2p^5 3s^2)$ projectiles are e.g. predominantly produced in a process leading to Ar^{4+} target ions, whereas electrons above 210 eV from $Ar^{7+}(2p^5 3\ell 5\ell')$ are coincident with Ar^{3+} target ions only. Pure two-electron transfer collisions apparently lead only to electrons below 100 eV.

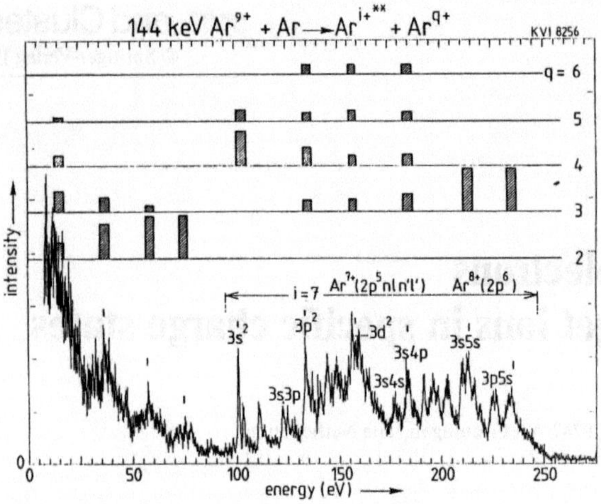

Fig.3. Energy spectrum of electrons emitted from projectile ions after electron capture in collisions of Ar^{9+} on Ar. The bar-graphs on top of several electron peaks indicate the charge state of the target ion, i.e. the number of electrons which were transferred to the projectile

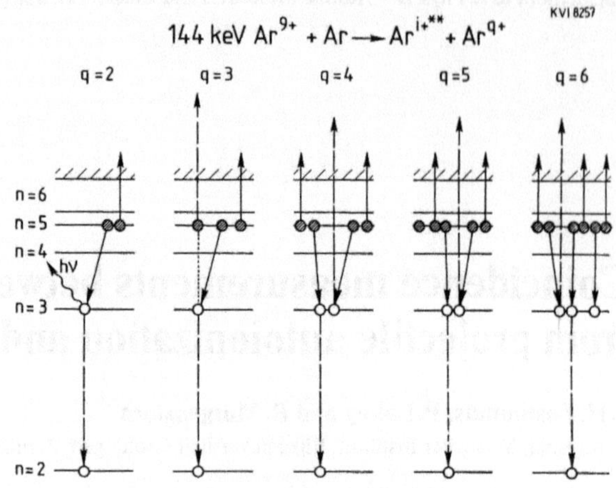

Fig.4. Schematic view of primary and secondary Auger processes following capture of q electrons into the n=5 shell of the Ar^{9+} ion.

IV. Discussion

These features can be explained in the following way: electron capture predominantly takes place into the *n*=5 shell of the projectile. This is in fact predicted by the extended classical oberbarrier (ECB) model [2]. If only two electrons are captured, an MOO-Auger process follows, resulting in low energy electrons and leaving the primary L-shell vacancy untouched. This is shown schematically in fig.4 (q=2). A three-electron transfer into the *n*=5 O-shell leads to an Auger cascade of first an MOO-transition, again with low-energy electrons resulting, and a subsequent LMO- transition leads to electron energies above 210 eV (see fig.4 q=3). If four electrons are captured, even two primary MOO-transitions can take place before the 2p-vacancy is filled in a secondary Auger process (see fig.4 q=4). This implies that most of the electrons above 100 eV are due to secondary Auger transitions. Without the knowledge of the target charge state one could be tempted to ascribe these electrons to decay from directly populated states, with wrong implications for the electron capture process.

V. Conclusions

By analyzing Auger electrons in coincidence with charge state selected target ions we have for the first time shown directly the importance of Auger cascades for the production of high energy electrons. Obviously these cascades have to be taken into account for a proper interpretation of electron transfer dynamics.

References

1. M.Mack, J.H.Nijland, P.v.d.Straten, A.Niehaus, R.Morgenstern, Phys.Rev. A **39**, 3846 (1989)
2. A.Niehaus, J.Phys.B **19**, 2925 (1986)

Supplement to Z. Phys. D - Atoms, Molecules and Clusters 21, S 289-290 (1991)

Atoms, Molecules
Zeitschrift für Physik D and Clusters

Single and double electron capture in 105 keV Ar^{7+} - He (H_2) collisions

P. Boduch, M. Chantepie, D. Hennecart, X. Husson, H. Kucal, D. Lecler, N. Stolterfoht
Laboratoire de Spectroscopie Atomique CNRS - URA 19, ISMRa, Bd Mal Juin, F-14050 Caen - Cedex, France

B. C. Fawcett
Rutherford Appleton Laboratory, Chilton, Didcot, Oxfordshire, OX11 0QX, UK

Abstract

Ar^{7+} ion beam collisions at 105 kev on He and molecular hydrogen targets are studied by photon spectroscopy. New transitions of the Ar^{6+} and Ar^{5+} spectra due to single and double electron captures are observed and interpreted through ab initio calculations.

I - Introduction :

Visible and near UV spectroscopies are very good tools to study single and double electron capture of Ar^{8+} ions produced by an ECR source on neutral targets (1). In order to extend these observations we have study the collisions of an Ar^{7+} ion beam on Helium and molecular hydrogen targets.
The collisions give rise to $Ar^{(7-k)+*}$ excited ions following the reaction.

$$Ar^{7+} + He(H_2) \rightarrow Ar^{(7-k)+*} + He^{k+}(H_2^{k+})$$

where k=1 corresponds to single electron capture and k=2 to double electron capture. $\Delta n=0$ transitions show that single electron captures populate the $1s^2 2s^2 2p^6$ 3s $n\ell$ (He n=3, 4 ; H_2 n=5) Ar^{6+} configurations. $\Delta n=1$ transitions show that double electron captures lead to $1s^2 2s^2 2p^6 3s^2 n\ell$, configurations with large values of n (Rydberg states) and large value of ℓ.

II - Experiment :

The experimental set up is as follows. Fig. 1

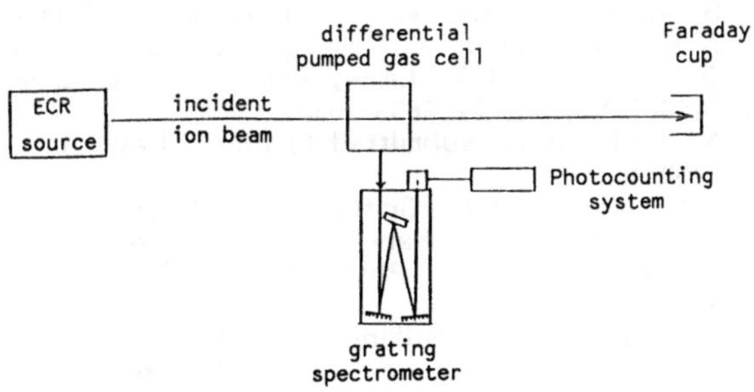

EXPERIMENTAL SET UP

A 105 kev Ar^{7+} ion beam provided by an ECR ion source of the GANIL test bench (Caen, France) is focused inside a differentially pumped gas cell. The vapor pressure in the cell is included between 5.10^{-5} and 2.10^{-4} mbar. The emitted photons are detected at right angle to the ion beam direction and analysed with a normal incidence grating spectrometer (1200 grooves/mm grating, 700nm blaze wavelength in the first order) fitted with photon counting photomultiplier. Fig. 2 and Fig. 2bis show examples of spectra obtained with the two targets.

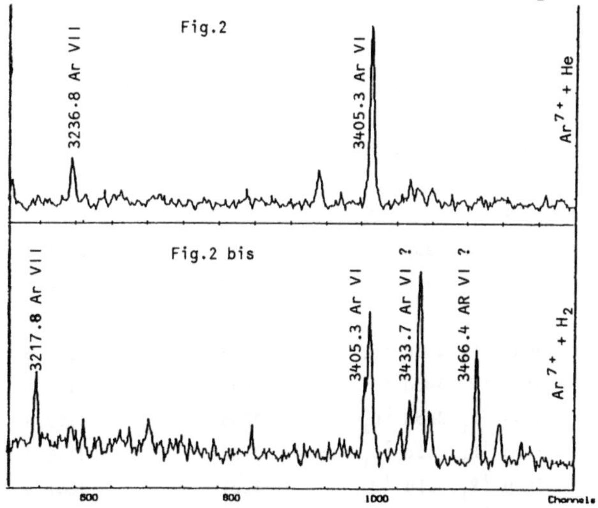

The lines due to electron capture are determined by comparing the spectra obtained successively with the Ar^{7+} and Ar^{6+} ion beam. Such selected lines are indicated in table I with their relative intensities.

III - spectral analysis :

According to the classical barrier model of Ryufuku et al [2] and in agreement with VUV observation [3] single electron captures in the ground state of the Ar^{7+} ion lead to $1s^2 2s^2 2p^6$ 3s $^{2S+1}L_J$ excited states of the Ar^{6+} ion with n=3 and 4 for He and n=5 for H_2. In the near UV and visible wavelength range only $\Delta n=0$ transitions with n=4 and n=5 are observable. Ab initio calculations have been performed with the Cowan code [4] on a CRAY XMP supercomputer at the Rutherford Laboratory. The theoretical predictions and the observed lines are indicated in table II.

Ar^{7+} + B		
Wavelength (Å)	B=He I	B=H_2 I
2250.2	22	
3116.7		51
3141.8	5	59
3217.8		44
3236.8	10	9
3402.7		56
3405.3	34	96
3422.4	2	22
3427.4	6	40
3433.7	4	129
3439.4	3	35
3466.4	2	71
3478.0		27
4117.8		80
4854.1		64
4919.1		46
5237.3	12	32
5262.9		32
5265.4	8	32
5271.9	28	55
5278.1		26
5288.4		25
5291.3		26
5296.9		18

λ in air

Δλ = ± 0.2Å

Table I

3s5l-3s5l' ArVII TRANSITIONS (Å)

Δn=0 TRANSITIONS	λ_{cal}	λ_{obs}
3s5s $^1S_0$ 3s5p $^1P_1$	4261	4117.8
3s5s $^3S_1$ 3s5p $^3P_1$	4855	4919.1
3s5s $^3S_1$ 3s5p $^3P_2$	4797	4854.1
3s5p $^1P_1$ 3s5d $^1D_2$	3208	3217.8
3s5p $^3P_1$ 3s5d $^3D_1$	3098	3117.8
3s5p $^3P_1$ 3s5d $^3D_2$	3096	3117.8
3s5p $^3P_2$ 3s5d $^3D_1$	3122	3141.8
3s5p $^3P_2$ 3s5d $^3D_2$	3121	3141.8
3s5p $^3P_2$ 3s5d $^3D_3$	3118	3141.8

3s4l-3s4l' ArVII TRANSITIONS (Å)

Δn=0 TRANSITIONS	λ_{cal}	λ_{obs}
3s4s $^1S_0$ 3s4p $^1P_1$	2244	2250.2
3s4d $^1D_2$ 3s4f $^1F_3$	3259	3236.8

Table II

Some lines which appear both with the helium and the molecular hydrogen targets cannot be interpreted in the Ar^{6+} spectrum. They originate from the Ar^{5+} ion produced by double electron capture. Such transitions have already been observed [5]. They coincide with ab initio calculated Rydberg transition --3s ^{2}n ℓ -3s^{2}n'ℓ' with large value of ℓ, and are indicated in table III.

$3s^2nl-3s^2n'l'$ ArVI TRANSITIONS (Angström)

Δn=1 TRANSITIONS	λ_{cal}	λ_{obs}
$3s^27i$ $^2I_{11/2}$ $3s^26h$ $^2H_{11/2}$	3412	3405.3
$3s^27i$ $^2I_{11/2}$ $3s^26h$ $^2H_{9/2}$	3412	3405.3
$3s^27i$ $^2I_{13/2}$ $3s^26h$ $^2H_{11/2}$	3412	3405.3
$3s^28k$ $^2K_{13/2}$ $3s^27i$ $^2I_{13/2}$	5285	5271.9
$3s^28k$ $^2K_{13/2}$ $3s^27i$ $^2I_{11/2}$	5285	5271.9
$3s^28k$ $^2K_{15/2}$ $3s^27i$ $^2I_{13/2}$	5285	5271.9
$3s^28i$ $^2I_{11/2}$ $3s^27h$ $^2H_{11/2}$	5250	5237.3
$3s^28i$ $^2I_{11/2}$ $3s^27h$ $^2H_{9/2}$	5249	5237.3
$3s^28i$ $^2I_{13/2}$ $3s^27h$ $^2H_{11/2}$	5250	5237.3

Table III

These double capture can be interpreted in terms of correlated electron capture [6]. The large values of ℓ of the final states are not clearly understood but have been already observed [7],[8].

Conclusion :

This experiment is an other proof that the photon spectroscopy can be a good mean to study charge exchange in collisions of multicharged ion beam on a neutral target. In the present case single and double electron captures are observed, they lead to new transitions of the Ar^{6+} and Ar^{5+} spectra and confirm some exchange mechanism already observed in other experiments.

References :

[1] P. Boduch, M. Chantepie, D. Hennecart, X. Husson, D. Lecler and I. Lesteven-Vaïsse, J. Phys. B (1989) 22L377-3800.

[2] H. Ryufuku, K. Sasaki and T. Watanabe, Phys. Rev. A (1980) 21-745.

[3] M. Druetta, S. Martin, T. Bouchama, C.Harel and H. Jouin, Phys. Rev. A (1987) 36 3071.

[4] F. Fawcett, Private Communication.

[5] S. Martin, A. Denis, Y. Ouerdane, A. Salmoun A.E. Motassadeq, J. Desesquelles, M. Druetta, D. Church and T. Lamy, Phys. Rev. Letters (1990) Vol 64, 22 2633-2636.

[6] N. Stolterfoht, submitted to Phys. Rev. A (1990).

[7] F.W. Meyer, D.C. Griffin, C.C. Havenen, M.S. Huq, R.A. Phaneuf, J.K. Sevensson and N. Stolterfoht, Phys. Rev. (1988) vol 60 18 1821-1824.

[8] E.M. Mack, J.H. Nijland, P.V.D. Straten and A. Niehaus, Phys. Rev. A 39, 8 3846-3854.

Supplement to Z. Phys. D - Atoms, Molecules and Clusters 21, S 291-292 (1991)

Atoms, Molecules and Clusters
Zeitschrift für Physik D

Electron transfer in nuclear reactions

D. H. Jakubaßa-Amundsen

Sektion Physik, Universität München, W-8046 Garching, FRG

Abstract

The influence of nuclear reactions on atomic charge transfer in energetic ion-atom collisions is studied within a two-channel model for the reactions and the impulse approximation for electron capture. Large interference structures in the capture probability are found for the test reactions $^{12}C(d,p)^{13}C$, $^{18}O(p,\alpha)^{15}N$, $^{19}F(p,\alpha)^{16}O$ and $^{32}S(p,p')^{32}S$.

I. Introduction

Interference effects from the interplay between nuclear reactions and electronic excitation can be understood in terms of a time-delay, caused by the finite lifetime of the transiently formed compound nucleus, between the atomic transition amplitudes for the incoming and outgoing parts of the collision. Energy matching between the energy ΔE transferred to the electron and the nuclear decay width Γ is a necessary criterion for the visibility of interference structures in the electronic transition probability. Theoretical and experimental investigations concerning ionisation and electron capture in the presence of elastic nuclear resonances have recently been reviewed /1/.

An extension of the theory to allow for nuclear reactions has been attempted some time ago in case of ionisation /2/. Nuclear reactions with their wealth of possibilities to probe electronic half-trajectory transition amplitudes, isotope or recoil effects have the advantage, compared to elastic resonances, that the reaction amplitude does not contain any Coulomb contribution which could veil the atomic interference effects.

II. Theory

The nuclear two-channel model considers only two channels for the decay of the compound nucleus, the initial (elastic) channel a and the reactive (inelastic) channel b. Correspondingly, the nuclear scattering state with incident momentum $\underline{K}_a$ is at internuclear distances R exceeding the nuclear interaction radius represented by

$$\psi^{(+)}_{a\underline{K}}(\underline{R}) = \left[e^{i\underline{K}_a\underline{R}_a} + A^{(+)}_{aa}(K_a,\Theta_a)\, e^{iK_aR_a}/R_a\right]\phi_a + \left[A^{(+)}_{ba}(K_a,\Theta_b)\, e^{iK_bR_b}/R_b\right]\phi_b \qquad (1)$$

where A_{aa} is the elastic scattering amplitude (including the Coulomb contribution) and A_{ba} the reaction amplitude. $\Theta_\lambda = \sphericalangle(\underline{K}_\lambda,\underline{R}_\lambda)$ and ϕ_λ (λ=a,b) denotes the internal states of the projectile and target nuclei which are assumed to form a complete and orthonormal set. The combination of the two-channel model with the impulse approximation (IA) for electron capture (which is the on-shell limit of the strong potential Born approximation (SPB)) proceeds in an analogous way as in the elastic (one-channel) case /3/. The different length scales of the atomic and nuclear processes allow an approximate representation of the capture amplitude in terms of products between the reaction amplitude and the partial atomic transition amplitudes /4/.

III. Results

We have investigated a series of nuclear reactions under the aspect of the criterion

$$\Gamma \approx \Delta E = \varepsilon_{fb} + \frac{1}{2} m v_b^2 - \varepsilon_{ia} \qquad (2)$$

where ε_{fb} and ε_{ia} denote the energies of the final and initial electronic states, and v_b is the relative projectile-target velocity in channel b. We have improved on the large-angle capture theory by applying the less restrictive transverse peaking approximation instead of the previously used full peaking approximation. Fig.1 demonstrates that in case of electron capture in the non-resonant p + C collision the present theory is much closer to the experimental data than the previous one /3/.

For the neutron transfer reaction, $^{12}C(d,p)$ at the deuteron energy E_d = 1.4495 MeV, the width Γ of the J=2 compound state is 7 keV as compared to ΔE = 2.6 keV, and the excursion of the capture probability P(Θ) across the resonance is a factor of 2 (at Θ = 60°). For the reaction $^{18}O(p,\alpha)$ at E_p = 2.363 MeV one has J=1/2, Γ= 4.3 keV, ΔE = 1.6 keV and an excursion of P(Θ) up to a factor of 2 for Θ < 90° /4/.

Fig.2 shows the energy variation of P(Θ) for the elastic and inelastic collision of p with ^{32}S near the resonance at E_p = 3.716 MeV with J = 5/2 and the partial widths Γ_a = 0.82 keV and Γ_b = 0.64 keV.

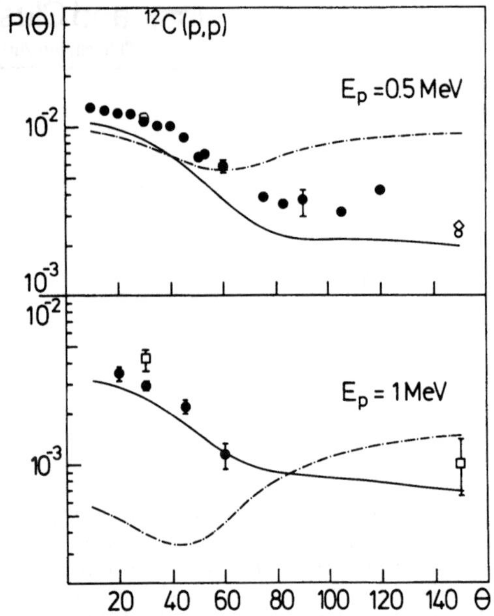

Fig.1. Capture probabilities for 0.5 MeV and 1 MeV p + C collisions versus scattering angle Θ. Theory (K-K capture): ——— transverse peaked IA, – – – – fully peaked SPB. Experiment: ●, Horsdal Pedersen /5/, open symbols, Meyerhof et al /6/ (o: E_p = 0.51 MeV, ◇ : E_p = 0.49 MeV)

For this reaction, ΔE = 3.23 keV and 4.48 keV for channels b and a, respectively, and large interference structures are found for channel b. In channel a, the excursion of P(Θ) is reduced due to the presence of the slowly varying Coulomb scattering amplitude in A_{aa}, but the interference structure is repeated (this is possible because $\Delta E > 2\Gamma$).

Pilot capture measurements have been made for the reaction $^{19}F(p,\alpha)$ at E_p = 0.34 MeV and 0.873 MeV /5/. For the higher energy, J=2, Γ = 4.3 keV and $\Delta E \leq$ 1.1 keV, and the excursion of P(Θ) is at most 50%. The capture probability has only been measured at Θ = 60°, 90° and 120°, without studying the dependence on the collision energy. Theory overestimates the data by a factor of 5, partly because capture from the target L shell is not included in the calculations.

IV. Conclusion

We have found interference structures in the capture probability when the collision energy is varied across the nuclear resonance, which are the more pronounced, the smaller Γ as compared to ΔE.

The angular dependence of P(Θ) near the resonance energy shows a monotonic decrease with Θ, in contrast to the oscillatory behaviour for elastic resonances which results from the influence of the Coulomb scattering amplitude.

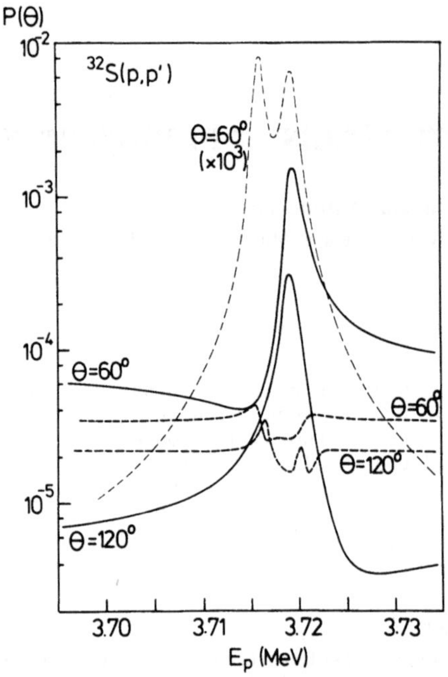

Fig.2. K-K capture probabilities at Θ = 60° and 120° in p + ^{32}S collisions as a function of proton energy. Theory: P(Θ) for inelastic nuclear scattering normalised to the nuclear reaction cross section (———) and to the elastic nuclear scattering cross section (–·–·–·–), P(Θ) for elastic nuclear scattering (– – – –)

References

1. Heinz,U.: Rep. Prog. Phys. 50, 145 (1987)
2. Anholt,R.: Atomic Inner-Shell Physics, ed. B.Craseman, p.581. New York: Plenum Press 1985
3. Jakubaßa-Amundsen,D.H., Amundsen,P.A.: J. Phys. B 18, 757 (1985)
4. Jakubaßa-Amundsen,D.H.: Phys. Rev. Lett., in print
5. Horsdal Pedersen,E.: Nucl. Instr. Meth. B 24/25, 130 (1987)
6. Meyerhof,W.E.: Private Communication; Scheurer, J.N., Baker,O.K., Meyerhof,W.E.: J. Phys. B 18, L85 (1985)

Supplement to Z. Phys. D - Atoms, Molecules and Clusters 21, S 293-294 (1991)

Atoms, Molecules
Zeitschrift für Physik D and Clusters

Inclusive probabilities for the scattering system 16 MeV-S^{16+} on Ar

P. Kürpick, W.-D. Sepp, B. Fricke
Fachbereich Physik, Universität Kassel, Heinrich-Plett-Straße 40, W-3500 Kassel, FRG

Abstract
We performed *ab initio* calculations of many particle inclusive probabilities for the scattering system 16 MeV-S^{16+} on Ar. The solution of the time-dependent DIRAC-FOCK-SLATER-equation is achieved via a set of coupled-channel equations with energy eigenvalues and matrix elements which are given by static SCF molecular many electron calculations.

PACS: 34.10.+x

During the last few years we developed a fully *ab initio* DIRAC-FOCK-SLATER (DFS) molecular code [1] which is able to provide energy eigenvalues and matrix elements between single particle levels. Using this input we are now able to perform coupled channel calculations. Knowing the transition amplitudes between the channels one is able to calculate inclusive probabilities [2, 3] for many electron transfer and excitation in many electron collision systems. Such a calculation is equivalent to the solution of the time dependent Dirac-Slater equation of the scattering problem with an adiabatic interaction potential. As a first example we have choosen the system 16 MeV-S^{16+} on Ar [4]. The experiment was carried out to answer the following questions: How large is the probability of finding:

1. One electron in Ar 1s and one electron in S 1s ?(P_{SA})
2. Two electrons in Ar 1s? (P_{SS})
3. Two electrons in S 1s? (P_{AA})

We performed a large number of selfconsistent static DFS calculations of the quasi molecular system S^{16+} - Ar at many internuclear distances to obtain the adiabatic energy eigenvalues as well as the d/dR and $d/d\omega$ coupling-matrix elements with an occupation according to the asymptotic occupation of the separated atoms in the incoming channel. This leads to the correlation diagram presented in Fig. 1. Due to the high degree of ionization the lowest level of this system for large internuclear distances is the 1s level of S although its Z value is smaller than the one for Ar.

In the actual coupled-channel calculation we included the 20 lowest relativistic channels (8 non relativistic

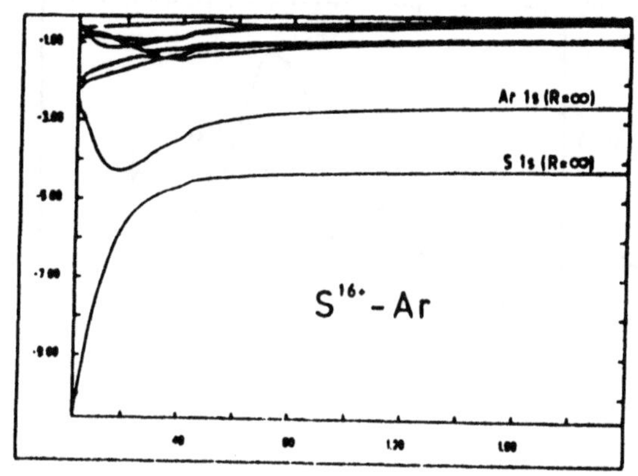

Figure 1: Correlation diagram : S^{16+}-Ar

channels) and calculated the amplitudes of the four lowest molecular levels (each doubly degenerate) which form the separated Ar 1s and S 1s levels after the collision. To evaluate the many particle probabilities we express the inclusive probabilities P_{SS}, P_{AA} and P_{SA} in terms of one electron amplitudes. The index S, resp. A, denotes a measured hole in the 1s shell of the Sulfur, resp. Argon. The 16 combinations to distribute zero to four electrons in the 1s shells of Sulfur and Argon are given in Tab. 1. The probabilities of each of these combinitions were calculated exclusive within the four lowest channels but inclusive with respect to the other channels. To get P_{SS}, P_{AA} and P_{SA} we sum up the probabilities of those combinations that contribute to the many particle probabilities considered according to the right side of Tab. 1. Probabilities of those combinations contributing to more than one many particle probability are distributed equally among them.

Figs. 2 and 3 show the inclusiv probabilities P_{AA}, P_{SS} and P_{SA} in comparison with the results of [4] i.e. P(b)-curves for the questions 1 to 3. The inclusive (four particle exclusive) probabilities will be published soon [5].

Occupation						
S 1s+	S 1s-	Ar 1s+	Ar 1s-	P_{SA}	P_{AA}	P_{SS}
○	○	○	○	◇	◇	◇
●	○	○	○	◇	◇	
○	●	○	○	◇	◇	
○	○	●	○	◇		◇
○	○	○	●	◇		◇
○	○	●	●			◇
○	●	●	○	◇		
●	●	○	○		◇	
●	○	○	●	◇		
○	●	○	●	◇		
●	○	●	○	◇		
●	●	●	○			
○	●	●	●			
●	○	●	●			
●	●	○	●			
●	●	●	●			

Table 1: The 16 possible ways of putting 0, 1, 2, 3, or 4 electrons into 4 levels and contribution to the inclusive probabilities. •: occupied; o: unoccupied

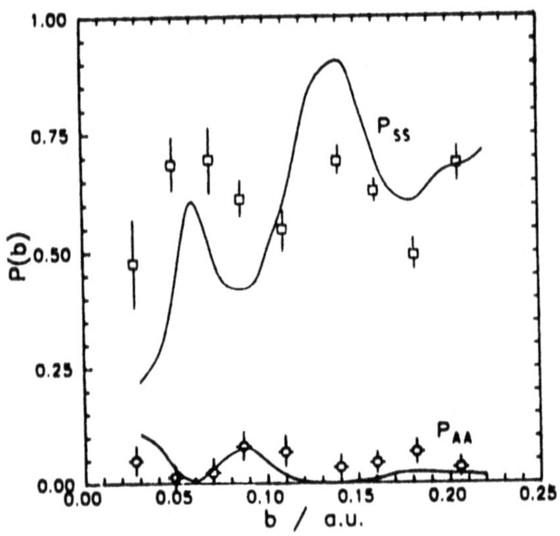

Figure 2: P_{AA} and P_{SS}

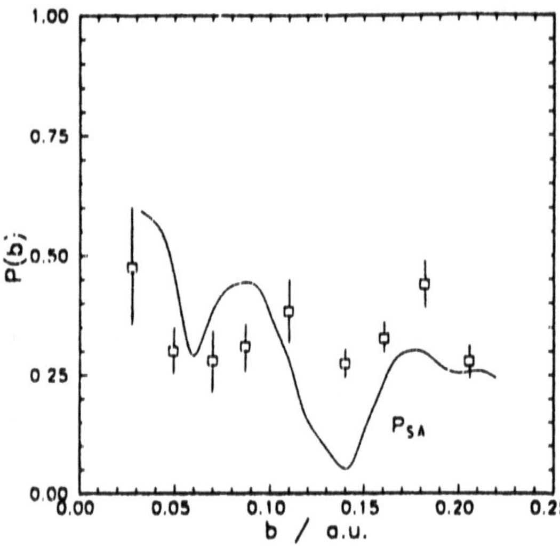

Figure 3: P_{SA}

References

[1] W- D Sepp et al., *Phys. Rev. A* **6** (1986) 3679

[2] H J Lüdde, R M Dreizler, *J.Phys. B:At. Mol. Phys.* **18** (1985) 107

[3] R L Becker, A L Ford, J F Reading *Phys. Rev. A* **6** (1984) 3111

[4] M Schulz, E Justiniano, J Konrad, R Schuch, A Salin, *J.Phys. B:At. Mol. Phys.* **20** (1987) 2057

[5] P. Kurpick et al. to be published

Supplement to Z. Phys. D - Atoms, Molecules and Clusters 21, S 295-296 (1991)

Atoms, Molecules
Zeitschrift für Physik D and Clusters

Azimuthal angular dependence of the K-x-ray emission in swift Ne^{7+}-Ne collisions*

T. Kambara[1], Y. Awaya[1], Y. Kanai[1], R. Dörner[2] and H. Schmidt-Böcking[2]
[1] RIKEN, Saitama-ken 353-01, Japan
[2] Institut für Kernphysik der Universität Frankfurt, W-6000 Frankfurt/M., FRG

* supported by DRG

In swift near symmetric heavy ion atom collisions the inner shell excitation is well described in the framework of electron promotion between quasi-molecular orbitals transiently formed during the collision process [1]. For K-shell excitation the vacancy transfer via $2p_\pi$-$2p_\sigma$-rotational coupling plays a dominating role and yields a characteristic dependence of the $2p_\sigma$-vacancy production probability $P_{2p\sigma}(b)$ on the impact parameter (b). The conservation of angular momentum, where angular momentum of the nuclear motion is transfered into electronic angular momentum, creates a significant shape of $P_{2p\sigma}(b)$. Systematic experimental investigations [1,2] have shown nice agreement with theory [3,4]. Because of the rotational coupling mechanism the $2p_\sigma$-electron should be promoted only into the $2p_{\pi x}$-orbital (its angular momentum vector is perpendicular to the scattering plane). If all other $2p_\pi$-states would be vacant, a subsequent x-ray emission from the $2p_{\pi x}$ into the isotropic 1s state should reflect by its dipole emission pattern directly this selective $2p_{\pi x}$-promotion. For heavy ions this alignment aspect has never experimentally been investigated. Such an experimental investigation would require that the emitted K-x-ray is measured in coincidence with the scattering plane of the colliding nuclear partners. Furthermore we must ensure that the L-shell of the emitting excited ion was completely vacant besides the one promoted $2p_{\pi x}$ electron. This is in general hard to control, however for 10 MeV Ne^{7+} on Ne by selection of the high projectile charge state exit channels, we are indeed able to separate such promotion channels. As expected for such exit channels a pronounced dipole emission pattern of K-x-ray emission with respect to the scattering plane has been observed.

The experiment was performed at the RIKEN heavy ion linear accelerator (RILAC). The charge state selected Ne^{7+} beam was well collimated by a triple set of collimators reaching an angular divergence of about 10^{-4} rad. The Ne^{7+} beam passed a thin Ne gas target viewed by a Si(Li) detector with a 12.5μ Be-window, to be able to detect the Ne-K-rays. The scattered projectiles were magnetically (25^{0}) charge-state selected and were detected by a two-dimensional position-sensitive parallel plate avalanche detector with a wedge and stip anode. The detector was placed about 6 m downstreams from the target. The overall position resolution (FWHM) including the beam collimation corresponds to a scattering angle uncertainty of $1.4 \cdot 10^{-4}$ rad.

Besides the x-ray energy, the coincidence time and three particle detector signals (3 charge devided branches) were recorded in list mode technique using standard fast/slow coincidence electronics. The obtained time resolution (for Ne-K-rays) was typically 90-110 nsec (FWHM), and the true to random ratio was always better than 5:1.

In figure 1 the measured impact parameter dependence of the Ne-K-x-ray emission for 10 MeV Ne^{7+} on Ne is shown. The upper part shows the data when the final charge state of the projectile is not determined, the lower part presents the data for the final charge state q_f=7. Both sets of $P_x(b)$ reflect nicely the expected $2p_\pi$-$2p_\sigma$-rotational coupling shape, the minimum towards small b and the so called adiabatic maximum somewhat outside the Ne-K-shell radius. Particulary in the region of the adiabatic peak and towards smaller b the $2p_\pi$-$2p_\sigma$-rotational coupling mechanism should dominate and populate only the $2p_{\pi x}$state. To ensure that only K-x-rays from one highly charged collision partner are measured we determined also the outgoing projectile charge state (e.g. q_f=7,8). Thus for incoming Ne^{7+} ($1s^2 2s^1$) the projectile L-shell before the x-ray decay can only be occupied by the promoted $2p_{\pi x}$ electron. The non distinguishable Ne-target K-rays are always strongly suppressed particulary for $q_f=7,8$. Here the mean final charge state q is between 4 and 5. For those charge states the Ne-target K-vacancies are predominately decaying by K-Auger emission. In figure 2 we present the measured angular azimuthal dependence. For $q_f=7$ and $b \approx r_{ad}$ indeed a dipole emission structure is observed.

In figure 3 for comparison for $q_f=4$ the measured azimuthal dependence is shown. For $q_f=4$ we expect several electrons in the projectile 2p state. The subsequent 2p-1s decay will be a superposition of the different dipole patterns thus smoothing out the dipole structure. Indeed this effect has been seen.

We conclude that in agreement with the quasi-molecular electron promotion picture in Ne^{7+}-Ne collisions for selected reaction channels a characteristic dipole pattern in the Ne-K-x-ray emission has been observed. Performing x-ray particle coincidence experiments with highly charged ions one has to aware that K-x-ray emission is not always an isotropic process.

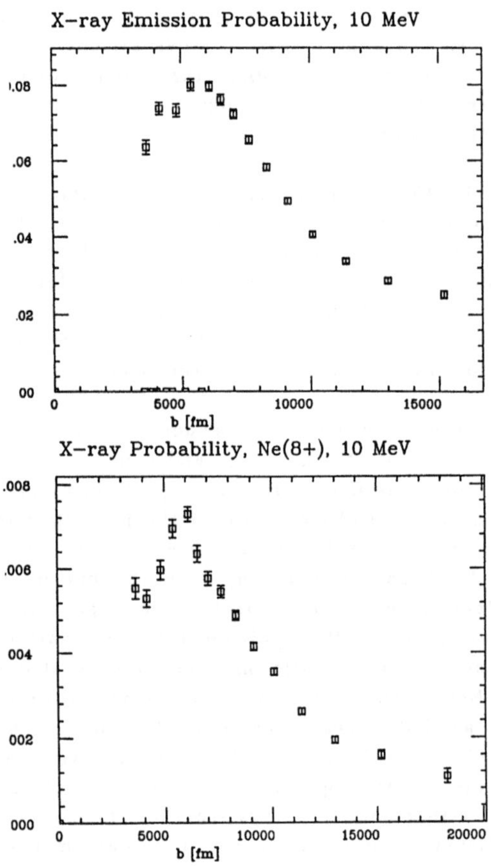

Figure 1

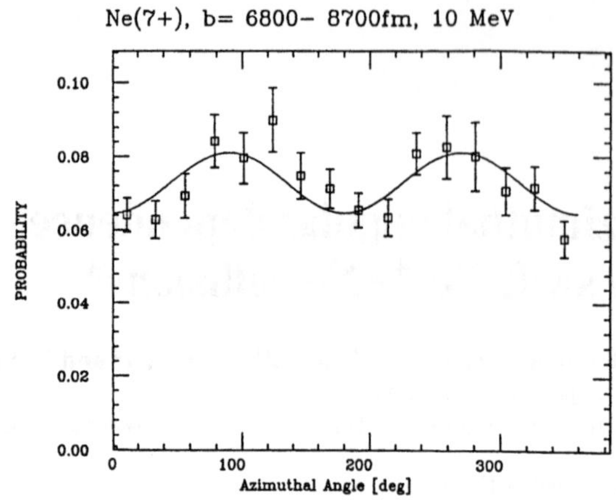

Figure 2

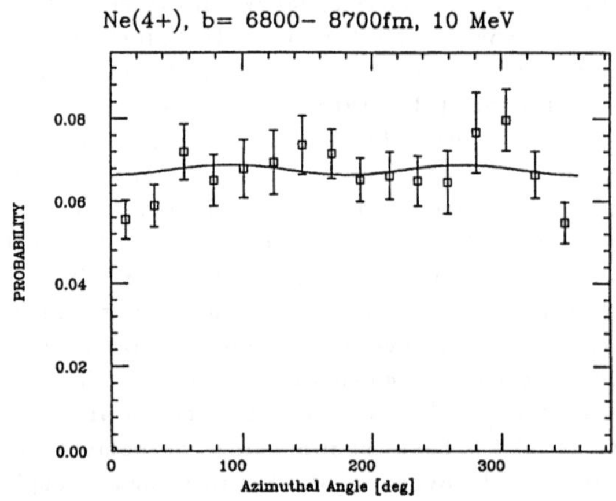

Figure 3

Acknowledgments:
Two of us (R.D. and H.SB) want to thank the RIKEN for the great hospitality they have obtained. We acknowledge further stimulating and helpfull discussions with P.Mokler.

References:

1. R.Schuch, R.Hoffmann, K.Müller, E.Pflanz, H.Schmidt-Böcking, H.Specht Z.Phys. A316 (1984) 5 and references therein
2. V.Dangendorf, H.Schmidt-Böcking, W.Fritsch, J.Ullrich, S.Kelbch, C.Kelbch Nucl.Instr.Meth. B23 (1987) 170
3. K.Taulbjerg, J.S.Briggs, J.S.Vaaben J.Phys. B9 (1976) 170
4. W.Fritsch and C.D.Lin Phys.Rev. A31 (1985) 1164

Supplement to Z. Phys. D - Atoms, Molecules and Clusters 21, S 297-298 (1991)

Atoms, Molecules
Zeitschrift für Physik D and Clusters

Study of the 4d-4f coherence of impact-excited He atoms by observing 1s3d-1s2p transitions

A. S. Aynacioglu, S. Heumann, G. von Oppen
Institut für Strahlungs- und Kernphysik, Technische Universität Berlin, Hardenbergstraße 36, W-1000 Berlin 12, FRG

Abstract:

Impact radiation of He atoms excited by $H_2^+(D_2^+)$ and $H_3^+(D_3^+)$ ions was investigated as function of electric fields parallel to the ion beam for the spectral lines at $\lambda(3^1D \rightarrow 2^1P) = 668$ nm and $\lambda(3^3D \rightarrow 2^3P) = 588$ nm. The intensity variations are explained by cascade feeding of the $1s3d$ states from the $1s4f$ states where the electric field causes both singlet-triplet demixing and $1s4d$ admixing.

I. Introduction

By investigating the $\lambda(4^3D \rightarrow 2^3P) = 447$ nm impact radiation of He atoms excited by collisions with 10-20 keV $H_2^+(D_2^+)$ ions as function of an electric field parallel to the ion beam, we recently demonstrated that electric dipole moments of impact-excited He atoms can be detected [1]. Here we present corresponding measurements on the electric-field dependence of the impact radiation at $\lambda(3^1D \rightarrow 2^1P) = 668$ nm and $\lambda(3^3D \rightarrow 2^3P) = 588$ nm emitted after $H_2^+(D_2^+)$ - He and $H_3^+(D_3^+)$ - He collisions. For these He measurements only weak electric fields $F \leq 2$ kV/cm were applied. Using stronger electric fields, similar experiments can be performed on energetic He-like ions which are less polarizable.

II. Experiment

The measurements were performed using the same experimental set-up as in [1]. Further details of the ion-beam apparatus have been described in [2] and [3]. 10, 15 and 20 keV $H_2^+(D_2^+)$ and $H_3^+(D_3^+)$ ions were used for excitation and the electric field was scanned between -2 kV/cm (directed upstream) and +2 kV/cm. Impact radiation emitted perpendicular to the ion beam and selected with interference filters was observed with a photomultiplier. Contrary to the $1s4d$ signals [1], the $1s3d$ signals have a pronounced broad structure with a maximum for the 588 nm signals (Fig. 1) and a minimum for the 668 nm signals (Fig. 2). Signals essentially symmetric with respect to the direction of the electric field were obtained after 10 keV D_3^+ excitation. At higher velocities of the ions the signals are asymmetric. Some signals, in particular those observed after $H_2^+(D_2^+)$ impact, have an additional narrow structure near $F = 0$: a sharp increase of the intensity in the field range $0 < F < 0.3$ kV/cm.

III. Discussion of results

Considering direct excitation of the $1s3d$ states, an electric-field dependence of the observed impact radiation can result at best from the Stark mixing of the $1s3d$ and $1s3p$ states. However, one estimates for the amplitude a_{3d}^{3p} of the $1s3p$ admixture to the $1s3d$ states $|a_{3d}^{3p}| < 1\%$ if $|F| \leq 2$ kV/cm. Therefore, the observed intensity variations can only be understood by assuming that cascade feeding from the $1s4f$ states contributes considerably to the population of the $1s3d$ states. The influence of the electric field on the $1s4f$ substates is twofold: (i) the $1s4f$ states acquire a $1s4d$ Stark admixture with $|a_{4f}^{4d}| \leq 20\%$ [1]; (ii) the singlet-triplet mixing in the $1s4f$ configuration is reduced [2]. Both effects are to be taken into account for explaining the broad structure of the 588 nm and 668 nm signals. For a preliminary quantitative analysis, we extended the two-state calculation used in [1] to a four-state calculation based on two singlet states * $|4^SF\rangle$ and $|4^SD\rangle$ and two triplet states $|4^TF\rangle$ and $|4^TD\rangle$. Due to Wigner's spin conservation rule [2], the excitation matrix $\rho_4 = \rho_4^S \otimes \rho_4^T$ of these four n = 4 states can be assumed to be diagonal with respect to the multiplicity. The 2 x 2 excitation matrices ρ_4^S and ρ_4^T refer to the singlet and triplet states, respectively. However, the evolution of the singlet and triplet states after excitation is coupled by the off-diagonal matrix elements of the spin-orbit coupling, which were incorporated into the analysis. Using this still simplifying approach for describing the electric-field dependence of the $3d-2p$ spectral lines, the observed broad signal structures could be modelled. The intensity

*The superscripts S and T are used to emphasize that we refer to Russell-Saunders basis states.

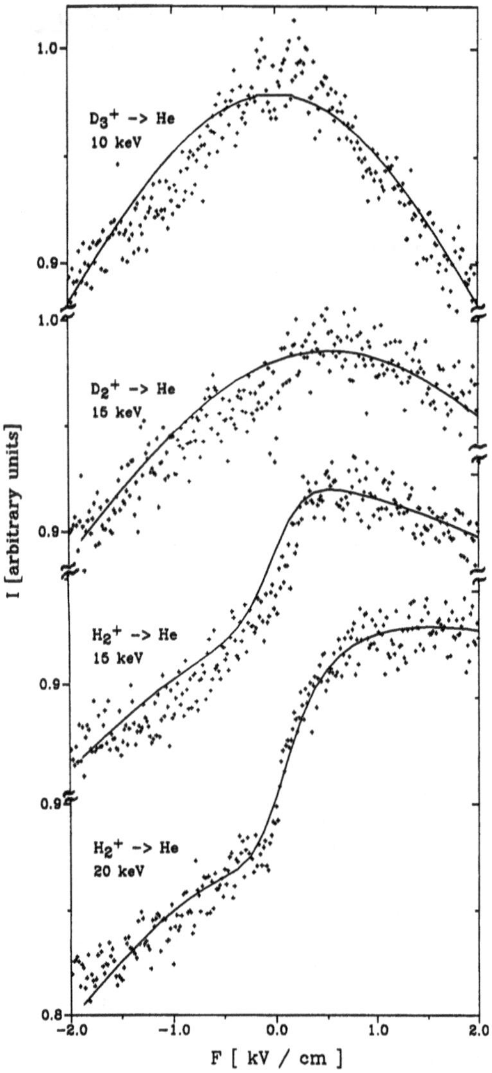

Fig. 1. Intensity I of the 588 nm line as function of the electric field F for the indicated collisions (solid lines: calculated signals).

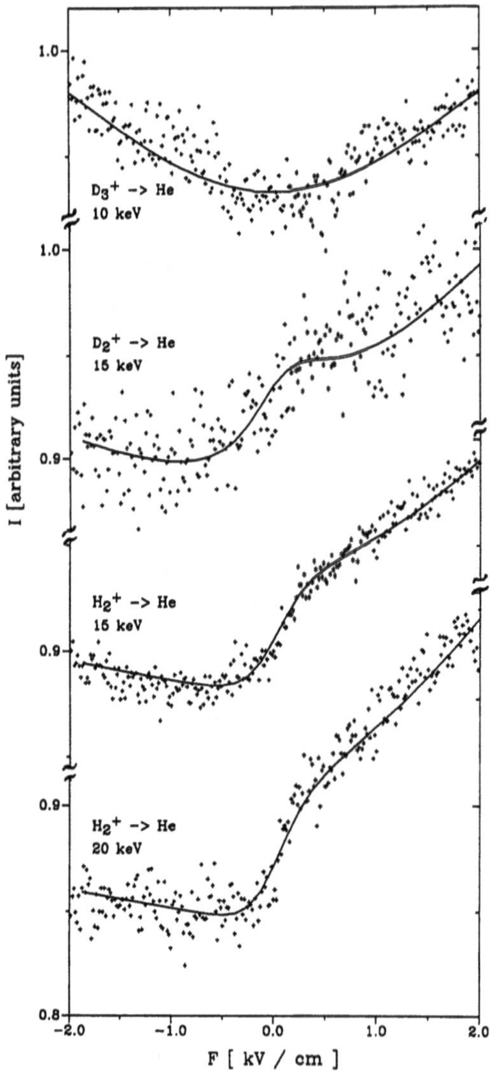

Fig. 2. Intensity I of the 668 nm line as function of the electric field F for the indicated collisions (solid lines: calculated signals).

maxima of I(588 nm) (Fig. 1.) and, correspondingly, the intensity minima of I(668 nm) (Fig. 2.) near $F = 0$ indicate that the excitation cross sections of the $4f$ singlet states are larger than those of the $4f$ triplet states. The asymmetric structures of both the triplet and the singlet signals indicate electric dipole moments of the impact excited He states directed upstream. The sharp increase of the $H_2^+(D_2^+)$ signals near $F = 0$ can only be explained by assuming a cascade feeding from the closely spaced $5f$ and $5g$ states in addition to the cascade feeding from the $4f$ states.

Acknowledgements: This work has been supported by the Deutsche Forschungsgemeinschaft.

References

1. A. S. Aynacioglu et al.: Phys. Rev. Lett. 64 (1990) 1879
2. A. S. Aynacioglu et al.: Z. Physik D6 (1987) 155
3. A. S. Aynacioglu et al.: Z. Physik D to be published

Supplement to Z. Phys. D - Atoms, Molecules and Clusters 21, S 299-300 (1991)

Atoms, Molecules and Clusters
Zeitschrift für Physik D

Strong directional out-of-plane scattering in multiple ionizing highly-charged ion-atom collisions

A. González[1,5], S. Hagmann[1], T. Quinteros[1,5], B. Krässig[2], R. Koch[3], A. Skutlartz[4], and H. Schmidt-Böcking[3]

[1] J. R. Macdonald Laboratory, Kansas State University, Manhattan, KS 66506-2604, USA
[2] Fakultät f. Physik, Universität Freiburg, FRG
[3] Institut f. Kernphysik, Universität Frankfurt, FRG
[4] East Carolina State University, Greenville, NC, USA
[5] Now at Manne Siegbahn Institut, Stockholm, Sweden

Abstract

The azimuthal (ϕ_P) and polar angle (θ_P) scattering of projectiles in coincidence with recoil ions has been studied for 0.53 MeV/u F^{8+} + Ne. For high degree of ionization of the target we find the resultant transverse momentum of all electrons emitted into the continuum to increase with the number of ejected electrons and to have a direction mostly not co-planar with the scattering plane.

I. Introduction

The mechanism by which many electrons are simultaneously emitted in comparatively slow collisions of highly charged ions with atoms has been of considerable interest, since experiments[1,2] characterizing the impact parameter dependence of the electron emission patterns in F^{8+} + Ne revealed strong angular dependences of the emission probabilities. Observation of very high longitudinal electron momenta [3] and strong polarization of target electron states [4] showed the strong influence of electron nucleus interaction.

We report here on inclusive measurements of multi-electron emission probabilities measured via a coincidence between recoil ions and scattered projectiles in well defined charge states q_f detected at scattering angle (θ_P, ϕ_P). The data show that the ejected electron momentum summed over all outgoing electrons increases with increasing degree of ionization and is mostly non co-planar to the scattering plane resulting in strong non co-planar nuclear scattering effects.

Experiment

A highly collimated beam of 0.53 MeV/u F^{8+} of the J.R. Macdonald Laboratories EN Tandem intersected with a Ne gas jet. The projectiles were charge state selected after the collision and then after 700 cm flight path detected with a 2-dimensional position sensitive parallel plate avalanche detector (PPAD) equipped with a wedge and strip anode. Recoil ions in different charge states q_R extracted electrostatically from the collision zone were detected in coincidence with scattered projectiles in cartesian coordinates (x, y), the recoil ion flight time identifying charge states q_R (see Fig 1). Due to the small extraction field the detection efficiency of the recoil spectrometer is maximal for recoil ions going parallel to the +x axis, i.e. with a vanishing y-component of the transverse momentum. A maximum in the coincident projectile azimuthal distribution thus will correspond to recoils with $(P_R^T)_Y = 0$. In off-line analysis for each polar angle θ_P recoil TOF spectra as function of the azimuthal angle ϕ_P were constructed from recoil projectile coincidence events recorded on-line with cartesian coordinates (x, y).

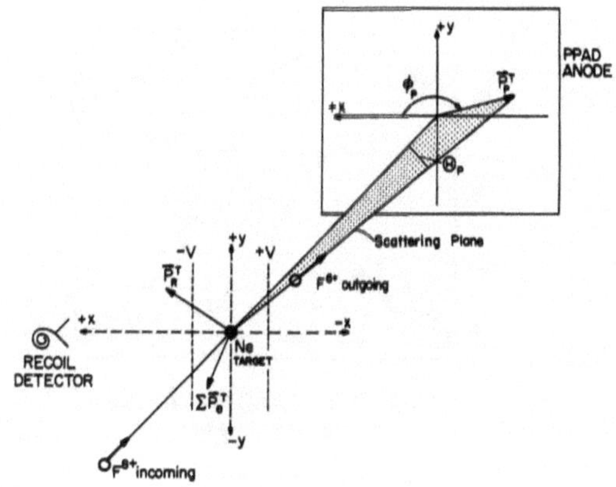

Fig 1. Experimental Set-up.

III. Results

Fig. 2 shows, for a fixed polar scattering angle θ_P=0.28 mrad, the azimuthal (ϕ_P) dependence of the recoil projectile coincident probability for selected recoil ion charge states $3 < q_R < 7$. We observe for a low recoil charge, q_R=3, i.e. 1 electron in the continuum and two electrons captured into the projectile a single peak for ϕ_P=180°. For higher recoil charge states, that is a higher multiplicity of electrons emitted into the continuum, we note that the single peak splits into two peaks moving away from 180°. For q_R=7+, that is for the emission of 5 electrons, the recoil projectile coincident rate peaks for ϕ_P=137° and ϕ_P=223°, which is clearly non-coplanar with the recoil ions. The analysis of the transverse momentum balance of scattered projectile and recoil ion allows to deduce the role which the momentum of electrons emitted into the continuum plays in the collision. Fig. 3 illustrates the transverse momentum balance. Recoil ions can only be detected if the transverse recoil momentum P_e^T lies within the acceptance cone of the TOF spectrometer. Then, for a fixed polar scattering angle $\theta_P \pm \Delta\theta_P$ (i.e. projectiles falling into the horizontally shaded ring) recoil ions can only be detected in coincidence with projectiles at azimuthal angles $\phi_P \neq 180°$, if $P_{Proj}^T + P_{Rec}^T + P_e^T = 0$, where P_e^T is the electron momentum vector summed over the transverse momenta of all emitted electrons.

The observation of maxima in the ϕ_P distribution of recoil projectile coincident events for fixed polar angles then imposes restrictions on the possible form of the momentum density distribution of electrons. If for a given degree $n_e = q_R - 2$ of ionization the probability distribution $n(|p_e(q_R)|)$ follow a monotonic decrease with a maximum at $|p_e|=0$ the recoil projectile coincidence yield as function of would always peak at $\phi_P=180°$. Only, with increasing mean value $<|p_e|>$ the observed peak width in the recoil projectile coincidence distribution as function of ϕ_P would increase too.

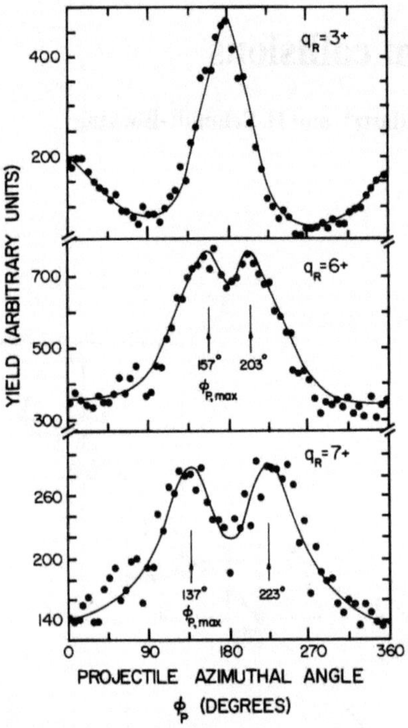

Fig 2. Recoil-projectile coincident distribution for different recoil charge states as function of ϕ_P for polar scattering angle $\theta_P=0.28$mrad of projectiles.

The observed maxima at $\phi_P \neq 180°$ can only be obtained if already the distribution $n_e(|p_e|)$ has a maximum at $|p_e| \neq 0$. The deviation of ϕ_P from 180° degree is therefore a direct measure of the mean electron momentum vector $|p_e|$.

An estimate for the magnitude of the transverse electron momenta can be derived as follows: at the azimuth ϕ_P where the coincident scattered projectile distribution has a maximum, we have

$$(1) \quad P_{Proj,x}^T + P_{Rec,x}^T + P_{e,x}^T = 0$$

$$(2) \quad P_{Proj,y}^T + 0 + P_{e,y}^T = 0.$$

Eq. (1) will not be used at this time due to large uncertainties in the determination of $P_{Rec,x}^T$ (see also Fig 3). Eq. (2) reads for small scattering angles

$$(3) \quad P_{Proj}^i * \theta_P * \sin \phi_{P,max} + P_{e,y}^T = 0,$$

where P_{Proj}^i is the incident projectile momentum. $P_{e,y}^T$ we then derive as lower bound for the electron transverse momentum $<|P_e^T|>$ [a.u.].

Table 1.

q_R	n_e	$\theta_P=0.28$mrad $<\lvert P_e^T\rvert>$	$\theta_P=1.1$mrad $<\lvert P_e^T\rvert>$
5	3	9	12
6	4	18	18
7	5	30	24
8	6	35	36

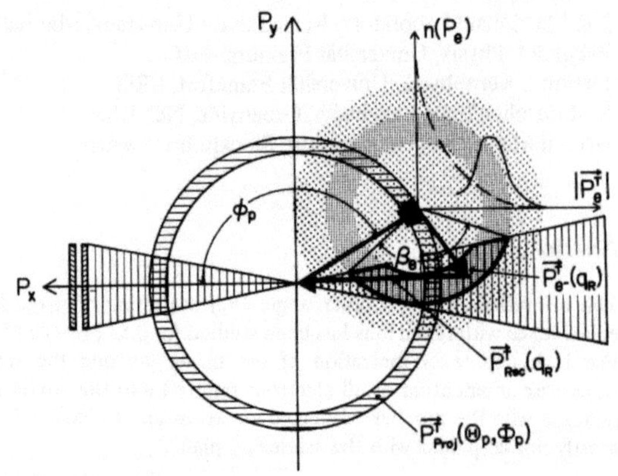

Fig 3. Transverse momentum balance for scattered projectile, recoil-ion and summed transverse electron momentum.

This leads to individual single electron energies between 122 and 463 eV in qualitative agreement with preliminary data from delta electron-recoil coincidence experiments [5].

IV. Summary

We have measured inclusive multielectron emission probabilities via recoil-scattered projectile coincidences with a defined scattering plane. We find that the total transverse electron momentum increases with the number of electrons emitted into the continuum. We also find that at very small scattering angles electrons are emitted non-isotropically with respect to the scattering plane giving rise to strong out of plane scattering of the projectile.

Acknowledgements

This work was supported by: the Division of Chemical Sciences, Office of Basic Energy Science, Office of Energy Research, U.S. Department of Energy; and by Deutsche Forschungsgemeinschaft DFG, Germany.

References

1. A. Skutlartz, S. Hagmann, Phys. Rev. A 28, 3268 (1983).
2. H. Schmidt-Böcking, A. Skutlartz, S. Hagmann, Phys. Lett. 122A, 421 (1987).
3. R. Schuch, H. Schöne, P.D. Miller, H.F. Krause, P.F. Dittner, S. Datz, Phys. Rev. Lett. 60, 925 (1988).
4. R. Doerner, J.Ullrich, H. Schmidt-Böcking, R. Olson, Phys. Rev. Lett. 63, 147 (1989).
5. B.Krässig, S.Hagmann, A.Gonzalez, and T.Quinteros (to be published).

Supplement to Z. Phys. D - Atoms, Molecules and Clusters 21, S 301-302 (1991)

Atoms, Molecules
Zeitschrift für Physik D and Clusters

δ-Electron emission in fast, highly-charged heavy ion-atom collisions*

C. Kelbch[1], S. Hagmann[2], R. Koch[1], G. Kraft[3], R. E. Olson[4], H. Schmidt-Böcking[1], J. Ullrich[3]

[1] Institut für Kernphysik, Universität Frankfurt, W-6000 Frankfurt/M., FRG
[2] Physics Department, Kansas State University, Manhattan, KS 66506, USA
[3] GSI Darmstadt, FRG
[4] Physics Department, University of Missouri-Rolla, Rolla, MO 65401, USA

* supported by BMFT and GSI

Systematic studies have been started to investigate the double differential cross sections for the emission of delta-electrons, differential in emission energy E_e and -angle ϑ_e in U - rare gases collisions. These studies have been performed in a projectile energy range between 0.4 MeV/u and 1.4 MeV/u. The energy spectra for different emission angles show unexpected structures in the region of the binary encounter peak [1], which vary strongly with the emission angle ϑ_e and projectile energy E_p. These structures appear e.g. at $\vartheta_e \approx 40°$ for an impact energy E_p/A of 1.4 MeV/u, at $\vartheta_e \approx 32.5°$ for E_p/A = 1 MeV/u and at $\vartheta_e \approx 22.5°$ for E_p/A = 0.5 MeV/u.

Figure 1 shows as an example the energy dependence of the double differential cross sections for delta-electron emission for different ejection angles in the 1.4 MeV/u U^{33+} on Ar system. Nearly identic structures have been observed for He, Ne and Xe as target and do not show any target dependence. The strong structures around $\vartheta_e \approx 40°$ are clearly observable. In a very similar scattering system, however, namely 1.3 MeV/u I^{28+} on Ar, where the effective charge is comparable to the U projectile, for our surprise no structures in the delta-electron spectra have been measured for this latter system. This indicates that obviously the screened nuclear charge rather than the ionic charge state plays an important role in the electron emission in such heavy ion-atom encounters. Such a structure is not at all expected in the framework of binary encounter electron emission in a point-charge like Coulomb potential.

Very recently, Olson et al [2,3] predicted for a screened Coulomb potential a strong enhancement of BE electron emission into forward angles. In this potential, BE-electrons from different trajectories are emitted into the same angle in the laboratory system. This different amplitudes might interfere and therefore might give rise to structures in the energy distribution of the electrons. In figure 2, for 1.4 MeV/u U on Xe, the integrated BE cross section is given in relative units as a function of the observation angle ϑ_e. As predicted by the CTMC-calculations, there is indeed a strong enhancement for anlgles $\vartheta_e < 40°$. The interesting fact is, that the structures in the energy spectra occur at exactly the same angle where the cross section starts to increase. This is an indication, that the discussed structures and the enhanced forward emission of BE-electrons might be related. To gain more insight into the delta-electron emission process, first electron-electron coincidence experiments have been performed. The electron energy spectra under $\vartheta_e = 0°$ coincident with

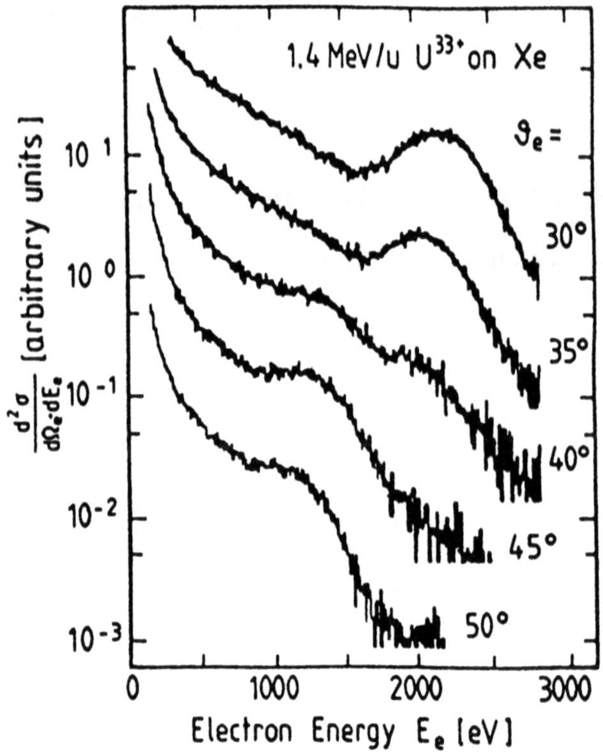

Figure 1 See Text

electrons emitted under $\vartheta_e \approx 60°$ show a strongly enhanced contribution at emission energies E_e between cusppeak and binary encounter peak. This broad peak shifts to higher electron energies at $\vartheta_e = 0°$ when higher electron energies under 60° are detected. This result indicates that "hot" electrons at $\vartheta_e = 0°$ and $\vartheta_e = 60°$ (where about 80% of all emitted electrons are found [4]) are resulting from the same collision.

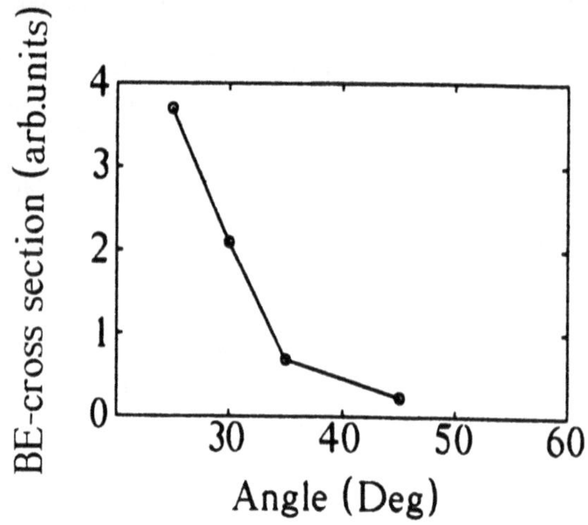

Figure 2 See text

[1] Kelbch C, Hagmann S, Kelbch S, Mann R, Olson RE, Schmidt S,Schmidt-Böcking H Phys. Lett. A 139 304 (1989)

[2] Olson RE, J. Phys. B 12 1843 (1979)

[3] Olson RE, Reinhold C, Schulz D J. Phys. B 23 (1990) L455

[4] Kelbch C, Olson RE, Schmidt S, Schmidt-Böcking H, J. Phys. B 22 2121 (1989)

Supplement to Z. Phys. D - Atoms, Molecules and Clusters 21, S 303-304 (1991)

Atoms, Molecules and Clusters
Zeitschrift für Physik D

EUV spectra from Ne and Ar recoil ions induced by 1.4 MeV/u heavy ion beams

B. Kraus* and K.-H. Schartner[1], F. Folkmann[2], A. E. Livingston[3], P. H. Mokler

[1] I. Physikalisches Institut, Universität Giessen, FRG
[2] Institute of Physics, University of Aarhus, Denmark
[3] Department of Physics, Notre Dame University, Indiana, USA
[4] GSI Darmstadt and Universität Giessen

Abstract

V^{15+}-, Ni^{17+}-, Xe^{25+}- and U^{32+}-ions at 1.4 MeV/u were used to ionize and excite Ne and Ar atoms. Spectra with wavelengths between 8 nm and 25 nm were recorded. Using tabulated wavelength values, the wavelength scale of the used grazing incidence spectrometer could be calibrated in situ within 0.01 nm.

Introduction

Spectroscopic investigations of few-electron ions are of interest in order to test fundamental atomic structure calculations, especially experiments with highly charged heavy ions for tests of the higher order terms of the relativistic and QED contributions. $\Delta n = 0$ ($n = 2$) transitions within heavy He- or Li - like ions, which are of special interest /1/, ask for spectroscopic techniques in the spectral range of the extreme ultraviolet (EUV). Within this context, first experiments in the field of recoil ion spectroscopy were carried out with the GSI 5m grazing incidence spectrometer /2/. They aim at an investigation of line identification and line intensities in the EUV. An accurate wavelength calibration in the 10 nm range is in our opinion still difficult.

Experimental set up

The GSI 5m grazing incidence spectrometer is a Rowland circle instrument designed by R. J. Speer /3/. It is a prefocussed modular system consisting of a one meter long beam and three basic modules: an entrance slit, a diffraction grating and a photoplate or double scanning detector. These three modules can be positioned and clamped anywhere on the rails of the chassis beam. Weak signals expected with beams of very highly charged ions made it necessary to equip the instrument with a position sensitive detector. This detector is based on channelplates followed by a wedge and strip anode for read out. It covers a wavelength range of about 3 nm in one position. More specific data of the spectrometer and the detector have been published in /2/.

* Part of thesis

In a first application at the UNILAC of the GSI we used V^{15+}-, Ni^{17+}-, Xe^{25+}- and U^{32+}-ions at 1.4 MeV/u to ionize and simultaneously excite Ne and Ar as target gases. Beam currents of 300 nA, 400 nA, 150 nA and 30 nA were available for the mentioned projectiles, respectively. The target pressure was 150 mTorr. The accumulation periods at one specific detector position amounted on the average to 1000 s.

Recorded spectra

Optical recoil ion spectra of Ne and Ar are displayed in Fig. 1 and 2. Slit widths amounted to 50 μm for Ne and the V^{15+}-Ar spectrum, while for Ar excited by U^{32+} the slit width had to be increased to 200 μm. With V^{15+} a grazing angle of 3° was used, with Ni and Xe of 2.6° and with U of 1.4°. The spectra in fig. 1 and 2 are composed from spectra at 9 successive positions of the detector.

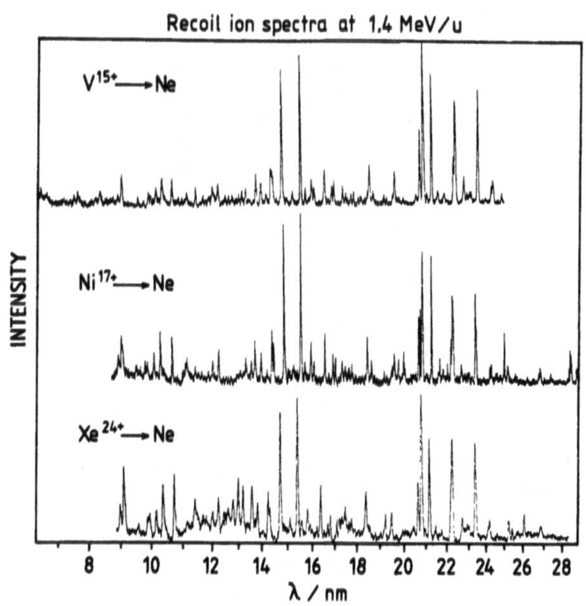

Fig. 1 Ne recoil ion spectra

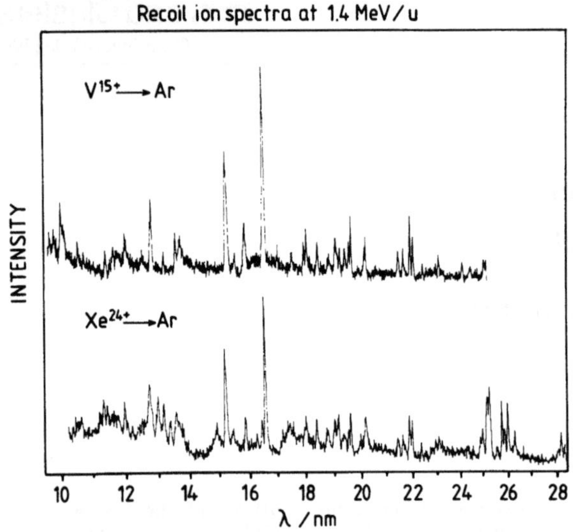

Fig. 2 Ar recoil ion spectra

Wavelength calibration procedure

The general features of the recorded spectra strongly resemble spectra accumulated earlier with 1.15 MeV/u Kr^{20+} /4/ or with 0.7 MeV/u Cl^{11+} /5/. Therefore we present no listing of the identified spectral lines. No dramatic change of the relative line intensities with increasing projectile charge was observed, though Xe^{24+} impact produces more lines from the more highly charged components of the target. For the calibration of the wavelength scale we proceeded in the following way, using the Ni^{17+}→Ne spectrum: The channel position in the original readouts were determined by a fitting program. Using a spectrometer simulation program with the known and measured geometrical quantities of the slit, grating and detector positions as input data, wavelengths were calculated and, for each detector position, compared with tabulated values /6/. The obtained differences are plotted in fig. 3 as a function of the tabulated wavelength. Two main features are present in fig. 3: Each detector position shows a systematic relation between measured and tabulated wavelength. Additionally a systematic correction seems to depend on the single detector position. The further analysis of the detector behaviour lead to a common quadratic relation between $\Delta\lambda$ and λ for each detector position. It can be explained by small deviations between the amplification factors of the three amplifiers used for the position analysis and by a slightly curved channelplate. The underlying overall nonlinearity, probably of mechanical origin, was additionally corrected by a quadratic function. The final residua are plotted in fig. 4 as function of the tabulated wavelength. As in fig. 3, the numbers denote the successive detector settings. In the short wavelength range from 85 to 130 Å we obtained an inaccuracy of the wavelength calibration of 0.1 Å. The inaccuracy increases for the longer wavelengths to 0.25 Å. Errors of the tabulated wavelengths range from 0.02 Å to 0.1 Å and are incorporated in the quoted uncertainties. The statistical errors of the line fits are also in the range 0.02 Å to 0.1 Å. The increase of the error towards larger wavelengths is due to an inaccurate read out of the detector position but is not of fundamental character.

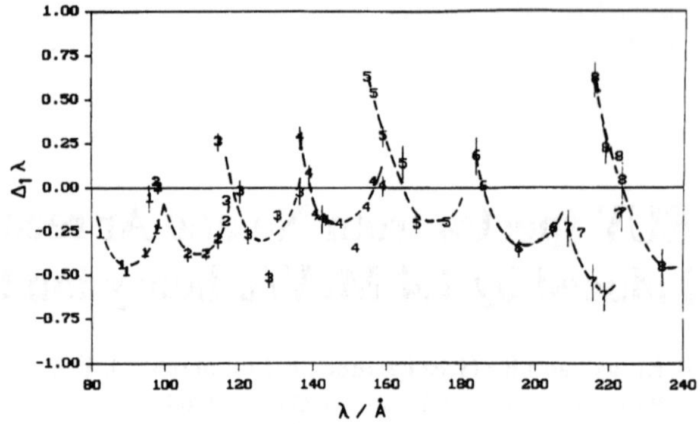

Fig. 3 Wavelength differences as function of tabulated wavelengths. Numbers correspond to successive detector positions, error bars mark uncertainties of the fitting procedure.

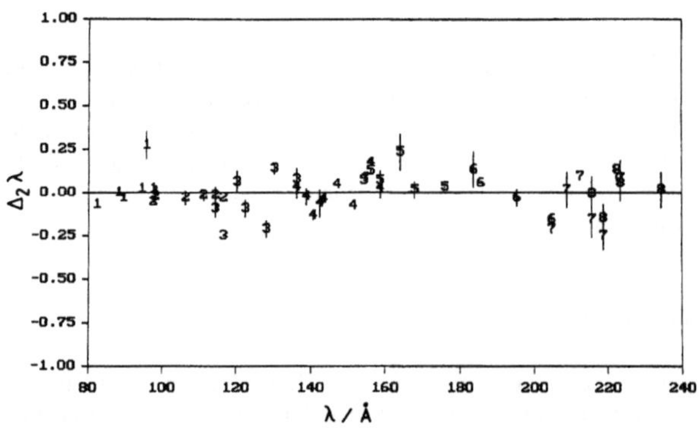

Fig. 4 Wavelength residues after compensation of the nonlinearities in fig. 3 by two quadratic functions.

Summary

Ne and Ar recoil ion spectra recorded with 1.4 MeV/u heavy ion beams were found to contain enough lines suitable for an in situ wavelength calibration of a grazing incidence spectrometer. Such a calibration could be carried out with an inaccuracy of 0.1 Å in the wavelength range from 85 Å to 130 Å. A further reduction of the calibration uncertainty is certainly of great interest. To achieve progress in this direction a stationary line source for the EUV would be of great value which is on the other hand not easy to realize.

References

1. A. E. Livingston; J. de Physique, 50, Cl-255 (1989)
2. B. Kraus, K.-H. Schartner, F. Folkmann, A. E. Livingston and P. H. Mokler, Proc. SPIE 1159, 217 (1989)
3. R. J. Speer, Space Sci. Instrum. 2, 463 (1976)
4. I. Lesteven-Vaisse, M. Chantepie, J. P. Grandin, D. Hennecart, X. Husson and D. Lecler, Physica Scripta 34, 138 (1986)
5. N. H. Eisum, F. Folkmann and P. H. Mokler, Nucl. Instr. and Meth. B23, 215 (1987)
6. R. L. Kelly, J. Phys. Chem. Ref. Data, 16, Suppl. 1 (1987)

Supplement to Z. Phys. D - Atoms, Molecules and Clusters 21, S 305-307 (1991)

Atoms, Molecules
Zeitschrift für Physik D and Clusters

Strong contributions from Rydberg transitions in EUV radiation from beam-foil excited heavy ions at 1.4 MeV/u

B. Kraus* and K.-H. Schartner[1], F. Folkmann[2], A. E. Livingston[3], P. H. Mokler[4]

[1] I. Physikalisches Institut, Universität Giessen, FRG
[2] Institute of Physics, University of Aarhus, Denmark
[3] Department of Physics, Notre Dame University, Indiana, USA
[4] GSI Darmstadt and Universität Giessen

Abstract

Beam-foil spectra from 1.4 MeV/u Ne^{6+}-, Ni^{17+}- and Xe^{24+}-ions transversing thin C-foils were recorded in the wavelength range from 40 Å to 280 Å. A 5 m grazing incidence spectrometer equipped with a position-sensitive detector was used. For Ne and Ar the excitation of low lying levels was observed while for Ni and especially for Xe strong Rydberg transitions ($n \leq 11$) dominate the spectra.

Introduction

A study of beam-foil spectra emitted from 1.4 MeV/u Ne-, Ni- and Xe-ions after passage of thin carbon foils was carried out. The spectra were recorded at wavelengths from 40 Å to 280 Å. The intention of the study was twofold. With an intense Ne beam the spectrometer performance in the EUV-range of 40 Å to 100 Å was to be tested. Published beam-foil spectra have documented strong lines from Ne X to VIII in this wavelength range /1/. With the heavier and more highly charged projectiles the contribution of Rydberg transitions in the spectra were of main interest. They are important as calibration lines for an in situ wavelength calibration of the spectrometer used. The experiments are part of a program to measure wavelengths of 2s-2p-transitions in He-, Li- and Be-like heavy ions with an accuracy sensitive to higher order relativistic and QED contributions.

Experimental set-up

The experiments were carried out at the UNILAC of the GSI. The GSI 5m grazing incidence spectrometer was used. It is designed for photon spectroscopy in the extreme ultraviolet (EUV). Details of the spectrometer and of the wavelength calibration procedure can be found in an accompanying contribution in this volume /2/.

In the experiments Ne^{6+}-, Ni^{17+}- and Xe^{24+}-ions of 1.4 MeV/U specific energy transversed 25 $\mu g/cm^2$ C-foils. The ion currents amounted to 6 μA, 400 nA and 300 nA, respectively. Accumulation times were in the order of 1000 s in each position of the position sensitive detector, which covers approximately 30 Å.

Recorded spectra

Fig. 1, 2 and 3 reproduce beam-foil spectra for Ne-, Ni- and Xe-ions. Fig. 1 contains 6 single spectra according to 6 subsequent detector positions. Fig. 2 and 3 are composed from 9 spectra. Slit widths of 70 μm, 80 μm and 200 μm were used for Ne, Ni and Xe, respectively. The angles of grazing incidence were 4.3° for Ne and 2.6° for Ni and Xe. The spectra of Fig. 1-3 are not corrected for the wavelength dependence of the spectrometer sensitivity.

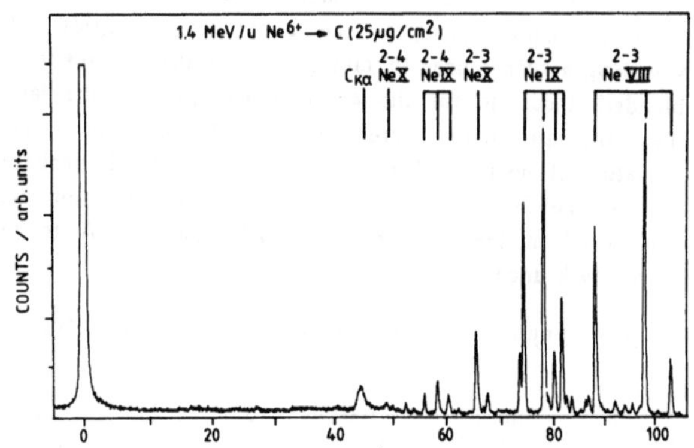

Fig. 1 Beam foil spectrum of 1.4 MeV/u Ne

Discussion

The lines in fig. 1 belong to $n = 2 - n = 3$ transitions in Li-, He- and H-like ions, with the strongest contributions being from Ne VIII. At the shorter wavelengths $n = 2 - n = 4$ transitions in Ne IX and X are seen. The spectrum resembles the spectrum observed by Watson et al. who also give a listing of the transitions /1/.

* Part of thesis

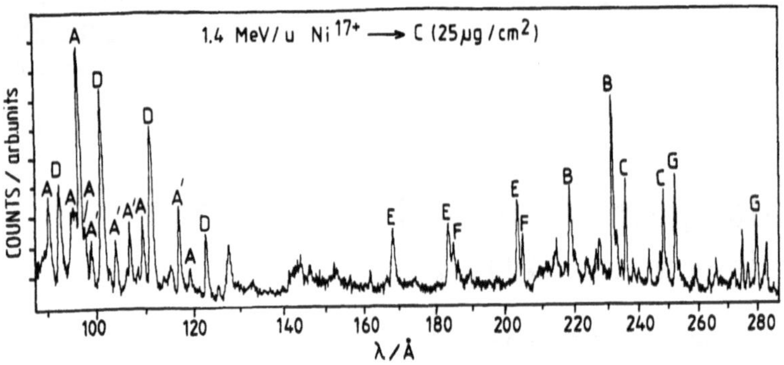

Fig. 2 Beam foil spectrum of 1.4 MeV/u Ni

A,A'	$2s^22p^n$ - $2s2p^{n+1}$		Q = 19 - 21
B	$2p^63p$ - $2p^63d$		Q = 17
D	4f - 5g		Q = 17 -20
E	5 - 6		Q = 18-20
F	6 - 8		Q = 18, 19
C	unclear		
G	6 - 7		Q = 20, 21

For Ni, only one published example is known to us, covering the wavelength range from 210 Å to 290 Å /3/. In the analysis of fig. 2 we concentrated on the short wavelength range using the wavelength calibration procedure described in /2/. This calibration is based on known transitions in Ne recoil ion spectra /4, 5, 6/ and could be carried out with an inaccuracy of 0.1 Å in the wavelength range from 85 Å to 130 Å. Due to mechanical problems the inaccuracy at the larger wavelengths of fig. 2 amounts to 0.25 Å. The Ni beam foil spectrum shows strong 2s-2p-transitions in Ni XXI and Ni XXII (A and A' in fig. 3), resulting from the excitation of a 2s electron into a 2p vacancy. This channel is closed for Ne-like Ni, while F-like Ni XX offers only one doublet which is weak and blended in fig. 3. The D-marked lines in fig. 2 are n = 4 from n = 5 transitions. While 2s-2p-transitions dominate for Ni^{20+}, 4 - 5-transitions dominate for Ni^{19+}. This could indicate a capture process for the population of the n = 5 states though conclusions in this direction would need a sensitivity calibration.

Table 1 shows a comparison of the measured and tabulated wavelengths /6/. The 4f-5g-values have been calculated by ourselves using hydrogen like Dirac term-energies. The wavelength differences in tabele 1 are < 0.1 Å except for blended lines and for the 4f-5g-transitions. For the latter they increase with decreasing ion charge. The largest deviation of nearly 1 Å is present for Na-like Ni, probably due to the 3s electron. A doppler effect correction was considered in the comparison contributing 0.05 Å to the wavelength uncertainty.

One should notice that for Ni^{17+} the 4f-5g-transition was identified and its wavelength has been tabulated /6/. In this case we find good agreement between our measured and the tabulated value. For Ni^{17+} also the 4d-5f-transition has been identified /6/ with a wavelength at 114.74 Å. We observe strong blendings in this wavelength range which nevertheless allow the statement that 5g is much stronger populated that 5f.

Strong Rydberg transitions are also observed at larger wavelengths (marked E, F and G). 5-6-transitions and 6-8-transitions nearly coincide. The calculated wavelengths deviate less than 0.4 Å from the measurement, which is slightly larger than the quoted calibration uncertainty of 0.25 Å. The B-marked lines agree within 0.1 Å with the 3p-3d transitions in Na-like Ni, the C-marked lines were so far not identified unambiguously.

Most of the observable Xe-ions will have a partially empty 3d-shell. Δn = 0 (n = 3,4) transitions occur at lower wavelengths than the investigated wavelengths /7, 8/. To the contrary Rydberg transitions with Δn = 1 (see fig. 4) and with Δn = 2 can be expected and are indeed dominating the displayed spectrum (fig. 3). Transitions from n as high as 11 were observed for a broad charge spectrum. Its width corresponds qualitatively to the width of the ion charge spectrum after foil passage, being 5 at FWHM. The latter peaks at Q = 28. Again, for a quantitative comparison of the charge state distribution with the intensities of Rydberg transitions, having the same Δn but different charge, the

Wavelength λ Kelly	exp.	delta λ	Q	Configuration	Terms	J - J	
88.81	88.80	-0.01	20	$2s^22p^4$ - $2s2p^5$	^3P - ^3P	2 - 1	
91.20	91.25	0.05	21	$2s^22p^3$ - $2s2p^4$	^2P - ^2P	1/2 - 3/2	
91.78*	91.25	-0.53	20	4f - 5g			
94.50	94.32	-0.18	19	$2s^22p^5$ - $2s2p^6$	^2P - ^2S	1/2 - 1/2	(bl)
95.85	95.77	-0.08	20	$2s^22p^4$ - $2s2p^5$	^3P - ^3P	2 - 2	
96.79	96.70	-0.09	20	$2s^22p^4$ - $2s2p^5$	^3P - ^3P	0 - 1	
98.16	98.04	-0.12	21	$2s^22p^3$ - $2s2p^4$	^2D - ^2D	3/2 - 3/2	
101.19*	100.49	-0.70	19	4f - 5g			(bl)
103.31	103.23	-0.08	21	$2s^22p^3$ - $2s2p^4$	^4S - ^4P	3/2 - 1/2	
106.04	105.98	-0.06	21	$2s^22p^3$ - $2s2p^4$	^4S - ^4P	3/2 - 3/2	
109.29	109.44	0.15	20	$2s^22p^4$ - $2s2p^5$	^3P - ^3P	1 - 2	
112.14*	111.46	-0.68	18	4f - 5g			
117.91	117.89	-0.02	21	$2s^22p^3$ - $2s2p^4$	^4S - ^4P	3/2 - 5/2	
120.33	120.26	-0.07	20	$2s^22p^4$ - $2s2p^5$	^1D - ^3P	2 - 2	
124.050	123.99	-0.06	17	4f - 5g	^2F - ^2G	7/2 - 9/2	
124.95*	123.99	-0.96	17	4f - 5g			

Table 1 Comparison of tabulated and experimental wavelengths for the Ni-beam foil spectrum.

*Calculated from hydrogenlike Dirac term energies, bl indicates blending.

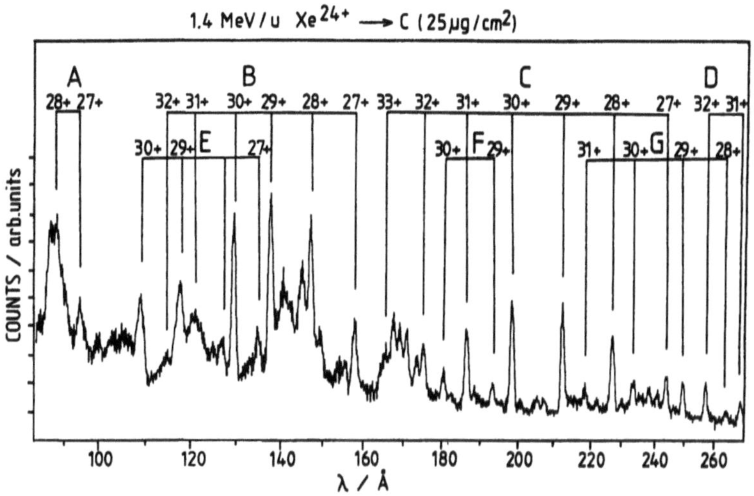

Fig. 3 Beam-foil spectrum of 1.4 MeV/u Xe A 5 - 6, B 6 - 7, C 7 - 8, D 8 - 9 E 7 - 9, F 8 - 10, G 9 - 11 Charges as indicated

relative detection sensitivity has to be known. We conclude from fig. 3 that also Xe ions carrying only one or two electrons should present strong Rydberg transitions, which can be used for calibration purposes. Respective observations have already been described /9/.

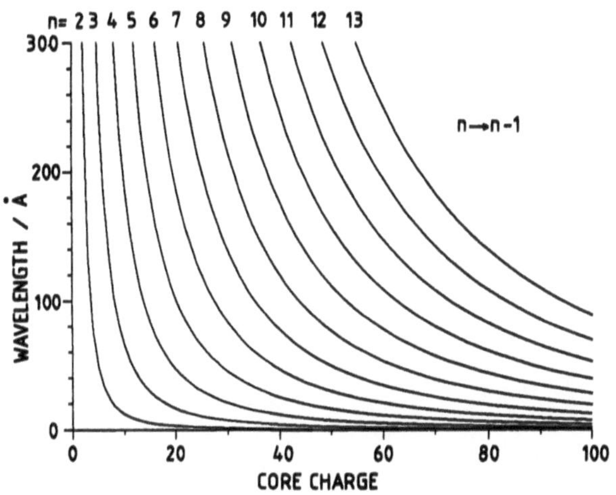

Fig. 4 Wavelengths for $\Delta n = 1$ transitions as function of the ion core charge, calculated from Dirac term-energies:

$$E_{nj=n-1/2} = \frac{m_0 c}{\sqrt{1 + \frac{(Z\alpha)^2}{n^2 - (Z\alpha)^2}}}$$

Summary

The described first application of the GSI 5m grazing incidence spectrometer in beam-foil spectroscopy has shown its utility for high resolution spectroscopy in the range of the EUV. While published Ne and Ar spectra could be reproduced new spectroscopic information for Ni and Xe could be obtained. Strong lines from Rydberg transitions were observed especially for the highly charged Ni- and Xe-ions.

References

1. J.A. Demarest and R.L. Watson, Physica Scripta 38, 670 (1988)
2. B. Kraus, K.-H. Schartner, F. Folkmann, A.F. Livingston, P.H. Mokler this volume
3. R. Hutton, N. Reistad and I. Martinson, Physica Scripta 35, 300 (1987)
4. N.H. Eisum, F. Folkmann, P.H. Mokler, Nuclear Inst. and Meth. B 23,(1987)
5. I. Lesteven-Vaisse, M. Chantepie, J.P. Grandin, D. Hennecart, X. Husson and D. Lecler, Physica Sripta 34, 138 (1986)
6. R.L. Kelly, J.Phys. Chem. Ref. Data, 16 Suppl. 1(1987)
7. J.F. Wyart, C. Bauche-Arnoult, E. Luc-Koenig, Physica Scripta 32, 103 (1985)
8. C. Breton, C. DeMichelis, W. Hecq, M. Mattioli, J. Ramette, B. Saoutic, Physica Scripta 37 , 33 (1988)
9. S. Martin, J.P. Buchet, M.C. Buchet-Poulizac, A. Dennis, J. Desesqelles, M. Druetta, J.P. Grandin, D. Hennecart, X. Husson, and D. Lecler, Europhysics Lett., 10, 645 (1989)

Supplement to Z. Phys. D - Atoms, Molecules and Clusters 21, S 309-311 (1991)

Atoms, Molecules
Zeitschrift für Physik D and Clusters

Absolute state selective cross sections for ionization-excitation and excitation of helium in $e^- + He$, $H^+ + He$, $H_2^+ + He$ and $H_3^+ + He$ collisions

S. Fuelling[1], R. Bruch[1], G. Liu[1*] M. Bailey[1], E. A. Rauscher[1], E. Träbert[2***] and P. H. Heckmann[3]**

[1] Department of Physics, University of Nevada, Reno, Reno, NV 89557, USA
[2] Harvard Smithonian Center for Astrophysics, Cambridge, MA 02139, USA
[3] Institut für Experimentalphysik III, Ruhr-Universität Bochum, FRG

Abstract

We have measured absolute EUV emission cross sections for ionization-excitation of helium by H^+, H_2^+ and H_3^+ (v: 1.4 to 8 a.u.) and e^- (v: 3.5 to 8 a.u.) and the absolute emission yields for excitation of He. An interesting comparison is made based on three different mass and structural states for H^+, H_2^+ and H_3^+. Also examined is the comparison of H^+ and e^- projectiles which involves a large mass difference and opposite charge state.

I. Introduction

In our experiments using the H^+, H_2^+, H_3^+ and e^- projectiles we have observed the EUV target emission for HeII ($1s$-np) and HeI $(1s^2)\,^1S$ - $(1snp)\,^1P^o$ transitions. The aims of the present research are (i) measurements of absolute cross sections for ionization plus excitation of helium versus velocity of the projectile, (ii) experiments to obtain absolute EUV emission yields as a function of the projectile velocity and target pressure and (iii) calculations of the cross section ratio, $\sigma_{e^-}/\sigma_{H^+}$ associated with the ionization-excitation process dependence on the projectile velocity. These results allow us to examine contributions to the scattering process from mass contributions, charge states, and projectile-structural effects as well as possible contributions of coherent superposition of scattering amplitudes.

II. Experiment

The experiments were conducted at the University of Nevada, Reno, using the 2 MV Van de Graaff accelerator of the Physics Department. We utilized intense (1-20 μA), tightly collimated beams of H^+, H_2^+ and H_3^+ and e^- which were passed through a differentially pumped gas cell. By measuring the beam current as a function of the target pressure for different energies we were able to deduce the absolute incoming particle flux. EUV photons were detected under an observation angle 90° with a 1.5m grazing incidence monochromator. This monochromator has been absolutely calibrated by electron impact studies of He and Ar [1,2].

III. Experimental Results and Discussion

A comparison of our experimantal cross sections for ionization-excitation of He by proton and electron impact is presented in Fig. 1. We can deduce from this figure that the trend of the cross sections versus velocity does not strongly depend on the specific $He^+(np)$ states and the electron cross sections lie consistently above the corresponding proton values for $v/v_0 \geq 3$. In Fig. 2 we display the absolute EUV emission cross section for single charged state projectiles differing by both one and by two atomic mass units. This data gives us an excellent opportunity to study the absolute cross section dependence on the projectile mass and structure for various velocities. It appears that the cross section is predominantly dependent on the projectile mass difference rather than the detailed molecular structure.

In addition we present our absolute emission yields for excitation of He as shown in Fig. 3 and 4. Threshold effects are evident for velocities smaller than 3 a.u. In the region for velocities between about 2.5 and 4 a.u. the H^+ and e^- data coincide. In Fig. 4 the emission yields for all three projectiles, H^+, H_2^+ and H_3^+ peak at about the same velocities and in all cases of the various target states the H_3^+ yield is larger than the H_2^+ or H^+ results. Similar effects have been reported by R. Hippler *et al.* [3] for protons and by F. G. Donaldson *et al.* [4] for electron impact on helium. Since we measured at relative high target gas pressures ($p = 50$ mTorr), we only give absolute yields for the excitation of helium, where resonant trapping of the emitted light and secondary collisions may appear. This does not occur, however, for ionization-excitation of He and we verified that the emission yield for Lyman-α of He^+ increases linearly with the target gas pressure.

IV. Conclusion

In our ongoing research, our data gives us the opportunity to examine mass, charge, and structural effects. From Fig. 2 preliminary analysis indicates that the cross sections depend on the mass and most probably (to a

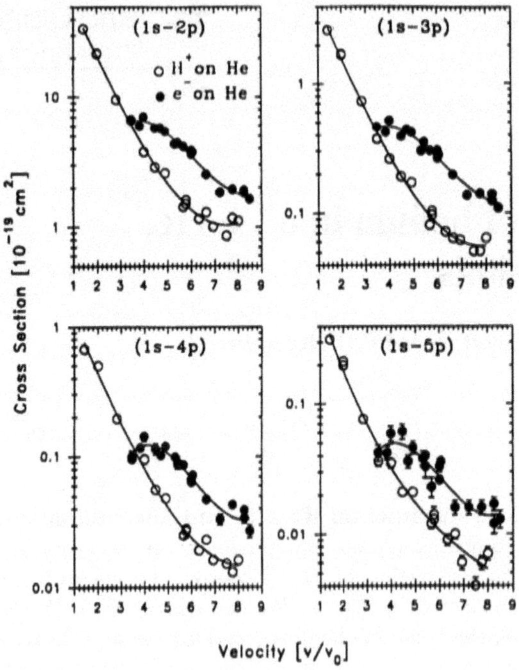

Fig. 1. Absolute EUV emission cross sections for ionization plus excitation of helium by e^- and H^+ impact. The cross sections have been derived from the Lyman series of He II ($1s$-np) for $n \leq 5$.

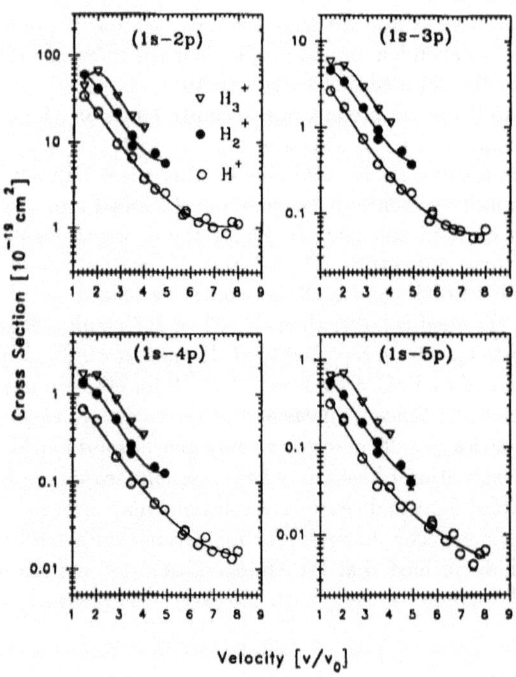

Fig. 2. Absolute EUV emission cross sections for ionization plus excitation of helium by H^+, H^{2+} and H^{3+} impact. The cross sections have been derived from the Lyman series of He II ($1s$-np) for $n \leq 5$.

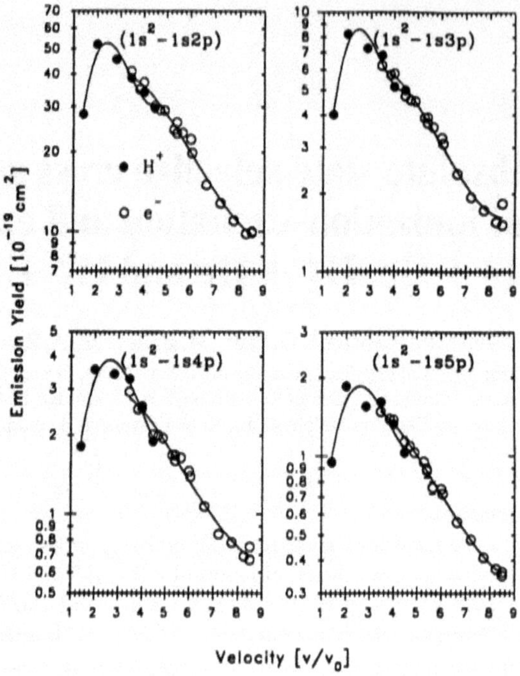

Fig. 3. Absolute EUV emission yields for excitation of helium by e^- and H^+ impact. The yields have been derived from the He I $(1s^2)\,^1S$ - $(1snp)\,^1P^o$ Rydberg series for $n = 2$ to 5.

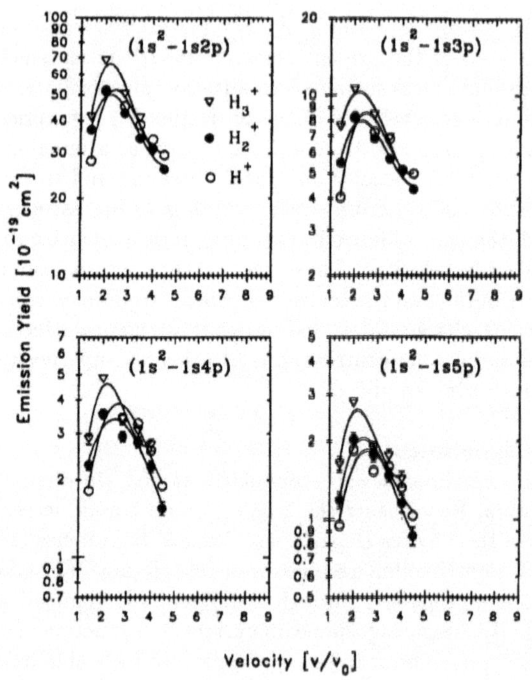

Fig. 4. Absolute EUV emission yields for excitation of helium by H^+, H_2^+ and H_3^+ impact. The yields have been derived from the He I $(1s^2)\,^1S$ - $(1snp)\,^1P^o$ Rydberg series for $n = 2$ to 5.

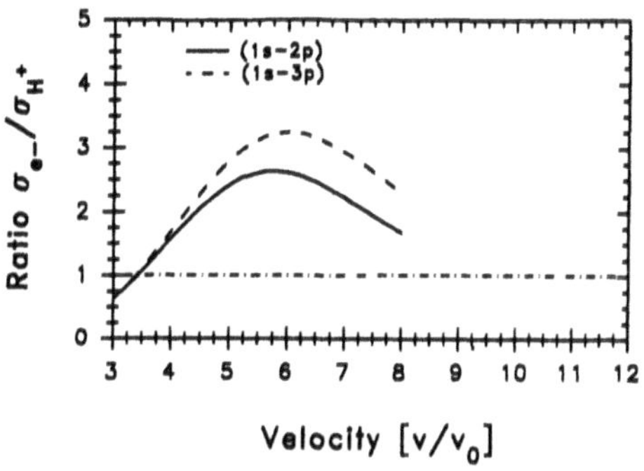

Fig. 5. Cross section ratio $\sigma_{e^-}/\sigma_{H^+}$ versus projectile velocity (a.u.) determined from Fig. 1.

much lesser degree) on the structure. The data displayed in Fig. 1 allows us to analyze a very interesting case where we have an extremely large mass ratio of 1836 and opposite charge states. In Fig. 5 we have plotted the electron to proton cross section ratio for n=2 and n=3 as a function of velocity. For n=2 our results are very similar to those of Pedersen and Folkmann [5]. The ratio for n=2 and n=3 both peak at about $v \sim 6.4$ a.u. From our extensive data base,we will proceed with the theoretical analysis utilizing coupled-channel models as well as R-matrix theory.

References

* Washoe Medical Center, Radiation Therapy, Reno, NV 89503 USA

** Theoretical Physics Division, Tecnic Research Laboratories, 7685 Hughes Dr, Golden Valley, NV 89506, USA

*** Permanent address: Institut für Experimentalphysik III, Ruhr-Universität Bochum, FRG

[1] J. L. Forand, K. Becker and J. W. McConkey, J. Phys. **B18**, 1409 (1985)

[2] K-H. Tan and J. W. McConkey, Phys. Rev. **A10**, 1212 (1974)

[3] R. Hippler and K-H. Schartner, J. Phys. **B7**, 618 (1974)

[4] F. G. Donaldson , M. A. Hender and J. W. McConkey, J. Phys. **B5**, 1192 (1972)

[5] J. O. Pedersen and F. Folkmann, J. Phys. **B23**, 441 (1990)

Supplement to Z. Phys. D - Atoms, Molecules and Clusters 21, S 313-314 (1991)

Atoms, Molecules
Zeitschrift für Physik D and Clusters

Absolute state selective cross sections for ionization-excitation of helium in fast H^+ + He and C^{6+} + He collisions: Z_p-dependence

S. Fuelling[1], R. Bruch[1], E. A. Rauscher[1*], G. Liu[1], E. Träbert[2***], and P. H. Heckmann[3]**

[1] Department of Physics, University of Nevada, Reno, Reno, NV 89557, USA
[2] Harvard Smithonian Center for Astrophysics, Cambridge, MA 02139, USA
[3] Institut für Experimentalphysik III, Ruhr-Universität Bochum, FRG

Abstract
Absolute EUV emission cross sections for simultaneous ionization and excitation of helium by H^+ and C^{6+} projectiles have been measured for impact energies between 0.83 and 2.3 MeV/u. In this energy range the mass, charge and velocity dependence of the ionization-excitation cross sections is compared for the $He^+(np)$ Rydberg states for $n \leq 5$. It is found that the experimental cross sections vary approximately as Z_p^3.

I. Introduction

Currently there exists a great interest in many-body collision theories based on time dependent perturbation methods [1] which overcome the inadequacies of previous ion-atom collision theories, earlier work in this area has been done by Rauscher *et al.* [2]. Consequently, we have studied the dynamic aspect of electron correlation due to the Coulomb interaction of a bare ion such as H^+ and C^{6+} with a two electron Fermi system like He. In particular, absolute state selective cross section measurements of ionization plus excitation of He by fast particle impact are very sensitive to small variations in the electron-electron, projectile-electron and projectile-nucleus potentials and hence are an excellent probe to test many body pertubation theories as a function of the projectile charge.

II. Experiment

Experimentally, we have examined the following reaction:

$$X^{q+} + \mathrm{He}(1s^2) \longrightarrow X^{q+} + \mathrm{He}^+(np) + e^- \\ \longrightarrow \mathrm{He}^+(1s) + h\nu,$$

where X^{q+} is either H^+ or C^{6+}. We have selected the impact energies (E: 0.83, 1.1, 1.6 and 2.3 MeV/u or $v \simeq 5.8$, 6.6, 8.0 and 9.6 a.u.) such that the velocities are large enough for perturbative many-body theories [1] to be applicable. In particular, He has been chosen as a target gas because it represents the simplest many-electron system and the deexcitation of the $He^+(np)$ Rydberg states via the ($1s$-np) Lyman series allows state selective measurements of the collisional process. Experiments were performed over a period of seven years on charge transfer [3] and multiple ionization plus excitation [4] using the Dynamitron Tandem Laboratory (DTL) facility at the University of Bochum, which provides intense, fast, multiply charged ion beams. In this work, H^+ and C^{6+} projectiles were passed through a differentially pumped gas cell and were collected in a Faraday cup for normalization purposes. The emitted EUV photons of the hydrogenic helium ions were analyzed by means of a 2.2m grazing incidence monochromator. The deduced ionization-excitation cross section of $He^+(np)$ were placed on an absolute scale by additional studies of electron and proton impact on He and Ar carried out with the 2 MV Van de Graaff accelerator at the Physics Department at the University of Nevada, Reno.

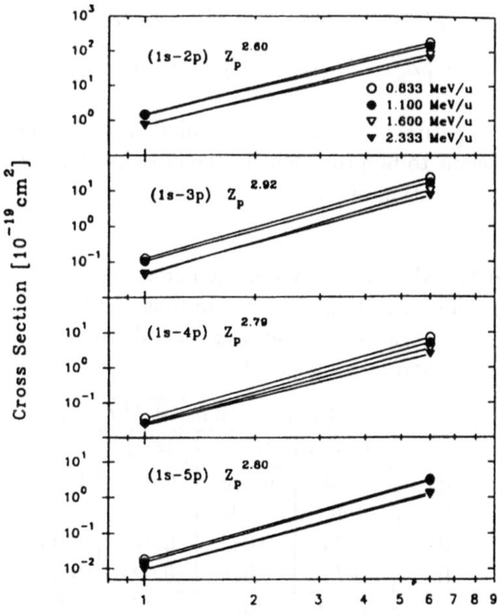

Fig. 1. Absolute EUV emission cross sections for ionization plus excitation of He by H^+ and C^6+ impact for varies projectile energies. The cross sections have been extracted from the Lyman series of He II ($1s$-np) for n=2 to 5.

III. Experimental Results and Discussion

Our absolute EUV emission cross sections for ionization plus excitation of He by H^+ and C^{6+} impact are plotted in Fig. 1 as a function of the projectile charge Z_p. The cross sections are listed in Table 1. It is interesting to study the cross section dependence on projectile charge Z_p. For H^+ and C^{6+} the charge difference is $\Delta Z_p = 5$. The variation of the measured cross sections on Z_p can be examined by taking the ratios

$$m = \ln\{\sigma_{C^{6+}}/\sigma_{H^+}\}/\ln\{Z_C/Z_H\},$$

which manifests the charge-state dependence $\sigma \sim Z_p^m$ between the endpoints for $Z_p = 1$ and $Z_p = 6$. It is evident from Fig. 1 and Table 2 that the cross sections vary between $Z_p{}^{2.6}$ and $Z_p{}^{3.06}$ for the observed velocity range. This looks almost like a $Z_p{}^3$ dependence. However, we need to explore intermediate cross section values for $Z_p = 2$ (He^{2+}), $Z_p = 3$ (Li^{3+}) and $Z_p = 5$ (B^{5+}) to confirm the $\sigma \sim Z_p{}^3$ variation. It is also interesting to note that the experimental $Z_p{}^m$ dependence is nearly the same for all $He^+(np)$ Rydberg levels.

According to McGuire [1], the scattering amplitude a at high velocities v for two electron transitions can be written as

$$a = -ic_1 Z_p - (\bar{c}_2 - ic_2) Z_p{}^2,$$

where c_1, c_2, and $\bar{c}_2$ are real coefficients which are independent of Z_p. For the cross section σ we find

$$\begin{aligned} \sigma &= \int |a|^2 d\mathbf{B} \\ &\simeq \int (c_1^2 Z_p^2 - 2c_1c_2 Z_p^3 + c_2^2 Z_p^4 + \bar{c}_2^2 Z_p^4) d\mathbf{B} \\ &\equiv C_1^2 Z_p^2 - 2C_{12} Z_p^3 + C_2^2 Z_p^4 + \bar{C}_2^2 Z_p^4 + O(Z_p^5). \end{aligned}$$

From experimental results ([6] and this work) the Z_p^3-term seems to be prominent in the cross section for ionization-excitation of He.

Table 1. Absolute EUV emission cross sections for ionisation plus excitation of He by protons and C^{6+} impact. The cross sections have been extracted from the He II ($1s$-np) Lyman series for n=2 to 5.

v/v_0 [a.u.]	Proj.	Cross Sections [10^{-19} cm^2] ($1s$-$2p$)	($1s$-$3p$)	($1s$-$4p$)	($1s$-$5p$)
5.8	H^+	1.50	0.123	0.037	0.0189
6.6	H^+	1.40	0.104	0.027	0.0155
8.0	H^+	0.73	0.043	0.024	0.0097
9.6	H^+	0.73	0.048	0.023	0.0092
5.8	C^{6+}	182.4	23.9	6.95	3.20
6.6	C^{6+}	146.6	17.5	5.03	2.89
8.0	C^{6+}	84.1	10.4	3.42	1.30
9.6	C^{6+}	64.3	7.3	2.33	1.16

Table 2. Z^m dependence of the EUV emission cross sections for ionization plus excitation of helium. The value m is listed for a variety of projectile velocities and for the different members of the He II ($1s$-np) Lyman series (n=2 to 5).

v/v_0 [a.u.]	$m = \ln\{\sigma_{C^{6+}}/\sigma_{H^+}\}/\ln\{Z_C/Z_H\}$ ($1s$-$2p$)	($1s$-$3p$)	($1s$-$4p$)	($1s$-$5p$)
5.8	2.68	2.94	2.92	2.86
6.6	2.60	2.86	2.91	2.92
8.0	2.65	3.06	2.76	2.73
9.6	2.50	2.81	2.57	2.70
Average	2.60	2.92	2.79	2.80

IV. Conclusion

In this work we have considerably extended experimental cross sections on ionization plus excitation of helium to Rydberg states $He^+(np)$ up to n=5 in H^+ on He and C^+ on He single collisions when compared to previous results from Pederson and Folkmann [6]. We also plan to perform coupled-channel calculations [7] and to examine potential well models with the R-matrix approach [2] to produce a more satisfactory agreement between experiment and theory.

References

* Theoretical Physics Division, Tecnic Research Laboratories, 7685 Hughes Dr, Golden Valley, NV 89506, USA
** Washoe Medical Center, Radiation Therapy, Reno, NV 89503 USA
*** Permanent address: Institut für Experimentalphysik III, Ruhr-Universität Bochum, FRG

[1] J. H. McGuire, Multiple Electron Excitation Ionization, Transfer in High Velocity Atomic and Molecular Collisions, Advances in Atomic, Molecular and Optical Physics, to be published
[2] K. Harada and E. A. Rauscher, Phys. Rev. **169**, 818 (1968)
[3] L. J. Dubé, U. Will, R. Bruch, E. Träbert and P. H. Heckmann, Nucl. Instr. Meth. **B9**, 408 (1985)
[4] E. Träbert, P. H. Heckmann, R. Bruch and S. Fuelling, Nucl. Instr. Meth. **B23**, 153 (1987)
[5] R. Bruch, L. Kocbach, E. Träbert, P. H. Heckmann, B. Raith and U. Will, Nucl. Instr. Meth. **B9**, 438 (1985)
[6] J. O. Pedersen and F. Folkmann, J. Phys. **B23**, 441 (1990)
[7] E. A. Rauscher, J. O. Rasmussen and K. Harada, Nuclear Phys. **A94**, 33 (1967)

Supplement to Z. Phys. D - Atoms, Molecules and Clusters 21, S 315-316 (1991)

Atoms, Molecules
Zeitschrift für Physik D and Clusters

A computer-aided crystal diffraction spectrometer for precision measurements of characteristic X-rays from highly-charged heavy ions

G. Zschornack, G. Musiol, V. Pohl, A. Reichmann and M. Schiekel
TU Dresden, Sektion Physik, WB Angewandte Kernphysik, Mommsenstraße 13, O-8027 Dresden, FRG

Abstract

A computer-aided crystal diffraction spectrometer for precise X-ray spectroscopy is described. The operation principle of the spectrometer allows to use plane crystals as well as bended ones. The obtained reproducibility of diffraction reflex positions under working conditions was 2 arcsec and 4 arcsec after mechanical modifications of the spectrometer kinematics.

I. Introduction

For investigations of basic properties of highly charged heavy ions and for ion source or plasma X-ray diagnostics it is often favourable to use wave - length dispersive methods, because the wavelengths of spectral lines from different atomic configurations and ionization stages are often very closed to each other /1/. To resolve such spectra from different highly charged ions we have developed a crystal diffraction spectrometer working in Bragg geometry as plane crystal or as Johann- or Johansson-spectrometer with crystals of different bending radius. In contrast to other spectrometers /2-5/ the goal is here, that in dependence of the experimental conditions one can in a comfortable manner select the operation principle, crystal properties and influence acting geometrical aber - ations /6/.

II. Operation principle

The operation principle is shown in Fig.1. A stepping motor drives a spindle for positioning the beam admission diaphragm. The position of the diaphragm is hereby controlled by an incremental transmitted light length measurement device. Analogous, the detector and the detector diaphragm are moved by corresponding spindles and stepping motors

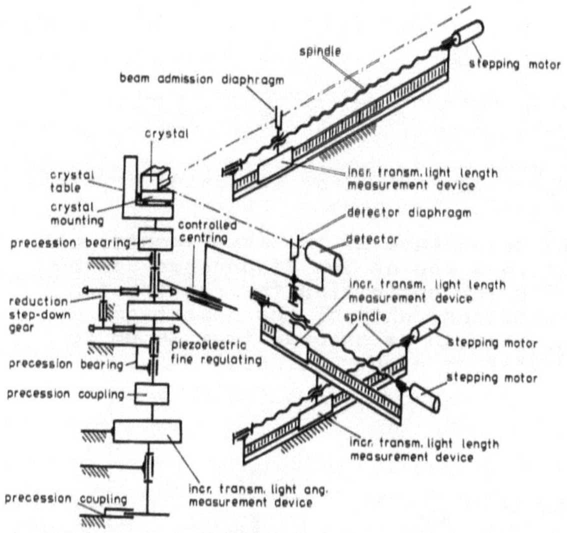

Fig.1
Operation principle of the crystall diffraction spectrometer

and the actual position is also controlled by incremental transmitted light length measurement devices. A change in the crystal angular position leads simultaneous to a reorientation of the detector through the controlled centring and the reduction step-down gear. The precise screw-down movement of the crystal is realized through a coupling and -if required- through a piezoelectric fine regulating device. An exact guidance of the rotation axis of the crystal, the crystal table and of the crystal mounting is obtained by a precession coupling. Forces from unwanted deviations of crystal rotation axis relative to the driving axis are compensated by couplings. All movable parts of the spectrometer are carried and mounted on a plate with special processed surface. An on-line control of the spectrometer operation allows to realize a detector motion to any position within the boundaries given by the mechanical construction.

III. Spectrometer parameters

The operation range for bended crystals is shown in Fig.2.

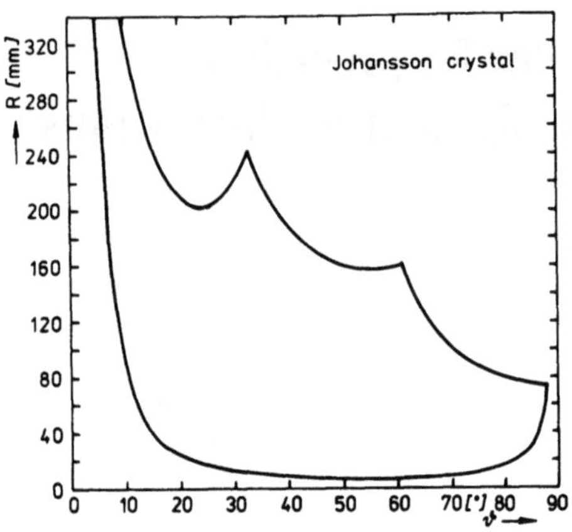

Fig.2
Operation range for focussing crystals.

In dependence of the experimental conditions one or two diaphragms can be used for optimizing the ray tracing conditions whereby the diaphragms widths can be choosen from 4o μm to different higher values.

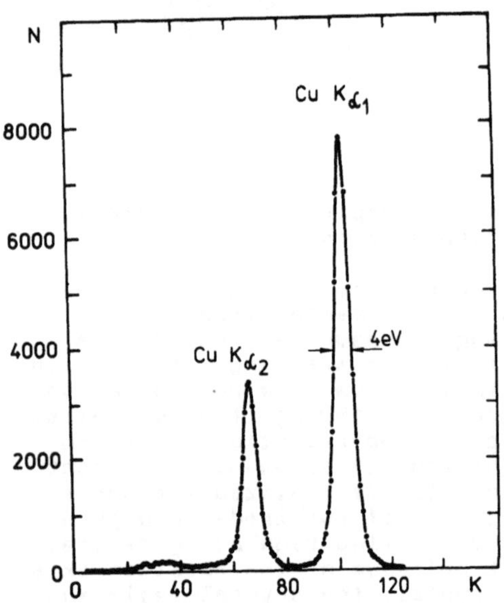

Fig.3
Diffraction reflexes of the $CuK_{\alpha1,2}$-dublett.

As basic operating principles it is possible to work in a sequential measurement regime with measuring times of up to 10^4 sec/step and in a continuous angular scanning mode with an angular velocity of 1/16,1/8,1/4,1/2,1,2 and 6 degrees per minute. For some applications it is essential to stabilize the temperature on the analyzer crystal. For this behaviour we have applied a discontinuous optimized controller to guarantee at the crystal a long-time temperature constancy of ± 0.04 K /7/.

The obtained reproducibility of the diffraction reflex position under working conditions was 2 arcsec and 4 arcsec after mechanical modifications of the spectrometer kinematics. Fig.3 shows the $K_{\alpha1,2}$-dublett of copper. The measurement was here provided with a plane SiO_2 (13$\bar{4}$0)-crystal and with a position sensitive X-ray detector (CCD-matrix). The luminosity of the spectrometer depends on the actual experimental conditions and varies in the range between 10^{-5} up to 10^{-8}.

IV. Conclusion

We have developed a computer-aided crystal diffraction spectrometer for a wide range of applications and with different operation regimes. As a next step it is foreseen to combine the spectrometer with different position sensitive X-ray detectors to increase the luminosity in experiments on heavy ion sources and with beams of highly charged heavy ions.

References

/1/ Zschornack G.,Musiol G., Wagner W.: Report ZfK-574, Rossendorf (1986)
/2/ Leonhardt G., Kopczynski A., Ehrhardt H., Meisel A.: Exp.Technik d.Physik XX, 221 (1972)
/3/ Morita S., Kadoto K., Fujita J.: Jap.Journ.of Appl.Phys. 22,1283(1983)
/4/ Jitschin W., Wisotzki B., Werner U., Lutz H.D.: J.Phys.E: Sci.Instrum. 17, 137 (1984)
/5/ Beyer H.F.: GSI 79-6, Darmstadt (1979)
/6/ Zschornack G., Müller G., Musiol G.: Nucl.Instr.Methods 200, 481 (1982)
/7/ Kreiseler D., Zschornack G.: Exp.Technik d.Physik 35, 111 (1987)

Supplement to Z. Phys. D - Atoms, Molecules and Clusters 21, S317-318 (1991)

Atoms, Molecules
Zeitschrift für Physik D and Clusters

Status of the EBIT project at the Institute of Spectroscopy

P. S. Antsiferov, V. G. Movshev
Institute of Spectroscopy, Troitzk, Moscow region, 142092, USSR

Abstract.
This communication presents the project and status of a simple EBIT-type device, intended for the investigation of n=2⇒n=3 transitions in ions with open L-shell around nuclear charge $Z_n \simeq 40$.

1. Introduction

The experimental study of the energy structure of atoms and their ions is far from complete now [1]. The main difficulties in the investigation of the different isoelectronic sequences towards increasing nuclear charge are the limited ionization capabilities of traditional spectral sources (vacuum sparks, laser produced plasma) and the large ($\Delta\lambda/\lambda \simeq 10^{-3}$) line widths. The use of an EBIT-type source can provide considerable progress. This is the main point of the EBIT project of the Institute of Spectroscopy. Apart from the usual goals of few-electron and extremely highly charged ions we are interested in a systematic study of comparably moderate charge states of ions with complex energy structure. A short review of the last works, dealing with n=2⇒n=3 transitions in ions with an open L-shell (ground configurations $2s^l2p^k$ l=1÷2, k=1÷5) is given in the table.1.

Table 1. The status of the study n=2⇒n=3 transitions in ions with the ground state configuration $2s^l2p^k$ (l=1÷2, k=1÷5)

Iso-electronic sequence	[Li]	[Be]	[B]	[C]	[N]	[O]	[F]
Max nuclear charge	32	34	24	29	29	29 54	38 54
Reference	[2]	[3]	[4]	[5]	[5]	[5] [6]	[6] [7]

It can be seen, that the region of $Z_n \simeq 40$ is quite interesting to investigate the Li-like ÷ F-like isoelectronic sequences. The needed parameters of the EBIT for this aim are: $j\tau \simeq 10 \div 100$ Ω/cm², $U \simeq 10$ keV. The spectral region of the transitions we are interested in is λ<10 Å. Here high-quality quartz-crystal X-ray spectrometers can be used. Since the EBIT is an extremely weak source of X-rays, we have analyzed the questions of efficiency and resolving power of curved-crystal spectrometers intended for the work with EBIT [8]. In principle the Johann type spectrograph with spectral resolution $\Delta\lambda/\lambda \simeq 10^{-4}$ allows one to store a spectral interval of $(\lambda_{max}-\lambda_{min})/\lambda \simeq 10^{-2}$ during a time interval of 10^2 to 10^3sec.

2. The EBIT design

Fig.1 demonstrates the adopted scheme of the ion trap unit.

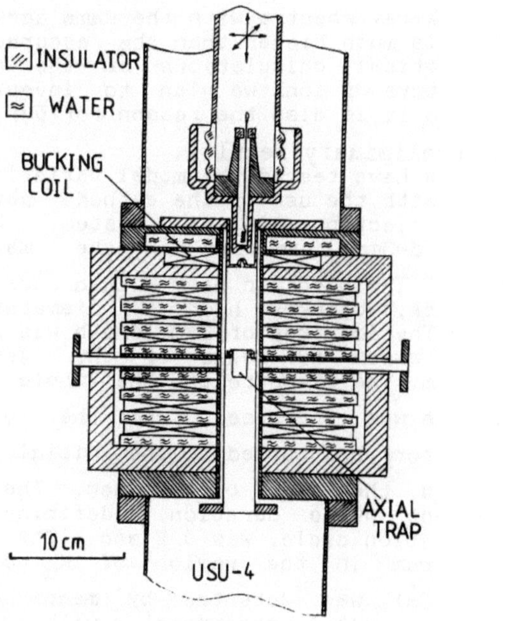

Figure 1: The ion trap unit on the entrance flange of USU-4.

The base of our construction is the commercially available high-vacuum device USU-4 (5×10^{-10}torr) and the conventional water-cooled solenoid, which allows one to obtain an axial field $B\simeq0.5T$ in the pulsed mode. The Pierce-type electron gun, adjusted by means of a bucking coil to the main solenoid, must deliver the electron beam with the current of $I\simeq0.1$ A and diameter of $d\simeq0.5$ mm. The electron-gun volume is differentially pumped by means of the sputter-ion pump NORD-100.
The injection of the element under investigation into the trap volume is done in two ways -a gas-puff system for work with **Kr** (Z_n=36),and a vacuum-arc system for work with **Y** to **Mo** (Z_n=39 to 42). We plan to use characteristic X-rays of single - ionized atoms for wavelength calibration of the spectra of multi charged ions. The same systems of substance injection will be used in this regime, but the quantity of substance, injected in the trap region, may be considerably greater, than in the multi charged ions regime.Also, the axial potential profile must be inverted, i.e. the ions must be deleted from the trap volume in order to obtain characteristic X-ray lines free from satellite lines. Certainly, here arises the problem of shifts of the characteristic lines in the vapor in comparison with solid-state characteristic lines, because practically all the available measured X-ray wavelengths are given for solid-state characteristic radiation. To date the accuracy of the measured experimental wavelengths of the characteristic X-ray lines is of the order of $\Delta\lambda/\lambda\simeq10^{-6}$. The use of such X-ray lines as a reference allows one to obtain in principle an absolute wavelength calibration of the EBIT X-ray spectra with the same accuracy. This is much higher than the accuracy of theoretical calculations of the energy structure of ions we plan to investigate and so it is also the reason for our work.

3. Preliminary results.

Now we have tested the model variation of EBIT with the use of the cathode part of the electron-welding system EOSS-2 (LaB_6,d=3mm) immersed in the magnetic field. The electron beam with a current of I=0.15A, U=10keV had a diameter of $\simeq$1mm.The ion trap of 2cm length was filled with argon by means of the gas-puff system. The pressure in the whole USU-4 volume was increased from the 10^{-8} to $\simeq10^{-6}$torr and turned to the initial value during the time of 10^{-2}sec. The beam current pulse duration, defining the ionization cycle, was 0.2 sec. The X-ray spectrum in the region of K_α of **Ar** (λ=4.2Å) was detected by means of a curved-crystal monochromator with scanning system [8], quartz,r_d=6.68 Å, R=520 mm. The results are shown on the fig.2.

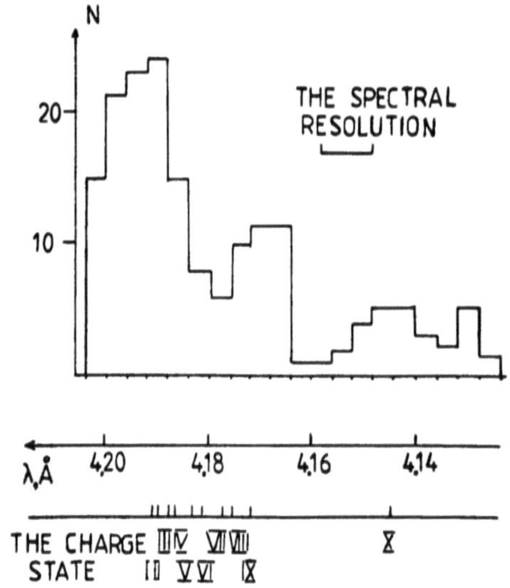

Figure 2: The K_α-X-ray spectrum of **Ar**, emitted by EBIT. The positions of K_α-transitions for the different charge states are taken from [9].

Each experimental point was stored during 32sec. The apparatus resolution was fully determined by the beam diameter. The shifted position of the K_α- transitions in the ions of different stages are taken in accordance with [9]. It seems, that we can speak about reaching **Ne**- like charge stage of ionization of **Ar** in this EBIT system, and,even probably **F**- like.

References

1. Kononov E.Ja., Ryabtsev A.N., Izv.Acad. of Sci, ser. phys., **48**, 689 (1984) (in russian)
2. Behring W.E., Seely J.F., Brown G.M., Feldman U., Knawer J.P., JOSA B6, 5319 (1989)
3. Boiko V.A., Pikuz S.A., Safronova U.I., Faenov A.Yu., J. Phys. B10, 1253 (1977)
4. Burkhalter P.G., Charatis G., Rockett P.D., Newman D.A., JOSA B1, 155 (1984)
5. Swartz M., Kasuter S., Rothe E., Newpert W., J. Phys B4, 1747 (1971)
6. Conturie Y., Yaakobi B., Feldman U., Doschek C.A., Cowan R.D., JOSA 71, 1309 (1981)
7. Hutcheon R.J., Bromage C.E., Cooke R.L., Key M.H., Physica Scripta 21, 89 (1980)
8. Antsiferov P.S., Movshev V.G., Preprint No 4, ISAN, Troitzk, 1990
9. House L.L., Astrophys. J. Suppl., **155**, 21 (1968)

Supplement to Z. Phys. D - Atoms, Molecules and Clusters 21, S 319-321 (1991)

Atoms, Molecules
Zeitschrift für Physik D and Clusters

Elastic ion collisions in multiply-charged ion sources

G. D. Shirkov
Joint Institute for Nuclear Research, Head Post Office, P. O. Box 79, 101000 Moscow, USSR

Elastic Coulomb electron-ion and ion-ion collisions in electron beams and plasma are studied. The use of low charged light ions for cooling multiply charge ions in Electron Beam Ion Source (EBIS) [1], Electron Ring Ion Source (ERIS) [2] and Eletron Cyclotron Resonance Ion Source (ECRIS) [3,4] is substantiated.

1. Introduction. EBIS, EBIT[5] and ECRIS are successfully used now for ion production. Recently, a new type of ion source (ERIS) was proposed. A lot of articles and reviews were devoted to the problem of ionization and charge exchange processes last yeas [6]. The elastic Coulomb collisions of charge particles in plasma have been studied well in connection with the thermonuclear fusion researches. It is described in the textbooks now [7,8]. But these processes are not yet taken into account in the calculations of ion appearance and accumulation in ion sources. The elastic Coulomb collisions can explain some observed effects and show the way to increase the ion sources effectiveness.

2. General equations. Elastic scattering of an electron beam on stored ions, residual gas or solid target is a cause of transverse beam dimension and emittance increase [3,9]. The ions receive an additional momentum in the Coulomb collisions. The energy increase of ions with charge state **Z** and mass **AM** is determined from the equation [3]:

$$dE/dt = r_e^2 Z^2 N_e m c^2 \Lambda / a_e^3 \beta A M \quad , \qquad (1)$$

where $\beta = v_e/c$; m, r_e, v_e are the mass, the classical radius and the velocity of electrons; c is the velocity of light; a-RMS small halfdimention of electron beam; Λ=10-20 is so called Coulomb logarithm.

In the electron beam ions are produced as the result of electron impact ionization of the neutral atoms or molecules in the beam. The energy of a single charged ion equals to the electric field potential in the place of ion appearance and as a rule is much more than the energy of a neutral. The ions remained in the potential well of the beam and begin to oscillate. They undergo further ionization and increase the charge and the energy. The average total energy of the ion transverse oscillations with the charge $Z \gg 1$ is defined as[3,10]

$$E = \sqrt{Z/\pi}\, N_e r_e m c^2 \qquad (2)$$

The energy of the ion is approximately equal to the value which does not depend on the charge in the electron beam with the Gaussian density in cross section[3]:

$$E = 2 N_e r_e m c^2 \qquad (3)$$

The formulas (2) and (3) were obtained without taking into account the ion neutralization of the electron beam charge and without probable electron heating or electron-ion instabilities. When the ion energy increases, the oscillation amplitudes rise and can become larger than the electron beam size. It was shown, that in this case the ions were outside the beam most of the time, they don't depend much on the electron component and we consider these ions to be the lost particles [11]. Let us assume that the energy

$$E_1 = Z N_e r_e m c^2 \qquad (4)$$

is the limit energy for ions. The ions with $E > E_1$ have oscillation amplitudes in constant density beam of more than the beam size and in Gaussian density beam of more than $1.65a_e$. The ions collide in the electron beam. The variation rate of the one kind particle energy (under the index "1") caused by the collisions with another kind of particles (under the index "2") is [8]

$$\frac{dE_1}{dt} = 4\pi Z_1^2 Z_2^2 n_2 m^2 c^4 r_e \frac{\sqrt{A_1 A_2}\,(E_2 - E_1)\Lambda}{\sqrt{M}\,(A_1 E_2 + A_2 E_1)^{3/2}} \qquad (5)$$

The same formula is used for E_2. When the particle lifetime $\tau \gg E_1/(dE_1/dt), E_2/(dE_2/dt)$, then the ion distribution function becomes three-dimensional Maxwell distribution with the common temperature and the equal energy for three degrees of freedom for all kind of ions. The ions with the higher temperature will cool off and with the lower will heat up. Let's analyze some concrete examples.

3. EBIS. EBIS is a linear continuous electron beam with high constant density. The highly charged ion losses are the main insufficiency of EBIS ionization times about seconds. Only 10^{-3}-10^{-4} part of the initial ion number are left in the beam. The formula (1) gives the heating time for Xe^{50+} for the KRION-2 ionizer parameters [1] (E_1=20keV, I=0.2A, a=0.15mm). The ion energy increase up to the limit energy (6) and the ion are lost out of the beam. The instabilities can be the cause of ion losses, too.

One can use the injection of low charged light ions in the beam for cooling the highly charged ions [11]. The ion energy in the constant density beam depends on the ion charge (4) and moreover the electrons heat the highly charged ion more rapidly (1). As a result, the highly charged ions have higher energy than the light ions. The ion collisions provide for equal energy for all kinds of ions: $\bar{E}=(E_1N_1+E_2N_2)/(N_1+N_2)$. If one takes a definite number and kind of light ions so, that $\bar{E}>E_{12}$, where E_{12} is the limit energy of heavy ions (4), then the light ions will cool the heavy ions, get lost and take away the energy from the beam. At the same time a new portion of light ions must be injected into the beam to compensate the ion losses. The energy of the lost ions equals the heating energy in the static case.

Let us take for example the source with parameters: E_e=70keV, I=0.5A, a=0.1mm. The equation (1) gives the time for highly charged Xe heating till E_1 (4) less than 1s. The alignment temperature time with C^{3+} ions is about 0.1ms. One must inject 3 10^7 ions of C^{3+} every 10 ms if the number of Xe ions is 2 10^{10}.

Donets has made the first successfully experimental cooling of Kr ions at KRION-2 ionizer in Dubna [13].

4. ERIS. ERIS construction is beginning in Dubna now. The electron rings of collective accelerator of heavy ions are used as a source of highly charged ions in ERIS [3,4]. There are plansto obtain about 10^{10} ion per cycle up to Xe^{48+} in electron rings with radius R=2cm, a=0.1cm, number of electrons about 5 10^{12} and life time about 0.1s. The ions in the electron rings will be used for X-ray spectroscopy of highly ionized atom electron shells. But the ion oscillation in a small cross section of a ring with electron current 1-2kA can be the cause of Doppler broadening of X-ray lines and will decrease the accuracy of measurement. The electron density in the beam cross section is near to Gaussian distribution. From (2) the mean energy of highly charged Xe ions is about 50keV. Doppler broadening of $K\alpha_1$ and $K\beta_1$ lines is about 25eV and less than the difference between the energy of line for the nearest next highly charge states.

Really, there are some possible processes for ion heating in the electron ring. The instabilities increase ion energy, except the collisions with electron, which do not influence much at times smaller than 0.1s. Some part of the ions can be lost and the remained ions can fill all the surface in the potential well of the ring. In this case the mean ion energy will be $\bar{E}=E_1/2= ZN_er_emc^2/2$. The broadening will reach the value about 100eV and it will be difficult to separate the lines of neighbouring charge states.

For ion cooling one can inject, for example, He atoms from the residual gas in the vacuum chamber or by a special neutral jet into the electron ring with Xe ions. The He^{2+} ions will be generated after a time of about 0.1ms. It is possible to vary the ion number by changing either the He gas pressure, or density and duration of the neutral jet. From (5) one can obtain that if the number of Xe^{48+} is 10^{10} and number of He^{2+} ions is 5 10^{11} the energies of all kind of ions will be equal to 25-35keV after the time about 10-20ms and broadening will decrease to 15eV.

5. ECRIS. ECRIS is a plasma in the open magnetic trap. The microwave heats the electrons to a keV temperature. The ions receive the energy (some eVs) in collisions with electrons and slowly diffuse from the trap. The ion energy depends on its charge

(1).The difference between electron and ion diffusion rates results the appearance of electric charge and a potential of the plasma.The confinement time and the charge state of ions are determined by the plasma potential and the ion energy [16].

It was experimentally discovered last years that the addition of light ions increases the extraction of multiply charged ions in ECR sources. The elastic ion-ion collisions and the redistribution between the light ions with low energy and the heavy multiply charged ions with high energy can explain this effect. The light ions take the energy of heavy ions and go away from the trap. The cooled heavy ions diffuse more slowly and have time for higher charge states.The energy redistribution time from (5) has a microsecond time scale for the typical ECR plasma parameters: n_e=3 10^{11}-3 $10^{12}cm^{-3}$, n_i=3 10^{10}-3 $10^{11}cm^{-3}$, $E_i \approx 10eV$, Z_1=10, Z_2=3. The most light ions have a low cooling rate and the ions with the similar masses have a similar diffusion rate and are not recommended for cooling. The decrease of ion energy causes the energy spread and emittance reduction in ECR sources.

Some gas mixing experiments in ECRIS were done in Holland [14]. It was shown here that the nitrogen addition to a plasma was more effective to increase the extraction of multiply charged argon ions, than the addition of helium. But the He was better for Ne ions. The addition of the heavy elements to the light ion plasma causes the decrease of the mean charge state for the light ions. The results obtained in Michigan University [15] showed that Ar ion temperature was decreasing from 11eV to 5eV when He ions were injected and to 3eV with O and N ions. The energy straggling reduction was registered too. All these facts are in good agreement with our model.

The calculations of Kr ion charge state distributions with N ions for ECRIS with n_e=3 $10^{12}cm^{-3}$ and E_e=1keV have been carried out. The charge state distribution of Kr ion is shown in the figure. The large points correspond to pure Kr, the small points correspond to the calculation when the number of Kr ions is equal to the number of N ions and the crosses for the number of Kr ions in twenty times less than the number of N ions. The calculations show, that the addition of N ions in the plasma with separated parameters reduce the ion temperature from 15eV to 6eV and the extraction of multiply charged ions increases by several ten times. These results are in good agreement with the experiments, too.

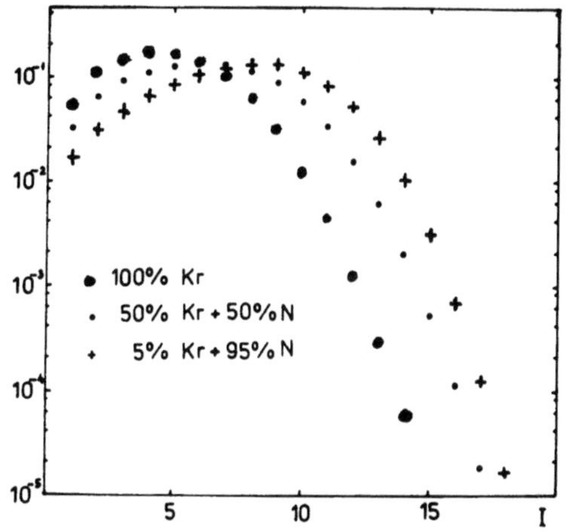

Fig.

6.Conclusions. The elastic collisions among the ions are more intensive than between electron and ion components in the plasma.This is connected with the high charge of the ion and the small relative velocities of heavy particles. The addition of light ions with low charge states permits one to cool heavy ions, to increase the effective confinement time and to reduce the energy straggling.

References.

1. Donets,E.D.: Phys. Scripta **T3**,11 (1983)
2. Geller,R., Jacquot,B.: NIM **184**,293 (1981)
3. Perelstein,E.A., Shirkov,G.D.: Particles & Nuclei **18**,68 (1987)
4. Perelstein,E.A., Shirkov,G.D.: In Proc. of Summer School and Int. Symp. on the Phys. of Ionized Gases, Saraevo 1988, pp.148-151
5. Levine,M.A.et al.:Phys.Scrip. **T22**,157 (1988)
6. Janev,R.K., Presnyakov,L.P.:Phys. Rep. **70**, 1 (1981)
7. Spitzer,L.: Physics of Fully Ionized Gases. Interscience Publ.,New York 1962
8. Longmire,C.L.: Elementary Plasma Physics. Interscience Publ.,New York 1863
9. Lee,E.P., Cooper,R.K.:Particle Acceler. **7**, 83 (1976)
10. Laslett,L.S.: Preprint ERAN-218, LBL, Berkeley 1972
11. Bulyak,E.V., Kurilko,V.I.: TsIIatominform, Moskow 1987
12. Donets,E. D., Shirkov,G.D.: USSR Patent Nr. 1225420, Bull. OIPOTZ **44** (1989)
13. Donets,E.D.:In Proc.of Int.Conf.on Ion Sources;Rev.of Scient.Instr.,**64**,N1,part 2,225 (1990)
14. Mack,M. et al.: In Contr.Papers of the 7th Workshop on ECR Ion Sources, p152, Julich 1986
15. Antaya,T.A.: J. de Physique, **50**, Colloque C1, p. C1-707 (1989)
16. Jongen,Y.: LC8001, Univer.Catholique de Lonvain (Belgium) 1980

Supplement to Z. Phys. D – Atoms, Molecules and Clusters 21, S 323–325 (1991)

Atoms, Molecules and Clusters
Zeitschrift für Physik D

Calculation of ion charge distribution in the Dubna ECRIS-14

V. B. Kutner and G. D. Shirkov
Joint Institute for Nuclear Research, Head Post Office, P. O. Box 79, 101000 Moscow, USSR

A self-consistent model for the calculation of ion charge state distribution (ICSD) in electron cyclotron resonance ion sourses (ECRIS) is proposed. The calculation of ICSD for neon, argon and krypton ions in Dubna ECRIS-14 have been carried out.

The creation of multiply ion source based on electron cyclotron resonance has begun now in Dubna [1]. The project DECRIS-14 is aimed for great expansion of the range and energy of ion beams from Dubna cyclotrons. The frequency 14.5Ghz will be used in DECRIS-14. The optimization of all ion source characteristics is very important in connection with the problem to obtain the beams of rare and expensive isotopes. The models of ion appearance and loss in ECR plasma can help to create the effective ion sources for cyclotrons.

The first calculations of ICSD for ECRIS ware made by Jongen [2], then by Bliman & Chan-Tung [3] and later by West [4]. In these papers ICSD were defined by solving the algebraic balance equations of ion charge states in plasma. The balance equations for neutrals and single charged ions in [2] and [4] and for the highest charge states in [3] were expelled from the consideration to simplify the problem. The densities of electrons n_e and ions n_i in plasma and temperatures of electrons T_e and ions T_i were used as initial parameters.

The mathematical methods used in the proposed calculations make possible to solve the set of equations for all ion charge states and neutrals in plasma taking into account any collision processes. Here we consider the double electron ionization and charge exchange processes. T_i is defined by the heating rate of electrons. T_e and n_e or the density of neutrals in plasma n_o are used as initial values.

The set of non-linear differential equations describe the ion formation and its transition from one charge state to another in the general case. ECRIS is an apparatus of continuous action and all the processes are in equilibrium. That is why the static solutions of the set should be searched, as a rule. In this case, these equations are reduced into the set of non linear algebraic equations:

$$
\begin{aligned}
& S v_o (n-n_o)/V - n_o\Big(\sum_{i=2}^{z} n_i \sigma_i^{ex} v_i + \sum_{i=3}^{z} n_i \sigma_i^{2ex} v_i + (\sigma_1^{i}+\sigma_1^{2i}) n_e v_e\Big) = 0 \\
& n_o\Big(\sigma_1^{i} v_e n_e + \sum_{k=2}^{z} n_k \sigma_k^{ex} v_k + n_2 \sigma_2^{ex} v_2 + n_3 \sigma_3^{2ex} v_3\Big) - n_1\big((\sigma_2^{i}+\sigma_2^{2i}) v_e n_e + 1/\tau_1\big) = 0 \\
& (n_1 \sigma_2^{i} + n_o \sigma_1^{2i}) v_e n_e + \Big(\sum_{k=3}^{z} n_k \sigma_k^{2ex} v_k + n_3 \sigma_3^{ex} v_3 + n_4 \sigma_4^{2ex} v_4\Big) n_o - n_2\big((\sigma_3^{i}+\sigma_3^{2i}) v_e n_e + n_o \sigma_2^{ex} v_2 + 1/\tau_2\big) = 0 \\
& (n_{i-1} \sigma_i^{i} + n_{i-2} \sigma_{i-1}^{2i}) v_e n_e + (n_{i+1} \sigma_{i+1}^{ex} v_{i+1} + n_{i+2} \sigma_{i+2}^{2i} v_{i+2}) n_o - n_i\big((\sigma_{i+1}^{i}+\sigma_{i+1}^{2i}) v_e n_e + (\sigma_i^{ex}+\sigma_i^{2ex}) v_i n_o + 1/\tau_i\big) = 0, \quad (i=3,4,\ldots z-2) \qquad (1) \\
& (n_{z-2} \sigma_{z-1}^{i} + n_{z-3} \sigma_{z-2}^{2i}) v_e n_e + n_z \sigma_z^{ex} v_z n_o - n_{z-1}\big(\sigma_z^{i} v_e n_e + (\sigma_{z-1}^{ex}+\sigma_{z-1}^{2ex}) v_{z-1} n_o + 1/\tau_{z-1}\big) = 0 \\
& (n_{z-1} \sigma_z^{i} + n_{z-2} \sigma_{z-1}^{2i}) v_e n_e - n_z\big((\sigma_z^{ex}+\sigma_z^{2ex}) v_z n_o + 1/\tau_z\big) = 0
\end{aligned}
$$

Here n is the density of neutrals outside the plasma; S and V – the area of surface and volume of the plasma; τ_i – the mean life time of ions with charge i; v_o, v_i, v_e are mean velocities of neutrals, ions and electrons; $\sigma_i^{i}, \sigma_i^{2i}$ are single and double ionization cross sections of the ions with charge i-1; $\sigma_i^{ex}, \sigma_i^{2ex}$ are single and double charge exchange cross sections of the ions with charge i.

The condition of charge neutralization of the plasma demands an additional equation:

$$\sum_{i=1}^{z} i\, n_i = n_e \qquad (2)$$

There are a lot of models for determination of the ionization cross section by electron impact with different precision in the ECR range of electron energies. The relatively simple formula by Lotz [5] was used in our case:

$$\sigma^{i} = (4.5\ 10^{-14}/T_e) \sum_{k=1}^{m} (n_k/I_k) \log(T_e/I_k)\ cm^2 \quad (3)$$

where m is the number of subshells in the ion; n_k is the number of equivalent electrons in the subshell. Muller got a cross section with a similar form by analyzing the experimental data for the double ionization [6]:

$$\sigma^{2i}_{i+1} = 2.6\ 10^{-14} \ln(T_e/(I_i+I_{i+1}))/(T_e(I_i+T_{i+1}))\ cm^2 \quad (4)$$

Muller & Salzborn's empirical formulas were used for charge exchange cross sections [7]:

$$\sigma^{ex}_{i} = 1.43\ 10^{-12} i^{1.17}/I_o^{2.76}\ cm^2$$
$$\sigma^{2ex}_{i} = 1.08\ 10^{-12} i^{0.71}/I_o^{2.80}\ cm^2 \quad (5)$$

where I_o is the ionization energy of neutrals.

Jongen's method was used to determine the electron and ion diffusion times. The longitudinal ambipolar diffusion in the magnetic mirror is the main process. Energy and collision rates determine the diffusion rates of the ions. The main ion losses happen along the magnetic field through the butt-end of the mirror. The magnetic field keeps the light electrons more effectively. Only the electrons with a velocity vector near the direction of the axis can leave the trap. The electrons can obtain this velocities in collisions. Hence, the electron life time is determined by the collision rate of the electrons with the particles in the plasma.

The microwave heats the electrons in the conditions of electron cyclotron resonance. In our model T_e is an extra parameter. The ions receive their energy in the collisions with hot electrons in the absence of any instabilities in strong magnetic field. When $T_e \gg T_i$ we may take $T_i = \tau_i dT_i/dt$. The ion heating by fast electrons was considered in [8,9]:

$$dT/dt = (4\sqrt{2\pi} m_e Z^2 r_e^2 m^{5/2} c^4 \Lambda_{ei})/(AMT_e^{1/2}) \quad (6)$$

where $\Lambda_{ei} \approx 10$. At the same time the energy exchange rate among the ions is some orders more effective [10]. The relaxation time of the temperatures for different kinds of ions is much less than the life time of the ions in the plasma. So we can suppose that ions have equal temperature:

$$T_i = \sum_{k=1}^{z} \frac{dT_k}{dt} \tau_k n_k \Big/ \sum_{k=1}^{z} n_k \quad (7)$$

The ion current density through the butt-end of the source can be valued as [4]:

$$J = 1.6\ 10^{-19}\ i\ n_i\ l/\tau_i\ Acm^2 \quad (8)$$

The mean life time of the ions in the set (1) is determined by the ion densities and ICSD. It is the reason for strong nonlinearity of the set. Some iteration methods were examined to solve the set. The best result was obtained by Zeidel method. The main results of calculations are shown in the figures. The ICSD for Ne, A and Kr ions in DECRIS-14 plasma were determined. It was supposed that ions are formed in the plasma with the size l=30cm and d=6.4cm from the residual gas.

In the fig.1-2 we present the ICSD dependence on the T_e and n_e for extraction of Kr ions from the source. The ion current densities for every charge states are in relative units. The total ion current density in these units is equal to 1. Dependences of Ti for Ne, A and Kr ions in source plasma on T_e and n_e are shown in the fig.3-4. The dependences of total ion current density I on T_e and n_e are shown in the fig 5-6. These results make possible to evaluate the ion extraction current from the source.

The results obtained in this work, give the opportunity to conclude that the ICSD in ECR plasma and the output of multiply charged ions depend much on the electron temperature. The T_e increase is the cause of the increase of the mean charge and the output current of multiply charged ions to increase sharply. These results are in agreement with [2] and [4]. At the same time the total ion current is decreasing. This fact is due to the reduction of the collision probability of the hot electrons in the plasma and thus the number of electrons in the cone of losses is reduced too. The total ion currents from the source must be equal to the electron current due tothe condition of plasma neutrality.

In [2] and [4] it was concluded, that electron density increasing is a means of rising of the numbers multiply charged ion. In our case, we have considered the balance equations for neutrals and singly charged ions and, so we have a selfconsistent problem. The total ion number (or ion density) depends on the neutral and electron densities. The condition of charge neutralization establishes the relationship between electron and ion total charges. Thus, the densities of electrons and ions are dependend values. Increasing of the neutral density in the plasma n_o (and thus the density outside the plasma n) demands the reducing of electron density

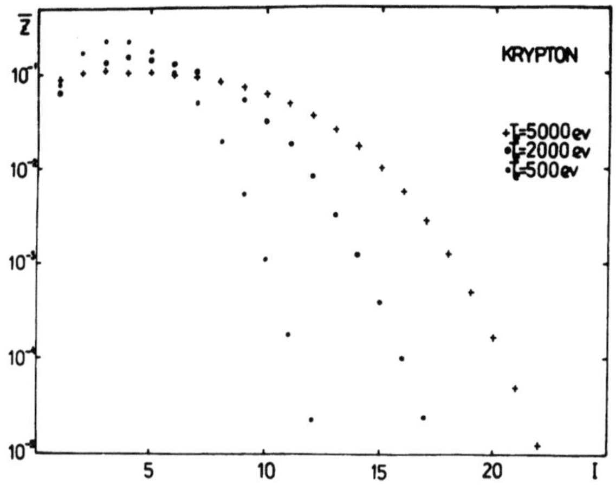

Fig. 1

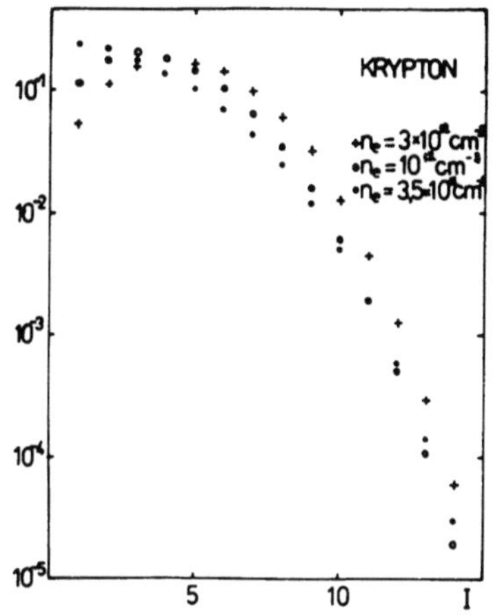

Fig. 2

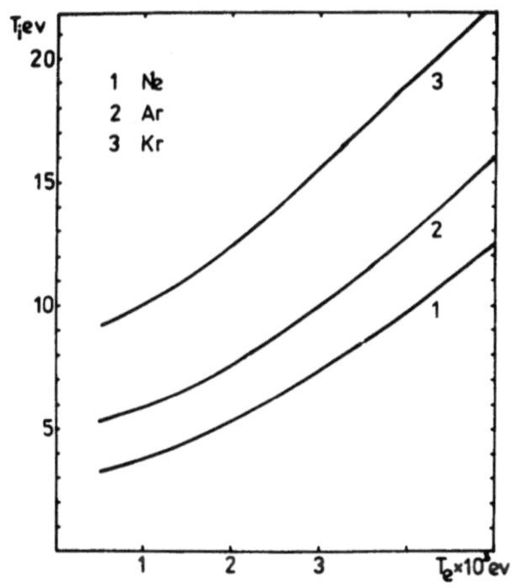

Fig. 3

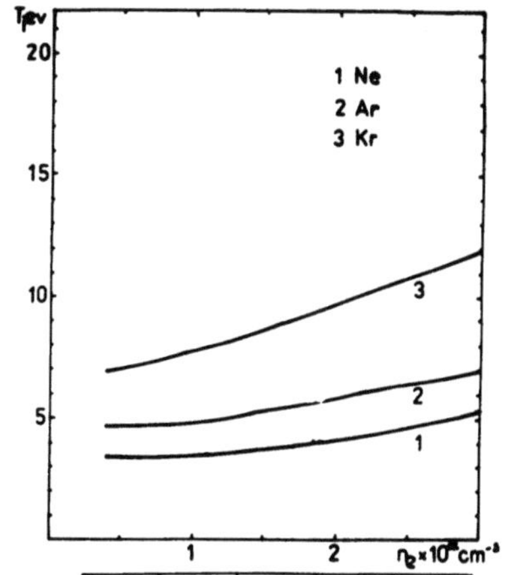

Fig. 4

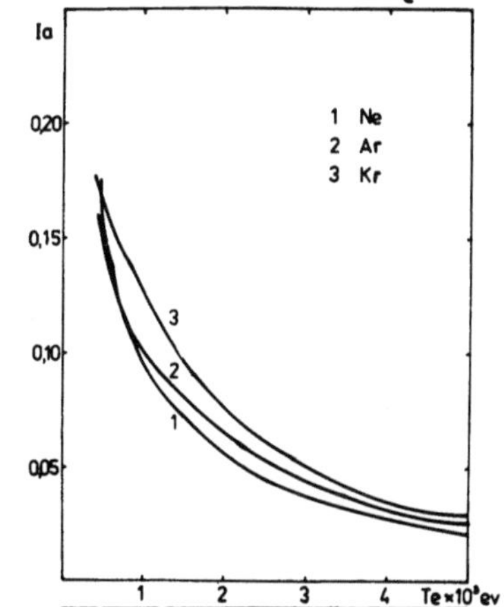

Fig. 5

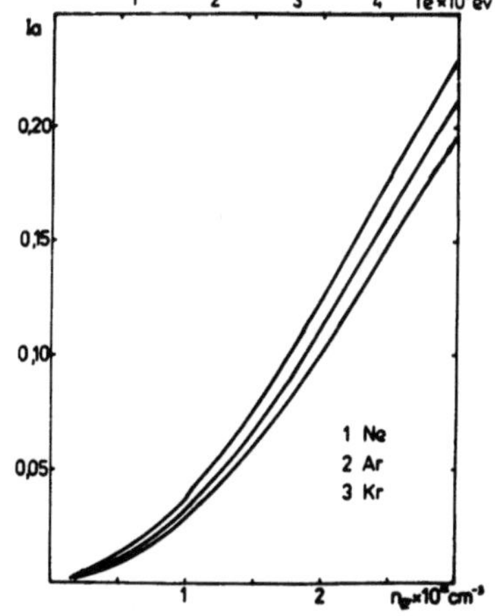

Fig. 6

n_e for self-consistent solution of the problem and vice versa. From the other hand, the increasing of neutral density decreases the mean charge of the ions and as a result, ICSD practically does not depend on electron density (fig.2,6), but the total ion current increases very much. Another way to increase the output of the multiply charged ions is the light ion cooling [10].

References.

1. Kutner,V.B.et al.:Prep.JINR R7-85-362,Dubna 1985
2. Jongen,Y.:LC-8001,LC-8003;Univer. Catholique de Louvain,Belginm 1980
3. Bliman,S.& Chan-Tung,N.:J.Physique **42**,1247(1981)
4. West,H.I.:UCRL-53391,Lawrence LNL,Califor.1982
5. Lotz,W.et al.:Z.fur Phys.**206**,205;**216**,241 (1968)
6. Muller,A.& Frodl,R.:Phys.Rev. **44**, 29 (1980)
7. Muller,A.& Salzborn,E.:Phys.Lett. **59A**,19 (1976); **62A**,39 (1977)
8. Perelstein,E.A.& Shirkov,G.D.:Particles & Nuclei **18**,68 (1987)
9. Wiesemann,K.:In Proc.of Int.Conf.on Mult.Charged Heavy Ion Sources,IEEE Trans. NS-19,**2**,156 (1972)
10. Shirkov,G.D.:JINR R9-89-600,Dubna 1989; GSI-tr-89-09,Darmstadt 1989

Supplement to Z. Phys. D – Atoms, Molecules and Clusters 21, S 327–328 (1991)

Atoms, Molecules
Zeitschrift für Physik D and Clusters

A 10 GHz electron-cyclotron-resonance (ECR) ion source for ion-electron and ion–atom collision studies

M. Liehr, M. Schlapp and E. Salzborn
Institut für Kernphysik, Universität Giessen, W-6300 Giessen, FRG

Abstract

In order to extend our investigations in the field of electron-ion collision studies involving ions in high charge states, we decided to build a newly designed 10 GHz Electron-Cyclotron-Resonance Ion Source. The new ion source is expected to produce ion beams comparable to those of the Minimafios ECR ion sources with respect to intensities and maximum charge state.

Mechanical Assembly

A schematic overview of the mechanical assembly is shown in Fig. 1. Both stages are made of stainless steel tubes with an inner diameter of 80 mm. The first stage, which will be operated at high gas pressure, is equipped with remote controlled valves for gas inlet and an axial microwave horn antenna.

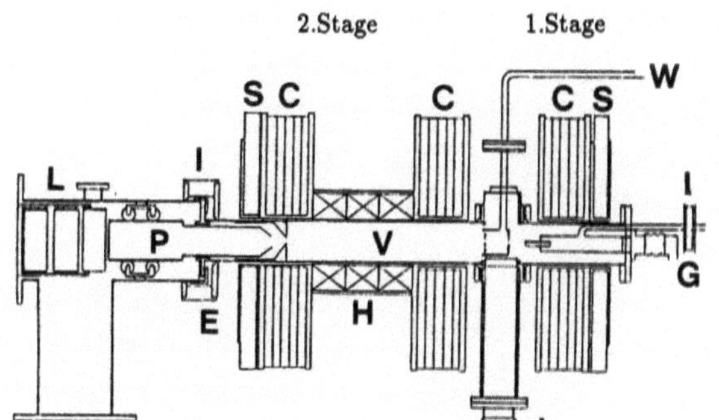

L	:	Puller-Electrode (movable)	H	:	Hexapole Magnet (FeNdB)
P	:	Einzel-Lens	V	:	Vacuum Chamber
I	:	Insulation	W	:	Waveguide (X-Band)
E	:	Extraction	G	:	Gas Inlet
C	:	Solenoid Coil	P1	:	Pump 700l/s
			P2	:	Pump 280l/s

Fig 1. Schematic set-up of the 10 GHz ECR ion source

Furthermore, an oven for metallic ion production can be installed at the rear flange. The difference in gas pressure is sustained by a ceramic window with an 8 mm hole. The "cold" plasma can drift through this hole into the second stage which will operate at a gas pressure lower to the order of two magnitudes with respect to the first stage. In this stage, the microwave-power will be fed by a H-horn antenna to a region with a strong magnetic field for hot plasma electron confinement. This is achieved mainly by a radial hexapolar magnetic field which is produced by FeNdB permanent magnets. Its cross section was derived from the AECR project (LBL Berkeley) and was optimized by computer calculations (see Fig. 2). The axial confinement is obtained by solenoid coils which provide a double magnetic mirror with a mirror ratio of 1:2.

To increase the ratio of the plasma density to that of residual gas particles, the second stage is equipped with an oil diffusion pump.

The extraction region is followed by a long puller electrode which can be adjusted with respect to the extraction hole.

The microwave system operates with one 10 GHz Klystron amplifier which yields a maximum power of 2.5 kW. The waveguide system supplies both ion source stages with microwave power independently (see Fig. 3).

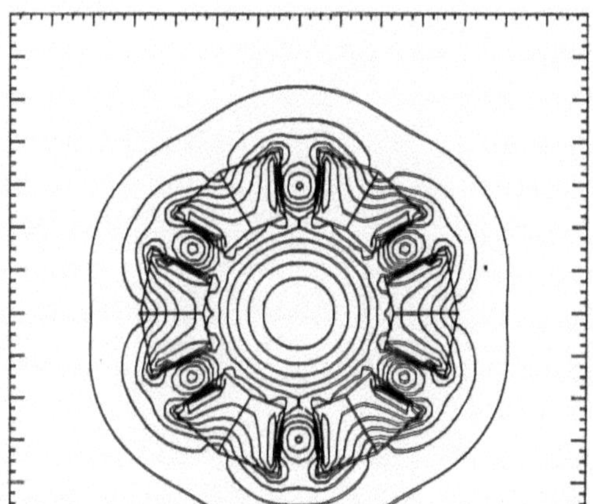

Fig 2. Cross sectional view of the hexapole structure with lines of constant B (mean values, calculated with PANDIRA-Code, A.M.Winslow, J.of Comp.Phys., (1957)149),the angle between direction of magnetization and the pole axis was chosen to be 26°.

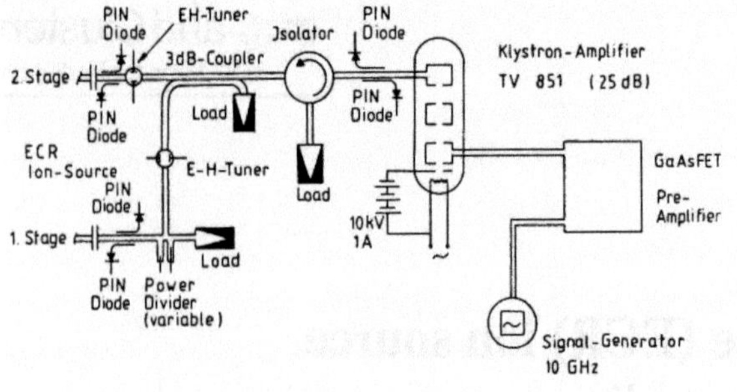

Fig 3. The microwave system supplying both stages with microwave power.

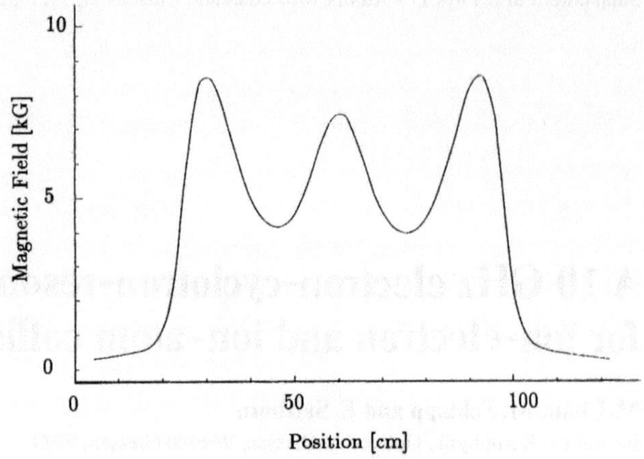

Fig 4. Longitudinal distribution of magnetic field measured along the axis of symmetry.

Dimensions and parameters

Electrical power supplies for longitudinal magnetic field

600 A	100 V
600 A	80 V
500 A	50 V

for 15 double layer pancake coils assembled in three groups

Permanent magnetic structure for radial hexapole field

FeNdB anisotropic material

Remanence:	1.2 T
Coercitive Force:	920 kA/m
Maximum field at pole bar:	0.85 T

Length of main plasma stage: 400 mm

Inner diameter: 80 mm

Supplement to Z. Phys. D - Atoms, Molecules and Clusters 21, S 329-330 (1991)

Atoms, Molecules
Zeitschrift für Physik D and Clusters

Charge state distributions of 44 MeV/amu $^{129}Xe^{44+}$ and 29 MeV/amu $^{208}Pb^{56+}$ after various solid targets

A. Cassimi, J. P. Grandin, M. G. Suraud
CIRIL, rue Claude Bloch, B.P. 5133, 14040 Caen Cedex, France

Abstract

Emerging charge state distributions of 44 MeV/amu $^{129}Xe^{44+}$ and 29 MeV/amu $^{208}Pb^{56+}$ transmitted through various solid targets have been measured and analyzed using the GANIL facility.

I. Introduction

The charge state of ions passing through matter undergoes changes due to electron capture and loss processes. There are practical demands for the prediction of charge state distributions. One of them, which comes from ion accelerator physicists, concerns the production of highly stripped heavy ion beams in a given charge state. Charge state distributions can be determined by ab initio calculations, such as those performed by J.P. ROZET et al [1], starting from the electron capture, loss, and excitation cross sections. The experimental study of these charge states is thus also of fundamental interest since it gives informations about atomic collision processes in complex atomic systems through the electron capture and loss cross sections.

The present work was devoted to the recording of experimental charge state distributions of 44 MeV/amu $^{129}Xe^{44+}$ and 29 MeV/amu $^{208}Pb^{56+}$ behind various solid targets (Be, C, Al) as a function of the target thickness. These results will increase the experimental data available for the charge state prediction when stripping and slowing down ion beams.

II. Experiment

The 44 MeV/amu $^{129}Xe^{44+}$ and the 29 MeV/amu $^{208}Pb^{56+}$ beams were delivered by the GANIL accelerator in CAEN(FRANCE). Measurements were performed on line LISE[2]. After passing

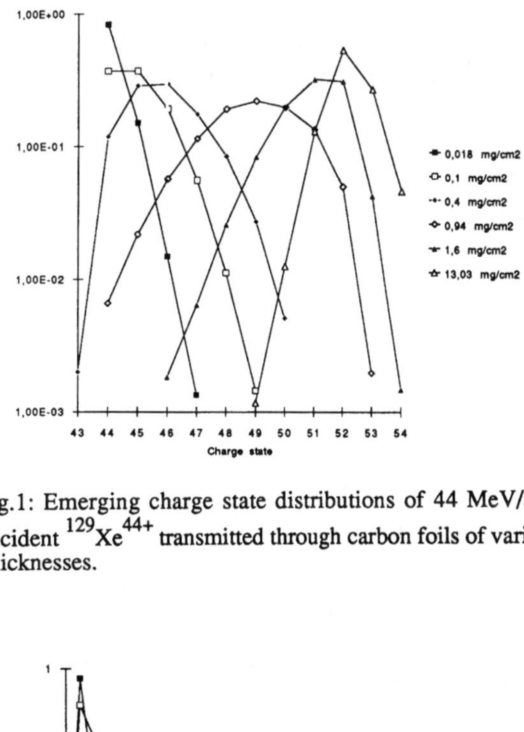

fig.1: Emerging charge state distributions of 44 MeV/amu incident $^{129}Xe^{44+}$ transmitted through carbon foils of various thicknesses.

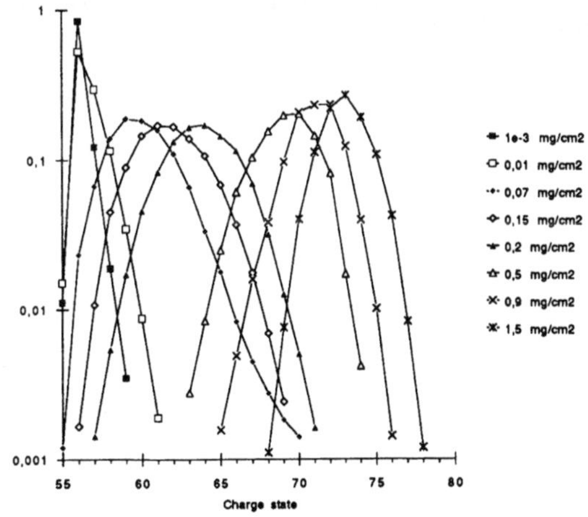

fig.2: Emerging charge state distributions of 29 MeV/amu incident $^{208}Pb^{56+}$ transmitted through carbon foils of various thicknesses.

through the foil, the beam was charge analyzed by a dipole magnet. Relative intensity of each produced charge state was measured using an Ar-CO_2 filled multiwire proportional chamber which allows to detect beam intensities varying between 10^2 and 10^8pps.

III. Results

The targets we used were C foils (from 0,001 to 37 mg/cm^2), Be foils (from 2,3 to 40 mg/cm^2) and Al foils (from 0,22 to 26 mg/cm^2). For each target we measured the charge state distribution. Some of these distributions are given in fig.1 and fig.2. Charge state evolution of 44 MeV/amu ^{129}Xe and 29 MeV/amu ^{208}Pb ions in C targets as a function of thickness has been determined. Evolution of the mean charge state with thickness is given in fig.3 and fig.4.

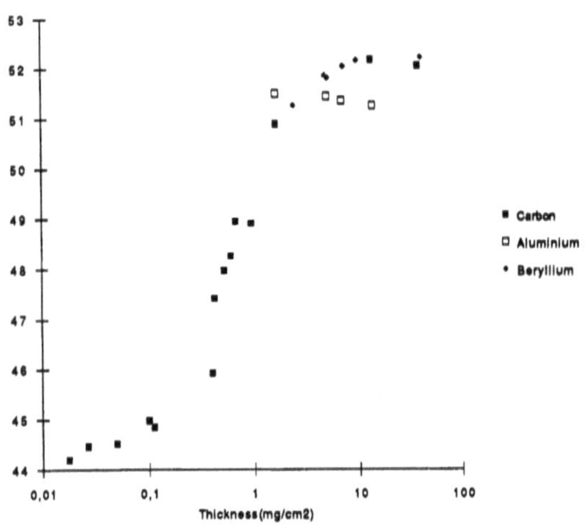

fig.3: Mean charge state of 44 MeV/amu ^{129}Xe as a function of target thickness.

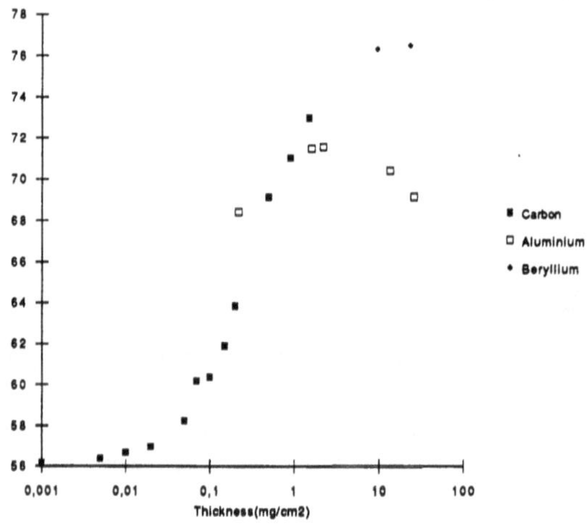

fig.4: Mean charge state of 29 MeV/amu ^{208}Pb as a function of target thickness.

Fig. 1 and 2 are very helpful since one can deduce the target thickness to use in order to center the distribution on a given charge state. It appears on fig.3 that the mean charge state evolutions for C and Be targets are very similar and, as expected, the equilibrium charge is smaller with Al than with C and Be targets. Fig. 4 shows that for C and Be the equilibrium is hardly reached while for Al the mean charge starts to decrease due to the slowing down of the ions.

References

[1]: J.P. ROZET, A. CHETIOUI, P. PIQUEMAL, D. VERNHET, K. WOHRER, C. STEPHAN and L. TASSAN-GOT, J. Phys. B 22 (1989) 33

[2]: R. ANNE, D. BAZIN, A.C. MUELLER, J.C. JACMART and M. LANGEVIN, Nucl. Instrum. Methods A 257 (1987) 215

Supplement to Z. Phys. D - Atoms, Molecules and Clusters 21, S 331-332 (1991)

Atoms, Molecules
Zeitschrift für Physik D and Clusters

Evolution of fine structure populations as a test of solid target effects

J. P. Rozet, A. Chetioui, P. Nicolaï, M. Chabot, M. F. Politis, A. Touati, D. Vernhet, K. Wohrer, C. Stephan and J. P. Grandin*
LPI2M2C, Université P. M. Curie, CNRS URA 1379 and Institut Curie, Section de Physique et Chimie,
11 rue P. M. Curie, F-75231 Paris Cedex 05, France
*CIRIL, 14040 Caen, France

Abstract

The evolution of fine structure components of Balmer α transition following electron capture by fully stripped 33 MeV/u Kr^{36+} ions in carbon targets has been studied in the target thicknesses range 10 - 200 $\mu g/cm^2$. Our results cannot be explained by a rate-equations model (binary collisions including ℓ-changing process), but constitute a first indication of a possible wake-field induced Stark mixing of the substates.

I. Introduction

Energy loss of swift heavy ions in condensed matter results from a collective response of the medium ("wake") as well as from a succession of binary ion-atom collisions along the path of the projectile. As both contributions follow the same scaling law, energy loss measurements cannot be used as a test of their calculated relative values. However, the evolution with target thickness of relative n, ℓ, m populations in hydrogenlike heavy ions under near single collision conditions can similarly be expected to result from both binary ion-atom collisions (collisional ℓ mixing) and a wake-field induced Stark mixing of the substates. In a recent paper [1], we have studied the evolution with target thickness of Lyman lines emission for the same collision partners as in the present work. Our results showed that using ℓ-mixing cross sections calculated in the Plane Wave Born Approximation (with screening and antiscreening effects accounted for), poor agreement was obtained between experiment and model calculations. On the other hand, a set of fitted ℓ-changing cross sections has been found to reproduce our data, these last cross sections being a factor 3-50 larger than the calculated ones. We suggested that this discrepancy may be an indication of a specific effect of free electrons in solid target, and that future experiments may help clarifying the problems. We report here on the first results of such an experiment.

II. Experiment

The experiment was conducted at the GANIL heavy-ion accelerator using the LISE line. The experimental set-up has been described in ref. [1]. Briefly, fully stripped 33.2 MeV/u Kr^{36+} ions were directed onto thin solid carbon foils. Lyman and Balmer lines from the projectile were detected simultaneously by a Si-Li detector positioned at a polarization insensitive angle of 120° with respect to the beam direction, and a high - resolution high - transmission crystal spectrometer set at a 30° angle with respect to the beam direction. Here, using a vertical reflection plane geometry results also in almost polarization insensitive intensities (the "magic" angle being 27.38° in the laboratory frame). Our spectrometer includes a flat HOPG graphite crystal of well known polarization factor and reflectivity [2] and a large (70 x 70 mm^2) position - sensitive proportional detector developed in our laboratory.

III. Results

A typical spectrum recorded with the crystal spectrometer is shown in figure 1. Absolute intensities of Balmer α and Lyman β lines have

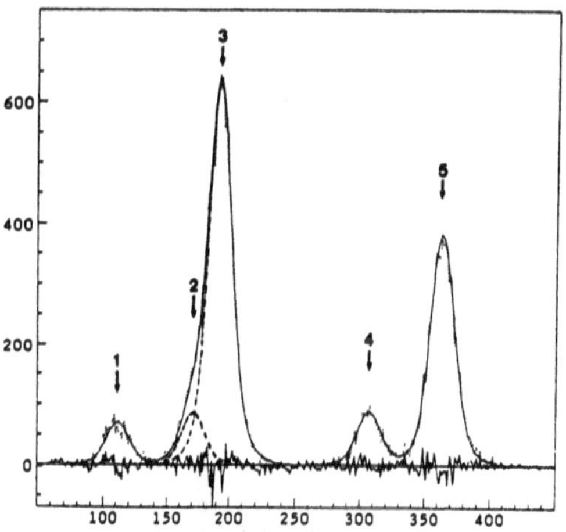

Figure 1 : Fine structure of Balmer α line observed for a 50 $\mu g/cm^2$ target. 1 channel corresponds to 0.5 eV (horizontal scale).
1 : $3s_{1/2} \rightarrow 2p_{3/2}$
2 : $3d_{3/2} \rightarrow 2p_{3/2}$
3 : $3d_{5/2} \rightarrow 2p_{3/2}$
4 : $3s_{1/2} \rightarrow 2p_{1/2} + 3p_{1/2} \rightarrow 2s_{1/2}$
5 : $3p_{3/2} \rightarrow 2s_{1/2} + 3d_{3/2} \rightarrow 2p_{1/2}$

been extracted from Si-Li spectra, and absolute intensity of total Balmer α as well as relative intensities of fine structure components from high resolution spectra. From these, we have extracted six *normalized* relative intensities $\overline{N}_R$ defined by :

$$\overline{N}_R = \frac{N_R(\text{exp.})}{N_R(\text{theo.})}$$

where

$$N_R(3\ell \to n'\ell') = \frac{N(3\ell \to n'\ell')}{\Sigma N(3\ell \to n'\ell')}$$

and where $N(3\ell \to n'\ell')$ is the experimental (or theoretical) intensity of $3\ell \to n'\ell'$ Balmer or Lyman β line. The theoretical values come from CDW calculations which proved to be very accurate when compared with experimental results in the case of gaseous targets [3,4].

IV. Discussion

A major consideration in performing the high resolution experiment was the ability to compare the evolution with target thickness of line intensities from $3d_{3/2}$ and $3d_{5/2}$ substates. Indeed, the rate equations model (with collisional ℓ mixing) predicts the evolution of the populations in these substates to be identical, whereas a very different behaviour must be expected to arise from a wake field-induced Stark mixing of the substates. In the first case, the $n\ell$-$n\ell'$ process (collisional ℓ mixing) leads to a statistical sharing of the populations between $3d_{3/2}$ and $3d_{5/2}$. In the second case, Stark mixing can only occur between near degenerate substates of same j and m_j, in view of the predicted wake field intensity of $\simeq 7\ 10^8$ V/cm [1,5]. Then, only the $3d_{3/2}$ substates can Stark mix with the $3p_{3/2}$ ones, whereas no such effect should occur in $3d_{5/2}$. The normalized relative intensities extracted from lines 2 and 3 (of high resolution spectra) from $3d_{3/2}$ and $3d_{5/2}$ substates respectively are plotted as a function of target thickness in figure 2, and compared with the predictions of the rate equations model. Model

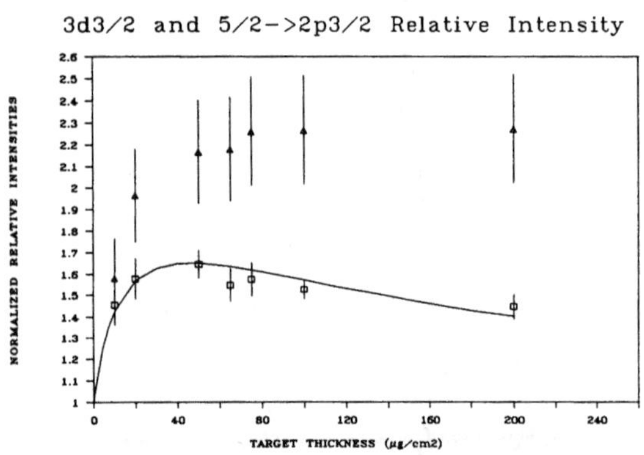

Figure 2 : Normalized relative intensities of Balmer lines from $3d_{5/2}$(□) and $3d_{3/2}$(Δ). The solid line is the prediction of the rate equation model (see text)

calculations have been performed using the same set of fitted cross sections used in [1] to reproduce the evolution of Lyman lines intensities.
The very good agreement obtained between theory and experiment for the $3d_{5/2}$ population is similar to the one observed in [1] for Lyman transitions. In both cases, cascades contributions are very large and originate from many substates, so that the general trends can be quite well reproduced by any model reproducing qualitatively the substates mixing, if ℓ mixing cross sections are properly adjusted. By contrast, the $3d_{3/2}$ level appears to be subject to much larger effects, in agreement with what is expected to occur from a Stark mixing between $3p_{3/2}$ and $3d_{3/2}$ sublevels : the initially much larger population in $3p_{3/2}$ is transferred to the $3d_{3/2}$ substates. From the calculated wake field intensity, the Stark splitting for substates with j=3/2 and m_j=3/2 is estimated to be 0.19 eV in Kr^{35+} ions, leading to a characteristic mixing time corresponding to a target thickness of $\simeq$ 85 $\mu g/cm^2$, which compares very well with experiment.

V. Conclusion

Using high resolution spectroscopy, we have studied the evolution with target thickness of relative populations in fine structure substates $3d_{3/2}$ and $3d_{5/2}$ of hydrogenlike krypton. Our results could indicate wake field induced substates mixing effects, and are in good agreement with wake field calculations [5]. For the future, we should be able to extract complementary informations from our results, since populations in any of the 5 fine structure substates in n=3 can be deduced from the six relative intensities we measured. The corresponding data analysis is under way, as well as calculations of the wake field effects, using a density matrix evolution formalism similar to the one used by Havener *et al.* [6] in the frame of their experimental n=3 density matrix determination.

References

1. Nicolaï, P., Chabot, M., Rozet, J.P., Politis, M.F., Chetioui, A., Stephan, C., Touati, A., Vernhet, D., Wohrer, K. : J.Phys.B 23 (1990) in press
2. Wohrer, K., Touati A., Chevallier, P., Chetioui, A., Rozet, J.P., Bouisset, P., Wang, J.X. : Nucl.Instr.Meth A274, 394 (1989)
3. Rozet, J.P., Chetioui, A., Bouisset, P. Vernhet, D., Wohrer, K., Touati, A., Stephan, C., Grandin, J.P. : Phys.Rev.Lett. 58, 337 (1987)
4. Wohrer, K., Ben Salah, F., Rozet, J.P., Chetioui, A., Touati, A., Bouisset, P., Vernhet, D. : Nucl.Instrum.Meth. B27 594 (1987)
5. Minchollé, J.L., Ugalde, J.M., Echenique, P.M. : Nucl.Instrum.Meth. B48, 21 (1990)
6. Havener, C.C., Rouze, N., Westerveld, W.B., Risley, J.S. : Phys.Rev. A33, 276 (1986)

Supplement to Z. Phys. D - Atoms, Molecules and Clusters 21, S 333-335 (1991)

Atoms, Molecules
Zeitschrift für Physik D and Clusters

Doubly differential electron emission cross sections for He^0 (0.5 MeV/amu) → He collisions*

O. Heil, R. Maier, R. D. DuBois#, M. Kuzel, K. O. Groeneveld
Institut für Kernphysik, J. W. G.-Universität Frankfurt/M., August-Euler-Straße 6, W-6000 Frankfurt/M. 90, FRG

Abstract

Absolute differential cross sections for electron emission occurring in fast (0.5MeV/amu) He^0-He collisions have been measured using standard non-coincidence as well as emitted electron- charge state analysed ejectile coincidence techniques. PWBA calculations for this system demonstrate good agreement with the experimental data for target and projectile ionization and indicate the importance of including simultaneous processes in the theoretical treatment.

I. Introduction

Measurements of doubly differential cross sections (DDCS) for electron emission test theoretical models of ionization in far greater detail than do total ionization cross section measurements. Collision systems where the projectile contains loosely bound electrons of its own allow to test target, projectile and simultaneous projectile-target ionization (1). Past DDCS studies have demonstrated that simultaneous projectile target ionization contributes significantly (≤50%) in the single (non-coincidence) electron spectra of simple H^0-He collision systems (2). In order to study simultaneous projectile target ionization in collisions where the projectile contains <u>two</u> loosely bound electrons of its own, experiments of fast (0.5 MeV/amu) neutral He^0-He collisons systems were performed.

II. Experiment

The experiments were conducted at the J.W.Goethe University 2.5 MV Van de Graaff accelerator in Frankfurt/M using a He^+- beam. The fast neutral He^0 beam was created via charge capture and electrostatically cleaned of any charged components before interacting with an effusive helium gas target. The metastable component of the He^0 beam at the target was estimated to be <35 to 40% (3). By using standard non-coincidence and emitted electron-charge state analysed ejectile coinicidence techniques, electron emission for projectile, target and simultaneous projectile-target ionization were obtained. The emitted electron spectra were measured at Θ =10° emission angle and between 20-1500 eV electron energy with a cylindrical mirror analyser (4). For the coincidence measurement, a post collision electric field deflected the stripped component of the ejectiles (e.g. He^+, He^{2+}) onto a secondary electron emission detector (5).

III. Results

Doubly differential cross sections (DDCS) for 0.5MeV/amu H^0 (6), He^0 impact on helium are presented in Fig. 1. In the upper portion of the figure, the non-coincidence DDCS(-- ;H^0, —;He^0 Single) for the sum of target and projectile emission are shown and, in addition, the emitted electron-charge state analysed projectile coincidence cross sections (□□;H^0->H^+, ●●;He^0->He^+, △△ ;He^0->He^{2+}). One observes clearly that the width of the peak structure in the electron spectra increases in the order of H^0 (□) impact, He^0 (●, single loss) and He^0 (△ , double loss). This broadening is better seen in the lower portion of Fig. 1; here the data are normalized to a common value at the peak maximum. For comparison, the non-coincidence He^0-He cross sections are also shown.

---

* supported by BMFT under contract no. 060F110/ Ti476Gr

perm. address: Pacific Northwest Laboratory, Richland, Washington 99352

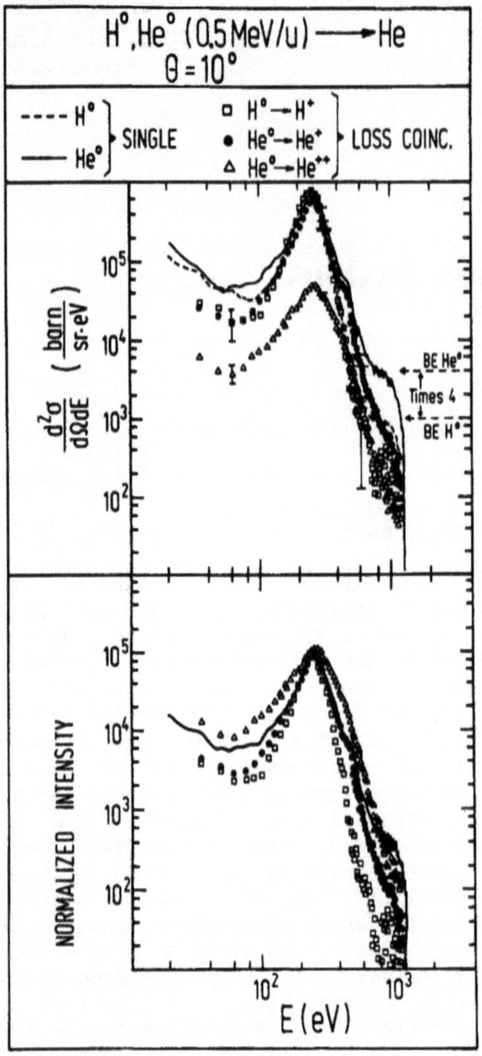

Fig.1. Doubly differential cross sections for 10° electron emission measured for 0.5MeV/amu H°, He° impact on helium (BE: Binary encounter peak; see text)

IV. Discussion

The dominant structures in the electron spectra are the well known electron loss peak, due to ionization of the projectile, and the continuum portion of the spectra, determined by the screening of the projectile nuclear charge by its electrons. For example, at high emitted electron energies (BE: Binary Encounter Peak), i.e. small impact parameter collisions, the electron screening is relatively ineffective and similar values for four times the H° cross sections and He° cross sections are observed. Here, the cross sections follow the Z_p^2- scaling of the unscreened nuclear charge (7). The width of the electron loss peak reflects the internal momentum distribution of the bound projectile electron, as clearly demonstrated for H°, He°-projectiles. For He° impact primarily single ionization (●) occurs roughly an order of magnitude more likely than double ionization (△). Suprisingly, the single H°, He° data and the coincident $H^0 \to H^+$ (□), $He^0 \to He^+$ (●) data show similar shape and magnitude for electron energies $E \ll E_{loss}$. However, at $E \gg E_{loss}$ the coincident $H^0 \to H^+$ and $He^0 \to He^+$ data deviate strongly from the single H°, He° data, but the coincident $H^0 \to H^+$ and $He^0 \to He^{2+}$ data have similar magnitudes. These coincidence cross sections below approximately 100 eV and above 600 eV are an indication of target ionization occurring in collisions where the projectile is also ionized. Thus simultaneous projectile-target ionization is a larger contributer to the non-coincidence DDCS for electron energies $E \ll E_{loss}$. In order to analyse projectile and target ionization in the electron spectra, PWBA calculations are performed for He°, He^+(0.5MeV/amu)-He collision systems and compared in Fig. 2 with the experimental data. Since the present theoretical model is for single transitions only, double electron loss by He° cannot be directly calculated.

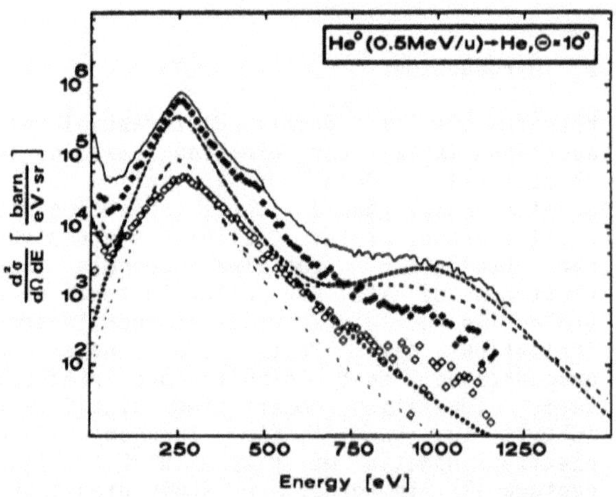

Fig. 2. Experimental and PWBA theoretical cross sections at 10° electron emission from 0.5MeV/amu He° impact on helium. Sum of target and projectile ionization: exp. He°, —; theory, He°, ▪▪▪; He⁺, - - -; Projectile and simultaneous target-projectile ionization: exp. He°, ♦ and ◊ for single and double projectile ionization. PWBA calculations for single projectile ionization: He°, ····, and He⁺, - - -.

However, simple projectile and target ionization is handled quite well by the PWBA indicating that the observed differences, between the He^{o} experimental and theoretical cross sections in the regions 0-200 eV and 350-600 eV, are caused by simultaneous target-projectile ionization events.

V. Conclusion

New measurements for ionization occurring in fast He^{o}-He collisions have been presented, which demonstrate the influence of two loosely bound projectile electrons on target, projectile, and simultaneous target-projectile ionization. Simultaneous projectile-target ionization for He^{o}-He collisions is dominated by one electron loss of the projectile. As a consequence cross section data for simultaneous target-projectile ionization show similar values at $E \ll E_{loss}$ for H^{o},He^{o}-He collision systems.

References

1. Manson. S.T., Toburen. L.H., Madison. D.H., Stolterfoht N.: Phys. Rev. A12, 60 (1975)
2. Heil. O., DuBois. R.D., Maier. R., Kuzel. M., Groeneveld. K.O.: Preprint, Phys. Rev. Lett. (1990)
3. Horsdal-Pedersen. E., Heinemeier. J., Larsen. L., Mikkelsen. J.V.: J. Phys. B13, 1167 (1980)
4. Bernardi. G., Suarez. S., Focke. P., Meckbach. W.: Nuc. Inst. and Meth. B33, 321 (1988)
5. Rinn. K., Müller. A., Eichenauer. H., Salzborn. E.: Rev. Sci. Instr. 53, 829 (1982)
6. Proceedings of the 4th Workshop on High-Energy Ion-Atom Collision Processes, Debrecen, Sep.1990 (to be published in Lecture Notes in Physics, Springer Verlag, Heidelberg)
7. Schader. J., Latz. R., Burkhard. M., Frischkorn. H.J., Hofmann. D., Koschar. P., Groeneveld. K.O., Berenyi. D., Kõver. A., Szabo. Gy.: Il Nuovo Cimento 7, 219 (1986)

Supplement to Z. Phys. D - Atoms, Molecules and Clusters 21, S 337-339 (1991)

Atoms, Molecules
Zeitschrift für Physik D and Clusters

Highly-charged ion-at-surface spectroscopy

E. D. Donets, V. G. Dudnikov, V. B. Dunin and S. V. Kartashov
Joint Institute for Nuclear Research, Dubna Head Post Office, P. O. Box 79, 101000 Moscow, USSR

Abstract

Hard X-rays arising from recombination of slow Kr^{35+} and Kr^{36+} ions at a solid surface were studied. Measurements were also carried out with highly charged xenon ions. The Z-dependence of K_βX-radiation probabilities for hydrogenlike ions is inferred from the data for Ar^{17+}, Kr^{35+} and Xe^{53+}. The binding energy of an electron in heliumlike krypton was also determined in the frame of ion-at-surface spectroscopy.

I. Introduction

Ion-at-surface spectroscopy is a method for investigating various features of atomic and ionic "superexcited states", which arise when slow highly charged ions recombine at the surface of condensed matter. The term "ion-at-surface" [1] was introduced by analogy with "beam-foil spectroscopy" and the term "superexcited" [2] means high excitation of a large number of electrons (in an atom or ion), so that in the limit all electrons of the atom are located near the continuum in levels with large main quantum numbers. The first results on Ar^{17+}-at-surface spectroscopy [3,4] were obtained by use of the electron beam ion source "Krion-2". Some new results of studies with highly charged ions of Kr and Xe are presented below.

II. Experimental Technique

The experimental techniques and procedures were not much different from those in the experiments with Ar^{17+} [3,4]. Highly charged ions, in particular Kr^{35+} and Kr^{36+} were produced in the electrostatic ion trap "Krion-2" [5,6] by energetic electron impact. Usually an electron beam of about 0.1 A current and about 400 A/cm^2 current density was used with energies sufficient to produce the desired ion charge states by successive ionization. Pulses of low-charged ions of the investigated elements were injected into the electron beam and then confined in the beam during variable periods of time (up to 40 s in case of production of $Xe^{53+,54+}$). Then the ions were extracted from the trap along the electron beam during approximately 3 ms and were directed at vertical incidence to the surface of a magnesium or carbon target with energy less than 12 keV/u. Hard X-rays were detected by a semiconductor germanium spectrometer situated behind a 50 μm mylar window of the reaction chamber under 5° to the target surface. The detection efficiency was about 10^{-5}. Calibration of the energy scale of the spectrometer is done with K_αX-rays of Br, Kr, Sr and Nb.

III. Experimental results

The spectra of KX-rays and K_βX-rays of Kr^{35+}-ions recombining at a surface of a carbon target are presented in Fig.1 and Fig.2. The shape of the Kr diagram lines is shown in Fig.1 for comparison. The KX-ray energy spectra of a mixture of Kr^{35+} and Kr^{36+}-ions were measured as well.

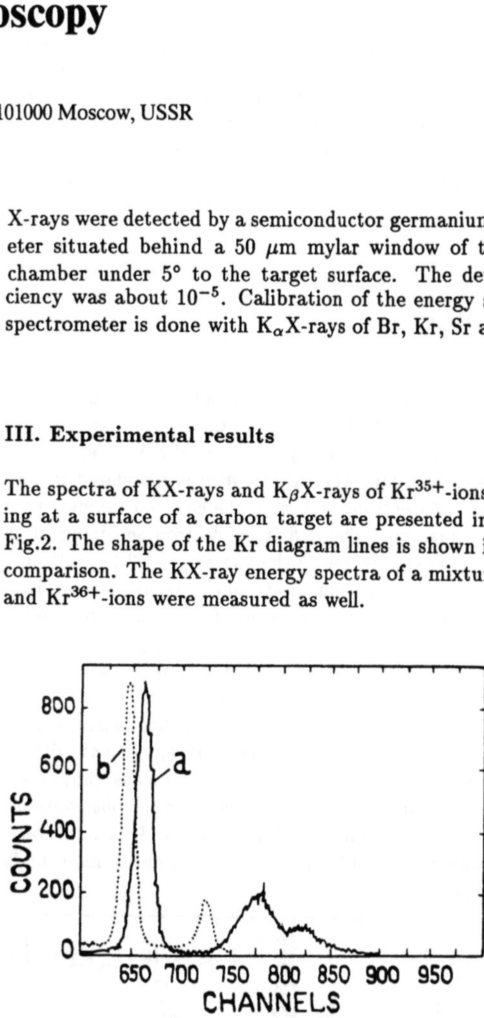

Fig.1. Spectrum of KX-rays of Kr^{35+}-at-surface (a) and spectrum of diagram KX-rays of neutral krypton (b).

For the most probable K_αX-ray energy for Kr^{35+} impact we obtained E_{K_α} = 12922.2±1.4 eV. For K_αX-rays following Kr^{36+} impact two different results in two different experiments were obtained for the most probable energy: E_{K_α} = 13397.1±32.0 eV and E_{K_α} = 13438.2±34.8 eV. In the case of a mixture of Kr^{35+} and Kr^{36+} the most probable K_αX-ray energy for Kr^{35+} was found to be a little less than for pure Kr^{35+}. The high-energy limit of K_βX-rays for Kr^{35+} is determined to be 17250 eV. The relative intensity of K_βX-rays for Kr^{35+} was determined versus full KX-ray intensity: $I_{K_\beta}/(I_{K_\alpha}+I_{K_\beta})$ = 0.39±0.02. The same relation was evaluated for Kr^{35+}: $I_{K_\beta}/(I_{K_\alpha}+I_{K_\beta}) = 0.61\pm0.06$. In the case of highly-charged xenon ions ($45\leq q\leq 52$) LX-rays with the maximum intensity in an energy region about 4.5 to 5.0 keV and an unresolved band of X-ray lines of higher energy up to 9.85 keV were observed. Sometimes sharp LX-ray peaks appeared on an almost smooth spectral distribution. Apart from these features very low intensity high-energy X-rays were observed in an energy region up to 18 keV.

When the high energy electron beam (about 55 keV) was used in the ion source and when a confinement time of xenon ions was chosen equal to about 40 s then very hard X-rays appeared in the spectra. Namely, a sharp line, with a foot on the right, of about 32 keV energy (xenon K_αX-rays) and many lines of higher energy (xenon K_βX-rays) were observed. This was an evidence for the production of hydrogenlike xenon in this experiment and of a small fraction of bare xenon in addition. Although the statistics were poor it was possible to evaluate the relative K_βX-radiation intensity for Xe^{53+}: $I_{K_\beta}/(I_{K_\alpha}+I_{K_\beta}) = 0.52\pm0.06$.

IV. Discussion

1. Kr^{35+}-at-surface.

The most probable value of the K_αX-ray energy (12922.2±1.4 eV) in comparison with the energy of the diagram krypton K_α-line (Fig.1) means that the population of the K-vacancy by an L-shell electron occurs when the L-shell is populated in average only by 2.7 electrons including the electron taking part in the L→K-transition. The deficit of electron screening of the krypton nuclear charge due to the absence of 5.3 L-electrons in average can provide approximately 300 eV energy shift of K_αX-rays. Let us consider the spectrum of K_βX-rays presented in Fig.2, where the theoretical energies of (3→1)-, (4→1)-, (5→1)- and (∞ →1)-transitions in heliumlike krypton [7] and energies of (3→1)- and (4→1)-diagram lines are given. First, the binding energy of 17296 eV of an electron in heliumlike krypton as calculated by Safronova [7] is in good agreement with the high-energy limit of the spectrum. This means that ion-at-surface spectroscopy provides a method for the observation of electron transitions from very high levels and, in addition, it allows measurements of binding energies of electrons in highly charged ions.

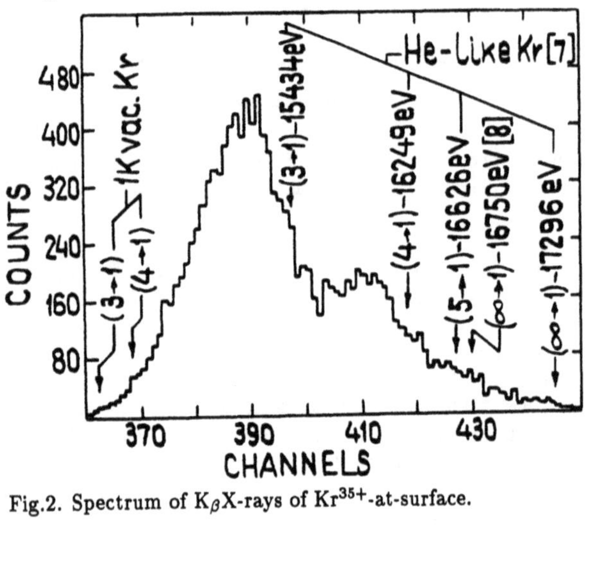

Fig.2. Spectrum of K_βX-rays of Kr^{35+}-at-surface.

The accuracy of these measurements can be estimated from an analysis of the spectral shape near the high-energy limit. One can see that the calculated electron binding energy [8] for Kr^{34+} is considerably less than the high energy limit of the spectrum: it is less than the actual value by about 0.5 keV.

As there is satisfactory agreement of the experimental and theoretical values of the binding energy of electrons in helium-like krypton we may use theoretical values for other transitions [7] to analyze the K_βX-ray spectrum. One can see that the energy spectrum of (3→1)-transitions is very asymmetric. The maximum (3→1)-transition energy in Kr^{34+} is only approximately 200 eV greater than the most probable energy of (3→1)-transitions in the process of neutralization of Kr^{35+}. But the radiation energy extends more than about 1.2 keV to lower energies. The shape of the spectrum of (3→1)-transition radiation can be explained as follows: The rapid growth of the (3→1)-transition probability at energies below 15434 eV is caused by population of the M-shell, electrons of which screen the (3→1)-transition relatively weakly. And then beyond the maximum the probability decreases slowly which can be explained by population of the L-shell. These electrons screen the (3→1)-transitions much stronger and they have greater probability to make a transition into the K-vacancy than do M-shell electrons. Under such conditions the K-vacancy does not live untill full population of the L-shell has occured as the intensity of X-rays in the region of the diagram line energy of (3→1)-transitions is equal to about zero. (By the way the same conclusion can be based on the position and the shape of the K_α-line). Similar conclusions can be drawn about (4→1)-transitions and about transitions from higher levels.

A small decrease of the most probable value of the K_αX-ray energy when the incident Kr^{35+} ion beam is mixed with Kr^{36+} can show that a Kr ion with one K-vacancy arising from Kr^{36+}-at-surface has -in the moment of the population of this vacancy- more L-electrons in average than Kr^{35+}-at-surface.

Thus the analysis of Kr^{35+}-at-surface KX-ray spectra confirms the idea [3,4], that the transformation of a highly charged ion into a neutral atom at a solid surface goes through the formation and decay of superexcited states of the ion (and atom, respectively) so that in the first stage high levels just below the continuum are populated and then more or less sequentially low lying levels are populated. This does not exclude of course direct transitions to the lowest vacancies. Namely, a definite probability of such transitions allows to see a picture of the transformation of highly charged ions through superexcited atoms to usual atoms.

2. Highly charged Xe-at-surface

Unfortunately in the experiments with highly charged xenon ions only mixtures of various charge states could be used. Nevertheless their results allow to make some conclusions. First of all, the higher the highest ion charge state in the ion beam, the higher will be the limit of LX-ray energies; i.e. the limit is equal to the electron binding energy in neon-like xenon for Xe^{45+}-at-surface and reach the electron binding energy in lithiumlike xenon for Xe^{52+}-at-surface. This also confirms that very high lying electron shells close to the continuum are populated by electrons and that this is the first stage of a neutralization of a slow highly charged ion at a surface.

Appearance of sharp peaks in the LX-ray spectra can be a proof of preferable probabilities of definite transitions. It would be very interesting to identify such transitions and to find the reasons for this preference. What we definitely know is that the low intensity and high energy component of LX-rays is not a result of any kind of pulse pile-up.

3. Probability of K_βX-radiation

The data for the relative probability of K_βX-radiation for hydrogenlike argon, krypton and xenon ions-at-surface and for bare krypton-at-surface as well are presented in Fig.3.

One can see that for hydrogenlike ions a significant increase of the probability against Z has been observed. This phenomenon does not contradict a decreasing population of the L-shell in the moment of a transition into the K-vacancy. The phenomenon is rather connected with an increase of transition energy as well as with an increase of the number of electrons in higher shells.

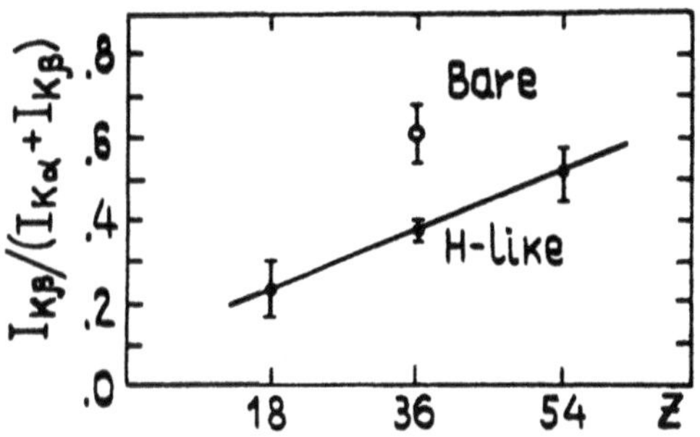

Fig.3 Z-dependence of K_βX-radiation probability.

The significant increase (approximately by a factor of 2) of the K_βX-radiation probability for Kr^{36+} in comparison with Kr^{35+} is caused by an increase of the transition probability for all higher levels as a consequence of the presence of two K-vacancies.

V. Conclusion

The present results confirm that the highly charged ion-at-surface spectroscopy provides possibilities for investigations in a new region of atomic spectroscopy: the spectroscopy of superexcited atoms.

References

1. Donets E.D.: Nucl. Instr. Meth. in Phys.Res. B9, 522 (1985)
2. Donets E.D., Kartashov S.V., Ovsyannikov V.P.: In: "Rapid Communication of JINR", N20-86, Dubna, p.27 (1986).
3. Donets E.D.: Physica Scripta, T3, 11 (1983).
4. Donets E.D., Trifonov V.A., Tutin G.A., Zschornack G., Eismont V.P.: Preprint of JINR, P7-83-627, Dubna (1983).
5. Donets E.D.: IEEE Trans. Nucl. Sci., NS-23, 897 (1976).
6. Donets E.D., Ovsyannikov V.P.: Preprint of JINR, P7-9799, Dubna (1976).
7. Vainshtein M.A., Safronova U.I.: Preprint of IS AN USSR, N2, Troitzk (1985).
8. Carlson et al.: Atomic Data, 2, 63 (1970)

Supplement to Z. Phys. D - Atoms, Molecules and Clusters 21, S 341-342 (1991)

Atoms, Molecules
Zeitschrift für Physik D and Clusters

K x-ray and Auger emission from H-like Ar ions slowly approaching a Ge surface

M. Schulz[1], C. L. Cocke, M. Stöckli, S. Hagmann, and H. Schmidt-Böcking[2]
Department of Physics, J. R. McDonald Lab., Kansas State University, Manhattan, KS 66502, USA
[1] Inst. für Kernphysik, Universität Frankfurt, W-6000 Frankfurt/M., FRG

[2] present address: Dept. of Physics, University of Missouri-Rolla, Rolla, MO 65401, USA

Abstract

We have studied K x-ray and Auger emission from Ar^{17+} ions interacting with a Ge surface. The x-ray yields were studied as a function of the vertical projectile velocity with respect to the surface.

I. Introduction

Highly charged ions have for a long time attracted a good deal of interest for a number of reasons. In a collision process the number of passive electrons which are only spectators for the process of interest is minimized for these ions which makes theoretical calculations and the interpretation of experimental data somewhat easier. Furthermore, in highly charged ions processes leading to a high degree of excitation involving a large number of active electrons can be studied under certain conditions. An example of such a condition is a slow highly charged ion interacting with a surface.[1] At sufficiently low collision velocity electrons are captured with an extremely large cross section predominantly into high n-states and a solid target surface offers a huge source of electrons. This can lead to very exotic ions with a large number of electrons in high n-states and empty inner shells.

Since powerful ion sources were developed a number of studies on highly charged ions interacting with a surface were carried out.[2-6] From these studies the following physical picture of the neutralization and deexcitation process of an ion near a surface has developed: At relatively large distances above the surface (typically a couple of nm) the ion starts drawing electrons from the surface which are captured into high n-states of the ion. The electrons then cascade down to lower n-states predominantly by Auger transitions before they start filling the innermost vacancy. Since the filling of the innermost vacancy is in general preceded by many cascading steps, it is expected to take a relatively long time. Depending on the time the ion spends above the surface the final deexcitation step takes place above the surface or inside the solid.

In this work we studied K x-ray yields as a function of the projectile velocity component perpendicular to the surface v_p in collisions of H- like Ar ions with a Ge target. Furthermore, the spectral shape of K Auger spectra were studied as a function of v_p.

II. Experiment

The experiment was performed at the CRYEBIS at Kansas State University. A 51 keV Ar^{17+} beam was extracted and collided with a single crystal Ge target. v_p was varied by rotating the target by an angle θ with respect to the beam axis. An angular range of 3° (near grazing incidence) to 60° was covered. At very small angles the vertical projectile energies may be strongly effected by the image potential the ions see from the surface. If the image potential is neglected the covered angular range corresponds to vertical energies of 140 eV to 38.2 keV. The x-rays were detected with a Si(Li) detector with an energy resolution of 180 eV at 5.9 keV. The electrons were energy analyzed with a spherical electron spectrometer.

III. Results

In fig. 1 x-ray spectra are shown for angles of 3° (top) and 60° (bottom). The K_α line at around 3.1 keV is resolved from a second line at around 3.5 keV which is mainly due to K_β transitions. Significant differences between these two spectra can be observed. At 3° the total x-ray yield is increased by about 40% compared to 30°. Furthermore, The K_β line in the 3° spectrum is shifted towards higher energies by about 150 eV compared to its counterpart in the 60° spectrum. The corresponding shift in the K_α line is about 30 eV. The centroid energies of the K_α and K_β lines are consistent with an average L-shell population of 3 to 4 electrons at 3° and approximately 5 electrons at 30°.

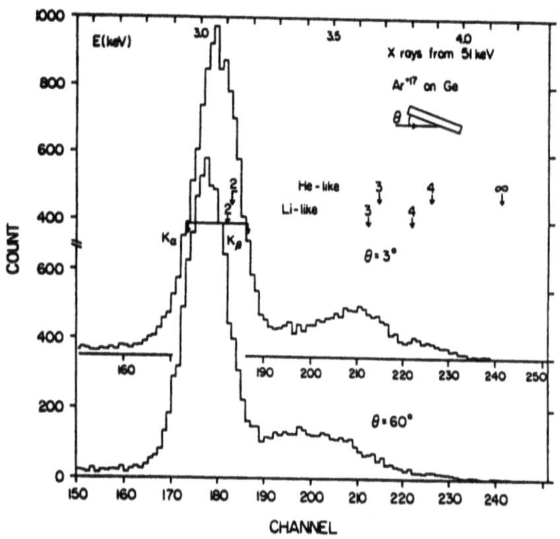

Fig. 1. Ar K x-ray spectra at angles of 3° (top) and 60° (bottom).

A smaller average L-shell population at smaller angles is also indicated by the electron spectra shown in fig. 2 for 3° (top) and 60° (bottom). Here the KLL Auger line at around 2.2 keV is well separated from the KLM line at about 2.4 keV. The energy calibration of the electron spectrometer is not straight forward and the uncertainty in the assigned energy scale is much larger than for the x-ray spectra. However, it can be seen that at 3° the KLL line is shifted towards lower electron energies compared to the 60° spectrum, whereas the KLM line is not shifted at all. In contrast to the x - ray energies, the KLL Auger energies are reduced with decreasing L-shell population, in agreement with our data. The Rydberg series limit of the KLn Auger series occurs at the same energy as the K_α x - ray transition. This means that the dependence of the KLn Auger energy on the L-shell population must turn over at some n, for which the energy would be almost independent on the L-shell population. This may just be the case for n = 3, which explains that the KLM line is not shifted.

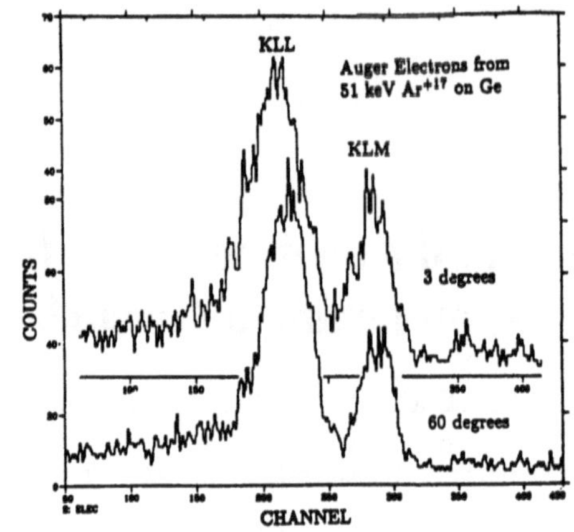

Fig. 2. Ar K Auger spectra at angles of 3° (top) and 60° (bottom).

These results we explain by an increasing fraction of K vacancies being filled above the surface with decreasing θ. Since electrons are more efficiently captured to low n-states inside the solid, this explains the large L-shell population for large θ. The increase in the x-ray yield with decreasing θ is due to a larger fluorescence yield for configurations with a small L-shell population.

IV. Conclusions

We have studied x - ray emission from H- like Ar ions slowly approaching a Ge surface. The data indicate an increasing fraction of x-rays emitted above the surface with decreasing vertical collision velocity. This work was supported by the Division of Chemical Sciences, Office of Basic Energy Sciences, Office of Energy Research, U.S. Department of Energy.

References

[1]H.J. Audrä, Nucl. Instr. Meth. B43, 306 (1989)
[2]E.D. Donets, Nucl. Instr. Meth. B9, 522 (1985)
[3]M. Delaunay, M. Fehringer, R. Geller, D. Hitz, P. Varga, and H. Winter, Phys. Rev. B35, 4232 (1987)
[4]S.T. de Zwart, Nucl. Instr. Meth. B23, 239 (1987)
[5]F.W. Meyer, J. de Phys. 50, C1 263 (1989)
[6]J.P. Briand, L. de Billy, P. Charles, S. Essabaa, P. Briand, R. Geller, J.P. Desclaux, S. Bliman, and C. Ristori, Phys. Rev. Let. 65, 159 (1990)

Supplement to Z. Phys. D - Atoms, Molecules and Clusters 21, S 343-345 (1991)

Atoms, Molecules
Zeitschrift für Physik D and Clusters

High resolution X-ray study of multi charged ions in collisions with surfaces

J. J. Bonnet, A. Fleury, M. Bonnefoy, M. Chassevent
Laboratoire de Physique des Collisions Atomiques et Moléculaires, CNAM 292, rue St. Martin, F-75141 Paris Cedex 03, France

T. Lamy, A. Brenac, A. Simionovici, H. J. Andrä
LAGRIPPA, CENG, BP 85X, F-38041 Grenoble Cedex, France

S. Andriamonje
CEN Bordeaux-Gradignan, Le Haut-Vigneau, F-33175 Gradignan Cedex, France

Abstract

High resolution X-ray spectra of the K_α satellites emitted from Multi Charged Ions (MCI) in collisions with surfaces have been measured as a function of the vertical velocity towards the surface. The results allow for the first time to differentiate between X-rays emitted from ions inside the bulk or ions experiencing a pure surface interaction.

Introduction

When hydrogenic ions of medium Z approach a surface or penetrate into the solid, np → 1s K X-ray satellites with n = 2,3,4... can be observed in competition with KLY-Auger transitions, where Y = L,M,N.. [1]. Donets et al. [2] were the first to observe these X-ray transitions with a Ge detector in the interaction of Ar^{17+} with a beryllium target at 0.42 keV/u. The analysis [3] showed that these X-rays were emitted from Ar atoms with one K-hole, four L-holes, and two or three M-holes. This strong population of the L and M shells of initial Ar^{17+} is in contradiction with the Resonance Neutralization (RN) - Auto-Ionization (AI) - cascade model in front of the surface [1,4,5]. In this model, resonant charge exchange dominates the transfer of electrons from the conduction band of the metal target to high lying Rydberg levels. These levels Auger-cascade down to the L-shell but the probability of another L-shell electron to be present when the Ly_α X-ray radiation is emitted is very small. This model applies well to the Auger emission from Ar^{9+} approaching a surface at very low energy (≈ 1.1 eV/u) [6]. Therefore, it has been suspected that the strong population of the L-shell observed with X-rays from ions of relatively high energy has been due to emission from the bulk. In order to prove this assumption, we present the first high resolution K X-ray measurements of Ne^{9+} emitted from ions approaching a surface at systematically varied velocity, i.e. with variable probability of penetration into the bulk.

Experiment

Ne^{9+} ions from the NEW LAGRIPPA FACILITY [7] have been impinging in a vacuum of 10^{-5} Pa and at energies of E_0 = 140 and 35 keV on Ta and doped Si targets at systematically varied angles of incidence θ_{in} in order to vary the "vertical energy" $E = E_0 \sin^2\theta_{in}$ of the projectile. The X-rays have been observed with a Bragg spectrometer [8] supplied with a flat RbAP crystal (2d = 2.612 nm). The Soller slit angular divergence has been 0.11° to yield an experimental resolution of E/ΔE ≈ 150 around E = 900 eV. The photons have been detected by a conventional flowing gas proportional counter filled with propane at a regulated pressure of 2 10^{-4} Pa and supplied with a 0.4 µm thick Makrofoil window. The spectra have not been corrected for the variation of the crystal reflectivity with the wavelength.

Results

In Fig. 1 the K X-ray spectrum in the wavelength range of 1.1 to 1.5 nm is shown for Ne^{9+} impinging at 35 keV on the highly polished polycrystalline Ta surface at θ_{in} = 50°. It exhibits the K_α satellite group KL^0 to KL^7 ($^1P_1$) with a maximum of intensity corresponding to a completely filled L-shell particularly when considering the monotonic increase of the fluorescence yield from KL^0 (0.013) to KL^7 (≈ 0.37) when taking into account KLM-Auger transitions.
The spectrum obtained for a θ_{in} = 5° is presented in Fig.2.It exhibits a striking modification of the relative intensities of the whole K_α satellite group

KL^0 to KL^7 with a tremendous increase of the KL^7 emission corresponding to a single 2p electron in the L-shell.

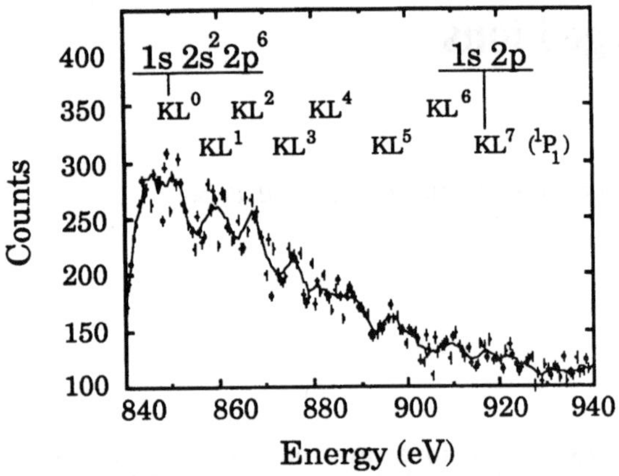

Fig.1. K X-ray spectrum in the wavelength range of 1.3 to 1.5 nm emitted from Ne^{9+} impinging at 35 keV and θ_{in} = 50° on a Ta surface. The full line is a 5 point smoothing to guide the eye. The background is about 120 counts. The initial configurations for the KL^0 and KL^7 transitions are indicated.

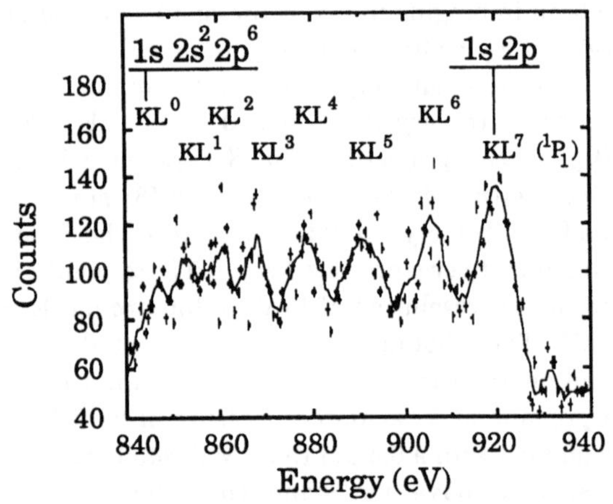

Fig.2. K X-ray spectrum in the wavelength range of 1.3 to 1.5 nm emitted from Ne^{9+} impinging at 35 keV and θ_{in} = 5° on a Ta surface. The full line is a 5 point smoothing to guide the eye. The background is about 50 counts. The initial configurations for the KL^0 and KL^7 transitions are indicated.

Discussion

The spectrum in Fig.1 can be interpreted in terms of an emission during the direct and rapid filling of the L and M shells from inner shells of the target atoms when Ne^{9+} is penetrating into the solid and slowing down inside the bulk at the given experimental conditions. The faster the filling, the stronger the K_α emission with a completely filled L-shell and the smaller the probability to emit with few electrons in the L-shell. The spectrum in Fig.2 has to be interpreted as a statistical blend of different processes. At a 265 eV "vertical energy", the time of approach for Ne^{9+} is sufficient to allow for the RN-AI cascade to supply with a finite probability one or two electrons to the L-shell for KL^7- or KL^6 emission, respectively. Such a "pure" spectrum is, however, not observed due to the large probability of the K-hole to survive up to the surface. There, the ion is either totally reflected or penetrates into the solid. At the distance of closest approach, the reflected ion may directly collect M shell electrons which LMM Auger-cascade very rapidly to the L-shell to yield Ly_α emission from the receding ion with about 1÷3 L-shell electrons, but still in greater favor of the KL^7 satellite, due to its greater fluorescence yield. The penetrating ion retains the same energy E_0 inside the bulk but enters into the solid with M-, L- and K-shell populations which vary as a function of E_0 and of θ_{in} and depend on their probability of survival at the surface boundary.
When this ion traverses the surface, the high rates of direct capture into the M- and L-shells set in and give rise to conditions for the emission of an intermediate spectrum between the one in Fig. 1 and a spectrum stemming from a pure surface interaction. Given a realistic surface with a distribution of θ_{in} around the nominal θ_{in}, one can readily explain the spectrum in Fig. 2 as a superposition of contributions from surface and solid interactions. In addition one has to take into account the emission from ions which penetrate and then re-escape after a short time T_s spent inside the bulk. The appearance of a dominant KL^7 peak is thus an unequivocal proof of a Ne^{9+}-surface interaction. This argument is not valid for the recent observation of the KL^7 line of Ar^{16+} [9] when Ar^{17+} is interacting with a solid. In Ar this line has a fluorescence yield 13 times higher than in Ne so that the KL^7 line shows up even during the rapid filling of the L-shell inside the bulk.

Conclusion

The systematic variation of the "vertical energy" of Ne^{9+} approaching a metal surface combined with a high resolution spectrometer has created experimental conditions which allow to identify for the first time X-rays emitted from ions purely interacting with the surface. All former observations of X-rays emitted from MCIs interacting with solids [9] can now undoubtedly be attributed to an emission from ions inside the bulk and as such are not related to the ion-surface interaction. A further refinement of the present experimental conditions is in preparation to test whether the X-ray spectroscopy of MCIs interacting with solids is a valuable tool for the investigation of the neutralization of MCIs in front of the surface or whether its preferential domain remains the interaction of MCIs with the bulk.

Acknowledgement

We would like to thank G. Lamboley for his ingenious mechanical solutions and for his continuous support. This work has been performed at LAGRIPPA, a joint laboratory of the CNRS and of the CEA.

References

1) H.J. Andrä, Nucl. Instr.and Meth. **B43**, 306 (1989).and references therein.

2) E.D. Donets, Phys.Scr. **T3**, 11 (1983); ibid. Nucl. Instr. and Meth. **B9**, 522 (1985).

3) G. Zschornak, G. Musiol, and W. Wagner, invited talk at the Workshop on "Energiereiche atomare Stöße" in Oberstdorf/FRG, Jan. 1985; and Phys.Scr. **T3**, 194 (1983).

4) H.D.Hagstrum, Phys.Rev.**96**, 325 (1954)

5) U.A.Arifov, L.M.Kishinevskii, E.S.Mukkamadiev, and E.S.Parilis, Zh.Tekh.Fiz. **43**, 181 (1973); Sov.Phys.-Tech.Phys. **18**, 118 (1973).

6) S.T. de Zwart, Thesis, Groningen (1987); ibid. Nucl.Instr. and Meth. **B23**, 239 (1987).

7) T. Lamy, G. Lamboley, D. Hitz, and H.J. Andrä, Rev.Sci.Instr. **61**, 336 (1990).

8) A. Fleury, J. Debernardi, M. Bonnefoy, S. Bliman, J.J. Bonnet, and M. Chassevent, Nucl.Instr.Meth **B14**, 353 (1986).

9) J.P. Briand, L. de Billy, P. Charles, S. Essabaa, P. Briand, R. Geller, J.P. Desclaux, S. Bliman, and C. Ristori, Phys.Rev.Lett. **65**, 159 (1990).

Supplement to Z. Phys. D - Atoms, Molecules and Clusters 21, S 347-348 (1991)

Atoms, Molecules
Zeitschrift für Physik D and Clusters

Two and three electron Auger transitions in collisions of highly-charged ions with surfaces

P. Moretto-Capelle, A. Bordenave-Montesquieu and P. Benoit-Cattin[1], S. Andriamonje[2], H. J. Andrä[3]
[1] Centre de Physique Atomique, Université Paul Sabattier, 118 route de Narbonne, F-31062 Toulouse Cedex, France
[2] Centre d'Etudes Nucléaires de Bordeaux-Gradignan, Le Haut-Vigneau, F-33175 Gradignan Cedex, France
[3] LAGRIPPA, DRF-CEN Grenoble BP 85X, F-38041 Grenoble Cedex, France

Abstract

The Auger electron spectra from Ar^{9+} approaching at 265 eV a Si or metal surface in vacua of 10^{-5} Pa or UHV are identical. Experiments on atomic physics in front of surfaces are thus possible in standard vacuum. N^{7+} approaching a surface at 1000 eV penetrates with great probability into the bulk and gives rise to K^2L^2L double Auger lines, observed for the first time with low energy highly charged ions.

Introduction

Whenever inner shell holes exist in highly charged ions approaching a metal surface at low normal velocities ($v \leq 10^5$ m/s) structures show up in the secondary electron spectra which can be attributed to Auger transitions which fill these holes. Taking Ar^{9+} as an example, these structures correspond to the 3s $nl \rightarrow$ 2pε LMn Auger transitions with n = 3,4,5,6. The well resolved spectra of these transitions in Fig.1 from the excellent work of S. deZwart [1] at ion energies as low as 45 eV, show that these transitions occur in ions in front of the surface.

Experiments

If this is indeed the case, the surface flatness and cleanliness should not play a significant role. Experiments concerning the atomic physics in front of the surface should thus be possible with moderate surface quality in a moderate vacuum.

In order to test this astonishing fact we have performed in a 10^{-5} Pa vacuum a test with Ar^{9+} ions impinging at 45 keV and an angle of incidence $\Phi = 5°$ on a doped Si single crystal surface without any further precautions. The electron spectrum observed in Fig.1 is an exact replica of the results obtained in [1] at the same vertical velocity or "vertical energy" = $E_0 \sin^2\Phi$ of the ion.

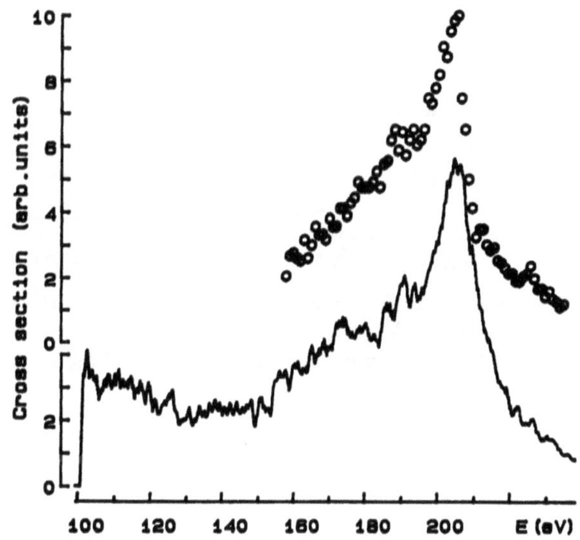

Fig. 1. 3s $nl \rightarrow$ 2pε LMn Auger spectrum of Ar^{9+} interacting with a conducting surface. Full line: Present experiment with conditions described in the text. Circles: From ref.[1].

We interpret this observation as further proof of the fact that these Auger transitions take place in ions in front of the surface and that the surface serves only as a supplier of electrons at binding energies of the conduction band plus or minus effects from adsorbates and surface corrugations. Taking N^{7+} as another example, it is the $2l\ nl' \rightarrow 1s\varepsilon$ K^2Ln hypersatellite and the $1s\ 2l\ nl' \rightarrow 1s^2\varepsilon$ KLn satellite Auger transitions with n = 2,3,4,5 which show up as a clearly resolved spectral feature between 280 and 500 eV. When analyzing more carefully the background above 500 eV another spectral feature shows up between 570 and 950 eV for ions impinging at 35 keV and at 10° on a Si single crystal. We interpret this feature as transitions involving three electrons of the type $2l\ 2l'nl'' \rightarrow 1s^2\varepsilon$ with a variable number (2 ÷ 4) of spectators in the L shell.

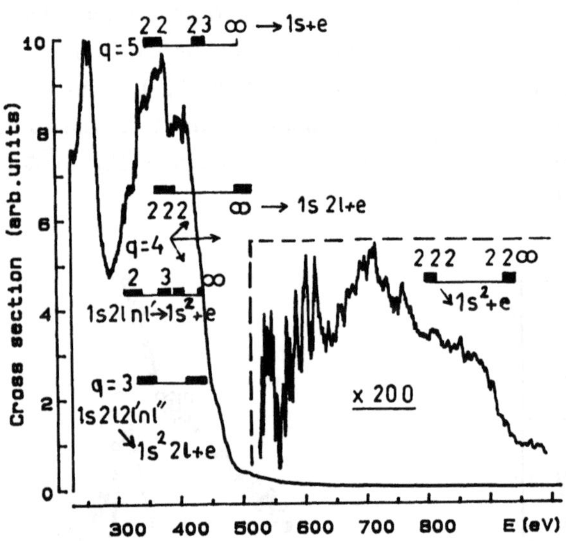

Fig. 2. $2l\ nl' \rightarrow 1s\varepsilon$ $K^2\ Ln$ hypersatellite and the $1s\ 2l\ nl' \rightarrow$ $1s\ 2\varepsilon$ KLn satellite Auger spectrum of N^{7+} interacting with a Si surface. In the insert is shown the appearance of the weak $K^2\ L^2\ L$ double Auger transitions.

While these K^2L^2L transitions have been observed earlier, after production of two K holes in energetic N^+ gas collisions [5], the original character of our experiment is characterized by an initially bare ion, the L and M shells of which have to be filled by its interaction with the solid. This filling has to be fast with respect to the KLL Auger transition rate in order to allow the accumulation of at least three L electrons before the transition takes place. Such conditions are only fulfilled inside the bulk where direct electron capture from inner shells of target atoms into the L and M shells of the projectile is fast enough. The choice of the 1000 eV "vertical energy" and of the Si target has thus been essential in ensuring a strong penetration of the ions into the solid and an efficient electron capture into N^{7+} in order to observe these K^2L^2L transitions.

Acknowledgement

It is a pleasure to thank the LAGRIPPA team composed of T. Lamy, A. Brenac, and G. Lamboley for the excellent beams and for the generous support during the runs. This work has been performed at LAGRIPPA, a joint laboratory of the CNRS and of the CEA in Grenoble, France

References

1) S. deZwart, Thesis, Groningen (1987)
2) Y.K. Ho, Phys.Rev. **A23,** 2137 (1981).
3) U.I. Safranova and V.S. Senashenko, J.Phys. **B11**, 2623 (1978).
4) L.N. Ivanov et al., J.Phys. **B11**, L175 (1978).
5) V.V. Afrosimov et al., Proceedings of the Xth ICPEAC, Paris (1977), p. 184.

Supplement to Z. Phys. D - Atoms, Molecules and Clusters 21, S 349-350 (1991)

Atoms, Molecules
Zeitschrift für Physik D and Clusters

Two electron-one photon X-ray emission in slow collisions between fully stripped ions and solids

S. Andriamonje[1], H. J. Andrä[2], A. Simionovici[2]

[1] CEN Bordeaux-Gradignan and IN2P3, Le Haut-Vigneau, F-33175 Gradignan Cedex, France
[2] LAGRIPPA/DRF-CEN Grenoble BP 85X, F-38041 Cedex, France

Abstract

Among the X-rays emitted in collisions between slow fully stripped ions with solids, a structure well above the limit of the Lyman series is observed. For Ne^{10+} ions this spectral feature has an average energy of 1.8 keV and thus evidently results from the deexcitation of two K shell vacancies via a two electron $2s2p \rightarrow 1s^2$ transition with the emission of a single photon.

Introduction

Two electron - one photon transitions are known to take place in energetic heavy ion - solid collisions [1-3] where two K-holes are created in the projectile via ionization while its L-shell is partially conserved. Among other dominant decay modes, two-electron $2s^n\,2p^m \rightarrow 1s^2\,2s^{n-1}\,2p^{m-1}$ dipole transitions under emission of a single photon may then occur with probabilities on the order of 10^{-3} at intermediate Z with respect to the Ly_α hypersatellite transitions [4]. These double K_α or $K_{\alpha\alpha}$ photons have energies well above the limit of the Lyman series, in a region where minute intensities have been observed recently in collisions between slow (< 10 keV/u) fully stripped ions with solids [5].

Experiment

Using the CAPRICE-ECR source at the NEW LAGRIPPA FACILITY in Grenoble [6], we have performed a systematic study of this spectral feature above the limit of the Lyman series.
Ne^{10+} ions have impinged on various solid targets (C, Si, Ta) at variable angles of incidence to yield "vertical energies" $E = E_0 \sin^2\theta_{in}$ ranging from 40 eV up to 4500 eV/u. The X-rays emitted in the collision have been detected by a Si(Li) detector placed at 90° with respect to the beam axis. The variation of the intensity of this spectral structure with respect to the Ly_α satellite intensity has been measured as a function of θ_{in} as shown in Fig.1, of different energies E_0, and of the thickness of aluminized mylar absorbers placed in front of the Si(Li) detector. The results clearly show that this feature can neither result from pile-up nor from two photons detected simultaneously in the detector.

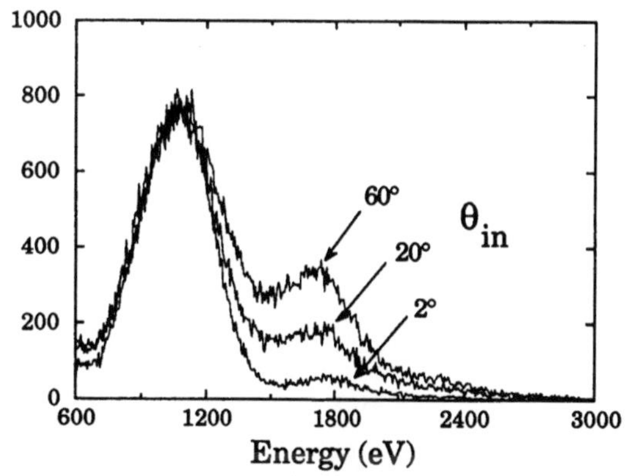

Fig. 1. X-ray spectra observed through 10 μm Mylar (25 μm Be window) from Ne^{10+} interacting at 140 keV and three angles of incidence with Ta. The spectra are normalized to the maximum of the residual Ly_α satellite peak.

Discussion

For Ne^{10+} ions, this spectral feature has an average energy of 1.8 keV. It is thus well placed between twice the energy of K_α and the sum of Ly_α (H-like) plus Ly_α (He-like). Its intensity increases significantly when the probability of penetration into the solid is increased by increasing θ_{in}. This is consistent with our original experimental situation for the observation of such $K_{\alpha\alpha}$ transitions : instead of ionizing the K shell, we start up with an initially bare ion, the L, M shells of which have to be filled by the interaction with the solid. This filling has to be fast with respect to the rates of the various L shell decay modes in order to allow the accumulation of at least two L electrons. Such conditions are only fulfilled inside the bulk where direct electron capture from inner shells of the target atoms into the L and M shells of the projectile is significant. The strong penetration of the ions into the solid has thus been essential for an efficient electron capture into Ne^{10+} in order to observe these two electron - one photon $2s^n\,2p^m \rightarrow 1s^2\,2s^{n-1}\,2p^{m-1}$ dipole transitions .

Acknowledgement

We would like to thank the LAGRIPPA team : T. Lamy, A. Brenac, and G. Lamboley for the excellent beams and for the continuous support. This work has been performed at LAGRIPPA, a joint laboratory of the CNRS and of the CEA in Grenoble, France.

References

1) W. Wölfli et al., Phys. Rev. Lett. **35** (1975) 656.
2) J.S. Greenberg, P. Vincent, and W. Lichten, Phys.Rev. **A16**, 964 (1977).
3) J. Auerhammer, H. Genz, A. Kumar, and A. Richter, Phys.Rev. **A38**, 688 (1988), and references therein.
4) T. Åberg, K.A. Jamison, and P. Richard, Phys.Rev.Lett. **37**, 63 (1976).
5) S. Andriamonje et al., 3th ECAMP Bordeaux 1989.
6) T. Lamy, G. Lamboley, D. Hitz, and H.J. Andrä, Rev.Sci.Instr. **61**, 336 (1990).

Supplement to Z. Phys. D - Atoms, Molecules and Clusters 21, S 351-352 (1991)

Atoms, Molecules
Zeitschrift für Physik D and Clusters

Neutralization of H-like ions at a W-surface: via Auger cascades?

L. Folkerts and R. Morgenstern

Kernfysisch Versneller Instituut, University of Groningen, Zernikelaan 25, 9747 AA Groningen, The Netherlands

Abstract

We present the electron spectra for O^{7+} on a polycrystalline tungsten surface at collision energies between 100 eV and 100 keV. We find clearly resolved structures from projectile KLL- and LMM Auger electrons. The relative intensities of these peaks show that the $n=2$ level is partly populated by 'side-feeding' processes. The time dependence of the KLL peak and the LMM peak can be described by a single exponential.

I. Introduction

An important neutralization and deexcitation mechanism in the interaction between highly charged ions and a metal surface is resonant neutralization (RN) into high Rydberg levels followed by a cascade of autoionization (AI) processes filling lower levels. With high resolution electron spectroscopy it should be possible to observe the corresponding Auger electrons of several cascade steps at the appropriate energies and with the appropriate intensities.

Until recently only those Auger electrons could be identified which correspond to the 'last' step in the cascade, leading to the filling of the innermost vacancy, and the ejection of a high energy electron [1,2,3]. However in our electron energy spectra, resulting from collisions of slow hydrogen-like ions (C^{5+}, N^{6+}, O^{7+}, Ne^{9+}) on a tungsten surface, we now succeeded also to identify electrons from the second last step, i.e. LMM Auger electrons. This allows conclusions about the validity of the cascade picture [4].

II. Experiment

The experiments were performed with multiply charged ion beams from the KVI-ECR ion source, colliding on a polycrystalline tungsten surface. The resulting electron energy distributions were measured with a rotatable 180° spherical electrostatic analyzer (ESA). The minimum electron energy for reliable measurements is about 25 eV due to residual magnetic fields.

A detailed description of the experimental setup is given elsewhere [1].

III. Results

Figure 1 shows the secondary electron energy spectra resulting from O^{7+} colliding on the W-surface with several collision energies. These spectra are characteristic for H-like ions on a metal surface.

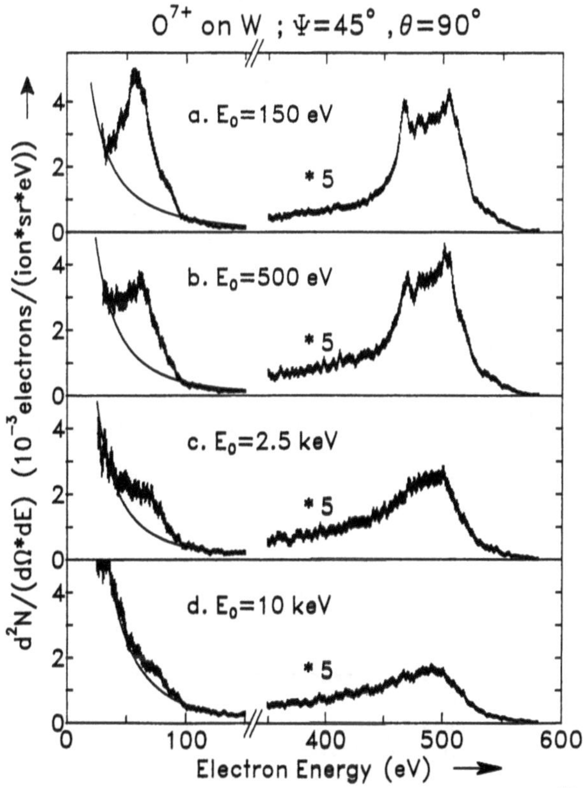

Fig.1. Secondary electron spectra for O^{7+} colliding on tungsten for several projectile energies. The spectral shape of the background fitted to the low energy peaks is given by eq.(2).

The main features of the spectra are two distinguished peak structures at low collision velocity, one at the low energy side and the other at the high energy side of the spectra.

Both peak intensities decrease with increasing collision velocity. Both peaks show the appropriate Doppler shift in energy with varying detection angle, indicating that the electrons are ejected from the projectile, while it is moving on the way in to the surface [4]. The high energy electrons are only present if the primary ions have a *1s*-vacancy, but the low energy peak also shows up in the electron spectra resulting from lower charged ions [4].
In accordance with these features and from the precise energy position and width's of the respective peaks, the high energy electrons are ascribed to KLL Auger processes from neutralized projectiles with the captured electrons in the *n=2* shell. The low energy electrons are ascribed to LMM Auger processes, also from neutral projectiles, with the captured electrons divided over the *n=3* and the *n=2* levels [4].
The LMM Auger peaks are superimposed on a continuous background which arises from other processes. We approximate the spectral shape by

$$y(E) = A\ E^{-1}\ \exp(-BE) \qquad (1)$$

where E is the electron energy, and A and B are constants which can be fitted to the background of the spectra (see figure 1). In this way is it possible to obtain an estimate of the LMM Auger electron intensities.

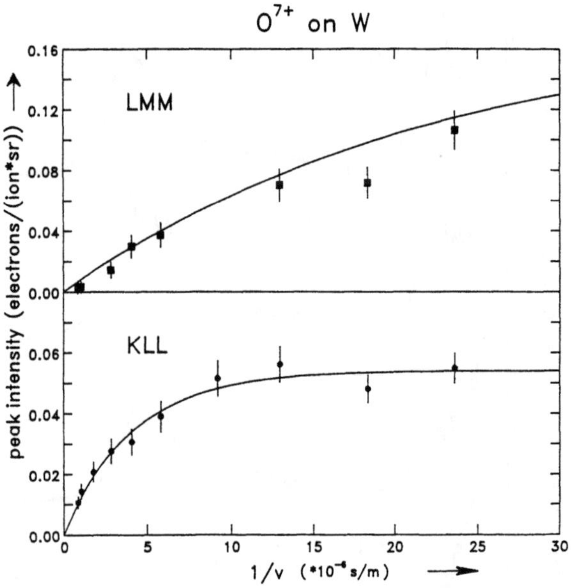

Fig.2. The peak intensities of the KLL and the LMM Auger electrons plotted against 1/v. The used energy range for the KLL Auger peak is 420 - 570 eV. The solid curves are fitted to the data points according to eq.(2).

Figure 2 shows the KLL-Auger and the LMM-Auger intensities plotted against $1/v$ (with v being the ion velocity, $1/v$ is a measure for the total interaction time). The time dependence of the KLL intensity is well fitted by a single exponential

$$I = I_{max}\ (1 - \exp(-\vartheta/v)) \qquad (2)$$

with a maximum of $I_{max} = 0.054$ KLL electrons per ion per sr, and a characteristic velocity $\vartheta = 2.4 \cdot 10^{5}$ m/s. The LMM intensity is also fitted by eq.(2), resulting in a significantly smaller characteristic velocity $\vartheta = 4.4 \cdot 10^{-4}$ m/s. The ratio of LMM to KLL intensities increases with increasing $1/v$ and has a maximum of 2 at low collision energy.

IV. Discussion

The absolute number of KLL electrons per ion can be estimated by integration over the solid angle, yielding $4\pi \cdot I_{max} = 0.6$ el/ion, which is quite close to the possible maximum of 1 el/ion (one *(1s)*-vacancy per ion).
The ratio of LMM to KLL intensities represents an upper limit to the number of LMM processes per ion and thus to the number of electrons entering the *n=2* level via such an LMM process. This ratio has a maximum of 2 LMM electrons per KLL electron at low collision energy (high $1/v$). However the position of the KLL-peaks indicate that a much larger number of electrons (most likely 7 L-electrons) is present in the *n=2* level when the KLL transition occurs. Therefore the LMM processes can only to a minor fraction contribute to the filling of the L-shell. Our conclusion is that the L-shell is partly filled by other processes, e.g. direct transitions from discrete target levels to the *n=2* level in the projectile, which we will refer to as 'side-feeding'.
The slow rise of the LMM intensity with decreasing velocity (increasing interaction time) shows that the characteristic time for the LMM Auger transitions is significantly larger than for the KLL Auger transitions. Assuming that the (one electron) 'side-feeding' processes are relatively fast we can estimate the KLL transition time by the characteristic velocity ϑ from eq.(2) and some typical ion-surface distance d at which the process start. From [5] we find d = 11 Å, which results in $\tau_{(KLL)} = 4.6 \cdot 10^{-15}$ s. For the LMM transitions this results in a characteristic time $\tau_{(LMM)} = 2.5 \cdot 10^{-14}$ s.

***

This work is part of the research program of the "Stichting voor Fundamenteel Onderzoek der Materie" (FOM) with financial support from the "Nederlandse Organisatie voor Wetenschappelijk Onderzoek" (NWO).

References

1. de Zwart, S.T., Drentje, A.G., Boers, A.L., Morgenstern, R.: Surf. Sci. **217**, 298 (1989)
2. Andrä, H.J.: Nucl. Instrum. and Meth. B **43**, 306 (1989)
3. Delaunay, M., Fehringer, M., Geller, R., Varga P., Winter, H.: Europhys. Lett. **4**, 377 (1987)
4. Folkerts, L., Morgenstern, R.: Europhys. Lett. in press
5. Snowdon, K.J.:Nucl. Instr. and Meth. B **34** 309 (1988)

Supplement to Z. Phys. D - Atoms, Molecules and Clusters 21, S 353-355 (1991)

Atoms, Molecules
Zeitschrift für Physik D and Clusters

Adiabatic orbitals of the ion-image-charge system in ion-metal-surface scattering

U. Wille
Bereich Schwerionenphysik, Hahn-Meitner-Institut Berlin, Postfach 39 01 28, W-1000 Berlin 39, FRG

Abstract

The electronic motion in the two-center potential generated by the core of a slow, highly charged ion and and by its (classical) image charge induced in a metal surface is studied. Adiabatic orbitals in this potential are calculated as a function of the distance of the ion from the surface. Critical distances at which the orbital-energy curves merge in the continuum are evaluated. A model is proposed for treating screening effects due to gradual neutralization of an ion approaching a metal surface.

I. Introduction

The interaction of atoms and ions with metal surfaces constitutes a subject of current interest [1]. The recent development of this field has been stimulated by the availability of novel ion sources providing intense beams of slow, highly charged ions [2,3], which allow the ion-surface interaction to be studied under near-adiabatic conditions [4].

The electronic potential in the ion-metal system reads

$$V = V_m + V_c + V_i \,, \qquad (1)$$

where V_m is the potential felt by the metal electrons in the conduction band, V_c is the potential of the ion core, and V_i is the sum of the potentials of the image charges induced in the metal by the ion core and by the valence electrons. The metal potential V_m is usually taken in the "jellium" approximation [5]. For not too small distances D of the ion from the surface, the image potential V_i can be approximated by its classical form [6]. The potential (1) exhibits a barrier between the metal surface and the ion core. For a step-like jellium potential and a Coulombic core potential with effective charge Z_c (the image potential of the valence electrons is disregarded here), the height of this barrier is independent of D, and its width at the Fermi level of the metal is

$$z_0 = D_0 \left\{ [1 + (D/D_0)^2]^{1/2} - 1 \right\} \qquad (2)$$

(unless stated otherwise we use atomic units), where $D_0 = Z_c/\Phi$, and Φ is the work function of the metal.

The present paper is devoted to a theoretical study of the influence which the image charge induced by the core of a slow, highly charged ion has on electronic processes in ion–metal-surface scattering. While the image-charge concept has been used since long in qualitative discussions, its fully quantitative elaboration has not been undertaken yet. Direct experimental information on image-charge effects is scarce. Recently, however, evidence has been presented [7] that even after complete neutralization of initially highly charged ions the image charge of the (incompletely screened) nuclear charge has an observable effect.

In our approach, the electronic motion in the combined field of the (positively charged) *ion core* and its (negative) *classical image charge* is viewed, for fixed position of the ion (and, accordingly, of its image charge), as an adiabatic two-center problem similar to the problem for two positively charged centers encountered in molecular physics and in the theory of slow ion-atom collisions [8]. By solving the adiabatic ion–image-charge problem exactly, we evaluate the distorting effect of the image charge on the ionic states in infinitely high order. In order to deal with the change in the ionic charge state and the corresponding change in screening due to gradual neutralization in a scattering event, we develop a simple model.

II. Theory

We assume the core potential to have the general form

$$V_c = -\frac{Z}{r}\, \chi(r; Z, q, \alpha) \,, \qquad (3)$$

where χ is some suitable screening function [8] depending on the nuclear charge number Z, the ionic charge state q, and a set of screening parameters denoted collectively by α. In the Coulombic approximation $\chi \equiv 1$, we write V_c as

$$V_c = -\frac{Z_c}{r} \,, \qquad (4)$$

with a q-dependent effective charge Z_c. The classical image potential V_i is related to the core potential V_c simply by

$$V_i(r) = -V_c(r) \,. \qquad (5)$$

Outside of the metal, the electronic Hamiltonian of the ion–image-charge system then reads

$$H = -\frac{1}{2}\nabla^2 + V_c(r_1; D) + V_i(r_2; D)$$
$$\equiv -\frac{1}{2}\nabla^2 - Z\left\{\frac{\chi(r_1; D)}{r_1} - \frac{\chi(r_2; D)}{r_2}\right\}, \quad (6)$$

where r_1 and r_2 are the distances of the electron from the nucleus of the ion and from the "image nucleus", respectively. The screening function χ as well as the effective core charge Z_c are now assumed to depend on the ion-surface distance D via a D-dependence of the parameters q and α:

$$\chi(r; D) \equiv \chi\{r; q(D), \alpha(D)\}, \quad (7)$$
$$Z_c(D) \equiv Z_c\{q(D)\}. \quad (8)$$

Note that the orbital angular momentum projection m along the surface normal is a good quantum number of the Hamiltonian (6), while parity is "maximally broken".

In order to solve the Schrödinger equation with the two-center Hamiltonian (6) at fixed distance D, we can exploit the formal similiarity between this Hamiltonian and the effective Hamiltonian frequently used [8] in the theory of slow ion-atom collisions to calculate quasimolecular electronic states at fixed internuclear separation. The common feature of these Hamiltonians is that they are in general *non-separable*, and the difference in sign of the charges attached to the two centers is formally of no significance. So we can readily solve the ion–image-charge problem by applying diagonalization methods developed to solve the quasimolecular problem [8].

In the purely Coulombic approximation $\chi \equiv 1$, the Hamiltonian

$$H = -\frac{1}{2}\nabla^2 - Z_c\left(\frac{1}{r_1} - \frac{1}{r_2}\right) \quad (9)$$

is separable in prolate spheroidal coordinates [9], and the associated Schrödinger equation can be solved directly by the same methods as the H_2^+-like molecular problem. Physically, the Hamiltonian (9) describes the motion of an electron in the field of a finite electric dipole with moment Z_cR, where $R = 2D$ for the ion–image-charge problem. The solution of the finite-dipole problem is known since long and has been applied in a variety of contexts [9,10].

III. The Ion–Image-Charge System in Coulombic Approximation

In Fig. 1, a selection of orbital-energy curves $\epsilon(D)$ corresponding to the Hamiltonian (9) is shown in a scaled plot $\epsilon(D)/Z_c^2$ vs. Z_cD. The curves are labelled by the set of quantum numbers $(n_\lambda, n_\mu, |m|)$, where n_λ and n_μ refer to the prolate spheroidal coordinates λ and μ, respectively [9], and $n_\lambda + n_\mu + |m| + 1 = n$, with n the principal quantum number of the unperturbed ionic state in the limit $D \to \infty$ (note the degeneracy of the orbitals with respect to the sign of m).

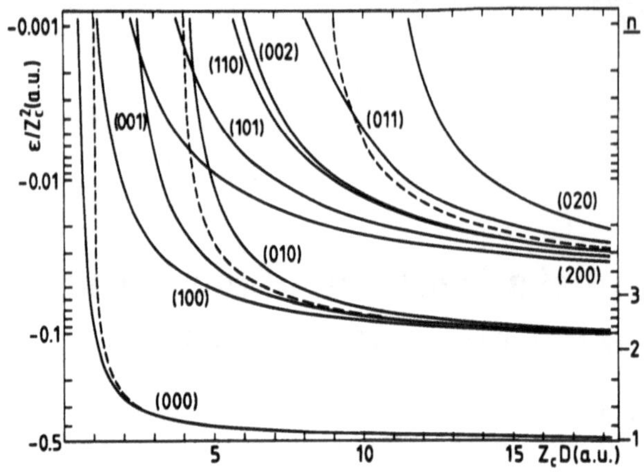

Fig. 1. Scaled adiabatic energy curves for the Coulombic ion–image-charge system. The broken curves are obtained by adding the asymptotically leading correction $Z_c/2D$ to the hydrogenic energies

While the monotonic increase of the energies with decreasing D is plausible, their behavior at small distances is most conveniently characterized by the "critical" distances D_{cr} at which the binding energy becomes zero, i.e., at which the orbitals merge in the continuum. These distances can be evaluated by applying the method of [11]. In Fig. 2, we show scaled D_{cr}-values as a function of m for various values of n_μ. The critical distance increases monotonically both with m (at fixed n_μ) and with n_μ (at fixed m), but is independent of n_λ.

The qualitative features exhibited by Figs. 1 and 2 allow a detailed discussion of resonance neutralization processes [4], in which conduction band electrons at the Fermi level are transferred into high-lying ionic Rydberg states. Here, we restrict ourselves to pointing out the effect which the image charge has on the density of the final ionic states

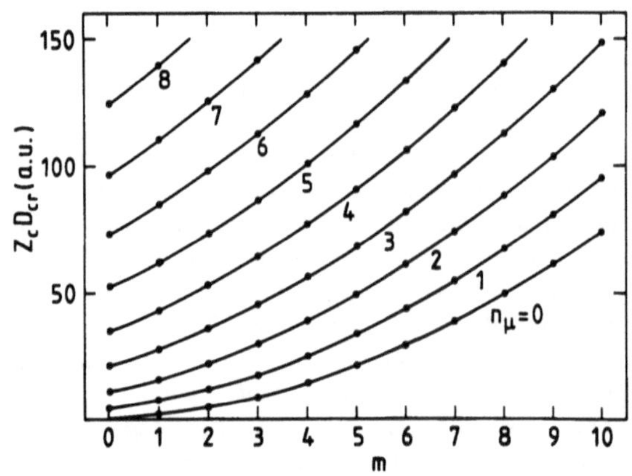

Fig. 2. Scaled critical distances D_{cr} as a function of orbital angular momentum projection m

available for the transfer processes. From Fig. 2, in particular, a lowering of the density of states with high m-values is anticipated. A more detailed discussion of this point will be presented elsewhere.

IV. Screening in the Ion–Image-Charge System

We now consider screening effects due to gradual neutralization of the incoming ion in an ion–metal-surface scattering event. In order to incorporate such effects into the calculation of the adiabatic orbitals of the ion–image-charge system, we parametrize the D-dependence of the charge state q and of the screening parameters α by smoothly interpolating their values between $D = \infty$ and some finite distance D_n.

For the charge state q, we assume that it decreases, on the incoming branch of the scattering process, monotonically from its initial value q_∞ at $D = \infty$ to a value q_n at $D = D_n$ (in specific cases, the value of D_n can be identified with the "neutralization distance" at which $q_n = 0$). The simplest function $q(D)$ which fulfils these conditions and has zero derivative at $D = 0$ and $D = \infty$ reads

$$q(D) = \begin{cases} (q_n + q_\infty x^2)/(1 + x^2) & , x \geq 0 \\ q_n & , x < 0 \end{cases} \tag{10}$$

where $x = (D - D_n)/\Delta$. The parameter Δ determines the center and the width of that D-range where the charge state changes most rapidly. Some experimental information on the neutralization process is required in order to assign meaningful values to Δ and D_n.

While a reliable parametrization of the D-dependence of the charge state q seems feasible, the D-dependence of the screening parameters α is very difficult to construct. The latter parameters depend on D not only via q, but also via the state of excitation of the ion. The state of excitation depends on the dynamics of the scattering process and accordingly is extremely hard to simulate in phenomenological screening parameters. The detailed investigation of this issue will be the subject of future work.

In order to give a simple illustration of our method, we consider a *neutral* Ar atom in its ground state. By diagonalizing the Hamiltonian (6) with a suitable screening function [8], we obtain the D-dependence of the image-charge induced level shift $\Delta\epsilon$ (i.e., the difference of the two-center energies and the unperturbed atomic energies) displayed in Fig. 3 for the occupied Ar orbitals. At distances smaller than about 2 a.u., the shifts are seen to acquire values of order 1 a.u. and larger. At larger distances, the calculated shift is consistent with the analysis of Ar LMM Auger electron spectra [7], which gives an image-charge induced shift of $\sim$ 2 eV in a neutral Ar atom at $D \simeq 9$ a.u. In any case, our result emphasizes the importance of taking into account image-charge corrections in the analysis of Auger electron emission following ion–metal-surface interactions.

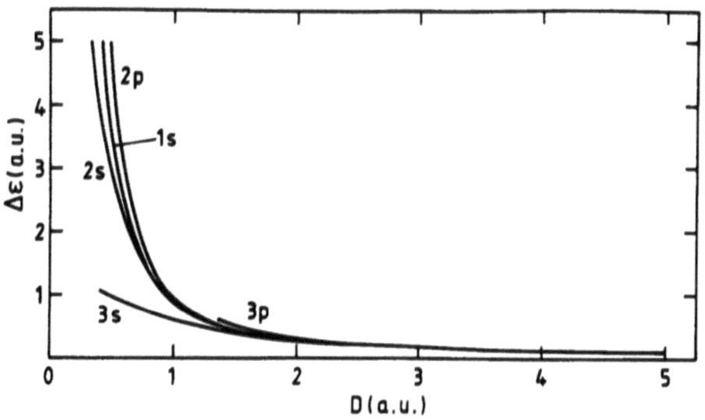

Fig. 3. Image-charge induced shift of the levels of the neutral Ar atom in its ground state

V. Summary and Conclusions

We have considered in this paper the electronic motion in the combined field of a highly charged ion core and its classical image charge induced in a metal surface when the ion is scattered by the surface. Adiabatic orbitals of the ion–image-charge system have been calculated for both the unscreened and the screened case. These orbitals may be of use in detailed investigations of the dynamics of resonance neutralization processes as well as in the analysis of decay processes in partly or fully neutralized ions. Further refinements of our model should include effects of the image potential of the valence electrons and an explicit treatment of the coupling of the image-charge distorted orbitals to the metal potential.

References

1. Andrä, H.J.: In: Fundamental processes of atomic dynamics. Briggs, J.S., Kleinpoppen, H., Lutz, H.O. (eds.), p. 631. New York, London: Plenum Press 1988
2. Proceedings of the Conference on the Physics of Multiply Charged Ions (Groningen, 1986). Morgenstern, R., Niehaus, A., de Heer, F.J., Drentje, A.G. (eds.). Published in: Nucl. Instr. and Meth. **B23**, no. 1,2 (1987)
3. Proceedings of the International Conference on the Physics of Multiply Charged Ions and International Workshop on E.C.R. Ion Sources (Grenoble, 1988). Bliman, S. (ed.). Published in: J. de Physique **50**, Colloque C1 (1989)
4. Varga, P.: Comments At. Mol. Phys. **23**, 111 (1989)
5. Zangwill, A.: Physics at surfaces, Chap. 4. Cambridge: Cambridge University Press 1988
6. Lang, N.D., Kohn, W.: Phys. Rev. B **7**, 3541 (1973)
7. Folkerts, L., Morgenstern, R.: In: Ref. 3, p. 541
8. Wille, U., Hippler, R.: Phys. Rep. **132**, 129 (1986)
9. Wallis, R.F., Herman, R., Milnes, H.W.: J. Molec. Spectrosc. **4** 51 (1960)
10. Turner, J.E.: Am. J. Phys. **45**, 758 (1977)
11. Coulson, C.A., Walmsley, M.: Proc. Phys. Soc. **91**, 31 (1967)

Supplement to Z. Phys. D - Atoms, Molecules and Clusters 21, S 357-358 (1991)

Atoms, Molecules
Zeitschrift für Physik D and Clusters

The average equilibrium charge-states of heavy ions with Z > 60 stripped in He and H_2

Yu. Ts. Oganessian, Yu. V. Lobanov, A. G. Popeko, F. Sh. Abdullin, Yu. P. Kharitonov, A. A. Ledovskoy and Yu. S. Tsyganov
Laboratory of Nuclear Reactions, Joint Institute for Nuclear Research, 141980 Dubna, USSR

Abstract

The equilibrium charges of heavy ions (61< Z < 101) with energies from 5 to 100 MeV stripped in He and H_2 have been measured. New empirical formulae for the average charge state are presented.

I. Introduction

Our interest in the determination of the average charge states of different heavy ions moving through gases is associated with the development of a gas-filled magnetic separator for the study of fusion reactions products [1]. Such a device separates particles according to their average magnetic rigidity, which depends on the momentum and the average charge of the ion. In recent years, similar devices have been designed in many laboratories [2,3,10].

II. Experiment

The separator of the Laboratory of Nuclear Reactions is installed on a cyclotron U-400 beam. It consists of a 23° dipole magnet (max. flux 1.5 T, radius 1.8 m, gap 60 mm) and a quadrupole doublet (max. - field gradient 12 T/m, yoke length 370 mm, aperture 100 mm). The target is placed in front of the dipole magnet.

The detection of ions in the focal plane of the separator is performed using an array of six surface-barrier detectors.

To produce heavy ions of different energies we have used the reactions of elastic target atoms· knock-out in interactions with projectiles, e.g.,Au + Ar, Cf + Ne..; complete-fusion reactions of a projectile and target nucleus, e.g., Pt + C → Po, E_{Po}=4 MeV ... Ba + Ni → Po, E_{Po}=66MeV. The ion energies were calculated according to kinematic characteristics of the reactions. We have used also the slowing down of ions in thin carbon, aluminum or gold - foils. For energy calculations we used tables [4].

The gas enters the target chamber and is pumped off at the detector position.

The movement of the ions through helium and hydrogen at a gas pressure of 1 Torr was studied in detail. In fig.1 the ions, for which the measurements have been performed, are marked.

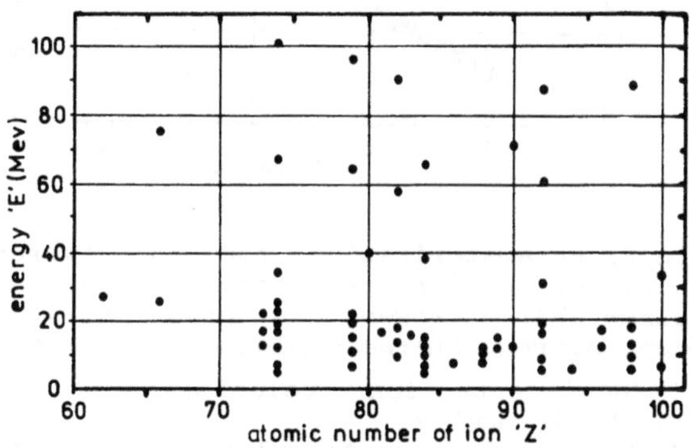

Fig.1. The ions with atomic numbers Z and energy E(MeV) for which we have measured the average charge.

III. Results

Several authors (see, e.g., refs. [5-11]) have obtained theoretical and empirical formulae for rather accurate calculations of the average charges of heavy ions passing through a dilute gas. A fit to our experimental data shows, that with accuracy of about 5% the average charge can be parameterized as follows: if $v*Z^{1/3}<2*10^7$:

$$\bar{q}=(1.8E-7)*v*Z^{1/3}+ 1.65 \text{ for helium,}$$

$$\bar{q}=(2.6E-7)*v*Z^{1/3}- 0.28 \text{ for hydrogen;}$$

if $v*Z^{1/3}>2*10^7$:

$$\bar{q}=(3.3E-7)*v*Z^{1/3}- 1.18 \text{ for helium,}$$

$$\bar{q}=(4.4E-7)*v*Z^{1/3}- 3.92 \text{ for hydrogen,}$$

where v(m/sec) is the ion velocity, Z is the ion atomic number.

Our experimental data are shown in fig.2a for helium and in fig.2b for hydrogen. Our parameterization allows one to get more confidence in the extrapolation of the $\bar{q}(v,Z)$- dependence beyond the known range.

Practically for all the complete-fusion reactions we need the interpolation of known data.

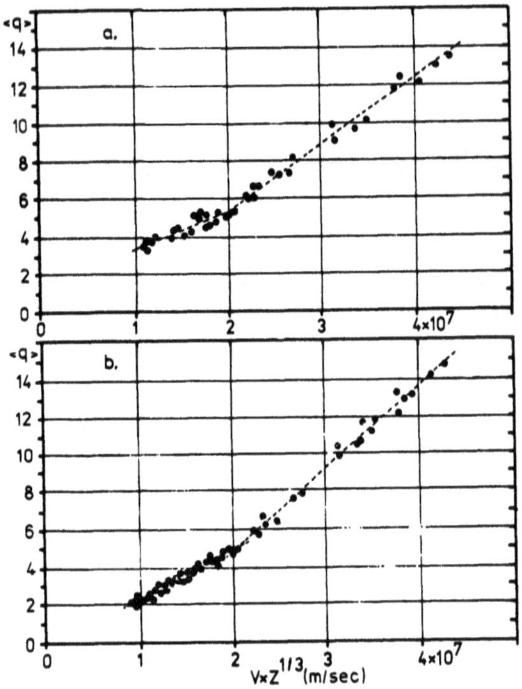

Fig.2. Parameterization for the average charge of the ion after traversal of a): helium, b): hydrogen. Gas pressure is 1 Torr. Average charge $\bar{q}$ plotted versus $v*Z^{1/3}$, v(m/sec) is the ion velocity, Z is the ion atomic number.

We have tested our formulae, e.g. for fermium-248 produced in the $^{235}U(^{18}O,5n)$ reaction.

The predicted average charge in hydrogen is $\bar{q}$=2.35 and the measured value is $\bar{q}$=2.29. We plan to investigate shell effects in the dependence of q more thoroughly.

We wish to thank the U-400 cyclotron crew for the excellent machine operation.

References

1. Yu.Ts.Oganessian, Yu.V.Lobanov, A.G.Popeko, J.Rigol in Abstr.of Int.School-Seminar on Heavy Ion Physics, Dubna, October 1989, Dubna, JINR D7-89-531,p.145.
2. G.Berthes, G.Muenzenberg, B.Langenbeck, H.Kiesewetter and P.Armbruster, GSI Scientific Report 1985, GSI 86-1, p.279.
3. H.Miyatake, T.Nomura, H.Kawakami, J.Tanaka, M.Oyaizu, K.Morita, T.Shinozuka, H.Kudo, K.Sueki and J.Iwata, Nucl. Instr.& Meth. B26,309(1987).
4. L.C.Northcliffe, R.F.Schilling, Nucl. Data Tables, A7,233(1970).
5. N.Bohr, Phys.Rev. 59,270(1941).
6. V.Nikolaev, I.Dmitriev, Phys.Lett. 28A,277(1968).
7. G.Sauter, S.Bloom, Phys.Rev, 6B,699(1972).
8. H.D.Betz, Rev.Mod.Phys. 44,465(1972).
9. J.F.Ziegler, Appl.Phys.Lett. 31,544(1977).
10. A.Ghiorso, S.Yashita, M.E.Leino, L.Frank, J.Kalnins, P.Armbruster, J.-P.Dufour and P.K.Lemmertz, Nucl.Instr.& Meth. A269,192(1988).
11. N.K.Skobelev, Phys. of Elementary Partic. and Atomic. Nucl., V.20, No.6, p.1439 (1989), Moscow, 1989.

Supplement to Z. Phys. D – Atoms, Molecules and Clusters 21, S 359–360 (1991)

Atoms, Molecules
Zeitschrift für Physik D and Clusters

The advanced light source – A new tool for research in atomic physics*

A. S. Schlachter

Advanced Light Source, Accelerator and Fusion Research Division, Lawrence Berkeley Laboratory, 1 Cyclotron Road, Berkeley, CA 94720, USA

* This work was supported by the Director, Office of Energy Research, Office of Basic Energy Sciences, Materials Sciences Division of the U. S. Department of Energy, under Contract No. DE-AC03-76SF00098

Abstract

The Advanced Light Source, a third-generation national synchrotron-radiation facility now under construction at the Lawrence Berkeley Laboratory in Berkeley, California, is scheduled to begin serving qualified users across a broad spectrum of research areas in the spring of 1993. Undulators will generate high-brightness, partially coherent, plane polarized, soft x-ray and ultraviolet (XUV) radiation from below 10 eV to above 2 keV. Wigglers and bend magnets will generate high fluxes of x-rays to photon energies above 10 keV. The ALS will have an extensive research program in which XUV radiation is used to study matter in all its varied gaseous, liquid, and solid forms.

I. Introduction

The availability of intense, tunable, collimated, polarized radiation in the x-ray and ultraviolet regions of the spectrum has driven the evolutionary development of dedicated facilities optimized for the generation of synchrotron radiation. The newest, third-generation synchrotron sources are based on the use of an electron or positron storage ring specifically designed to have very low emittance and several long straight sections containing insertion devices (wigglers and undulators). The combination of a very-low-emittance storage ring with optimized undulators makes possible the generation of radiation with a spectral brightness that is a factor of 20 or more over that of existing, second-generation sources, depending on the spectral range. Around the world, construction of several third-generation sources is either under way or planned. They include the Advanced Light Source (ALS) at the Lawrence Berkeley Laboratory [1]. The project is scheduled to be completed in April 1993.

II. Accelerator

An overall layout of the facility's accelerator complex, which consists of a 50-MeV linac, a 1-Hz, 1.5-GeV booster synchrotron, and an electron storage ring, is shown in Fig. 1. The ALS lattice has 12 long straight sections that are joined by 12 achromatic arcs, each containing three bending magnets. The ALS produces electron beams that are bunched, with a bunch spacing of 2 ns and a length is 35 ps. For particular experiments—for example, those involving time-of-flight measurements—it will be possible to have only one or a few circulating electron bunches in the storage ring.

III. Photon Sources

The ALS lattice is optimized for the use of insertion devices. The periodic magnetic field of an insertion device bends the electrons into an approximately sinusoidal trajectory in the horizontal plane, causing the emission of synchrotron radiation, as shown in

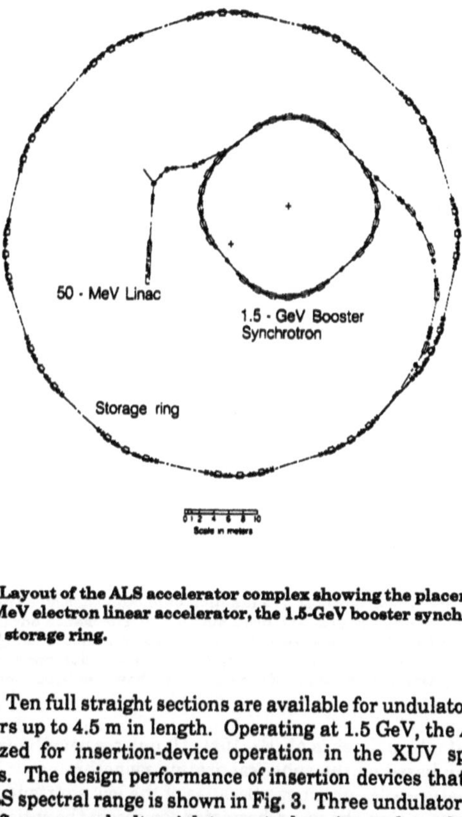

FIG. 1. Layout of the ALS accelerator complex showing the placement of the 50-MeV electron linear accelerator, the 1.5-GeV booster synchrotron, and the storage ring.

Fig. 2. Ten full straight sections are available for undulators and wigglers up to 4.5 m in length. Operating at 1.5 GeV, the ALS is optimized for insertion-device operation in the XUV spectral regions. The design performance of insertion devices that cover the ALS spectral range is shown in Fig. 3. Three undulators span the soft x-ray and ultraviolet spectral regions when the ALS operates at 1.5 GeV, and a wiggler extends spectral coverage into the x-ray region beyond 10 keV. Between them, the undulators will be able to excite the K shell of elements through silicon and the L shell of elements up to krypton, while the wiggler will be able to excite the L shell of nearly every element in the periodic table. The spectral range of the undulator is scanned by varying the undulator magnetic field. The wiggler, which operates with a high magnetic field, generates a broad continuous spectrum extending into the hard x-ray region near 10 keV. High-quality synchrotron radiation will be available from the 24 bend-magnet ports as well.

IV. Insertion-Device Beamlines

Low-emittance storage rings and insertion devices have created new challenges for designers of UV and soft x-ray optics. The attainment of higher resolution by use of small slits also becomes practical (the spectral-resolution goal of monochromators in

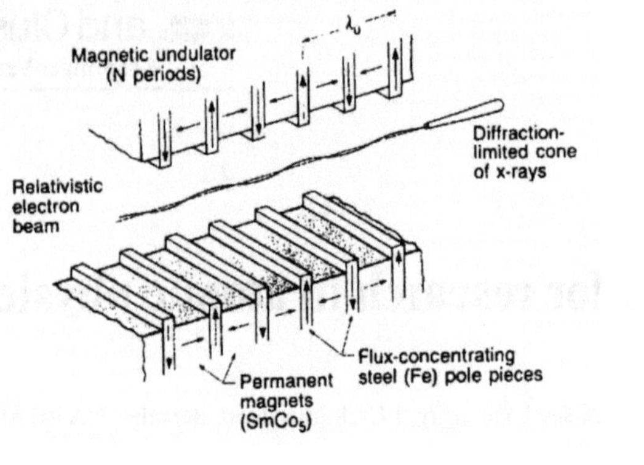

FIG. 2. Schematic drawing of a periodic magnet structure (an undulator) of period λ_u and with a number of periods, N. The oscillations of the electron beam passing through the structure produce ultraviolet and soft x-ray radiation (photons) of high spectral brightness.

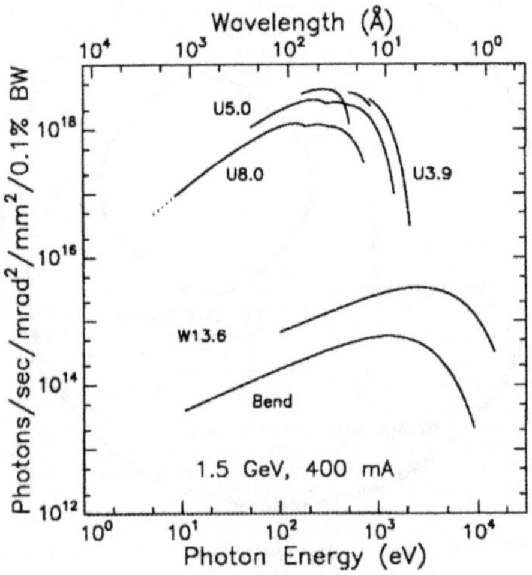

FIG. 3. Spectral brightness as a function of photon energy for proposed ALS undulators, wiggler, and bend magnet. Performance envelope for each insertion device is shown. This figure was calculated using $\varepsilon_x = 10^{8}$ m-radian; effects of electron energy spread and undulator errors are relatively small on the scale of the figure and have not been included.

undulator beamlines is $\Delta E/E \approx 10^{-4}$). Undulator beamlines at the ALS are based on the spherical-grating monochromator system with water-cooled gratings. Because of the low emittance of the ALS storage ring, the monochromator can accept the entire undulator beam in most cases, even at a slit-width of 10 μm.

V. Atomic Physics and Highly Charged Ions

The ALS will be a national user facility that is open to all qualified scientists and technologists. Participating research teams (PRTs) consisting of investigators with related research interests from one or more institutions will be assigned privileged access to the facilities it helps develop for its own research program. A substantial fraction of the beam time at every beamline will be allocated to independent investigators by means of a proposal-review process.

The ALS is particularly well suited for research on tenuous targets, as, for example, occurs in atomic physics and chemistry, in which targets are often gas-phase, beams, or an excited population. A few examples are photon-ion interactions, in which either the ion beam after the collision [2], or electrons emitted are detected. Other examples are coincidence experiments [3], where low collection efficiencies require intense photon beams; experiments with actinide targets, where material is either scarce or highly radioactive; and laser-excited targets [4]. Many other applications in atomic physics have been considered [5].

Highly charged ions can be produced by photon impact; these ions are very cold [6], and could be used for a variety of collisions and especially spectroscopy experiments. These ions could also be stored in a trap [7].

Scientists desiring to do research at the ALS can submit a proposal to form a new insertion-device or bend-magnet team, they can request to join an existing team, or they can submit a proposal to do research as an independent investigator. Interested persons should contact the author for additional information.

VI. Acknowledgments

This work was supported by the Director, Office of Energy Research, Office of Basic Energy Sciences, Division of Materials Sciences, of the U.S. Department of Energy under Contract No. DE-AC03-76SF00098.

References

1. *An ALS Handbook*, PUB-643 Rev. 2 (Lawrence Berkeley Laboratory, Berkeley, CA 1989).
2. I. C. Lyon, B. Peart, J. B. West, and K. Dolder, *J. Phys. B* 19, 4137 (1986).
3. J. C. Levin, C. Biedermann, N. Keller, L. Liljeby, C.-S. O, R. T. Short, I. A. Sellin, and D. W. Lindle, *Phys. Rev. Letter* 65, 988 (1990).
4. M. Ferray, F. Gounand, P. D'Oliveira, P. R. Fournier, D. Cubaynes, J. M. Bizau, T. J. Morgan, and F. J. Wuilleumier, *Phys. Rev. Lett.* 59, 2040 (1987).
5. M. O. Krause, D. W. Lindle, and R. C. C. Perera, LBL Pub-5207 Rev. (March 1988) [unpublished].
6. R. T. Short, C. S. O, J. C. Levin, I. A. Sellin, L. Liljeby, S. Huldt, S. E. Johansson, E. Nilsson, and D. A. Church, *Phys. Rev. Lett.* 56, 2614 (1986).
7. D. A. Church, K. W. Jones, B. M. Johnson, M. Meron, and J. A. Sellin, *J. Phys. B* 17, L401 (1984); D. A. Church, S. D. Kravis, I. A. Sellin, C. S. O, J. C. Levin, R. T. Short, M. Meron, B. M. Johansson, and K. W. Jones, *Phys. Rev. A* 36, 2487 (1987).

Author Index

MIX
Papier aus verantwortungsvollen Quellen
Paper from responsible sources
FSC® C105338
FSC
www.fsc.org

If you have any concerns about our products,
you can contact us on
ProductSafety@springernature.com

In case Publisher is established outside the EU,
the EU authorized representative is:
Springer Nature Customer Service Center GmbH
Europaplatz 3, 69115 Heidelberg, Germany

Printed by Libri Plureos GmbH
in Hamburg, Germany